CLINICAL CHEMISTRY

CLINICAL CHEMISTRY

David White

Nigel Lawson

Paul Masters

Daniel McLaughlin

NEW YORK AND LONDON

Garland Science
Vice President: Denise Schanck
Senior Editor: Elizabeth Owen
Assistant Editor: David Borrowdale
Senior Production Editor: Georgina Lucas
Illustrator: Matthew McClements, Blink Studio Ltd
Layout: Georgina Lucas
Cover Designer: Matthew McClements, Blink Studio Ltd
Copyeditor: Jo Clayton
Proofreader: Sally Huish
Indexer: Bill Johncocks

ISBN 9780815365105

Library of Congress Cataloging-in-Publication Data
Names: White, David, 1943- author. | Lawson, Nigel, 1954- author. | Masters, Paul, 1964- author. | McLaughlin, Daniel, 1965- author.
Title: Clinical chemistry / David White, Nigel Lawson, Paul Masters, Daniel McLaughlin.
Description: New York, NY : Garland Science/Taylor & Francis Group, [2017] |
Includes index.
Identifiers: LCCN 2016030252 | ISBN 9780815365105 (alk. paper)
Subjects: | MESH: Chemistry Techniques, Analytical | Chemistry, Clinical--methods
Classification: LCC RB37 | NLM QY 90 | DDC 616.07/56--dc23
LC record available at https://lccn.loc.gov/2016030252

Published by Garland Science, Taylor & Francis Group, LLC, an informa business,
711 Third Avenue, New York, NY 10017, USA, and 2 Park Square, Milton Park, Abingdon, OX14 4RN, UK.

Printed in the United States of America

15 14 13 12 11 10 9 8 7 6 5 4 3 2 1

Visit our web site at http://www.garlandscience.com

About the authors

David White was awarded a BSc in Chemistry and a PhD in Physiological Chemistry at the University of Birmingham. Following postdoctoral study at The Johns Hopkins Medical School in Baltimore, he pursued an academic career at the University of Nottingham where he was promoted to Associate Professor and Reader in Medical Biochemistry and awarded a DSc in 1998. Dr White has taught undergraduate courses to both bioscience and medical students, supervised many graduate PhD students, and been external examiner in Biochemistry at medical schools in the UK, the Republic of Ireland, and Libya. He was appointed Director of Academic Affairs at the University of Nottingham Graduate Entry Medical School in 2002.

Nigel Lawson graduated in 1976 (BSc Biochemistry, University of Sheffield) and 1980 (PhD, University of Nottingham). In 1982 he became a Clinical Biochemist at Heartlands Hospital, Birmingham. From 1991 to 2005 he worked as a Consultant Clinical Biochemist (Nottingham City Hospital), and from 2005 to 2013 the Royal Derby Hospital as Clinical Director for Pathology. Dr Lawson has taught Clinical Chemistry to undergraduates and postgraduates for many years and is a Fellow of the Royal College of Pathologists, and was an Examiner for the College. He established the Nottingham-Derby NHS Clinical Chemistry training scheme and is an Emeritus Member of the Association of Clinical Biochemistry.

Paul Masters graduated in Medicine from the University of Leeds in 1988, having done an intercalated BSc in Chemical Pathology. After house jobs, he did postgraduate training in Leeds and Nottingham. He took up the post of Consultant in Chemical Pathology and Metabolic Medicine in Chesterfield in 1997, later becoming Clinical Lead for Blood Sciences. He currently works as an NHS hospital consultant in both Chesterfield and Derby. As well as practising and teaching clinical chemistry, Dr Masters also runs outpatient clinics for lipid disorders, diabetes and metabolic bone disease.

Danny McLaughlin earned his BSc in Human Physiology from Glasgow University in 1986 and his PhD from the Council for National Academic Awards in 1991. Since 2003 Professor McLaughlin has taught medical students at the University of Nottingham and Durham University, where he is the Academic Director of Undergraduate Medicine. Professor McLaughlin is a Senior Fellow of the UK Higher Education Academy, a member of the Association for Medical Education in Europe, and a Fellow of Durham University's Wolfson Research Institute for Health and Wellbeing.

PREFACE

The discipline of Clinical Chemistry has many aliases throughout the world, including Chemical Pathology, Medical Biochemistry, and different permutations of the terms clinical, medical, and biochemistry. Even within the departments in which it is a subdiscipline it may be known variously as Clinical Pathology, Blood Sciences, or—increasingly—Laboratory Medicine. Despite this slight identity issue, Clinical Chemistry is a strong discipline that underpins much of the diagnosis and monitoring of patients in modern health care systems. A good understanding of clinical chemistry and an awareness of its use in the management of patients are essential for both users and providers.

This text evolved from our combined experiences in the postgraduate teaching of Clinical Chemistry, using material from courses in biochemistry, physiology, and sub-specialties of medicine. Content has been expanded to include core subjects from the syllabuses of examinations for the Royal Colleges, particularly of the Royal College of Pathologists, and other graduate courses. The book is thus a suitable text for both undergraduate students of medicine and biosciences as well as for postgraduates undergoing professional training.

From the start, we wanted the book to be relevant to clinical practice, but also to include the core knowledge that enables the reader to see the full picture from first principles. Our extensive combined experiences both as full-time academics at the University of Nottingham (David and Danny) and as hospital consultants in Clinical Chemistry at the Chesterfield Royal and Royal Derby Hospitals (Paul and Nigel) have been equally valuable in enabling us to achieve these aims. The book contains many analytical and clinical practice points, distilled as "pearls of wisdom," which are intended to be relevant on a day-to-day basis to people working (or wishing to work) in Clinical Chemistry. In addition, there are many case histories included that help to put the subject into its clinical context. We believe that inclusion of clinical cases in each chapter is essential to illustrate the way in which an understanding of the underlying basic science enables interpretation of laboratory data in real-life clinical practice. Each case is based on genuine data and scenarios that we have encountered in our laboratory and clinical services. These case histories have been found to be very popular with medical students, especially those undertaking case-oriented courses, such as the University of Nottingham's Graduate Entry Medicine program in the UK.

In selecting topics for the book, we have sought to provide a varied yet comprehensive approach to Clinical Chemistry, with a mixture of chapters; some describe individual analytes, while others emphasize tissue integrity and function. In addition, we have included chapters on therapeutic drug monitoring, poisons, pregnancy, inborn errors of metabolism, and, importantly, immediate assessment of the critically ill patient. Clinical Chemistry, like any other discipline, is always developing, and we have tried to ensure that the book reflects current practice.

We have been very fortunate to work with many excellent academic, scientific, and clinical staff from the University of Nottingham and the Royal Derby and Chesterfield Royal Hospitals. Such good working interactions are vital in running successful Clinical Chemistry departments. We believe those working relationships have been reflected in how we have put Clinical Chemistry into context with the rest of medicine. Countless numbers of students have been through our training courses, and many have gone on to become medical and scientific consultants. We hope that this book will enable a new generation of students to develop their careers, and hopefully some will become Clinical Chemists.

We are forever indebted to various members of staff at Garland Science for commissioning the book. The initial project would never have got off the ground without the continued "gentle" prompting and encouragement of our editor Liz Owen and, latterly, of our assistant editor David Borrowdale, whose tireless work, patience, and humor ensured the eventual delivery of the first set of draft chapters and who oversaw the rewriting following the comments of the reviewers. Matt McClements deserves credit for turning our self-made lecture slides into the full-color pictures presented here. We would also like to thank Georgina Lucas (Senior Production Editor) for typesetting of the book and delivery of the final product.

Thanks also go to many of our close colleagues, especially Julia Forsyth and John Monaghan in the Clinical Chemistry Department of the Royal Derby Hospital.

We would also like to thank Angela White for producing many of the anatomical drawings used in the book. Finally, we wish to thank our wives and families who have given us constant support throughout and have made it possible for us to write this book.

David White, Nigel Lawson, Paul Masters, and Danny McLaughlin

ACNOWLEDGMENTS

The authors and publishers have benefitted greatly from the many people who have reviewed the text and figures and made suggestions for improvements. We would like to thank the following for their contribution:

Sheila A. Alexander, University of Pittsburgh, USA; Jose A. L. Arruda, University of Illinois, USA; Ian Beales, University of East Anglia, UK; Michael J. Bennett, University of Pennsylvania, USA; Roger L. Bertholf, University of Florida College of Medicine, USA; Peter D. Brown, University of Manchester, UK; Nicola Brunetti-Pierri, University of Naples Federico II, Italy; Diane Davis, Salisbury University, USA; Christian Delles, University of Glasgow, UK; Kathy Dugan, Auburn University at Montgomery, USA; Karen J. Gibson, University of New South Wales, Australia; Aidar R. Gosmanov, Albany Medical Center, USA; David A. Grahame, University of the Health Sciences, USA; David G. Grenache, University of Utah, USA; Nancy Hakooz, University of Jordan, Jordan; Ibrahim A. Hashim, University of Texas Southwestern Medical Center, USA; Kenneth U. Ihenetu, Albany College of Pharmacy and Health Sciences, USA; Farideh Javid, University of Huddersfield, UK; Michael W. King, Indiana University School of Medicine, USA; Anders Larsson, Uppsala University, Sweden; Jaleel Miyan, University of Manchester, UK; Abhay Moghekar, Johns Hopkins University, USA; Michael L. Moritz, Children's Hospital of Pittsburgh, USA; Vinod Patel, University of Warwick, UK; Vivian Pijuan-Thompson, University of Alabama at Birmingham, USA; David Rees, King's College Hospital, UK; Nancy J. Rehrer, University of Otago, New Zealand; Ana Ricardo, University of Illinois College of Medicine, USA; Damian G. Romero, University of Mississippi Medical Center, USA; Audrey Skaggs, Arkansas State University, USA; Nelson Tang, Chinese University of Hong Kong, China.

CONTENTS IN BRIEF

CONTENTS

LABORATORY MEDICINE —AN INTRODUCTION

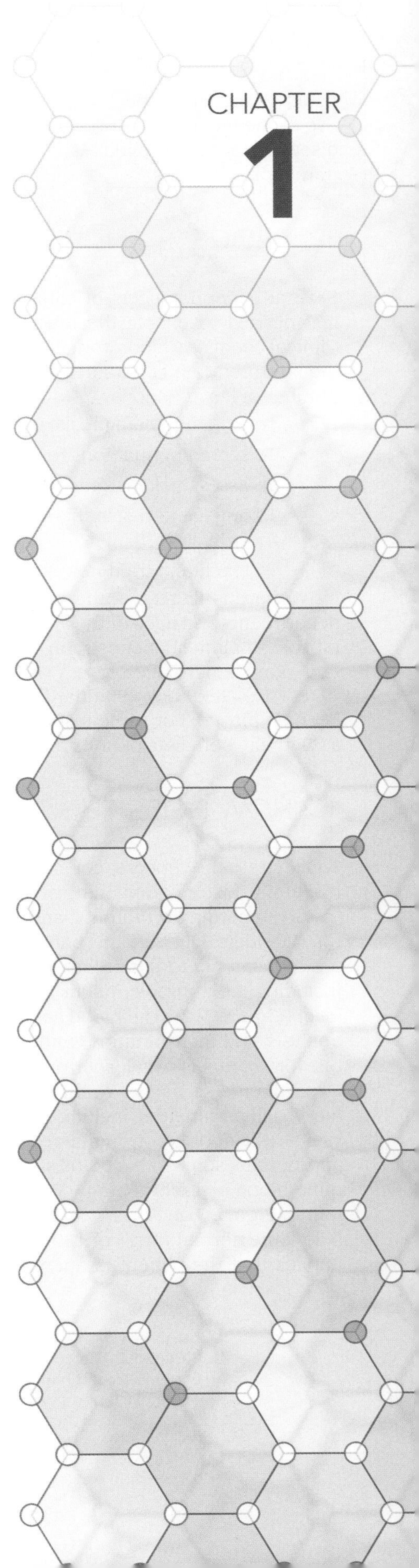

IN THIS CHAPTER

PROCESS AND TERMINOLOGY
HOW DOES THE SYSTEM WORK?
USING FLUIDS TO ASSESS TISSUE DISEASE
PRE-ANALYTICAL FACTORS
WHAT CONSTITUTES AN ABNORMAL RESULT?
WHAT IS A REFERENCE RANGE?
AN IDEAL DIAGNOSTIC TEST
USING REFERENCE RANGES
HOW DOES CLINICAL CHEMISTRY OPERATE?
QUALITY ASSURANCE
QUALITY CONTROL
ACCREDITATION

From a somewhat humble beginning using rudimentary chemical and biological tests and microscope slides, laboratory medicine has progressed to become an essential component in the diagnosis and management of disease. It has evolved into a group of subspecialties which includes Histopathology or Cell Pathology, Hematology, Microbiology, Clinical Biochemistry or Clinical Chemistry, and Immunology and these disciplines have specialized further as new areas of science and technology have been adopted. For example, the techniques of molecular biology and genetics have been embraced by all disciplines leading to the development of separate departments in some institutions. In other institutions, however, such developments have prompted a rethink of working practices and the combination of aspects of hematology, immunology, and biochemistry into all-embracing departments of Blood Sciences or Clinical Laboratory Sciences. Throughout these changes, the principles of clinical biochemistry have remained constant, relating and adapting human biochemistry to the development of diagnosis and the management of disease.

Although terminology may vary between institutions, the terms Chemical Pathology, Clinical Chemistry, Medical Biochemistry, and Clinical Biochemistry are often used interchangeably; the term Clinical Chemistry will be used in this text. The clinical biochemistry of tissues, organs, and molecules is discussed in the context of disease processes and related to the diagnosis, monitoring, and management of disease. This will include outlining the fundamental metabolism of biochemical processes and physiological interrelationships. Some space has been given to descriptions of analytical processes and theory, such as immunoassay, and how these relate to clinical practice. Although the emphasis of this book is clinical biochemistry, some chapters include sections on hematology, radiology, and microbiology as appropriate only where this helps in the understanding of disease processes. However, it is not intended to provide the detail expected in specialized textbooks on these disciplines. The increasing use of the techniques of molecular biology and genetics in the investigation of disease is acknowledged also by appropriate inclusion of these disciplines in a number of chapters.

1.1 PROCESS AND TERMINOLOGY

Several general aspects of clinical biochemistry process and terminology are first defined to enable the reader to better understand subsequent chapters. Clinical chemistry laboratories are engaged in the analysis of tissue samples and the interpretation of data derived from these analyses. In doing so, they have a number of key functions:

- To help in the diagnosis, monitoring, and management of disease
- To screen populations for disease
- To assess risk factors for disease processes
- To engage in research and development of analytical and diagnostic tools
- To teach and train new staff

While this illustrates the scope of the work of major clinical biochemistry departments and each of the above functions is important per se, the fundamental role of clinical chemistry in most hospitals is in the diagnosis, monitoring, and management of disease.

For this textbook, Standard International (SI) units of measurement have been employed. For some tests, where non-SI units are in common use as well as SI units, both sets of units are quoted.

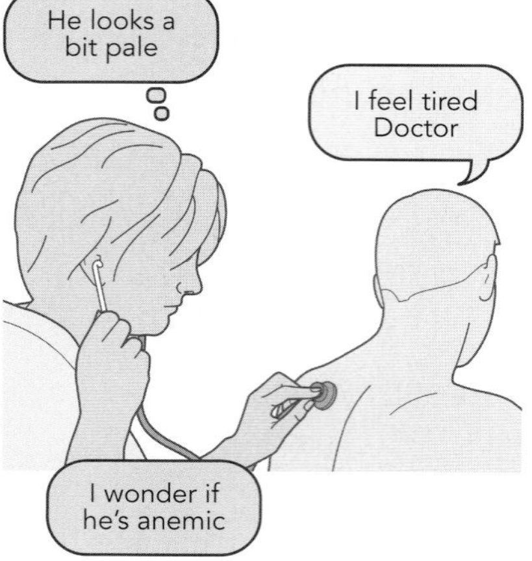

Figure 1.1 Before tests are performed.
Before a doctor considers requesting tests on a patient they will usually take a history, check for symptoms, and consider a possible diagnosis.

1.2 HOW DOES THE SYSTEM WORK?

To illustrate the process, consider this simple scenario where a patient attends his family doctor complaining of tiredness (**Figure 1.1**). After first taking a short history, recording symptoms, and carrying out a physical examination, the doctor considers whether the patient might be anemic. The next step is to take a blood sample to determine whether the tiredness and any other symptoms, including looking pale, might be due to anemia; that is, to a low hemoglobin level (**Figure 1.2**). The blood sample is transferred to a (hospital) laboratory for analysis and interpretation, and a report is sent to the requesting doctor. The test for anemia is usually performed by Hematology or Combined Blood Science Departments as part of a full blood count (FBC) which assesses a number of blood cell parameters, including hemoglobin concentration. Usually, such analyses are undertaken using large automated cell counters (instruments). Where appropriate, some of the blood sample is examined under a microscope to determine blood cell morphology. For the patient above, the following results were found (**Figure 1.3**). The hemoglobin concentration of 97 g/L (9.7 g/dL) is clearly below the normal range of 130–180 g/L (13.0–18.0 g/dL) for an adult male and implies that the patient is anemic. A decreased concentration of hemoglobin

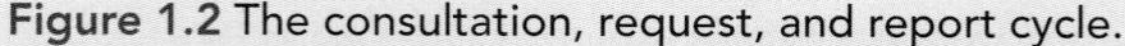

Figure 1.2 The consultation, request, and report cycle.
Once the doctor has made a potential diagnosis, in this case anemia, they will send a blood sample to the laboratory and request the sample is tested for anemia, and the result is reported back to the doctor. Anemia can be defined as having a low blood hemoglobin, which will lead to symptoms such as tiredness. Therefore the laboratory will measure the patient's blood hemoglobin concentration, as well as other red blood cell parameters.

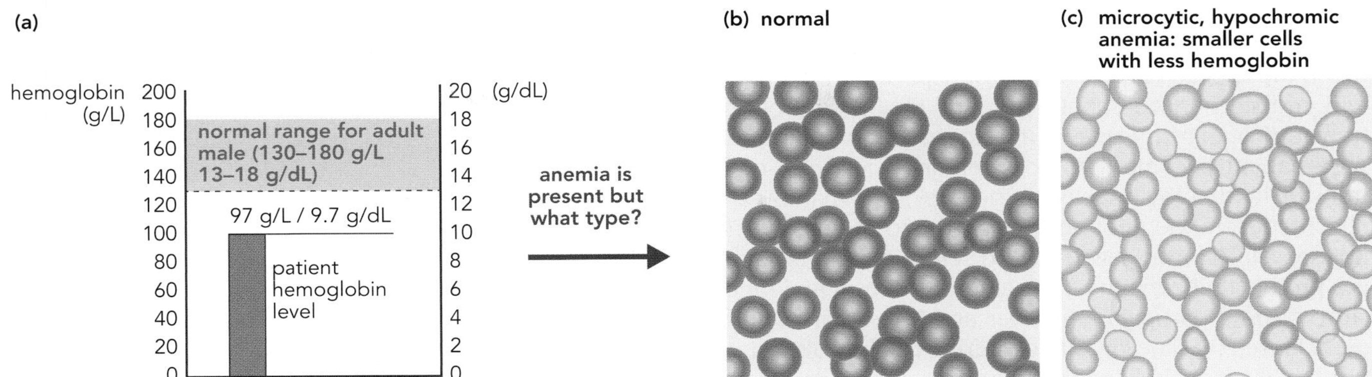

Figure 1.3 Does the patient have anemia? Hematology results. (a) The patient has a blood hemoglobin result of 97 g/L (9.7 g/dL) of blood, which is low compared to most adult males, who usually have a blood hemoglobin greater than 130 g/L (13.0 g/dL) in health; that is, the patient has anemia. Normal results for adult males are shown in parentheses. In addition, the patient's red blood cells will be examined by microscopy and compared to normal cells (b) to help to define what type of anemia is present. In this case, the patient's red blood cells (c) are smaller (microcytic) and have less pigment (hypochromic) than normal. The patient has a microcytic, hypochromic anemia. This type of anemia is often found in patients with iron deficiency.

lowers the capacity of the red cells to transport oxygen to the tissues throughout the body, giving rise to tiredness, and explains some of the patient's symptoms. However, there are a number of different types of anemia (see Chapters 10 and 16) and further tests are required to ascertain which is present in this case.

Microscopic examination of the patient's blood film (Figure 1.3c) indicates that the red cells of the patient have less color and are smaller than red cells from a healthy, adult male. In summary, the patient has small red cells lacking in pigmentation and a low hemoglobin content; that is, the patient has microcytic, hypochromic anemia. A common cause of this type of anemia is iron deficiency and further tests in the clinical biochemistry laboratory are required to investigate this.

Although some clinical biochemistry analyses are made on whole blood, most tests have been developed on the supernatant fluid remaining after the blood cells have been separated by centrifugation. If blood is taken into a plain glass or plastic tube and the sample is allowed to clot, the resulting supernatant is termed serum. However, if an anticoagulant is added to the tube to prevent clotting, the supernatant liquid is termed plasma. Measurements assessing iron deficiency may be made on either serum or plasma. In the current scenario, the patient's serum iron is indeed below the normal range for adult males (**Figure 1.4**), suggesting iron deficiency. This may be due to decreased

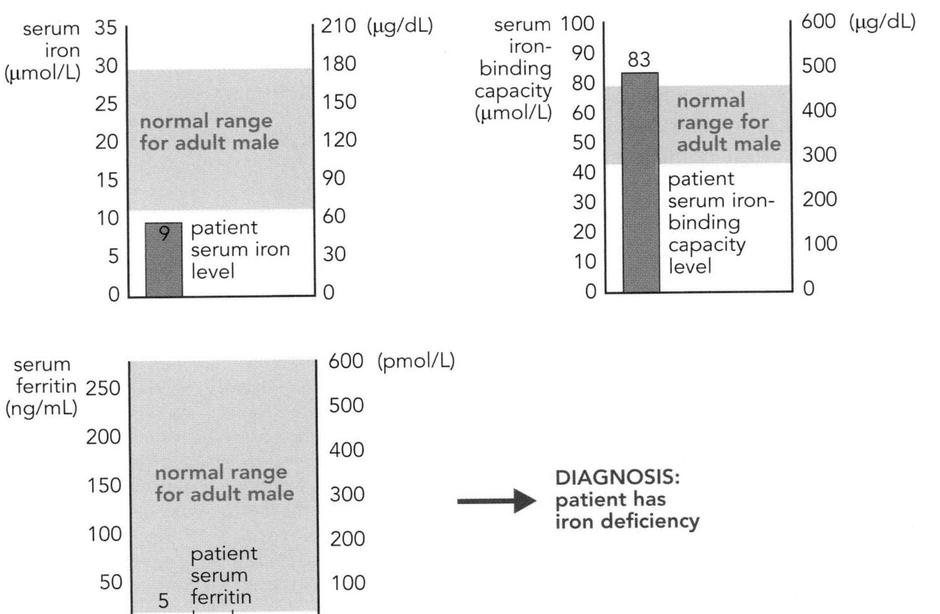

Figure 1.4 Clinical chemistry results. In order to confirm the diagnosis of iron deficiency the doctor will request further tests on the patient. With iron deficiency, a patient's serum iron would be expected to be low, as would an estimate of the patient's total body iron stores (serum ferritin). Conversely, the blood capacity to transport iron is increased in iron deficiency. Comparing this patient's results to normal (see Table 10.1), they have a low iron and ferritin but a high iron-binding capacity. They have iron deficiency.

intake into the body due to low dietary iron or malabsorption, increased loss through bleeding, or a combination of both. Further tests are required to investigate these possibilities. Iron is transported in the circulation bound to a protein, transferrin (see Chapter 10), and the amount of iron that can be bound to transferrin can be measured and expressed as total iron-binding capacity (TIBC). As a patient becomes progressively iron deficient, the liver synthesizes more transferrin, thereby increasing the TIBC, to maximize the amount of iron being absorbed from the gut and transported to the tissues. This response in itself is not specific for iron deficiency, so serum ferritin is routinely measured. Ferritin is a protein that can bind large amounts of iron and functions as an intracellular store of iron, particularly in liver and bone marrow. Small amounts of ferritin are released into the circulation during cell turnover and measurement of serum ferritin gives an indication of the iron stores in these tissues. The lower the serum ferritin level is, the lower the tissue ferritin stores and thus whole body iron stores. The results for the patient shown in Figure 1.4 indicate that all three additional parameters measured are below the normal range for adult males, confirming that the patient has iron deficiency. This is a diagnosis and helps to direct investigations into the cause of iron deficiency (see Chapter 10). Where possible, the underlying cause is treated and the patient is given iron supplements to return his iron and hemoglobin concentrations to within the normal range. Hematology and clinical biochemistry tests are used to monitor the effectiveness of treatment.

1.3 USING FLUIDS TO ASSESS TISSUE DISEASE

Although several diseases are due to abnormalities in the blood, for example abnormal red blood cells, most diseases are characterized by abnormalities in specific tissues. Since it is difficult to routinely take biopsy material from the affected tissue, blood and urine are frequently examined to show whether tissue damage or disease is present. Many compounds are released into the blood by different organs and have a specific function within the bloodstream, for example hormones. These are routinely measured to assess disease. However, there are also many compounds circulating in the bloodstream which have no known function there. They may be being transported to and from other tissues, or alternatively they are present due to the constant turnover of cells. Most of these compounds are transported to various organs, metabolized, and then recycled or excreted. Such compounds can also be measured in urine if they are excreted by the kidneys.

Enzymes are typical examples of such compounds, being used extensively as specific markers of organ damage in clinical biochemistry. The plasma level of an enzyme reflects the general turnover of cells that contain significant concentrations of that particular enzyme. As a cell breaks down, proteins (including enzymes) will appear in the bloodstream, and their level in plasma will be dependent upon the amount of enzyme in the organ or tissue, the rate at which it appears in the plasma, and the rate at which it is cleared from the plasma. This can be seen for many enzymes, for example lactate dehydrogenase (LDH), an enzyme found in the cytoplasm of most human cells. The measured value in the plasma reflects a steady state which will exist in the normal, non-diseased state, where there is a balance between production and elimination. This will be equivalent to the reference range. Changes to this equilibrium can occur due to more enzyme being present in the tissue or organ, which may be simply due to more cells being present; for example, in some lymphoma cases, higher LDH values are seen as the tumor mass increases. Alternatively, higher values can occur due to reduced clearance of the enzyme from the plasma. In the case of many diseases, increased enzyme values occur due to damage to the cells of the tissue or organ (**Figure 1.5**). The increase in the plasma enzyme is related to the

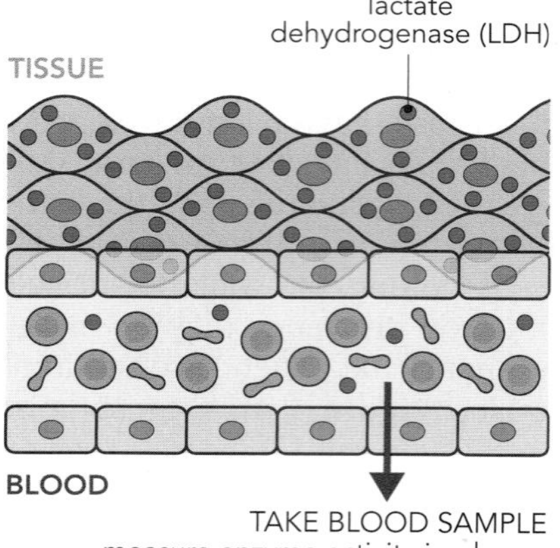

Figure 1.5 Enzymes in plasma. Enzymes, for example lactate dehydrogenase, are frequently measured in plasma or serum to help to diagnose and monitor disease. The plasma level is dependent upon the rate at which the enzyme is released into the plasma from its tissue of origin, and how quickly it is cleared from the plasma by various mechanisms, including removal by the kidneys and liver. Clearly an increased release of enzymes into the plasma, for example by damage to the tissue, will lead to increased levels in the plasma. Similarly, reduced clearance may also lead to raised levels in the plasma. Most of these processes are dynamic, with the enzyme levels rising and falling in disease dependent upon the degree and duration of tissue damage and the rates of clearance.

TABLE 1.1 Enzyme activities in tissues in comparison to plasma

Tissue	Aspartate transaminase (AST) relative to plasma	Alanine transaminase (ALT) relative to plasma
Heart	7800	450
Liver	7100	2850
Skeletal muscle	5000	300
Kidney	4500	1200
Pancreas	1400	130
Spleen	700	80
Lung	500	45
Erythrocytes	15	7
Plasma	1	1

Table shows the relative concentrations of two enzymes, aspartate transaminase (AST) and alanine transaminase (ALT), compared to their concentrations in plasma, where plasma concentration has been defined as 1 arbitrary unit. Clearly, the enzymes have much higher concentrations in tissues than plasma, and there is a degree of tissue specificity, with much higher ALT values in liver compared to other tissues.

degree of damage. More damage means higher plasma enzymes. In addition, the intracellular location of the enzyme could determine how much is seen in plasma. The level of a cytoplasmic enzyme is likely to rise quicker in the bloodstream than is that of a mitochondrial enzyme.

Although LDH is used to monitor certain diseases it is not very organ specific. Other enzymes have more specificity and this is reflected in their clinical use. Both aspartate transaminase (AST) and alanine transaminase (ALT) are found in many tissues as well as liver, but in the case of ALT the liver has the highest concentration. In addition, there is a huge difference in the concentration of ALT in the liver and plasma, allowing small amounts of damage to be reflected in changes in the plasma concentration (**Table 1.1**).

Since the enzyme level reflects production and clearance, raised values are dependent on whether the increased damage is a chronic effect or not, and how quickly the enzyme is cleared from the plasma. In acute tissue damage, the enzyme levels will rise and fall, and the speed and magnitude of these rises depend upon the plasma half-lives of each enzyme (**Table 1.2**).

TABLE 1.2 Half-lives of enzymes in plasma

Enzyme	Half-life
Lactate dehydrogenase	3–7 days
Alkaline phosphatase	3–7 days
Gamma-glutamyl transferase	3–4 days
Alanine transaminase	1.5–2.5 days
Creatine kinase	12–24 hours
Amylase	3–6 hours

1.4 PRE-ANALYTICAL FACTORS

Whilst the examples above illustrate the role of clinical chemistry in diagnosis and monitoring of therapy, there are a number of factors which need to be considered in obtaining appropriate samples for analysis. These are often termed pre-analytical factors and include an awareness of the patient's medical history and the preparation of a patient for taking samples, as well as the procedures of sample collection, processing, and storage. It is essential that these are carried out correctly to provide valid information for diagnosis and treatment.

Analytical practice point 1.1

Whether an electronic requesting system or handwritten forms are used, it is essential to clearly label all samples with the correct patient identifiers (name, date of birth, and so on). It does not matter how accurate and precise the analysis has been, sending the result out to the wrong patient can have disastrous consequences.

Patient preparation

Consider as an example blood glucose measurement, where correct patient preparation has a major influence on the interpretation of results and diagnosis of disease. Patients with poorly controlled type 2 diabetes mellitus have a raised fasting blood glucose concentration of >7.0 mmol/L (126 mg/dL) and this is diagnostic for the condition (see Chapter 2). However, the blood glucose concentration is raised significantly in healthy individuals following a meal, particularly one enriched in carbohydrate. It is essential therefore that subjects be fasted fully prior to taking a blood sample for glucose analysis to ensure a meaningful result. Fully fasted requires that the subject ingests only water for 12 hours prior to sampling. Other pre-analytical factors include dietary and drug histories, consideration of diurnal and seasonal variations, and the effects of stress, exercise, and posture on analysis. These are discussed in context in later chapters where relevant.

Taking samples

Once a patient has been prepared correctly, samples may be taken. The majority of clinical biochemical analyses are performed on whole blood, serum, or plasma (separated blood), but many other sample types may be used, including:

- Urine
- Cerebrospinal fluid (CSF)
- Sweat
- Saliva
- Feces
- Pleural, ascitic, and other fluids
- Tissue biopsies

Of course, the quality of the sample taken is dependent often on the skill of the health care professional taking the sample, but other samples, such as 24 h urine collection, require an understanding of sample collection by the patient. In such cases, understanding sampling protocols is essential since the discarding of preservative from collection bottles or incomplete urine collection, for example, may lead to incorrect diagnosis despite accurate sample analysis in the laboratory.

Blood sampling tubes

Selection of the appropriate tube for blood collection is extremely important and will depend on whether or not the blood sample may be left to clot prior to measurement of a particular analyte. If the analyte is stable, blood may be collected into a tube containing no anticoagulant and allowed to clot. Since this may take some time, in excess of 30 minutes at ambient temperature, some collection tubes contain clot activators to speed up the process of coagulation. On the other hand, if whole blood or plasma is required, anticoagulant is added to the collection tube prior to collection of blood. The choice of anticoagulant will depend on which parameters are being assayed; potassium-EDTA (ethylenediaminetetraacetic acid) is the anticoagulant of choice for full blood counts. However, potassium-EDTA tubes are obviously inappropriate for measurements of plasma potassium and, since EDTA is a very good chelator of divalent cations, the use of these tubes is also precluded in the measurement of calcium and magnesium. A display of tubes currently available commercially is shown in **Figure 1.6**. Since there is no universal color coding for different tubes, it is

Analytical practice point 1.2

Correct selection of blood tube and anticoagulant is essential. The wrong tube (or contamination by mixing the contents of two different tubes) can give spurious results.

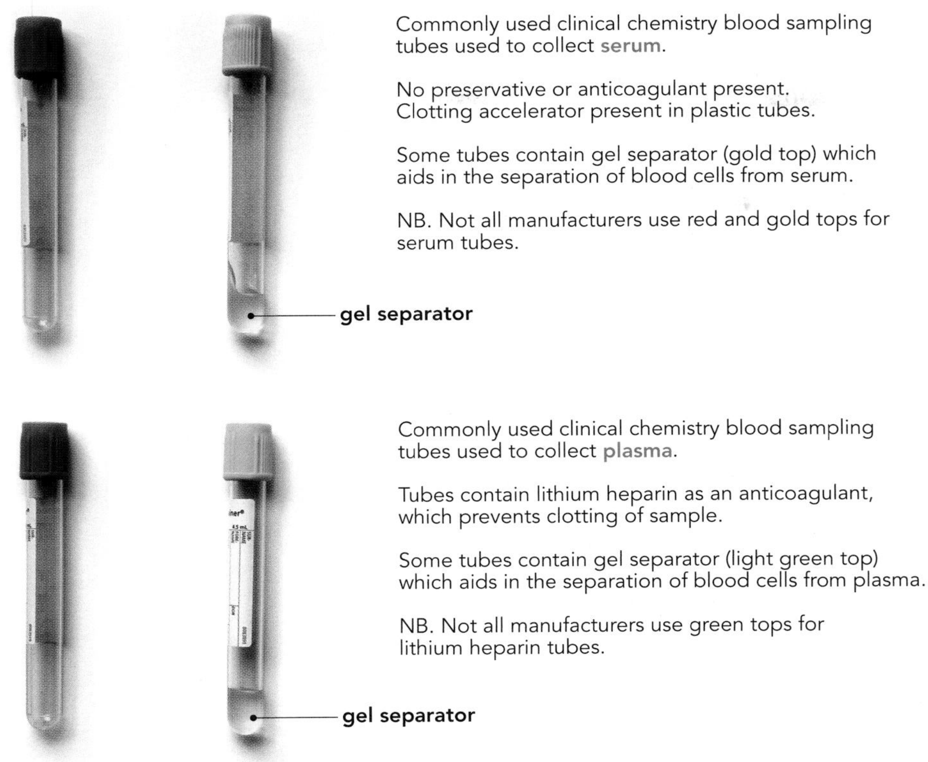

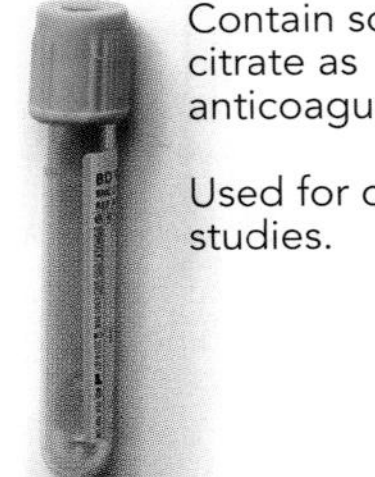

Figure 1.6 Blood collection tubes. Different types of blood tubes are used depending upon what tests are required. The simplest tubes are just straightforward plain tubes to collect clotted samples; that is, there is no anticoagulant or preservative present. Different anticoagulants can be used, usually lithium heparin, EDTA, oxalate, or citrate. Occasionally preservatives are needed, for example fluoride to inhibit glycolysis when measuring glucose.

essential to read the label on tubes before blood sample collection to ensure that the tube chosen is fit for purpose.

Processing and storage of samples

The processing of a blood sample depends on the stability of the analyte to be measured. It should be remembered that red cell metabolism, particularly that of glucose, does not stop when blood is collected into a tube and, unless steps are taken to remove the red cells from the sample by centrifugation or to inhibit this metabolism by the use of metabolic poisons, erroneous results may be obtained. For example, if blood samples are kept at room temperature, anaerobic glycolytic activity in red cells will reduce the blood glucose concentration at a rate of 5–10% per hour with a concomitant rise in lactate concentration. Also, potassium ions leak from red cells when blood is kept at ambient temperature for a few hours giving rise to elevated plasma potassium levels. Paradoxically, potassium also leaks from red cells when blood samples are cooled, due to slowing of the activity of the sodium-potassium pump, and this again produces an elevated plasma potassium level. In both these cases, the measured potassium level is in excess of the true plasma concentration in the patient. Thus separation of cells from plasma is a key process in sample preparation and storage. Centrifugation is usually carried out at 3000*g* for 5–10 minutes, taking care not to damage cells. Some collection tubes contain a gel that creates a barrier between plasma and blood cells (see Figure 1.6). If the analyte to be assayed is labile, the time between phlebotomy and centrifugation should be minimized. Further knowledge of the stability of the analyte will dictate at what temperature a sample is centrifuged and stored, and whether preservatives and/or metabolic inhibitors such as fluoride (to inhibit red cell glycolysis) are added to the sample. Particular instances of sample processing and storage which impact on results and may affect their clinical interpretation are included in individual chapters.

1.5 WHAT CONSTITUTES AN ABNORMAL RESULT?

The normal range for any given parameter or analyte defines the range of values found in a normal healthy adult population. It is against this range that any value for a given patient is compared. Simplistically, it may be thought that an abnormal result is one that lies outside of the normal range, being lower or higher. To some extent, however, the concept of a normal range is somewhat artificial because in many cases (diseases) there may be considerable overlap between values in the normal and diseased states. Furthermore, because of a number of confounding factors, it is often difficult to decide on what is normal since what may be normal in one population group may be distinctly abnormal in another. Thus, when an analyte is measured, the result is compared with a reference range for that analyte taking into account any potential confounding factors.

1.6 WHAT IS A REFERENCE RANGE?

When the concentration of an analyte in a given population is normally distributed, the reference range is calculated from the mean and standard deviation of results from that population. The reference range is defined as the mean ± two standard deviations (mean ± 2SD); this will include 95% of the results from the population group and exclude the top and bottom 2.5% (**Figure 1.7**). Results outside the reference range (mean ± 2SD) are considered abnormal but do not indicate the presence of disease per se merely because the result lies outside of the reference range. However, the likelihood of disease being present usually increases with the distance of the measured value from the mean.

Analytical practice point 1.3

Any result has a 5% chance of being outside the reference range and the more tests are done, the greater the chance of at least one result being abnormal. The chance is $1 - 0.95^n$, so that if a profile of 10 tests is done, the chance of at least one abnormal result is 40%.

The creation of a reference range needs to take into account potential confounding factors that might influence the concentration of a given analyte in a patient. For example:

Time of day

Measurement of blood glucose may be affected by several factors and the sources of variation must be minimized prior to taking a sample for analysis. Since blood glucose concentration rises after a meal, the reference range of 3.5–5.8 mmol/L (63–104 mg/dL) may be quoted for fasting subjects. Values that are slightly above or below the reference range are not necessarily indicative of disease, and there may be no obvious symptoms or signs in these patients

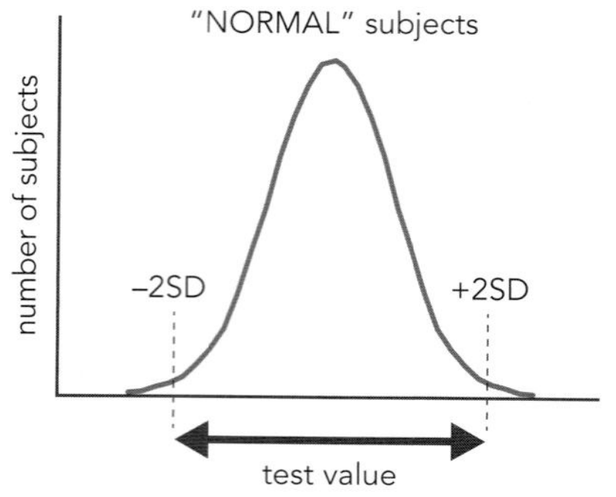

Figure 1.7 The reference range.
The range of results found in the general healthy population when they are distributed normally. This reference range in a normally distributed population would consist of all results which are ±2 standard deviations (SD) of the mean. That refers to 95% of the population.

compared to those whose blood glucose concentrations lie within the reference range. Should the results exceed three standard deviations from the mean, the probability of processes and symptoms of disease being present increases considerably. A blood glucose concentration below 2.5 mmol/L (45 mg/dL), for example, is insufficient to support normal brain function and symptoms of hypoglycemia (tremor and eventually loss of consciousness; see Chapter 2) are manifest.

Gender
Testosterone concentrations would be expected to be much higher in men than in women.

Time of year
Vitamin D concentrations in blood would be higher in summer compared to winter due to exposure of skin to sunlight.

Age of patient
There is usually less lean body mass (muscle) in the young and in the elderly and so blood parameters which are derived from muscle, creatinine for example, will be lower in these two age groups compared with a healthy adult group.

The number of factors that have the potential to influence the reference range makes it difficult if not impossible to generate a generic reference range for all subject groups. It is relatively straightforward for instance to collect blood from, say, 200 healthy medical students and to establish reference ranges for a number of analytes in the student group. These reference ranges are unlikely, however, to be applicable to the majority of patients who are elderly, or to young children and babies. Indeed, the establishment of reference ranges for normal healthy children creates problems of its own because of ethical concerns in collecting blood and CSF, both painful procedures from which the child derives no obvious benefit.

1.7 AN IDEAL DIAGNOSTIC TEST

In terms of usefulness in diagnosis, an ideal test is one where the reference range for a normal population is normally distributed and quite separate from the reference range derived from the disease group, also normally distributed; that is, there is no overlap of the two reference ranges (**Figure 1.8**). This is rarely the case, however, and overlap of the two ranges occurs due to false positives—individuals who have no disease but have a positive result (a value within the disease range)—and false negatives, patients who have disease but whose test result lies within the normal reference range (**Figure 1.9**). Knowledge of

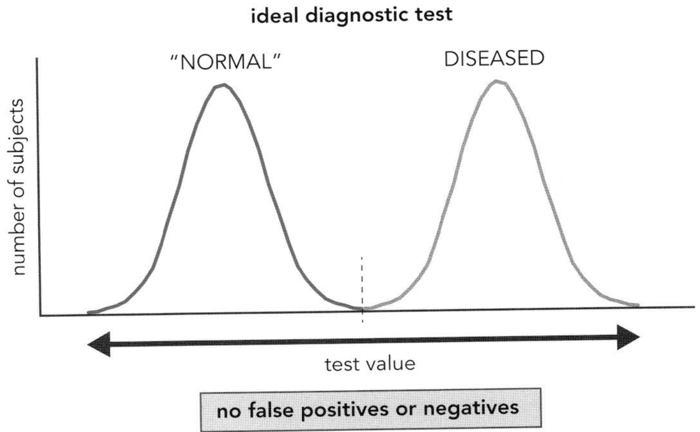

Figure 1.8 Ideal diagnostic test. Hypothetical results from an ideal diagnostic test where the test result is increased in disease (orange line) compared to the non-disease state (blue line). There is no overlap between values found in patients with disease compared to those without. There would be no false positive results or false negative results.

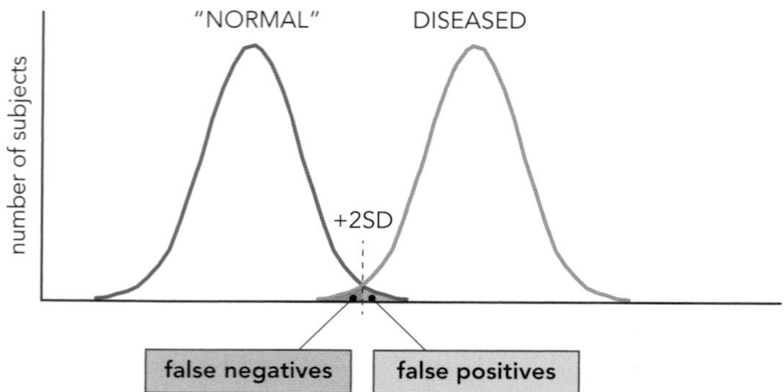

Figure 1.9 Uncertainty of results.
In the vast majority of pathology tests there is overlap between results found in the diseased and non-diseased states. In this hypothetical case, where the diseased state would consist of raised results, an abnormal result would be considered as any value greater than two standard deviations (2SD) above the mean. Of course this would include 2.5% of the normal population without disease; therefore at least 2.5% of a normal population would give a false positive result. Dependent upon the disease and test, patients with a disease may produce results that are within the reference range; that is, false negative results.

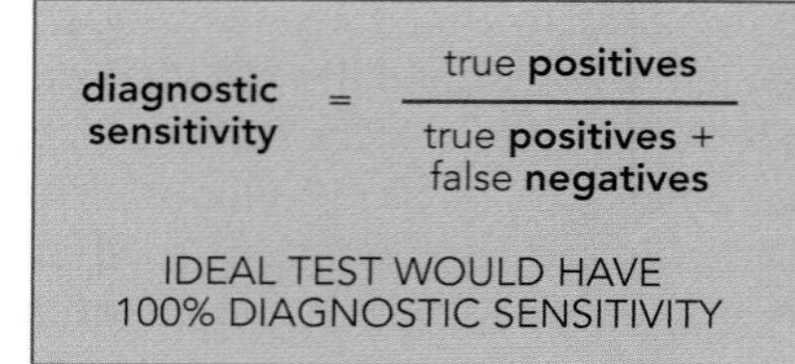

diagnostic specificity = true **negatives** / true **negatives** + false **positives**

IDEAL TEST WOULD HAVE 100% DIAGNOSTIC SPECIFICITY

Figure 1.10 Diagnostic sensitivity and specificity.
Simple equations can determine the diagnostic sensitivity and specificity of a test. True positives are defined as people who have disease and produce abnormal results, whilst true negatives are defined as people who do not have disease and have normal results. False positives are defined as people who do not have disease but produce abnormal results, whilst false negatives are defined as people who have disease but produce normal results. Sensitivities and specificities are expressed as percentages.

the numbers of true and false positives and negatives allows the determination of the diagnostic sensitivity and specificity of a test (**Figure 1.10**). A test of high diagnostic sensitivity has a low number of false negative results, while high diagnostic specificity is associated with a low number of false positives. In an ideal test, both sensitivity and specificity are 100%. Unfortunately, in real life this seldom occurs and although some test ranges are normally distributed (for example, blood glucose in normal and diabetic individuals) many are not and the use of parametric statistics is precluded. Many analytes, for example prolactin, have skewed reference ranges (**Figure 1.11**) and the use of the mean ± 2SD would indicate many normal results to be abnormal; thus, nonparametric statistics are used to determine 95% confidence intervals and reference ranges.

1.8 USING REFERENCE RANGES

As mentioned earlier, a number of factors other than disease may impact on the concentration of a particular analyte and these should be borne in mind in determining reference ranges and interpreting test results. However, as is shown in

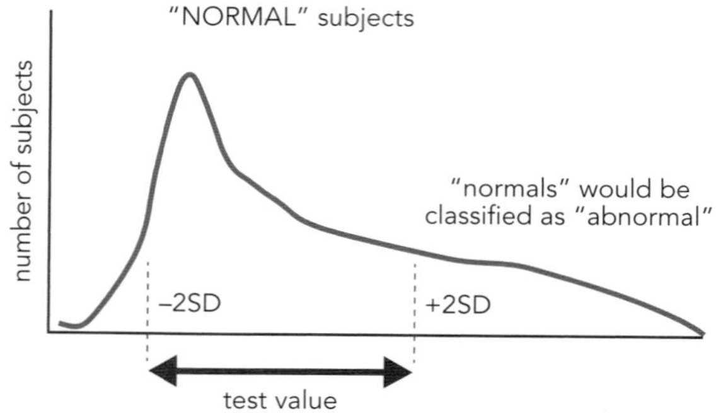

Figure 1.11 Skewed reference ranges.
The diagram shows that when the results in a healthy population have a nonparametric distribution, in this case a positive skew, the reference range cannot be determined using a simple mean ± two standard deviations (2SD). In these cases, the true 95% confidence limits have to be determined to calculate the reference range.

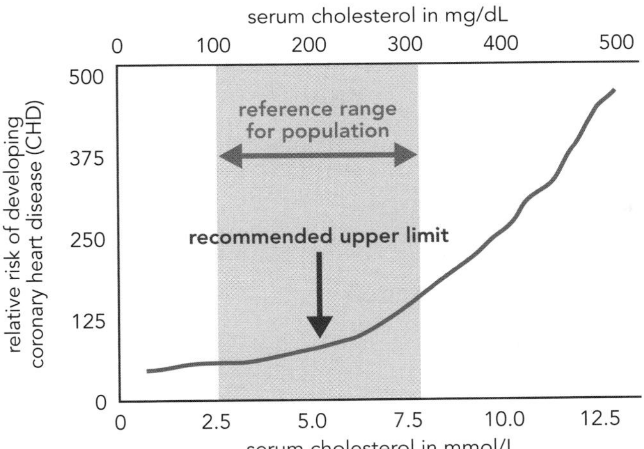

Figure 1.12 Not using the reference range.
The figure shows the relative risk of developing coronary heart disease (CHD; *y* axis) compared to the serum concentration of cholesterol (*x* axis). Statistically, the higher the concentration of serum cholesterol, the greater the risk of developing CHD. Certain populations have reference ranges where the upper 97.5% confidence interval is much higher than the recommended upper level for cholesterol, due to diet and other factors. In this example, an individual with a serum cholesterol at the upper end of the reference range (7.8 mmol/L or 300 mg/dL) would have a much higher risk of developing CHD than with one of the quoted recommended upper levels (5.2 mmol/L or 200 mg/dL). In these situations, quoting a clinical reference range is not clinically useful, and referral to an ideal range or referral to guidelines would be preferable.

later chapters, the use of reference ranges has proved extremely useful in the investigation of disease, particularly using serum samples. Consider, for example, the determination of tissue damage in acute myocardial infarction. Cardiac-specific proteins, troponins, are released from heart cells into the circulation in low concentration as part of the homeostatic process of cell turnover. When cell damage occurs, as during a myocardial infarct, cell proteins are released into the blood and there is a transient abnormal rise in troponins which exceeds the background normal range and is diagnostic of infarction. Perversely, there are also cases where the reference range for an apparently healthy population (one showing no obvious signs or symptoms of disease) may not be ideal and a value in the normal range might indicate a risk factor for disease. For example, the normal range for blood cholesterol concentration in Western societies probably constitutes a risk factor for coronary heart disease and reference ranges for cholesterol are now rarely quoted; rather, ideal target ranges and relative risk factors are proposed (**Figure 1.12**).

1.9 HOW DOES CLINICAL CHEMISTRY OPERATE?

The range of institutions involved in the delivery of clinical chemistry is extremely broad. State-of-the-art research laboratories involved in the development of new assays are quite different from routine hospital laboratories providing a day-to-day service. Even hospital laboratories vary considerably, from large factory-like units with wide analytical repertoires processing millions of tests per year to smaller units processing a few tests per day. In addition of course, some automated tests may be carried out at the bedside, in the ward, or at home by a doctor or nurse or sometimes the patients themselves (so-called point-of-care testing).

A typical large, 1000-bed general hospital in Europe or North America processes over 500,000 samples and carries out over 5,000,000 tests per year. This can be achieved only through the use of large-scale automation, robotics, and sophisticated computing techniques. Many such hospital laboratories are able to perform over 300 different tests using a wide variety of analytical techniques including:

- Photometrics (simple color changes and absorption in the ultraviolet range)
- Enzyme assays
- Ion-selective electrodes (ISE)

- Gas chromatography (GC)
- Mass spectrometry (MS)
- High performance liquid chromatography (HPLC)
- Electrophoresis
- Immunoassays
- Polymerase chain reaction (PCR)
- Atomic absorption

Many of these techniques have been automated using large-scale analyzers, which combine sample processing with analysis to produce total laboratory automation systems.

1.10 QUALITY ASSURANCE

The concept of quality assurance is embodied in two questions related to accuracy and precision: Are you measuring what you think you are measuring? If you measured it again tomorrow, would the result be the same as today? The ability to ensure that the correct result is reported on the correct patient in the right place at the right time is key to the working practice of a clinical chemistry laboratory. Failure to comply with good quality assurance may lead to disastrous consequences including misdiagnosis, mistreatment, and, in the worst-case scenario, death of a patient.

Accuracy

Accuracy concerns the specificity of a test for a given analyte and the reliability of that test when used in different laboratories. For example, in the measurement of blood glucose, a high degree of accuracy implies that the test measures only glucose and not potential interfering molecules such as other reducing sugars of similar structure. Furthermore, when the test is used in different laboratories, the same result is found for a given standard glucose solution. There is a high degree of accuracy in tests for many of the common analytes and reference ranges for these analytes are common across a number of laboratories. However, for more complex assays, particularly immunoassays, methods may vary between laboratories and result in different reference ranges between individual laboratories. In such cases, a result from a patient must be compared with the reference range from the hospital in which the analyte was measured. Occasionally, this may be the case for a common analyte such as creatinine. It could be argued that accuracy might not be critical if a routine method in widespread use generated results for a standard reference sample that were consistent and for which there was good agreement between laboratories. However, comparison of accuracy between laboratories becomes more of a problem for assays for which there is no standard reference material and for which interfering molecules have not been identified, for example assays for some peptides and tumor markers. In these cases, misleading results may be produced and inappropriate treatment may ensue.

Precision

The practice of clinical chemistry involves the measurement of a particular analyte and comparison of the result with a reference range or with a measurement made hours, days, months, or even years earlier. The precision of a test is a measure of the reliability of the test from day to day and can be determined by repetitive measurements of a reference standard assayed with the patient samples. Obviously, for diagnosis and monitoring response to treatment, high precision is of the essence.

1.11 QUALITY CONTROL

Since the production of a wrong result can have serious consequences, it is essential that results are as accurate and precise as possible, and much time and expense is spent on quality control—both internal (IQC) and external quality control (EQC)—to monitor this. Commercial quality control samples of known composition are available for many common analytes, which allows individual laboratories to determine their own accuracy and precision. With less common analytes, such material may not be available and laboratories may obtain reference material and prepare their own in-house quality control standards or reach a consensus on values with other laboratories.

Analytical variation

Most IQC checks on precision are based upon the analytical variation of a test. This is often quoted as an analytical coefficient of variation (CV) which is expressed as

$$\mathbf{CV} = \frac{\mathbf{SD \times 100}}{\mathbf{Mean}} \qquad \text{(Equation 1.1)}$$

where the mean and standard deviation are derived from multiple assays of samples from a reference standard. Traditionally, where the assays are performed on a sample from a single batch, the CV is referred to as an intrabatch CV; when samples are taken from different batches, the CV is referred to as an interbatch CV. With the increasing use of auto-analyzers and other automated techniques, where analyses are performed under constant conditions, it is more relevant to refer to within-day and between-day CVs.

The performance of any individual assay can then be monitored regularly using simple IQC rules. Although there are many excellent quality control packages in use, most work on the basis that essentially batches would be accepted if the IQC samples fell within two standard deviations of the expected mean value. Such a value can be achieved by chance. If the IQC result falls outside of this range the batch can be rejected, and corrective action such as recalibration can be taken. The IQC shows the degree of imprecision and the degree of bias away from the true result.

Many routine assays have a very low CV, less than 1%. So, for example, in the case of calcium where the control value was 2.00 mmol/L (8.00 mg/dL), the batch would be considered to be acceptable if the result for the IQC sample was between 1.96 and 2.04 mmol/L (7.84 and 8.16 mg/dL). This degree of precision is very important for analytes like calcium, where the physiological concentration is kept within a very tight range and where small changes can be important clinically. Many other assays in routine use are more imprecise, with a CV in excess of 5% or even 10%. Imagine trying to use a calcium assay with a CV of 10% with a control value of 2.50 mmol/L (10.00 mg/dL). Acceptance limits of ±2SD would mean that IQC values ranging between 2.00 and 3.00 mmol/L (8.00 and 12.00 mg/dL) could be accepted. Considering the fact that the reference range for calcium is often given as 2.20 to 2.60 mmol/L (8.80 to 10.40 mg/dL), this degree of imprecision is not acceptable.

External quality assurance

Although internal quality control enables laboratories to keep their laboratory tests performing within certain limits of precision and bias, it by no means enables total quality assurance, nor does it enable ready comparison with other laboratories. The majority of clinical laboratories participate in external quality assurance (EQA) schemes for most of the tests that they undertake. Essentially, these schemes send samples to participating laboratories; these samples have predetermined concentrations of specific analytes, but the participating laboratories do not know the values. After the samples have been analyzed, the quantitative and/or qualitative results, along with interpretations in some cases, are sent back to the coordinating center for the scheme. Laboratories can then be scored for accuracy and precision, and ranked for performance. Laboratories that perform poorly over time can thus be identified and action taken to improve performance. This process, when performed correctly, should assure the quality of a laboratory and should give reassurance to the physician and patient that the correct results and interpretative information are being produced.

1.12 ACCREDITATION

As pathology services have developed in general, there has been an increasing awareness of the need for formal assessment of laboratories to ensure they are operating appropriately. Early schemes for pathology were Clinical Laboratory Improvement Amendments (CLIA) and Clinical Pathology Accreditation (CPA), which both were operating along the lines of International Organization for Standardization (ISO), Good Laboratory Practice, and Food and Drug Administration (FDA) guidelines.

GLUCOSE

CHAPTER

2

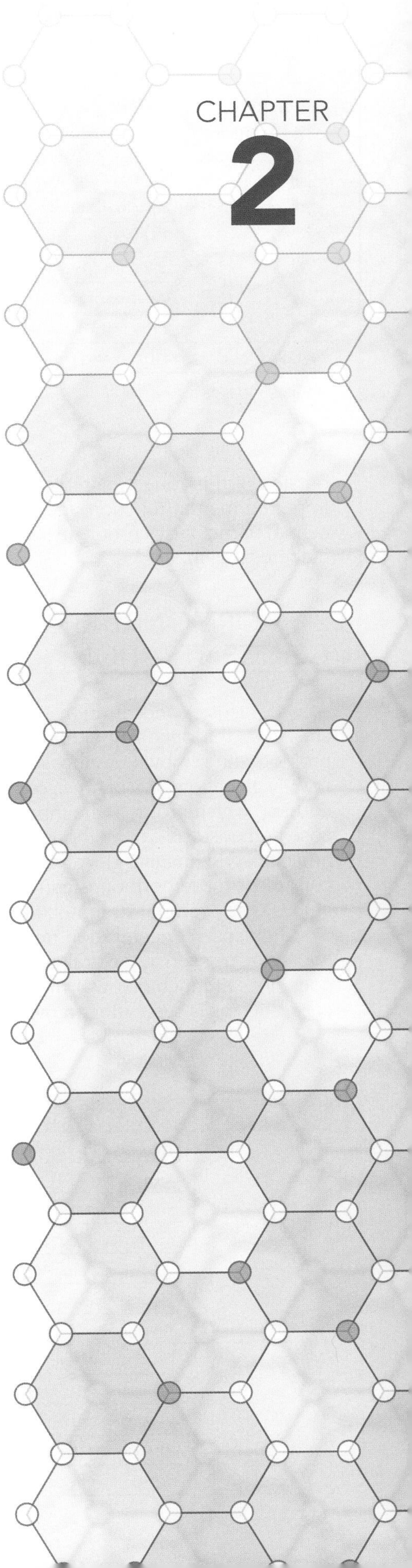

IN THIS CHAPTER

Glucose is the single common substrate that can be used by all tissues as an energy source. Although most tissues can also use fatty acids for energy, nervous tissue, red cells (erythrocytes), kidney, and, in normal circumstances, brain have an obligatory requirement for glucose (**Table 2.1**). It is for this reason that the maintenance of plasma glucose within a narrow concentration range (**Table 2.2**) is fundamental to health. The normal distribution range of blood glucose in the UK population is 4.1–5.9 mmol/L (74–106 mg/dL) with a slightly higher range for adults of 60 years and above of 4.4–6.4 mmol/L (80–115 mg/dL). Pathological consequences ensue when the glucose concentration lies

TABLE 2.1 Major fuel sources consumed by tissues after a meal

Tissue	Immediately postprandial	2–5 hours postprandial
Brain	Glucose	Glucose
Red blood cells	Glucose	Glucose
Nervous tissue	Glucose	Glucose
Kidney	Glucose	Glucose (fatty acids)
Skeletal muscle	Glucose	Fatty acids (glucose)
Liver	Glucose	Fatty acids (glucose)
Heart	Glucose	Fatty acids (glucose)
Lung	Glucose	Fatty acids (glucose)
Gastrointestinal tract	Glucose	Fatty acids, glucose
Adipose	Glucose	Fatty acids (glucose)

Substrates in parentheses are secondary and minor fuel sources 2–5 hours postprandial. During longer-term fasting, all tissues except liver, kidney, adipose, and red blood cells adapt to using ketone bodies as the major metabolic fuel for energy production.

TABLE 2.2 Normal ranges for glucose, HbA1c, and associated parameters

		SI units	Conventional units
Plasma	Glucose (fasting)	3.9–6.0 mmol/L	70–110 mg/dL
	Insulin (fasting)	14–140 pmol/L	2.0–20 μIU/mL
	C-peptide (fasting)	0.17–0.90 nmol/L	0.5–2.7 ng/mL
	3-Hydroxybutyrate	<0.1 mmol/L	<1 mg/dL
	Acetoacetate	<0.1 mmol/L	<1 mg/dL
	Free fatty acids	0.3–0.9 mmol/L	8–25 mg/dL
Whole blood	HbA_{1c}	<48 mmol/mol Hb	<6.5% of Hb

HbA_{1c}, glycated hemoglobin (Hb).

outside of this range for prolonged periods; hypoglycemia is characterized by impaired mental function and severe hypoglycemia by coma, while hyperglycemia, particularly chronic hyperglycemia, promotes the insidious pathology of diabetes mellitus.

2.1 HORMONAL REGULATION OF BLOOD GLUCOSE CONCENTRATION

Blood glucose concentration is maintained within the normal range (homeostasis) by the opposing actions of anabolic hormones, mainly insulin, which promote the removal of glucose from the circulation, and catabolic hormones, mainly glucagon and, in times of stress, also epinephrine, growth hormone, and cortisol, which promote the input of glucose into blood. For the most part, glucose homeostasis can be thought of as a balance between the actions of insulin and glucagon (insulin–glucagon ratio; **Table 2.3**). The effectiveness of the acute control by these hormones can be seen from the observation that while blood glucose may reach >10 mmol/L (>180 mg/dL) after a carbohydrate-rich meal, blood glucose concentration falls to less than 6.7 mmol/L (121 mg/dL) within two hours as seen in the oral glucose tolerance test. Furthermore, blood glucose does not fall much below 4 mmol/L (72 mg/dL) in the absence of dietary glucose, such as is seen after an overnight fast.

TABLE 2.3 Actions of insulin and glucagon on carbohydrate metabolism

Tissue	Metabolic pathway	Insulin effect	Glucagon effect
Liver	Glycolysis	Stimulatory	Stimulatory
	Glycogen synthesis	Stimulatory	Inhibitory
	Gluconeogenesis	Inhibitory	Stimulatory
	Lipolysis	Inhibitory	Stimulatory
	Lipogenesis	Stimulatory	Inhibitory
	Ketogenesis	Inhibitory	Stimulatory
Muscle	Glucose uptake	Stimulatory	Limited or no action
	Glucose metabolism	Stimulatory	Limited or no action
	Glycogen synthesis	Stimulatory	Limited or no action
	Amino acid uptake	Stimulatory	Limited or no action
	Protein synthesis	Stimulatory	Limited or no action

2.2 GLUCOSE STORAGE AND METABOLISM

Glucose is stored as glycogen, a macromolecular branched polymer, primarily in liver and muscle. Storing glucose units as a macromolecule eliminates the problem of increased osmotic pressure that would arise if glucose were stored as individual molecules in the cell. The highly branched glycogen molecule consists of chains of glucose monomers linked α1:4, with branch points linked α1:6, and a single reducing terminal glucose (**Figure 2.1**). Such a structure presents a multitude of substrate sites for the enzymes that are involved in both its hydrolysis (by glycogen phosphorylase) and its synthesis (by glycogen synthase). The amount of glycogen present in the tissue at any one time represents a balance between the activities of these two enzymes, both of which are under hormonal control.

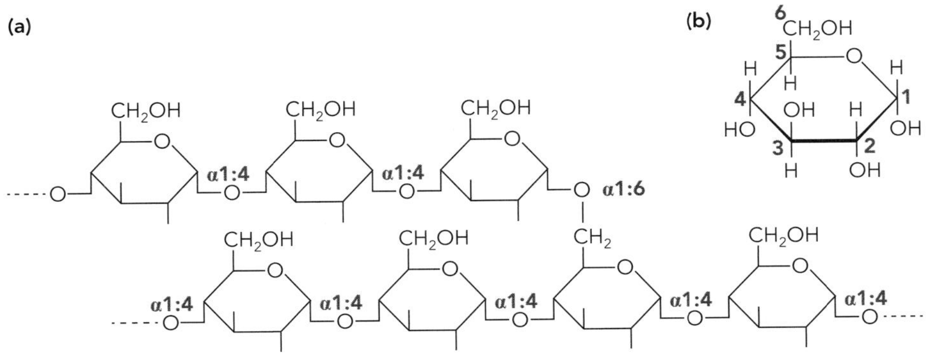

Figure 2.1 Partial structure of glycogen showing a branch point. (a) Glucose monomers are linked α1:4 intrachain and α1:6 interchain (branch point). (b) Structure of glucose with carbon numbering in red; the reducing carbon is C1. Hydrogen atoms are included in the monomeric structure but have been omitted from the polymer for clarity.

Quantitatively, muscle stores more glycogen *in toto* than liver—approximately 250 g in muscle and approximately 75 g in liver—reflecting the relative tissue masses, but the concentration is greater in liver (0.5–1% in muscle, 3–5% in liver). However, only liver can generate free glucose from the breakdown of stored glycogen; muscle lacks the key enzyme, glucose 6-phosphatase, required for this. Thus liver alone is capable of releasing glucose into the circulation for use by other tissues. Its glycogen content will thus fluctuate throughout the day as it releases glucose to maintain homeostasis and replenishes this lost glucose from dietary intake. Glucose enters and leaves the liver via a high-capacity, low-affinity transporter, GLUT2. In contrast, muscle glycogen is used solely to provide energy for contraction and its concentration will only fall during exercise and again be replenished from dietary glucose during the recovery period.

The five metabolic pathways involving glucose are described in **Table 2.4** and shown diagrammatically in **Figure 2.2**. The two pathways regulating the

TABLE 2.4 Pathways involving the metabolism of glucose

Metabolic pathway	Role of pathway in metabolism
Glycogenesis	Storage of glucose as glycogen in liver and muscle; glucose is first activated to UDP-glucose
Glycogenolysis	Hydrolysis of glycogen producing glucose 1-phosphate (G1P). G1P is converted to glucose 6-phosphate (muscle and liver) and then glucose (liver)
Glycolysis	Metabolism of six-carbon glucose to three-carbon pyruvate (oxidized in the mitochondrion to CO_2 and water)
Pentose phosphate pathway	Metabolism of glucose to five-carbon sugars (for nucleic acids) plus generation of reducing equivalents (NADPH) for anabolism
Gluconeogenesis	Synthesis of glucose from non-carbohydrate precursors: glycerol, lactate, and amino acid carbon skeletons

UDP-glucose, uridine diphosphate glucose.

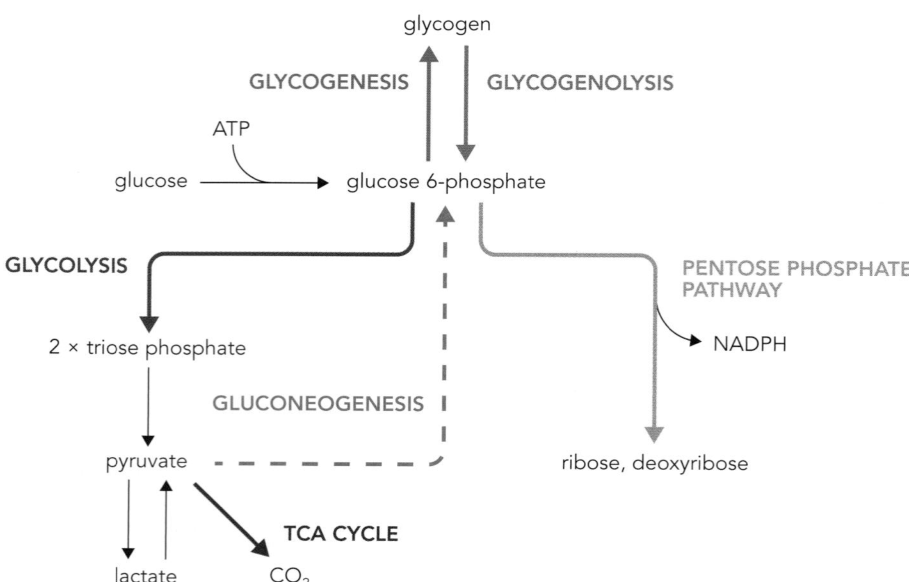

Figure 2.2 Pathways of glucose metabolism in liver.
Free glucose entering liver from blood is phosphorylated to glucose 6-phosphate by hexokinase before further metabolism via one of three pathways to: (i) pyruvate (by glycolysis), which is reduced to lactate anaerobically or to carbon dioxide (CO_2) aerobically via the tricarboxylic acid (TCA) cycle; (ii) the five-carbon sugars ribose and deoxyribose (by the pentose phosphate pathway), which are essential for nucleotide synthesis; or (iii) glycogen (by glycogenesis). Glucose 6-phosphate may also be synthesized from three-carbon precursors (gluconeogenesis) and from stored glycogen (glycogenolysis).

amount of glucose stored as glycogen in liver and muscle, glycogenesis and glycogenolysis, exhibit a reciprocal relationship which is controlled mainly by insulin and to a lesser extent by glucagon in liver and epinephrine in muscle. Thus the plasma insulin–glucagon ratio is a key factor in controlling the two processes.

Glycogenesis

The pathway of glycogenesis (**Figure 2.3**) is prominent following ingestion of a carbohydrate-rich meal. Diet-derived glucose and amino acids and also gastrointestinal peptides stimulate the release of insulin from the β cells of the pancreas, and insulin promotes the synthesis of energy stores including glycogen and triacylglycerol. Glycogenesis begins with the phosphorylation of free glucose by the enzymes glucokinase and hexokinase in liver and hexokinase in muscle, using adenosine triphosphate (ATP) as the phosphate donor. The product of this reaction is glucose 6-phosphate, which inhibits hexokinase but not glucokinase, and this difference has implications in the resynthesis of glycogen from dietary glucose following the depletion of hepatic glycogen stores brought about by fasting. Glucokinase (K_m 5–7 mmol/L) is a glucose-specific, inducible enzyme whose activity increases when the portal blood glucose concentration rises above 5 mmol/L (90 mg/dL).

Glucose 6-phosphate is converted to glucose 1-phosphate by the enzyme phosphoglucomutase before being activated to uridine diphosphate glucose (UDP-glucose) by uridine diphosphate pyrophosphatase. UDP-glucose is the donor of glucose to the growing glycogen molecule in a reaction catalyzed by glycogen synthase. The ability of liver to store glucose as glycogen is primarily due to the rapid activation of this enzyme. Glycogen synthase adds glucose in

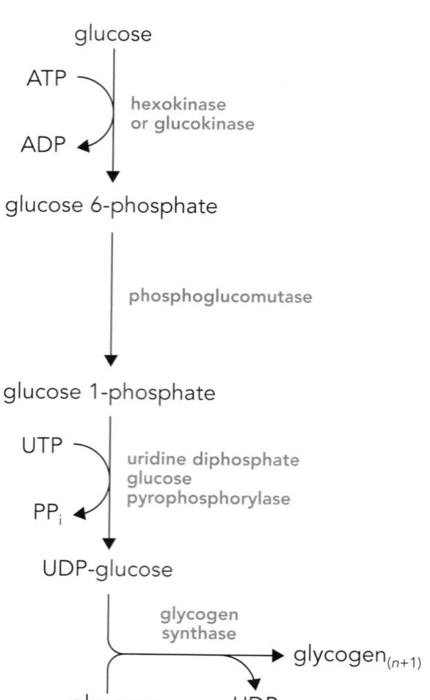

Figure 2.3 Glycogenesis in liver.
Glucose is activated to glucose 6-phosphate which undergoes isomerization to glucose 1-phosphate. The donor of glucose to the growing glycogen chain is uridine diphosphate glucose (UDP-glucose) formed from UTP and glucose 1-phosphate. PP_i, pyrophosphate; *n*, the number of glucose monomers in the polymeric glycogen molecule.

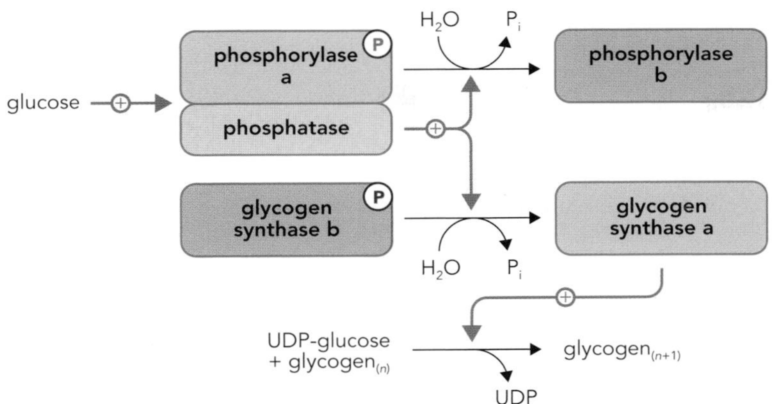

Figure 2.4 Stimulation of hepatic glycogen synthase by glucose. After a meal, the hepatic portal vein carries a high concentration of glucose to the liver. This glucose is rapidly incorporated into glycogen through activation of glycogen synthase. Glycogen synthase, which is inactive in the phosphorylated form (glycogen synthase b), is activated by the same phosphatase that inactivates glycogen phosphorylase a by converting it to inactive phosphorylase b. The phosphatase itself is inactive when complexed with phosphorylase a, but at intracellular concentrations of glucose >5 mmol/L (>90 mg/dL), glucose binds to the phosphorylase a–phosphatase complex and allows the phosphatase to dephosphorylate phosphorylase a, forming inactive phosphorylase b; this also releases the phosphatase in order to activate glycogen synthase b to glycogen synthase a. Free glucose can thus stimulate its own incorporation into glycogen. P_i, inorganic phosphate; *n*, the number of glucose monomers in the polymeric glycogen molecule.

an α1:4 linkage and gradually builds a chain of α1:4-linked glucose units. When the chain length reaches eight glucose units, a branching enzyme, a transglycosylase, transfers some of the chain to form an α1:6 linkage, now providing two chains for subsequent addition of glucose. In this way, the multibranched nature of the mature glycogen molecule is achieved. Glycogen synthase is the key regulatory enzyme of glycogenesis and its activity is increased by insulin released from the pancreas in response to a rise in blood glucose. Insulin increases the expression of genes coding for enzymes of carbohydrate storage as well as protein tyrosine phosphorylation which, by activation of GTPase, phosphodiesterase, and phosphoprotein phosphatases, inhibits the cyclic adenosine monophosphate (cAMP)-activated pathway of glycogenolysis. Insulin also inhibits the pancreatic secretion of glucagon, the hormone that activates glycogen phosphorylase.

Glycogen synthase exists in an inactive phosphorylated form, glycogen synthase b, and activation involves dephosphorylation to glycogen synthase a by a specific phosphatase, the same enzyme that inactivates glycogen phosphorylase a by a similar dephosphorylation. In the liver, the phosphatase forms an inactive complex with phosphorylase a; as the intracellular concentration of glucose rises above 5 mmol/L (90 mg/dL), glucose binds to phosphorylase a which acts as an intracellular glucose receptor, changing its shape such that it becomes susceptible to the action of the phosphatase (**Figure 2.4**). Dephosphorylation of phosphorylase a to inactive phosphorylase b releases the phosphatase, which can now activate glycogen synthase b by converting it to glycogen synthase a. Free glucose in the liver is thus able to stimulate its own incorporation into glycogen. This is enhanced by raised insulin and the stimulatory action of glucose on hepatic glucokinase. A similar cAMP-independent mechanism involving activation of the phosphatase that is common to the glycogen synthase–phosphorylase system exists in muscle. Insulin also stimulates the uptake of glucose into muscle and adipose tissue via the GLUT4 transporter system. It has no effect on glucose uptake into liver, however, where the rate of uptake reflects the glucose concentration gradient across the plasma membrane. Interestingly, the concentration of insulin required to stimulate glycogenesis is lower in muscle than in liver.

Glycogenolysis

Glycogenolysis (**Figure 2.5**) involves the release from glycogen of terminal glucose (nonreducing) units as glucose 1-phosphate; this is termed phosphorolysis and is catalyzed by the action of glycogen phosphorylase a. The phosphate donor is inorganic phosphate present in the cytosol. Different phosphorylases are present in liver, kidney, and muscle. Under catabolic physiological conditions where plasma epinephrine is raised, glucose is mobilized rapidly

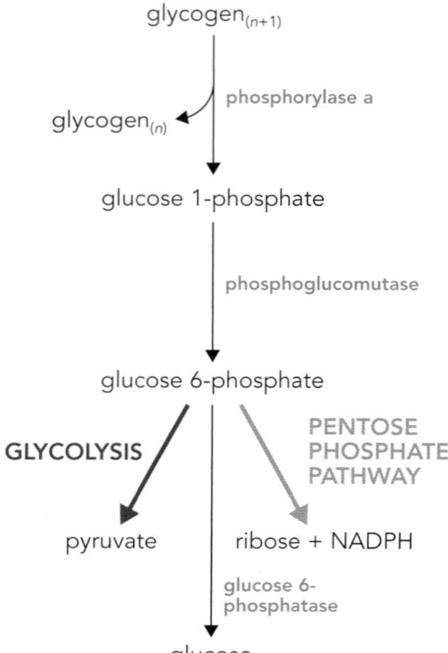

Figure 2.5 Glycogenolysis and the possible fates of glucose 6-phosphate in liver.
Glycogen phosphorylase hydrolyzes glucose monomers as glucose 1-phosphate from the nonreducing ends of the glycogen macromolecule. Glucose 1-phosphate is converted to glucose 6-phosphate before metabolism via glycolysis or the pentose phosphate pathway. Only liver contains glucose 6-phosphatase which hydrolyzes glucose 6-phosphate to free glucose for the circulation in times of decreased intake of dietary glucose. *n* is the number of glucose monomers in the polymeric glycogen molecule.

from glycogen in skeletal muscle and also in the liver when epinephrine levels are very high, whilst raised glucagon stimulates glycogenolysis only in liver (**Table 2.5**). The molecular mechanism of the stimulation of glycogen phosphorylase by epinephrine in muscle has been studied in detail and a similar mechanism obtained for the action of glucagon in liver. A rise in the plasma concentration of glucagon increases the binding of the hormone to its specific receptor on the plasma membrane of the target tissue (liver) causing activation of the intracellular synthesis of cAMP via the adenylate cyclase system, shown in outline in **Figure 2.6**. Raised intracellular cAMP initiates a phosphorylation cascade resulting eventually in the phosphorylation of the less active form of phosphorylase, phosphorylase b, to the more active phosphorylase a. Thus, an increase in the intracellular concentration of cAMP simultaneously increases glycogen breakdown and decreases glycogen synthesis (by inactivating glycogen synthase).

The effect of epinephrine on glycogenolysis in liver is not mediated by cAMP but via a rise in intracellular free calcium concentration which effects a direct activation of phosphorylase b kinase, the enzyme that phosphorylates inactive phosphorylase b to active phosphorylase a. A similar mechanism is found in skeletal muscle, where a rise in cytosolic calcium as a result of neural stimulation can activate phosphorylase b kinase and promote glycogen breakdown independently of stimulation by epinephrine. In this case, Ca^{2+}-induced glycogenolysis depends on calcium binding to the calmodulin subunit of the kinase and is linked to contraction. The simultaneous phosphorylation of glycogen synthase under these conditions causes its inactivation and minimizes the futile cycling of glucose. The active conformation of phosphorylase is stabilized by phosphorylation even in the absence of an allosteric activator and only glucose is a significant inhibitor of the phosphorylated form.

Glucose 1-phosphate released from glycogen by the action of glycogen phosphorylase a then isomerizes to glucose 6-phosphate under the action of

TABLE 2.5 Actions of hormonal changes on glycogenolysis and glycolysis in liver and muscle

Tissue	Hormone change	Glycogen to glucose	Glucose to pyruvate
Liver	Insulin decrease	No effect	No effect
	Glucagon increase	Major stimulus	Inhibition
	Epinephrine increase	Some effect at high concentration of hormone	No effect
Muscle	Insulin decrease	No effect	No effect
	Glucagon increase	No effect	No effect
	Epinephrine increase	Major stimulus	Major stimulus

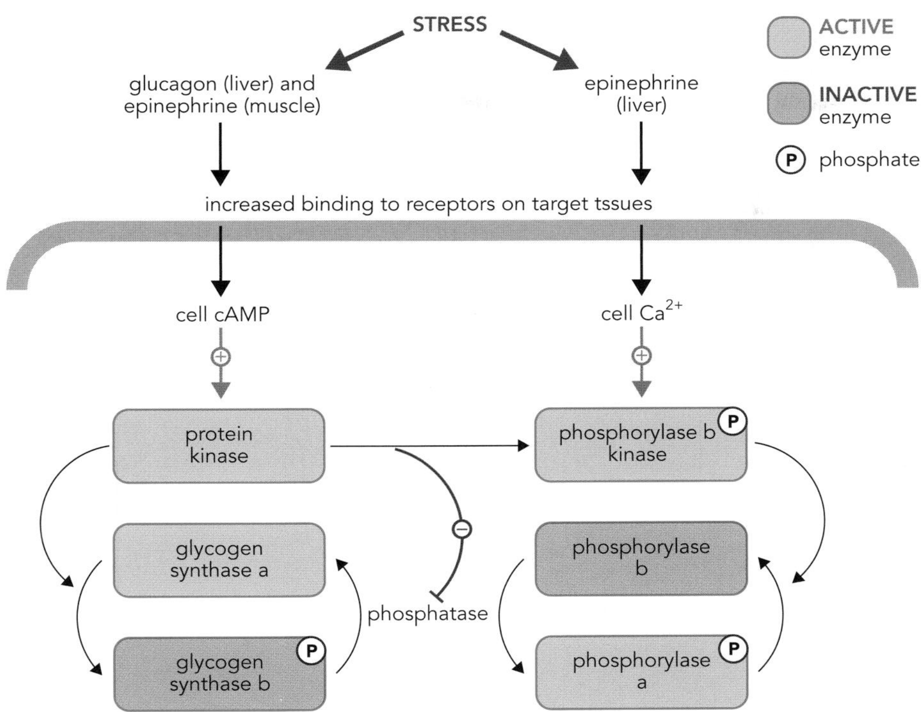

Figure 2.6 Activation of glycogen phosphorylase and inhibition of glycogen synthase by glucagon (liver) or epinephrine (muscle). A rise in the plasma concentration of glucagon during stress increases its binding to receptors on target tissues, activating the synthesis of intracellular cAMP. This molecule initiates a phosphorylation cascade that eventually results in the net phosphorylation of less active glycogen phosphorylase b to more active glycogen phosphorylase a, and of active glycogen synthase a to less active glycogen synthase b. Epinephrine has similar actions in skeletal muscle. The action of epinephrine in liver, however, is not mediated via cAMP but via a rise in intracellular Ca^{2+}.

phosphoglucomutase (see Figure 2.5). This enzyme is present predominantly in liver, kidney, and muscle. Release of free glucose from glucose 6-phosphate is mediated by the enzyme glucose 6-phosphatase, which is present only in liver and kidney. The lack of this enzyme in muscle prevents release of free glucose from the tissue. Phosphorylase a hydrolytic activity is specific for α1:4 links and the enzyme does not hydrolyze glucose monomers linked α1:6. It will hydrolyze α1:4-linked glucose residues along a glycogen branch to within 3–4 residues from an α1:6 branch point, where it is unable to reach the bond to be cleaved. A debranching enzyme (transglycosylase) cleaves the short sequence at the α1:6 site and transfers it to the end of a near-neighbor α1:4 chain, leaving a single glucose at the branch. This residue is cleaved as free glucose by an α1:6-glucosidase, leaving another α1:4 chain for phosphorylase activity. The transglycosylase and glucosidase activities, which release small amounts of free glucose from exposed α1:6 branch points during glycogen breakdown, are not regulated directly by hormone action. Approximately 90% of glucose stored in glycogen is released as glucose 1-phosphate with the remainder as free glucose.

Glycolysis

Glycolysis (**Figure 2.7**), in which six-carbon glucose is metabolized to two molecules of three-carbon pyruvate, occurs in all tissues and can be thought of as anabolic for lipid synthesis in liver and white adipose tissue or catabolic for energy production in tissues such as exercising muscle, brain, and red cells (erythrocytes). The activity of the glycolytic pathway is high in those tissues that depend on glucose as their major energy source (for example brain, erythrocytes, and kidney) although most tissues will also metabolize glucose via this route after a carbohydrate-rich meal. Only small amounts of ATP can be generated from glycolysis but for mature erythrocytes, which lack mitochondria, this represents their only means of generating ATP. In other tissues, much more energy is generated by subsequent metabolism of pyruvate in mitochondria, as shown in the overview of oxidation in **Figure 2.8**. The entry of glucose into resting muscle and also white adipose tissue is insulin dependent, but glucose

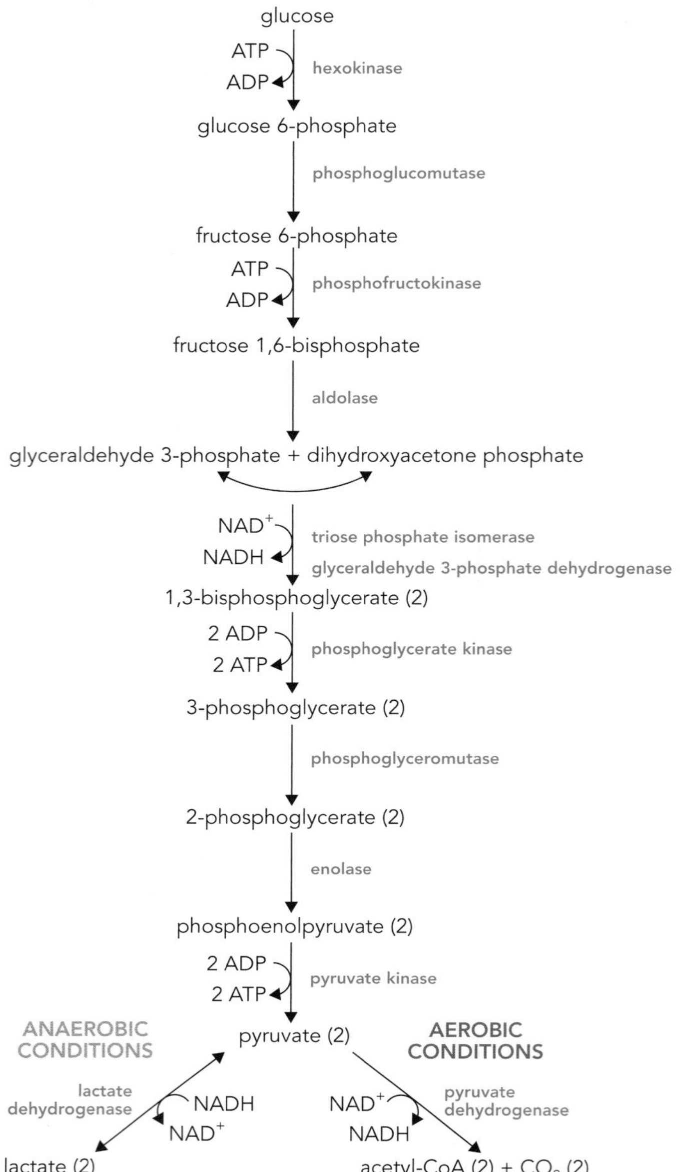

Figure 2.7 The glycolytic pathway. The metabolism of six-carbon glucose to three-carbon pyruvate occurs in the cell cytosol. Two molecules of the intermediates from 1,3-bisphosphoglycerate to lactate are formed per molecule of glucose entering the pathway. Under aerobic conditions pyruvate enters the mitochondrion and is oxidized to CO_2 whilst under anaerobic conditions it is reduced to lactate.

entry into working muscle is less dependent on insulin and glycolysis can proceed from exogenous glucose during exercise, even under catabolic conditions.

The major regulatory step of the glycolytic pathway is at the level of phosphofructokinase (PFK; **Figure 2.9**), which catalyzes the step prior to the cleavage of six-carbon sugar to two three-carbon molecules. This enzyme is inhibited by allosteric effectors that reflect the energy charge ratio (ATP–AMP ratio) of the cell. For example, with a plentiful supply of energy substrate, the level of ATP will rise relative to AMP leading to an inhibition of PFK and, as a consequence, inhibition of glycolysis. Raised levels of ATP also slow the tricarboxylic acid (TCA) cycle, raising the cytosolic concentration of citrate, another allosteric inhibitor of PFK. Inhibition of PFK will in turn lead to an increase in glucose 6-phosphate, an inhibitor of hexokinase, and this will decrease the rate of glucose entry into the glycolytic pathway.

The most potent activator of glycolysis in the liver is fructose 2,6-bisphosphate, a molecule formed when circulating glucose is abundant, after a meal for example. This molecule also acts allosterically to activate PFK by antagonizing

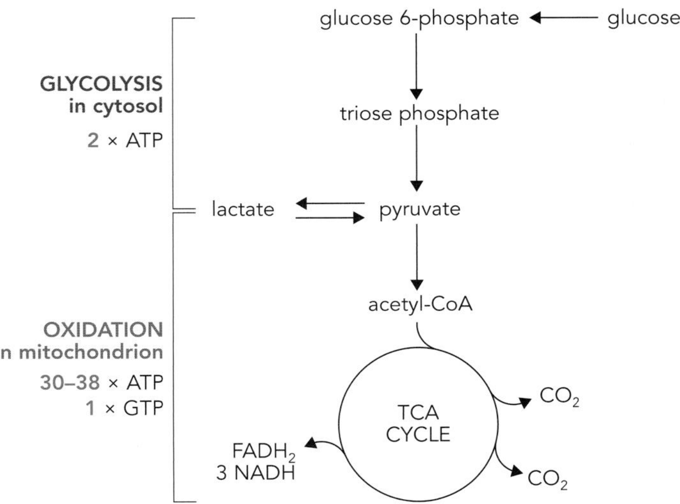

Figure 2.8 Overview of glucose oxidation.
There is a net formation of two molecules of ATP per molecule of glucose entering the glycolytic pathway. ATP is formed in two reactions—catalyzed by phosphoglycerate kinase and pyruvate kinase (four ATP in total per molecule of glucose)—while one molecule of ATP is consumed by each of the hexokinase and phosphofructokinase steps. On entering the mitochondrion, pyruvate is oxidized to acetyl-CoA which in turn is oxidized to CO_2 in the TCA cycle with the formation of reducing equivalents in the form of reduced co-factors NADH and $FADH_2$. These molecules, NADH and $FADH_2$, donate electrons to the electron transport chain on the inner mitochondrial membrane and drive the synthesis of ATP by oxidative phosphorylation: three molecules of ATP are generated per pair of electrons from NADH and two molecules per pair of electrons from $FADH_2$. Red numbers indicate the yield of ATP molecules per molecule of glucose metabolized: two in the cytosol versus 30–38 via mitochondrial oxidation of pyruvate. A single molecule of GTP is generated by substrate-level phosphorylation at the succinyl CoA synthase step of the cycle.

the inhibitory effects of ATP and citrate. Hepatic glycolysis is inhibited at two points in the pathway by the raised circulating glucagon concentration of the catabolic state: (i) at pyruvate kinase and (ii) at PFK. Glucagon action decreases the intracellular concentration of fructose 2,6-bisphosphate thereby decreasing PFK activity; glucagon also stimulates cAMP-activated phosphorylation of pyruvate kinase, producing an enzyme which is inhibited by the prevailing concentration of phosphoenolpyruvate.

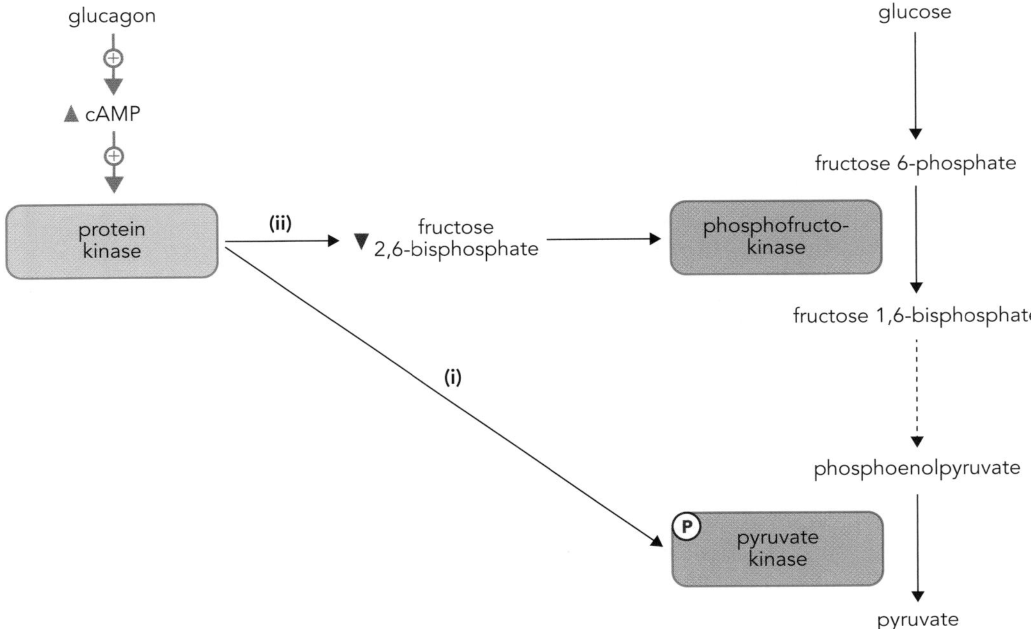

Figure 2.9 Inhibition of hepatic glycolysis by glucagon.
The glucagon-stimulated elevation of cAMP in liver activates protein kinases which are responsible for the inhibition of glycolysis at two sites in the pathway: (i) by phosphorylating pyruvate kinase so that it is inhibited by the prevailing concentration of phosphoenolpyruvate and (ii) by decreasing the intracellular concentration of fructose 2,6-bisphosphate. Phosphofructokinase is a major regulatory enzyme of the glycolytic pathway and fructose 2,6-bisphosphate is the most potent activator of this enzyme, antagonizing the allosteric inhibition exerted by ATP and citrate. A decreased fructose 2,6-bisphosphate concentration reduces the activity of phosphofructokinase.

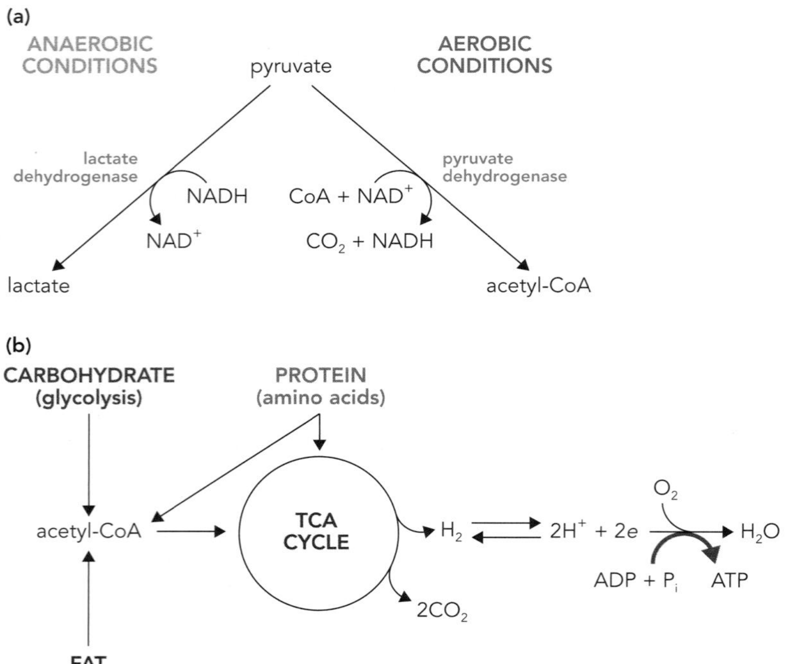

Figure 2.10 Metabolism of pyruvate under aerobic and anaerobic conditions and oxidation of acetyl-CoA in the TCA cycle. (a) Pyruvate is oxidized to acetyl-CoA and CO_2 aerobically and reduced to lactate anaerobically. NADH is derived from glycolysis. Pyruvate dehydrogenase is inhibited by acetyl-CoA. (b) Acetyl-CoA formed from oxidation of pyruvate is further oxidized to CO_2 and H_2O in the TCA cycle in mitochondria. Some of the energy released from this exothermic process drives the synthesis of ATP.

Further metabolism of pyruvate

Under aerobic conditions, where the oxygen supply to tissues is not limited, pyruvate enters the mitochondrion and is eventually oxidized via the TCA cycle to carbon dioxide and water (**Figure 2.10**). The initial step in this process, the oxidative decarboxylation of pyruvate to acetyl-CoA, is irreversible and, because human cells cannot synthesize glucose from two-carbon molecules such as acetate, this represents a net loss to the body of potential carbohydrate reserves. Conservation of three-carbon molecules is essential when the exogenous supply of glucose is limited and under such conditions, fasting for example, the oxidation of pyruvate is tightly regulated. The enzyme that catalyzes this step, pyruvate dehydrogenase (PDH), undergoes a cAMP-independent phosphorylation-dephosphorylation cycle catalyzed by a specific kinase and phosphatase, respectively. In the fed state, raised plasma insulin and raised cytosolic calcium stimulate the phosphatase, thereby increasing the amount of active enzyme and pyruvate oxidation. On the other hand, the kinase that phosphorylates and inactivates PDH is activated by acetyl-CoA and NADH which are produced when tissues, particularly liver, switch to metabolism of fatty acids for energy. In this way, pyruvate oxidation is inhibited by both the decreased insulin concentration and by the increased β-oxidation activity of the catabolic state.

Under anaerobic conditions where the oxygen supply is limited, such as in muscle during high-intensity exercise, or where no mitochondria are present, as in mature erythrocytes, pyruvate is reduced to lactate by the enzyme lactate dehydrogenase. Lactic acid is a waste end product in muscle and erythrocytes and, being a strong acid, is toxic to these tissues. It diffuses out of muscle and erythrocytes and is carried in the blood to the liver. The liver is well-oxygenated through its blood supply via the hepatic artery (see Chapter 3) and converts the lactate back to pyruvate. Pyruvate is either oxidized to CO_2 and water in mitochondria or used as a substrate for glucose synthesis via gluconeogenesis. The process whereby glucose from muscle glycogen is metabolized to lactate and this lactate is used for glucose synthesis in liver is known as the Cori cycle (**Figure 2.11**), named for the husband and wife team who first described it and who were awarded the Nobel Prize in Medicine in 1947 for their work on glycogen metabolism.

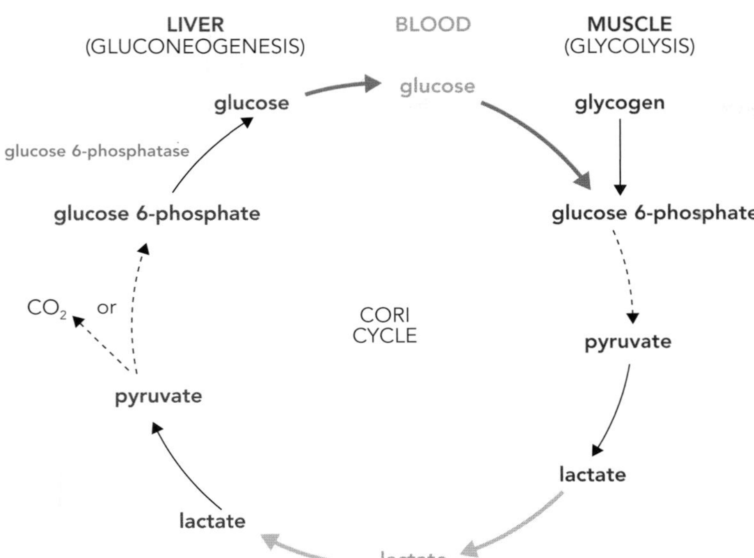

Figure 2.11 The Cori cycle: recycling of muscle glucose. Lactate produced via anaerobic metabolism of glucose derived from muscle glycogen breakdown during exercise, for example, is transported via the blood to liver. Liver, being well vascularized, oxidizes lactate to pyruvate, which can be oxidized to CO_2 or used as a substrate for gluconeogenesis through production of glucose 6-phosphate. Liver also possesses the enzyme glucose 6-phosphatase and can thus release free glucose to the circulation. Glucose can then enter the circulation and return to muscle.

Pentose phosphate pathway

The pentose phosphate pathway (**Figure 2.12**), also known as the hexose monophosphate shunt, is particularly active in rapidly dividing tissues, including activated lymphocytes, and in tissues engaged in reductive biosynthesis, particularly of fatty acids, such as lactating mammary gland and adipose tissue. As its name implies, in this pathway glucose 6-phosphate is metabolized to five-carbon sugars, which may be used as precursors of the ribose and 2-deoxyribose components of nucleic acids. The pathway is also a major source of reducing equivalents in the form of NADPH (reduced nicotinamide adenine dinucleotide phosphate) which is required for the reductive biosynthesis of macromolecules such as fatty acids and cholesterol. NADPH is also the electron donor to the

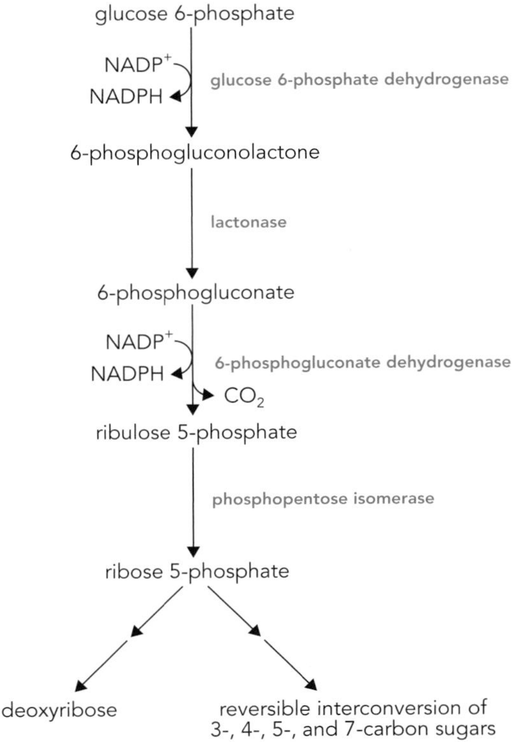

Figure 2.12 The pentose phosphate pathway. In this pathway, also known as the hexose monophosphate shunt or phosphogluconate pathway, six-carbon glucose is converted to five-carbon sugars such as ribose with the release of CO_2. The pathway is also an important source of NADPH (at steps 1 and 3) required for the reductive biosynthesis of fatty acids. Five-carbon sugars are components of ribonucleotides and deoxyribonucleotides.

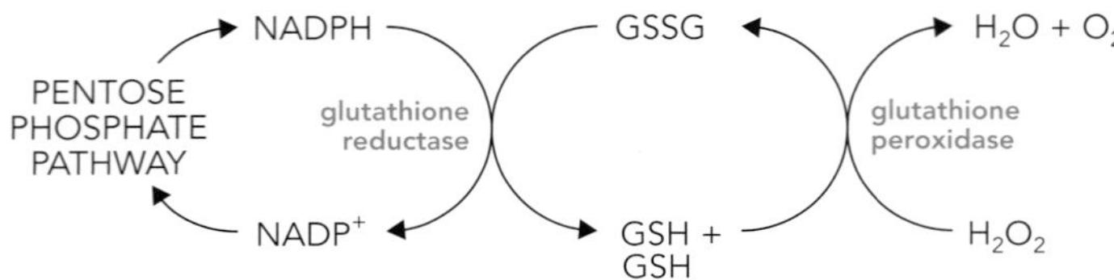

Figure 2.13 The role of the pentose phosphate pathway and glutathione in the reduction of oxidative stress in erythrocytes. The presence of oxygen and reactive oxygen species such as peroxide in erythrocytes causes a stress that promotes the denaturation of membrane proteins and other proteins. Protection against this stress is afforded by the alternative oxidation of reduced glutathione (GSH) to its oxidized form (GSSG). With only a finite amount of glutathione in the cell, this protection is lost when all the glutathione is oxidized. Glutathione reductase, with its co-factor NADPH derived from the pentose phosphate pathway, converts oxidized glutathione to its reduced form to allow continued protection from oxidative stress. Glutathione peroxidase reduces peroxide to water.

hepatic drug-metabolizing system described in Chapter 3. Importantly, high activity of the pentose phosphate pathway is found in activated macrophages where NADPH provides electrons required for the generation of reactive oxygen species, such as superoxide and hydroxyl radical, which are involved in killing bacteria and degradation of macromolecules. In erythrocytes, an important role of NADPH produced from pentose phosphate pathway activity is the maintenance of glutathione in its reduced form to counter the oxidative stress put upon the cells (**Figure 2.13**).

Gluconeogenesis

Gluconeogenesis (**Figure 2.14**) describes the synthesis of new glucose from non-carbohydrate precursors such as lactate (via the Cori cycle), glycerol, and amino acids. Its name suggests that it is an anabolic process but it occurs only when the body is in an overall catabolic state. Gluconeogenesis is stimulated by raised plasma concentrations of glucagon (acutely) and glucocorticoid (chronically) and inhibited by raised plasma insulin. The minimal structural requirement for a molecule to serve as a precursor for gluconeogenesis is three carbons. This precludes fatty acids since they are degraded into two-carbon acetyl-CoA and, unlike plants, humans lack the enzymes required to synthesize glucose from acetate. The major gluconeogenic tissue is liver although kidney and, to some extent, small intestine also possess gluconeogenic activity. Gluconeogenesis in the liver provides a mechanism for synthesis of glucose from lactate, derived from exercising muscle or anaerobic glycolysis in

acetyl-CoA
lactate
aspartate & glutamate
alanine
pyruvate
pyruvate kinase
oxaloacetate
phosphoenolpyruvate carboxykinase
phosphoenolpyruvate
glucagon
glycerol
fructose 1,6-bisphosphate
▼fructose 2,6-bisphosphate
phosphofructokinase
fructose 1,6-bisphosphatase
fructose 6-phosphate
glucose 6-phosphate
hexokinase
glucose 6-phosphatase
GLUCOSE

Figure 2.14 Gluconeogenesis: synthesis of new glucose. Gluconeogenesis, the synthesis of glucose from non-carbohydrate precursors (lactate, glycerol, and amino acids), occurs mainly in the liver but also in kidney. It is essentially the reverse of the glycolytic pathway apart from at three key, energetically unfavorable steps: conversion of phosphoenolpyruvate to pyruvate (catalyzed by pyruvate kinase); phosphorylation of fructose 6-phosphate (by phosphofructokinase); and phosphorylation of glucose (by hexokinase). These steps are overcome by the actions of phosphoenolpyruvate carboxykinase, fructose 1,6-bisphosphatase, and glucose 6-phosphatase, respectively. A decrease in fructose 2,6-bisphosphate stimulates the action of fructose 1,6-bisphosphatase. The minimum structural requirement for a substrate for gluconeogenesis is three carbons. Gluconeogenic precursors are shown in green. Red arrows show reactions inhibited by the catabolic state. Dashed arrows indicate intermediate steps that are not shown.

erythrocytes, and from glycerol released from hydrolysis of adipose tissue triglyceride. It is particularly important during the early stages of fasting when, in the absence of dietary glucose, hepatic glycogen levels become depleted and the liver synthesizes glucose to satisfy the needs of tissues that have an obligatory requirement for glucose as an energy source, particularly brain and erythrocytes. In this case, the carbon skeletons of amino acids derived from hydrolysis of muscle protein are the precursors for hepatic glucose synthesis. The acute stimulation of gluconeogenesis by glucagon is due to inhibition of synthesis of fructose 2,6-bisphosphate, which thereby inhibits the metabolism of glucose via glycolysis (by reducing PFK activity), and stimulation of fructose-1,6-bisphosphatase activity, a key step on the gluconeogenic pathway. The lowered plasma insulin level seen in the fasting state also promotes the phosphorylation and thus the inhibition of pyruvate dehydrogenase, which conserves three-carbon pyruvate for gluconeogenesis. Chronic hormonal changes in the prolonged catabolic state, particularly raised glucocorticoid, lead to increased synthesis of key enzymes of the gluconeogenic pathway.

The rate of gluconeogenesis is governed by substrate supply and is inhibited by high redox states (high NADH–NAD^+ ratios) in the hepatocyte, such as is seen in excess alcohol ingestion (see Chapter 22). Even under optimal conditions of substrate supply, however, the liver is incapable of meeting the needs of glucose-dependent tissues and the adaptation of the brain to the use of fatty acid-derived ketone bodies is described in Chapter 3. Indeed, the rate of hepatic gluconeogenesis falls during prolonged fasting as the tissue engages in ketone body synthesis and, under this condition, kidney becomes the major gluconeogenic tissue.

2.3 INSULIN

The actions of insulin are key to the whole-body metabolism of glucose and lack of insulin action results in the hyperglycemia of diabetes mellitus and severe, pathological consequences. The insulin gene is located on the short arm of chromosome 11 and encodes an initial translation product, a 109-amino-acid peptide called preproinsulin, from which an N-terminal signal peptide sequence of 23 amino acids is cleaved co-translationally to yield proinsulin (86 amino acids). Proinsulin undergoes post-translational cleavage involving excision of a C-peptide to yield A and B chains, which are linked via disulfide bridges to form the parent hormone comprising 51 amino acids. Insulin is thus a dimeric protein linked through both intra- and interchain disulfide bridges (**Figure 2.15**). The

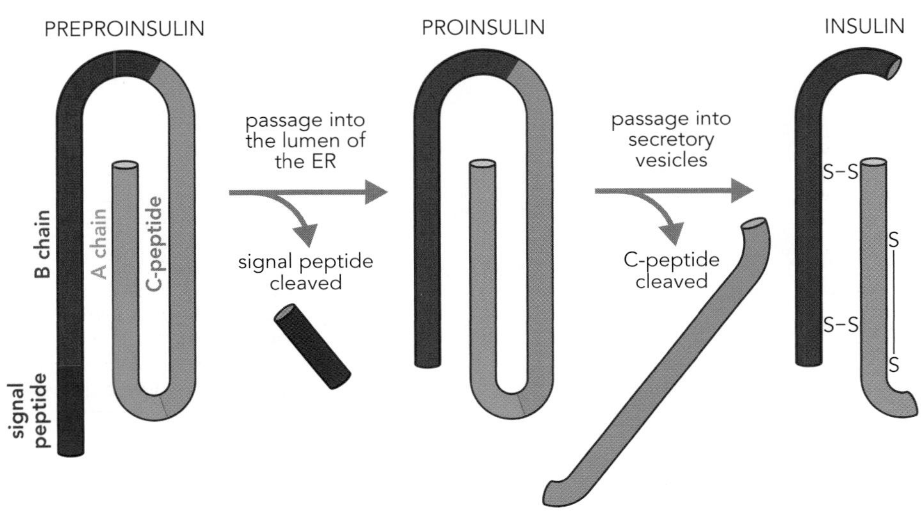

Figure 2.15 Structure of insulin. The nascent insulin polypeptide (preproinsulin, 109 amino acids) is translated as a monomeric species from its messenger RNA on the rough endoplasmic reticulum (ER) of pancreatic β cells. Its passage into the lumen of the ER is directed by an N-terminal signal peptide (23 amino acids) and this is cleaved co-translationally. The signal peptide is not found in the mature hormone. Further post-translational processing of the insulin polypeptide occurs during its passage through the secretory pathway into secretory vesicles. This involves excision of the connecting C-peptide and production of the A and B chains. Disulfide bond formation, both intra- and interchain, then occurs to form the dimeric parent hormone. The C-peptide is released with the parent hormone from secretory vesicles.

parent hormone and the C-peptide are stored in secretory vesicles complexed with zinc and both are released into the circulation by a hyperglycemic signal in the portal circulation or in response to an intestinal peptide that promotes insulin secretion (see Section 2.4 below). Thus, when the plasma glucose rises above the homeostatic concentration of about 4.5 mmol/L (81 mg/dL), glucose enters the pancreatic β cell via the GLUT2 transport protein and is metabolized via glycolysis; this initiates a signal transduction system that closes ATP-sensitive potassium channels thereby depolarizing the cell and stimulating calcium entry into the cytoplasm. The subsequent rise in cytoplasmic calcium triggers fusion of insulin-containing secretory vesicles and exocytosis of vesicle contents into the bloodstream. The magnitude of insulin secretion is dependent on the plasma glucose concentration. Both insulin and C-peptide are released into the portal circulation and approximately 50% of the parent hormone is metabolized in the first pass through the liver, such that the concentration of insulin in the portal vein is two- to fourfold greater than in the peripheral circulation. The C-peptide, however, does not undergo a first-pass effect and measurement of C-peptide rather than of insulin itself is a more precise marker of endogenous insulin secretion.

Metabolic actions of insulin

Insulin has both acute anabolic effects on metabolism and, via somewhat complex signaling to the nucleus, more chronic effects on protein synthesis, cell growth, and replication. Only the acute effects are relevant to the current text.

Following a rise in plasma glucose, after a meal for example, the primary actions of insulin are to promote glucose uptake into liver, muscle, and adipose tissue and glucose storage as glycogen in liver thereby reducing the rise in plasma glucose concentration.

Insulin acts by binding to its specific receptors on target cells (**Figure 2.16**). The receptor is heterodimeric, consisting of two surface-bound α-subunits and two membrane-spanning β-subunits linked via disulfide bonds. The β-subunits have intrinsic tyrosine kinase activity and can autophosphorylate on serine, threonine, and tyrosine residues. Insulin binding to the α-subunits situated on the extracellular surface of the plasma membrane promotes autophosphorylation of the β-subunits and initiation of signal transduction pathways that promote the short- and longer-term effects of insulin mentioned above. One such pathway involving 1-phosphatidylinositol 3′-kinase leads to translocation of the

Figure 2.16 Overview of insulin action on glucose transport into muscle and adipose tissue. Insulin binding to its receptor stimulates receptor dimerization and activates a protein tyrosine kinase cascade. This in turn promotes the translocation of GLUT4 (glucose transporter 4) from intracellular sites on the endoplasmic reticulum to the cell surface thereby increasing glucose uptake into the tissues. In the unstimulated, basal state, shown as (i), greater than 95% of the GLUT4 glucose transporters are in intracellular vesicles with only a few in the plasma membrane. Dimerization of the insulin receptor effected by the binding of insulin initiates a tyrosine kinase-mediated signaling cascade, which ultimately leads to the translocation of the intracellular GLUT4 molecules to the plasma membrane (ii), thereby increasing glucose transport into the cell by facilitated diffusion. Transport of glucose out of the cell is prevented by its rapid phosphorylation to glucose 6-phosphate (iii). The translocation of GLUT4 is readily reversible and the basal state (i) is re-attained within an hour of removal of the insulin stimulus.

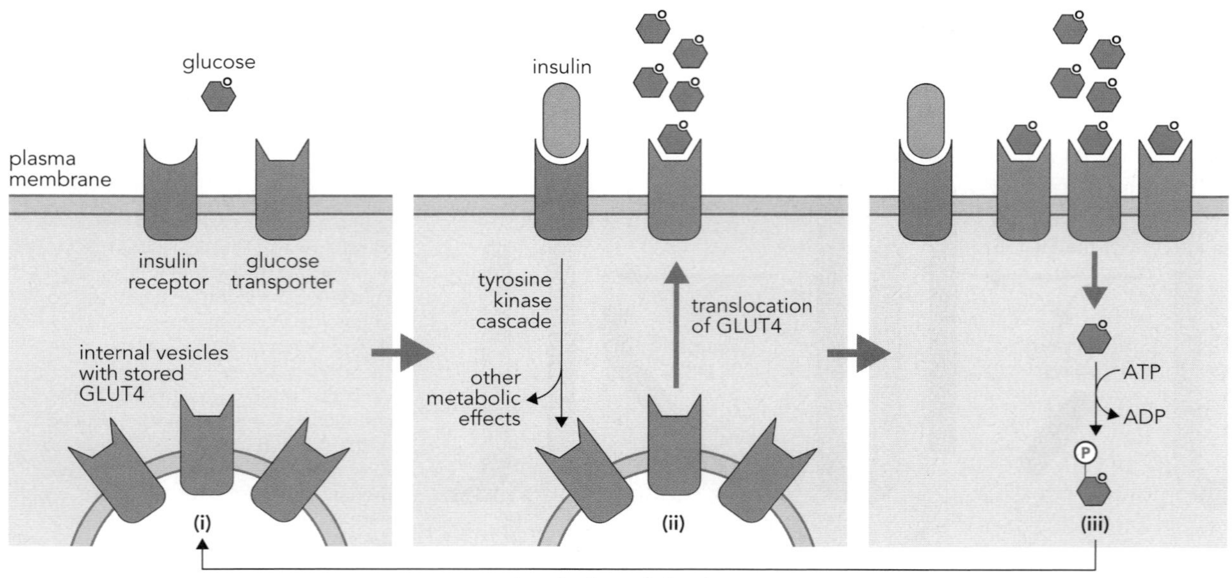

glucose transport protein GLUT4 to the cell surface and facilitation of glucose uptake into muscle and adipose tissue. Uptake of glucose into liver, however, is not facilitated by GLUT4 and is not stimulated by insulin; glucose enters the liver down a concentration gradient. The process of glycogenesis, however, is stimulated by insulin and this is responsible for increased glucose uptake by the liver. Approximately 30% of the glucose ingested during a meal is stored in the liver as glycogen, replacing that lost during the previous fasting period. Insulin also stimulates triglyceride synthesis in liver and, as a consequence, promotes very-low-density lipoprotein (VLDL) synthesis and secretion. The insulin-stimulated uptake of glucose into muscle promotes glycogen synthesis and storage by the tissue, and increased glucose uptake into adipose tissue leads to storage of triglyceride.

Hepatic actions of insulin

The overall action of insulin in the liver is anabolic and it regulates the supply of fuels, such as glucose and ketone bodies, to the systemic circulation. Under normal dietary conditions of moderate glucose consumption, the major hepatic actions of insulin are to decrease glucose production and promote glucose storage by inhibiting glycogenolysis and gluconeogenesis and promoting glycogenesis, processes which are sensitive to small increases in insulin concentration. It also stimulates glucose metabolism via glycolysis and the TCA cycle to supply energy to the tissue and inhibits β-oxidation of fatty acids and ketogenesis. The liver hydrolyzes 80% of insulin arriving from the portal circulation such that, under normal conditions, the concentration of insulin in the systemic circulation is considerably lower than that of the portal system.

Extrahepatic actions of insulin

The principal sites of insulin action in the systemic circulation are skeletal muscle and adipose tissue. Here, its major action is to increase the recruitment of insulin-sensitive glucose transport proteins (GLUT4) to the plasma membrane of both tissues and thereby increase the rate of glucose uptake and supply of metabolic substrate. In skeletal muscle, the increased supply of glucose and elevated insulin concentration promote glucose storage as glycogen as a result of activation of glycogen synthase and inhibition of glycogenolysis. This storage is an important energy source for muscle not only for acute exercise but also during long-term fasting; its mobilization is considerably slower than for hepatic glycogen reserves. In addition, insulin stimulates amino acid uptake into muscle and, by also inhibiting proteolysis, it promotes protein accumulation. An important anabolic action of insulin is to promote the storage of triglyceride in adipose tissue. Dietary fatty acids are transported in the systemic circulation as the triglyceride component of chylomicrons, and activation by insulin of lipoprotein lipase in adipose tissue results in hydrolysis of the triglyceride and uptake of free fatty acids into the tissue. Activation of glycolysis in the tissue also provides energy and glycerol phosphate for triglyceride synthesis. In addition to this action, insulin inhibits the hormone-sensitive lipase of adipose tissue required for mobilization of fatty acids from adipose-tissue triglyceride, such that the overall effect of the hormone is to promote fat storage.

When dietary carbohydrate exceeds total daily energy expenditure, a raised insulin concentration stimulates the conversion of glucose to fatty acids in liver and formation of triglyceride. This is exported as a component of very-low-density lipoprotein and can also donate its fatty acids to adipose tissue by the same route as for chylomicrons. In this way, excess dietary glucose is converted to stored fat.

Two final important actions of insulin are (i) inhibition of ketogenesis in liver and (ii) promotion of the uptake of K^+ ions by cells through stimulation of the plasma membrane Na^+/K^+-ATPase; insulin may be used to correct hyperkalemia, as described later. Hypokalemia may also be present in patients with certain types of pancreatic tumors (ectopic insulin).

Thus insulin is primarily an anabolic hormone that regulates the supply of glucose into the systemic circulation, facilitates the synthesis of energy stores in the form of glycogen and triglyceride, and promotes protein deposition.

2.4 GLUCAGON

Glucagon may be considered metabolically as an insulin antagonist, its actions for the most part being the opposite of those of insulin. The gene for glucagon is located on the long arm of chromosome 2 and encodes a large precursor molecule, preproglucagon, which is expressed mainly in the α cells of the pancreas and to a lesser extent in the duodenum and brain. The glucagon gene is a member of a multigene superfamily that includes secretin, vasoactive peptide, and gastric inhibitory peptide (GIP). In the pancreatic α cells, the initial translation product, preproglucagon, undergoes intracellular processing to yield the parent 29-amino-acid hormone and two further peptides, glucagon-like peptides 1 and 2 (GLP-1, GLP-2; **Figure 2.17**). Proteolytic processing of preproglucagon in the L-cells of the intestine yields peptides that serve to promote insulin secretion by pancreatic β cells (GLP-1 and glicentin) and which may suppress appetite following a meal (oxyntomodulin). For the purposes of this text, only the metabolic actions of glucagon are considered.

Glucagon secretion increases rapidly in response to a fall in plasma glucose and glucagon acts acutely to provide glucose to the circulation. It has a very short half-life of about 5 minutes and is destroyed rapidly when carbohydrate is consumed. The level of glucagon in the blood rises between meals as the glucose concentration falls and is raised chronically during fasting or on consumption of a diet low in carbohydrate. Glucagon is also a stress-response hormone and is released through direct sympathetic stimulation of the pancreas during times of acute psychological or physical stress.

Metabolic actions of glucagon

Glucagon acts by binding to its trimeric, G-protein-linked receptor on the surface of the target tissue, mainly liver and adipose tissue, and signaling via a cAMP-activated phosphorylation cascade to activate glycogen phosphorylase and inhibit glycogen synthase in liver, and activate hormone-sensitive lipase in adipose tissue. In this way, it mobilizes readily metabolizable fuels—glucose and fatty acids—for use by tissues when the dietary supply of these fuels is reduced or at zero. Activation by glucagon of the hormone-sensitive lipase in adipose tissue results in the release of free fatty acids and glycerol. The free fatty acids are transported in the circulation bound to albumin and provide fuel to most tissues. Glycerol is returned to the liver and used as a substrate for gluconeogenesis. In

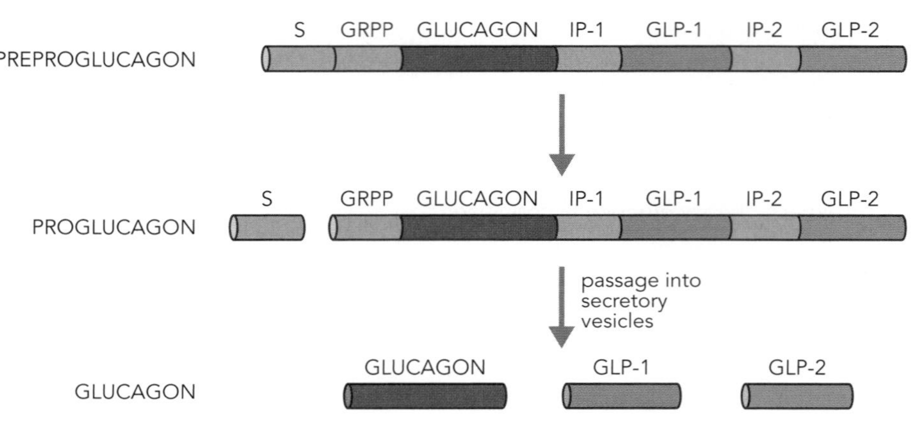

Figure 2.17 Structure of glucagon. Glucagon is formed from the translation product of the GCG gene, preproglucagon, which is translated as a monomeric polypeptide on the rough endoplasmic reticulum (ER). An N-terminal signal peptide (S) directs the passage of the nascent polypeptide into the lumen of the ER, after which it is cleaved by a signal peptidase. Further tissue-specific processing in the secretory pathway yields other biologically active peptides including, in pancreatic α-cells, glucagon. GLP, glucagon-like peptide; GRPP, glicentin-related peptide; IP, intervening peptide.

the longer term, glucagon also promotes the process of ketogenesis, key to the provision of ketone bodies as an alternative fuel for the brain during prolonged fasting (see Table 2.3).

2.5 MAINTENANCE OF GLUCOSE HOMEOSTASIS IN THE FED AND FASTING STATES

Blood glucose concentration is maintained through the opposing actions of insulin and counter-insulin hormones—glucagon, epinephrine, cortisol, and growth hormone—which combine to ensure that the amount of glucose entering and leaving the bloodstream is balanced in both the fed and fasting states. The glucose supply to the circulation during a 24-hour period is shown in **Figure 2.18**. Insulin concentration will rise as the blood glucose level rises after each meal, while levels of the counter-insulin hormones rise as glucose concentration falls. Of paramount importance is the supply of glucose to the brain, other nervous tissue, and erythrocytes.

The rise in blood glucose following digestion of a carbohydrate-enriched meal is detected by glucose receptors in the β cells of the pancreas and triggers the release of insulin. At the same time, glucagon release from the pancreatic α cells is suppressed. By increasing the number of glucose-specific transporters (GLUT4) on the cell surface in peripheral tissues, especially muscle and adipose tissue but not liver and brain, insulin promotes the uptake of glucose into these tissues. In addition, a raised concentration of glucose arrives at the liver via the hepatic portal circulation and glucose enters the tissue via insulin-independent, low-affinity, high-capacity transporters (GLUT2). The increase in the insulin-to-glucagon ratio promotes hepatic glycogen synthesis and inhibits glycogenolysis and gluconeogenesis such that there is an increase in glucose uptake by the liver. The condition of raised insulin and intracellular glucose concentrations promotes the synthesis and activity of glucokinase, which converts glucose to glucose 6-phosphate. Since this enzyme is not product-inhibited, the hepatic concentration of glucose 6-phosphate rises and the flux of glucose 6-phosphate into glycolysis, the pentose phosphate pathway, and glycogenesis is increased. Thus, enhanced hepatic metabolism of glucose also contributes to the clearance of blood glucose following a meal. The rise in blood glucose will be maintained until the clearance of glucose into liver and peripheral tissues is greater than the rate of glucose released from the splanchnic bed (arising from ingested food and endogenous production). At this point, the glucose concentration begins

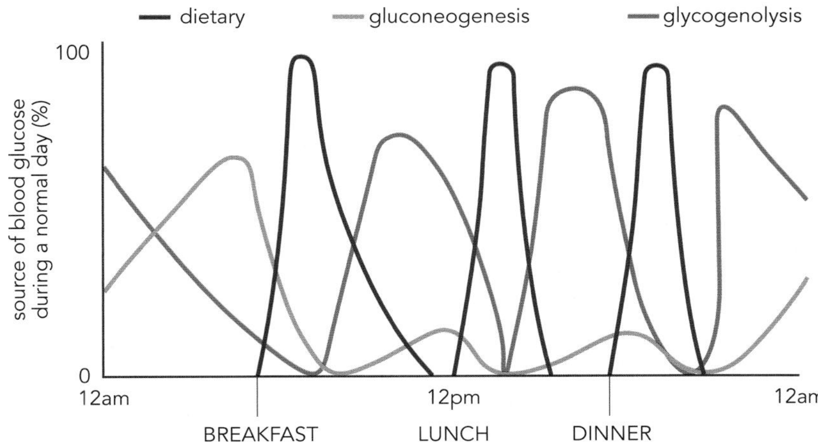

Figure 2.18 Contributions to the maintenance of blood glucose through a period of 24 hours. The figure assumes consumption of three meals (breakfast, lunch, and dinner) during the period. Sources of glucose are diet (exogenous), hepatic gluconeogenesis (endogenous), and hepatic glycogenolysis (endogenous). Between meals, hepatic glycogenolysis becomes the major contributor. Only overnight with a longer interval between meals (just prior to the next breakfast) does hepatic gluconeogenesis play any significant role.

to fall toward its preprandial level, with a concomitant fall in insulin and a rise in glucagon, such that the basal level is reached 4–6 hours after the meal. The various interacting factors that limit the rise in glucose concentration following a meal and give rise to a smooth return to homeostasis are:

- The rate of glucose absorption from the gut
- The relative timing and amounts of insulin and glucagon secreted by the pancreas
- The ability of the liver to store and release glucose
- The responses of the liver, muscle, and adipose tissue to insulin and counter-insulin hormones

The fasted state

As the blood glucose falls slightly overnight, glucose is released from the liver (**Figure 2.19**), initially from glycogen breakdown and later through gluconeogenesis from glycerol, lactate, and amino acids, which occurs mainly in liver but also a little in kidney. The overall rate of glucose production at this stage is about 2 mg/kg/min (11 μmol/kg/min).

In the absence of ingested carbohydrate, hepatic gluconeogenesis becomes a progressively more and more important source of blood glucose as the fast continues and hepatic glycogen is used up. This is to meet the needs of the glucose-dependent tissues. Obviously, the rate of hepatic glycogen depletion will depend on the previous nutritional state of the subject (that is, having a full store of glycogen to start with) and factors such as exercise. The major need now is to reduce the demand for glucose, particularly by the major consumer, the brain. Most tissues, including liver, muscle, and adipose tissue, will already be using fatty acids as their major respiratory substrate and the brain will also gradually adapt to using fatty acid-derived substrates—ketone bodies (acetoacetate and 3-hydroxybutyrate)—for energy. A reduction in blood glucose concentration reduces insulin and increases glucagon, cortisol, and growth hormone, and the swing in favor of the action of the counter-insulin hormones increases lipolysis in adipose tissue and ketogenesis in liver. The use of ketone bodies by the brain reduces the overall glucose demand and thereby the rate of hepatic gluconeogenesis from carbon skeletons of amino acids; in this way, vital protein is conserved. This is extremely important since even at its maximal rate (approximately 40 g/day), hepatic gluconeogenesis cannot supply sufficient glucose to

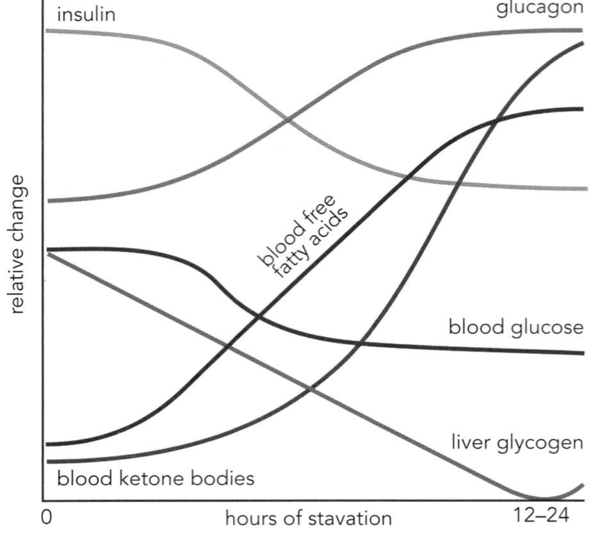

Figure 2.19 The effect of fasting over a period of 24 hours on blood insulin and glucagon concentrations and consequent changes in metabolic parameters. The change from an anabolic to a catabolic state results in a total loss of liver glycogen and a marked reduction in blood glucose as tissues apart from brain, other nervous tissue, and kidney switch to the use of fatty acids and ketone bodies for metabolic fuel. The concentrations of free fatty acids and ketone bodies rise dramatically over the time period.

satisfy the needs of the brain. The overall switch of the body from glucose-based metabolism to fatty acid-based metabolism allows the glucose concentration to fall to 2–3 mmol/L (36–54 mg/dL) without inducing the symptoms commonly seen in hypoglycemia.

As may be deduced from this discussion, defects in enzymes of the glycogenesis, glycogenolysis, or gluconeogenesis pathways, lack of substrates, or abnormal levels of insulin or counter-insulin hormones will prevent the maintenance of normoglycemia.

2.6 GLYCATION OF PROTEINS

Glucose is an aldohexose, a six-carbon sugar with a functional aldehyde group at carbon 1 as illustrated in **Figure 2.20**, which depicts both the Haworth ring and Fischer open-chain representations of the molecule. The aldehyde group allows it to react nonenzymatically with a free amino group, of a polypeptide for example, to form a Schiff base (an aldimine), which can undergo an Amadori rearrangement to a ketimine (**Figure 2.21**) where the double bond at carbon 1 moves to carbon 2 and forms a fructosamine derivative. The formation of the Schiff base is reversible but the rearrangement to the stable ketimine is irreversible. As with all chemical reactions, the rate of formation of product is proportional to the concentration of reactants and so Schiff base formation is proportional to the concentration of glucose and/or the amino group. Glycation of proteins occurs even at concentrations of glucose within the normal physiological range and assumes importance pathologically as the concentration of plasma glucose rises. The extent of glycation in plasma can be measured as glycated hemoglobin (HbA_{1c}); in this case, the free amino group of the N-terminal valine of the β-chain of hemoglobin reacts with glucose to form a Schiff base initially and this undergoes rearrangement to form the stable fructosamine conjugate. Measurement of HbA_{1c} is used to assess glycemic control in diabetic patients.

HAWORTH

6 CH_2OH 5 O 4 OH 1 HO 3 2 OH OH

α-D-glucopyranose

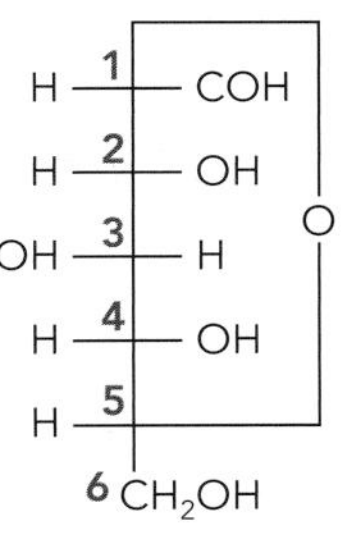

FISCHER

Figure 2.20 Haworth and Fischer projection formulas of glucose (α-D-glucopyranose).
The aldehyde reducing carbon is C1 on both structures, which are shown in the lactone form. Brown numbers show the numbering of carbons.

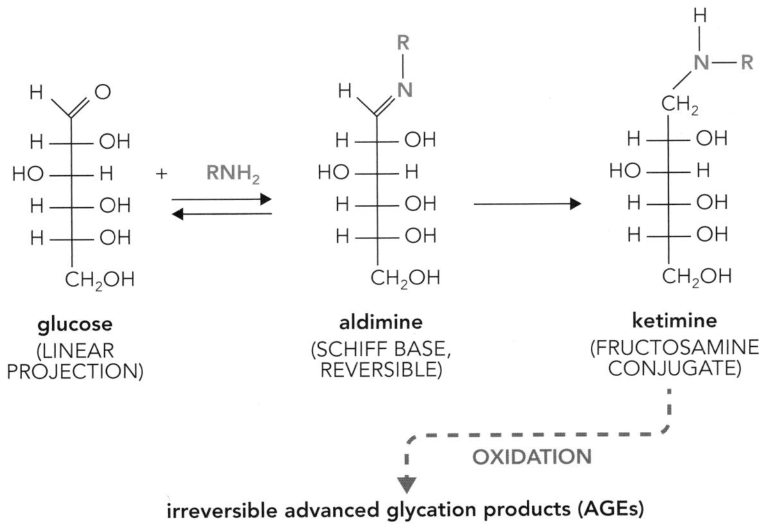

H O / H OH / HO H / H OH / H OH / CH_2OH + RNH_2 ⇌ R N H / H OH / HO H / H OH / H OH / CH_2OH → H N—R / CH_2 / H OH / HO H / H OH / H OH / CH_2OH

glucose (LINEAR PROJECTION)

aldimine (SCHIFF BASE, REVERSIBLE)

ketimine (FRUCTOSAMINE CONJUGATE)

OXIDATION

irreversible advanced glycation products (AGEs)

Figure 2.21 Glycation of proteins leading to formation of advanced glycation end products.
Glucose is shown here in the open chain form to illustrate the reversible formation of a Schiff base between the aldehyde moiety on C1 and a free amino group on proteins, RNH_2 (internal lysine residues and N-terminal amino acids). The further formation of the fructosamine conjugate (ketimine) and subsequent oxidation to advanced glycation end products (AGEs) are irreversible.

TABLE 2.6 Definitions of various glycated hemoglobins

Hemoglobin	Definition
Glycated hemoglobin	Carbohydrate, usually glucose, is bound to any free amino group (of the N-terminal amino acid or internal lysine) in the hemoglobin chains
HbA_1	Carbohydrate is bound to the N-terminal valine of hemoglobin β-chains
HbA_{1a1}	Fructose 1,6-bisphosphate is bound to N-terminal valine
HbA_{1a2}	Glucose 6-phosphate is bound to N-terminal valine
HbA_{1b}	Unknown carbohydrate is bound to N-terminal valine
HbA_{1c}	Glucose is bound to N-terminal valine

In each hemoglobin subspecies, the carbohydrate is covalently bound to the hemoglobin chain via a Schiff base linkage to a free amino group.

While other sugars, including fructose 1,6-bisphosphate and glucose 6-phosphate, with a functional keto group can also form glycated proteins with free amino groups on hemoglobin (**Table 2.6**), only HbA_{1c} is measured in clinical practice.

The ε-amino groups of internal lysine residues on proteins also present opportunities for Schiff base formation and eventual formation of glycated proteins. Thus, virtually any protein can undergo glycation, including albumin and plasma proteins, and measurement of plasma fructosamine (glycated plasma proteins, mainly glycated albumin) provides an index of short-term (2–3 weeks) glycemic control in individuals.

Glycation leads to changes in the three-dimensional shape of a protein such that it exhibits an altered antigenic profile and may no longer be recognized by its receptor. For example, glycation of apoprotein B100, the only protein of low-density lipoprotein (LDL), yields a product that is no longer recognized by the LDL receptor and is cleared by unregulated scavenger receptors, leading to cholesterol deposition in extrahepatic tissues. In addition, in the longer term, oxidation of the fructosamine conjugates can lead to further cross-linking of proteins, including collagen, and the formation of advanced glycation end products (AGEs; see Figure 2.21) with subsequent changes in protein function and turnover. In the case of collagen and other structural proteins, this increases the rigidity of the cytoskeletal network and probably contributes to circulatory, joint, and vision problems, particularly in patients with diabetes.

2.7 HYPERGLYCEMIA AND DIABETES MELLITUS

Transient hyperglycemia may be seen in normal individuals after ingestion of a carbohydrate-enriched meal and poses no clinical problems. However, chronic hyperglycemia gives rise to disease, the treatment of which is a major cost to national health budgets.

Diabetes mellitus

Diabetes mellitus was defined by the World Health Organization in 2000 as a metabolic disorder of multiple etiology characterized by chronic hyperglycemia with disturbances of carbohydrate, fat, and protein metabolism resulting from defects in insulin secretion, insulin action, or both. Thus, although the diagnosis of diabetes depends upon demonstration of hyperglycemia, it is important to remember that other metabolic processes are also affected.

Diabetes mellitus is a chronic, noncommunicable condition which, if untreated, proceeds to micro- and macrovascular disease. The trademark hyperglycemia of diabetes mellitus arises from an inability to clear glucose into insulin-sensitive tissues, particularly skeletal muscle and adipose tissue, due to either (i) decreased insulin production by the β cells of the pancreas (type 1; juvenile-onset, insulin-dependent diabetes mellitus) or (ii) decreased sensitivity of tissues to the prevailing insulin concentration (type 2; adult-onset, non-insulin-dependent diabetes mellitus).

Diabetes was recognized as a metabolic disease by the ancient Greeks and the term diabetes (Greek for siphon) was used in the second century by Aretaeus of Capodocia, who observed that patients with the disease "suffered liquefaction of the flesh and bones into urine such that kidneys and bladder do not cease emitting urine." Avicenna (980–1037) observed that the urine of diabetic patients was "wonderfully sweet" and the Latin term mellitus ("honey sweet") was added by Thomas Willis (1621–1676) who noted that diabetic patients "piss a great deal" whilst "suffering from a persistent thirst." It is now evident that the diuresis and sweet-tasting urine are caused by glycosuria.

The early descriptions of the wasting disease and associated thirst, polyuria, and early death clearly refer to the type 1, insulin-dependent form. This insulin-dependent form was shown in the 1950s to be an autoimmune condition that gives rise to the highly selective destruction of the insulin-producing β cells of the pancreas. It was in the 1880s that the French physiologist Lanceaux differentiated between this maigre (thin) presentation and the gras (fat) presentation characterized by corpulence and stupor, which is now associated with type 2, non-insulin-dependent diabetes mellitus.

Since the discovery of insulin by Banting and Best in Canada in the early 1920s, early death from acute insulin deficiency has become avoidable, and insulin-dependent diabetic patients can lead relatively normal lives with daily injections of the hormone. However, this increase in life-expectancy is associated with long-term complications of diabetes which have significant effects on the overall morbidity and mortality of the disease. Such complications, described later, are also associated with the later-onset type 2 form of the disease. This form accounts for about 95% of patients presenting with diabetes in Western societies and is becoming an increasing problem in developing countries. Such patients present with variable combinations of insulin resistance and β-cell dysfunction leading to defects in insulin secretion. It is now apparent that type 2 diabetes mellitus is a syndrome with many different causes including a more sedentary lifestyle, obesity, and dietary factors imposing upon an innate genetic susceptibility.

Diagnosis of diabetes

Until 2009, diagnosis of diabetes was based entirely on blood or plasma glucose concentrations exceeding thresholds specified by the World Health Organization. These concentrations are different for samples taken in the fasting state (**Table 2.7**), at random times during the day, and after a standard 75 g oral glucose load (oral glucose tolerance test [OGTT]; **Figure 2.22** and **Table 2.8**). The use of HbA_{1c} for monitoring of long-term glucose control has been well established for almost two decades, but its use as a diagnostic test for diabetes is controversial. The reasons for this include differences in standardization of measurement, lack of availability in poorer countries, variable correlation with average blood glucose, and the effect of variant hemoglobins and states which alter the red cell survival time. Nevertheless, a number of diabetes organizations have recently agreed that a confirmed HbA_{1c} value of ≥6.5% (48 mmol/mol) is

TABLE 2.7 Ranges of fasting blood glucose concentration in normal, glucose-intolerant, and diabetic subjects

Subject	Range (mmol/L)	Range (mg/dL)
Normal	Up to 6	Up to 108
Impaired glucose tolerance	6–7	108–126
Diabetic	7 and above	126 and above

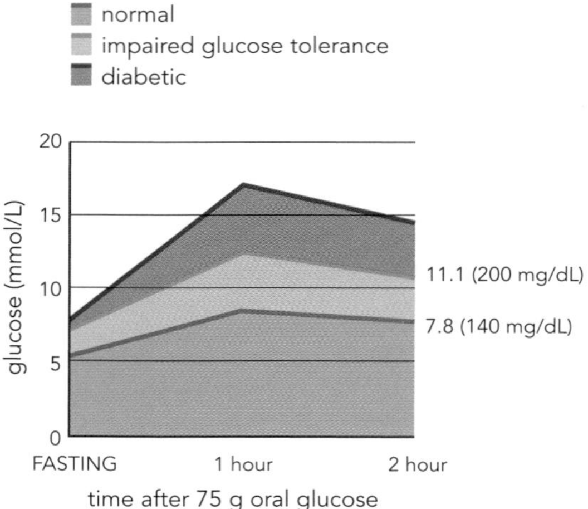

Figure 2.22 Oral glucose tolerance tests in normal subjects and patients with diabetes mellitus types 1 and 2.
An oral glucose tolerance test involves ingestion of 75 g of glucose dissolved in 100 mL of water and measurement of blood glucose over the following two hours. In normal, non-diabetic subjects, blood glucose rises during the first hour after ingestion, peaking at about 7.8 mmol/L (140 mg/dL), and then falls as glucose is cleared by liver, adipose tissue, and muscle in response to a rise in insulin. This contrasts with the situation in patients with diabetes, where insulin is either absent (type 1) or tissues are less responsive to the prevailing insulin concentration (type 2); in these patients, glucose may rise to concentrations >15 mmol/L (270 mg/dL) and fall slowly subsequently, remaining at >11.1 mmol/L (200 mg/dL) after two hours. Blood glucose levels tend to be higher in untreated type 1 than in untreated type 2 diabetes because patients with type 2 diabetes retain some insulin activity. Some subjects with impaired glucose tolerance have raised glucose levels but these do not exceed values used to diagnose the presence of diabetes mellitus.

sufficient evidence to diagnose diabetes without glucose testing. This approach has not been universally accepted and it remains to be seen whether misdiagnosis is a significant problem.

Analytical practice point 2.1

Glucose decreases rapidly in whole blood samples, due to glycolysis, and must be collected into tubes containing fluoride/oxalate unless analyzed immediately.

TABLE 2.8 World Health Organization (WHO) criteria for the diagnosis of diabetes mellitus following an oral glucose tolerance test

Subject	Fasting glucose		Blood glucose after 2 h	
	(mmol/L)	(mg/dL)	(mmol/L)	(mg/dL)
Non-diabetic	<6.0	<108	<7.8	<140.4
Impaired glucose tolerance (IGT)	<6.7	<120.6	7.9–11.0	142.2–198
Impaired fasting glucose	6.1–6.9	109.8–124.2	>7.8	>140.4
Diabetic	>7.0	>126	>11.1	>199.8

Subjects are given 75 g of glucose in water (100 mL) and the glucose concentration in capillary whole blood is measured over a period of two hours.

Types of diabetes

Diabetes is classified into several Types (**Table 2.9**) with the largest numbers of patients having either type 1 or type 2. About 90% of diabetes worldwide is type 2. Diabetes of other types should not be overlooked but make up a very small fraction of cases.

TABLE 2.9 Classification of diabetes mellitus and probable causal mechanisms

Classification	Causal mechanism
Type 1	Insulin deficiency
Type 2	Relative insulin deficiency/peripheral insulin resistance
Gestational diabetes	Increased insulin resistance during pregnancy
Maturity-onset diabetes of the young	Single gene defect affecting glucose metabolism, for example, glucokinase
Secondary diabetes due to	
• endocrine disease	Increased concentration of hormones antagonistic to insulin*
• pancreatic disease	Relative to absolute insulin deficiency
• malnutrition	Relative to absolute insulin deficiency
• drugs	Impaired insulin secretion or increased insulin resistance

*For example, increased thyroid hormones in thyrotoxicosis.

TABLE 2.10 Major clinical differences between patients with type 1 and type 2 diabetes mellitus

Clinical feature	Type 1	Type 2
Usual age at onset	<40 years	>40 years
Obesity	Uncommon	Usual
Prone to ketosis	Yes	No
Insulin secretion	Absent	Present
Insulin resistance	Rare (unless obese)	Usual
Genetic factors*	+	++
HLA association	Yes	No
Islet cell antibodies	Yes	No

*There appear to be genetic influences on both disorders which are more pronounced in the case of non-insulin-dependent diabetes mellitus.

Pathogenesis and clinical aspects of diabetes mellitus type 1 and type 2

The major differences between diabetes mellitus types 1 and 2 are shown in **Table 2.10** and the clinical presentations of each in **Table 2.11**. The most striking difference is the inability to synthesize and secrete insulin in type 1 diabetes. Polyuria, thirst, and polyphagia are features common to both types and their causes are explained in **Table 2.12**. In type 1 diabetes there is destruction of pancreatic β cells, due to an autoimmune process, leading to insulin deficiency. This usually presents as an acute illness with weight loss, symptoms of hyperglycemia, and sometimes ketoacidosis. Type 2 diabetes is a more slowly progressive condition, often asymptomatic for many years, characterized by peripheral resistance to insulin and a compensatory increase in β-cell secretion of insulin. As the disease progresses there is a gradual failure of β cells superimposed on the insulin resistance, but this is partial rather than total, and ketoacidosis rarely occurs. Insulin resistance is strongly correlated with obesity and lack of physical activity and its prevalence is increasing across the globe in parallel with the growth in unhealthy lifestyle. The development of hyperglycemia is often the final step in the progression of insulin resistance, which may exist for many years in association with cardiovascular risk factors including dyslipidemia and hypertension. This combination has been termed the metabolic syndrome and is considered to be a major contributor to cardiovascular disease.

Whilst most patients with type 1 or type 2 diabetes conform to the expected phenotypes listed in Table 2.10, it should be realized that exceptions occur and the conditions are not mutually exclusive. Thus, increasing obesity amongst the

TABLE 2.11 Clinical presentation of diabetes mellitus

Diabetes mellitus type	Clinical signs and symptoms
Type 1 (usually acute onset)	Polyuria
	Thirst
	Acute visual changes
	Weight loss
	Ketosis
Type 2 (usually slow onset)	Tiredness, mood changes, fungal infections
	Incidental finding in routine examination for other diseases
	Polyuria, thirst, visual changes if severe hyperglycemia
	Late presentation may be with complications, for example sight loss, foot ulcer

TABLE 2.12 Clinical symptoms and their causes in diabetes mellitus

Symptom	Cause
Polyuria	Retention of glucose in renal tubule as glucose load exceeds the reabsorptive capacity of the kidney. Glucose acts as an osmotic diuretic causing the production of large volumes of urine
Thirst	CNS-driven response to dehydration; may be mediated by angiotensin secreted in response to hypovolemia
Polyphagia	Hunger stimulated by nonutilization of dietary glucose
Weight loss	Increased catabolism of all stored metabolic fuels; muscle glycogen and protein, and adipose tissue triglyceride
Tiredness	Muscular weakness due to (i) proteolysis and mobilization of muscle protein and (ii) reduced availability to muscle of metabolic substrate (glucose)
Blurred vision	Systemic dehydration of the lens and aqueous and vitreous humors, thereby reducing visual acuity
Vomiting	CNS-driven response to ketones stimulating the area postrema in the floor of the fourth ventricle
Hyperventilation (Kussmaul breathing)	Respiratory compensation to metabolic acidosis due to raised concentrations of lactic acid and keto acids in plasma
Itching	Impaired humoral immunity leading to increased risk of skin infections

CNS, central nervous system.

young is increasing the prevalence of type 2 diabetes amongst the under 40s, even affecting children. As people with type 1 diabetes conform to the population trend in obesity, they may become insulin resistant as well as deficient. Equally, there is no age beyond which β-cell destruction cannot occur and cases of type 1 diabetes appearing in the ninth decade of life have been described. Slow onset of type 1 diabetes is also seen on occasion with patients initially classified as type 2. It is more accurate to consider diabetes as a spectrum with pure insulin deficiency at one end and insulin resistance at the other. The acute effects of insulin deficiency are shown in **Figure 2.23**.

Clinical practice point 2.1

Type 1 and type 2 diabetes are not mutually exclusive conditions, and elements of both can occur in the same individual.

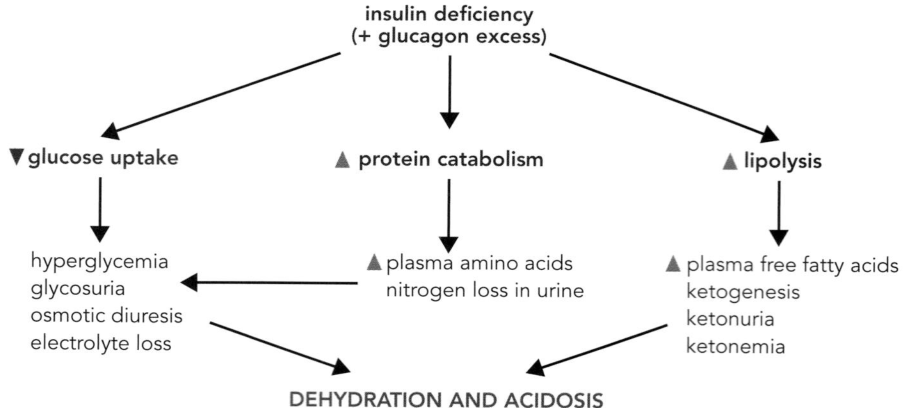

Figure 2.23 Acute metabolic effects of insulin deficiency. Decreased glucose uptake by tissues leads to hyperglycemia and increased excretion of glucose in urine (glycosuria) accompanied by osmotic diuresis and dehydration. The catabolic state, stress for example, increases the blood concentrations of amino acids and fatty acids, which are mobilized by hydrolysis of proteins and stored fat, and is responsible for increased nitrogen loss in urine and formation of ketone bodies and acidosis.

Diabetes mellitus type 1

The biochemical signs and their causes in type 1 diabetes are listed in **Table 2.13**.

The propensity to develop ketosis is one of the most important distinguishing features of insulin deficiency and its presence usually means that lifelong insulin treatment is mandatory. However, some patients with type 2 diabetes may develop ketosis when ill but may not always need long-term insulin therapy. The major clinical aspects of ketoacidosis are as follows:

- It generally occurs in known diabetic patients
- It is often precipitated by intercurrent illness, especially infections

TABLE 2.13 Biochemical signs and their causes in insulin-dependent diabetes mellitus

Biochemical sign	Cause
Hyperglycemia	(1) Decreased uptake of glucose into peripheral tissues (2) Increased mobilization of hepatic glycogen (3) Increased hepatic gluconeogenesis
Glycosuria	Glucose load exceeds capacity for reabsorption in renal tubule
Ketoacidosis	Increased β-oxidation of adipose tissue-derived fatty acids in liver; leads to raised hepatic acetyl-CoA concentration and ketone body synthesis
Ketonuria	Loss of ketone bodies to urine via renal tubule
Hyperlactatemia	Mobilization and metabolism of muscle glycogen to lactate, a precursor for hepatic gluconeogenesis (Cori cycle)
Hyperlipidemia	Free fatty acids derived from increased lipolysis in adipose tissue
Hypertriglyceridemia	Increased synthesis of triglycerides in liver, secreted as component of VLDL
Hypovolemia/hyperosmolarity	Excessive loss of body water as urine due to glucose acting as an osmotic diuretic
Hyponatremia	Loss of body sodium due to glucose-induced osmotic diuresis

- It develops relatively slowly compared to hypoglycemia, which is immediate
- Anorexia, nausea, vomiting, polyuria, and thirst are associated symptoms
- Untreated patients present with stupor progressing to coma; dehydration, shock, air hunger, and acetone in breath are features
- Urine is typically markedly positive for glucose and ketones

Pathogenesis of diabetic ketoacidosis

The pathogenesis of the diabetic ketoacidosis seen in poorly controlled type 1 diabetes mellitus is shown in **Figure 2.24** and may be deduced from the lack of action of insulin (see Figure 2.23).

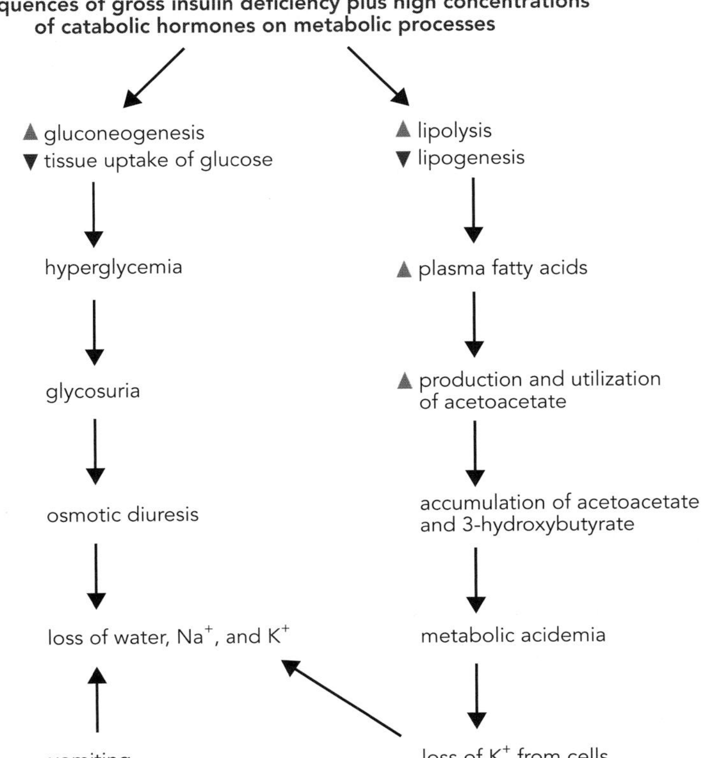

Figure 2.24 Pathogenesis of diabetic ketoacidosis. Diabetic ketoacidosis is seen only in patients with poorly controlled diabetes mellitus type I where insulin is absent. The metabolic sequelae are similar to those seen in Figure 2.23 and can only be prevented/reversed by administration of insulin to the patient.

Hyperglycemia

With no insulin to support glycogen synthesis and inhibit gluconeogenesis, the partition between the portal and systemic glucose concentrations no longer exists and so all glucose derived from the diet moves directly to the systemic circulation, producing a marked hyperglycemia. With a total lack of insulin, hepatic carbohydrate metabolism is responsible for glucose production from the breakdown of stored glycogen (glycogenolysis) and glucose synthesis (gluconeogenesis) and this glucose is released directly to the systemic circulation, contributing to the hyperglycemic state. The hyperglycemia is also exacerbated to some extent by the reduced uptake of glucose by adipose tissue and skeletal muscle. Even during starvation—the major therapy for type 1 diabetic patients prior to the discovery of insulin in 1923—the high production of glucose and reduced utilization of glucose is sufficiently high to maintain a state of hyperglycemia. The presence of a raised blood glucose concentration has a number of physiological sequelae including glycosuria and dehydration.

Glycosuria and dehydration

Glucose is normally completely reabsorbed in the proximal renal tubule such that it is undetected in urine. However, once the capacity for reabsorption is exceeded (10mmol/L; 180mg/dL), glucose is lost into the urine. The presence of glucose in the loop of Henle and the distal tubule acts as an osmotic diuretic giving rise to the production of large volumes of dilute urine (polyuria). Such increased production of urine shortens dramatically the tubular transit time of urine and decreases the ability of the kidneys to reabsorb water. The combined effects of these processes can lead to gross diuresis and severe dehydration.

Sequelae of dehydration

Dehydration and an increased serum osmolarity activate the thirst center located in the hypothalamus and promote polydipsia, a marked drinking response. Failure to compensate adequately for dehydration results in depletion of intravascular volume, hypotension, and reduced peripheral circulation. Furthermore, a reduction in peripheral perfusion reduces the oxygen supply to tissues leading to increased anaerobic metabolism of glucose and production of lactic acid, thereby creating a metabolic acidosis.

Catabolism in skeletal muscle and adipose tissue

The absence of insulin in the systemic circulation promotes catabolism in muscle and adipose tissue. In muscle, gluconeogenic precursors are released into the circulation following the catabolism of stored glycogen to pyruvate and lactate, and of protein to amino acids, mainly alanine and glutamine. These gluconeogenic precursors are transported to the liver where they are converted, somewhat inappropriately, to glucose and serve to perpetuate the hyperglycemic state. In adipose tissue, the breakdown of stored triglyceride to free fatty acids and glycerol leads to an increase in plasma free fatty acid and glycerol concentrations. Fatty acids bind to albumin in the blood and are transported to the liver where they undergo mitochondrial β-oxidation to acetyl-CoA. With no insulin to control catabolism, the rate of production of acetyl-CoA exceeds the capacity of the normal route of metabolism to CO_2 and water via the TCA cycle and acetyl-CoA is converted to ketone bodies (acetoacetate and 3-hydroxybutyrate). This process of ketogenesis is normally inhibited by even low levels of insulin. Both ketone bodies are strong acids and their release into the circulation contributes to the worsening metabolic acidosis. The condition of diabetic ketoacidosis now exists. Free fatty acids are toxic to the liver and if the rate of ketogenesis is insufficient to reduce their concentration in the liver, the tissue responds by converting them to triglycerides and exporting them as part of very-low-density lipoproteins, giving rise to hypertriglyceridemia.

Acid–base balance

A fall in blood pH as a consequence of metabolic acidosis causes an efflux of potassium ions from cells via the K^+/H^+ antiport system in their plasma membranes in an attempt to maintain acid–base balance. Much of this potassium is lost due to the severe diuresis and significant depletion of the total body pool of potassium ensues. Paradoxically, however, the initial rapid efflux of potassium from the tissues can lead to a potentially dangerous hyperkalemia. Another characteristic of patients with diabetic ketoacidosis is Kussmaul breathing, or air hunger. In this situation, the respiratory drive is increased, giving rise to rapid, shallow breathing, to blow off CO_2 in a further attempt to alleviate the metabolic acidosis.

The pathological changes described above are self-perpetuating and, unless the patient is treated, will continue to deteriorate, progressing to coma and eventually death of the patient.

Laboratory investigations in ketoacidosis

The biochemical hallmarks of diabetic ketoacidosis (**Table 2.14**) are hyperglycemia (plasma glucose typically 20–40 mmol/L [360–720 mg/dL]) plus ketones detectable in either serum or urine and evidence of systemic acidosis. Acidosis may be diagnosed by a venous blood gas analysis showing decreased pH (increased hydrogen ion concentration) with low CO_2 content due to compensatory hyperventilation. Blood gas analysis is not always essential, however, if the clinical picture is sufficiently convincing. A low bicarbonate or total CO_2 in the urea and electrolyte profile is also used as evidence of a metabolic acidosis. Serum potassium is often elevated due to the effects of insulin deficiency and acidosis allowing movement of potassium from the intracellular to the extracellular space. However, total body potassium tends to be low due to increased renal losses. This results in a risk of hypokalemia when insulin therapy is started, and so potassium is added to the intravenous fluid until serum potassium reaches the reference range. Serum sodium may be low, normal, or high in ketoacidosis. This is because several factors operate in different directions. Ketones are anions and require a balancing cation when excreted in urine. This may be sodium or potassium. The inability to reabsorb water in the distal nephron, despite high levels of antidiuretic hormone, is due to unabsorbed glucose reducing the transmembrane osmotic gradient (osmotic diuresis). This net loss of free water tends to cause hypernatremia. Finally, the hyperglycemia itself has an osmotic effect across cell membranes, pulling water out and diluting the sodium. The net effect on serum sodium thus depends on the relative magnitudes of these different effects.

Serum urea and creatinine are elevated, reflecting decreased glomerular filtration due to intravascular fluid loss. Urea may be disproportionately higher as it is able to diffuse back from the tubular fluid when the flow rate falls, whereas creatinine cannot. In some assays for creatinine, ketones may positively interfere, giving falsely high results.

Analytical practice point 2.2

Blood gas analysis can be performed on venous blood in diabetic ketoacidosis. Arterial samples are unnecessary.

TABLE 2.14 Clinical chemistry of diabetic ketoacidosis

Patient group	Typically in patients with diabetes mellitus type 1
Biochemical basis of disorder	Diminished glucose utilization and excessive lipolysis with ketogenesis
Clinical chemistry features	Hyperglycemia; plasma glucose usually >40 mmol/L (720 mg/dL)
	Plasma and urine positive for ketones
	Low plasma bicarbonate
	Plasma sodium usually low but may be normal or high
	Plasma potassium high or high-normal

There are situations where detection of ketones in urine is unreliable. Dipsticks detect acetoacetate rather than 3-hydroxybutyrate. In situations of tissue hypoxia, relatively more 3-hydroxybutyrate is formed than acetoacetate, thus potentially giving false negative results when using dipsticks. In addition, excretion of ketones requires an adequate glomerular filtration rate (GFR). As this rate diminishes, the renal threshold rises and ketonuria ceases. The ketones are still being produced and retained in the plasma, however. Hence it may be wrongly deduced that the patient is getting better because the ketones are no longer detected in urine, when in fact this just reflects deteriorating renal function.

Other findings in these patients include hypertriglyceridemia, due to increased synthesis of VLDL, and sometimes raised plasma amylase. When associated with abdominal pain, which is not uncommon in ketoacidosis, this may lead to an incorrect diagnosis of acute pancreatitis.

Analytical practice point 2.3

Ketones are not detected in urine if renal function is significantly reduced.

Diabetes mellitus type 2

The pathogenesis of grossly uncontrolled non-ketotic diabetes mellitus is shown in **Figure 2.25**. In contrast to the situation described above for type 1 patients, hepatic function is relatively normal in non-insulin-dependent diabetic patients and this has marked consequences for the presentation and progression of the disease. For example, the synthesis of ketone bodies is extremely sensitive to insulin and even low levels of insulin in the portal circulation will inhibit ketogenesis and protect against ketoacidosis. Thus, type 2 diabetic patients do not develop ketosis. In these patients, the major metabolic consequences arise from effects on muscle and adipose tissue, particularly those resulting from insulin resistance and the increased action of catabolic hormones. Decreased glucose uptake into both tissues leads to hyperglycemia, and increased lipolysis in adipose tissue causes a raised plasma free fatty acid concentration. In the absence of ketogenesis, fatty acids arriving at the liver must be oxidized or converted to triglyceride for export as VLDL and contribute to the hyperlipidemia. The reduced peripheral response to the prevailing insulin concentration will also decrease the activity of lipoprotein lipase and slow the clearance of VLDL. The hyperglycemia in type 2 patients may go undetected for months or even years and, without intervention, plasma glucose concentrations *in extremis* can often reach very high levels of >40 mmol/L (>720 mg/dL). Even though such patients do not develop ketoacidosis, they are still at risk from severe dehydration and increased serum osmolarity and have a marked increase in the risk of developing a major vascular (arterial and venous) thrombosis. If untreated, these patients can enter a diabetic, hyperglycemic, hyperosmolar, non-ketotic coma and die. The clinical features of grossly uncontrolled diabetes mellitus are:

- Patients are usually elderly and often not previously known to be diabetic.
- It is often precipitated by intercurrent illness, especially infection.
- It is often associated with arterial and venous thrombosis.
- The condition develops over several days.
- Features include polyuria, thirst, and dehydration; stupor progressing to coma; no air hunger.
- High mortality related to age of patient and arterial thrombosis.
- It may be diagnosed from glucose in CSF and serum osmolarity.
- There are occasionally trace of ketones in urine.

The major therapeutic requirement for patients is rehydration.

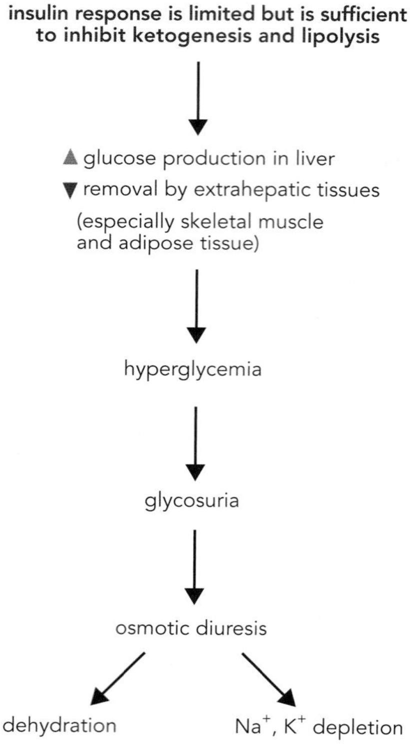

Figure 2.25 Pathogenesis of grossly uncontrolled non-ketotic diabetes mellitus. Lipolysis and ketogenesis are very sensitive to insulin and in non-ketotic diabetes mellitus (type 2), although the insulin response is blunted, it is sufficient to inhibit these catabolic processes.

The hyperglycemic, non-ketotic, hyperosmolar state

In type 2 diabetes, severe, decompensated hyperglycemia is not usually associated with significant ketone production (**Table 2.15**). If acidosis is present it is

TABLE 2.15 Clinical chemistry of hyperosmolar non-ketotic crisis

Patient group	Typically seen in older patients with diabetes mellitus type 2
Biochemical basis of disorder	Diminished glucose utilization, severe dehydration, and lactic acidosis
Clinical chemistry features	Severe hyperglycemia; plasma glucose often >40 mmol/L (720 mg/dL)
	Plasma and urine negative for ketones
	Plasma bicarbonate normal or slightly low
	Plasma sodium high
	Plasma potassium normal or high

usually due to lactic acid, reflecting the tissue hypoxia caused by the reduced circulating volume. There is polyuria and polydipsia, which may continue for several days and be exacerbated by the patient having sugary drinks. As a result of free water loss, hypernatremia develops, despite the osmotic draw of water from the intracellular compartment. When insulin treatment is initiated, the fall in glucose may cause the hypernatremia to worsen, as water moves back into cells.

Complications of diabetes

Diabetic complications may be divided into short-term (acute) and long-term (chronic) problems. The acute complications are hypo- and hyperglycemic crises due to short-term metabolic decompensation. The chronic complications (**Figure 2.26**) are manifestations of vascular disease and may be subdivided into microvascular and macrovascular. Microvascular complications are unique to diabetes and are caused to a large extent by hyperglycemia itself. The small vessels supplying blood to the retina, nerves, and kidney tissues are affected causing retinopathy, neuropathy, and nephropathy, respectively. Diabetic nephropathy is the most common cause of renal failure. Tight control of blood glucose (as shown by low HbA_{1c}) has been shown to reduce these complications significantly. Macrovascular complications are ischemic heart disease (angina and myocardial infarction), stroke, and peripheral vascular disease. These occur in the general population as well as in diabetic patients, but hyperglycemia is an additional risk factor and these complications often affect younger individuals.

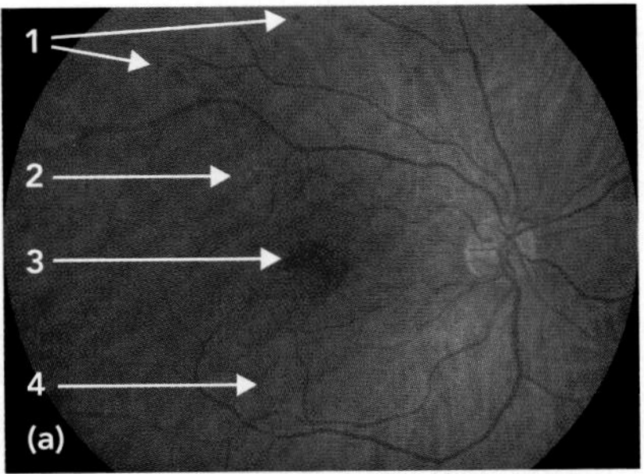

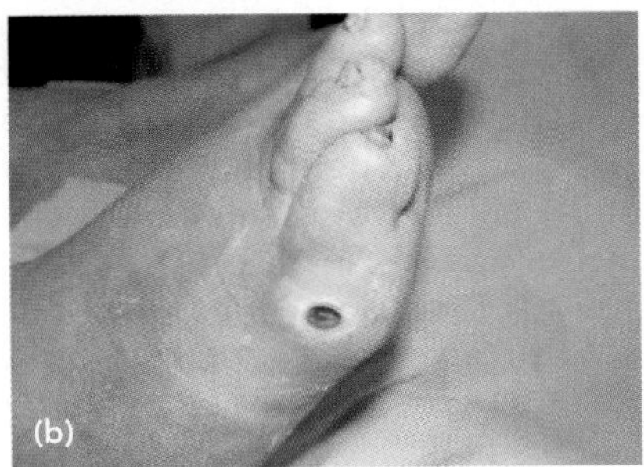

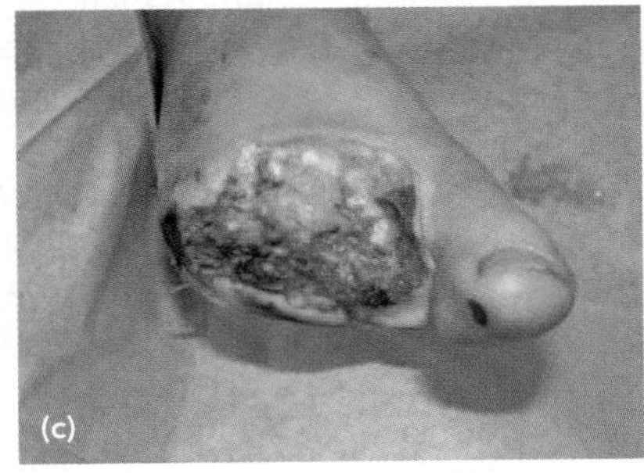

Figure 2.26 Long-term clinical complications of diabetes mellitus. (a) Diabetic retinopathy and diabetic maculopathy. Indicated by arrows are: (1) dot blot hemorrhages in retinal periphery; (2) hard semicircinate exudates superior to macula; (3) diabetic maculopathy with hemorrhages; and (4) inferior cotton wool spot indicating retinal inflammation. (b) A small, clean, innocuous-looking foot ulcer must be treated aggressively to prevent progression to amputation. (c) Ulceration 8 weeks after initial presentation as a small area of redness. No treatment was sought during this period. Amputation is now essential. (a, courtesy of David Bennett. b and c, courtesy of Vinod Patel, Warwick Medical School.)

2.8 CLINICAL CHEMISTRY MARKERS OF GLYCEMIC CONTROL

Self-monitoring of blood glucose

The explosion in the prevalence of diabetes has created a huge market for patient-held blood glucose meters. These have become progressively smaller and more convenient to use, producing accurate results within seconds from as little as 1 μL of capillary blood. Whilst self-monitoring is extremely useful in many patients, it may be unnecessary and a waste of resources in others. The key is whether the patient is able to act on the results obtained, and if they are not then collecting the data is not worthwhile. Where patients are on an insulin regime requiring variable doses or where hypoglycemic awareness is impaired, self-monitoring of glucose is mandatory. However, for patients who manage their diabetes with dietary restriction alone, self-monitoring is rarely necessary.

Glycated hemoglobin: HbA_{1c}

The proportion of hemoglobin that is glycated during its circulation in the vascular system has been used as a surrogate marker of average blood glucose since

TABLE 2.16 Assumptions and pitfalls in the measurement of HbA_{1c}

Assumptions	Blood glucose is the only variable
	Red cell survival time is constant
Pitfalls	Shortened red-cell survival due to iron deficiency, hemolytic anemias, or venesection
	Prolonged red-cell survival due to drugs such as dapsone or erythropoietin
	Interference in assay due to hemoglobin variants such as HbS, HbD, or HbC (>900 variants identified)

the 1980s. Its clinical value has been shown by two landmark studies published during the 1990s, which demonstrated the beneficial effects of improved glycemic control as measured by HbA_{1c}. The Diabetes Control and Complications Trial (DCCT) looked at type 1 diabetes, whilst the United Kingdom Prospective Diabetes Study (UKPDS) was a study of type 2. Techniques for measuring HbA_{1c} have become better standardized following the DCCT and UKPDS, allowing clinicians and patients to aim for the same target values.

There are a number of factors to be considered in the interpretation of HbA_{1c} measurements (**Table 2.16**). As hemoglobin glycation is nonenzymatic, the two main factors determining %HbA_{1c} are the prevailing concentration of glucose and the length of time the red cells (and hence the hemoglobin) are present in the circulatory system. Most clinicians focus on the first factor and overlook the second. In the majority of cases this does not matter, but in conditions where red-cell survival is longer or shorter than average, the HbA_{1c} will be higher or lower, respectively, for the same average blood glucose concentration. In addition, the presence of hemoglobin variants may affect the way HbA_{1c} is calculated, and this depends on the exact laboratory method used. Most commonly, a chromatographic technique is used to separate hemoglobin types and the relative proportions can then be calculated. If a variant co-migrates with either the HbA_{1c} (numerator) or the HbA_0 (denominator), the HbA_{1c} expressed as a percentage will be falsely high or low.

Analytical practice point 2.4

Glycation of hemoglobin is affected by red blood cell survival time as well as average blood glucose concentration. HbA_{1c} is lowered when red-cell turnover is high.

An international initiative to improve further the standardization of HbA_{1c} resulted in a change of the units in which it is reported (**Table 2.17**). Since 2011, HbA_{1c} has been reported as mmol HbA_{1c} per mole of total Hb (mmol/mol). The numerical results thus appear very different from those that had become familiar and so, for a period of two years from 2009, dual results in both % and mmol/mol were reported.

Universal HbA_{1c} targets, however, are controversial. A comparison of diagnostic criteria for diabetes mellitus using HbA_{1c} measured as % or concentration is shown in **Table 2.18**. Nevertheless, there are clinical situations in which these figures do not apply:

- Conditions with abnormal red-cell turnover such as anemias from hemolysis, spherocytosis, or iron deficiency (for example, in pregnancy)
- Hemoglobinopathies; HbS, HbC, HbF, and HbE, for example, may interfere with the measurement of HbA_{1c}, depending on the method used.
- In rapid-onset diabetes, such as in type 1 diabetes, the HbA_{1c} can be within the normal range despite marked hyperglycemia
- Near-patient testing using current HbA_{1c} tests is not deemed to be sufficiently accurate for diagnosis

In these and other cases where there is doubt as to the use of HbA_{1c}, the glucose criteria shown in Table 2.8 must be used. Concerns regarding renal failure can be overcome if specific assays are used.

TABLE 2.17 Changes in the units used in reporting HbA_{1c}

HbA_{1c} %	HbA_{1c} mmol/mol total Hb
6.0	42
7.0	53
8.0	64
9.0	75

HbA_{1c} has been reported as mmol/mol total Hb since June 1st 2009. The new target range for diabetic patients on treatment is 48–59 mmol HbA_{1c} /mol total Hb. New units and old were reported together until 2011.

Although it is accepted that good glycemic control reduces complications, this comes at the expense of more frequent hypoglycemia, which is unpleasant

TABLE 2.18 Diagnostic criteria for diabetes mellitus using glycated hemoglobin (HbA_{1c}) data

	HbA_{1c} (%)*	HbA_{1c} (mmol/mol total Hb)**
Diabetes mellitus	≥6.5	≥48
High-risk diabetes	6.0–6.4	42–47
Normal	≤6	≤42

*Data from Diabetes Control and Complications Trial; **Data from International Federation of Clinical Chemists.

and potentially dangerous. Most diabetes societies recommend an HbA_{1c} of around 6.5 to 7.5% (47.5–53.0 mmol HbA_{1c}/mol Hb) as the best compromise, but this range has been extrapolated from DCCT and UKPDS data. Recent trials aiming for lower values have not demonstrated any overall benefit and some individuals cannot achieve even the current targets with available treatment. When HbA_{1c} is compared to average blood glucose concentration measured at several points throughout the day, there is a good correlation at a population level. However, there is a large scatter around the regression line, meaning that two individuals may have identical HbA_{1c} results but very different average glucose concentrations. Thus, one may have regular hypoglycemic episodes whilst the other has daily hyperglycemia. This may reflect inter-individual differences in average red cell survival time or other factors affecting the rate of glycation or de-glycation.

Despite the growing use of HbA_{1c} as a diagnostic tool for diabetes, there are a number of clinical situations in which the use of HbA_{1c} is inappropriate and these include:

- All children and young adults
- Patients of any age suspected of having type 1 diabetes
- Patients with symptoms of diabetes for less than two months
- Patients at high diabetes risk who are acutely ill (for example, those requiring hospital admission)
- Patients taking medication that may cause a rapid glucose rise (for example, steroids or antipsychotics)
- Patients with acute pancreatic damage, including pancreatic surgery
- In pregnancy
- Presence of genetic, hematologic, and/or illness-related factors that influence HbA_{1c} and its measurement

However, use of the HbA_{1c} criteria shown in Table 2.18 should reduce the need to carry out oral glucose tolerance testing in the bulk of cases, particularly in the elderly and some ethnic minority groups where fasting glucose concentrations may not be a reliable indicator of diabetes.

Clinical practice point 2.2

HbA_{1c} may be used for the diagnosis of diabetes as well as monitoring, but there are a number of caveats.

Laboratory markers of complications

Glycemic control is not the only modifiable risk factor for diabetic complications. Both microvascular and macrovascular complications are more likely to occur in the presence of hypertension, hyperlipidemia, and smoking. Management of these risk factors, monitoring their levels, and looking for emerging complications is a key part of long-term diabetes care. The clinical laboratory contributes to this in a large part by measuring serum lipids and creatinine (as a marker of renal function). It also measures urinary albumin as a marker of nephropathy, the only complication that can be monitored biochemically.

When glomeruli are damaged, the barrier that prevents albumin and other plasma proteins from entering the renal filtrate is compromised and proteinuria results. This is detectable on urine dipsticks. It was recognized during the 1970s and 1980s that more sensitive assays for albumin could detect more subtle

TABLE 2.19 Clinical chemistry of microalbuminuria	
Pathological basis of disorder	Glomerular damage prevents albumin and plasma proteins entering renal filtrate thereby increasing urinary protein
Timing of sampling of urinary albumin	Measure on first morning urine as albumin–creatinine ratio (ACR)
Values of ACR indicating microalbuminuria	**UK and Europe:** Males: >2.5 mg albumin/mmol Females: >3.5 mg albumin/mmol
	USA: Within the range 30–299 μg albumin/mg creatinine
Repeat measurements	Should be confirmed in 2 out of 3 samples

degrees of proteinuria; that is, above normal but below the measuring range of conventional dipsticks. This more subtle proteinuria was given the term microalbuminuria, which is a misnomer as it implies a very small form of albumin, when it really refers to a very small amount. Over the last three decades, urinary albumin has proved to be a useful marker of diabetic nephropathy as it is detectable many years before GFR becomes impaired. This gives time for intensive treatment of glycemia, high blood pressure, and other risk factors, which can slow the rate of progression of nephropathy. In some cases, microalbuminuria is reversible. However, once it has progressed to a level which is dipstick-positive it is termed proteinuria (or macroproteinuria, again a misnomer), and this is almost always progressive. The presence of microalbuminuria is also a predictor of cardiovascular disease and death. Urinary albumin was originally measured on 24-hour or overnight collections, but this is impractical and has been replaced by measurement on early morning urine (when it is most concentrated; **Table 2.19**). The results are expressed as albumin–creatinine ratios (mg/mmol or μg/mg, depending on geography).

Analytical practice point 2.5

Tests for microalbuminuria are extremely sensitive and may be falsely raised by infection or inflammation anywhere in the genitourinary tract.

2.9 HYPOGLYCEMIA

Hypoglycemia is a clinical condition in which the concentration of glucose in extracellular fluid (ECF) is so low as to deprive the brain of its obligatory source of energy. It occurs when the rate of glycogenolysis and/or gluconeogenesis is insufficient to correct the low glucose concentration. Particularly significant conditions in which hypoglycemia may arise are:

- Complications of insulin treatment
- Ketotic hypoglycemia
- Insulinoma
- Ethanol-induced hypoglycemia
- Neonatal hypoglycemia

As described previously, the glucose concentration is maintained through the balanced actions of anabolic and catabolic hormones with insulin as the most important regulator, modulating glucose clearance from ECF and glucose release into ECF from body stores. Insulin, with an equimolar amount of C-peptide, is released from proinsulin in the pancreas in response to hyperglycemia, and plasma C-peptide is a marker of insulin secretion. In health, insulin secretion is controlled by the glucose concentration in ECF, and is amplified by GIP and other gut hormones. Conversely, when hypoglycemia exists, the secretion of insulin is decreased and the corrective action of catabolic hormones, particularly glucagon, epinephrine, and cortisol, stimulates glycogenolysis and gluconeogenesis in liver to restore plasma glucose concentration to the normal range. A further consequence of the actions of catabolic hormones is to stimulate lipolysis and ketogenesis and this occurs in starvation, for example, where insulin concentration falls to very low levels. However, hypoglycemia due to insulin excess is not associated with ketosis due to the strong antilipolytic and antiketotic actions of insulin which antagonize the weaker lipolytic action of glucagon and epinephrine.

Pathological causes of hypoglycemia

Pathological causes of hypoglycemia are listed in **Table 2.20**.

Hypoglycemia arising from treatment of diabetes mellitus

In practice, insulin treatment of type 1, insulin-dependent diabetic patients is, by far, the commonest cause of hypoglycemia in adults and older children and illustrates the relative inflexibility of exogenous administration of the hormone. Thus hypoglycemia may arise in such patients as an unwanted consequence of a number of situations:

- Use of an inappropriate preparation
- Missed meals
- Unusual heavy exercise
- Maintaining a particular dose when requirement falls (for example, during recovery from acute illness or weight loss)
- Use of a new injection site
- Attempting to abolish glycosuria
- Willful overdose of insulin

Induced hypoglycemia

Hypoglycemia may also arise (be induced) when patients are removed from a situation where they are exposed to hyperglycemia. For example (i) patients coming off a glucose drip (infusion) and (ii) neonates of hyperglycemic, diabetic mothers. In both of these cases, the level of insulin in the patient/baby will be raised as a response to the original hyperglycemia and will remain raised temporarily after removal of the hyperglycemic stimulus (glucose infusion/maternal circulation) thereby inducing hypoglycemia.

Hypoglycemia may also be induced by alcohol, particularly in conjunction with a poor diet, and misuse of drugs including insulin and sulfonylureas (occasionally). Consumption of the unripe West Indian ackee fruit will also give rise to hypoglycemia. This fruit contains toxins that inhibit the β-oxidation of fatty acids, which leads to overconsumption of glucose.

TABLE 2.20 Pathological causes of hypoglycemia

Cause	Examples
Excess plasma insulin	Insulin overdose (exogenous)
	Insulinoma (endogenous)
Increased insulin-like activity	Secretion of insulin-like growth factor 2 (IGF2)
Deficiency of catabolic hormone	Glucagon
	Cortisol
Liver failure	End-stage
	Fulminant
	Reye's syndrome
Specific inherited enzyme defects	Glycogen storage diseases
Failure of gluconeogenesis	Chronic alcoholism
Low carbohydrate intake	Anorexia
	Prolonged starvation

CASE 2.1

A 34-year-old woman presents to the emergency department with a collapse at home. She is found to have low blood glucose on a capillary blood sample taken in the ambulance and is given intravenous glucose. She gives a history of several similar collapses. Her 8-year-old son has type 1 diabetes. A blood sample taken before the glucose was given shows the following results.

	SI units	Reference range	Conventional units	Reference range
Plasma				
Glucose	1.8 mmol/L	3.9–6.1	32 mg/dL	70–110
Insulin	300 pmol/L	14–140	43 μIU/mL	2–20
C-peptide	0.09 nmol/L	0.17–0.90	0.26 ng/mL	0.5–2.7

- **What are the possible causes of hypoglycemia with raised insulin?**
- **How does the C-peptide level help with the diagnosis?**
- **Why is it important to take blood samples before giving glucose?**
- **How is the family history relevant?**

Hypoglycemia may be caused by insulin excess if the insulin is produced endogenously from an islet cell tumor (insulinoma) or if it is administered by injection. Every endogenous insulin molecule that is secreted is accompanied by one molecule of C-peptide. However, pharmaceutical insulin has no C-peptide. Therefore, raised insulin with low C-peptide indicates an exogenous source. This may be given by the patient injecting themselves or given by another person, for example a carer of a child or an elderly person. In this case, the son having diabetes means the patient will have access to insulin and this may be a factitious, self-inflicted illness. It should be noted that modern insulin analogs used to treat diabetes may not be detected in all insulin immunoassays, due to small but important changes in the amino acid sequence. Samples in hypoglycemic patients must be taken before glucose is administered, otherwise an appropriate secretion of insulin and C-peptide may be stimulated.

Spontaneous hypoglycemia

The group of conditions associated with spontaneous hypoglycemia includes:

- Endogenous hyperinsulinism as a result of an insulinoma (nesidioblastosis, insulin-binding antibodies)
- Severe liver disease which eliminates the capacity of the liver to carry out carbohydrate metabolism (glycogenolysis and gluconeogenesis); fulminant hepatic failure, end-stage liver failure, or Reye's syndrome
- Non-pancreatic tumors such as mesothelioma, hepatoma, and adrenal carcinoma
- Severe nutritional failure
- Type I glycogenosis and fructose 1,6-bisphosphatase deficiency

Ketotic hypoglycemia

Ketotic hypoglycemia occurs in children and particularly in underweight boys aged 2–4 years. It is a self-limiting condition and episodes of variable severity are precipitated by missed meals and intercurrent illness. As the name implies, it presents with low plasma glucose and ketosis, a situation akin to accelerated starvation. The plasma insulin is low and high levels of ketones are present in both plasma and urine. Causative factors include decreased mobilization of alanine from muscle to provide substrate for gluconeogenesis between meals and decreased responses in the adrenal gland, both cortex and medulla, to the prevailing hypoglycemia.

Ethanol-induced hypoglycemia

The oxidized form of NAD (NAD^+) is a key co-factor in the process of gluconeogenesis. NAD^+ is also a co-factor for the two enzymes that catalyze the oxidation of ethanol to ethanal (acetaldehyde) and subsequently to acetic (ethanoic) acid (**Figure 2.27**). Oxidation of ethanol depletes the cytosolic concentration of NAD^+ in the hepatocyte and thereby impairs gluconeogenesis. Hypoglycemia, which may be severe, occurs during fasting when the store of glycogen has been exhausted, often several hours after ingestion of alcohol. Subjects who present in clinical practice are often children who drink surreptitiously and adults, particularly vagrants who drink heavily but do not eat. Such subjects do not elicit a rise in plasma glucose on injection of glucagon. A mild, clinically unimportant hypoglycemia may arise via potentiation of insulin secretion when ethanol is taken with a small amount of carbohydrate, such as gin and tonic, when fasting.

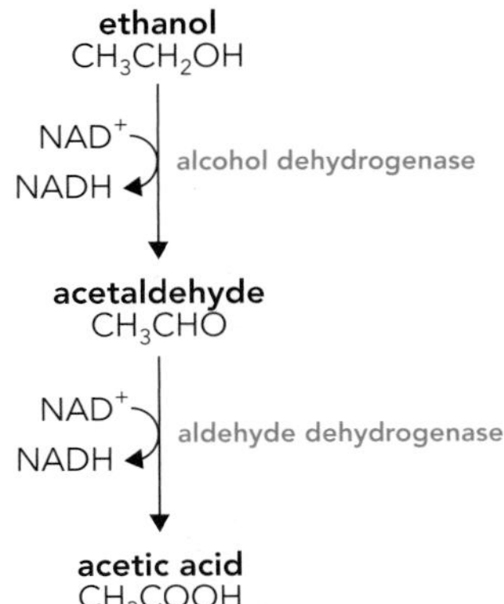

Figure 2.27 Metabolism of ethanol. Ethanol is oxidized sequentially to acetic acid (ethanoic acid) by the action of two enzymes, alcohol dehydrogenase and aldehyde dehydrogenase, both of which require NAD^+ as a co-factor. In the first reaction, ethanol is oxidized to acetaldehyde by alcohol dehydrogenase (co-factor NAD^+). In the second reaction, acetaldehyde is oxidized to acetic acid by aldehyde dehydrogenase (co-factor NAD^+).

Insulinoma

The diagnosis of insulinoma depends on the demonstration of autonomous hypersecretion of insulin. Insulinomas are rare, usually small, single, benign tumors occasionally associated with other endocrine tumors (for example, multiple endocrine neoplasia type 1). The hypoglycemia resulting from secretions by the tumor is of variable severity and often precipitated by fasting and exercise but sometimes also by food.

Hypoglycemia in neonates

Neonatal hypoglycemia is characterized by a plasma glucose concentration <2.0 mmol/L (<36 mg/dL) within the first 72 hours after delivery and <2.5 mmol/L (<45 mg/dL) following this period. The major causes of neonatal hypoglycemia are included in **Table 2.21**.

Laboratory investigation of hypoglycemia

Clinically, the diagnosis of hypoglycemia depends upon the demonstration of Whipple's triad: (i) symptoms consistent with hypoglycemia; (ii) low blood glucose at the time of symptoms; and (iii) resolution with the provision of glucose. This may be difficult to fulfill—in particular, there may be no measurement of

TABLE 2.21 Major causes of neonatal hypoglycemia

Incidence	Causes
Common: mild, short-lived hypoglycemia	Intrauterine malnutrition due to immaturity of glucose homeostatic mechanisms in pre-term and small-for-dates babies
	Serious illness including infection and hypoxia
	Poorly controlled diabetes mellitus in mother
	Abrupt cessation of intravenous glucose
Less common: severe, persistent hypoglycemia	Inborn errors of metabolism: for example, galactosemia, type I glycogenosis, organic acidemias especially methylmalonic, tyrosinemia type I, fatty acyl-CoA dehydrogenase deficiency
	Hyperinsulinism including nesidioblastosis, Beckwith–Wiedemann syndrome, or severe Rhesus incompatibility (erythroblastosis fetalis)
	Hormone deficiency including congenital adrenal hyperplasia and congenital hypopituitarism

blood glucose during an episode and a patient may require hospital admission to observe for symptoms and obtain the crucial sample. If hypoglycemia is demonstrated, it is also critical that blood samples and further investigations are performed before the patient is given glucose. It is very difficult, if not impossible, to interpret results in the presence of rising plasma glucose.

The most important measurements are insulin and C-peptide. These should be suppressed in hypoglycemia. If not, they indicate inappropriate endogenous insulin secretion, such as due to an insulinoma. If the insulin is raised but the C-peptide is suppressed, this indicates exogenous insulin administration, either by the patient or a third party. Pharmacological insulin used to treat diabetes has no C-peptide as this is removed during manufacture. If insulin and C-peptide are both appropriately suppressed, this indicates either failure to mobilize fatty acids or inability to metabolize them for energy. Measurement of plasma free fatty acids and ketones (3-hydroxybutyrate) can distinguish these two possibilities. Endocrine deficiencies can cause hypoglycemia and measurement of plasma cortisol and growth hormone are useful. These are usually stimulated to high levels by hypoglycemia and their failure to rise points to pituitary or adrenal insufficiency.

An unusual cause of hypoglycemia is excessive production of insulin-like growth factor 2 (IGF2), which can be measured by reference laboratories. IGF2 is an embryonic growth factor, the gene for which is repressed after birth. In certain tumors, in particular retroperitoneal fibrosarcomas, the gene is derepressed and IGF2 is released into the circulation. As the name suggests, it has insulin-like effects, including causing hypoglycemia.

Clinical practice point 2.3

Insulin and C-peptide are the first-line tests for spontaneous hypoglycemia and can distinguish between endogenous and exogenous sources of insulin.

Analytical practice point 2.6

Modern insulin analogs may not be detected in routine insulin immunoassays.

2.10 METABOLISM OF FRUCTOSE AND GALACTOSE

Sucrose (cane sugar) and lactose (milk sugar) are important disaccharides that may contribute appreciably to carbohydrate intake. The structures of the two disaccharides are shown in **Figure 2.28**. Hydrolysis of sucrose yields fructose and glucose, while lactose is hydrolyzed to galactose and glucose. Metabolism of both fructose and galactose involves converting them into intermediates of the glycolytic pathway.

Fructose

Fructose is initially phosphorylated to fructose 1-phosphate in a reaction catalyzed by fructokinase, and fructose 1-phosphate undergoes cleavage by a specific aldolase, fructose 1-phosphate aldolase, to dihydroxyacetone phosphate (a glycolytic intermediate) and glyceraldehyde (**Figure 2.29**). Phosphorylation of glyceraldehyde by triose kinase yields another glycolytic intermediate, glyceraldehyde 3-phosphate.

The affinity of fructose for hexokinase is an order of magnitude lower than of glucose and so little fructose 6-phosphate (a glycolytic intermediate) is formed in liver under the prevailing intracellular glucose concentration. The competition for hexokinase in adipose tissue, however, is not as great because glucose is not present intracellularly in sufficiently high concentration so appreciable amounts of fructose 6-phosphate may be formed from fructose in this tissue.

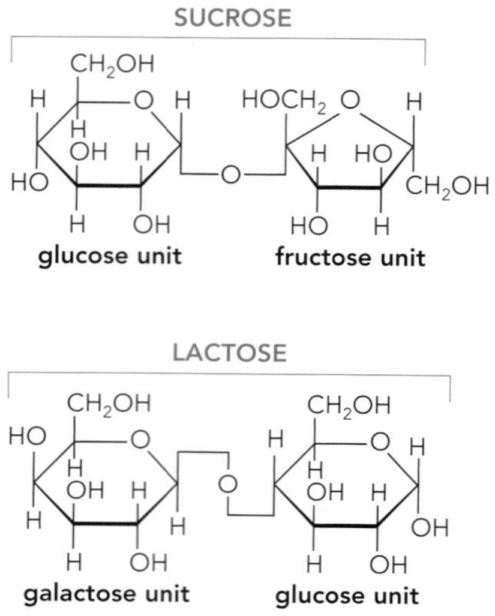

SUCROSE

CH_2OH H O H H OH H HO H OH $HOCH_2$ O H H HO CH_2OH HO H

glucose unit — fructose unit

LACTOSE

CH_2OH HO O H OH H H H H OH O CH_2OH H O H H OH H OH H OH

galactose unit — glucose unit

Figure 2.28 Chemical structure of sucrose (glucose α(1→2)β fructose) and lactose (galactose β(1→4) glucose).
Sucrose, from cane and beet, is a major dietary source of glucose. It is hydrolyzed to its constituent monomers, glucose and fructose, by sucrases on the intestinal brush border prior to absorption. Lactose, from milk, is hydrolyzed to its monomers, galactose and glucose, by lactase, secreted from the intestinal villi.

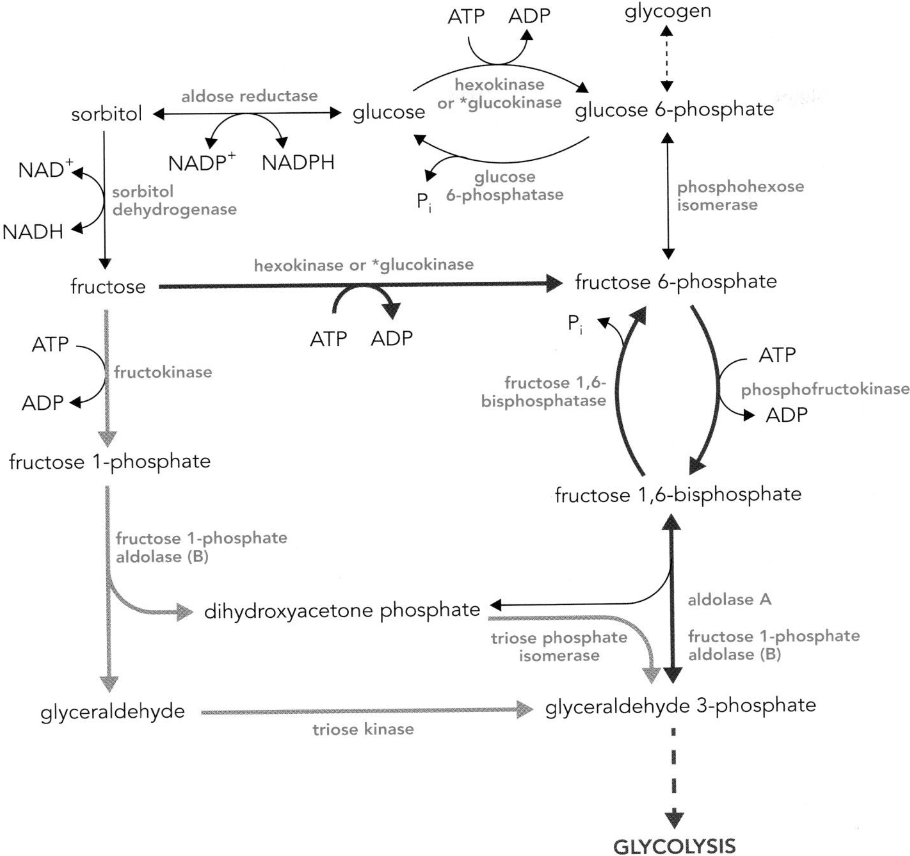

Figure 2.29 Metabolism of fructose.
The affinity of fructose for hepatic hexokinase is twentyfold lower than of glucose and thus in the liver (orange arrows) fructose is metabolized mainly to fructose 1-phosphate by fructokinase; fructose 1-phosphate is in turn hydrolyzed to glyceraldehyde and dihydroxyacetone phosphate by a specific hepatic fructose 1-phosphate aldolase. Triose kinase phosphorylates glyceraldehyde to glyceraldehyde 3-phosphate, and both glyceraldehyde 3-phosphate and dihydroxyacetone phosphate can enter the glycolytic pathway. In tissues such as adipose (blue arrows), where intracellular glucose concentration is much lower than in liver, fructose is a substrate for hexokinase and can enter glycolysis at the level of fructose 6-phosphate. Dashed arrows indicate intermediate steps not shown in the diagram.
* The affinity of glucose and fructose for glucokinase is much lower than for hexokinase. Glucokinase action becomes significant only at high substrate concentrations.

Galactose

Three enzymes are required to convert galactose to glucose 6-phosphate (**Figure 2.30**). Galactose is first phosphorylated to galactose 1-phosphate by galactokinase, and UDP-galactose is formed by UDP transfer from UDP-glucose

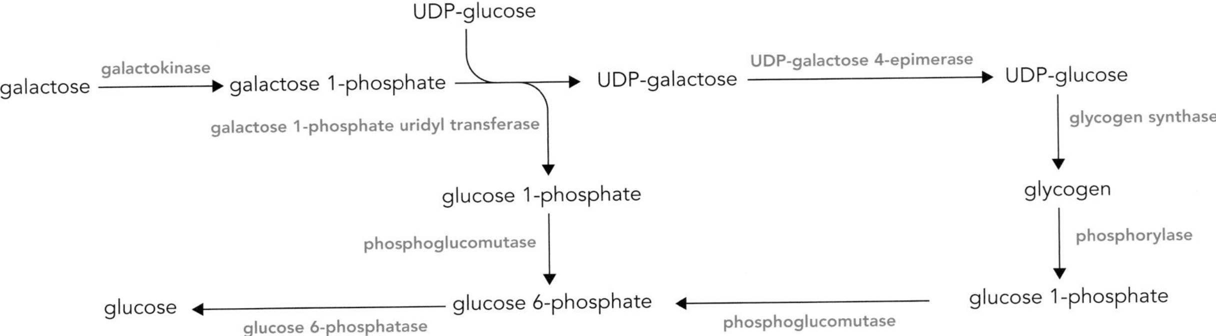

Figure 2.30 Metabolism of galactose.
Galactose, derived from dietary lactose, is phosphorylated in liver by galactokinase to galactose 1-phosphate, which in turn is converted to uridine diphosphate galactose (UDP-galactose) by transfer of a uridyl group from UDP-glucose. This reaction is catalyzed by galactose 1-phosphate uridyl transferase and glucose 1-phosphate is another product of the reaction. This product is converted to glucose 6-phosphate by phosphoglucomutase. Epimerization, by UDP-galactose 4-epimerase, of the hydroxyl group of UDP-galactose at C4 of the galactose moiety yields UDP-glucose, which can add glucose to stored glycogen.

in a reaction catalyzed by galactose 1-phosphate uridyl transferase. The other product of this reaction is glucose 1-phosphate, which can be converted to glucose 6-phosphate by phosphoglucomutase. The reversible epimerization of UDP-galactose, catalyzed by UDP-galactose 4-epimerase, yields UDP-glucose which in turn enters the glycolytic pathway.

2.11 INBORN ERRORS OF CARBOHYDRATE METABOLISM

There are many inborn errors of carbohydrate metabolism and some of these, including disorders of glucose, fructose, galactose, and glycogen metabolism, are described in Chapter 23.

LIVER

CHAPTER
3

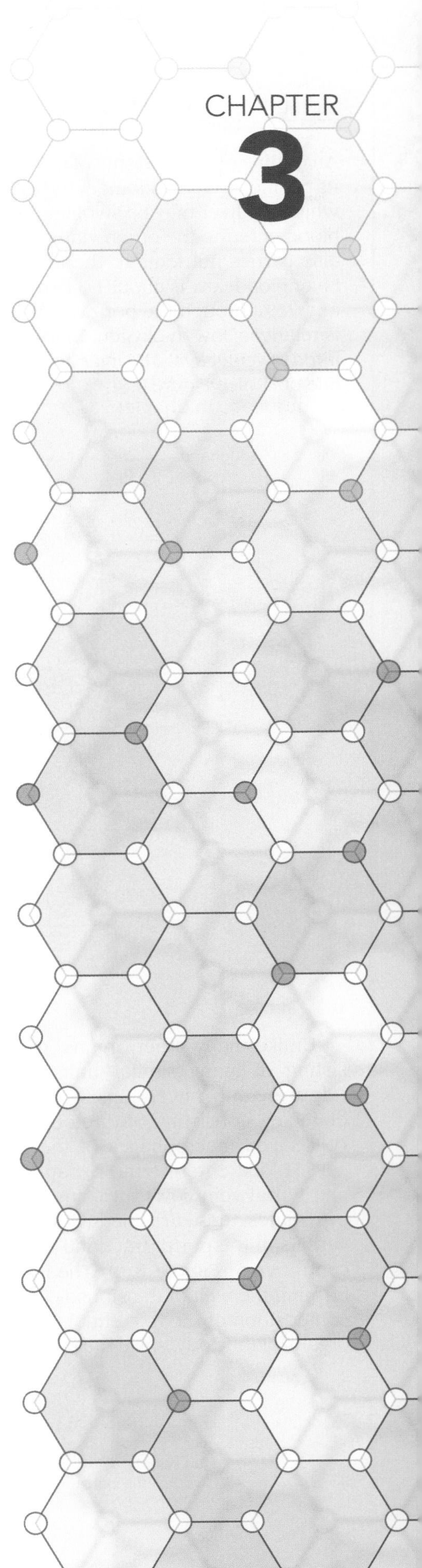

IN THIS CHAPTER

LIVER ANATOMY AND STRUCTURE

FUNCTIONS OF LIVER

MAJOR CAUSES OF LIVER DISEASE AND THEIR INVESTIGATION

The liver is the largest organ in the body, weighing approximately 1.5 kg in the adult human. As a proportion of total body weight, it is relatively larger in infants. The liver has a large metabolic capacity and plays a central role in whole-body energy metabolism. This large reserve of metabolic capacity ensures that even when two-thirds of the tissue is removed or damaged there may be little loss of overall metabolic function; that is, it can still carry out essential metabolic functions of plasma protein synthesis and carbohydrate and fat metabolism. It also means that mild liver disease may present with no symptoms other than a propensity to tiredness and general malaise with diagnosis arrived at only after phlebotomy and blood analysis by liver function tests.

Alone amongst the major organs, the liver has a remarkable ability to regenerate, even if partially removed by surgery. Any acute liver injury, if not severe enough to cause fatal liver failure, may completely resolve without long-term damage. In some situations, however, continuing injury leads to chronic liver disease. The liver's response to ongoing damage is a combination of the regeneration of hepatocytes plus scarring with fibrotic tissue. This in turn alters the flow of blood through the lobules, leading to further hypoxic cell death and increasing back pressure in the portal vein (portal hypertension). The end stage of this process is called cirrhosis. Eventually, the number of viable hepatocytes may fall to the point where the liver can no longer maintain its metabolic function and decompensation occurs, causing liver failure. At this point, portal hypertension has often caused the development of varices (enlarged and thin-walled veins) at the lower esophagus, which can bleed catastrophically. Regenerating nodules in cirrhosis increase the risk of hepatocellular carcinoma, which is another potentially fatal complication.

This chapter describes the major functions of the liver and how the loss of these functions relates to symptoms and findings in disease states. The major emphasis is on bilirubin, bile salts, the urea cycle, and drug metabolism. Other functions including carbohydrate and fat metabolism, plasma proteins (including clotting factors), and digestion are described in other chapters.

3.1 LIVER ANATOMY AND STRUCTURE

An outline of the anatomy of the liver is shown in **Figure 3.1**. As can be seen by its position in the abdomen, the edge of the organ can be palpated in diseases in which the liver may be swollen (hepatomegaly). It has a dual blood supply, with blood entering the organ via both the hepatic artery and the portal vein, which also carries nutrients to the liver from the gastrointestinal tract (**Figure 3.2**). Liver blood flow is about 1.3 L/min, with 25% flowing through the hepatic artery and 75% through the portal vein. This means that the blood supply to the liver is relatively low in oxygen. Blood drains into the hepatic veins and then into the inferior vena cava. The liver also has bile ducts that drain into the main common bile duct (see Figure 3.2).

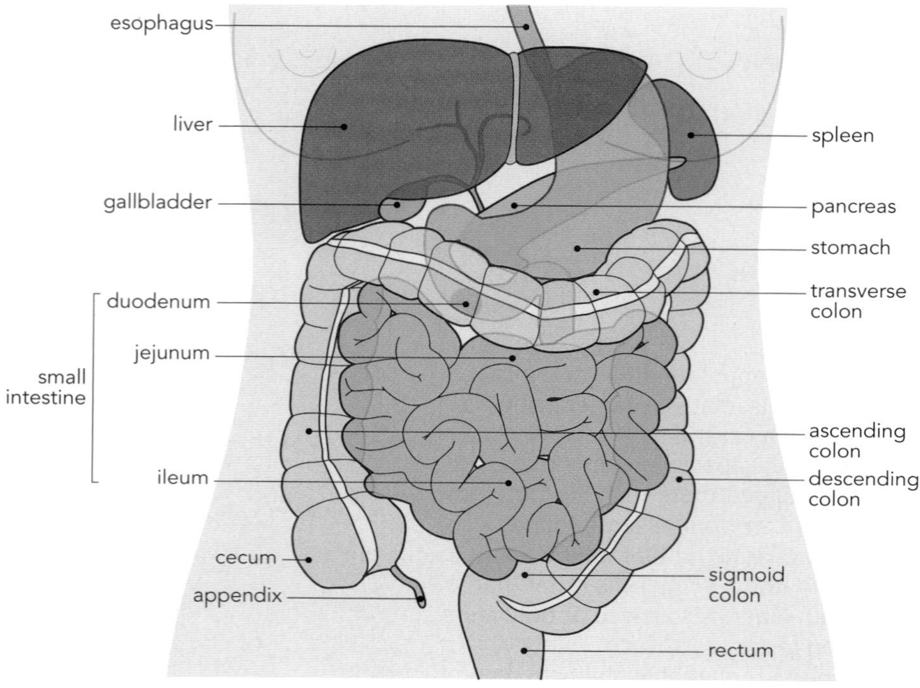

Figure 3.1 Anatomy of the liver. The figure shows the liver in relation to the other main organs in the abdomen.

Unlike many other organs, the liver is very homogeneous in nature, consisting of many similar units of processing power called hepatic lobules (**Figure 3.3**). A liver consists of about 50,000 lobules, each about 1–2 mm in diameter, consisting of three cell types: hepatocytes (true liver cells), Kupffer cells (fixed macrophages of the reticuloendothelial system), and endothelial cells. Blood entering the hepatic lobules from branches of the hepatic artery and portal vein flows into sinusoids, capillary-like channels between hepatocyte plates; these are lined with endothelial cells. Blood enters the lobules from vessels in the portal tract and flows centripetally through the sinusoids to the central vein (**Figure 3.4**). The interstitial space between the endothelial cells forming the wall of the sinusoid and the hepatocytes is the space of Disse. The composition of the interstitial fluid in the space of Disse is unlike that of other organs of the body, since the protein concentration is approximately 90% that of

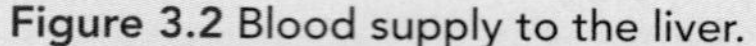

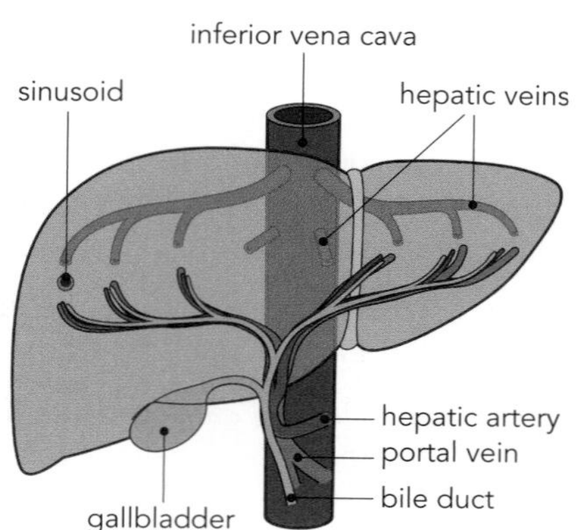

Figure 3.2 Blood supply to the liver. The liver has a dual blood supply. Oxygenated blood enters the liver via the hepatic artery whilst blood draining from the intestinal tract enters the liver via the portal vein.

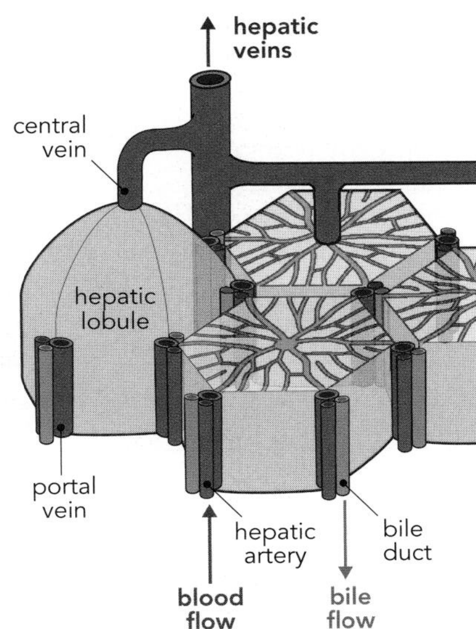

Figure 3.3 Lobular structure of the liver.
Most of the main functions of the liver can all be carried out in each lobule. Each lobule has the dual blood supply, venous outflow, and bile duct.

plasma. This means that the hepatocytes are readily able to absorb bloodborne proteins and secrete proteins into the blood.

The filtering of splanchnic blood from the gastrointestinal tract and spleen by the liver allows it a measure of pharmacodynamic control over ingested drugs, and also provides a mechanism for detoxification of any poisonous material absorbed from the gastrointestinal tract. The pattern of delivery of blood causes the hepatocytes of the various parts of the hepatic lobule to be prone to different insults and has implications for the pattern of damage that can be seen in the liver when a biopsy is taken. If the liver is subject to an ischemic insult, hepatocytes closer to the central vein of the hepatic lobule are more prone to damage because these cells receive relatively less oxygen than those near the portal tract. On the other hand, if the liver is damaged by a hepatotoxic substance carried in the blood or arriving from the gastrointestinal tract, the hepatocytes closer to the portal tract are more prone to damage because these cells are exposed to a higher dose of the toxin. Blockage of hepatic blood flow leads to shunting of mesenteric blood around the liver with consequent encephalopathy and hemorrhage from esophageal and gastric varices.

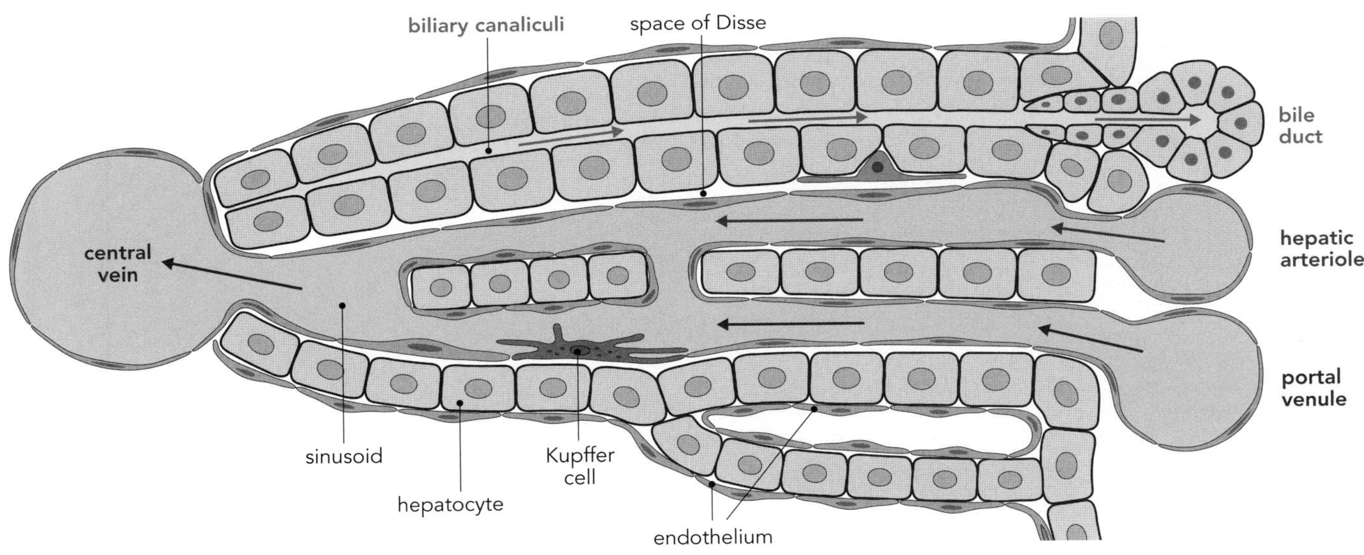

Figure 3.4 Acinar structure of the liver.
Cellular-level view of the mixing of the arterial and portal blood, and the outflow of the bile via the biliary canaliculi.

3.2 FUNCTIONS OF LIVER

Serum reference ranges in the assessment of liver function are shown in **Table 3.1** and the functions of the liver are shown in **Table 3.2**. As these functions become impaired, symptoms of liver disease present, with jaundice being the most obvious example. Where relevant, these functions will be discussed in more detail in this chapter, and extra information can be found in other appropriate chapters.

TABLE 3.1 Serum reference ranges in the assessment of liver function

Analyte	Serum reference range	
	SI units	Conventional units
Total bilirubin	<21 μmol/L	<1.3 mg/dL
Alanine transaminase (ALT)	5–40 U/L	5–40 U/L
Aspartate transaminase (AST)	5–40 U/L	5–40 U/L
Alkaline phosphatase (ALP)	30–120 U/L	30–120 U/L
Gamma-glutamyl transferase (GGT)	5–45 U/L	5–45 U/L
Albumin	32–52 g/L	3.2–5.2 g/dL
Total protein	60–80 g/L	6.0–8.0 g/dL

TABLE 3.2 Functions of the liver

Energy metabolism	Major site of oxidative metabolism for generation of: (1) ATP and reduced co-factors for chemical reactions (2) Heat; approximately 25% of total heat production at rest
Intermediary metabolism	Glycogen synthesis, storage, and mobilization
	Gluconeogenesis
	Ketogenesis
	Urea synthesis
	Interconversions of sugars; for example, galactose and fructose to glucose
Plasma protein synthesis	Synthesis of most plasma proteins (except immunoglobulins) Angiotensinogen
Lipid metabolism	Formation of bile; synthesis of bile acids and bile salts
	Synthesis and excretion of cholesterol via bile
	Synthesis and secretion of lipoproteins VLDL and nascent HDL
	Removal of LDL from circulation
Detoxification	Metabolism of endogenous and exogenous molecules (xenobiotics) including drugs
	Solubilization by hydroxylation and conjugation of excretory products; for example, bilirubin
Precursor activation	Activation of precursor molecules; for example, cholecalciferol to 25-hydroxy cholecalciferol
Storage	Cobalamins (vitamin B_{12}), iron, and fat-soluble vitamins

VLDL, very-low-density lipoprotein; HDL, high-density lipoprotein; LDL, low-density lipoprotein.

Energy metabolism

The ability of the liver to maintain its temperature at 37°C is dependent on the generation of heat (thermogenesis) through metabolic processes, particularly oxidation and the hydrolysis of energy-rich nucleoside triphosphates such as ATP. The splanchnic region, which includes the gastrointestinal tract, liver, and spleen, is a major site of thermogenesis, and metabolism in the liver may account for up to 25% of the body's heat production at rest.

Carbohydrate metabolism

Detailed information about carbohydrate metabolism can be found in Chapter 2. However, it is important to emphasize the unique role of the liver in the maintenance of blood glucose concentration. It is able to respond to a rise in plasma glucose, after a meal for example, by storing it as glycogen (glycogenesis), and respond to a fall in plasma glucose by releasing glucose into the blood, initially from the stored glycogen (glycogenolysis) and later from the synthesis of new glucose (gluconeogenesis).

Fat metabolism

The liver is one of the major sites for the processing of lipids, including the synthesis and degradation of fatty acids and triglycerides, and the synthesis of cholesterol and its degradation and excretion. Details of some of the basic processes of lipid metabolism can be found in Chapter 18. It is worth noting here, however, those specific aspects of hepatic fat metabolism which, when impaired, produce symptoms of disease. For example, there are two possible routes of fatty acid metabolism in liver—oxidation to CO_2 in mitochondria or incorporation into triglyceride for very-low-density lipoprotein (VLDL) assembly—and the relative flow into each is dependent on the hormonal state. Thus in catabolic states (for example, fasting) the bulk of the fatty acids are oxidized, but in more overwhelmingly catabolic states, particularly where there is peripheral resistance to insulin (for example, uncontrolled diabetes mellitus, trauma, or acute alcohol intoxication), hepatic synthesis of triglyceride becomes significant.

Increased lipolysis may be accompanied by ketonemia, as the liver synthesizes ketone bodies—acetoacetate and 3-hydroxybutyrate—when β-oxidation proceeds at a rate which exceeds that at which acetyl-CoA enters the tricarboxylic acid (TCA) cycle. Ketogenesis is the synthesis of ketone bodies (ketones) from fatty acid-derived acetyl-CoA. This process occurs only in the liver since it is the only tissue with the enzymes required for ketogenesis and has a limited TCA cycle activity. The switch from glucose to ketone body utilization in the brain during starvation reduces considerably the body's daily requirement for glucose and, by reducing the need for gluconeogenesis, helps to conserve lean body mass.

Protein and amino acid metabolism

Nitrogen balance

Although the lean body mass (muscle mass) of healthy adults changes little from day to day, the tissue proteins, including those of muscle, turn over at a rate of between 300 g and 400 g of protein per day and this involves degradation and resynthesis of protein. Healthy adults are said to be in nitrogen balance, where the amount of nitrogen entering the body (mainly dietary protein) is balanced by the amount of nitrogen excreted each day (mainly as urea in the urine).

Amino acids are also an important source of metabolic energy, providing substrates for gluconeogenesis (see Chapter 2). However, even in excessive catabolic states where gluconeogenic rates are high (for example, major burns and infection), amino acids derived from protein degradation (mainly muscle) never provide more than about 20% of the metabolic fuel requirements.

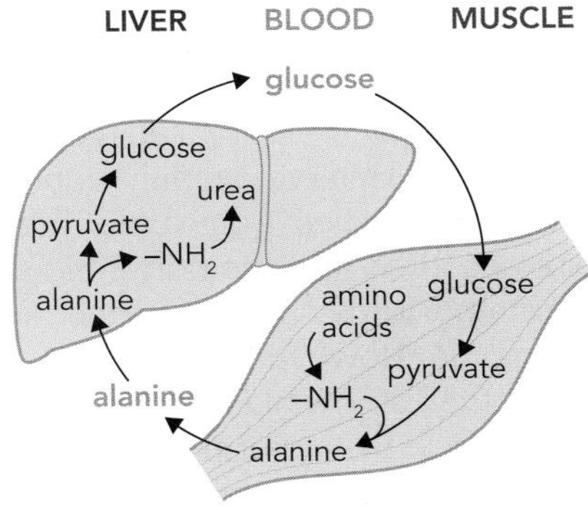

Figure 3.5 The glucose–alanine cycle.
Liver is the only tissue capable of synthesizing urea from the α-amino groups of amino acids derived from muscle proteins. The α-amino groups of muscle amino acids are transported to liver after transamination of pyruvate to form alanine, which exits the muscle and is carried to the liver. Once in the liver, the alanine amino group is incorporated into urea and excreted via the urine, whilst pyruvate is used to form glucose via gluconeogenesis. Glucose can, in turn, be transported back to muscle and undergo glycolysis to pyruvate.

Protein and amino acid catabolism

Both glucocorticoids and glucagon stimulate proteolysis in muscle and amino acid catabolism in liver by increasing the expression of the enzymes involved in catabolism. Glucagon also stimulates the uptake and subsequent metabolism by the liver of circulating amino acids, and alanine in particular (**Figure 3.5**); this is important in the catabolism of muscle-protein-derived amino acids (see Chapter 2). The initial reaction in the breakdown of many amino acids involves the removal of the α-amino group, by pyridoxal phosphate-dependent transamination to a ketoacid, and catabolism of the remaining carbon skeleton. The nitrogen of amino acids is a metabolic waste product. The plasma activities of two of these aminotransferases, alanine transaminase (ALT) and aspartate transaminase (AST), are measured in liver function tests since they are released into the circulation when the hepatocyte plasma membrane is damaged.

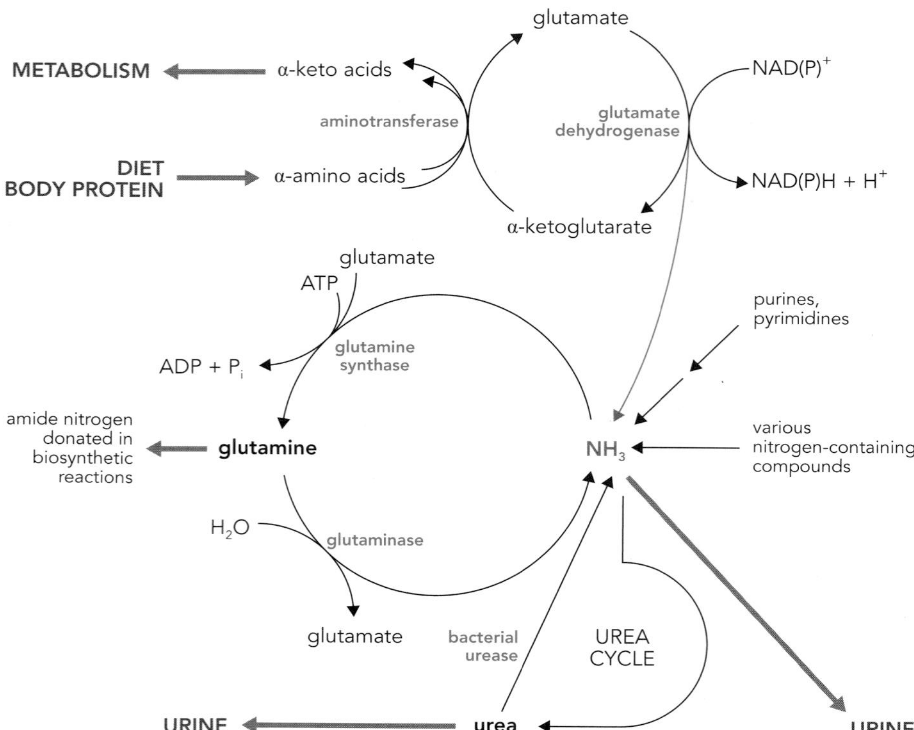

Figure 3.6 Glutamine synthase and glutaminase.
The formation of glutamine from glutamate serves as a way of storing nitrogen for anabolic reactions and for removal of excess ammonia. Its synthesis from glutamate and ammonia is catalyzed by glutamine synthase in a reaction requiring energy from the hydrolysis of ATP. Glutamate itself may be synthesized by transamination of amino groups from amino acids to α-ketoglutarate, which is formed in the TCA cycle. Glutamate can also provide ammonium ions by the action of glutamate dehydrogenase (red arrow).

Amino acid catabolism and ammonia production

Ammonia is produced during the catabolism of many nitrogenous molecules, the most important quantitatively being amino acids. Ammonia production from amino acids derived from catabolism of dietary and endogenous proteins is approximately 1 mol/day. The production of ammonia in the liver from muscle-derived amino acids requires transamination steps in both tissues followed by oxidative deamination of glutamate (**Figure 3.6**). Small amounts of ammonia formed from amino acids by bacterial enzyme action in the colon are absorbed into the portal blood and also cleared by the liver. Deamination of amino acids such as serine, leucine, and glycine in muscle also produces ammonia. However, little nitrogen is excreted as ammonia, since ammonia is toxic even at low plasma concentration. This problem of toxicity is overcome in the muscle by the formation of glutamine from glutamate and ammonium ions by the enzyme glutamine synthase. Glutamine, which is water soluble, enters the circulation and transports nitrogen to the gut or the liver. In these tissues, which have active glutaminase enzymes, glutamine is converted back to glutamate and ammonium ions (see Figure 3.6). In the gut, a rapidly dividing tissue, ammonium ions provide nitrogen for nucleic acid synthesis, while in the liver they are converted into urea.

Urea cycle

The incorporation of ammonia into urea in the urea cycle is quantitatively the most important mechanism of elimination of ammonia from the body. Liver is the only tissue capable of synthesizing urea from the α-amino groups of amino acids, derived from proteins in the liver itself or from muscle proteins and delivered to liver as alanine or glutamine (see above); glucose–alanine cycle). Urea is small, extremely water soluble, and electrically neutral, with a high proportion of nitrogen per molecule; these are properties which enable it to be handled easily by the kidney. The nitrogen atoms of urea are derived by transfer of α-amino

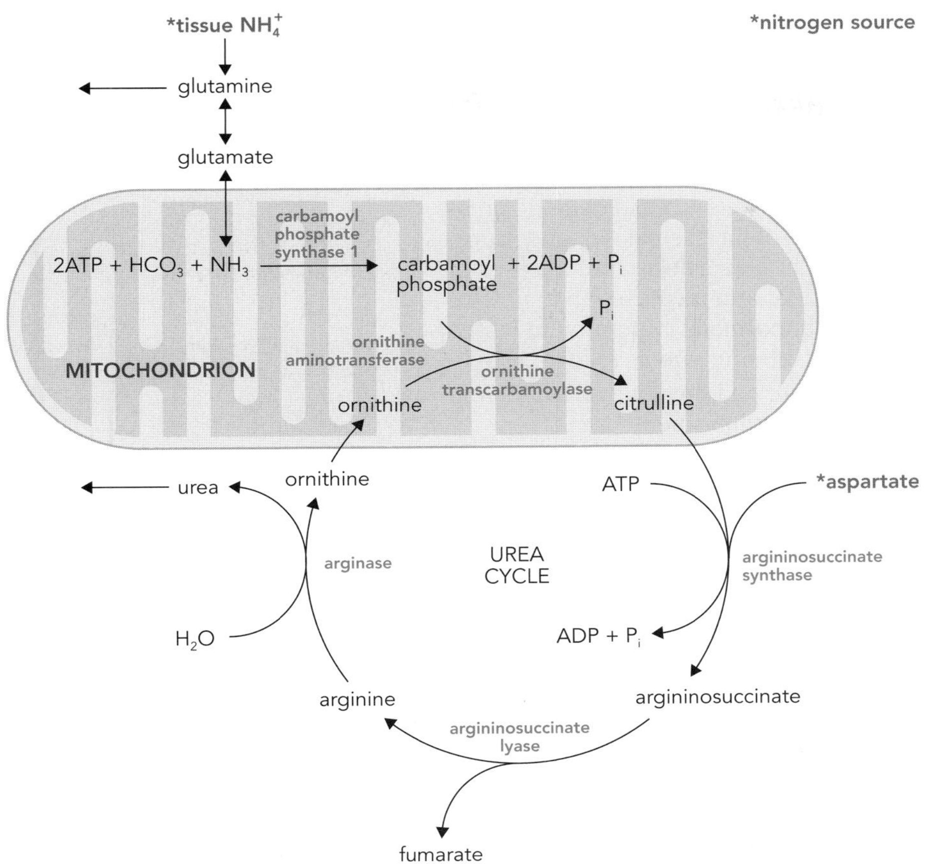

Figure 3.7 The urea cycle. Ammonia is a highly toxic metabolic product and may cause symptoms at concentrations above 50 µmol/L. The urea cycle of five enzymes, present in liver, serves as a mechanism to prevent accumulation of ammonia by the formation of a soluble, neutral, nitrogen-rich molecule (urea), which is excreted in the urine. The first two enzymes of the cycle, carbamoyl phosphate synthase 1 and ornithine transcarbamylase, are present in the mitochondrion, while argininosuccinate synthase, argininosuccinate lyase, and arginase are found in the cytoplasm. *N*-acetyl glutamate, whose synthesis is catalyzed by *N*-acetyl glutamate synthase, is a potent activator of carbamoyl phosphate synthase 1 and modulates the activity of the cycle and the mitochondrial transporters, ornithine transport protein and aspartate carrier protein (SLC25A13), not shown here.

groups of amino acids to α-ketoglutarate or oxaloacetate to form glutamate and aspartate, respectively. Oxidative deamination of the glutamate yields ammonium ions, which are converted to carbamoyl phosphate for entry into the urea cycle. Aspartate donates the other nitrogen atom later in the cycle (**Figure 3.7**). Carbamoyl phosphate synthase 1 catalyzes the formation of carbamoyl phosphate and requires the presence of *N*-acetyl glutamate as a co-factor. Carbamoyl phosphate condenses with ornithine to form citrulline, a reaction catalyzed by ornithine transcarbamylase. The final reaction of the cycle catalyzed by arginase liberates urea for excretion and ornithine for another round of the cycle. In addition to being incorporated into urea, excretion of ammonia as ammonium ions (NH_4^+) in urine is used as a normal means of eliminating hydrogen ions from the body and may reach as much as 100 mmol/day in a healthy adult. In this case, ammonium ions are generated from glutamine by glutaminase activity in the renal tubules rather than from general catabolism of amino acids. Hyperammonemia results from a failure to synthesize urea, such as in acute liver failure or through genetic defects in enzymes of the cycle.

Protein synthesis

The liver is the major site of synthesis of plasma proteins. Apart from γ-globulins (antibodies) and von Willebrand factor, which are synthesized in B lymphocytes and endothelial cells, respectively, most plasma proteins are synthesized in the liver. These include albumin, coagulation or clotting factors, complement proteins, and transport proteins. The details of protein synthesis per se are not pertinent to this text and the reader is referred to a textbook of biochemistry for this. Failure of the overall process of hepatic protein synthesis has profound consequences, however.

Albumin

The plasma concentration of albumin is between 32 and 52 g/L (3.2–5.2 g/dL) and this level represents a balance between the rates of hepatic synthesis and degradation. Albumin is synthesized only in liver—about 15 g of albumin per day (100–200 mg/kg body weight/day)—and has a relatively long plasma half-life of about 20 days. The half-life may be affected by a number of factors including nutritional state, thyroid hormone and cortisol levels, colloid osmotic pressure, disease states (it is a negative acute-phase protein; see below), and drugs, including alcohol. Thus the rate of synthesis of albumin is decreased in cases of malnutrition such as starvation and malabsorption, and increased in nephrotic syndrome in an attempt to compensate for the abnormally large protein losses that occur via the kidney. The major functions of albumin are as follows.

- Major contributor to plasma oncotic pressure and thus of distribution of water between plasma and interstitial fluid compartments
- Carrier of bilirubin, free fatty acids, thyroxine (T_4) and tri-iodothyronine (T_3), and aldosterone
- Binds metals (for example calcium, zinc) and drugs (for example penicillin, salicylate)

Although the clearance of albumin is not understood completely, most albumin is removed by uptake into tissues and degradation by intracellular proteases, with smaller amounts being filtered through the glomeruli and degraded in renal tubules (5 g/day) or lost into the small intestine and degraded by proteases (approximately 2 g/day). The liver has a great capacity for albumin synthesis (and protein synthesis in general) and will maintain plasma levels of albumin even when a large fraction of the organ is damaged. Furthermore, plasma levels are maintained for some time even when the rate of hepatic synthesis falls, due to decreased rates of clearance. Thus a decreased plasma albumin level (hypoalbuminemia) is usually a late feature of advanced liver disease. Besides liver disease, plasma albumin levels may also be decreased due to losses in cases of nephrotic syndrome, protein-losing enteropathy, severe burns, exfoliative dermatitis, and major gastrointestinal bleeding (**Table 3.3**).

Clotting factors

Liver is the site of synthesis of most of the blood coagulation proteins, including fibrinogen (Factor I), prothrombin (Factor II), and Factors V, VII, IX, X, XI, and XIII. All of these proteins are synthesized and secreted into the plasma as inactive precursor proteins which undergo proteolytic activation at the coagulation site. The antithrombotic proteins, protein C and protein S, are also synthesized by the liver. Hepatic syntheses of Factors II, VII, IX, X, protein C, and protein S

TABLE 3.3 Causes of hypoalbuminemia

Decreased albumin production	Acute-phase response
	Nutritional failure
	Chronic liver disease
Increased albumin loss	Nephrotic syndrome
	Protein-losing enteropathy
	Burns
	Hemorrhage
	Catabolic state
Others	Translocation from plasma in acute illness
	Hypervolemia

are vitamin K dependent. The calcium-activated proteins undergo post-translational γ-carboxylation of specific glutamic acid residues in the molecule in a reaction requiring vitamin K and which is inhibited by warfarin. Failure of this modification results in ineffective clotting factors and increased clotting time. The synthesis of these proteins thus requires intraluminal bile and micelle formation to allow absorption of dietary vitamin K. A single intravenous dose of vitamin K would relieve dietary deficiency but have no effect on other clinical causes of increased clotting time.

Transport proteins

Liver is the site of synthesis of a number of proteins that transport molecules of low water solubility through the aqueous environment of the plasma. Such proteins include the hormone-binding proteins—cortisol-binding protein, sex hormone-binding globulin, thyroxine-binding globulin, and transthyretin (thyroid hormone-binding prealbumin)—and the proteins responsible for the transport of iron (Fe^{3+}; transferrin) and copper (Cu^{2+}; ceruloplasmin), both of which are acute-phase proteins. The protein-bound lipophilic hormones are biologically inactive and serve as a reservoir for the active free form with which they are in equilibrium.

Complement proteins

The complement system consists of about twenty interacting soluble proteins (for example, C_1–C_9), the majority of which are synthesized in the liver. Like the coagulation proteins, these are synthesized as inactive precursors which undergo proteolytic activation during infection.

Acute-phase proteins

Acute-phase proteins are proteins whose concentration in serum alters nonspecifically in response to tissue insult due to trauma, necrosis, inflammation, or infection. Their synthesis in liver is modulated by pro-inflammatory cytokines, particularly interleukins-1, -2 and -6 and tumor necrosis factor released by macrophages at the site of tissue damage or infection. The magnitude of the acute-phase response is related to the severity of trauma or infection and the change in the serum concentration of such proteins may be positive or negative. Examples of proteins that increase in plasma concentration during an acute phase include:

- C-reactive protein
- α_1-Antitrypsin
- α_1 Acid glycoprotein
- α_2-Macroglobulin
- Haptoglobins
- Ceruloplasmin
- Fibrinogen

Examples of proteins that decrease in plasma concentration during an acute phase include:

- Albumin
- Prealbumin
- Transferrin

C-reactive protein (CRP), so-called because of its ability to bind to pneumococcal C polysaccharide, is synthesized in the liver. It is probably the most useful acute-phase protein for detection and monitoring of disease for a variety of reasons, including:

- Serum concentration often rises tenfold
- Response is seen in all age groups
- Rapid response to tissue insult (< 8 hours)

- Prolonged response (half-life of 24 hours)
- Remains raised in active, chronic illness
- Turnover not affected directly by steroids and other drugs

Measurement of CRP is indicated in the detection of organic disease or of inapparent infection in neonates and immunodeficient patients, for example. However, little or no change in serum CRP concentration is found in a number of chronic diseases, including:

- Connective tissue disorders (systemic lupus erythematosis, dermatomyositis, systemic sclerosis)
- Ulcerative colitis
- Leukemia
- Rejection of bone marrow transplant

Biologically, CRP appears to bind to macromolecules released by damaged tissue and infective agents and promote their phagocytosis by macrophages.

The protease inhibitors α_1-antitrypsin and α_1-antichymotrypsin are glycoproteins and, as their name implies, inhibit several proteases including those synthesized endogenously and those produced by bacteria. Their serum concentration rises after tissue trauma and their action is thought to promote the repair of damaged tissue. Serum activity of less than 30% of normal is associated with early-onset emphysema, particularly in smokers. α_1-Antitrypsin (AAT) deficiency is a well-recognized determinant of inherited susceptibility to liver disease and to emphysema. AAT is encoded by the *Protease Inhibitor* (*PI*) gene, known also as *Serpin Peptidase Inhibitor A1* (*SERPINA1*), on chromosome 14. A range of electrophoretic variants of the AAT protein have been described since the early 1960s and over 90 allelic variants of the AAT gene have been identified to date. AAT variants include null mutations (that is, complete absence of protein), deficiency mutations, and mutations giving rise to altered AAT function. For example, the *PiZ* allele is a deficiency mutation associated with the secretion of approximately 15% of normal AAT amounts in homozygous individuals (PiZZ genotype), while the remaining 85% of AAT protein accumulates in the endoplasmic reticulum of hepatocytes, resulting in hepatic damage. Only 12–15% of individuals with the rare PiZZ genotype progress to liver disease, suggesting that other genetic variants may influence susceptibility. Of more importance from a population perspective are individuals who are heterozygous for the *Z* allele (PiMZ genotype) or the less severe *S* deficiency allele (PiMS genotype) who have an elevated lifetime risk of developing emphysema. The Pi *S* and *Z* alleles are relatively recent and are restricted to European populations and populations of European ancestry.

Ceruloplasmin synthesis is dependent on the availability of copper which binds to it irreversibly. Conditions causing raised serum concentrations of ceruloplasmin include:

- Acute-phase reaction
- Induction by estrogen (including in pregnancy)

Decreased serum concentrations of ceruloplasmin are seen in:

- Wilson's and Menkes' diseases
- Severe liver failure
- Protein-loss states
- Acquired copper deficiency

Detoxification

The liver is the tissue responsible for detoxification and excretion of a wide variety of both endogenous (for example peptide and steroid hormones, catecholamines) and exogenous (for example drugs and toxins) molecules. While peptide hormones are degraded to amino acids by hepatic proteases, more hydrophobic molecules, such as steroid hormones and drugs, undergo reactions that make them more polar (phase I reactions) and conjugation with groups such as sulfate, glucuronic acid, or glutathione (phase II reactions). These latter reactions serve to increase the water solubility of the substrates and facilitate their excretion via the bile. This is exemplified by the hepatic metabolism of bilirubin described later in this chapter. Here, insoluble bilirubin is conjugated with glucuronic acid to form mainly bilirubin diglucuronide, and this hydrophilic conjugate is excreted into bile (**Figure 3.8**). Phase I reactions generally produce a more hydrophilic product which is less active than the parent drug but this is not always the case. Some pro-drugs for instance are converted to their active form by hepatic metabolism and the particular case of acetaminophen (paracetamol) is described later. Given that the drug-metabolizing system is capable of handling an almost infinite variety of molecular structures, it might be expected that the substrate specificity of the enzymes involved in the phase I reactions is rather low.

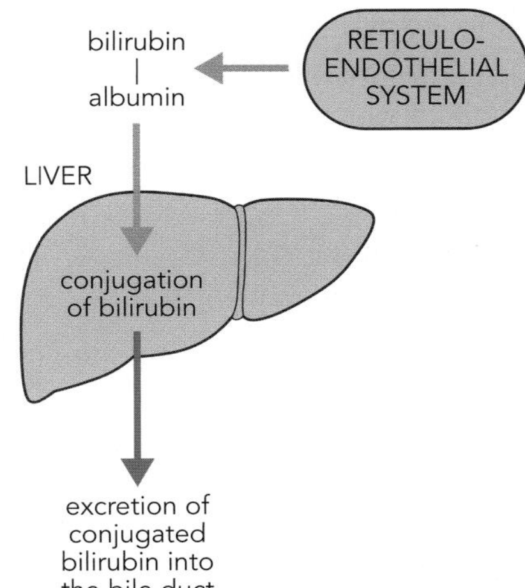

Figure 3.8 Bilirubin metabolism. The figure shows how the insoluble bilirubin produced by the reticuloendothelial system is transported to the liver in the blood bound to albumin. In the liver it is converted to a soluble form via two conjugation steps. The orange arrows refer to unconjugated bilirubin whilst the blue arrow refers to conjugated bilirubin.

Phase I reactions

These reactions are catalyzed by members of the microsomal cytochrome P450 mixed function oxidases (CYPs) and serve to increase the hydrophilicity or polarity of their substrate by oxidation or hydroxylation. They require NADPH as a source of reducing equivalents (electrons) and molecular oxygen. During the reaction, one atom of molecular oxygen is introduced into the substrate while the other is reduced to water. The major cytochrome P450-linked oxidases for drug detoxification are CYP1, CYP2, and CYP3, which are heme-containing proteins forming part of the microsomal electron-transport system which includes NADPH–cytochrome P450 reductase and cytochrome b_5. The activity of this hydroxylating system may be induced by certain drugs including phenobarbital and phenytoin, and concomitant administration of such drugs may increase the rate of metabolism and elimination of other drugs such as oral steroids. Considerable natural variation is observed among the cytochrome P450 enzymes such that, based upon DNA sequence homology, multiple gene families and subfamilies have been identified. Furthermore, significant levels of polymorphism have been observed for individual cytochrome P450 genes within the human population. For example the *CYP2D6* gene, which codes for the enzyme debrisoquine 4-hydroxylase, is known to have more than seventy variant alleles associated with different levels of enzyme activity. The phenomenon of polymorphism among cytochrome P450 enzymes has direct clinical consequences. Individuals whose CYP genotype makes them poor metabolizers of specific drugs will be at substantially increased risk of adverse drug reactions; additionally, they may show reduced or absent therapeutic effect for drugs that require metabolic activation, such as codeine. Conversely, individuals may be ultrarapid drug metabolizers due to the presence of CYP gene duplication; these patients will not respond to normal therapeutic doses of drugs. As a result, many clinical laboratories provide polymerase chain reaction (PCR)-based CYP genotyping assays so that patients can be pre-screened for the presence of the more common CYP gene variants associated with aberrant drug metabolism.

Phase II reactions

Phase II reactions are carried out in the main by transferases, which conjugate glucuronic acid, sulfate, or glutathione to the functional group introduced by the phase I reaction. The glutathione conjugate loses glutamic acid and glycine such that the eliminated product (a mercapturic acid) contains only cysteine from the glutathione added in the conjugation process. The addition of these hydrophilic entities increases the water solubility of the drug and enhances its excretion from the body. Overwhelming the conjugation pathway, as occurs in certain drug overdoses, can lead to hepatic toxicity.

Bile synthesis and excretion

A major function of the liver is the production and excretion of bile into the bile canalicular system (see Figure 3.4). Bile consists of:

- Cholesterol
- Bile salts
- Bilirubin
- Phospholipid
- Salts including potassium chloride and sodium chloride

Bile is secreted into the common hepatic duct and stored in the gall bladder. On contraction of the gall bladder, the bile is excreted into the cystic duct and common bile duct and thence into the second part of the duodenum via the duodenal papilla. The common bile duct passes posterior to the head of the pancreas on its way to the duodenal papilla. Bile flow can therefore be obstructed by stones in the gall bladder or the biliary tree, or by cancer of the head of the pancreas (see the section below [obstructive jaundice] for the signs of biliary obstruction). Bile, formed in the hepatocytes, passes into the canaliculi between hepatocytes and into the biliary ductules in a nearby portal tract. Thus, in the portal tract, bile flows in the opposite direction to that of blood. The functions of the biliary system enable:

- Secretion and storage of bile salts
- Excretion of cholesterol to maintain cholesterol homeostasis
- Excretion of bilirubin (mainly as the diglucuronide conjugate)
- Excretion of organic ions
- Excretion of drug detoxification products
- Emulsification, digestion, and subsequent absorption of dietary lipids in the gastrointestinal tract

The concentration of canalicular bile is more dilute than that stored in the gall bladder and concentration occurs during passage through the biliary tract. For example, the total lipid concentration in canalicular bile is about 3 g/L and this is increased to about 10 g/L in the gall bladder. The detergent action of bile salts is critical in maintaining lipid in solution here, particularly cholesterol. Contraction of the gall bladder by the action of the hormone cholecystokinin causes the ejection of bile into the duodenum through the cystic and common bile ducts. This occurs after a meal, when the presence of food in the duodenum stimulates the mucosal cells to secrete cholecystokinin.

Bilirubin production, metabolism, and excretion

Bilirubin is the main breakdown product of heme moieties in the body. The daily bilirubin production in adults is approximately 300 mg (approximately 4 mg/kg body weight) derived mainly from degradation, in fixed macrophages including Kupffer cells, of the heme moiety of hemoglobin, with smaller contributions from the heme groups of myoglobin and cytochromes. Approximately 85% is derived from effete erythrocytes (average lifetime 120 days), 10% from ineffective erythropoiesis, and 5% from degradation of cytochromes.

The degradation of spent erythrocytes in reticuloendothelial cells of the spleen releases heme from hemoglobin. Heme is metabolized to biliverdin, a green pigment, and then to bilirubin, a brown pigment, by heme oxygenase (think of color changes in a bruise). Unconjugated bilirubin is extremely hydrophobic and fat soluble but is insoluble in water. Free (unconjugated) bilirubin released into plasma is thus transported to the liver from the spleen bound to plasma albumin. Although unconjugated bilirubin cannot be excreted via the glomerulus or bile directly, its hydrophobic nature allows it to cross the blood–brain barrier and placenta. Free bilirubin also dissolves in lipid-rich

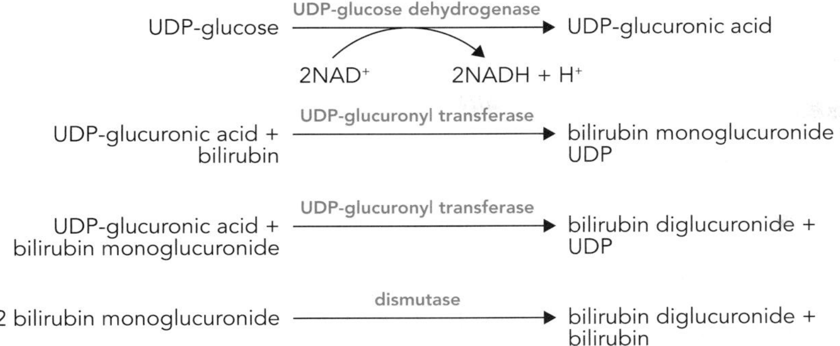

Figure 3.9 Bilirubin conjugation. Bilirubin conjugation in the liver, where two molecules of glucuronic acid are added to each molecule of bilirubin, making the bilirubin more soluble in water.

environments such as plasma membranes of tissues and in fat. The metabolism of bilirubin proceeds in a stepwise fashion, as follows.

1. Uptake into liver

Bilirubin dissociates from albumin in the space of Disse and free bilirubin is taken up by liver via a specific transport system (rifampicin competes at this site). It is stored intracellularly bound to proteins known as ligandins.

2. Conjugation

The liver processes bilirubin to make it more water soluble by a process of conjugation (phase II reaction) with one or usually two molecules of glucuronic acid donated by UDP-glucuronic acid; the reactions are catalyzed by UDP-glucuronyl transferase (**Figure 3.9**). This enzyme is expressed in the fetal liver during the third trimester of pregnancy.

3. Excretion

The major conjugation product, bilirubin diglucuronide (**Figure 3.10**), is excreted into bile canaliculi by an active transport process. When this is impaired, conjugated bilirubin will be forced from hepatocytes back into plasma causing a rise in conjugated bilirubin levels. In this situation, conjugated bilirubin, being water soluble and less tightly bound to albumin, is filtered by the kidney and appears in the urine giving it a dark color (choluria). Conjugated bilirubin normally enters the intestine from bile and is hydrolyzed by bacterial glucuronidases to free bilirubin. This is further metabolized by gut bacteria to urobilinogens (colorless) which are excreted in the feces. Urobilinogens are oxidized on exposure to air to the stercobilins which give feces their brown color. About 20% of urobilinogen is reabsorbed from the gut and undergoes enterohepatic circulation. Urobilinogens are low-molecular-weight, water-soluble molecules that are able to cross the glomerulus, and small amounts are normally found in the urine of healthy individuals.

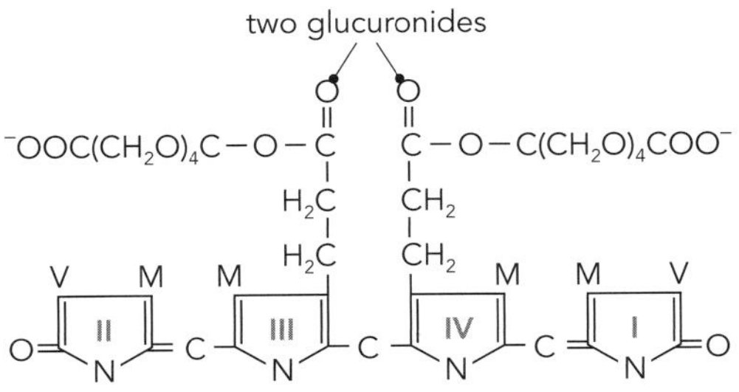

Figure 3.10 Structure of conjugated bilirubin (bilirubin diglucuronide). Unconjugated bilirubin is insoluble in water, whilst the addition of two molecules of glucuronic acid makes it soluble in water. The roman numerals I–IV refer to the pyrrole rings of heme, which is degraded to bilirubin. M, methyl; V, vinyl.

Synthesis of bile acids and bile salts

Bile acids are synthesized in the liver from cholesterol, the rate-limiting step being the production of 7α-hydroxy cholesterol by the 7α-hydroxylase, a cytochrome P450-linked enzyme which requires NADPH as an electron donor. The synthesis of the two primary bile acids, cholic acid and chenodeoxycholic acid, requires further hydroxylations, involving cytochrome P450-linked enzymes, at carbons 3, 7, and 12 (cholic acid) and at 3 and 7 (for chenodeoxycholic acid). Conjugation of bile acids with the amino acids glycine or taurine forms the major bile salts glycocholate and taurocholate in the ratio of 3:1, and these are excreted into bile where they help to solubilize the biliary lipids, mainly free cholesterol (**Figure 3.11**), and prevent precipitation of cholesterol as gallstones during storage in the gall bladder.

The amphipathic nature of bile salts enables them to function as detergents in the emulsification of dietary lipids in the duodenum and small intestine. The action of peristalsis mixes bile salts with the lipids, gradually dispersing the lipid into smaller and smaller droplets which eventually become mixed micelles of bile salts, triglycerides, cholesterol, phospholipids, and fat-soluble vitamins. The lipid emulsion is optically clear at this stage. Bile salts, phospholipids, and cholesterol are present on the surface of the micelle with their hydrophilic moieties in contact with the aqueous environment of the gastrointestinal tract and their hydrophobic moieties pointing inward toward the oily center of the micelle, which contains the bulk neutral lipid, triglycerides, cholesterol ester, and fat-soluble vitamins. This action of emulsification increases the surface area of the dietary lipid mass several thousand fold and presents a huge increase in the number of substrate sites for the action of pancreatic, lipolytic enzymes secreted into the duodenum.

Secondary bile acids (deoxycholic acid and lithocholic acid) are formed in the gut by the action of bacteria in the colon; this involves deconjugation and removal of the hydroxyl group at carbon 12 of the primary bile salts.

Figure 3.11 Bile acids and bile salts. The primary bile acids are cholic acid and chenodeoxycholic acid. The secondary bile acids, lithocholic acid and deoxycholic acid (not shown), are derived from the primary bile acids by bacterial action in the gut. Amidation of bile acids with glycine or taurine results in the formation of bile salts: glycocholate and taurocholate from cholic acid, and glycochenodeoxycholate and taurochenodeoxycholate from chenodeoxycholate. These salts are preferentially secreted into bile. Bile salts are amphipathic molecules synthesized from cholesterol via the cholesterol 7α-hydroxylase pathway and their synthesis accounts for approximately 50% of the hepatic metabolism of cholesterol.

BILE ACIDS

BILE SALTS

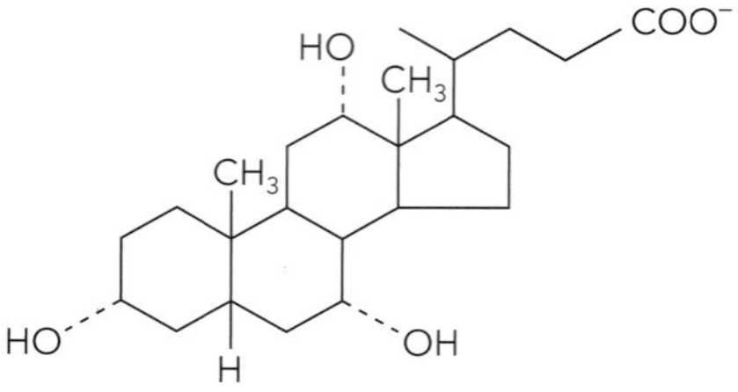

cholic acid

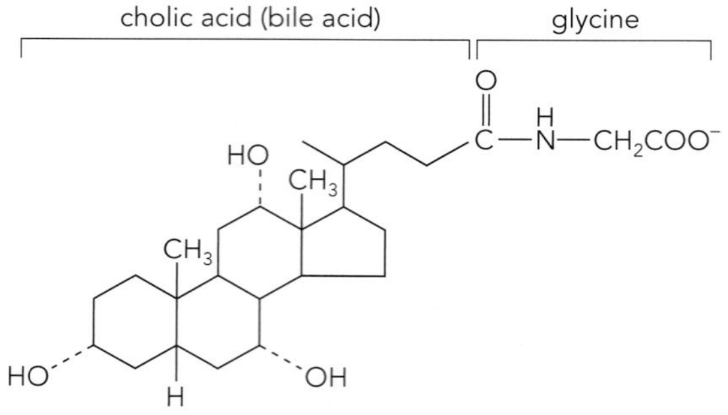

glycocholic acid (bile salt)

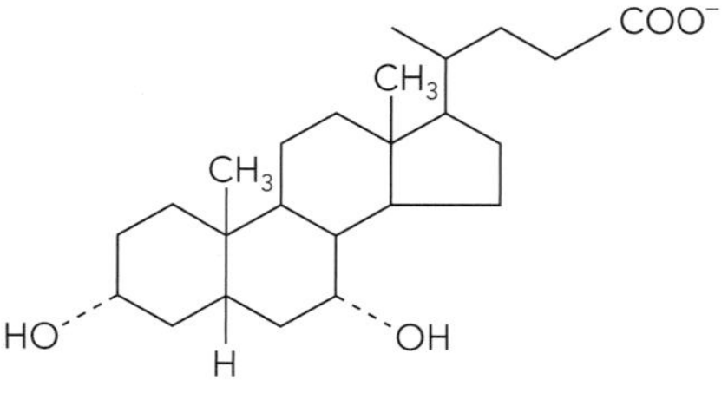

chenodeoxycholic acid

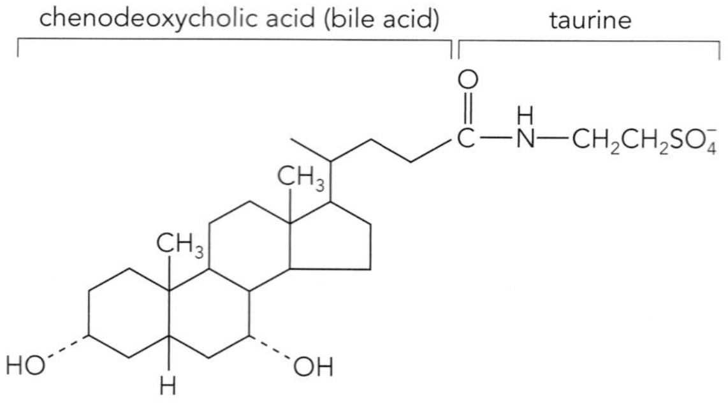

taurochenodeoxycholic acid (bile salt)

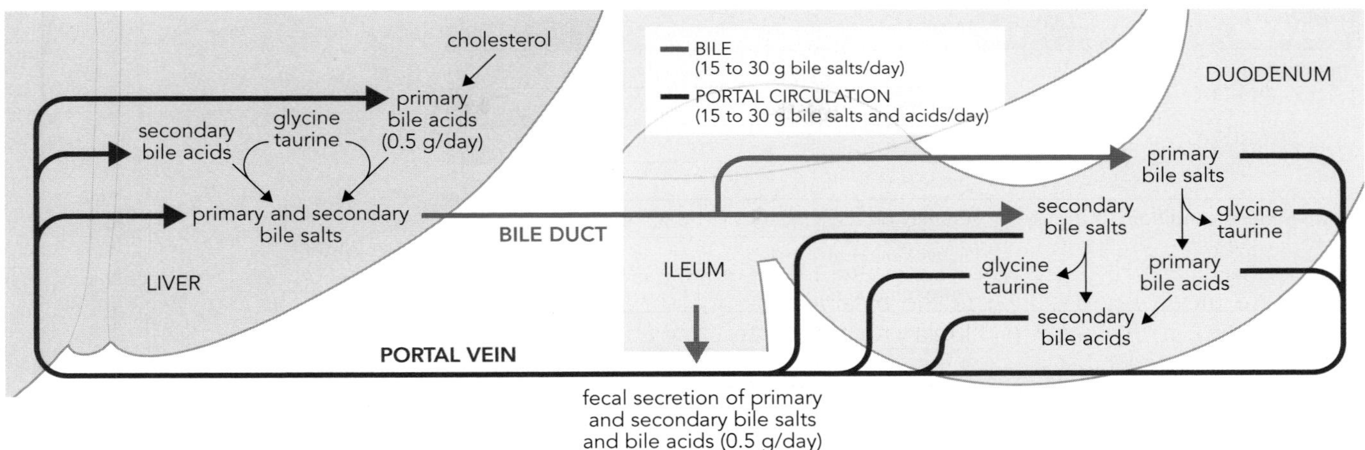

Figure 3.12 Enterohepatic circulation of bile acids. Bile acids synthesized in the liver are converted to bile salts and secreted into bile along with free cholesterol and phospholipids. Hydrolysis of bile salts occurs in the small bowel and approximately 90–95% of the secreted bile acids are reabsorbed via a sodium-dependent transporter in the distal ileum into the enterohepatic circulation, being delivered to the liver via the portal vein. The total bile acid pool circulates in this way between 5 and 6 times per day with only a small loss (approximately 0.5 g/day) to the feces.

Metabolism of bile salts

Bile acids are reabsorbed from the terminal ileum and undergo enterohepatic circulation such that the body's bile acid pool of 3–5 g circulates between 5 and 6 times daily (**Figure 3.12**). Less than 10% (approximately 0.5 g) of the pool is excreted into the feces daily (mainly as lithocholic acid), and this amount is replaced by daily synthesis. The rate-limiting enzyme of bile acid synthesis, cholesterol 7α-hydroxylase, is subject to feedback inhibition by bile acids, and enterohepatic recycling of bile acids limits the daily conversion of cholesterol to bile acids. Interruption of this enterohepatic circulation by oral administration of bile-acid-binding resins, such as cholestyramine, relieves the inhibition of cholesterol 7α-hydroxylase and promotes hepatic synthesis of bile acids from cholesterol.

3.3 MAJOR CAUSES OF LIVER DISEASE AND THEIR INVESTIGATION

Liver disease may arise from an extremely diverse range of pathological causes (**Table 3.4**). Different patterns of liver injury or impairment may arise depending on the cause, which in turn give different clinical presentations and blood test results.

Clinical presentation of liver disease

A patient may be investigated for a range of presenting symptoms which may of themselves suggest liver disease. In other cases, the presence of liver disease may be established by routine biochemical tests when nonspecific symptoms of disease are present. In yet other cases, abnormal liver function tests may be found by chance when no obvious symptoms are apparent. Most symptoms are related to disruption of the various biochemical functions of the liver, and usually the severity of the symptoms is related to the degree of liver damage (**Figure 3.13**); examples follow below.

TABLE 3.4 Causes of liver disease

Infection	Viruses (hepatitis A–E; non-hepatitis viruses)
	Bacteria
	Protozoa
	Parasites
Autoimmune conditions	Primary biliary cirrhosis
	Sclerosing cholangitis
	Autoimmune hepatitis
Tumors	Benign
	Primary hepatocellular carcinoma
	Secondary (metastatic) tumors
Biliary obstruction	Gallstones
	Extrahepatic tumors
	Strictures
Metabolic causes	Non-alcoholic fatty liver disease
	Hemochromatosis
	Wilson's disease
Toxins	Alcohol
	Drugs

(i) Bile acid metabolism

One of the more common signs of liver disease is jaundice, arising from an inability of the liver to process and/or excrete bilirubin. Also, as discussed earlier, the synthesis of bile salts by the liver is essential for the digestion and absorption of dietary fat (including fat-soluble vitamins) in the small intestine. Hepatic diseases that impair bile-salt synthesis often present with malabsorption of fat and deficiency of the fat-soluble vitamins A, D, E, and K. Fat malabsorption is a feature in patients with cholestatic liver disease, and night blindness may arise from a persistent lack of vitamin A absorption. The role of vitamin D in calcium

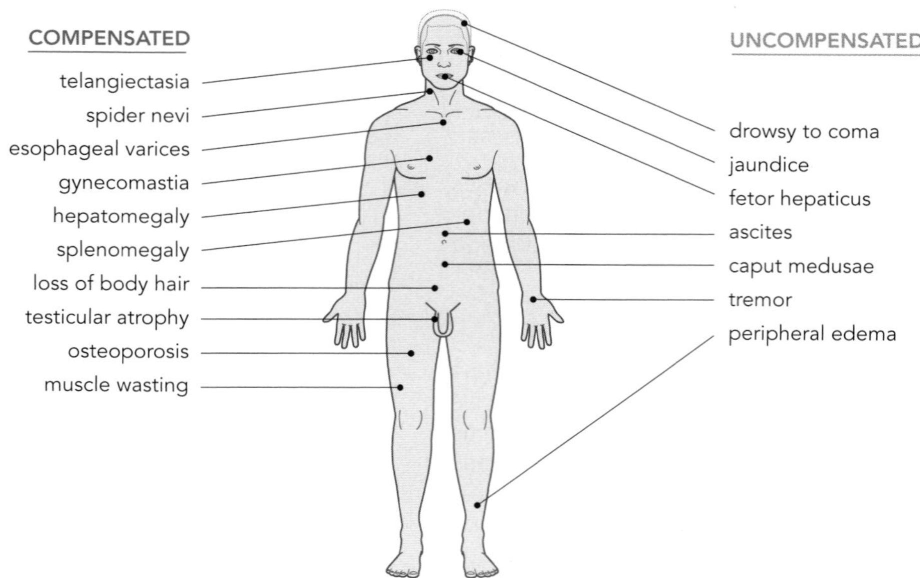

Figure 3.13 Signs of severe liver disease.
Signs and symptoms in a patient with sever chronic liver disease. The body tries to adapt to the consequences of loss of hepatic function; that is, a compensated state. When the changes are even more severe, the body cannot compensate enough, and further signs and symptoms become apparent; that is, the uncompensated state. The change from compensated to uncompensated state is more of a gradient, with different stages dependent on the stage of the disease.

and bone metabolism is described in Chapter 9, but it is worth noting at this point that fat malabsorption may lead to deficiency of parent vitamin D. Severe liver disease may also impair one of the steps in the synthesis of active vitamin D (1,25-dihydroxy vitamin D), namely the 25-hydroxylation step. This in turn leads to a decrease in the level of active vitamin, increased parathyroid hormone drive, and osteoporosis (see Chapter 9).

(ii) Cholesterol metabolism

Cholesterol is normally excreted via the bile, but may accumulate in liver disease, especially in cholestatic disease, leading to elevated cholesterol levels in the blood and the deposition of cholesterol under the skin (xanthoma, xanthelasma). In some diseases, the liver may be swollen (hepatomegaly) and palpable just below the rib margin.

(iii) Protein synthesis

As discussed above, the liver is involved in the synthesis of many plasma proteins, and decreased hepatic protein synthesis may lead to a fall in the concentration or total absence of these proteins in the plasma and characteristic clinical symptoms. In severe liver disease, for example, plasma albumin levels can be very low (see Table 3.3) and lead to a decrease in oncotic pressure in the blood and consequent peripheral edema and the presence of excess fluid in the peritoneal cavity (ascites). Ascites is caused when blood flow through the liver is affected by disease. Blood backs up into the mesenteric vascular beds, causing increased capillary hydrostatic pressure and fluid movement from the vascular to the peritoneal space.

Decreased hepatic synthesis of other plasma proteins such as clotting factors may present with an increased tendency to bruise, while differences in the synthesis of sex-hormone-binding globulins, which result in changes in the concentrations of biologically active free sex hormones, may result in changes in the male–female sex hormone ratio. In males, a decrease in the male–female sex hormone ratio may lead to breast development (gynecomastia), testicular atrophy, and to muscle wasting and loss of body hair. The relative increase in levels of estrogens is thought to cause changes in capillaries on the skin surface leading to spider nevi and telangiectasia. These symptoms are not specific for liver disease but are found frequently in severe chronic liver disease (see Figure 3.13).

(iv) Intermediary metabolism

The central role of the liver in intermediary metabolism of fat, carbohydrate, and amino acids has been mentioned earlier, and damage to the liver may impair essential metabolic processes; for example, hypoglycemia resulting from inhibition of gluconeogenesis (see Chapter 2) is a feature of cirrhosis and other causes of liver failure. Similarly, fatty liver is a common presentation in alcoholic liver disease and, in some liver pathologies, there is an increase in secretion of bile salts into the blood, a situation which is thought to contribute to the itching and scratch marks found in many patients. Importantly, in terms of amino acid metabolism, the liver is the site of urea synthesis, and disruption of this process may lead to elevated levels of potentially toxic ammonia (hyperammonemia), which contributes to the drowsiness and coma that is associated with liver failure.

(v) Drug metabolism

The liver is also responsible for the metabolism and excretion of many drugs and toxins, and patients with liver disease may have a reduced ability to activate pro-drugs or to eliminate drugs and toxins. They may thus become more sensitive to normal prescribed drug doses. The impaired metabolism of drugs may mean that such patients require less frequent and smaller doses of drugs, especially those which are administered orally and which are subject to significant amounts of first-pass metabolism in the liver.

(vi) Anatomical features

Some of the symptoms observed in severe liver disease (see Figure 3.13) are related to the anatomy of the liver. Portal hypertension, which is caused by obstructed blood flow due to fibrosis, forces splanchnic blood through limited connections with the systemic circulation. These are the portal–systemic anastomoses where varices may form. The most important of these are in the lower esophagus, but they are also found in the lower rectum and around the umbilicus, where they are known as *caput medusae* (Medusa's head). Initially minor bleeds produce iron-deficiency anemia but the potential for acute massive bleeds can lead to the death of the patient.

As always, a thorough clinical history of the patient is most important and may often point to a likely cause of disease and guide the choice of appropriate investigations. For example, patients with gallstones classically have upper abdominal pain after eating a fatty meal, whereas patients with a malignancy may have weight loss and little pain, at least initially. History of alcohol consumption and drug use, contact with hepatitis carriers, and previous blood transfusions are amongst the vital items of information to seek. A family history will also be very important in patients with defects arising from inborn errors of metabolism.

Laboratory assessment of liver disease

In some cases of symptomatic liver disease, for example different types of jaundice and hepatocellular disease, liver function tests (LFTs) have proved extremely useful in diagnosis and monitoring of treatment. However, initial investigation of clinical liver disease or of persistently asymptomatic abnormalities of LFTs (for example, early alcoholic liver disease, fatty liver, and cirrhosis) consists of a limited number of first-line investigations followed as necessary by second-line tests (**Figure 3.14**). The test considered to be the gold standard for diagnosis is histological examination of a piece of liver removed by needle biopsy. This may have the advantage of providing prognostic information as well in some cases. However, although generally safe, the procedure is invasive and does have a finite complication rate; it can be fatal on occasion. If possible, it is preferable to reach a likely diagnosis by noninvasive tests. First-line tests usually consist of immunoglobulins and autoantibodies (for autoimmune causes), hepatitis viral serology, transferrin saturation or ferritin (for hemochromatosis), and ultrasound scanning of the liver. The last of these can show the presence of gallstones, tumors, and fatty infiltration. Second-line tests for more rare conditions include markers of copper metabolism (to diagnose Wilson's disease) and

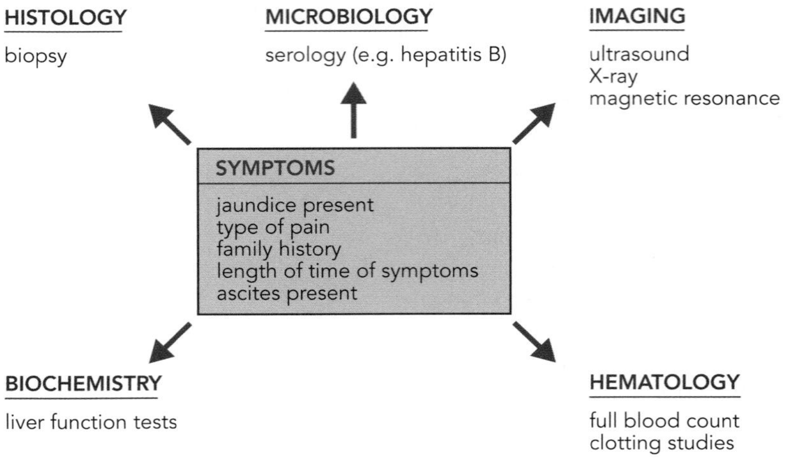

Figure 3.14 Testing strategies for investigating liver disease. When investigating liver disease, various approaches will be used dependent upon the initial history and symptoms of the patient. The diagnostic approaches range from simple blood tests (for example, liver function tests) through X-ray, ultrasound, and a biopsy and full histological classification.

α_1-antitrypsin concentration and phenotyping. A good illustrative example of the use of LFTs in helping diagnosis is in distinguishing between the different causes of jaundice.

Liver function tests

Clinically, a routine LFT consists of a number of core analyses, some of which are markers of liver function while others give some idea of the extent, and occasionally the type, of liver damage. LFTs are therefore a panel of tests. This panel may include some or all of the following:

- Total bilirubin
- ALT
- AST
- Alkaline phosphatase (ALP)
- Gamma-glutamyl transferase (GGT)
- Albumin
- Total protein

Analytical practice point 3.1

The exact components of a liver function test (LFT) panel may vary between laboratories.

CASE 3.1

A 53-year-old man is seen by a hepatologist due to abnormal liver function tests (LFTs). He admits to drinking no more than two units of alcohol per day on average. He has no risk factors for bloodborne viruses, has not traveled abroad, and his body mass index is 22 kg/m^2. On examination he has spider nevi and the physician notices that the patient's breath smells of alcohol. Although he denies having drunk that day, he consents to blood alcohol measurement and LFTs. The results show the following.

	SI units	Reference range	Conventional units	Reference range
Plasma				
Bilirubin	25 μmol/L	< 21	1.5 mg/dL	< 1.3
ALT	126 U/L	5–40	126 U/L	5–40
AST	243 U/L	5–40	243 U/L	5–40
GGT	987 U/L	5–45	987 U/L	5–45
ALP	180 U/L	40–120	180 U/L	40–120
Albumin	37 g/L	32–52	3.7 g/dL	3.2–5.2
Ethanol	70 mmol/L	<4	322 mg/dL	<20

- **What pattern of LFTs is present?**
- **What is the cause of these results?**
- **How do the other results support this diagnosis?**

The very elevated ALT and AST with only minor increase in ALP is a hepatocellular pattern. This indicates hepatocyte injury, which may be caused by many things, including viruses, toxins, metabolic disorders, or ischemia. In this case, the very high ethanol level points to alcoholic hepatitis. Other evidence in favor of this is disproportionately high GGT, which is often seen with chronic alcohol excess, and an AST–ALT ratio that is approaching 2. In most causes of liver disease, the AST is lower than ALT, but the ratio is classically reversed when alcohol is the cause. At present, the normal albumin and bilirubin show that the liver is able to maintain its metabolic functions, but if the patient continues to drink there is a high risk of liver failure.

Each individual test may be affected by conditions unrelated to the liver, and the diagnostic value of LFTs usually lies in particular patterns which suggest a likely cause of disease. However, the lack of both sensitivity and specificity for liver disease is important but often forgotten. Better markers of liver disease have been sought for many years but none has yet proved to be sufficiently useful to displace traditional LFTs as initial tests. The precise analyses constituting an LFT vary between laboratories, often for historical or economic reasons rather than sound clinical ones. For example, some laboratories might not offer GGT on all patient samples, and the choice of transaminase (ALT or AST or both) may differ, but generally ALT is preferred to AST. Despite any reservations from the forgoing discussion, however, many hundreds of thousands of LFTs are performed each year in most large hospital laboratories worldwide. The reference ranges are given at the beginning of this chapter (see Table 3.1). This battery of tests constitutes a routine LFT, and the unconjugated and conjugated fractions of bilirubin can be measured.

Bilirubin

Total bilirubin can be measured in many ways, including spectrophotometric methods using dual absorbance at 455 nm and 575 nm. Such simple methods can be adapted for use in point-of-care testing (POCT) settings, such as neonatal baby units. However, the majority of laboratories use modifications of the original chemical reaction with Ehrlich's reagent, leading to the formation of azobilirubin, which can be measured colorimetrically. Since unconjugated bilirubin is not very soluble and is normally bound to albumin in plasma, it has to be stripped from the protein before it can be measured. This process is accomplished by the use of an accelerator (for example, salicylate). When measuring bilirubin with and without the accelerator, an estimate of the unconjugated and conjugated bilirubin can be determined. The measurement that is made before the accelerator has been added is called the direct bilirubin, and refers to the soluble or conjugated bilirubin. By subtracting the direct bilirubin from the total bilirubin, an estimate of the less soluble or unconjugated bilirubin can be determined; that is, the indirect bilirubin. This is how the abbreviated request of D and I bilirubin is derived. The normal total bilirubin concentration is usually less than 21 μmol/L (1.3 mg/dL).

Analytical practice point 3.2

Direct bilirubin measures conjugated bilirubin, which is water soluble. Unconjugated (indirect) bilirubin may be calculated by subtraction of direct from total bilirubin.

Liver enzymes

The enzymes used as part of an LFT are large intracellular molecules and an increased concentration in plasma implies liver cell damage rather than providing a measure of liver cell function. However, because of the different distributions of these enzymes both anatomically and intracellularly, their measurement in plasma gives an indication of the type and degree of liver disease present. Thus ALT, found predominantly in the cytoplasm of the hepatocyte, is a marker of hepatocellular damage, while ALP, found predominantly in the biliary aspects of the liver, is a marker of biliary damage. GGT is an inducible enzyme and an increase in plasma levels, particularly when other liver enzymes are normal, is an early sign of liver insult, particularly by drugs such as anticonvulsants and especially by alcohol.

Albumin and total protein

Total protein and albumin are usually both measured. The difference between the two measures is the level of globulins, which consist mainly of immunoglobulins. High total protein is usually a reflection of raised immunoglobulins in infective or inflammatory (such as autoimmune) disorders. Total protein may be normal if there is a simultaneous decrease in serum albumin. Since albumin is the major protein synthesized by the liver, its plasma concentration is often considered to give an estimation of the liver's synthetic capacity. However, because of the huge reserve capacity of the liver, plasma albumin levels may be maintained until the late stages of liver disease. In addition, factors that increase the loss of albumin from the circulation have a greater and more rapid effect on serum albumin than does decreased synthesis.

Major examples of liver disease

Jaundice

Jaundice, from the French "jaune" meaning yellow, is used to describe conditions in which there is a defect in the metabolism or excretion of bilirubin and the pigment accumulates in blood and tissues. The yellow hue of the skin, sclera, and mucous membranes is due to the color of the breakdown products of bilirubin; this is the same color as that seen during the healing of a bruise. Jaundice is usually classified into three main types—pre-hepatic (hemolytic), hepatic, and post-hepatic (cholestatic)—although minor variants of these also exist and in clinical practice a mixture of types is often present. The causes of jaundice can be:

- Overproduction of bilirubin
  - Hemolysis
  - Ineffective erythropoiesis
- Impaired hepatic uptake of bilirubin
- Impaired hepatic conjugation of bilirubin with glucuronic acid
- Impaired hepatic excretion
- Biliary tract obstruction
  - Intrahepatic
  - Extrahepatic

In the first instance it is perhaps simpler to consider each type individually. The synthesis of bilirubin and its metabolism to urobilinogen and urobilin are shown in **Figure 3.15**. Some urobilinogen is reabsorbed into the bloodstream from the small intestine, and since this is a water-soluble molecule of low molecular weight, small amounts are present in the urine of normal healthy people and can be measured by simple qualitative or semiquantitative methods. In contrast, the somewhat larger conjugated bilirubin molecule is not reabsorbed from the intestine and does not appear in urine.

The plasma total bilirubin concentration in a healthy adult is usually less than 21 μmol/L (1.3 mg/dL). However, when the ability of the liver to conjugate and excrete bilirubin is compromised due to disease, the bilirubin concentration

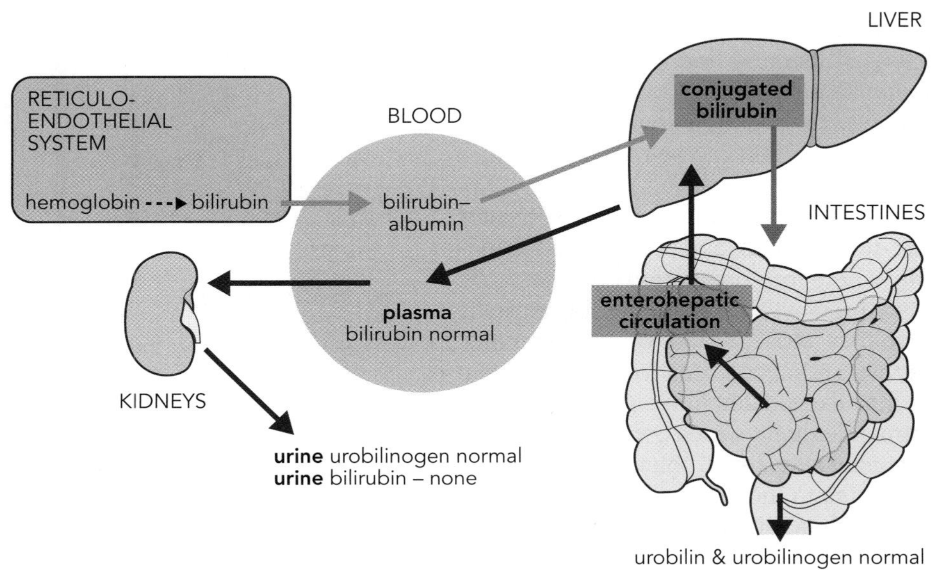

Figure 3.15 Normal bilirubin metabolism. In a healthy person, in whom the metabolism and excretion of bilirubin is unimpaired, the concentration of plasma bilirubin is within the reference range. In a healthy individual, the bulk of circulating bilirubin is unconjugated and hence bound to albumin; therefore no bilirubin should be present in the urine. Since the enterohepatic circulation is working normally, the bilirubin breakdown products from the intestines, urobilinogen and urobilin, will be found in the blood, and since urobilinogen is water soluble, it will also be found in the urine. Conjugated bilirubin will be totally excreted into the intestine and should not be found in the blood. The orange arrows refer to unconjugated bilirubin, green arrows to conjugated bilirubin, and blue arrows to urobilinogen and urobilin.

in the plasma increases. As values rise above 30 μmol/L, bilirubin leaks from the vasculature into the tissues and deposits as a yellow pigmentation, known as icterus, in the skin and sclera of the eyes; this is commonly observed as jaundice (**Figure 3.16**). The level of jaundice increases with the increase in plasma bilirubin. The yellow-brown pigmentation can be clearly seen in icteric plasma (**Figure 3.17**) characterized by pigmentation being seen in the bubbles that form as the plasma sample is gently shaken. Occasionally the precursor to bilirubin, biliverdin (which is green in color), builds up in the plasma (biliverdinemia; see Figure 3.17).

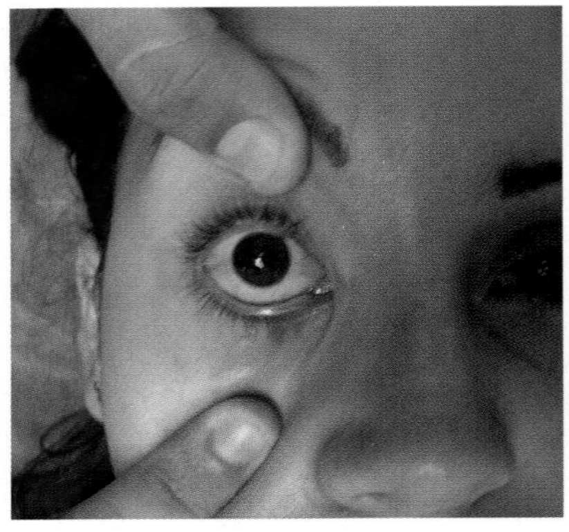

Figure 3.16 Jaundice.
Yellow pigmentation seen in the eye, which is characteristic of jaundice. (Courtesy of Bob J Galindo under CC BY-SA 3.0.)

Jaundice caused by unconjugated hyperbilirubinemia (pre-hepatic jaundice)

Raised plasma concentrations of bilirubin may arise as a result of (i) oversaturation of the hepatic conjugation mechanism, (ii) inability to transport bilirubin from the bilirubin–albumin complex in plasma into the liver, or (iii) partial or complete absence of the enzymes of hepatic conjugation.

The capacity of the liver to conjugate and excrete bilirubin into bile is finite and, when this capacity is exceeded, bilirubin accumulates in the plasma without impairment of liver function. Thus, in conditions in which excessive intravascular hemolysis occurs with consequent greatly increased degradation of hemoglobin (for example, ABO incompatibility, sickle cell crisis, and multiple blood transfusions), the supply of unconjugated bilirubin into the bloodstream increases markedly (**Figure 3.18**). Although the liver can deal with some of the increased supply of bilirubin, the unconjugated bilirubin builds up in the plasma and leads to jaundice. This is known as hemolytic jaundice and moderately raised levels of plasma bilirubin can occur, but rarely greater than 100 μmol/L (5.85 mg/dL). The increased supply of bilirubin to the liver leads to higher rates of conjugation and increased excretion of conjugated bilirubin into the gut. Here, deconjugation and metabolism produce increased amounts of urobilinogen, some of which undergoes enterohepatic circulation, resulting in increased excretion of urobilinogen in urine (see Figure 3.18). Despite the increased concentration of unconjugated bilirubin in plasma, it is not found in urine, since the unconjugated bilirubin–albumin complex is too big to pass through the glomerulus. The small increases in unconjugated bilirubin found in hemolytic jaundice do not usually lead to any pathological damage. In most situations, the liver enzymes assayed in the LFT will all be within reference range. The excess unconjugated bilirubin bound to albumin in plasma is eventually metabolized and excreted by the liver. However, situations can arise where the ability to conjugate bilirubin is massively exceeded, and in such cases the concentration of bilirubin in plasma may rise to levels where the albumin binding capacity is exceeded (>200 μmol/L or 11.7 mg/dL). The resulting free unconjugated bilirubin will be taken up rapidly in the lipophilic environment of soft tissues (for example, brain and nervous tissue) leading to severe pathological damage, or kernicterus. This free fraction of unconjugated bilirubin can be measured, and is sometimes termed δ-bilirubin.

Figure 3.17 Hyperbilirubinemia and biliverdinemia.
Plasma samples from a healthy individual ("normal") and patients with hyperbilirubinemia (jaundice) and biliverdinemia. Normally the plasma is straw colored, whilst in jaundice the plasma is brown, and this change in color can also be seen in the bubbles that form when the sample is shaken. Occasionally a green plasma can be seen when the normal conversion of biliverdin to bilirubin is impaired. This is called biliverdinemia.

A not uncommon situation in which free bilirubin accumulates in soft tissues is in some newborn babies, especially pre-term infants, and presents as neonatal jaundice. This is due to liver immaturity where the conjugation enzymes have not yet fully developed; the hepatic UDP-glucuronyl transferase is not normally expressed until the third trimester of pregnancy. Most of these cases of jaundice resolve after a few days and the bilirubin rarely reaches toxic levels. However, in some infants the bilirubin concentration may exceed 200 μmol/L (11.7 mg/dL). In such cases, exposure of the baby to ultraviolet light, which converts the insoluble bilirubin IX isomer to a soluble form, alleviates the problem and increases the rate of degradation of pigment accumulated in the tissue.

A benign form of unconjugated hyperbilirubinemia leading to mild jaundice can also occur in the common disorder Gilbert's disease. Affecting some 5%

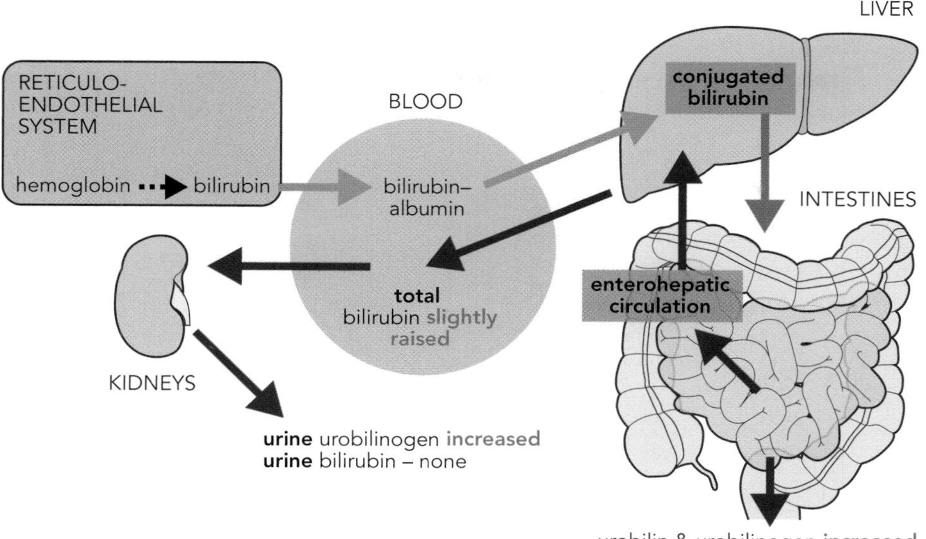

Figure 3.18 Pre-hepatic jaundice. In hemolytic or pre-hepatic jaundice the primary defect is a vast increase in the supply of bilirubin, which results in a slight rise in the plasma concentration of bilirubin (usually to less than 100 μmol/L [5.85 mg/dL]). It is the insoluble, unconjugated bilirubin that will be increased, which is bound to albumin in the plasma, and as such will not be excreted into urine. Since there is normal enterohepatic circulation, the intestinal production of urobilinogen and urobilin will be increased, and the urine levels of urobilinogen will be increased. The orange arrows refer to unconjugated bilirubin, green arrows to conjugated bilirubin, and blue arrows to urobilinogen and urobilin.

of the population, it is named after the French-Canadian physician Augustin Gilbert, who first described the syndrome. Patients with this condition have an impaired ability to transport unconjugated bilirubin from the bloodstream into the liver. This is due to mutations affecting the coding region or the promoter sequence of the UDP-glucuronyl transferase (*UGT1A1*) gene on chromosome 2, resulting in reduced enzyme activity. Normally, these patients do not present with jaundice, but in times of physiological stress (for example, fasting) or in response to various stimuli (for example, alcohol or drugs), serum levels of bilirubin may rise to 50 μmol/L (2.9 mg/dL) leading to a mild jaundice. As in hemolytic jaundice, the liver enzymes in such patients are normal. Gilbert's disease and hemolytic jaundice may be distinguished by measurement of plasma haptoglobin concentration. Haptoglobin is a protein that transports hemoglobin in the circulation and is cleared rapidly during intravascular hemolysis as the body seeks to remove hemoglobin released from red cells. Plasma haptoglobin remains above 500 mg/L (50 mg/dL) in Gilbert's disease while it falls to levels below 200 mg/L (20 mg/dL) in cases of hemolytic jaundice.

Clinical practice point 3.1

Gilbert's disease is a very common cause of mild unconjugated hyperbilirubinemia, but hemolytic disease must be excluded before making the diagnosis.

Another well-defined inherited disorder that gives rise to unconjugated hyperbilirubinemia is Crigler–Najjar disease, a more serious condition in which the liver bilirubin conjugation enzymes are defective or absent. Although sometimes classified as pre-hepatic, the defect is in the hepatocyte itself and thus could be thought of as hepatic jaundice. This condition is characterized by complete absence (type I) or partial absence (10% of normal activity; type II) of the hepatic UDP-glucuronyl transferases and leads to accumulation of bilirubin in plasma. Both types are due to mutations in the *UGT1A1* gene, just as in Gilbert's disease. The absence of conjugated bilirubin in type I results in a pale bile, while in type II the amount of monoglucuronide in bile is raised. Treatment of patients with phenobarbital would be expected to raise the hepatic phase I enzymes (for example, microsomal cytochrome P450-linked hydroxylases, described earlier) and consequently increase the activity of UDP-glucuronyl transferases by an increase in substrate supply. This happens to some extent in type II disease, with increased synthesis of bilirubin monoglucuronide and subsequent reduction in plasma bilirubin. In contrast, this cannot happen in type I where the active enzyme is not expressed at all and consequently phenobarbital treatment has no effect on plasma bilirubin. Thus, whereas type II is usually benign, the prognosis for type I is poor and the accumulation of unconjugated bilirubin leads to extensive brain damage in neonates, the condition known as kernicterus.

CASE 3.2

A 53-year-old woman sees her primary care physician with a history of generalized itching, worsening over the last year. There is no evidence of a rash on examination.

	SI units	Reference range	Conventional units	Reference range
Plasma				
Bilirubin	15 μmol/L	< 21	0.9 mg/dL	< 1.3
ALT	31 U/L	5–40	31 U/L	5–40
AST	26 U/L	5–40	26 U/L	5–40
GGT	245 U/L	5–45	245 U/L	5–45
ALP	1020 U/L	40–120	1020 U/L	40–120
Albumin	36 g/L	32–52	3.6 g/dL	3.2–5.2

- **What are the possible sources of the high ALP?**
- **What is the most likely cause for these results?**
- **What autoantibody test would confirm the diagnosis?**

There are several isoforms of alkaline phosphatase. The predominant sources of ALP are liver and bone. The placenta also produces ALP, with the plasma level increasing through pregnancy. Smaller increases may occur due to the presence of an intestinal isoform in some individuals. ALP isoforms can be distinguished by differences in heat stability and electrophoretic mobility. In a middle-aged woman, the combination of itching and very high ALP (in this case from the liver, as suggested by raised GGT) points to primary biliary cirrhosis (PBC), which is associated with a high titer of anti-mitochondrial antibodies. There is progressive destruction of the small bile ducts, which leads to jaundice and cirrhosis.

Hepatocellular or cholestatic jaundice

When jaundice is caused by a liver problem, the type of jaundice may be described as being either hepatocellular or cholestatic, but the latter term is often used loosely. Cholestasis implies a blockage of biliary secretion and this may be due to an excretory defect in the liver itself, in which case it is hepatic cholestasis, or to a defect in the route between the liver and gall bladder (biliary tree); that is, post-hepatic. In both cases the picture is one of obstructive jaundice (**Figure 3.19**). In clinical practice, many patients present with a mixed picture, where the jaundice may be predominantly cholestatic or predominantly hepatocellular in nature. In the cholestatic form, the liver has the capacity to conjugate bilirubin but there is a problem in excreting the conjugated bilirubin into the gut, for example a blockage caused by gallstones. This is one of a number of causes of obstructive jaundice, which include:

- Gallstones in the common bile duct
- Tumors of the head of the pancreas
- Tumors of the gallbladder
- Sclerosing cholangitis
- Primary biliary cirrhosis
- Biliary atresia

Since reduced amounts of conjugated bilirubin reach the gut, metabolism to urobilinogen and production of urobilin is decreased and there is a reduction in pigmentation of feces; pale stools thus are a characteristic of cholestatic jaundice.

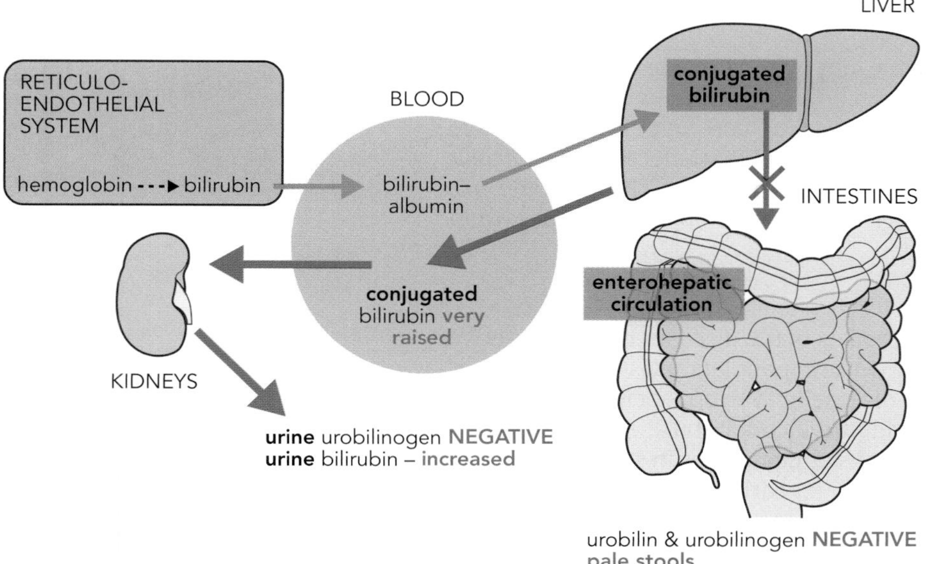

Figure 3.19 Obstructive jaundice. In obstructive jaundice the main defect is a blockage in the excretion of conjugated bilirubin into the intestines. The liver can still conjugate the bilirubin but it cannot excrete it; therefore the conjugated bilirubin leaks into the blood, giving rise to high plasma bilirubin concentrations. Most of this raised bilirubin is conjugated and, since it is readily soluble in water, will readily pass into urine, giving rise to increased urine bilirubin levels and making the urine look dark brown. Since there is a block in the excretion of conjugated bilirubin, enterohepatic circulation is impaired and therefore fecal excretion of the colored compounds (such as urobilin) will be markedly decreased, leading to pale stools. In addition, no urobilinogen will be found in the urine. The orange arrows refer to unconjugated bilirubin and green arrows to conjugated bilirubin.

The inability to move conjugated bilirubin from the biliary tree causes it to back up in the liver and eventually conjugated bilirubin flows from the liver into the blood. The raised concentration of conjugated bilirubin in the blood results in jaundice. Conjugated bilirubin, a small, water-soluble molecule, is able to pass through the glomerulus and thus appears in urine causing a brown-yellow darkening in its color. A further consequence of the reduction in urobilinogen synthesis in the gut in obstructive jaundice is the reduced enterohepatic circulation of urobilinogen and reduced urinary urobilinogen. Typical clinical chemistry results for patients with obstructive jaundice are given in **Table 3.5**. It should be noted that ALP levels are higher in children than in adults; levels of up to 400 U/L can be seen due to growth spurts.

In all cases of conjugated hyperbilirubinemia there is a defect or obstruction in the excretion of conjugated bilirubin. As mentioned earlier, problems in the biliary tree or with the process of biliary secretion would be seen as post-hepatic jaundice; for example, where gallstones or a tumor of the bile duct (cholangiocarcinoma) block bile flow, or where a tumor on the head of the pancreas blocks the common bile duct. However, abnormalities that damage the liver itself can also lead to conjugated hyperbilirubinemia and present as hepatocellular or hepatic cholestatic jaundice; for example, (i) the autoimmune disorder primary biliary cirrhosis (PBC), a relatively rare disorder more common in women than in men, and (ii) biliary atresia seen in newborn babies. Many patients with PBC have measurable levels of mitochondrial antibodies in their serum, which can

TABLE 3.5 Typical LFT results in obstructive jaundice (adult)

Analyte	Reference values*	Expected results in obstructive jaundice*
ALP	40–120 U/L	Raised, up to and above 1000 U/L
GGT	5–45 U/L	Raised, up to and above 1000 U/L
ALT	5–40 U/L	Slightly raised, above 100 U/L
Bilirubin	<21 μmol/L (1.3 mg/dL)	Raised up to above 100 μmol/L (5.9 mg/dL)
		In severe cases can be higher than 500 μmol/L (29.2 mg/dL)
Albumin	32–52 g/L (3.2–5.2 g/dL)	Often normal
Total protein	65–80 g/L (6.5–8.0 g/dL)	May be raised in autoimmune disease, but not usually greater than 100 g/L (10 g/dL)

*Values are given in SI units with conventional numbers in brackets where they differ.

be used in the initial diagnosis of the disease. A firm diagnosis is made by histology on a liver biopsy.

In addition, Dubin–Johnson syndrome and Rotor syndrome are two relatively rare benign disorders of conjugated bilirubin excretion where one of the hepatic transporter proteins in the bile canaliculi is defective; mutations have been identified in Dubin–Johnson syndrome patients that affect the *Canalicular Multispecific Organic Anion Transporter* (*CMOAT*) gene on chromosome 10. Such patients may present with very high serum bilirubin levels greater than 500 μmol/L (29.2 mg/dL).

Hepatocellular liver disease

While the biochemical and pathological profiles of the diseases described above are well characterized, their incidence in Western populations is somewhat low. By far the most common causes of liver disease in Western societies, and to an increasing extent in the developing world, are alcohol abuse and infection by hepatitis viruses, particularly from the use of shared needles in subjects injecting drugs. In the early stages of direct liver damage, early hepatitis for example, a somewhat mixed picture of clinical chemistry changes may present, often with raised plasma levels of transaminases rather than ALP (**Figure 3.20**). However, in chronic hepatocellular damage caused by prolonged infection or by drugs, more consistent changes are found; typical results are shown in **Table 3.6**. Infectious causes of liver damage commonly include hepatitis A and B, and less commonly hepatitis C and Weil's disease.

Clinical practice point 3.2

Hepatocellular disease is suggested by transaminases (ALT and AST) raised out of proportion to ALP.

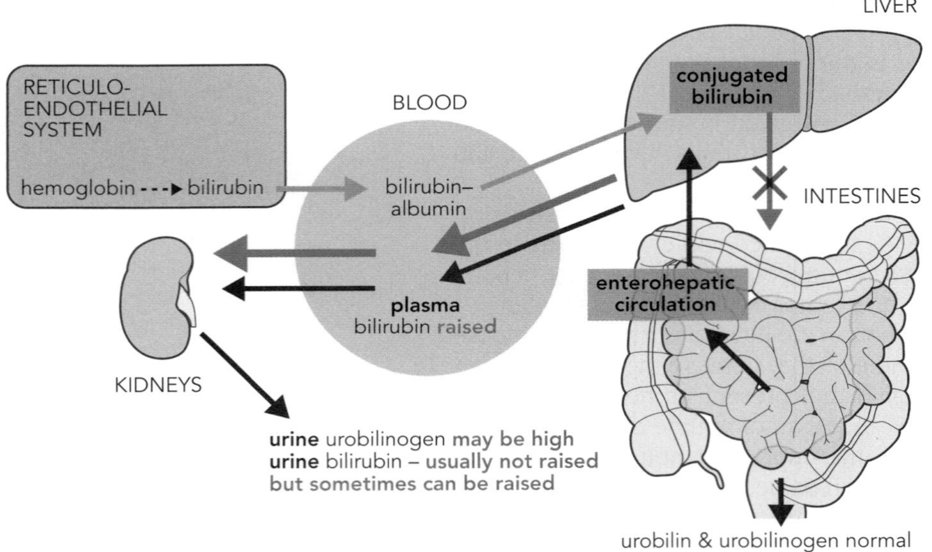

Figure 3.20 Early hepatitis.
In the jaundice associated with early hepatitis there may be a variety of findings dependent on the extent and nature of the damage. Invariably there will be raised bilirubin but this will be mixed, with conjugated and unconjugated levels raised. The amount of conjugated and unconjugated bilirubin present will depend on which of the steps in bilirubin metabolism and excretion have been impaired, and by how much, in the hepatitis. The orange arrows refer to unconjugated bilirubin, green arrows to conjugated bilirubin, and blue arrows to urobilinogen and urobilin.

TABLE 3.6 Typical LFT findings in cases of chronic hepatocellular jaundice (adult)

Analyte	Reference values*	Expected results in chronic hepatocellular jaundice*
ALP	40–120 U/L	Can be normal or slightly raised, up to 150 U/L
GGT	5–45 U/L	Slightly raised, up to 100 U/L
ALT	5–40 U/L	Markedly raised, can be above 1000 U/L
Bilirubin	Less than 21 μmol/L (1.3 mg/dL)	Raised, up to above 100 μmol/L (5.9 mg/dL)
Albumin	32–52 g/L (3.2–5.2 g/dL)	Often normal
Total protein	65–80 g/L (6.5–8.0 g/dL)	Often raised, due to antibody increases or acute-phase response, but not usually greater than 100 g/L (10 g/dL)

*Values are given in SI units with conventional numbers in brackets where they differ.

Hepatic damage due to acetaminophen overdose

Although many toxins and drugs can cause hepatocellular damage, some of the most severe cases of hepatocellular damage, with dramatic rises in LFTs, occur with acute acetaminophen toxicity. Accidental or deliberate acetaminophen overdose causes an enormous rise in ALT (to >5000 U/L) due to acute centrilobular damage and release of the enzyme into the blood (**Figure 3.21**). In severe cases, almost every hepatocyte may be damaged. Acetaminophen is taken up by hepatocytes and metabolized to a free radical, *N*-acetyl-*p*-benzoquinone imine (NAPQI). Scavenging of the toxic free radical by glutathione eventually depletes the cell of this essential antioxidant and leads to cell death. LFTs become abnormal a couple of days after overdose and death from liver failure may occur after two weeks. Treatment for acetaminophen overdose is relatively simple and involves oral administration of other free-radical scavengers such as *N*-acetylcysteine or methionine. This helps to conserve endogenous glutathione for normal hepatocyte function and thereby to decrease cell damage. Effective treatment requires administration of the free-radical scavengers within 12–18 hours of the overdose. There are a number of drugs that can cause hepatocellular toxicity, including:

- Anesthetics
  - Halothane
- Anticonvulsant drugs
  - Phenytoin
  - Valproate
  - Carbamazepine
- Antibiotics and antifungal agents
  - Erythromycin
  - Penicillins
  - Ketoconazole
  - Antituberculous drugs (rifampicin, isoniazid)
- Cardiovascular drugs
  - Amiodarone
  - Methyldopa
  - Lipid-lowering agents (statins, fibrates)
- Immunosuppressive drugs
  - Methotrexate
  - Azathioprine
  - Cyclosporine
- Psychotrophic drugs
  - Phenothiazines
  - Benzodiazepines
  - Tricyclic antidepressants

Clinical practice point 3.3

ALT may be extremely high following acetaminophen overdose, but prothrombin time is a better index of liver function and prognosis.

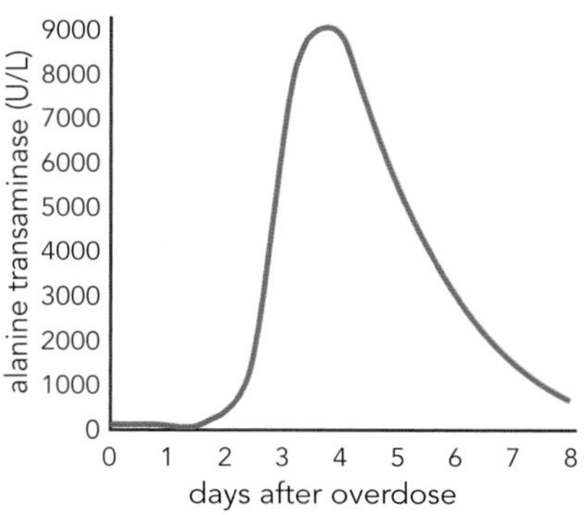

Figure 3.21 Kinetics of alanine transaminase release in acetaminophen overdose. The increase in alanine transaminase starts to occur two days after the initial overdose of acetaminophen, and the very high levels of alanine transaminase found peak at 4–6 days after overdose.

Alcoholic and non-alcoholic liver disease

Alcohol, or more probably its metabolite, acetaldehyde, is toxic to the liver. Persistent excess intake of alcohol, above the recommended weekly number of units, can lead to cirrhosis. Increases in the rate of alcoholic liver disease can be seen in many countries around the world. On an individual level, genetic susceptibility also plays a part and not everyone will develop liver disease at the same level of intake. Thus, women are more susceptible overall than men and hence there are different recommended weekly limits for the two sexes. A very high, short-term intake of alcohol (binge drinking) may cause an acute hepatitis with both a toxic and inflammatory component. This may be fatal but, in those who survive, the liver usually recovers if drinking is discontinued. The general pattern of hepatic pathology in response to alcohol abuse follows three stages: (i) fatty liver (steatosis); (ii) alcoholic hepatitis (inflammation); and (iii) cirrhosis (fibrosis). Damage caused in stages (i) and (ii) is potentially reversible once the insult, in this case alcohol, is removed. However, cirrhosis, where fibrosis occurs as part of the healing response to major tissue injury, causes irreversible damage to the liver and major loss of function. Typical biochemical changes seen in alcoholic liver disease are as follows: GGT raised out of proportion to the other liver enzymes (although a high GGT is not of itself diagnostic of alcoholism); raised AST–ALT ratio (typically greater than 2); and raised immunoglobulins (especially immunoglobulin A). Other common laboratory abnormalities are raised plasma triglycerides and HDL-cholesterol, raised mean red cell volume (macrocytosis), and decreased serum vitamin B_{12} and folate. The last of these may not be seen with heavy drinkers of beer, as this is a source of folic acid.

CASE 3.3

A 19-year-old woman with a history of depression is investigated by a neurologist as she has recently developed slurred speech and a tremor. She appears to be slightly jaundiced. On examination there is a brown ring present around the iris of her eyes. She has LFTs and copper measurements performed.

	SI units	Reference range	Conventional units	Reference range
Plasma				
Bilirubin	35 µmol/L	< 21	2.1 mg/dL	< 1.3
ALT	97 U/L	5–40	97 U/L	5–40
AST	85 U/L	5–40	85 U/L	5–40
GGT	150 U/L	5–45	150 U/L	5–45
ALP	190 U/L	40–120	190 U/L	40–120
Albumin	37 g/L	32–52	3.7 g/dL	3.2–5.2
Copper	8 µmol/L	11–22	51 µg/dL	70–140
Ceruloplasmin	100 mg/L	200–400	10 mg/dL	20–40
Urine				
Copper	10.2 µmol/24 h	0–0.8	650 µg/24 h	0–50

- **How do her neurological and liver abnormalities relate to her copper results?**

The patient has Wilson's disease, a genetic disorder of a copper-exporting protein of the same name. Copper accumulates in the tissues, predominantly the liver and the brain, giving rise to neuropsychiatric symptoms and chronic liver disease, which may present as liver failure. Most cases present before the age of 20. Accumulation of copper in the cornea gives rise to the physical sign of Kayser–Fleischer rings. Ceruloplasmin is the main copper-transporting protein in plasma. Levels of this are reduced in Wilson's disease, as is total copper, although free copper levels are increased. This is demonstrated by increased excretion of copper in the urine. Treatment consists of chelation therapy (usually penicillamine) to deplete copper stores by increasing urine copper excretion still further, followed by zinc supplements, which reduce the intestinal absorption of copper, and a low-copper diet.

Non-alcoholic fatty liver disease

It is possible to progress through the stages of liver disease described above—steatosis, inflammation, fibrosis, and cirrhosis—in the absence of excessive alcohol intake. Indeed, this so-called non-alcoholic fatty liver disease (NAFLD) or non-alcoholic steatohepatitis (NASH) is now recognized as the most common cause of cirrhosis previously labeled as of unknown origin (termed as idiopathic or cryptogenic). The underlying cause is related to insulin resistance, type 2 diabetes, and hypertriglyceridemia and is a manifestation of the metabolic syndrome (or syndrome X). This in turn is driven largely by the prevalence of obesity in the population, a growing public health problem in many countries.

Clinical practice point 3.4

Liver disease due to excessive food intake has been underrecognized. It may go through the same stages as that caused by alcohol. In some cases cirrhosis and malignancy may occur.

Liver tumors

The most common liver cancers in many countries are secondary tumors from colonic cancer, which frequently metastasize into the liver. The predilection for the liver is explained by the drainage of blood from the gut into the portal vein, allowing malignant cells to be carried directly to the liver. In the early stages of the disease all of the LFTs can be normal. However, raised liver enzymes without elevated bilirubin is a characteristic picture of a space-occupying lesion and should prompt a request for imaging with an ultrasound scan. Primary hepatocellular carcinomas, or hepatomas, are usually found in patients who have had chronic cirrhosis, such as with alcoholic liver disease or some cases of viral hepatitis. The most useful test for detecting the development of hepatoma is alpha-fetoprotein (AFP), which is elevated in up to 80% of cases. The concentration of AFP, a glycoprotein structurally related to albumin, is high in fetal blood but falls rapidly after birth when hepatic synthesis of the protein is switched off. A high plasma concentration of AFP in an adult with cirrhosis is almost always diagnostic of hepatoma, although smaller increases can be seen with benign liver disease.

Liver cirrhosis and failure

As discussed earlier, the huge metabolic overcapacity of the liver allows it to compensate for the loss of a proportion of its cells through damage, such that patients with liver disease may be asymptomatic. However, when liver disease becomes extensive and few functioning cells remain, severe symptoms may develop and may result in death. Often the functioning tissue is replaced by nonfunctioning, fibrotic tissue and continuing damage yields the pathological processes leading to cirrhosis (**Figure 3.22**). Since there are few functioning cells in the cirrhotic liver, plasma enzyme levels are low apart from the readily inducible GGT. Hyperbilirubinemia and jaundice arise from the marked reduction in the ability of the liver to conjugate bilirubin, and albumin levels are frequently low due to reduced capacity for protein synthesis. Hypoalbuminemia

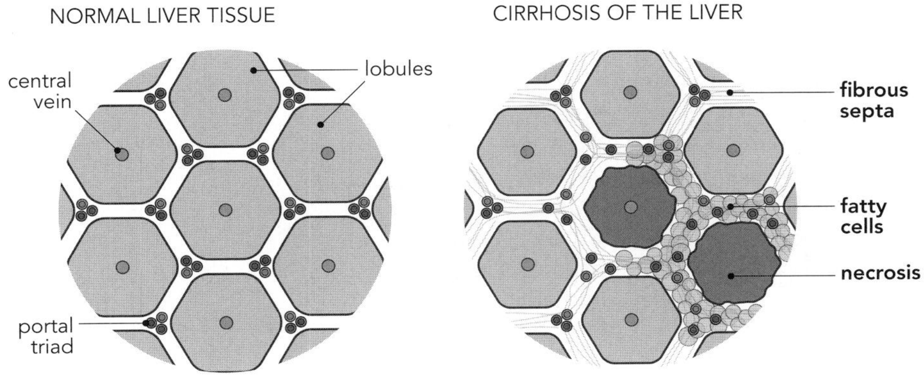

Figure 3.22 Liver cirrhosis. The replacement of normal healthy liver cells with fibrotic, nonfunctioning tissue.

TABLE 3.7 Typical LFT findings in cases of liver cirrhosis

Analyte	Reference values*	Expected results in chronic hepatocellular jaundice*
ALP	40–120 U/L	Slightly raised or often normal, usually less than 150 U/L
GGT	5–45 U/L	Usually raised, above 100 U/L
ALT	5–40 U/L	Slightly raised, usually below 60 U/L
Bilirubin	Less than 21 µmol/L (1.3 mg/dL)	Very high, greater than 300 µmol/L (17.5 mg/dL)
Albumin	32–52 g/L (3.2–5.2 g/dL)	Low, less than 30 g/L (3.0 mg/dL)
Total protein	65–80 g/L (6.5–8.0 g/dL)	Can be low, normal, or high

*Values are given in SI units with conventional numbers in brackets where they differ.

in turn contributes to the ascites frequently seen in patients with liver failure. Typical LFT results for patients with cirrhosis are shown in **Table 3.7**. Other results will also become abnormal; for example, as the urea cycle is disrupted, plasma ammonia will become high and the urea will become low in the absence of renal impairment.

Clinical practice point 3.5

Transaminases may not be elevated in cirrhosis as relatively few hepatocytes remain, having been replaced by fibrous tissue.

Measurement of a marker protein of fibrosis, PIIINP, has become available more recently. This protein is released into the blood from active fibrotic tissue. Measurement of PIIINP has proved useful in monitoring patients being treated with potentially hepatotoxic drugs such as methotrexate, but it is not yet in widespread use.

Urea cycle disorders

The urea cycle disorders (UCDs) are a group of inherited metabolic disorders characterized by the accumulation of ammonia (hyperammonemia) and other metabolites of the urea cycle during the early neonatal period. UCDs are caused by mutations of the genes coding for six key enzymes of the urea cycle, resulting in partial or complete loss of enzyme activity. The six UCDs are:

- Carbamoyl phosphate synthase 1 (CPS1) deficiency
- Ornithine transcarbamoylase (OTC) deficiency
- Argininosuccinate synthase (ASS) deficiency (or classic citrullinemia)
- Argininosuccinate lyase (ASL) deficiency (or argininosuccinic aciduria)
- *N*-acetyl glutamate synthetase (NAGS) deficiency
- Arginase (ARG) deficiency

A protein load will result in increased flux through the urea cycle and, when enzyme activity is deficient, ammonia and precursor metabolites accumulate. Five enzymes of the urea cycle (CPS1, OTC, ASS, ASL, and ARG) are involved in catalyzing the individual reactions, while the sixth, *N*-acetyl glutamate synthetase (NAGS), produces *N*-acetyl glutamate, an allosteric activator of CPS1. Additionally, two mitochondrial transport proteins—citrin and ornithine transporter protein-1 (ORNT1)—are involved in moving citrulline and ornithine, respectively, between the mitochondrial and cytoplasmic compartments (**Figure 3.23**). Most of the UCDs are inherited as autosomal recessive disorders, the exception being ornithine transcarbamoylase deficiency, which is an X-linked recessive condition. Patients with complete enzyme deficiencies present as newborns, while those with partial defects tend to have a later, often atypical presentation. In the classic presentation, the newborn appears normal for the first 24 to 48 hours until feeding begins and ammonia accumulates; this in turn causes cerebral edema with accompanying signs including vomiting,

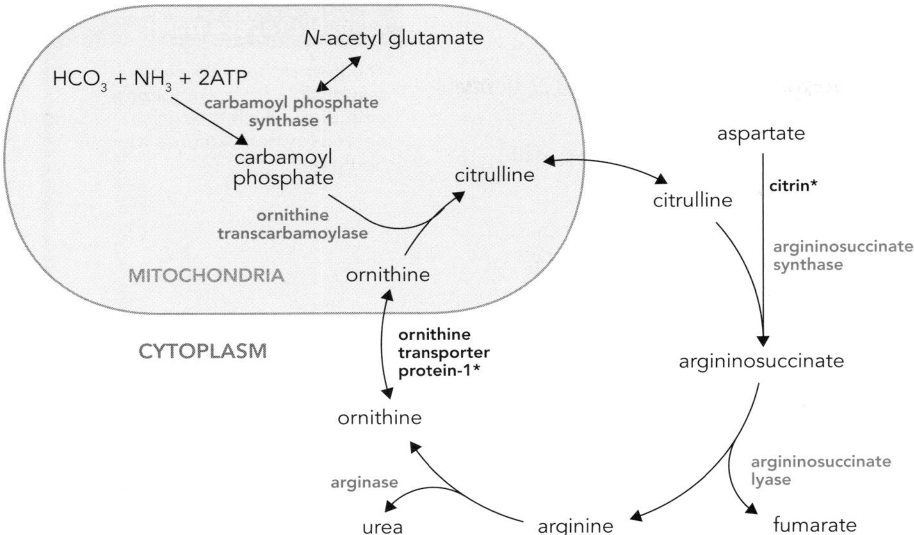

Figure 3.23 Enzyme deficiencies in the urea cycle disorders. Disorders of the urea cycle occur at a rate of about 1 per 25,000 live births. Complete enzyme deficiency of the first four enzymes of the cycle—carbamoyl phosphate synthase 1, ornithine transcarbamylase, argininosuccinate synthase, and argininosuccinate lyase—leads to early presentation (within 2–4 days of birth) of severe hyperammonemia. Arginase deficiency is of later onset, and leads to hyperarginemia with physical signs apparent after 2–3 years. Deficiency of two mitochondrial transporter proteins (marked by the asterisks), ornithine transporter protein-1 (ORNT1) and the aspartate transporter SLC25A13 (trivially known as citrin), also gives rise to hyperammonemia.

lethargy, and coma. Hyperammonemia in patients with a partial enzyme defect is a component of metabolic decompensation triggered usually by infection, stress, surgery, or trauma and can occur at any time in life. Because OTC deficiency is an X-linked disorder, a proportion of heterozygous (carrier) females can present with hyperammonemia and the severity of presentation can range from mild to severe as a result of variable X-chromosome inactivation. The investigation of an infant presenting with hyperammonemia is complex and involves an array of analyses including quantitative analyses of plasma amino acids and of urinary organic acids. Classical newborn screening (for a limited spectrum of disorders) does not include the UCDs. More modern, extended newborn screening based upon tandem mass spectrometry is of limited value given that it cannot detect metabolite patterns diagnostic of CPS1, OTC, or NAGS deficiency. DNA-based diagnostic testing can provide confirmatory diagnoses in UCD cases and can be used in antenatal testing situations.

Hyperammonemia

Ammonia is found in the plasma at concentrations of less than 32 μmol/L (<45 μg/dL). Hyperammonemia is a condition in which the plasma ammonium ion concentration is raised above 100 μmol/L (140 μg/dL) in neonates and 40 μmol/L (56 μg/dL) in older children and adults. It arises when there is an imbalance between ammonia production and its elimination from the body. This may be due to failure of urea synthesis, to increased amino acid metabolism from protein hypercatabolism, or to bypassing of liver by ammonia formed in the colon or occasionally the bladder. The principal causes of hyperammonemia are as follows.

- In neonates:
  - Inborn errors of metabolism, especially of the urea cycle, and organic acidemias
  - Acute liver failure
  - Nonspecific, severe illness, especially infections and with hypoxia
  - Some pre-term neonates
  - Excessive nitrogen intake via parenteral nutrition

- In other children:
  - Less severe forms of inborn errors of metabolism, as for neonates
  - Valproate administration, which reduces the concentration of *N*-acetyl glutamate
  - Nonspecific, severe illness, especially infections and with hypoxia
- In adults:
  - Acute or end-stage liver failure
  - Urinary stasis with infection by an organism synthesizing urease (ammonia absorbed from bladder)

Clinical practice point 3.6

The liver is the main organ responsible for metabolizing ammonia, which is toxic to the central nervous system, to urea which is nontoxic.

KIDNEY

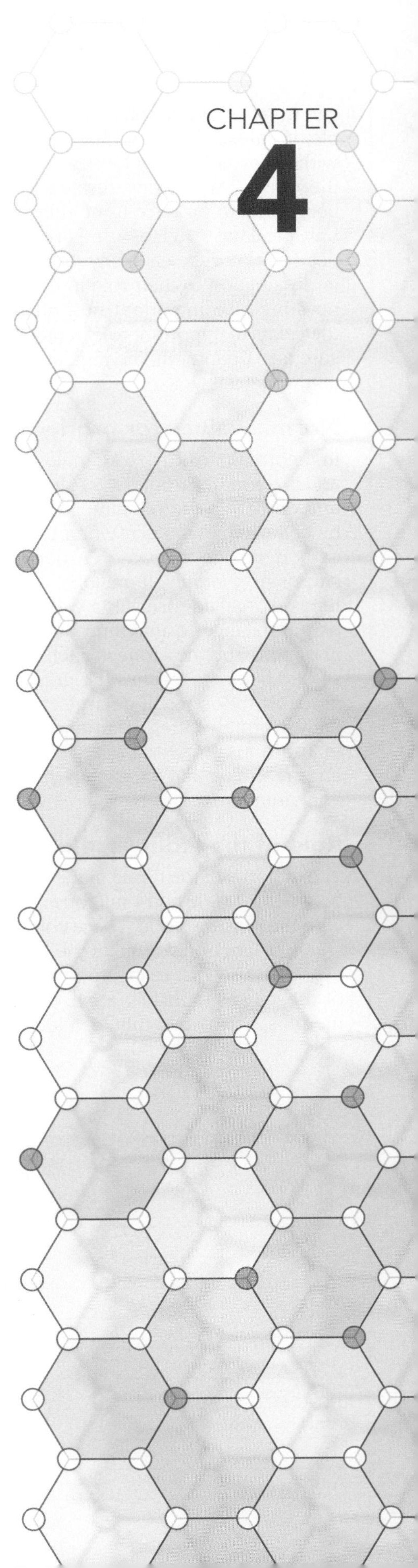

IN THIS CHAPTER

The renal system has evolved to ensure that the body eliminates various waste products of metabolism, in the form of nitrogen-containing compounds like urea, acids like phosphoric acid, and potentially toxic metabolites of many exogenous chemicals, including drugs. At the same time, the renal system must control the volume and concentration of urine to balance the variable daily intake of fluids and electrolytes. Although we tend to think that the main product of the kidneys is urine, it is in fact more accurate to say it is extracellular fluid, with urine being merely the waste output. Measurement of urinary composition can be useful in determining the underlying kidney function, especially when combined with simultaneous measurements in plasma. Failure of renal function is associated with severe metabolic derangement and the various forms of renal failure comprise a major cause of ill health and death.

4.1 PHYSIOLOGY OF THE RENAL SYSTEM

The kidneys lie on the posterior abdominal wall at the level of the lowest extent of the ribcage (the lowest two ribs on either side are close anatomical relations) and are highly innervated. Their blood supply comes from the renal arteries, which branch off from the abdominal aorta at an angle of roughly 90°, resulting in a degree of turbulent blood flow in this region. This means that the junction of the abdominal aorta and the renal arteries is a prime site for the development of atheroma and thrombosis. The renal arteries and their branches are end-arteries, meaning that there is no collateral circulation from other vessels, and so occlusion or clamping of the renal artery rapidly results in necrosis of the kidney.

Between 20 and 25% of cardiac output perfuses the kidneys, so normal renal blood flow is approximately 1.1 L/min. With a normal hematocrit of 0.45, this corresponds to renal plasma flow of 600 mL/min (1.1 L/min × 0.55). On passing through the kidneys, 20% of the plasma (120 mL/min) is filtered, meaning that >170 L of plasma are filtered by the kidneys in a day. Only around 1% of this volume is actually lost in the urine, the remainder being reabsorbed in the kidneys after filtration. In addition to adjusting the volume of urine produced, the kidneys can make dilute or concentrated urine to suit the needs of the body. For example, if an excess of fluid has been ingested, then dilute urine will be produced with an osmolality as low as 50–100 mOsm/kg H_2O, but with fluid deprivation the urine can be concentrated to an osmolality of up to 1200 mOsm/kg. By being able to produce urine that is significantly hypertonic to plasma, we need only produce 0.75 L of urine in order to excrete the 900 mOsm of solute that we need to eliminate each day (to balance intake and metabolism).

In addition to their role in the excretion of waste products in urine, the kidneys are also important in a number of other bodily processes. For example, they play an important role in medium- to long-term maintenance of blood pressure (see Chapter 8), the regulation of calcium homeostasis (see Chapter 9), and erythropoiesis.

Normal values for production of urine

In a normal, healthy 70 kg male, daily water loss is matched to daily water gain and is normally around 2.5 L. In terms of intake, around 2.2 L is ingested in food and drink and the remaining 0.3 L comes from oxidation of energy substrates by cellular metabolism when fats and carbohydrates are oxidized to produce carbon dioxide and water. Around 0.2 L of water is lost in the feces daily and a surprisingly large volume (0.8 L) is lost via skin secretions like sweat and via the lungs (so-called insensible water loss). This leaves 1.5 L of water in the urine as the balance to make up the total of 2.5 L. Approximately 170 L of plasma are filtered by the kidneys each day, so more than 99% of filtered fluid is reabsorbed before it reaches the urine, and only small alterations (for example, 1%) in reabsorption result in large changes (100%) in the amount of urine produced daily. Normal values for urinary composition in a healthy 70 kg male are given in **Table 4.1**. Note the relatively large amounts of waste products such as urea and creatinine that are excreted, compared to nutrients such as glucose, protein, and amino acids.

Role of the kidneys in filtration of plasma

The majority of the tissue in the kidneys is epithelial; each kidney is composed of between 200,000 and 1 million extensively coiled tubes, each in potential contact with the outside world. These coiled epithelial tubes are known as nephrons, the smallest functional units of the kidneys (**Figure 4.1**). The nephron is the site of blood filtration, of reabsorption of substances from this filtrate, and of addition of substances to this filtrate to make urine. Each nephron is composed of a glomerulus, a proximal tubule, a loop of Henle, a distal tubule, a collecting tubule,

TABLE 4.1 Normal urine composition in humans

Parameter	Reference values	
	Conventional units	SI units
Urinary volume	800–2500 mL/day (dependent on fluid intake)	
Urinary osmolality	80–1200 mOsm/kg	
Urinary urea (nitrogen)	12–20 g/day	200–330 mmol/day
Urinary creatinine	15–25 mg/kg/day	133–221 mmol/kg/day
Urinary urobilinogen	<2.5 mg/day	<4.2 μmol/day
Urinary protein	<100 mg/day	
Urinary amino acids	200–400 mg/day	14–29 nmol/day
Urinary glucose	<300 mg/day	<1.6 mmol/day
pH	4.5–8	

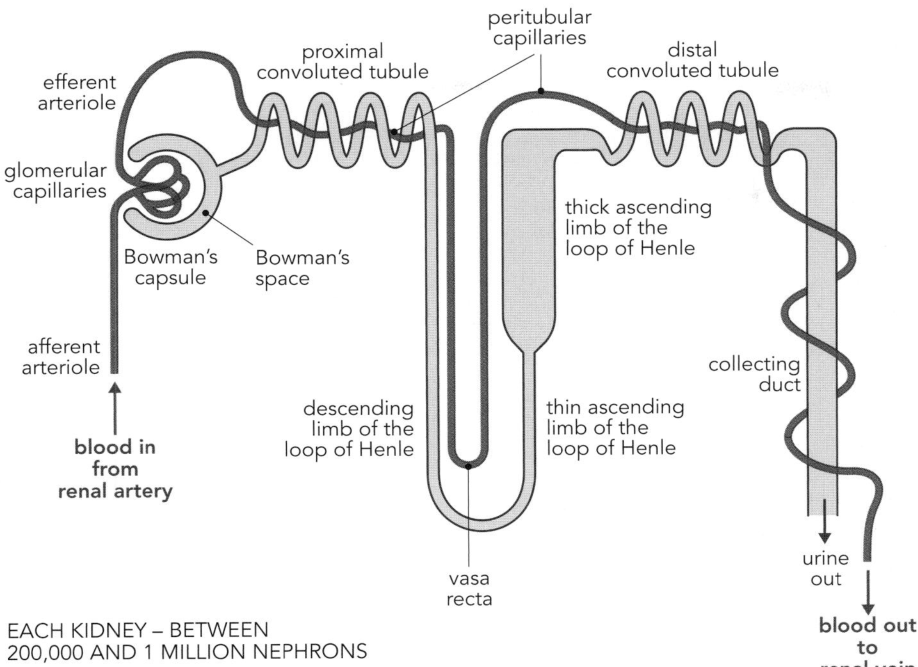

Figure 4.1 Schematic diagram of the nephron, showing its constituent parts.

and a collecting duct. Each of these parts of the nephron has a particular role in the formation of an appropriate volume of urine of the appropriate composition, as outlined in **Table 4.2**. At all points along the nephron there exists the tubular space (containing filtered fluid that will become urine), epithelial cells (of which the tubule is composed), the interstitial space (between epithelial and other cells in the kidney), and blood vessels. More detail on the function of these various parts of the nephron follows in later sections.

The site of filtration of the blood is known as the renal corpuscle and the remainder of the nephron is known as the renal tubule. **Figure 4.2** shows what the nephron would look like were it to be unraveled.

At the renal corpuscle, bundles of leaky capillaries allow some components of the blood plasma to enter Bowman's space. The first stage in urine production is filtration of the plasma at the glomerulus; as the filtered fluid subsequently passes along the renal tubule, many substances are reabsorbed into the blood and some substances are added, so that by the end of the renal tubule (collecting duct), the filtered and modified fluid has become urine.

TABLE 4.2 Major functions of the parts of the nephron

Part of the nephron	Major role
Glomerulus	Filtration of blood to form glomerular filtrate
Proximal tubule	Reabsorption of the majority of filtered fluid, electrolytes, and nutrients (for example glucose, amino acids); secretion of organic ions
Loop of Henle	Formation of medullary concentration gradient
Distal tubule and collecting tubule	Final modifications to urine: reabsorption of certain electrolytes (for example Ca^{2+} ions), acidification of the urine
Collecting duct	Formation of variably concentrated urine; important in K^+ balance

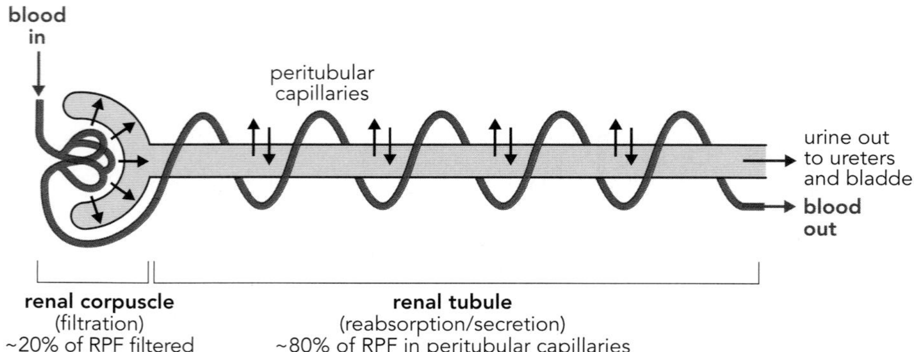

Figure 4.2 Events in the renal corpuscle (glomerular capillaries and Bowman's capsule) and the renal tubule (remainder of the nephron).
RPF, renal plasma flow.

The glomerular filter

Under normal circumstances, the glomerular filtrate has a composition much like plasma but without the plasma proteins; it has the same concentration of ions, glucose, and amino acids and has the same pH. Proteins and blood cells are normally excluded from the glomerular filtrate because they are too large to pass through the glomerular filtration barrier (**Figure 4.3**).

The glomerular filtration barrier is composed of three layers:

- 1. The endothelial cells of glomerular capillaries;
- 2. The foot processes of podocytes (the epithelial cells lining Bowman's space); and
- 3. Between these two layers, the fused basement membrane of both these cellular layers.

The endothelial cells lining the glomerular capillaries are fenestrated. Not only are there small gaps between the cells, as in most capillaries in the body, but the cells have pores through them, 50–100 nm in diameter. Thus, glomerular capillaries are extremely leaky; plasma proteins, with a much smaller molecular size, would be able to pass through these pores were it not for some of the other features of the glomerular filter. The glomerular basement membrane (between the endothelial cells and the podocytes) is composed of layers of interwoven proteins, with small spaces in between. The epithelial cells lining Bowman's space (podocytes) interdigitate with one another through their foot processes. Between these foot processes there exists a filtration slit diaphragm, made up of specific protein elements that act as a molecular filter, allowing molecules of no more than 20 nm diameter through into Bowman's space.

The likelihood of a molecule passing from the plasma into the glomerular filtrate is determined by two factors: molecular size and molecular electrical charge. Molecules with a molecular mass less than 10 kDa can pass through the glomerular filter provided that they are not bound to plasma proteins. Above this

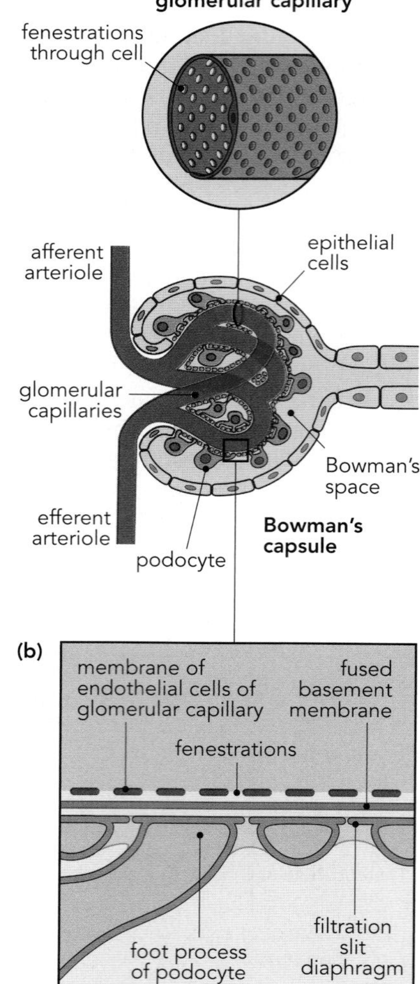

Figure 4.3 Diagram showing (a) fenestrations in the endothelial cells making up glomerular capillaries and (b) the layers of the glomerular filtration barrier.
Approximately 20% of renal plasma flow is filtered in the kidney glomeruli. Fenestrations in the endothelial cells making up the glomerular capillaries allow more fluid to leave the plasma here than in other tissues of the body. Under normal circumstances, the combination of fenestrations, the glomerular basement membrane, and the foot processes of the podocytes making up Bowman's capsule allows fluid, electrolytes, glucose, amino acids, and some other small molecules to pass from the capillaries into Bowman's space, whilst retaining plasma proteins and blood cells in the capillaries.

molecular mass, molecules find it more difficult to pass through the barrier, until at around 100 kDa none can pass through. The various studies that have been conducted into glomerular filtration have estimated that, if the glomerular filter functioned on the basis of molecular size alone, the pores should have a size of between 7 and 10 nm in diameter, but no such small-diameter pores have ever been seen. Thus, much of the sieving properties of the glomerular filter must be conferred in other ways.

All elements of the glomerular filter contain fixed negative charges (the glycocalyx on the cellular elements and sialic acid on the basement membrane) such that, above a given molecular mass, negatively charged (anionic) molecules are less likely to be filtered than neutral or cationic molecules. This is the major reason why serum albumin, despite its molecular diameter of approximately 7.5 nm, is not present in glomerular filtrate under normal circumstances; at physiological pH, most serum proteins are negatively charged and are repelled by the fixed negative charges on the glomerular filtration barrier. Small anions such as Cl^- and HCO_3^- are completely filterable because of their extremely small size. When the glomerular filter is damaged by disease, proteinuria can result.

As with movement of fluid out of capillaries in other tissues, Starling's forces are important in determining the ultrafiltration pressure (UP) at the glomerulus:

$$\mathbf{UP = (P_{cap} + \pi_{BS}) - (P_{BS} + \pi_{cap})} \qquad \text{(Equation 4.1)}$$

Glomerular capillary hydrostatic pressure (P_{cap}) and Bowman's space hydrostatic pressure (P_{BS}) oppose each other. Blood colloid osmotic pressure (π_{cap}) and Bowman's space colloid osmotic pressure (π_{BS}) also oppose each other but because blood contains far more protein than the glomerular filtrate in Bowman's space, π_{BS} can be discounted under normal circumstances. Thus the equation that describes glomerular UP normally takes the form:

$$\mathbf{UP = P_{cap} - P_{BS} - \pi_{cap}} \qquad \text{(Equation 4.2)}$$

Glomerular filtration rate

In a normal, healthy 70 kg male, renal plasma flow is approximately 600 mL/min. Normally, around 20% of this renal plasma flow is filtered at the glomerulus, meaning that glomerular filtration rate (GFR) is normally 120 mL/min. One of the ways in which a substance is removed (cleared) from the plasma is via filtration. The equation for calculating the clearance of substance X (the volume of plasma completely cleared of the substance in a given period of time, C_X) can be derived in the following manner:

1. Measure the concentration of substance X in the plasma (P_X)
2. Measure the concentration of substance X in the urine (U_X)
3. Measure urine output over a period of 24 hours (V)

Since, in the case of the kidneys, substances are cleared into the urine, the amount cleared should equal the amount that is found in the urine. Thus the relationship between C_X, P_X, U_X, and V takes the form:

$$\mathbf{C_X \times P_X = U_X \times V} \qquad \text{(Equation 4.3)}$$

Equation 4.3 can be rearranged to form an equation for calculating C_X:

$$\mathbf{C_X = (U_X \times V) / P_X} \qquad \text{(Equation 4.4)}$$

Creatinine is one of the by-products of protein catabolism. Creatinine is freely filtered at the glomerulus but on the whole is neither secreted by the renal tubule nor reabsorbed as it moves along the tubule. This means that the volume of plasma completely cleared of creatinine in a given period of time (the creatinine clearance) is an estimate of GFR. In actual fact, to accurately calculate GFR, one should infuse a subject with the plant polysaccharide inulin until plasma inulin concentrations reach equilibrium, then calculate inulin clearance (because inulin is freely filtered at the glomerulus and is never secreted or reabsorbed by the renal tubule).

Note that if a given substance is secreted by the renal tubule in addition to being freely filtered (for example H^+ ions), then the numerator on the right of Equation 4.4 will increase and clearance will be greater than for a substance like creatinine or inulin. If the substance is reabsorbed by the renal tubule in addition to being freely filtered (for example glucose), then the numerator on the right of Equation 4.4 will be lower and clearance will be lower than for creatinine or inulin.

As an example, using Equation 4.4, one can calculate $C_{creatinine}$ to obtain an estimate of GFR if one knows $P_{creatinine}$ (for example 80 μmol/L), $U_{creatinine}$ (for example 7.1 mmol/L), and V (for example 2 L/day):

$$\mathbf{C_{creatinine} = (U_{creatinine} \times V) / P_{creatinine}}$$

$$\rightarrow \mathbf{C_{creatinine} = (7100\ \mu mol/L \times 2\ L/day) / 80\ \mu mol/L}$$

$$\rightarrow \mathbf{C_{creatinine} = 177.5\ L/day = 0.123\ L/min = 123\ mL/min}$$

(Equation 4.5)

Thus a reasonable estimate of GFR from the data given above is 123 mL/min. It is important to realize that by calculating $C_{creatinine}$ one is only estimating GFR. This is because some creatinine is secreted by the renal tubule. Therefore, $U_{creatinine}$ can be as much as 20% higher than if no creatinine were secreted by the renal tubule. However, the traditional methods for measuring $P_{creatinine}$ are not entirely accurate either (being nonspecific and detecting non-creatinine substances), so that the errors in both the numerator and the denominator on the right side of Equation 4.5 cancel each other out over a wide range of GFR. In modern clinical practice, $P_{creatinine}$ (measured from a blood sample) combined with age, gender, and race is used routinely to estimate GFR, as described in Section 4.2.

Low GFR might indicate a reduced number or function of glomeruli or, since UP is largely dependent upon P_{cap}, lowered kidney perfusion. Thus it is worthwhile considering whether low GFR is symptomatic of glomerular damage or of low kidney perfusion, and estimating renal plasma flow is an important diagnostic tool in some circumstances. The distinction is important in understanding two of the types of kidney failure discussed later: namely pre-renal and intrarenal.

Measurement of renal plasma flow

Renal plasma flow is normally around 600 mL/min, of which 120 mL/min is filtered at the glomerulus. The remaining 80% of renal plasma flow then makes up the blood that perfuses the rest of the kidney in the peritubular capillaries (see Figure 4.2). If a substance is secreted into the urine from the plasma as it passes along the peritubular capillaries, its clearance will be greater than GFR; if it is reabsorbed from the urine into the peritubular capillaries, its clearance will be lower than GFR. This principle can be applied to calculating renal plasma flow.

Many molecules absorbed from the lumen of the nephron into the blood, or secreted into the lumen from the blood, are transported across cell membranes via specific transport proteins. However, the rate of transport of substances into or out of the lumen of the nephron is limited by the transport maximum (T_m) for the molecule in question (usually quoted in mmol/min). T_m is determined partly by the kinetics of the transporter and the number of transporter molecules available. Para-aminohippuric acid (PAH) is a substance that, in addition to being freely filtered at the glomerulus, is secreted into the renal tubule from the peritubular capillaries. The plasma concentration of PAH is usually low enough that all of the remaining nonfiltered PAH in the plasma can subsequently be secreted into the renal tubule from the peritubular capillaries. Hence C_{PAH} is a measure of the total renal plasma flow.

Reabsorption of electrolytes

As is the case in many cells (for example, in nerve and muscle cells where the extracellular concentration of Na^+ ions is high and the intracellular concentration of Na^+ ions is low), the function of a renal tubule cell is dependent upon the cell's ability to regulate the concentration of various ions on either side of the cell membrane. However, because renal tubule cells are epithelial and form a surface that is in potential contact with the outside world, this arrangement of ions is made more complicated; one part of the cell (the luminal surface) is in contact with the tubular fluid, whereas other parts of the cell (the basolateral surfaces) are in contact with the interstitial fluid between the cell and the bloodstream. This cellular arrangement allows the cell to preferentially send different ion channels, ATPase pumps, and exchanger proteins to different cell surfaces in order to affect absorption of substances from, or secretion of substances into, the tubular fluid.

An example of the ability of renal tubule cells to reabsorb substances that have been filtered at the glomerulus is the reabsorption of NaCl in the thick part of the ascending limb of the loop of Henle (**Figure 4.4**). The epithelial cells of the thick part of the ascending limb of the loop of Henle are important in reabsorption of around 20% of the Na^+ ions filtered at the glomerulus. These cells send more Na^+/K^+-ATPase pumps and a K^+/Cl^- co-transporter to the basolateral surface (blood side) of the cell than to other parts, and send more $Na^+/K^+/2Cl^-$ co-transporters and Na^+/H^+ exchanger proteins to the luminal surface (urine side) of the cell. K^+ and Cl^- ion channels are also placed on specific cell surfaces. Active pumping of Na^+ and K^+ ions against their concentration gradients (out of and into the cell, respectively) at the basolateral surfaces results in a high concentration of K^+ ions inside the cell and a high concentration of Na^+ ions in the interstitial fluid. Some K^+ ions are able to leak back out of the cell through K^+ ion channels on the basolateral membrane and others drag Cl^- ions with them through the K^+/Cl^- co-transporters; this process maintains the function of

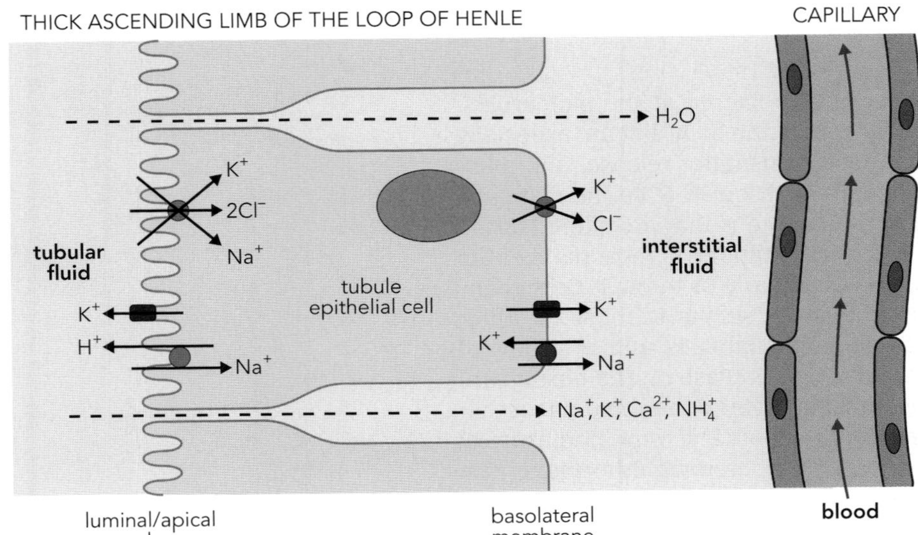

Figure 4.4 Cell model for reabsorption of solute in the thick ascending portion of the loop of Henle.

the Na^+/K^+-ATPase, which would stop working if K^+ ion concentrations fell to too low a level. The function of all tubular epithelial cells requires the formation of large amounts of ATP and so most such cells have a high number of mitochondria. The pumping of Na^+ ions out of the cell at the basolateral side also sets up a low concentration of Na^+ ions intracellularly, creating a concentration gradient for Na^+ ions to move into the cell. Na^+ ions enter the cell at the luminal surface, each bringing a single K^+ ion and two Cl^- ions with them through the $Na^+/K^+/2Cl^-$ co-transporter. Other Na^+ ions enter the cells in exchange for H^+ ions, using the Na^+/H^+ exchanger on the luminal membrane. Some K^+ ions also leak out through K^+ ion channels on the luminal surface, helping to maintain the function of the $Na^+/K^+/2Cl^-$ co-transporter.

These various ionic movements lead to a slightly more electropositive environment in the tubular fluid than in the interstitial fluid, resulting in some passive movement of small cations such as Na^+, K^+, Ca^{2+}, and NH_4^+ from the tubular fluid into the interstitial fluid via tiny gaps between the epithelial cells. Water also moves from the luminal fluid to the interstitial fluid, following the net movement of solute. In fact, the major portion of fluid reabsorption from the tubular fluid goes via this mechanism in all parts of the nephron. The luminal $Na^+/K^+/2Cl^-$ co-transporter proteins are blocked by loop diuretic drugs such as furosemide, and if reabsorption of Na^+ ions falls, so does the net movement of water, resulting in production of greater volumes of urine (diuresis).

These epithelial cells utilize a wide range of membrane transport processes in a unique way in order to affect reabsorption of Na^+ and other ions and excretion of acid. The movement of substances between the tubular fluid and the interstitial fluid is controlled by similar yet subtly different mechanisms in different parts of the nephron, in order to bring about reabsorption or secretion of different substances at different parts of the tubule. See later in this chapter for an example of secretion of H^+ ions.

Na^+ and K^+ balance

Na^+ and K^+ ions exist in the plasma at concentrations of approximately 140 mmol/L and 4.2 mmol/L, respectively; this means that approximately 25,200 mmol of Na^+ ions and 760 mmol of K^+ ions are filtered at the glomeruli each day. In order for our bodies to function effectively, almost all of these filtered ions have to be reabsorbed by the kidney following filtration; this is done by active transport processes in the renal tubular epithelial cells. In general, almost all of the Na^+ and K^+ ions that we ingest via our diet (around 200 mmol and 100 mmol daily, respectively) is balanced by Na^+ and K^+ excretion in the urine. In the case of Na^+, this is normally done as part of our body's normal mechanism for maintaining plasma osmolality.

Ca^{2+} balance

Ca^{2+} ions, like all small inorganic ions, are freely filtered at the glomerulus. Ca^{2+} ions are essential for the function of all cells in the body, but are especially important in the mineralization of bone, neurotransmitter release, and muscle cell contraction. Only around 3 mmol Ca^{2+} is absorbed from the diet each day, so it is important that the majority of the Ca^{2+} ions that are filtered at the glomerulus (360 mmol) are reabsorbed in the renal tubule. In most parts of the renal tubule, Ca^{2+} ions move by diffusion paracellularly or through Ca^{2+} channels on renal tubule cells, via mechanisms that are similar to those for other ions. However, in the late distal tubule, the epithelial cells utilize a luminal Ca^{2+} channel and Na^+/Ca^{2+} exchangers and a Ca^{2+}-ATPase on the basolateral membrane. Within the cytoplasm, Ca^{2+} is transported by a Ca^{2+}-binding protein known as calbindin. The action of calbindin is dependent on 1,25-dihydroxy vitamin D, the formation of which is controlled by parathyroid hormone (see Chapter 9). When serum free Ca^{2+} levels fall below the normal physiological range and parathyroid hormone (PTH) is secreted in increased amounts, some of the effect of PTH to increase serum free Ca^{2+} levels is mediated via the action of 1,25-dihydroxy vitamin D in the kidney. PTH also inhibits the reabsorption of phosphate ions in the proximal convoluted tubule, resulting in increased phosphate excretion in the urine, in an effort to limit the effect on serum phosphate concentrations of the bone resorption that is increased by PTH (see Chapter 9). Calcium ions and phosphate ions can combine and crystallize in alkaline urine to form calcium phosphate kidney stones in conditions such as hyperparathyroidism and renal tubular acidosis.

Carrier-mediated reabsorption of glucose

Glucose is an uncharged monosaccharide sugar with a molecular mass of 180 Daltons and a molecular radius of <0.4 nm. These three characteristics (charge, mass, and size) mean that glucose is freely filtered at the glomerulus. Plasma glucose concentrations are normally around 6 mmol/L and so with a GFR of 120 mL/min, the filtration rate for glucose is 0.72 mmol/min. With normal renal function, all of this glucose would usually be absorbed well before reaching the urine (usually by the end of the proximal tubule) because the T_m for glucose is normally around 1.8 mmol/min. Thus, with normal renal function and a GFR of 120 mL/min, plasma glucose levels would have to rise to above 15 mmol/L before any glucose would find its way into the urine. This is the reason why people with diabetes mellitus (see Chapter 2) usually have trace amounts of glucose in their urine.

Over the course of a day, our kidneys must reabsorb around 1 mole of filtered glucose; this is equivalent to 180 g of glucose. This is an amazing feat of cell biology, especially when one realizes that the plasma glucose concentration in the peritubular capillaries is roughly the same as that of the glomerular filtrate and there is therefore no concentration gradient that would naturally drive glucose reabsorption. In order to allow this to take place, reabsorption of glucose in the proximal part of the nephron is linked to the reabsorption of Na^+ ions (**Figure 4.5**). The epithelial cells of the proximal tubule have large numbers of Na^+/K^+-ATPase pumps on their basolateral membranes, actively pumping out Na^+ ions and pumping in K^+ ions. The K^+ ions are able to leave the basolateral surface of the cell and keep the Na^+/K^+-ATPases supplied with K^+ via leak

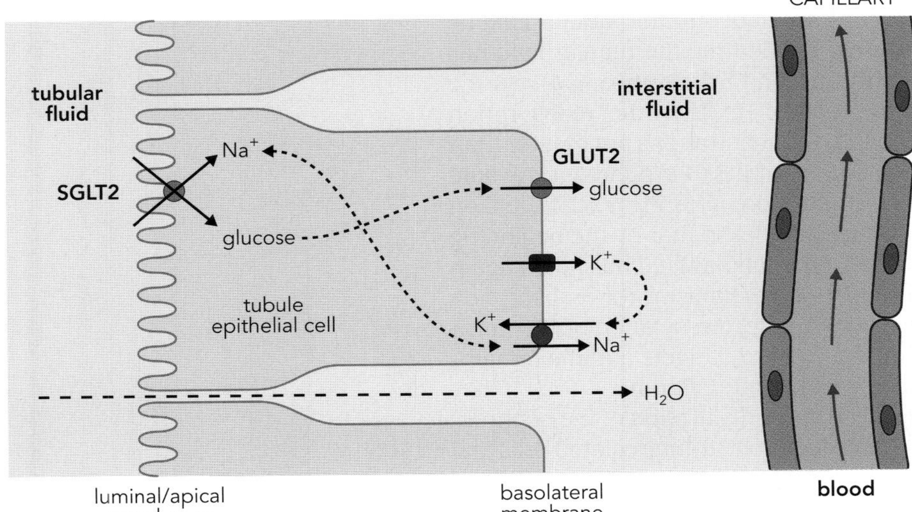

Figure 4.5 Cell model for glucose reabsorption in the proximal parts of the nephron.

K^+ channels. This creates a low Na^+ ion concentration in the proximal tubule epithelial cell. The proximal tubule cells also have large numbers of glucose/Na^+ co-transporters (SGLT2 transporters) on their luminal membrane and these transporters use the large concentration gradient for Na^+ ions to drag glucose into the proximal tubule cell (the glomerular filtrate will have a Na^+ concentration like that of plasma, around 140 mmol/L). This movement of glucose into the proximal tubule cell is a form of secondary active transport, using the energy stored in the Na^+ concentration gradient to move glucose against its concentration gradient. Glucose is then able to move from the intracellular space to the interstitial space between the proximal tubule cells and the blood via another glucose transporter protein (GLUT2) on the basolateral membrane.

Role of the kidneys in acid–base balance

Cellular metabolic processes result in the production of CO_2 and, in some cases, nonvolatile acids such as lactic acid. CO_2 can be blown off in the lungs by increasing the rate of ventilation, but in order to maintain plasma pH between 7.45 and 7.35 (H^+ concentration of 35–45 nmol/L), other mechanisms must exist to rid the body of the other acids. The acid–base status of blood plasma is monitored by chemoreceptors in the carotid and aortic bodies, which sense pH and PCO_2 (they also monitor PO_2, but that is largely irrelevant in discussion of acid–base balance). When PCO_2 rises or pH falls, respiratory areas in the brain stem produce an increased drive to ventilation, helping to blow off CO_2 and bring pH and PCO_2 back within the normal range (negative feedback). However, for a number of acid–base imbalances, CO_2 cannot be formed from the nonvolatile acids, or respiratory compensation mechanisms are insufficiently powerful. In such cases, alteration of the rate of excretion of H^+ and/or HCO_3^- by the kidneys comes into play (renal compensation).

Acidification of the urine

The glomerular filtrate produced in the renal cortex is more or less iso-osmotic with the plasma and has the same pH of 7.4. Urine produced and stored in the urinary bladder has an average pH of 6.0. Under normal conditions, the kidneys excrete the same amount of H^+ ions as are produced by cellular respiration. This is normally around 70 mEq/day. If the kidneys produce 1.5 L urine/day and the urinary pH is 6.0, then only 1.5 μEq are free in the urine (since the pH scale indicates free H^+ ions). The majority of H^+ ions excreted by the body in the urine (70 mEq) are combined with anions in the form of titratable acid ($H_2PO_4^-$) and with ammonia (NH_3) as ammonium (NH_4^+) ions.

The majority of H^+ ions that are secreted into the urine enter the tubular fluid in the proximal parts of the renal tubule (up to the end of the loop of Henle). In the proximal convoluted tubule (PCT), H^+ ions are added to the tubular fluid by a luminal Na^+/H^+ exchanger and H^+-ATPases. By the time the tubular fluid has reached the end of the PCT, pH has fallen to around 6.7 (**Figure 4.6**). As the fluid flows into the descending limb of the loop of Henle, water reabsorption increases until, at the bottom of the loop, HCO_3^- concentrations are greater than at the top; this results in an increase in the luminal pH to around 7.4. As fluid flows up the ascending limb, it is acidified by the action of Na^+/H^+ exchangers on the luminal surface of the epithelial cells of the tubule, so that by the beginning of the distal convoluted tubule (DCT), the pH of the tubular fluid has fallen again to around 6.7. Additionally, there are a number of mechanisms for reabsorption of NH_4^+ ions in the ascending limb of the loop of Henle; dissociation of NH_4^+ provides H^+ ions to serve as a substrate for the Na^+/H^+ exchangers and the remaining NH_3 is secreted into the collecting duct from the interstitium.

Further acidification of the urine takes place in the distal parts of the nephron. Two types of intercalated cells of the collecting duct exist: α-intercalated cells and β-intercalated cells. The α-intercalated cells secrete H^+ and the β-intercalated cells secrete HCO_3^-. In the collecting duct, H^+ ions secreted by α-intercalated

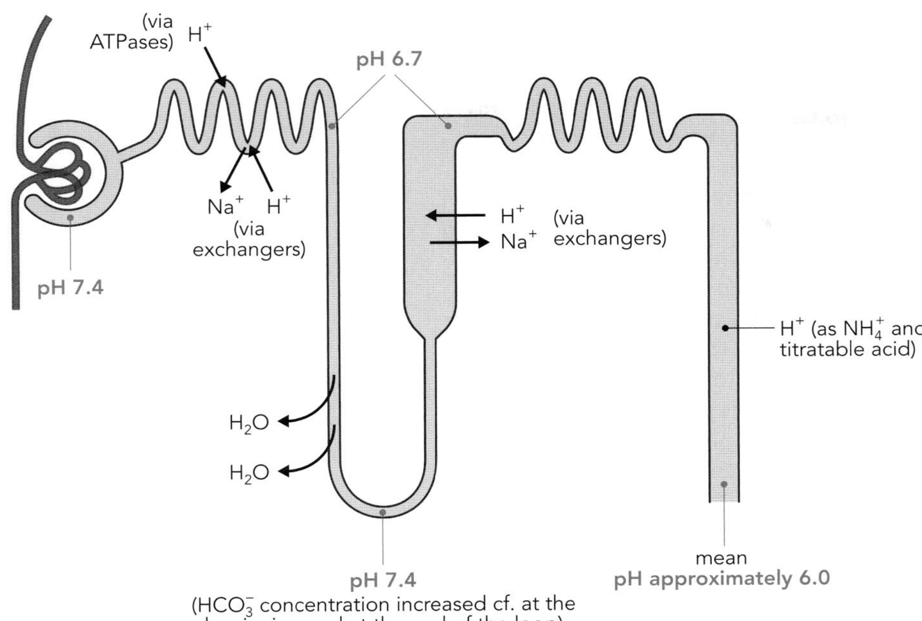

Figure 4.6 Schematic diagram of the nephron, showing sites of acidification and alkalinization of the tubular fluid.

cells can combine with the NH_3 that diffuses into the ductal fluid from the interstitium, leading to the formation of NH_4^+ in addition to titratable acid.

Modulation of HCO_3^- excretion

Virtually all of the HCO_3^- ions filtered at the glomerulus (>4000 mmol/day) must be reabsorbed by tubular epithelial cells if blood pH is to maintain much of its buffering capacity, because HCO_3^- is the major buffer in the blood. Some processes in the nephron allow filtered HCO_3^- ions to be reclaimed, but the formation of new HCO_3^- ions is a central part of the process for production of H^+ ions and acidification of the urine when acidemia (blood pH <7.35) cannot be corrected by respiratory mechanisms. The model for the reclamation of HCO_3^- ions from the tubular fluid in the proximal tubule is shown in **Figure 4.7**.

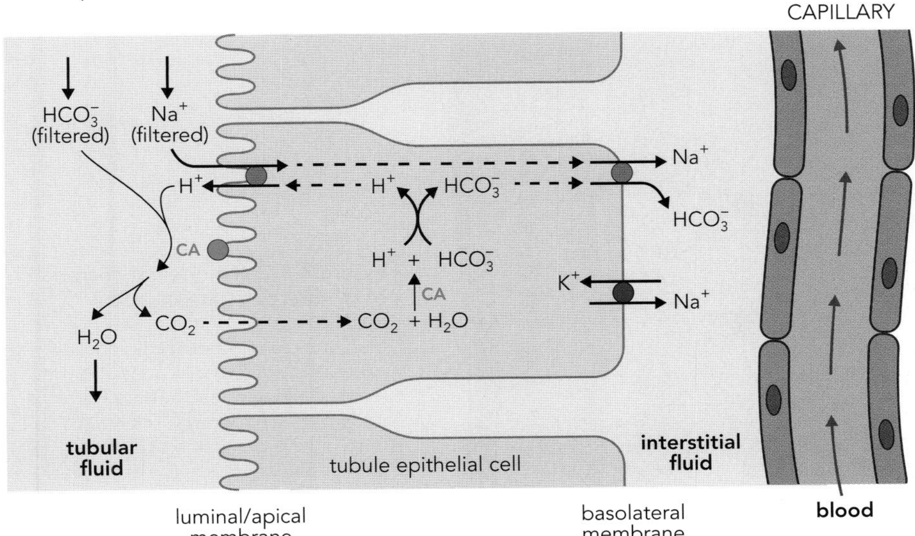

Figure 4.7 Cell model for reclamation of HCO_3^- ions in the proximal parts of the nephron. Carbonic anhydrase (CA), expressed on the apical surface of the epithelial cells, converts filtered HCO_3^- to H_2O and CO_2. The CO_2 diffuses into the cell, where CA catalyzes the formation of HCO_3^- and H^+ ions. HCO_3^- ions can then be transported into the interstitial space by the action of a Na^+/HCO_3^- co-transporter and H^+ ions can be secreted into the tubule via a Na^+/H^+ antiport. The driving force for Na^+ reabsorption is the Na^+/K^+-ATPase pumps on the basolateral surfaces of the epithelial cells.

Within the cells of the proximal tubule, the enzyme carbonic anhydrase converts water and carbon dioxide to carbonic acid (H_2CO_3). This carbonic acid rapidly dissociates to form H^+ ions and HCO_3^- ions (let us leave the HCO_3^- ions for the moment). The H^+ ions are exchanged for Na^+ ions, which have been filtered at the glomerulus, via a Na^+/H^+ exchanger on the luminal surface of the epithelial cell. The H^+ ions (now in the lumen of the tubule) can then combine with filtered HCO_3^- ions to form carbonic acid, which is broken down by carbonic anhydrase associated with the brush border of the epithelial cell to form water and carbon dioxide. The water remains in the tubular fluid, but the carbon dioxide can diffuse into the epithelial cell, where it becomes one of the substrates (along with water) for formation of carbonic acid via cellular carbonic anhydrase. The HCO_3^- ions formed from dissociation of this carbonic acid leave the cell via a Na^+/HCO_3^- co-transporter on the basolateral surface of the cell, to add a reclaimed HCO_3^- ion to the interstitial fluid. It is important to note that no new HCO_3^- ions are formed by this process in the proximal tubule; filtered HCO_3^- ions are converted to carbon dioxide, which is then used by the cell to form intracellular HCO_3^- ions that can only leave via the basolateral membrane.

Modulation of H^+ excretion

An example of the ability of renal tubule cells to secrete substances into the tubular fluid, the secretion of H^+ ions as titratable acid ($H_2PO_4^-$), is shown in **Figure 4.8**. Acid secretion is a property of many of the epithelial cells forming the walls of the nephron, but is most apparent in the proximal tubule, the ascending limb of the loop of Henle, and the collecting duct. Epithelial cells lining the tubule use a number of different methods for secreting acid into the lumen, of which Figure 4.8 shows only one.

Na_2HPO_4 filtered at the glomerulus flows along the nephron. Within the epithelial cells, the enzyme carbonic anhydrase converts water and carbon dioxide to carbonic acid (H_2CO_3). This carbonic acid rapidly dissociates to form H^+ ions and HCO_3^- ions and the H^+ ions are exchanged for one of the two Na^+ ions associated with the HPO_4^{2-} ions flowing in the tubular fluid, via a Na^+/H^+ exchanger on the luminal surface of the epithelial cell. The H^+ ion is then excreted in the urine associated with the HPO_4^{2-} ions as NaH_2PO_4 (a form of titratable acid). The HCO_3^- ions formed by the dissociation of carbonic acid in the cell leave the cell

passive transport
ATP-dependent active transport
passive diffusion

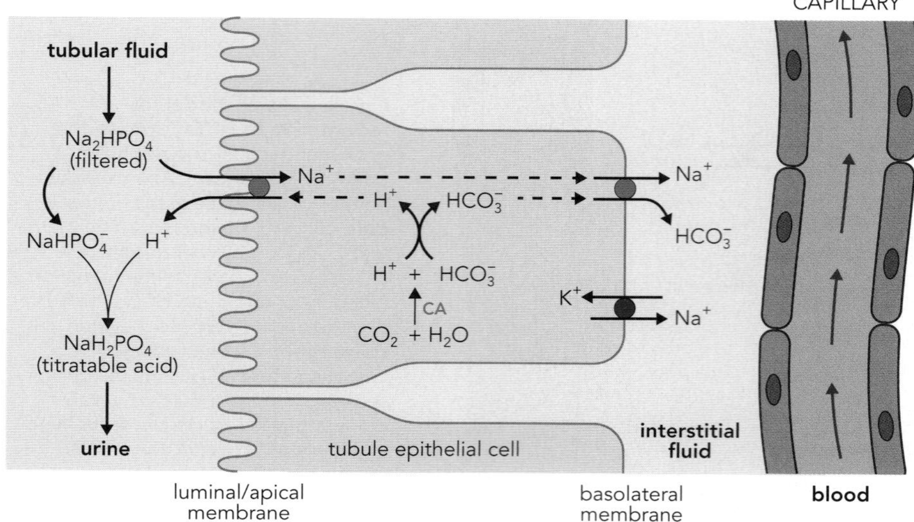

Figure 4.8 Cell model for secretion of H^+ ions as titratable acid (NaH_2PO_4). Many tubular epithelial cells are capable of secreting H^+ ions into the renal tubule and reabsorbing Na^+ ions, via Na^+/H^+ antiporters. The source of H^+ ions is the action of carbonic anhydrase (CA) inside the epithelial cell, which catalyzes the conversion of H_2O and CO_2 into H^+ and HCO_3^- ions.

on the basolateral side, via a Na^+/HCO_3^- co-transporter, to add a new HCO_3^- ion to the interstitial fluid. Thus, on average, for each H^+ ion that is secreted into the urine by this process, a new HCO_3^- ion is added to the interstitial fluid. It is worth mentioning that for the kidneys to excrete large amounts of acid, induction of ammonia production must occur, and NH_4^+ ions must be formed, because the supply of HPO_4^{2-} ions is limited.

Fluid balance

In an average 70 kg male, 60% of body weight is taken up by water, amounting to 42 kg. This body water is distributed in three major compartments: intracellular space; interstitial and lymphatic fluid; and blood plasma. Plasma water (3.5 L) makes up only around 5% of body weight; bear in mind that the fluid inside blood cells is not included in this measure. The interstitial fluid, which bathes the majority of the cells of the body, can be considered a filtered form of plasma. The electrolyte composition of the interstitial fluid normally mirrors that of the plasma as there is free exchange of dissolved electrolytes between the plasma and the interstitial compartment. Free exchange of proteins between the plasma and the interstitial fluid does not usually occur, so the interstitial fluid has a very low protein content compared to plasma. For these reasons, plasma electrolyte composition is normally considered to be a good marker of the composition of interstitial fluid and clinical analysis of blood samples relies on this fact. Plasma Na^+ ion concentration is held to be indicative of the concentration of Na^+ ions in the larger, but inaccessible, interstitial fluid compartment. This free exchange of fluid and electrolytes between the plasma and interstitial fluid is also important in the reabsorption of substances into the blood from the renal tubule.

Reabsorption of fluid in the renal tubule is obviously very important in maintaining water balance. If only 98% of filtered fluid were reabsorbed, compared with the normal 99%, this would mean that we would need to ingest 1.8 L more fluid than normal every day. The converse is also true; by drinking 0.9 L less in a day, we would only have to reabsorb 98% of the filtered fluid, rather than the normal 99%. One of the major ways in which fluid balance is maintained is by monitoring the osmolality of the plasma; this is done by specialized osmoreceptor cells in the anterior parts of the hypothalamus, which signal to other hypothalamic neurons to promote drinking and also to release or retain various hormones (particularly antidiuretic hormone [ADH]). Another method for maintaining fluid balance or fluid load is by monitoring blood pressure; activation of baroreceptors in the heart and the arteries also influences the secretion of hormones that act on the kidney. Understanding how these various hormones have their effect on plasma volume is dependent upon understanding the normal mechanisms by which the kidneys reabsorb fluid.

Role of the loop of Henle

Around 70% of the fluid and electrolytes filtered at the glomerulus is reabsorbed in the PCT. Urinary osmolality can be as high as 1200 mOsm/kg if very little fluid is available through lack of intake or excessive losses in the sweat. The kidneys are able to produce such concentrated urine because of the arrangement of the loop of Henle and vasa recta in the renal medulla and the differing permeability of various parts of the loop of Henle to water.

NaCl reabsorption in the ascending limb of the loop of Henle follows the model shown in Figure 4.4. In the ascending limb, water is unable to follow solute into the interstitial space because this part of the loop is impermeable to water. Imagine beginning from a starting point (**Figure 4.9a**) where fluid entering the loop of Henle is iso-osmotic with plasma (since water reabsorption in the PCT follows solute) and imagine fluid being added in stages to the descending limb. As NaCl is reabsorbed from the ascending limb into the interstitium (**Figure 4.9b**) and because water cannot follow the solute, fluid in the ascending limb becomes hypo-osmolar. The interstitium would then become

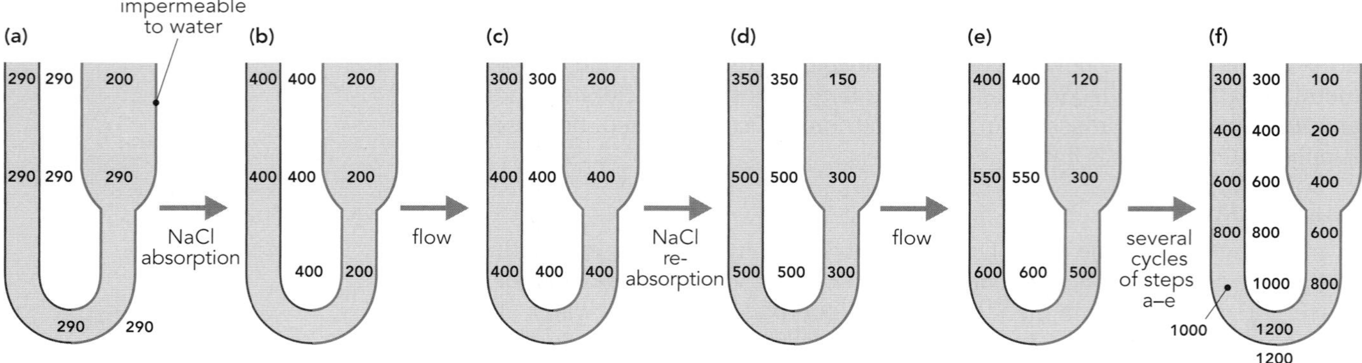

hyperosmolar and the fluid in the descending limb would then also become hyperosmolar over time as water left the descending limb to equilibrate with the interstitial fluid. On the next introduction of fluid and solute to the descending limb (**Figure 4.9c**), more solute would be pumped out of the tubule in the ascending limb, water would not follow, and the tubular fluid in the ascending limb would become more hypo-osmolar still (**Figure 4.9d**). The fluid in the descending limb and the interstitium would then become more hyperosmolar to balance the solute concentration in the interstitium.

Figure 4.9 (a)–(f) Schematic diagrams showing the method by which hyperosmolar conditions are set up in the renal medullary interstitium.
Values shown are osmolality of tubular fluid or interstitial fluid, in mOsm/kg.

If this chain of events were continued many times over, the situation could arise where interstitial fluid osmolality would become progressively greater as one moved deeper into the medulla (**Figure 4.9e and f**), although the degree of hyperosmolality of the interstitial fluid would depend on how much NaCl were being actively reabsorbed from the ascending limb.

Variations in solute reabsorption in the loop of Henle and variations in the permeability of the collecting duct epithelium to water mean that we can make hyperosmolar or hypo-osmolar urine to suit our requirements for fluid and electrolyte balance. The $Na^+/K^+/2Cl^-$ co-transporter on the luminal side of the cells in the thick portion of the ascending limb of the loop of Henle provides the major source of the solute that is pumped out into the interstitium. Hence, factors that change the expression of the genes encoding this co-transporter or the activity of the transporter (such as the hormone aldosterone) can have a major effect on this mechanism in the loop of Henle.

The purpose of setting up this osmotic gradient (from 100 to 1200 mOsm/kg) as one moves deeper into the medulla from the cortex is to make it possible to make more concentrated urine as the tubular fluid passes down the collecting duct toward the pelvis of the kidney. Movement of water from the tubular fluid in the collecting duct takes place via specific water pores formed by the protein aquaporin-2. If, due to the action of circulating ADH, the cells of the collecting duct have a large number of aquaporin-2 molecules on their cell membrane, then it is easier for water to leave the tubular fluid under the osmotic drive of the hyperosmolar medullary interstitial fluid. The converse is true if fewer aquaporin-2 molecules are present on the luminal side of these cells. Hence, by altering the number of aquaporin-2 molecules available to the cells of the collecting duct, and by actively inserting more or fewer aquaporin molecules in the cell membrane, the osmolality of the urine being produced can be altered, as a direct consequence of varying circulating ADH levels in response to plasma osmolality and/or volume.

Role of urea

Urea is the major nitrogenous compound in human urine and is formed from metabolism of amino acids and purines. The pathway for urea production is shown in **Figure 4.10** (see also Chapter 3). Uric acid can also be formed from purine metabolism, but uric acid, ammonia (as NH_3 and NH_4^+), and creatinine normally make up <20% of the nitrogen-containing compounds in the urine.

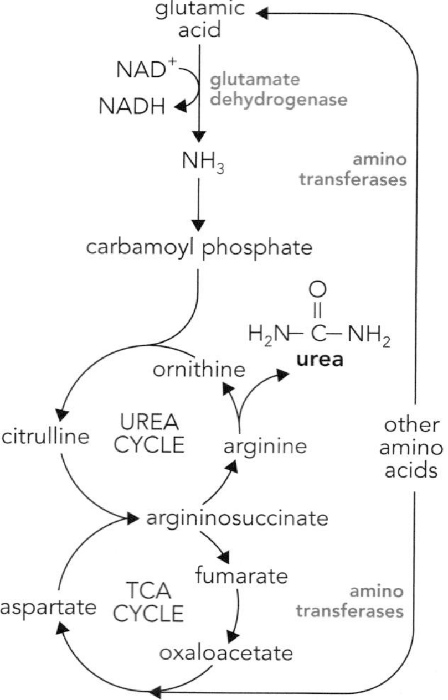

Figure 4.10 Biochemical pathways for the formation of urea.
NH_3, ammonia.

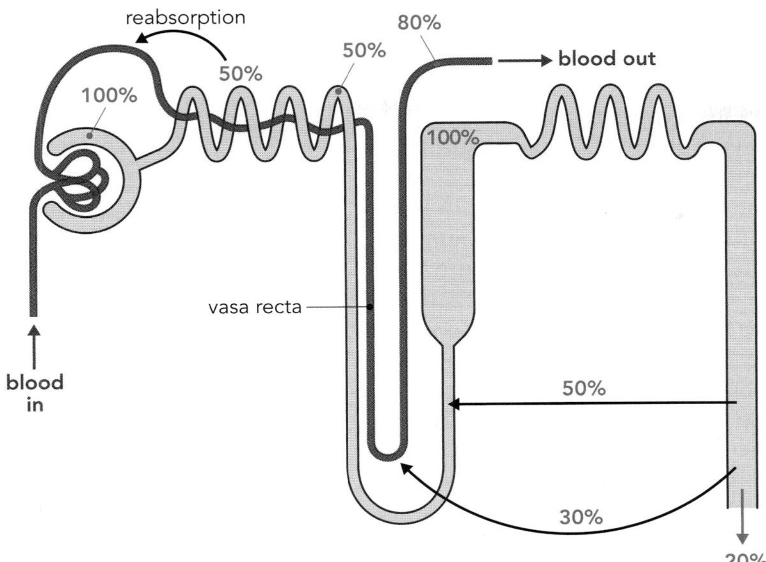

Figure 4.11 Urea recycling in the nephron.
Of the urea filtered, only around 20% finds its way into the urine. Around 50% is utilized in the osmolar gradient in the inner medulla, helping to concentrate the urine. Values presented are % of filtered load.

Significant amounts of uric acid are only found in the urine when there are problems with metabolism. Under those circumstances, uric acid crystals may form in the body and this can lead to the painful condition known as gout.

Urea is also very important in the urinary concentrating mechanisms of the kidney. This was first hinted at when it was noticed that people who had diets low in protein, leading to decreased urea production, could not produce maximally concentrated urine. Urea is recycled from the collecting duct to the loop of Henle by a mechanism demonstrated in **Figure 4.11**.

Urea is freely filtered at the glomerulus, but because the PCT is permeable to urea, by the time the filtrate has reached the top of the loop of Henle, around 50% of this urea has been reabsorbed. The cells of the loop of Henle secrete urea so that by the time the tubular fluid has reached the DCT, the amount of urea present in the tubular fluid is similar to that in the glomerular filtrate. The thick portion of the loop of Henle and the rest of the renal tubule up to and including the outer medullary parts of the collecting duct are impermeable to urea, so little urea is present in the outer medulla. However, the inner medullary parts of the collecting duct are highly permeable to urea via a urea transporter (activated by ADH), so urea leaves the tubular fluid in the inner medulla, where it contributes to the osmotic potential of the medullary interstitial fluid, playing an important role in the kidney's major urinary concentrating mechanism. Once this mechanism reaches equilibrium, the urea being recycled between the collecting duct and the loop of Henle amounts to about 50% in excess of the filtered load, and of the filtered load, 50% is reabsorbed in the PCT, 30% finds its way into the vasa recta leaving the medulla, and 20% is excreted in the urine.

Fluid reabsorption in the distal convoluted tubule

The DCT is a further site of fluid reabsorption in the nephron. In the epithelial cells of the DCT, basolateral Na^+/K^+-ATPases pump Na^+ ions out of the cell and K^+ ions into the cell. K^+ ions are able to leak back out of the basolateral side of the cell via K^+ channels, to maintain function of the Na^+/K^+-ATPase. The consequent low Na^+ concentration inside the cells of the DCT creates a concentration gradient for Na^+ ions to enter the DCT cells from the luminal fluid, a process facilitated by a Na^+/Cl^- co-transporter on the luminal membrane. Cl^- ions are able to leave the cell on the basolateral side via Cl^- channels. This DCT transport of NaCl facilitates the movement of water from the luminal fluid into the interstitial space. The permeability of the DCT to water is low and so the osmolality of the fluid in the DCT can remain low. Despite this, any factor that affects transport of NaCl in the DCT could have some limited effect on fluid reabsorption.

Hormonal influences on fluid balance

Three major hormones are involved in the regulation of fluid balance in the kidney. These are ADH (also known as vasopressin, in view of its effect of increasing the tone of blood vessels), aldosterone, and atrial natriuretic peptide (ANP). All three hormones act on various parts of the renal tubule to alter fluid absorption.

ADH, as its name suggests, promotes absorption of fluid by the kidney. ADH is a peptide hormone released from neuronal processes that extend down from the supraoptic nuclei and the paraventricular nuclei of the hypothalamus, via the infundibulum, to the posterior pituitary. Circulating levels of ADH are normally low, but are increased by raised plasma osmolality when osmoreceptor cells in the hypothalamus are activated. Above the normal plasma osmolality of around 280 mOsm/kg, there is a direct linear relationship between plasma osmolality and ADH secretion. The major effect of ADH is on the epithelial cells of the kidney collecting duct; ADH acts on these cells to increase the trafficking of aquaporin-2 to the luminal membrane and to increase the expression of the gene for aquaporin-2. Increasing the permeability of the collecting duct to water in this way facilitates the concentrating effect of the high medullary osmolality set up by the loop of Henle, resulting in increased water reabsorption and concentrated urine. Under normal circumstances, this increased water reabsorption should correct the initial increase in osmolality of the plasma that stimulated release of ADH. ADH also activates the urea transporter molecules in the cells of the inner medullary collecting duct, thereby increasing the osmotic potential in the medulla. Circulating levels of ADH are extremely low in the disease diabetes insipidus, in which either the infundibulum or the posterior pituitary have been damaged. As one might expect, this leads to production of copious amounts (5–20 L/day) of dilute urine, accompanied by excessive drinking in order to offset this level of fluid loss.

Aldosterone is the major mineralocorticoid in the body and is released from the adrenal cortex. Its secretion is controlled in part by angiotensin II, one of the hormones of the renin–angiotensin–aldosterone system (**Figure 4.12**). Renin is an enzyme released from cells in the wall of the glomerular afferent arteriole in response to two stimuli: reduced kidney perfusion pressure and decreased glomerular filtration rate (sensed as reduced delivery of Na^+ ions to the DCT). Such changes could be brought about by a reduction in blood pressure. The interaction between the glomerulus and the DCT has led to this area of the nephron

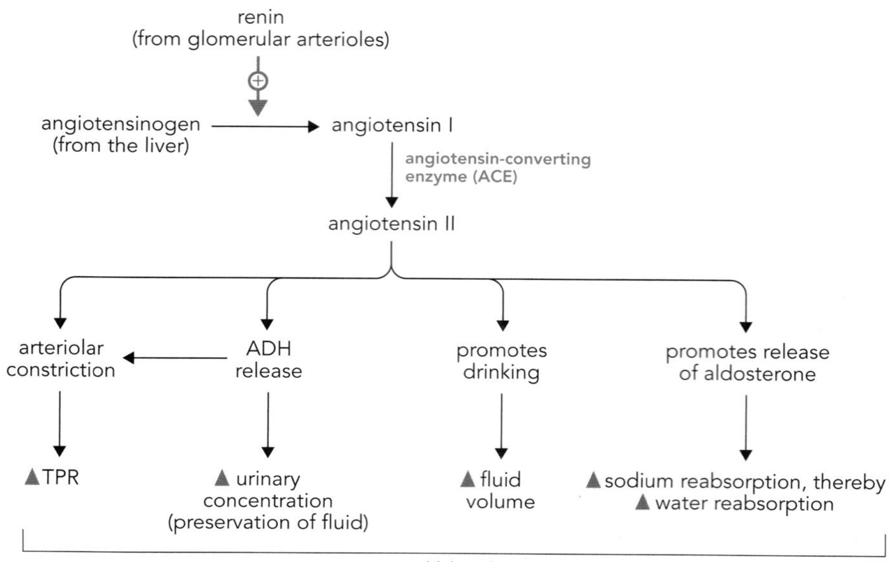

Figure 4.12 The renin–angiotensin–aldosterone system in fluid balance and long-term regulation of blood pressure.
ADH, antidiuretic hormone; TPR, total peripheral resistance.

being named the juxtaglomerular apparatus. Renin is a small peptide enzyme that acts in the bloodstream to convert angiotensinogen (produced by the liver) to angiotensin I. Angiotensin I is inactive, but can be converted to an active form, angiotensin II, by the action of angiotensin-converting enzyme (ACE). ACE is found in multiple sites throughout the body, but is present at high levels in the vascular endothelial cells of the lungs. Angiotensin II has four major effects, mediated by specific angiotensin II receptors: it constricts arterioles, promotes the release of ADH, stimulates release of aldosterone from the adrenal cortex, and promotes drinking through an action on the hypothalamus. All four of these effects are consistent with an attempt to raise blood pressure and/or blood volume. Aldosterone increases the expression of the $Na^+/K^+/2Cl^-$ co-transporter and other transporters in the ascending limb of the loop of Henle, which increases the amount of solute that is transported into the interstitium from the tubular fluid. As we have seen, this effect, coupled with an increase in ADH secretion, can result in the production of more concentrated urine and an increase in plasma volume.

Members of another hormone family, the natriuretic peptides, are released from the atria and ventricles of the heart in response to stretch. The nomenclature of these hormones can be misleading. Atrial natriuretic peptide (ANP), as suggested, is released from the cardiac atria. The ventricular equivalent was first discovered in the brain, and hence was originally called brain natriuretic peptide (BNP). This is now referred to as B-type natriuretic peptide and can be measured using commercial assays as a marker of heart failure. These hormones act on the kidney to reduce Na^+ reabsorption, hence their name. As we have seen, if more Na^+ is excreted in the urine, water follows. Thus ANP and BNP can cause both natriuresis and diuresis and reduce or limit the expansion of blood volume. The natriuretic peptides also have negative effects on renin and aldosterone secretion and relax vascular smooth muscle.

Diuretic drugs

In view of the close relationship between fluid balance, blood volume, and blood pressure, a number of different drugs that have diuretic effects are used to treat hypertension and edema. These include loop diuretics such as furosemide, thiazide diuretics such as bendroflumethiazide, and aldosterone receptor antagonists such as spironolactone.

Furosemide blocks the action of the $Na^+/K^+/2Cl^-$ co-transporters on the cells of the thick ascending limb of the loop of Henle that are so important in concentration of the urine, leading to increased fluid loss in the urine. Bendroflumethiazide blocks the Na^+/Cl^- co-transporter on the epithelial cells of the DCT and hence is a less potent diuretic than furosemide, since only a small amount of filtered fluid is reabsorbed in the DCT. One potential adverse effect of these diuretics is that they promote loss of K^+ in the urine, which can lead to hypokalemia in some cases. This effect is due to increased presentation of Na^+ to the aldosterone-sensitive site at which exchange for K^+ occurs, and also to the increased tubular flow rate causing K^+ washout.

Spironolactone reduces the renal loss of potassium by blocking the action of aldosterone on its (mineralocorticoid) receptors. Spironolactone is one of a group of drugs known as K^+-sparing diuretics and, in some circumstances, is preferred to or combined with loop diuretics like furosemide.

4.2 CLINICAL ASSESSMENT OF GLOMERULAR FUNCTION

GFR may be assessed very accurately by the use of radiolabeled tracers, which are injected into the patient and their rate of disappearance is measured in nuclear medicine departments. However, these techniques are not practical for use in the vast number of patients in whom renal function needs to be assessed and monitored. For many decades the concentration of creatinine in the plasma has been used as an index of renal function, supplemented in some cases by the measurement of creatinine clearance as a measure of GFR. Creatinine clearance requires a 24-hour urine collection to measure creatinine excretion, plus a sample for plasma creatinine taken during the same period. Creatinine is assumed to be produced at a constant rate, to be fully filtered at the glomerulus, and neither secreted nor reabsorbed in the renal tubules. These assumptions are not entirely true. Creatinine, as a product of muscle, is present in meat and so is increased in the plasma following a meal containing cooked meat. In addition, it is secreted by the renal tubules. However, traditional chemical methods for measuring serum creatinine using the alkaline picrate (Jaffe) reaction are nonspecific and so tend to overestimate its concentration. This has the effect of canceling out the overestimation of creatinine excretion. These problems, together with the recognized difficulties in obtaining genuinely complete 24-hour urine collections, have led to measurement of creatinine clearance being discouraged by clinical laboratories and expert guidelines.

Plasma creatinine alone is an insufficiently reliable index of renal function as its production is proportional to muscle mass. This means that a young bodybuilder with normal GFR may have high plasma creatinine, whilst an amputee or frail elderly woman with a low GFR may have normal plasma creatinine. This is particularly important in the elderly, where drugs that rely on renal clearance may be given in excessively high doses if the reduced GFR is not appreciated.

In order to make plasma creatinine measurement more useful, it is current practice to use it with other data to calculate an estimate of the GFR (eGFR). The most widely used formulas for estimating GFR are those of Cockcroft and Gault (which, strictly, is an estimate of the creatinine clearance, itself an indirect measure of the GFR) and from the Modification of Diet in Renal Disease (MDRD) trial.

The Cockroft and Gault formula for estimated creatinine clearance (eCCr) is as follows:

For plasma creatinine in mg/dL:

$$\mathbf{eCCr = (140 - age) \times mass\ (kg) \times 0.85\ (if\ female) / 72 \times creatinine}$$

(Equation 4.6)

For plasma creatinine in μmol/L:

$$\mathbf{eCCr = (140 - age) \times mass\ (kg) \times constant) / creatinine}$$

(Equation 4.7)

where the constant is 1.23 (males) or 1.04 (females).

The Modification of Diet in Renal Disease (MDRD) formula is as follows:

For plasma creatinine in mg/dL:

$$\mathbf{GFR\ (mL/min/1.73\ m^2) = 175 \times (creatinine)^{-1.154} \times (age)^{-0.203} \times (0.742\ if\ female) \times (1.212\ if\ African\ American)}$$

(Equation 4.8)

For plasma creatinine in μmol/L:

$$\mathbf{GFR\ (mL/min/1.73\ m^2) = 175 \times (0.0113 \times creatinine)^{-1.154} \times (age)^{-0.203} \times (0.742\ if\ female) \times (1.212\ if\ African\ American)}$$

(Equation 4.9)

The MDRD formula has the advantage of only requiring knowledge of age, sex, and racial origin in addition to the plasma creatinine. The Cockroft and Gault formula also requires weight, which is not information routinely provided to clinical laboratories. Both formulas can only be used in the steady state, so are not applicable in acute renal failure, nor are they validated in children or pregnant women. In addition, creatinine-based estimates are derived for the average muscle mass of the reference population and so are not valid if an individual has a greater or smaller amount of muscle than the average person of the same age. The examples of the bodybuilder or amputee, mentioned previously, would still have their eGFR underestimated and overestimated, respectively. However, the average elderly patient would have their eGFR correctly assigned. There is also the underappreciated effect of the secular trend for populations to become increasingly obese, resulting in a greater proportion of body mass being due to adipose tissue and a correspondingly smaller contribution from muscle. This will, over time, invalidate the assumptions about average population muscle mass, which may have been derived from studies in previous decades. The ingestion of cooked meat prior to measurement of plasma creatinine will also continue to cause underestimation of eGFR and so ought to be avoided. For all its drawbacks, creatinine remains the most widely used index of renal function, mainly because it is historically so embedded in clinical practice and also because it is so inexpensive to measure. In recent years, the nonspecific nature of the Jaffe reaction used to measure creatinine has led to the recommendation that creatinine be measured instead by enzymatic methods as a matter of routine. Such methods are far more specific and avoid overestimation in the presence of interfering substances such as glucose, ketones, and bilirubin. They are relatively more expensive, however, which has prevented universal uptake. The improvement in creatinine methodology does not, however, remove its inherent limitation for assessing glomerular filtration rate; that is, its dependence on muscle mass.

Analytical practice point 4.1

Use of estimated GFR improves detection and monitoring of kidney disease compared to plasma creatinine, but is based on population averages and does not account for extremes of muscle mass.

Analytical practice point 4.2

The alkaline picrate (Jaffe) method is prone to interferences and tends to overestimate creatinine. Enzymatic methods avoid many of these issues but are more expensive.

Creatinine measurement by enzymatic assays may utilize either creatininase (creatinine aminohydrolase) or creatinine deaminase. The measurable products of these assays are, respectively, creatine and ammonia. An example of the enzymes that may be used in series to measure creatinine, using creatininase initially, is shown below. The hydrogen peroxide (H_2O_2) formed can be measured with a final reaction that forms a colored product. An alternative approach is to use creatine kinase as the second enzyme to produce phosphocreatine, a reaction which can be monitored by the change in NADH.

$$\textbf{creatinine} + \textbf{H}_2\textbf{O} \xrightarrow{\text{[creatininase]}} \textbf{creatine}$$

$$\textbf{creatine} + \textbf{H}_2\textbf{O} \xrightarrow{\text{[creatinase]}} \textbf{sarcosine} + \textbf{urea}$$

$$\textbf{sarcosine} + \textbf{O}_2 + \textbf{H}_2\textbf{O} \xrightarrow{\text{[sarcosine oxidase]}} \textbf{glycine} + \textbf{formaldehyde} + \textbf{H}_2\textbf{O}_2$$

An alternative to using creatinine-based GFR estimates is the use of serum cystatin C measurement. This is a low-molecular-weight protein (mass approximately 13 kDa) that is freely filtered by the kidney; plasma concentration rises as GFR decreases, as with creatinine. However, cystatin C production is much less affected by age, gender, race, and muscle mass than creatinine, making it theoretically a more useful marker of renal function. Its major drawback is cost, since immunoassays required to measure cystatin C are significantly more expensive than even enzymatic creatinine assays (which are themselves several times the cost of Jaffe reagents). There are also clinical situations where cystatin C synthesis is altered, including pregnancy, malignancy, thyroid disease, and human immunodeficiency virus (HIV) infection. Whilst it is unlikely to completely replace creatinine measurement in the short term, it is a useful alternative in specific patients and its use will undoubtedly grow.

4.3 PROTEINURIA

Normal urine, as an ultrafiltrate of plasma, is essentially protein free—certainly when assessed by insensitive methods such as urine dipsticks. The presence of dipstick-detectable protein is an important sign of urinary tract disease anywhere from the kidneys to the urethra. Quantitation of urine protein excretion and identification of the type of protein present are useful for diagnosis and monitoring of several diseases, which include not only primary renal disorders but also diseases such as diabetes mellitus and multiple myeloma.

Quantitation of proteinuria has traditionally required a 24-hour urine collection. However, as with creatinine clearance, these samples are often poorly collected by patients and provide logistical problems for the laboratory, the samples being bulky and difficult to transport and handle. In recent years there has been a move toward using random urine samples and expressing the protein as a ratio to urine creatinine concentration. There is generally a good correlation between protein–creatinine ratios (PCR) and 24-hour urine protein. The albumin–creatinine ratio (ACR) is advocated by some authorities as a more precise and specific measurement, although it is considerably more expensive since the measurement of low levels of albumin requires immunoassay. Grading of proteinuria using dipsticks, 24-hour urine, and spot urine indices is summarized in **Table 4.3**.

Analytical practice point 4.3

Protein–creatinine or albumin–creatinine ratio has superseded 24-hour urine protein estimation in most situations.

Damaged or injured glomerular capillaries are excessively permeable to proteins because of increased pore size and/or reduction of the negative charge on the basement membrane. This latter cause is particularly important in so-called minimal change disease, where there is no visible injury on microscopy. Proteins are lost into the urine when the amount in the glomerular filtrate exceeds the reabsorptive capacity of the renal tubules. When glomerular injury is relatively minor, as in minimal change disease, loss into the urine is largely confined to proteins of low molecular weight, such as albumin and transferrin (selective proteinuria). With more severe glomerular injury, larger proteins such as immunoglobulins (for example, IgG) also appear in the urine (nonselective proteinuria).

Hypoalbuminemia occurs when the rate of albumin synthesis in the liver, although greater than normal, is insufficient to replace albumin lost in the urine plus that lost through normal protein turnover and thus is insufficient to maintain a plasma albumin concentration within the normal range. Because the filtered load falls in parallel with plasma albumin concentration, a new equilibrium is established at a lower plasma concentration.

Hypoalbuminemia reduces the plasma oncotic pressure, thus allowing the accumulation of fluid and electrolytes, especially sodium and chloride, in the interstitial compartment; that is, there is development of edema, most noticeably at the ankles and sacrum where the effect of gravity is greatest, but also around the eyes and fingers.

In addition, in many patients a fall in effective plasma volume stimulates the release of renin, causing secondary hyperaldosteronism, which is another

TABLE 4.3 Gradation of urine dipstick for proteinuria

Designation	Approximate amount	
	Concentration	Daily excretion
Trace	5–20 mg/dL	
1+	30 mg/dL	Less than 0.5 g/day
2+	100 mg/dL	0.5–1 g/day
3+	300 mg/dL	1–2 g/day
4+	More than 300 mg/dL	More than 2 g/day

mechanism causing the retention of water, sodium, and chloride. The pathogenesis of edema is summarized in **Figure 4.13**.

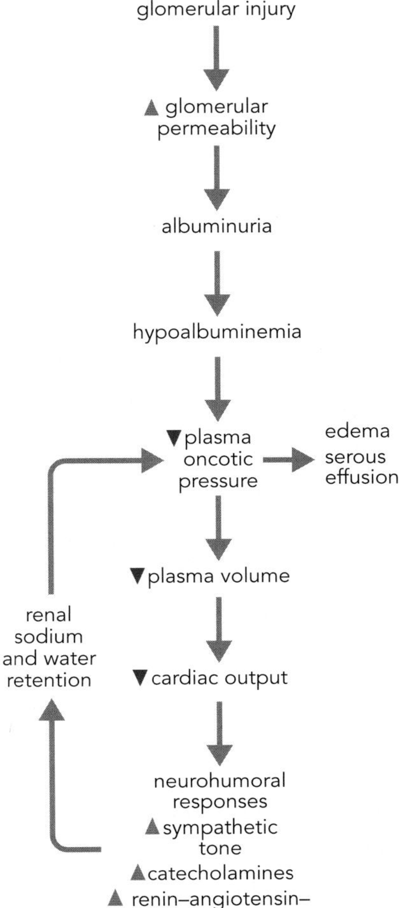

Figure 4.13 Pathogenesis of edema.
Increased permeability of the glomerulus results in plasma proteins like albumin entering the urine, which in turn reduces the oncotic pressure of the plasma. This allows more fluid than normal to enter body tissues (most often in the legs). In serious cases, the resultant hypovolemia triggers a range of neurohumoral responses that compound this problem.

Nephrotic syndrome

The triad of albuminuria, hypoalbuminemia, and edema is called the nephrotic syndrome. Renal excretory function (as measured by GFR) is initially intact but in some patients glomerular lesions progress and chronic renal failure ensues.

The degree of proteinuria required to diagnose nephrotic syndrome has historically been 3.5 g or more per 24 hours. However, 24-hour collections are avoided for practical reasons, where possible, and random urine PCR or ACR may be used instead. A PCR of >300 mg/mmol (3 mg/mg) or ACR of >250 mg/mmol (2.5 mg/mg) is consistent with nephrotic-level proteinuria.

Nephrotic syndrome is caused by glomerular disease, with different patterns of injury (glomerulonephritis) being identifiable on histological examination of a renal biopsy. The major patterns are described as minimal change, focal segmental, membranous, mesangial proliferative, and rapidly progressive. There may be an identifiable underlying cause for the histological appearance; otherwise the diagnosis is primary glomerulonephritis. For example, minimal change disease may be triggered by drugs or allergic reactions, focal segmental by HIV, and membranous by infections, diabetes mellitus, or systemic lupus erythematosis.

Other laboratory abnormalities that may be found in cases of nephrotic syndrome include hypercholesterolemia due to increased hepatic synthesis of apolipoprotein B, and reduced antithrombin III due to urinary loss, which predisposes to thrombosis.

Microalbuminuria

Some diseases, in particular diabetes and hypertension, may cause more subtle but progressive glomerular damage. Albumin leaks into the urine but at low concentrations and its detection requires more sensitive immunoassay techniques. This degree of albuminuria is termed microalbuminuria, a term which was originally defined as an amount of albumin that is greater than normal but below the detection limit of conventional dipsticks. The name is misleading as it does not refer to a small form of albumin but to a low-yet-abnormal concentration of normal albumin molecules. Microalbuminuria is discussed in greater detail in relation to diabetes in Chapter 2.

Another mechanism by which urine protein is increased is when small proteins, which can pass through the glomerular filter, are produced in excess amounts. The classical example of this is multiple myeloma, where a clonal proliferation of plasma cells produces a large amount of immunoglobulin light chains. This is termed Bence Jones protein and is discussed further in Chapter 19.

4.4 RENAL FAILURE

Kidney failure can be divided into acute and chronic, depending on the rapidity of onset: acute failure occurs over the course of hours to days and chronic over months or years. It is a major burden on health services in the developed world, being a relatively common condition and becoming more common with the increase in diabetes, which is the leading cause of chronic kidney disease. At the end stage of renal failure, the kidneys are unable to carry out their excretory function, to the point where death will occur unless this function is replaced either by dialysis or transplantation. Renal dialysis services are expensive and have a finite capacity. Renal transplantation depends on an adequate supply of donated organs and requires lifelong immunosuppressive drugs.

Acute kidney injury can often recover fully, although temporary support with dialysis may be required in the short term to correct potentially fatal metabolic derangement, particularly hyperkalemia. Chronic kidney failure (more recently termed chronic kidney disease, CKD) occurs progressively over years and may never require renal replacement if the trajectory is slow and the patient is elderly. The eGFR may be followed over time to predict when dialysis or transplantation is likely to be considered. Complications of CKD include anemia, bone disease, and increased risk of cardiovascular disease, particularly heart attack. In the last decade there has been increased awareness of chronic kidney disease in the population by the reporting of eGFR, as well as plasma creatinine, by clinical laboratories. This has allowed classification of CKD into categories, which enables treatment priorities to be set for each stage. CKD classification is given in **Table 4.4**.

TABLE 4.4 Stages of chronic kidney disease (CKD)

Stage	Estimated GFR	Description
1	>90 mL/min/1.73 m^2 with other evidence of kidney damage	Kidney damage with normal or high eGFR
2	60–89 mL/min/1.73 m^2 with other evidence of kidney damage	Mild CKD
3a 3b	45–59 mL/min/1.73 m^2 30–44 mL/min/1.73 m^2	Moderate CKD
4	15–29 mL/min/1.73 m^2	Severe CKD
5	<15 mL/min/1.73 m^2	Established renal failure

The addition of the suffix p at any stage indicates proteinuria.

The biochemical effects of kidney failure are summarized in **Table 4.5**. Although effects are similar in acute and chronic disease, they have different patterns, with raised potassium and acidosis being more important in acute renal failure, and disordered calcium and phosphate metabolism being more prominent in chronic failure.

TABLE 4.5 Biochemical changes in renal failure

Change in serum concentration	Major mechanism	Acute	Chronic
Sodium decrease	Extracellular fluid water retention	++	+
Potassium increase	Reduced clearance	+++	+
Urea and creatinine increase	Reduced clearance	++	+++
pH decrease	Reduced clearance of fixed acids	+++	+
Calcium decrease	Decreased activation of vitamin D	+	+++
Phosphate increase	Reduced clearance	+	+++
PTH increase	Stimulated by low calcium	+	+++
Magnesium increase	Reduced clearance	+	++

Relative severity: + mild, ++ moderate, +++ severe.

Acute renal failure (acute kidney injury)

Rapid loss of renal function over a few days carries a high morbidity and mortality. It may occur in many settings and is often seen in hospitalized patients as a complication of illness, dehydration, and medications. The importance of recognizing this has led to worldwide initiatives in the definition and diagnosis of acute renal failure. The preferred term is now acute kidney injury (AKI) and is based on arbitrary increases in serum creatinine within 7 days. Definitions are being revised by expert groups as more evidence is gathered over time regarding the prognostic values of different scores. Kidney Disease: Improving

TABLE 4.6 Kidney Disease: Improving Global Outcomes (KDIGO) staging of acute kidney injury		
AKI stage	**Serum creatinine**	**Urine output**
1	1.5–2 times baseline OR ≥27 μmol/L (0.3 mg/dL)	<0.5 mL/kg/h for 6–12 h
2	2–3 times baseline	<0.5 mL/kg/h for ≥12 h
3	3 or more times baseline OR ≥354 μmol/L (4.0 mg/dL) OR starting renal replacement therapy	<0.3 mL/kg/h for ≥24 h OR anuria for ≥12 h

Global Outcomes (KDIGO) has defined AKI as an increase in serum creatinine of 26.5 μmol/L (0.3 mg/dL) within 48 hours, an increase of 1.5 times baseline in 7 days, or a decrease in urine output to <0.5 mL/kg/h for 6 hours. From these definitions, laboratory alerts of increasing severity have been developed corresponding to AKI 1 (creatinine increase of 1.5 to 2 times baseline), AKI 2 (increase of 2 to 3 times), and AKI 3 (over 3 times baseline). The full definitions are shown in **Table 4.6**.

The underlying causes of AKI are traditionally categorized as pre-renal, intrinsic renal, and post-renal in origin (**Table 4.7**). The term pre-renal refers to the situation where poor renal blood flow, due to decreased circulating vascular volume (for example, caused by a large hemorrhage or loss of fluid from the gut), results in reduced GFR. This situation is potentially treatable by volume expansion (for example, a blood transfusion or saline infusion). Biochemical pointers to pre-renal failure are low urinary sodium (less than 10 mmol/L) and high urine osmolality (>750 mOsm/kg). These are, respectively, markers of activation of the renin–angiotensin–aldosterone axis and ADH, as well as the kidneys' ability to respond to these hormonal stimuli. In the intrinsic renal form of AKI, some of the causes of which are listed in **Table 4.8**, there is either prolonged

Clinical practice point 4.1

Acute renal failure is now called acute kidney injury (AKI) and is staged 1 to 3 by the degree of increase in serum creatinine.

TABLE 4.7 Classification of acute kidney injury	
Classification	**Cause**
Pre-renal	Decreased perfusion
Intrinsic renal	Primary renal disease or systemic disease
Post-renal	Obstruction of urinary tract

TABLE 4.8 Causes of intrinsic acute kidney injury	
Causes	**Examples**
Acute tubular necrosis	Prolonged poor perfusion (pre-renal failure)
Disorders affecting glomeruli and small vessels	Glomerulonephritis
	Connective tissue diseases (for example, systemic lupus erythematosis)
	Accelerated hypertension
Nephrotoxins	Hypercalcemia
	Liver failure
	Drugs
Tubular obstruction	Urate nephropathy
	Myeloma (light chains)
	Calcium oxalate from ethanediol poisoning
	Hemoglobinuria
	Myoglobinuria

hypoperfusion, causing tubular necrosis, or direct injury to the kidney and there is no response to the hormonal stimuli. Hence the urine has high sodium (>20 mmol/L) and the osmolality is similar to plasma (around 300 mOsm/kg). Post-renal AKI refers to obstruction of the urinary tract from the ureter downward and is due to the back-pressure causing loss of tubular function.

In all types of AKI the creatinine, by definition, rises rapidly. However, this is only a marker of decreasing GFR and is not in itself toxic. The major hazard lies in the rapid rise in plasma potassium, which affects the membrane potential of muscle cells. This includes skeletal, smooth, and, most importantly, cardiac muscle. Cardiac arrhythmias and abnormalities of the electrocardiogram (ECG) occur and eventually lead to cardiac arrest (see Chapter 6). Metabolic acidosis, due to inability to clear the normal products of metabolism, is another serious biochemical complication. This is also caused by reduced urinary phosphate (due to low clearance) and ammonia (from tubular synthesis), which are the major hydrogen-ion acceptors at the distal tubule. Hyperkalemia and acidosis reinforce each other due to the competition between H^+ and K^+ for Na^+ at the sodium/potassium pump. Severe AKI is a medical emergency requiring immediate treatment to control hyperkalemia and institute renal replacement in the

Clinical practice point 4.2

Rapidly rising plasma potassium is the most serious consequence of AKI as it may cause cardiac arrest.

CASE 4.1

A 75-year-old man is on a surgical ward following abdominal surgery. He is having twice-weekly blood tests to monitor his kidney function. On his last test 3 days ago his creatinine was 75 μmol/L (0.85 mg/dL). It had been similar prior to his surgery 2 weeks ago.

	SI units	Reference range	Conventional units	Reference range
Plasma				
Sodium	137 mmol/L	136–142	137 mEq/L	136–142
Potassium	7.2 mmol/L	3.5–5.0	7.2 mEq/L	3.5–5.0
Bicarbonate	13 mmol/L	21–28	13 mEq/L	21–28
Urea (blood urea nitrogen, BUN)	25 mmol/L	2.9–8.2	70 mg/dL	8–23
Creatinine	235 μmol/L	53–106	3.4 mg/dL	0.6–1.2

- **What does his current creatinine indicate?**
- **What is the most life-threatening abnormality?**
- **Why is the estimated GFR not stated?**

The rise in his creatinine is more than three times the baseline and therefore indicates acute kidney injury stage 3 (AKI 3). The cause may be pre-renal (due to hypovolemia from extracellular fluid loss with inadequate replacement), intrinsic renal (for example, a reaction to nephrotoxic drugs), or post-renal (for example, an obstructed urethra causing urinary retention). The most serious abnormality is hyperkalemia. Rapidly rising serum potassium, caused by reduced renal clearance in this case, affects cardiac muscle function and may cause potentially fatal arrhythmias and cardiac arrest. This must be treated as a medical emergency. The other biochemical abnormalities shown are low bicarbonate, indicating metabolic acidosis, and raised urea. The estimated glomerular filtration rate is not reported here as it is invalid in rapidly changing renal function and should only be used for assessing chronic kidney disease.

form of dialysis. However, lesser degrees of AKI may be managed by careful fluid and electrolyte balance, avoidance of nephrotoxic drugs, and, in some cases, targeting the underlying cause, for example with immunosuppressive therapy or relief of an obstructed urinary tract.

The prognosis for survivors of AKI is extremely variable. In many cases there may be full recovery of renal function, since the renal tubules have the ability to regenerate. However, in other cases there may be irreversible damage resulting in chronic kidney disease, which may be progressive and ultimately require renal replacement in the form of dialysis or transplantation.

Chronic renal failure (chronic kidney disease)

Chronic renal failure refers to slowly decreasing kidney function over months or years and is a complication of underlying disease: 75% of cases are due to diabetes mellitus, hypertension, or glomerulonephritis. Like AKI, chronic renal failure (or CKD) has in recent years undergone redefinition based on specific levels of serum creatinine. Thus, CKD may be classified in order of increasing severity as stages 1 to 5 (see Tables 4.4 and 4.5).

Clinical practice point 4.3

Chronic kidney disease (CKD) is staged 1 to 5, depending on the estimated GFR, which is calculated from serum creatinine and demographic factors (for example, age and sex).

The major causes of CKD are:

- Glomerulonephritis
- Interstitial disease
- Vascular causes, especially hypertension and/or diabetes mellitus
- Cystic disease
- Renal obstruction
- Drugs

The underlying disease results in the destruction (usually progressive) of nephrons with reduction of GFR. The surviving nephrons operate at maximal capacity and lose their functional flexibility. This increases the risk of acute (but potentially reversible) worsening of renal function in situations such as hypovolemia and hypercatabolism. In other words, patients are more vulnerable to AKI with intercurrent illness or certain drugs. As renal failure progresses, the composition of urine increasingly approaches that of glomerular filtrate; that is, there is less modification that the nephron can make in order to match changing demands.

In CKD, there is not usually a rise in serum potassium until the very late stages. However, other metabolic effects occur, particularly the loss of the abilities to produce the hormone erythropoietin (EPO), to activate vitamin D, and to clear phosphate. The first of these effects causes anemia due to suppression of bone marrow. It is characteristically normochromic, normocytic anemia and does not respond to hematinic replacement (iron, folic acid, and vitamin B_{12}). It is treated with EPO replacement.

The reduced activation of vitamin D follows from the lack of the enzyme 1α-hydroxylase, which converts 25-hydroxy vitamin D to 1,25-dihydroxy vitamin D, the main active form (see Chapter 9). As a result, there is a reduced absorption of dietary calcium from the gut and a fall in serum calcium. This is exacerbated by an increase in serum phosphate, due to its reduced renal clearance. The fall in serum calcium stimulates the secretion of PTH, which may reach extremely high levels. This secondary hyperparathyroidism stimulates osteoclastic resorption of bone and is a major contributor to renal bone disease (osteodystrophy), which is a common problem in stage 5 (and to a lesser extent stage 4) CKD. The consequence is an increased rate of fractures. Renal osteodystrophy may be difficult to treat and nephrologists generally aim to keep PTH levels below four times the upper reference limit by using vitamin D analogs that do not require 1α hydroxylation. Lowering PTH excessively is associated with another form of renal osteodystrophy, namely adynamic bone disease.

In the latter stages of CKD, the increased serum phosphate tends to cause deposition of calcium and phosphate in soft tissue. This includes the walls of arteries and contributes to the increased risk of heart disease, which is the major cause of death in end-stage renal failure. The conventional cardiovascular risk factors do not appear to account for this increased risk and the situation has been referred to as reverse epidemiology, since factors such as obesity appear to be relatively protective. In contrast, the management of CKD stages 3 and 4 includes the strict control of blood pressure and serum lipids to reduce the risk of atherosclerotic vascular disease.

CASE 4.2

A 68-year-old woman has chronic kidney disease due to diabetes and hypertension, and has been on hemodialysis three times a week for two years.

	SI units	Reference range	Conventional units	Reference range
Plasma				
Sodium	136 mmol/L	136–142	136 mEq/L	136–142
Potassium	5.3 mmol/L	3.5–5.0	5.3 mEq/L	3.5–5.0
Bicarbonate	20 mmol/L	21–28	20 mEq/L	21–28
Urea (BUN)	15.0 mmol/L	2.9–8.2	42 mg/dL	8–23
Creatinine	612 μmol/L	53–106	6.9 mg/dL	0.6–1.2
Estimated GFR (eGFR)	6 mL/min/1.73 m^2	>90	6 mL/min/1.73 m^2	>90
Adjusted or corrected calcium	2.15 mmol/L	2.20–2.60	8.6 mg/dL	8.8–10.4
Phosphate	1.85 mmol/L	0.74–1.52	5.73 mg/dL	2.3–4.7
Alkaline phosphatase	356 U/L	30–120	356 U/L	30–120
PTH	51 pmol/L	1.1–6.8	486 pg/mL	10–65

- What stage of chronic kidney disease (CKD) is this?
- Why is the potassium not raised to the same extent as in Case 4.1?
- Why is the corrected calcium low and the PTH raised?

This is CKD stage 5 as indicated by eGFR <15 mL/min/1.73 m^2 and renal replacement by dialysis. Unlike in AKI, in CKD the kidney is able to adapt over a period of time to increase average potassium excretion per nephron and maintain serum potassium in the reference range, until eGFR is extremely low. However, in CKD 5 potassium is liable to increase between dialysis sessions and potassium-rich foods should be avoided. Care must also be taken with drugs that are likely to increase potassium, both in CKD 5 and in less severe CKD. Disturbance of calcium metabolism occurs in the latter stages of CKD (stages 4 and 5) due to impaired ability to activate vitamin D from the 25-hydroxy form to the 1,25-dihydroxy form using the renal enzyme 1α-hydroxylase. PTH rises in response to low serum calcium and increases calcium and phosphate liberation from bone by osteoclasts. Serum phosphate increases in part due to this but mainly due to impaired renal clearance. Calcium phosphate precipitation in soft tissues may lower serum calcium further. In the long term, renal osteodystrophy increases the risk of fractures. Prevention and management include the use of calcium, vitamin D analogs, and phosphate binders.

4.5 DISORDERS OF TUBULAR FUNCTION

Kidney diseases usually affect both glomerular and tubular function. However, some disorders predominantly affect tubular function, whilst overall renal function (as measured by plasma creatinine and eGFR) is relatively spared. Some specific disorders that affect the renal tubules include:

- Renal glycosuria
- Nephrogenic diabetes insipidus
- Renal tubular acidosis
- Hypophosphatemic rickets
- Aminoacidurias including cystinuria and Hartnup disease

Some of these are single-gene defects involving individual transport proteins, but more commonly they are secondary to other diseases. These are many and diverse and include immunological, infective, vascular, and metabolic disorders. Lesions affecting the proximal and distal tubules and the loop of Henle result in different patterns of abnormalities. Proximal tubular injury causes inability to reabsorb small molecules and ions that pass freely through the glomerulus, such as bicarbonate, amino acids, glucose, phosphate, and urate. All of these are present in large amounts in the urine. Generalized proximal tubular leakage is known as the Fanconi syndrome. Injury to the loop of Henle or collecting duct, as may occur with predominantly renal medullary disease, impairs the ability to concentrate urine and thereby conserve water. Distal tubular lesions impair secretion of hydrogen ions and the ability to exchange sodium for potassium under the control of aldosterone.

Renal glycosuria

Renal glycosuria due to impaired tubular reabsorption in the presence of normal plasma glucose may occur as an isolated defect or in association with multiple defects of tubular function, including impaired absorption of amino acids, phosphate, and urate in the Fanconi syndrome. When the defect is isolated, renal glycosuria occurs because individuals have a proportion of tubules with lower-than-usual ability to reabsorb glucose. **Figure 4.14** shows the relationship between glucose filtration (measured in mg/min or mmol/min) and plasma glucose. The dotted line in the figure indicates that the filtered load at the glomerulus is proportional to the plasma glucose. However, as mentioned above, a threshold is reached (T_m) at which all the glucose transporters become saturated, and above this plasma glucose level glucose begins to be excreted in the

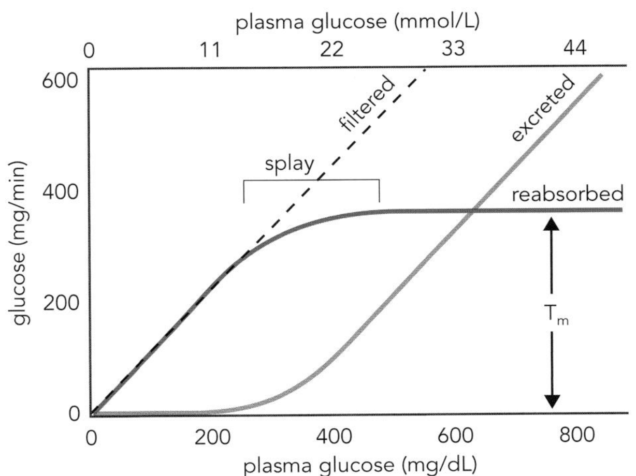

Figure 4.14 Renal handling of glucose.
The filtered load of glucose increases with plasma glucose, as shown by the dotted line. Glucose transport in the proximal tubule also increases, as shown by the first part of the blue line, and no glucose appears in the urine. However, when the glucose transporters become saturated (shown as T_m), any further increase in the plasma glucose results in glycosuria. The orange line shows that glucose appears in the urine at increasing concentration above the renal threshold (here set at 200 mg/dL [11.1 mmol/L]). The slope of the line is the same as the glomerular filtration rate (GFR) and the threshold can therefore also be expressed as T_m/GFR. The splay portion of the line is due to differences in the glucose transport capacity of individual nephrons.

urine, as shown by the orange line in Figure 4.14. Due to individual differences in the reabsorptive capacity of different nephrons, the transition from full reabsorption to some glucose being excreted is seen as splay rather than an abrupt change.

The slope of the orange line beyond the splay portion is given by the excreted glucose divided by the plasma glucose (P). The excreted glucose is calculated from the urine volume (V) in a measured time multiplied by the urine glucose concentration (U). Hence the slope equals UV/P, which we have seen is the same as the GFR. The x-axis intercept of the orange line—that is, the threshold concentration at which glucose first appears in the urine—can therefore be expressed as T_m/GFR. Interpolation of the part of the orange line with the slope UV/P back to the y-axis gives the y-intercept as $-T_m$.

Two types of renal glycosuria due to low renal threshold may occur: either T_m is smaller or the slope of the orange line is steeper (the interpolated y-axis intercept remains the same, as it is $-T_m$). Low T_m results from defects in SGLT2 and is clinically benign, despite severe glycosuria. A steeper gradient causes less severe glycosuria and may be seen with SGLT1 defects. Because this transporter is shared with the small intestine, the main consequence is sugar malabsorption. Increased GFR will also increase the slope and explains the frequent finding of glycosuria in pregnancy. The expansion of extracellular volume increases renal blood flow and GFR, causing the renal threshold for glucose to decrease.

Analytical practice point 4.4

Glycosuria occurs if the renal threshold is exceeded. This may be due to defects in tubular mechanisms responsible for glucose reabsorption or an increase in GFR.

Renal glycosuria is diagnosed by default when hyperglycemia has been eliminated as the cause of glycosuria (see Chapter 2). No further investigation is necessary in older children or adults but in young children tests should be performed for multiple tubular defects, including chromatography for unselected aminoaciduria and demonstration of hypophosphatemia with reduced renal threshold for phosphate (T_mP/GFR; see below).

In many cases it may not be possible to identify an underlying cause for Fanconi syndrome; it is then termed idiopathic. However, it may be secondary to another metabolic disorder, including cystinosis, galactosemia, hereditary fructose intolerance, or Wilson's disease.

Nephrogenic diabetes insipidus

Damage to the loop of Henle compromises urinary concentrating ability and hence can cause nephrogenic diabetes insipidus. This presents as polyuria with very dilute urine, which in turn causes polydipsia. If insufficient fluids are consumed to compensate for renal water losses, then hypernatremic dehydration develops. Diagnosis of diabetes insipidus requires simultaneous measurement of the osmolalities of plasma and urine. Further details are given in Chapter 5.

Renal tubular acidosis

Injury to the distal tubule causes inability to excrete acid and generate bicarbonate, leading to systemic acidosis, termed renal tubular acidosis. There may also be impaired ability to respond to aldosterone, causing hyperkalemia and urinary salt wasting. Investigation of renal tubular acidosis, which is described in detail in Chapter 7, requires measurement of urine acidification during acidosis, either spontaneous or provoked by ingestion of, for example, ammonium chloride.

Hypophosphatemic rickets

Renal tubular handling of phosphate is important, as excessive urinary loss of phosphate causes hypophosphatemia and bone disease. Again, this is most commonly part of the Fanconi syndrome, although specific phosphate loss may occur in rare conditions. A phosphaturic hormone, known as phosphatonin, had been postulated for many years, and was finally identified in 2000. It is a protein from the fibroblast growth factor (FGF) family known as FGF23. It is secreted by osteocytes in response to 1,25-dihydroxy vitamin D and acts to decrease phosphate transport in the proximal tubule. Thus, more phosphate is excreted and serum phosphate falls. Excess FGF23 may occur in both congenital and acquired disorders. The condition known as autosomal dominant hypophosphatemic rickets (formerly called vitamin D-resistant rickets) results from a mutation in FGF23 that renders it more resistant to proteolytic breakdown. An X-linked condition caused by overexpression of an FGF23 promoter causes a similar clinical picture, with short stature, bowed legs, and other features of rickets. Tumors that secrete FGF23 may occur in later life and may present with unexplained hypophosphatemia. Such tumors, which can be relatively small, are of mesenchymal origin, hence their ability to secrete fibroblast growth factors.

Assessment of phosphate handling may be made by the calculation of the parameter T_mP/GFR. This refers to the maximum phosphate reabsorption in the tubules; that is, the concentration in plasma at which phosphate starts to appear in the urine. This may also be referred to as the renal threshold for phosphate. This is entirely analogous to the T_m/GFR for glucose. T_mP/GFR is lowered by PTH as well as by FGF23.

Aminoacidurias

Separate mechanisms exist for renal tubular absorption of groups of amino acids and a number of transporters with varying degrees of specificity have been described (**Table 4.9**). In some cases there are related transport mechanisms in the upper small intestine, and both renal and intestinal mechanisms may be defective, as in cystinuria and in Hartnup disease, for example.

TABLE 4.9 Amino acid transport systems in the kidney

Specific transport system	Amino acids transported
Dibasic	Cysteine, ornithine, arginine, lysine
Dicarboxylic	Aspartate, glutamate
Neutral I	Proline, hydroxyproline, glycine
Neutral II	Alanine, serine, threonine, valine, leucine, isoleucine, phenylalanine, tyrosine, tryptophan, histidine
β-Amino acids	β-Alanine, β-amino-isobutyrate, taurine

Aminoacidurias may be selective, in which case urinary excess is confined to a single amino acid or to amino acids transported by the same mechanism, or generalized, in which there is excessive loss of all types of amino acid. Most or all mechanisms are defective in the Fanconi syndrome, in which there is generalized aminoaciduria.

Aminoacidurias may arise from an increase in amino acid concentration arriving at the kidney (overflow), decreased tubular reabsorption (leakage), or a mixture of both. Overflow occurs with an increase in plasma amino acid concentration; tubular reabsorption is intact. Examples include phenylalanine in phenylketonuria and the branched-chain amino acids valine, leucine, and isoleucine in maple syrup urine disease. Leakage is due to decreased tubular reabsorption; plasma amino acid concentrations are normal. Examples are dibasic amino acids (cysteine, ornithine, arginine, and lysine) in cystinuria and neutral amino acids in Hartnup disease. Mixed aminoaciduria is seen when an increased concentration of specific amino acids in plasma saturates tubular transport mechanisms, causing decreased reabsorption of other amino acids present in normal concentration in plasma that are transported by the same mechanism. For example, excretion of cysteine, ornithine, and lysine is increased in patients with the urea cycle disorder arginase deficiency, the primary effect being increased plasma arginine.

TABLE 4.10 Classification of cystinuria

Biochemical change	Type I	Type II	Type III
Increased urinary cysteine, ornithine, arginine, and lysine			
Homozygotes	+++	+++	+++
Heterozygotes	Unchanged	++	++
Absorption of cysteine by intestine	Decreased	Decreased	+/–

+++ large increase; ++ moderate increase; +/– minor increase or decrease.

Cystinuria

Cystinuria is an autosomal, recessively inherited disorder with defective transport of dibasic amino acids by renal tubules and, in most cases, by intestinal cells (**Table 4.10**). Urinary cysteine excretion, which in healthy adults is normally in the range 30–100 mg/day, is increased tenfold in patients with cystinuria (300–1000 mg/day) and gives rise to cysteine stones.

Hartnup disease

In this disease, named after the family in which it was first described, there is defective transport of neutral II amino acids by intestinal mucosa and renal tubules. Excess neutral II amino acids (see Table 4.9) and indoles are found in urine. Decreased absorption of tryptophan from the gut impairs the synthesis of nicotinamide and patients are treated with small doses of nicotinamide. Increased levels of amino acids in the intestine may lead to the production of toxins by the action of microorganisms. Clinical consequences include pellagra (photosensitive rash), ataxia, and emotional lability. Secondary nicotinamide deficiency may occur with metastatic carcinoid tumors.

4.6 RENAL STONES

The deposition of crystalline material (stones) in the urinary tract, a process called urolithiasis, is a common disease in industrialized societies. Between 5 and 12% of the population have at least one stone by the age of 70 years. As urine becomes concentrated in the nephron, the concentration of some less-soluble constituents may exceed their solubility product and form crystals, especially if a small particle is present to act as a nidus. The crystal may remain and grow within the renal calyx and cause gradual loss of renal function. Alternatively, it may partially or in its entirety move down the ureter. When the smooth muscle peristalsis of the ureter is unable to move the obstruction, it causes extremely severe pain, called renal colic. Some small renal stones may be passed spontaneously through the urethra, but larger ones may require surgery or lithotripsy using external sound waves to break them into smaller pieces before being passed in the urine. It is important to differentiate between nephrolithiasis (that is, kidney stones) and nephrocalcinosis, diffuse calcification in the kidneys.

It is also essential clinically to identify the type of stone, as the underlying metabolic causes and therefore treatments are different. The composition of urinary tract stones in the UK population is shown in **Table 4.11**. The figures for the USA are similar.

The commonest types of renal stones are composed of calcium oxalate and, along with calcium phosphate (hydroxyapatite), account for almost three-quarters of all stones. Oxalate is produced by hepatic synthesis and by conversion from ascorbic acid. Some edible plants are particularly rich in oxalate, which forms an insoluble precipitate with calcium. Usually, this remains unabsorbed

TABLE 4.11 Composition of urinary tract stones in the UK	
Component	**% Total by weight**
Calcium oxalate and/or phosphate	70
Magnesium ammonium phosphate	15
Uric acid	9
Mixed inorganic with uric acid	5
Others, chiefly cysteine	1

within the gut. However, oxalate absorption is increased if there is less calcium available. Thus, restriction of dietary calcium in dairy products to prevent renal stones may paradoxically increase the risk. Malabsorption syndromes are also associated with an increased risk of oxalate-containing stones. Unabsorbed fatty acids and bile acids bind to calcium in the gut lumen, increasing the amount of free oxalate available to be absorbed into the blood and cleared by the kidney.

Also common are stones consisting of uric acid (with hyperuricemia being a risk factor; see Chapter 17) or magnesium ammonium phosphate. The latter are sometimes referred to as struvite and are often seen in association with infection of the urinary tract with bacteria that can convert urea to ammonia (hence called urea-splitting organisms). The resulting alkaline pH promotes the precipitation of struvite. Anatomical defects that impair urine flow from the kidneys to the bladder are likely to result in bacterial colonization of the stagnant urine and a higher risk of struvite stones. Rarer types of stones include cysteine and xanthine. Pure oxalate stones are relatively small, while magnesium ammonium phosphate and cysteine form the largest stones. Chemical analysis of renal stones (if they can be collected) can be useful, particularly in the rarer types, but interpretation may be hindered by the presence of other components that crystallize on the primary stone. For example, cysteine stones may also contain calcium and magnesium phosphate. The major risk factors for stone formation are shown in **Table 4.12**.

Investigation of a single stone episode in an adult is rarely required. However, repeated episodes, especially in a younger person, is likely to have an underlying cause, either anatomical or metabolic. As well as imaging, biochemical investigations may be indicated. A number of metabolic disorders are associated with

TABLE 4.12 Risk factors for renal stones	
Abnormality	**Comment**
Low urinary volume and concentrated urine	Favors stone formation; stones are more common in hot than temperate climates
Urinary pH	pH conducive to stone formation: Calcium phosphate — High Magnesium ammonium phosphate — High Uric acid — Low Cysteine — Low
Urinary tract infection	Usually stone of magnesium ammonium phosphate (struvite) due to: Increased pH Decreased concentration of inhibitors of crystallization Cellular debris acting as nidus
Reduced citrate excretion	Citrate is an inhibitor of calcium salt crystallization. Excretion is reduced in acidosis and diets high in animal protein

renal stones (**Table 4.13**) and the tests may be selected according to clues from the clinical history. More commonly, the cause is multifactorial and a range of investigations are done to identify potential over- or underexcretion of promoters and inhibitors. The main example of the latter is citrate. Investigations performed on blood and urine samples to investigate recurrent stone formation are also shown in Table 4.13.

Analytical practice point 4.5

Chemical analysis of renal stones can identify the major components and may point to a treatable cause, but mixed stones are common.

TABLE 4.13 Metabolic disorders associated with renal stones

Condition	Clinical cause	Investigations
Hypercalciuria	Idiopathic hypercalciuria with normocalcemia: increased absorption of calcium from gut and/or defective renal handling of calcium	24-hour urine calcium Plasma calcium Plasma parathyroid hormone (PTH)
	Immobilization in patients with excessive bone turnover	
	Prolonged hypercalcemia (for example, primary hyperparathyroidism)	
Hyperoxaluria	Enteric, due to hyperabsorption of oxalate associated with impaired fat absorption in ileal disease or bowel resection	24-hour urine oxalate
	Rare inborn enzyme defects resulting in overproduction of oxalate	
Hyperuricosuria	Primary gout	24-hour urine urate Plasma urate
	Increased nucleoprotein turnover in malignant disease, especially with chemotherapy	
	Overproduction due to one of several rare inborn enzyme defects	
Cystinuria	Inborn defect of proximal renal tubular absorption of cysteine and other dibasic amino acids	Urine amino acids
Renal tubular acidosis	Systemic acidosis reduces urine citrate	24-hour urine citrate
Miscellaneous	Xanthinuria—a rare inborn defect of activity of xanthine oxidase Silica—in patients having prolonged treatment with magnesium trisilicate	Analysis of renal stone

SODIUM AND WATER

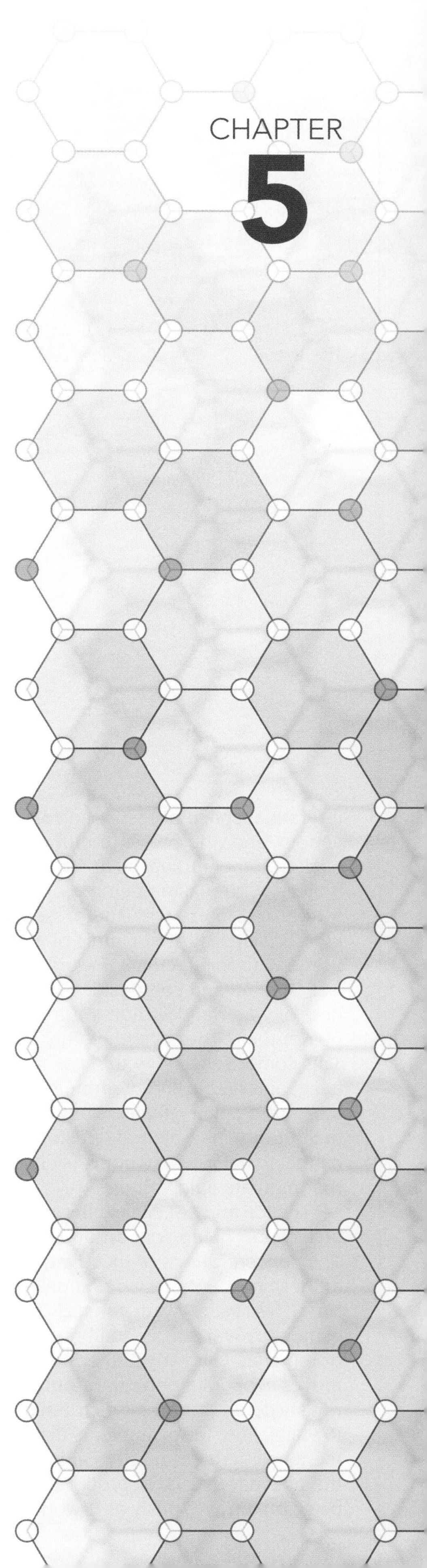

IN THIS CHAPTER

FLUID COMPARTMENT MODEL

REGULATION OF SODIUM AND WATER BALANCE

CLINICAL DISORDERS OF PLASMA SODIUM

Cellular function and physiological processes are critically dependent on a stable internal environment, or "milieu intérieur," as first described by the French physiologist Claude Bernard in the nineteenth century. This requires the maintenance of such parameters as temperature, pH, and osmotic strength of the extracellular fluid (ECF) within narrow and highly regulated ranges (**Table 5.1**). The major reason for this is to maintain protein conformation, particularly of enzymes. A protein's function is determined by its three-dimensional shape and any degree of change will dramatically alter its efficiency. The volume and osmolality of the ECF are regulated by overlapping mechanisms that control urine composition through effects on the kidney. Clinical abnormalities of plasma sodium arise when these mechanisms fail or are pushed beyond their normal limits by extreme loss or gain of water or solutes.

TABLE 5.1 Reference ranges for plasma and serum

Analyte	SI units	Conventional units
Sodium	136–142 mmol/L	136–142 mEq/L
Potassium	3.5–5.0 mmol/L	3.5–5.0 mEq/L
Urea (blood urea nitrogen, BUN)	2.9–8.2 mmol/L	8–23 mg/dL
Creatinine	53–106 μmol/L	0.6–1.2 mg/dL
Osmolality	275–295 mmol/kg	275–295 mOsm/kg

5.1 FLUID COMPARTMENT MODEL

The average adult human contains approximately 42 liters of water (**Figure 5.1**). This varies relatively little, with differences in weight between individuals consisting mainly of adipose tissue. Just over a half (23 L) of this water is within cells, with the remainder mainly residing in the extracellular space (12 L). ECF comprises the plasma within the arteries, veins, and capillaries (the intravascular space) and that fluid which immediately surrounds the cells (the interstitial fluid), taking nutrients to and waste products from cells. Plasma differs from interstitial fluid in that it contains proteins which, as macromolecules, cannot usually pass across the capillary walls. Interstitial fluid is, therefore, an ultrafiltrate of plasma in much the same way that glomerular filtrate is. At the arterial end of the capillary bed, the hydrostatic pressure forcing water into the

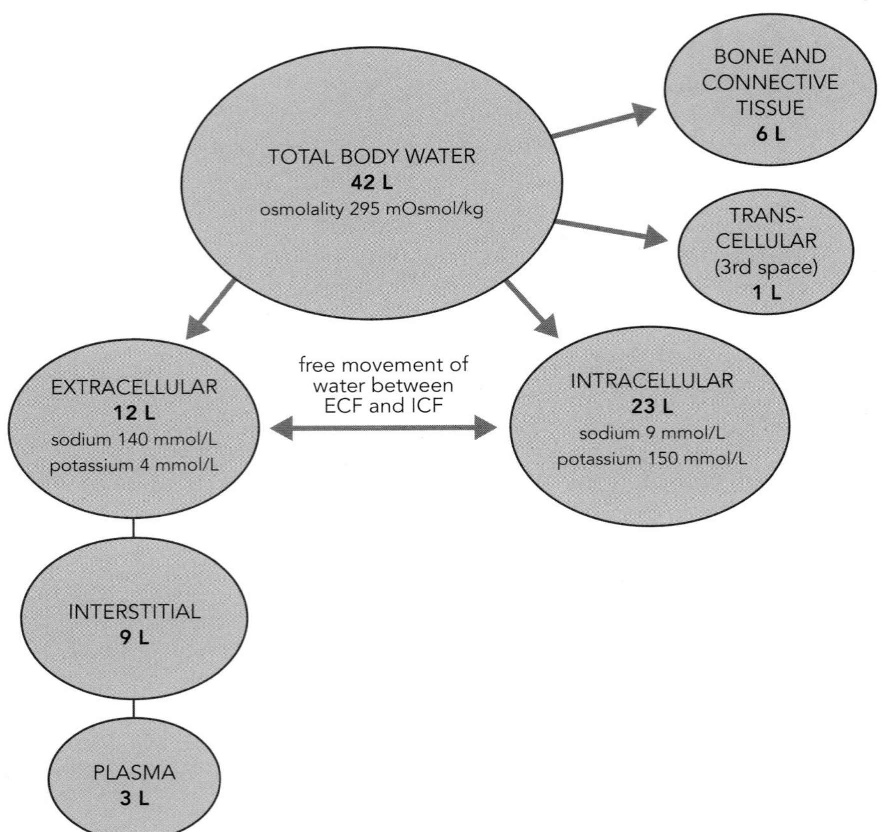

Figure 5.1 Distribution of water across the different fluid compartments.
Total body water is the sum of the extracellular (ECF) and intracellular (ICF) fluid volumes plus water locked within bone and connective tissue and around 1 liter within the lumen of the gut. The ECF volume is half that of the ICF and is further divided into the interstitial fluid (ISF), which bathes the cells, and the intravascular fluid or plasma. ISF and capillary plasma differ by only their protein concentrations, since ISF is an ultrafiltrate of plasma. Water can move freely between ECF and ICF compartments in order to maintain equal osmolality. Thus, the addition of 1 liter of water will expand the ICF by 667 mL (two-thirds of 1 L) and the ECF by 333 mL (one-third of 1 L). The water in bone, connective tissue, and the transcellular space equilibrates very slowly and can be ignored in this example. The differences in ECF and ICF sodium and potassium concentrations are maintained by Na^+/K^+-ATPase. A sodium-rich fluid will remain predominantly in the ECF.

interstitial space exceeds the colloid oncotic pressure pulling it back in. At the venous end, the pressure difference is reversed and interstitial fluid is pulled back into the intravascular space. The interstitial fluid volume is approximately 9 liters, with plasma being around 3 to 3.5 liters. Blood volume is made up to a total of 5 liters with the presence of cells, predominantly erythrocytes (red blood cells). A further 6 liters of water is held within bone and dense connective tissue. This is effectively locked in and responds only very slowly to changes in ECF volume and composition. It can be ignored in the context of clinical disorders of water and sodium. A final fluid compartment is referred to as transcellular or the third space (the other two being the intracellular fluid [ICF] and ECF). This comprises fluid within the gastrointestinal tract, such as saliva and gastric juice, and within the peritoneal and pleural spaces. Usually, this only amounts to around 1 liter of water, but can be significantly more in certain disorders of the gut or with a pleural effusion or ascites. Fluid in the third space, like that in connective tissue and bone, is effectively invisible to physiological mechanisms that regulate fluid volume.

The ICF and ECF have different compositions, with the predominant extracellular cation being sodium and the major intracellular cation being potassium. The concentrations of these are reversed across cell membranes. Maintenance of these concentration gradients is energy dependent and requires the action of Na^+/K^+-ATPase. Water can pass freely across cell membranes, however, and the osmolality will reach equilibrium across all fluid compartments if water is added or removed from any of them. As a consequence, cell volume changes as water moves in or out along osmotic gradients. The composition of fluids given intravenously determines their distribution across the different compartments. Normal (0.9%) saline, for example, is distributed across only the extracellular compartments, with three-quarters going into the interstitial space and one-quarter into the plasma; that is, in proportion to their relative volumes. Pure water cannot be given intravenously as it would cause localized osmotic swelling and lysis of red blood cells at the infusion site. The nearest equivalent fluid that can be used clinically is a solution of 5% glucose (dextrose). After metabolism of the glucose, the remaining water is distributed across all fluid compartments in proportion to their sizes. Thus, two-thirds goes into the intracellular space and one-third into the extracellular space, with only one-quarter of this going into the intravascular space (plasma). Thus, a liter of water theoretically expands the plasma by only 83 mL ($1/3 \times 1/4 \times 1000$), compared to 250 mL ($1/4 \times 1000$) for normal saline. This explains why saline is more effective than water (or a solution of isotonic glucose) for restoring intravascular volume.

Osmolality and tonicity

Osmolarity is the sum of all dissolved particles—of whatever size—in a unit volume of water. A related term, osmolality, refers to the total number of particles in a unit mass of water. Volume is temperature dependent and also itself affected by the presence of dissolved particles. However, for clinical purposes, the two terms are almost interchangeable. Osmolality can be measured in plasma and urine by the clinical laboratory. Osmometers most commonly work on the principle that the depression of freezing point is proportional to the concentration of particles present. It can also be estimated by adding together the molar concentration of the major solutes: in the case of plasma these being sodium, potassium, urea, and glucose. As sodium and potassium are cations, they must have corresponding anions and are therefore doubled for the purpose of calculating osmolality. Glucose and urea are the only other substances present at high enough concentrations to affect the plasma osmolality. Proteins are measured in g/L and their molar concentrations are small, around 1 mmol/L. Creatinine is present in micromolar concentrations and even with severe renal failure will not rise above 1–2 mmol/L. Osmolality is therefore estimated as follows:

Calculated (estimated) osmolality = 2 × (Na^+ + K^+) + urea + glucose (all in mmol/L) (Equation 5.1)

Another term, often used in relation to osmolality, is tonicity. This refers to the osmotic effect of a concentration gradient across a cell membrane, whereas osmolality is a physical property of the solution. Tonicity can therefore never exceed osmolality, but can often be much lower. For example, urea is freely permeable across most cell membranes whereas sodium is not. Thus, a solution of urea has a much lower tonicity than one of sodium chloride, even if they contain identical numbers of particles per kg of water and have the same measured and calculated osmolalities.

5.2 REGULATION OF SODIUM AND WATER BALANCE

Water intake varies greatly from day to day depending on availability, habit, and social factors. In order to remain in balance, the kidneys have to be able to vary the output of urine, in terms of both volume and concentration. This is explained in detail in Chapter 4. Insensible loss of water occurs due to evaporation from the skin and lungs and is increased in hot climates, exercise, and fever. Water balance is summarized in **Figure 5.2**. Sodium intake varies widely across the world, largely depending on intake of meat and processed foods. A typical Western diet may contain 100–200 mmol of sodium per day, whilst some isolated tribesmen may survive on less than this in a month. There are minimal non-renal losses of sodium, with almost all excretion being into the urine.

The two major homeostatic systems that are involved in the control of salt and water balance in the body are the renin–angiotensin–aldosterone (RAA) system and antidiuretic hormone (ADH). When these systems are fully stimulated, via such factors as hypovolemia, hypotension, decreased renal perfusion, and,

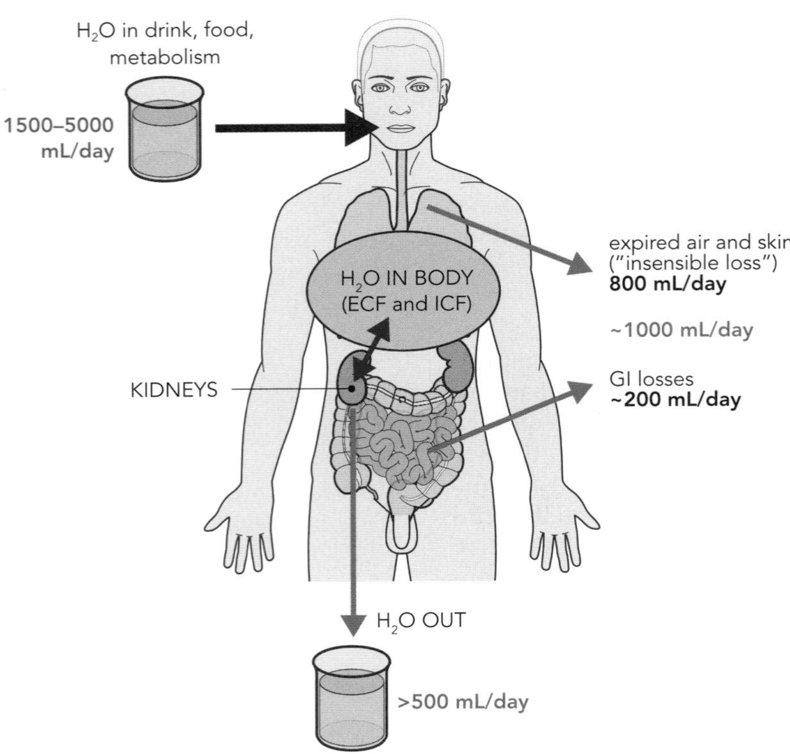

Figure 5.2 Water balance. Under normal conditions, water intake (blue arrow) and loss (green arrows) are in balance each day. Water balance is maintained by adjusting urine volume and concentration (red arrow) to match varying water intake and output from non-renal routes. Such losses are increased by evaporation from the lungs and skin (for example, in hot climates or fever) or by losses from the gastrointestinal tract (for example, vomiting, diarrhea, and fistula). If water intake cannot be increased to match these losses, renal conservation increases, resulting in reduced urine volume and increased urine concentration (osmolality).

in the case of ADH, hyperosmolality, they exert their separate effects on the kidneys. These result in reabsorption of sodium and water into the circulation. More detail of the nature of the action of these two hormone systems on renal function can be found in Chapter 4. A third mechanism that is often overlooked is thirst. Increased osmolality (or more accurately, tonicity) in the ECF exerts a powerful effect on thirst to the point that it overrides other thought processes. Fortunately, most people never experience this severe type of thirst because they will tend to drink in anticipation or according to habit.

Renin is a small enzyme that is released from cells in the afferent arterioles of the kidney in response to several factors, including reduced renal perfusion and reduced glomerular filtration rate. Renin cleaves the liver peptide angiotensinogen to form angiotensin I, which is further metabolized by angiotensin-converting enzyme to angiotensin II. Angiotensin II stimulates the secretion of the mineralocorticoid hormone aldosterone from the adrenal cortex.

The major effect of aldosterone is to promote the retention of sodium ions (in exchange for potassium ions) at the distal tubule. Reabsorbed sodium ions are accompanied by water and this has the effect of increasing the intravascular volume. This tends to increase blood pressure and renal perfusion toward normal levels. Urinary electrolytes reflect the action of aldosterone and provide clinical evidence of its secretion and renal response. Under conditions of maximal aldosterone secretion, the urinary sodium will be low; that is, <10 mmol/L. The potassium will usually be high, for example >50 mmol/L, but this is not used diagnostically as there is no specific concentration for determining maximal aldosterone action.

ADH (also called arginine vasopressin, AVP, or just vasopressin) is released in response to an increase in plasma osmolality, but is also released when significant reduction in effective circulating volume occurs in the intravascular space. Thirst is stimulated by the same mechanisms. In this situation, the need to reabsorb water overrides the need to maintain normal osmolality. Angiotensin II also promotes the release of ADH from the posterior pituitary. The major effect of ADH is to promote the retention of water by the kidneys, thereby potentially reducing plasma osmolality and increasing circulating blood volume in order to correct hyperosmolality and hypovolemia, respectively. Reabsorption of water alone is, however, relatively ineffective as it is distributed across all body fluid compartments and only approximately 10% remains within the intravascular space. As with aldosterone, the presence and action of ADH can be inferred from its effect on urinary parameters, in this case the urine osmolality. When ADH is maximally secreted, and the renal collecting ducts are able to respond, the urine becomes maximally concentrated; that is, the osmolality is >750 mmol/kg. In some extreme cases it may reach 1500 mmol/kg. The majority of the raised osmolality is due to the presence of urea and potassium, since the concentration of sodium is low due to the action of aldosterone.

5.3 CLINICAL DISORDERS OF PLASMA SODIUM

The osmolality of plasma, and hence ECF, is largely determined by the concentration of sodium, and is regulated by ADH and thirst. The total amount of sodium determines the volume of the ECF and cannot be estimated from the

CASE 5.1

A 79-year-old woman had hip-replacement surgery 48 hours ago. She had normal electrolytes, including sodium, preoperatively. She is now found to be unconscious with upgoing plantar reflexes (a sign of upper motor neuron lesion).

	SI units	Reference range	Conventional units	Reference range
Plasma				
Sodium	108 mmol/L	136–142	108 mEq/L	136–142
Potassium	3.6 mmol/L	3.5–5.0	3.6 mEq/L	3.5–5.0
Urea (BUN)	2.1 mmol/L	2.9–8.2	5.9 mg/dL	8–23
Creatinine	50 μmol/L	53–106	0.57 mg/dL	0.6–1.2
Osmolality	232 mmol/kg	275–295	232 mOsm/kg	275–295
Urine				
Sodium	5 mmol/L	Not applicable*	5 mEq/L	Not applicable*
Osmolality	95 mmol/kg	Not applicable*	95 mOsm/kg	Not applicable*

*Urine sodium and osmolality are interpreted as whether appropriate for plasma results and clinical status.

- **What are the most important features of this case?**
- **What further information is required to confirm the cause?**
- **How should this patient be treated**

She has rapidly falling sodium and neurological impairment; hence this is a medical emergency. This is iatrogenic acute water intoxication due to inappropriately large volumes of hypotonic fluid: 5% dextrose or 4% dextrose/0.18% saline. The most important information is fluid balance charts detailing intake and urine output. Aggressive treatment to rapidly raise sodium toward normal is required in an intensive care setting. Hypertonic (1.8%) saline may be given. Acute hyponatremic encephalopathy is the only situation where rapid correction of sodium is required. In other situations, the risk of osmotic demyelination is greater than the risk from hyponatremia. Another osmotically active substance, mannitol, is used to treat cerebral edema in other situations, for example brain tumors, and would also be effective here. However, it is not usually used because it would paradoxically lower the plasma sodium even more, due to osmotic pull of water from the ICF, and this seems counterintuitive to most clinicians.

concentration, which is merely the ratio of sodium to water. Abnormally high or low plasma sodium concentrations (hyper- and hyponatremia, respectively) can therefore be due to excess or deficiency of either sodium or water. It is far more common for gain or loss of water to be the primary pathology, rather than disorders of sodium. Because sodium is present in such a high concentration, relative to other solutes, its fall is more readily apparent when dilution of the ECF occurs. Dilution of the ECF by 10% will cause plasma sodium to fall by around 14 mmol/L, but potassium by only 0.4 mmol/L and calcium by 0.2 mmol/L. These are clinically insignificant changes and may be within the biological and analytical variation of the measurements.

If pure water is added rapidly to the ECF, it reduces the osmolality and, since the intracellular and extracellular spaces are in osmotic equilibrium, water will also move into the cells across their membranes. This will therefore increase their volume. Conversely, removal of water from the intracellular space will lead to cells shrinking. The only organ that is threatened by this change in volume is the brain, constrained as it is by the surrounding skull and vasculature. Symptoms of hypo- and hypernatremia are therefore predominantly neurological, and initial treatment is aimed at restoring normal brain volume.

Because of the interplay of different causes and compensating mechanisms, sodium disorders are often poorly understood by students. Specific examples are useful to demonstrate the relevant mechanisms. There are therefore a number of illustrative cases in this chapter.

Clinical practice point 5.1

Disorders of plasma sodium are more often due to gain or loss of water, rather than gain or loss of sodium.

Clinical practice point 5.2

Clinical effects of hypo- and hypernatremia are neurological, due to changes in brain volume.

CASE 5.2

A 72-year-old woman has had severe vomiting and diarrhea for 48 hours.

	SI units	Reference range	Conventional units	Reference range
Plasma				
Sodium	125 mmol/L	136–142	125 mEq/L	136–142
Potassium	2.8 mmol/L	3.5–5.0	2.8 mEq/L	3.5–5.0
Urea (BUN)	15 mmol/L	2.9–8.2	42 mg/dL	8–23
Creatinine	150 μmol/L	53–106	1.70 mg/dL	0.6–1.2
Osmolality	278 mmol/kg	275–295	278 mOsm/kg	275–295
Urine				
Sodium	5 mmol/L	Not applicable*	5 mEq/L	Not applicable*
Osmolality	810 mmol/kg	Not applicable*	810 mOsm/kg	Not applicable*

*Urine sodium and osmolality are interpreted as whether appropriate for plasma results and clinical status.

- **Why is the urinary sodium low and the osmolality high?**
- **Why is the plasma potassium low?**
- **Why is the patient hyponatremic even though she has lost ECF, which contains sodium?**

This is hyponatremia initiated by loss of sodium-containing fluid from the gastrointestinal tract. This causes reduced ECF volume and low circulating volume (hypovolemia). The plasma sodium concentration is primarily low because of the increase in water reabsorption in the renal collecting ducts, stimulated by ADH. The concentrated urine (osmolality >750 mmol/kg) is evidence of both ADH release and renal response. The simultaneous activation of the renin–angiotensin–aldosterone axis results in renal loss of potassium in exchange for sodium, and hence hypokalemia.

Hyponatremia

Hyponatremia, defined as plasma sodium concentration below the lower reference limit (around 135 mmol/L), is a common electrolyte abnormality, particularly in hospitalized patients, but clinical effects may not be apparent until the plasma sodium concentration reaches 125 mmol/L or lower. The rate of decrease is also important. Chronic hyponatremia may be present for several months, or even years, and be apparently asymptomatic, whilst the same degree of hyponatremia may be rapidly fatal if occurring over a few hours. This is because compensatory mechanisms are able to mitigate the effect of hyponatremia but take time to act. In response to the rise in cellular volume, cells are able to pump out (extrude) organic and inorganic particles (osmolytes), which take water with them and lower the intracellular volume. Rapid decreases in plasma sodium and osmolality overwhelm these mechanisms and result in cerebral edema. A range of symptoms may occur, ranging from nausea and confusion through to coma, and even death. Acute hyponatremia is therefore a medical emergency and must be treated early and aggressively. In contrast, chronic hyponatremia, which is far more common, may be surprisingly asymptomatic in some individuals, even with very low plasma sodium concentrations,

CASE 5.3

A 50-year-old man complains of swelling of his ankles, fingers, and face over several weeks. He is also known to have high cholesterol (9 mmol/L)

	SI units	Reference range	Conventional units	Reference range
Plasma				
Sodium	125 mmol/L	136–142	125 mEq/L	136–142
Potassium	2.9 mmol/L	3.5–5.0	2.9 mEq/L	3.5–5.0
Urea (BUN)	10 mmol/L	2.9–8.2	28 mg/dL	8–23
Creatinine	130 μmol/L	53–106	1.47 mg/dL	0.6–1.2
Osmolality	270 mmol/kg	275–295	270 mOsm/kg	275–295
Albumin	19 g/L	35–50	1.90 g/dL	3.5–5.0
Urine				
Sodium	5 mmol/L	Not applicable*	5 mEq/L	Not applicable*
Osmolality	800 mmol/kg	Not applicable*	800 mOsm/kg	Not applicable*
Albumin–creatinine ratio	325 mg/mmol	2.5–3.5	3900 μg/mg	30–300

*Urine sodium and osmolality are interpreted as whether appropriate for plasma results and clinical status.

- **What is the diagnosis, based on the clinical features and results?**
- **Is the total body sodium likely to be increased, normal, or decreased?**
- **Why is the patient hyponatremic even though he has edema?**

The triad of edema, hypoalbuminemia, and albuminuria is diagnostic of nephrotic syndrome. This is explained in greater detail in Chapter 4. Hypercholesterolemia is also a common feature. Due to low plasma oncotic pressure, fluid leaks from the intravascular space into the interstitial space and does not return at the end of the capillary bed, causing edema and reduced circulating volume. This stimulates ADH release and the RAA system as in Case 5.2. The only difference here is that fluid remains within the body, rather than being lost to the outside world, but is in the wrong space. Total body sodium is elevated and correlates with the degree of edema, reflecting expanded ECF volume.

for example 110 mmol/L. This indicates that compensation of cell volume has occurred. Aggressive treatment to rapidly raise the plasma sodium in this situation is not required and may cause serious, and even fatal, brain injury. A specific complication which may arise in the treatment of hyponatremia is central pontine myelinolysis (CPM) or osmotic demyelination, which can cause permanent paralysis and death. It is particularly likely to occur if sodium correction leads to an overshoot into hypernatremia.

Causes of hyponatremia may be divided into those where the primary disorder is gain of water and those where it is due to depletion of the ECF volume.

Clinical practice point 5.3

Rapid changes in plasma sodium concentration are more likely to cause clinical effects and rapid correction of sodium in chronic disorders can be dangerous.

Primary water excess

Most simply, this is due to input of water into the ECF through the oral or intravenous route at a rate that exceeds renal capacity to excrete it. This most commonly is seen in hospital patients who are given excessive volumes of dextrose-containing fluid. Although 5% dextrose has an osmolality similar to plasma (to avoid lysis of red blood cells at the infusion site) the dextrose is metabolized

CASE 5.4

A 45-year-old man has been unwell with weight loss for several months. He has a low blood pressure of 90/50 mmHg. On examination, skin and mouth pigmentation is noted.

	SI units	Reference range	Conventional units	Reference range
Plasma				
Sodium	125 mmol/L	136–142	125 mEq/L	136–142
Potassium	5.9 mmol/L	3.5–5.0	5.9 mEq/L	3.5–5.0
Urea (BUN)	10 mmol/L	2.9–8.2	28 mg/dL	8–23
Creatinine	130 μmol/L	53–106	1.47 mg/dL	0.6–1.2
Osmolality	278 mmol/kg	275–295	278 mOsm/kg	275–295
Cortisol	259 nmol/L	140–690	9.4 μg/dL	5–25
Urine				
Sodium	98 mmol/L	Not applicable*	98 mEq/L	Not applicable*
Osmolality	750 mmol/kg	Not applicable*	750 mOsm/kg	Not applicable*

*Urine sodium and osmolality are interpreted as whether appropriate for plasma results and clinical status.

- **What is the likely diagnosis and how would you confirm it?**
- **Why is the urinary sodium high in the presence of hyponatremia?**
- **Does the plasma cortisol result exclude adrenal insufficiency**

This is Addison's disease (primary adrenal insufficiency) causing deficiency of glucocorticoids and mineralocorticoids. This is explained in greater detail in Chapter 11. A random cortisol result in the reference range does not, however, exclude the diagnosis and a short Synacthen test is required. The urinary sodium is high because aldosterone is not produced. This also causes renal retention of potassium and hyperkalemia. Loss of sodium reduces ECF and intravascular volume. ADH secretion is stimulated, resulting in concentrated urine. Hyponatremia occurs due to water reabsorption from this appropriate secretion of ADH. Lack of cortisol impairs the excretion of free water, and hyponatremia can occur even if the secretion of mineralocorticoid is intact, as would be the case in secondary hypoadrenalism due to pituitary disease.

rapidly, leaving free water, which lowers the ECF osmolality. Careful monitoring of fluid balance and use of intravenous fluids of appropriate composition are required to avoid this situation.

Excess water intake by simply drinking too much fluid is uncommon, but is seen in some people with a psychiatric condition called psychogenic polydipsia. There are also cases described of hyponatremic encephalopathy—sometimes fatal—occurring in marathon runners, extreme dieters, and even water-drinking competitions.

More chronic hyponatremia occurs in people who drink large volumes of low-electrolyte liquid (beer is such a fluid) and eat very little food. This is sometimes called beer drinker's potomania and is due to an inability to clear the free water. The kidney is unable to excrete pure water: the minimum achievable osmolality is 50–100 mmol/kg Therefore the excretion of 3 liters of water a day requires a minimum of 150–300 mmol of solute. If this is not taken in the diet, free water will accumulate in all fluid compartments and dilute the contents of the ECF.

Syndrome of inappropriate ADH

The action of ADH is to increase water reabsorption in the collecting ducts of the kidney. If primary oversecretion of ADH occurs, this will cause hyponatremia. Such a situation may occur with ADH-secreting tumors of the lung or with excess secretion from the posterior pituitary due to certain brain, lung, or metabolic disorders or particular dugs. This secretion of ADH is not properly regulated and is called the syndrome of inappropriate ADH or SIADH. The causes of SIADH may be classified as follows:

- Ectopic secretion of ADH
  - Small-cell carcinoma of the bronchus
- Cerebral disorders
  - Tumor
  - Infection
  - Stroke
  - Trauma
- Pulmonary disorders
  - Pulmonary embolus
  - Pneumonia
  - Tuberculosis
- Drugs
  - Selective serotonin reuptake inhibitors (SSRIs)
  - Carbamazepine
  - Opiates
  - Oral hypoglycemics
  - Oxytocin
- Metabolic disorders
  - Acute porphyria

SIADH is a common cause for hyponatremia, especially in hospitalized patients. However, there is no single diagnostic test for it. Measurement of ADH is of little value, due to different patterns of secretion, and is not in the repertoire of any but the most specialized reference laboratories, especially within a clinically useful timescale. Instead, SIADH is a diagnosis of exclusion. In other words it is only diagnosed when other conditions that may cause hyponatremia have been excluded first. As well as hyponatremia, the serum osmolality must be low, with urine that is inappropriately concentrated. There must be no evidence of either ECF volume contraction (dehydration) or expansion (edema) on

CASE 5.5

A 63-year-old man has a long history of smoking and is known to have a bronchial carcinoma.

	SI units	Reference range	Conventional units	Reference range
Plasma				
Sodium	121 mmol/L	136–142	121 mEq/L	136–142
Potassium	3.5 mmol/L	3.5–5.0	3.5 mEq/L	3.5–5.0
Urea (BUN)	3.0 mmol/L	2.9–8.2	8.4 mg/dL	8–23
Creatinine	61 µmol/L	53–106	0.69 mg/dL	0.6–1.2
Osmolality	260 mmol/kg	275–295	260 mOsm/kg	275–295
Urine				
Sodium	96 mmol/L	Not applicable*	96 mEq/L	Not applicable*
Osmolality	573 mmol/kg	Not applicable*	573 mOsm/kg	Not applicable*

*Urine sodium and osmolality are interpreted as whether appropriate for plasma results and clinical status.

- **What is the likely diagnosis?**
- **Why are the urinary sodium and osmolality high in the presence of hyponatremia?**
- **Would restriction of fluid intake have any effect on the plasma osmolality?**

This is syndrome of inappropriate ADH (SIADH) due to ectopic secretion from a small-cell (oat-cell) carcinoma of the bronchus. Such tumors are of neuroendocrine origin and often secrete ADH and other hormones. This ADH secretion is not subject to negative feedback and is inappropriate for the osmolar and volume status. The urine osmolality is high. There is slight volume expansion, hence the RAA system is not activated and there is high urinary sodium. Restriction of water intake will result in a net loss of pure water due to insensible losses by evaporation from the lungs and skin.

clinical examination, as these would be associated with the appropriate secretion of ADH. Kidney disease, adrenal insufficiency, and hypothyroidism are all potential causes of hyponatremia, so must also be ruled out. Thus, the diagnostic criteria are:

- Hyponatremia and low serum osmolality
- Urine that is not maximally dilute (osmolality >100 mmol/kg)
- Normal ECF volume
- Normal kidney, adrenal, and thyroid function
- Patient is not on any drugs that may cause hyponatremia

Clinical practice point 5.4

SIADH is a diagnosis of exclusion: measurement of ADH is not required or helpful.

Primary extracellular volume depletion

Loss of sodium-containing fluid from the gut or kidney will deplete the ECF volume. If this loss is isotonic, there is no osmolar pull of water from the ICF. Falling ECF volume is detected by the juxtaglomerular apparatus in the kidneys, leading to activation of the RAA system. This results in the aldosterone-driven reabsorption of sodium and water (and excretion of potassium) with the effect of increasing ECF volume again. When the circulating fluid volume falls by around 10%, the baroreceptors in the great veins and heart are also activated and stimulate

CASE 5.6

A 48-year-old woman has a recent diagnosis of multiple myeloma.

	SI units	Reference range	Conventional units	Reference range
Plasma				
Sodium	125 mmol/L	136–142	125 mEq/L	136–142
Potassium	4.2 mmol/L	3.5–5.0	4.2 mEq/L	3.5–5.0
Urea (BUN)	12.0 mmol/L	2.9–8.2	33 mg/dL	8–23
Creatinine	120 µmol/L	53–106	1.36 mg/dL	0.6–1.2
Glucose	6.3 mmol/L	3.9–6.1	113 mg/dL	70–110
Osmolality	293 mmol/kg	275–295	293 mOsm/kg	275–295
Urine				
Sodium	20 mmol/L	Not applicable*	20 mEq/L	Not applicable*
Osmolality	300 mmol/kg	Not applicable*	300 mOsm/kg	Not applicable*

*Urine sodium and osmolality are interpreted as whether appropriate for plasma results and clinical status.

- **Why is the plasma osmolality not low when the patient is hyponatremic?**
- **What else can you measure to demonstrate your answer?**
- **How else can you assess the patient**

This is pseudohyponatremia due to the high serum protein caused by the monoclonal immunoglobulin secreted by the abnormal plasma cells in myeloma. Serum protein measurement, electrophoresis, and immunoglobulin assay will demonstrate this. Myeloma is explained in more detail in Chapter 19. Osmolality is normal because the protein molecules are large and only present at low molar concentration. Measurement of sodium by direct ion-selective electrode on a point-of-care analyzer will give the sodium as an activity in the aqueous component of plasma. This is more clinically relevant than the sodium in a unit volume of plasma (see Figure 5.3).

the release of ADH from the posterior pituitary. This results in increased free-water reabsorption and hence dilution of the ECF. This will be exacerbated if free water is drunk in response to the stimulation of thirst, which is also a baro-receptor-mediated effect in this situation. It is a greater priority for the body to restore circulating volume than it is to maintain osmolality. In this situation, the secretion of ADH is deemed to be appropriate.

Pseudohyponatremia

Artifactual hyponatremia or pseudohyponatremia may occur when the proportion of water in a unit volume of plasma is decreased because the nonaqueous proportion due to lipids and protein (usually 7%) is increased (**Figure 5.3**). The sodium is confined to the aqueous part and use of a method based on diluting a fixed volume of plasma before measuring the sodium will therefore underestimate the concentration. This is the situation for most laboratory methods (where an indirect ion-selective electrode or flame-emission photometry is used). Point-of-care methods, however, use direct ion-selective electrodes that measure the activity of sodium ions in the plasma water using undiluted samples and make no assumptions about plasma volume. These paradoxically give results that are clinically correct, compared to the central laboratory.

Analytical practice point 5.1

Pseudohyponatremia may occur with extreme hypertriglyceridemia and hyperproteinemia: use of direct ion-selective electrodes gives the correct result.

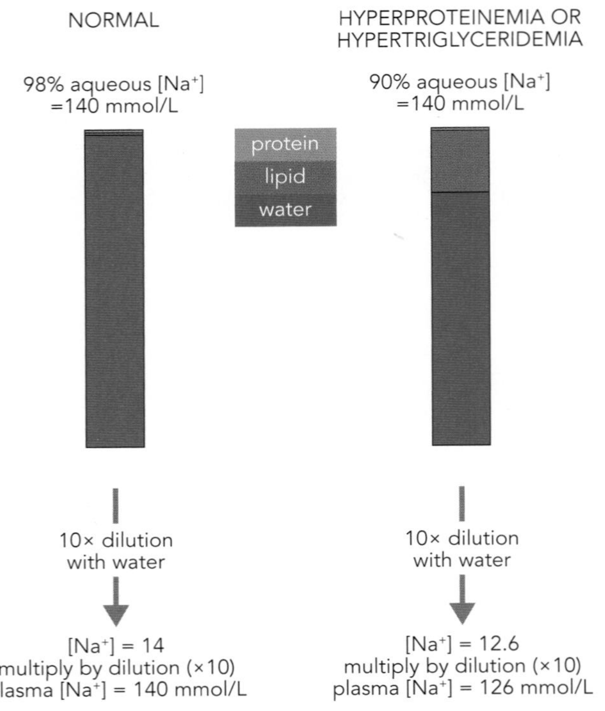

Figure 5.3 Pseudohyponatremia in plasma due to increased nonaqueous volume.
A unit volume of normal plasma (on the left) contains mainly water, with small amounts of protein and lipid. If either protein or lipid is significantly increased (on the right), the aqueous portion is decreased. The sodium concentration in the aqueous component is the same in both cases, but will be measured as falsely low if the sample on the right is diluted first and sodium concentration calculated assuming that a unit volume has normal composition. Routine laboratory measurements by indirect ion-selective electrode (ISE) or flame-emission photometry are affected by this. Measurement of sodium by direct ISE gives the result as an activity and is the same in both samples. This is the physiologically relevant measure. Note also that the osmolality of the two samples is the same. The molar concentrations of the lipid and protein particles are so small in comparison to sodium that they can be ignored.

CASE 5.7

A 75-year-old man is on a medical ward following a hemiplegic stroke 7 days ago.

	SI units	Reference range	Conventional units	Reference range
Plasma				
Sodium	155 mmol/L	136–142	155 mEq/L	136–142
Potassium	3.0 mmol/L	3.5–5.0	3.0 mEq/L	3.5–5.0
Urea (BUN)	25 mmol/L	2.9–8.2	70 mg/dL	8–23
Creatinine	220 μmol/L	53–106	2.49 mg/dL	0.6–1.2
Osmolality	345 mmol/kg	275–295	345 mOsm/kg	275–295
Urine				
Sodium	5 mmol/L	Not applicable*	5 mEq/L	Not applicable*
Osmolality	810 mmol/kg	Not applicable*	810 mOsm/kg	Not applicable*

*Urine sodium and osmolality are interpreted as whether appropriate for plasma results and clinical status.

- **Why is the urine osmolality high?**
- **Why is the urinary sodium concentration low?**
- **Is this patient's total body sodium likely to be increased, normal, or low?**

This is hypernatremic dehydration due to inability to drink. Insensible losses are not replaced and pure water loss occurs. This is predominantly from the intracellular space as this has the largest volume. Both ICF and ECF osmolality increase and this stimulates ADH, which causes renal conservation of water. With eventual decreased plasma volume, the RAA system is activated, causing urine sodium conservation. Total body sodium is normal (or low), but total body water volume is very low.

Hypernatremia

Hypernatremia occurs less commonly than hyponatremia and may be due to gain of salt or, more usually, loss of water. When water loss is the primary cause, ECF volume depletion occurs late as it is buffered by losses from the much larger ICF volume. On the other hand, when salt gain does occur, the volume excess it causes is clinically apparent much earlier, as it is confined to the ECF alone. The single most useful test for identifying the cause is the measurement of urine osmolality.

Primary water loss

Human beings cannot survive more than a few days without water intake. This comes mainly from drinks, but some foods also contain significant amounts of water. There is a continual loss of free water by evaporation from the lungs and skin, termed insensible loss. In temperate climates this may be around 800 mL per day. A further small loss occurs in the stools. Water losses are therefore accelerated by hot climates, fever, mechanical ventilation, and diarrhea. If these are not replaced by drinking fluids, there will be a gradual rise in plasma sodium and osmolality. This results in secretion of ADH, causing a small urine volume of very high osmolality (up to 1000 or even 1500 mmol/kg). Thus, anyone who is deprived of water due to lack of provision or physical inability to drink will

CASE 5.8

A 75-year-old man is on an orthopedic ward following elective surgery. He also has chronic bipolar disorder.

	SI units	Reference range	Conventional units	Reference range
Plasma				
Sodium	155 mmol/L	136–142	155 mEq/L	136–142
Potassium	3.0 mmol/L	3.5–5.0	3.0 mEq/L	3.5–5.0
Urea (BUN)	25 mmol/L	2.9–8.2	70 mg/dL	8–23
Creatinine	220 μmol/L	53–106	2.49 mg/dL	0.6–1.2
Osmolality	345 mmol/kg	275–295	345 mOsm/kg	275–295
Urine				
Sodium	5 mmol/L	Not applicable*	5 mEq/L	Not applicable*
Osmolality	200 mmol/kg	Not applicable*	200 mOsm/kg	Not applicable*

*Urine sodium and osmolality are interpreted as whether appropriate for plasma results and clinical status.

- How does this differ from Case 5.7?
- Is the urine osmolality high?
- Why is the urinary sodium concentration low?
- Is this patient's total body sodium likely to be increased, normal, or low?

The urine osmolality is inappropriately low in the presence of hypernatremia, indicating a defect in secretion or action of ADH (diabetes insipidus; see Figure 5.4). This is nephrogenic diabetes insipidus due to lithium treatment. Chronic treatment with lithium makes the collecting ducts unresponsive to ADH, and hence large volumes of dilute urine are passed. These must be replaced by an equivalent amount of oral fluid or hypernatremia will occur. The results are identical to Case 5.7 except the urine osmolality is inappropriately low (<750 mmol/kg). Total body sodium is normal to low, but total body water is very low.

develop hypernatremia. This does not just occur in desert explorers: cases of fatal hypernatremic dehydration in otherwise healthy patients are reported in twenty-first-century hospitals in developed countries, due to poor medical and nursing care. Patients at particular risk are those who are immobile, unable to swallow, or unable to voice their thirst. This often means the elderly in hot weather.

Hypernatremia may also occur when the secretion of ADH is impaired, or there is an inability of the collecting ducts to respond to it. A large volume of hypotonic urine is excreted, stimulating thirst. This is termed diabetes insipidus (DI) and may be divided into central (or cranial) DI (lack of secretion) or nephrogenic DI (lack of effect). The key distinguishing features from inadequate water intake are polyuria, as long as the patient has access to fluid, and an inappropriately low urine osmolality (that is, <750 mmol/kg) (**Figure 5.4**).

The final mechanism for excess water loss is osmotic diuresis, most commonly due to severe glycosuria in uncontrolled diabetes mellitus. The proximal tubular mechanism for glucose reabsorption is saturated, leading to increased glucose in the collecting ducts, which lowers the osmotic gradient between the tubular fluid and renal medulla (see Chapter 4). Thus, even with maximal

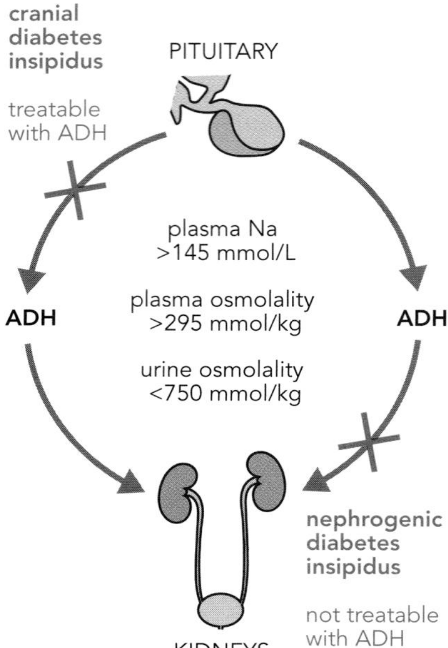

Figure 5.4 Types of diabetes insipidus. Diabetes insipidus (DI) may be due to either lack of ADH secretion from the posterior pituitary (left) or normal secretion but lack of action on the collecting ducts of the kidney (right). These are termed cranial (or central) DI and nephrogenic DI, respectively. The expected plasma and urine measurements, as shown, are the same in both conditions: they are distinguished by their clinical context, response to administration of ADH (in the form of arginine vasopressin or AVP), and occasionally by measurement of plasma ADH concentration.

CASE 5.9

A 79-year-old man has been unwell for a week with severe thirst and polyuria. He has been drinking large quantities of orange juice..

	SI units	Reference range	Conventional units	Reference range
Plasma				
Sodium	175 mmol/L	136–142	175 mEq/L	136–142
Potassium	4.0 mmol/L	3.5–5.0	4.0 mEq/L	3.5–5.0
Urea (BUN)	15 mmol/L	2.9–8.2	49 mg/dL	8–23
Creatinine	200 μmol/L	53–106	2.26 mg/dL	0.6–1.2
Osmolality	425 mmol/kg	275–295	425 mOsm/kg	275–295
Urine				
Sodium	5 mmol/L	Not applicable*	5 mEq/L	Not applicable*
Osmolality	810 mmol/kg	Not applicable*	810 mOsm/kg	Not applicable*
Glucose	+++ on dipstick	Negative	+++ on dipstick	Negative

*Urine sodium and osmolality are interpreted as whether appropriate for plasma results and clinical status.

- **Is the total body sodium likely to be increased, normal, or decreased?**
- **Why is the plasma osmolality so high?**
- **Why is the urinary osmolality not higher?**

This is a hyperglycemic, hyperosmolar non-ketotic state, as seen in type 2 diabetes mellitus. The severe hyperglycemia causes osmotic diuresis and a large water deficit, with severe thirst being stimulated. Consumption of sugary drinks, including fruit juices, aggravates the hyperglycemia and worsens the polyuria. The glucose in the glomerular filtrate overwhelms the proximal tubular reabsorption and is present in the excreted urine; hence the osmolality is similar to plasma. Urine cannot be concentrated any further, as despite maximal ADH secretion and opening of water channels in the collecting ducts, the osmolar gradient between the lumen and the renal medulla is reduced. Loss of water affects all fluid compartments, with the major volume deficit being in the ICF space.

ADH-driven opening of the water channels, little water is reabsorbed. Urine osmolality in this situation is similar to that of the plasma, which itself is higher than usual because of hyperglycemia. The most severe form of this is seen with hyperosmolar non-ketotic states in type 2 diabetes.

Clinical practice point 5.5

Urine osmolality is the most useful test for distinguishing causes of hypernatremia.

Primary salt gain

Gain of salt is an uncommon cause of hypernatremia. When it does occur, it is usually either iatrogenic (due to intravenous fluids containing high concentrations of sodium) or deliberate, due to child or elder abuse. Iatrogenic hypernatremia may occur when hypertonic saline is used to treat acute hyponatremia, or sodium bicarbonate is used to treat acidosis. Some antibiotics are sodium salts and the sodium content may be significant when given to small infants. Deliberate poisoning with salt is recognized, and has resulted in prosecutions of parents and carers, but it is often difficult to prove. To do so requires serial measurements of plasma and urine electrolytes and body weight, which falls rapidly as the kidneys remove the excess sodium and water.

POTASSIUM

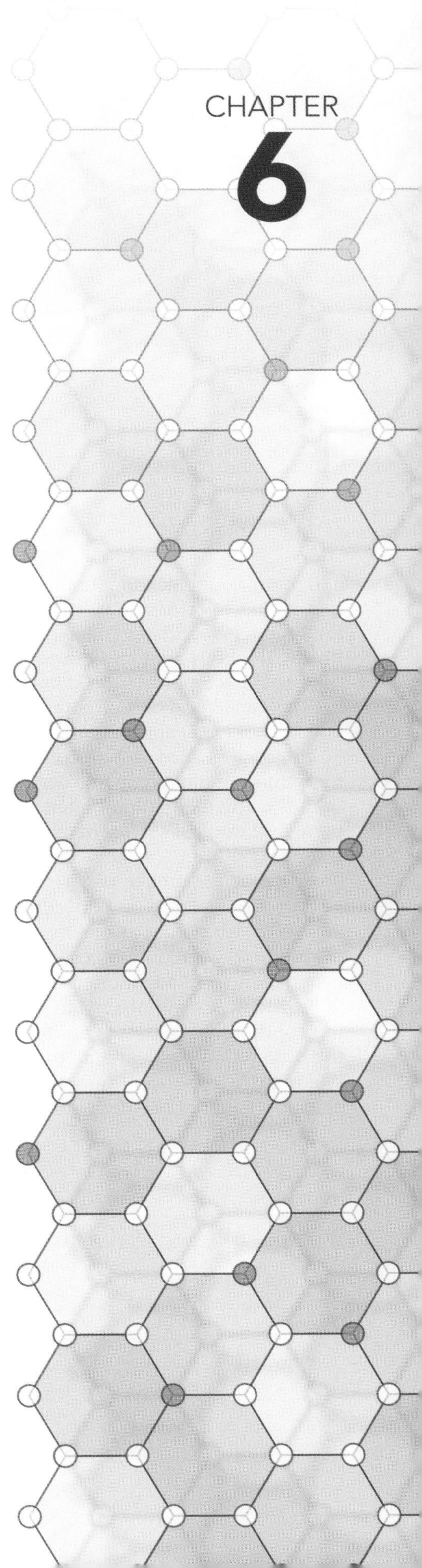

IN THIS CHAPTER

POTASSIUM BALANCE

HYPOKALEMIA

HYPERKALEMIA

Potassium and sodium are the major cations in body water, making significant contributions to the osmotic potential of the different fluid compartments that make up total body water. Throughout this chapter, potassium and sodium will imply their ionic species. The two cations have different roles in the body; sodium is involved particularly in the maintenance of extracellular fluid (ECF) volume while potassium plays a major role in neuromuscular activity. Reference ranges for potassium and related analytes are shown in **Table 6.1**.

The distribution of the two cations between the fluid compartments in the body is also quite different. The bulk (98%) of potassium is in intracellular compartments whereas most of the sodium is present in extracellular fluid compartments. Total body potassium is related to lean body mass rather than gross body weight and is approximately 10% higher in males than females (males ~45 mmol/kg; females ~40 mmol/kg). The approximate distribution of potassium in fluid compartments is shown in **Figure 6.1**. An assumed total body water volume of 42 L includes 12 L of ECF comprising 3 L of intravascular and 9 L of interstitial fluids. The homeostatic potassium concentration in these ECF compartments is between 3.6 and 5.0 mmol/L and contrasts with a sodium concentration of 136–142 mmol/L. The largest fraction of body water (23 L) is in intracellular fluid (ICF) and the concentration difference between potassium and sodium here is the reverse of what is seen in ECF; that is, potassium is

TABLE 6.1 Reference ranges for potassium and related analytes in serum and plasma

Analyte	Reference range in serum or plasma	
	SI units	Conventional units
Potassium	3.5–5.0 mmol/L	3.5–5.0 mEq/L
Sodium	136–142 mmol/L	136–142 mEq/L
Urea	2.9–8.2 mmol/L	8–23 mg/dL (expressed as urea nitrogen)
Creatinine	53–106 μmol/L (varies with age and sex)	0.6–1.2 mg/dL

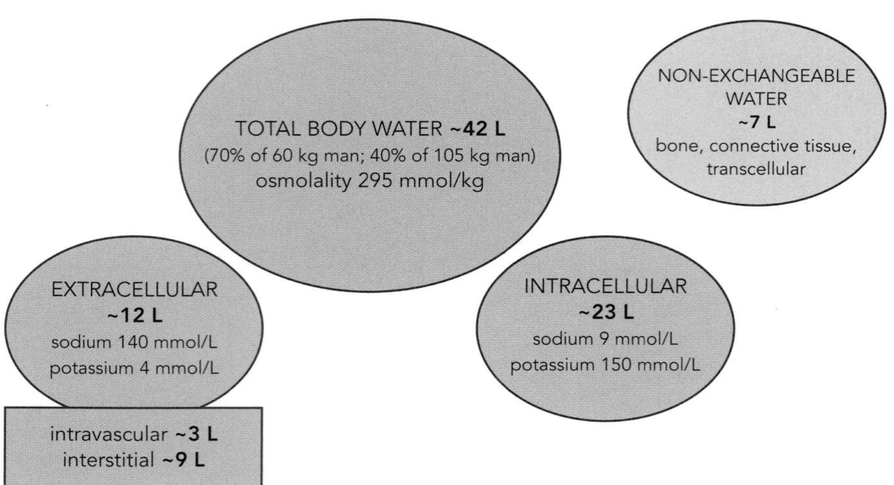

Figure 6.1 Fluid compartments in the body, showing the differential distribution of potassium. Sodium and potassium are mainly found as ions in water, which adds up to approximately 42 L in an adult male. The concentration of potassium is much higher within cells compared to the extracellular compartments—so much so, that 95% of the body's potassium is intracellular; the opposite to sodium. SI units are shown on the diagram; conventional units for sodium and potassium are mEq/L, and are mOsm/kg for osmolality. The actual amounts are identical using either SI or conventional units.

150 mmol/L and sodium 9 mmol/L. A further 6 L of water is found in the bone and connective tissue, with the remaining 1 L found in transcellular spaces. Potassium is found in varying concentrations with these two compartments. In health, the body content of the two cations is controlled mainly by modulation of urinary excretion via the kidney by aldosterone (see Chapter 4).

The concentrations of potassium in ICF and ECF imply that approximately 4000 mmol (98%) appear in the intracellular compartments and bone, while just 70 mmol (less than 2%) are present in extracellular fluids. The major effector mechanism maintaining this concentration difference across cell plasma membranes is the membrane-bound Na^+/K^+-ATPase (sodium pump) which actively pumps K^+ into cells in exchange for Na^+ to counterbalance a passive loss of K^+ from cells into ECF. Insulin and other drugs activate the sodium pump to promote the active uptake of K^+ into cells, while changes in the pH of ECF affect passive movement of K^+ between ECF and ICF, as discussed later.

It is worth noting at this point the distribution of potassium in body fluids (**Table 6.2**), since changes in any of these may contribute to hypo- and hyperkalemia. As mentioned earlier, the potassium concentration of plasma is approximately 4 mmol/L. That of whole blood is about 70 mmol/L and reflects the contribution of potassium from intracellular fluid (mainly red blood cells); the ratio of intracellular fluid to extracellular fluid in whole blood is about 1:1. The potassium concentration of bile and pancreatic juice is similar to that of plasma. Higher concentrations are found in gastric juice and sweat but, under

TABLE 6.2 Concentration of potassium in body fluids

Fluid	Approximate concentration	
	SI units	Conventional units
Plasma	4 mmol/L	4 mEq/L
Gastric juice	10 mmol/L	10 mEq/L
Bile, pancreatic juice	5 mmol/L	5 mEq/L
Sweat	10 mmol/L	10 mEq/L
Diarrhea	40 mmol/L	40 mEq/L

normal physiological conditions, losses of potassium via these routes are trivial. Also, little potassium is found in feces normally, but under conditions such as diarrhea associated with cholera, for example, this represents a pathological loss.

6.1 POTASSIUM BALANCE

A healthy individual on a normal Western diet is said to be in potassium balance whereby a daily intake of 60–150 mmol is balanced by an equivalent daily output, mainly via the urine. Enriched dietary sources of potassium are bananas, citrus fruits, watermelons, milk, and tomatoes, but as all animal and vegetable cells contain potassium, most if not all foodstuffs are sources of the cation. As implied above, the major route of potassium excretion is via the kidneys in urine, with only rather trivial amounts being excreted via the feces and skin. Potassium is filtered through the glomerulus and reabsorbed completely in the proximal tubules. It is excreted (secreted) into urine through the distal tubules linked with H^+ in exchange for sodium in a process stimulated by aldosterone. This is described in more detail in Chapter 4.

The maintenance of ECF potassium in the narrow range of 3.5–5.3 mmol/L (mEq/L) is vital to health for the maintenance of muscular function; this is illustrated perhaps best by the distinctive changes in the electrocardiogram (ECG) in hypo- and hyperkalemia (**Figure 6.2**), which reflect clinical changes in the way the heart works. *In extremis*, hyperkalemia can induce ventricular fibrillation and cardiac arrest, while hypokalemia results in pronounced muscle weakness.

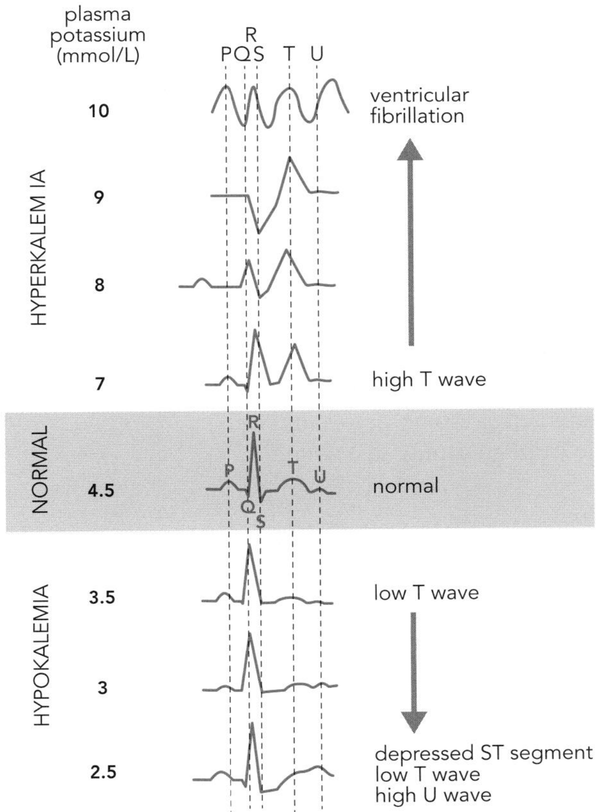

Figure 6.2 Potassium and the heart. The figure shows the effect of plasma potassium on the heart as reflected by changes in the ECG.

6.2 HYPOKALEMIA

When persistent hypokalemia is caused by a loss in potassium from the body, the actual body deficit can be very high. For example, a fall in plasma potassium below the reference range to1.5 mmol/L could reflect a loss of more than 1000 mmol of potassium from the body, equating to approximately 300 mmol per mmol fall in plasma potassium (**Figure 6.3**).

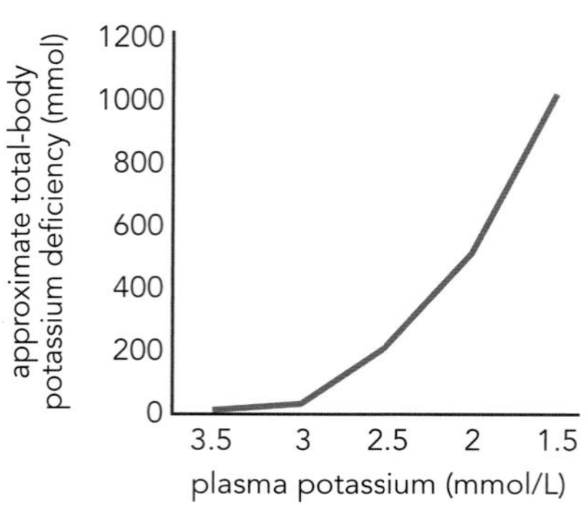

Figure 6.3 Whole-body potassium in hypokalemia.
Since the vast majority of potassium is intracellular, the low plasma potassium levels in hypokalemia (due to potassium deficiency) reflect very large deficits in total body potassium. Thus a plasma potassium of less than 2.0 mmol/L (2.0 mEq/L) may reflect a deficit of 1000 mmol.

There are a myriad of causes of hypokalemia, the more well-known ones being:

- Intravenous insulin administration
- Steroid use
- Laxative use
- Diarrhea
- Renal loss due to different causes, including diuretics
- Treatment of diabetic ketoacidosis
- Secondary to hypomagnesemia
- Cushing's syndrome
- Renal tubular acidosis (type I)

Most of the different causes arise from one or a combination of the following:

- Inadequate intake
- Excessive output, usually via the kidneys
- A shift from ECF to ICF

Since in the distal tubule sodium is reabsorbed in exchange for either potassium or hydrogen ions, the acid–base status of the patient can cause the plasma potassium to rise or fall. In alkalosis, where the concentration of hydrogen ions is low, potassium will be preferentially secreted, and therefore there will normally be a tendency toward hypokalemia. In acidosis, there is a high concentration of hydrogen ions, therefore hydrogen ions will be preferentially secreted, leading to hyperkalemia.

Inadequate intake

The major causes of inadequate intake can be summarized as:

- Prolonged starvation or dysphagia
- Prolonged vomiting
- Potassium-free or potassium-low fluids (oral or parenteral)

This list includes patients who have undergone prolonged starvation or who are unable to ingest oral fluids or solids (dysphagia). In the case of excessive vomiting, initially ingested potassium is obviously not available for absorption.

Other causes may be administration of intravenous fluids low in potassium or, the classical artifact, due to taking a blood sample from an arm with a saline or low-potassium drip.

Excessive output

Potassium is lost via the gut and particularly the urine. Increased losses via the gastrointestinal tract may result from persistent vomiting and diarrhea. Some potassium is lost in the vomit itself. However, these direct losses are usually quite small as gastric juice only contains up to 10 mmol/L potassium. Since there is a loss of hydrogen ion in vomit, there will be a tendency toward alkalosis, and hence hypokalemia. Furthermore, since vomiting is often associated with dehydration, the renin–angiotensin–aldosterone system will be activated in order to maintain sodium and fluid (secondary hyperaldosteronism), leading to more loss of potassium in exchange for sodium ions, further exacerbating the hypokalemia.

TABLE 6.3 Causes of excessive loss of potassium that can lead to hypokalemia

Site of loss	Mechanism
Gastrointestinal (GI) tract (all GI tract causes have renal secondary loss)	Vomiting
	Diarrhea
	Fistula
Urine	Drugs, especially diuretics and steroids
	Adrenal-linked: Cushing's syndrome Ectopic ACTH production Conn's syndrome
	Renal damage: Recovery from acute kidney injury Renal tubular acidosis Hypomagnesemia

ACTH, adrenocorticotropic hormone.

The major causes of excessive renal potassium loss are mostly due to factors that modulate secretion of the cation in the distal tubules.

Since the concentration of potassium in diarrhea can be quite high (30–40 mmol/L) compared to normal feces (**Table 6.3**), the direct loss of potassium is much higher in diarrhea than vomiting. Furthermore, since the sheer volume of fluid lost in patients with diarrhea can be high (4–5 L/day in acute cholera),

CASE 6.1

A 73-year-old woman presents to the emergency department after fainting at her daughter's house. She has a five-day history of diarrhea, possibly caused by food poisoning. Her long-term medication for mild hypertension includes a calcium-channel blocker (nifedipine) and a diuretic (bendroflumethiazide). Her results were as follows:

	SI units	Reference range	Conventional units	Reference range
Plasma				
Sodium	142 mmol/L	136–142	142 mEq/L	136–142
Potassium	2.5 mmol/L	3.5–5.0	2.5 mEq/L	3.5–5.0
Urea (blood urea nitrogen, BUN)	11.3 mmol/L	2.9–8.2	32 mg/dL	8–23
Creatinine	133 μmol/L	53–106	1.74 mg/dL	0.6–1.2

- **What is the significance of her urea and creatinine results?**
- **What are the contributing factors to her hypokalemia?**

On examination, the patient was clinically dehydrated, which would explain the urea and creatinine results, and this was mainly caused by the recent history of diarrhea. The hypokalemia is probably due to a combination of factors. First, she has been on a long-term thiazide diuretic, bendroflumethiazide, which causes potassium loss via the kidney, making patients potassium deficient. She then had a severe episode of diarrhea, which would cause further rapid potassium loss. Although some potassium is lost directly from the gut, the major source of loss is the kidney due to activation of the renin–angiotensin–aldosterone system in response to loss of extracellular fluid volume. Since she was already deficient in potassium, this extra loss would lead to profound hypokalemia, perhaps requiring intravenous replacement of potassium.

this will lead to severe hypokalemia. As with vomiting, diarrhea is often associated with dehydration, and the resulting secondary hyperaldosteronism will worsen the hypokalemia. Pancreatic cholera is the trivial name given to the condition caused by a rare pancreatic islet-cell tumor which secretes vasoactive intestinal polypeptide and also gives rise to a marked diarrheal potassium loss (Verner–Morrison syndrome). Excessive use of laxatives, a small but significant problem in weight-conscious societies, also promotes hypokalemia.

Excessive losses of potassium via the urine often arise from the use of diuretics that indirectly inhibit its reabsorption. The increased delivery of sodium to the distal tubules resulting from the action of diuretics prior to this site gives rise to a secondary hyperaldosteronism by stimulation of the renin–angiotensin system and release of aldosterone from the adrenal cortex with consequent increased reabsorption of sodium at the expense of potassium. In a similar manner, genetic disorders (such as Bartter's syndrome, a family of congenital disorders of chloride transport in the loop of Henle) that result in an increased delivery of sodium chloride to the renal collecting ducts also give rise to hypokalemia. Here, volume depletion (dehydration) stimulates the renin–angiotensin system and a secondary loss of potassium due to a rise in aldosterone. The main features of Bartter's syndrome are:

- Variable severity and age of diagnosis
- Secondary increase in renin and aldosterone
- Excessive renal loss of potassium
- Hypokalemia
- Alkalosis
- Normal blood pressure

Similar results are seen in patients with laxative and/or diuretic abuse. Less severe hypokalemia can be seen in the related inherited disorder, Gitelman's syndrome.

Increased endogenous production of mineralocorticoids as seen in Conn's syndrome, a tumor in the adrenal cortex producing aldosterone, can produce hypokalemia and usually hypertension. In health, the renal 11β-hydroxylase enzyme converts cortisol (which can bind to the cellular aldosterone receptor) to cortisone (which cannot bind to the cellular receptor) thereby stopping glucocorticoids from having mineralocorticoid effects. However, in disease states of excess glucocorticoid, such as Cushing's syndrome and particularly with ectopic production of adrenocorticotropic hormone (ACTH), the protective effect of the renal 11β-hydroxylase can be overwhelmed and a mineralocorticoid effect will occur, leading to hypokalemia. Interestingly, licorice extract produced in Europe contains glycyrrhizic acid, a molecule which inhibits the renal 11β-hydroxylase enzyme, and chronic ingestion of licorice induces a syndrome similar to endogenous hyperaldosteronism (pseudohyperaldosteronism).

There are several disorders in which there is an underlying loss of potassium but patients do not present initially with hypokalemia. For example, in diabetic ketoacidosis, patients frequently present with hyperkalemia, even though their body stores of potassium are often low due to renal losses from the osmotic diuresis. However, when the patient is treated with insulin, the potassium-deficient state will become unmasked and, unless the patient is given extra potassium, hypokalemia will result.

As discussed earlier, alkalosis is usually associated with hypokalemia. If a patient presents with acidosis and hypokalemia, a diagnosis of classical renal tubular acidosis is probable, where the normal secretion of hydrogen and potassium ions is out of balance. Hypokalemia also occurs in patients recovering from acute renal failure. Finally, because magnesium is required for the renal reabsorption of potassium, hypomagnesemia will lead to a refractory hypokalemia that can be reversed when the patient is made magnesium replete.

CASE 6.2

A 45-year-old woman has presented to her doctor on three occasions with marked hypertension (highest recorded value was 175/115 mmHg) over a twelve-month period. In between the hypertension episodes her blood pressure is normal. The doctor measures some basic blood tests during one of the hypertensive periods. The results were as follows:

	SI units	Reference range	Conventional units	Reference range
Plasma				
Sodium	143 mmol/L	136–142	143 mEq/L	136–142
Potassium	2.3 mmol/L	3.5–5.0	2.3 mEq/L	3.5–5.0
Urea (BUN)	2.1 mmol/L	2.9–8.2	5.9 mg/dL	8–23
Creatinine	50 μmol/L	53–106	0.57 mg/dL	0.6–1.2

When the patient was not hypertensive her results were within the reference ranges.

- **Aside from the profound hypertension, what are the two important features of this case?**
- **What further testing may be required?**

This case is unusual in that the symptoms are intermittent, and the marked hypokalemia only occurs when the patient is severely hypertensive. This picture could be explained by the patient having a mineralocorticoid-secreting tumor, such as Conn's syndrome, which is intermittently secreting large amounts of aldosterone. This would lead to sodium retention, urinary potassium loss, and hypertension. Plasma aldosterone would be raised, whilst the plasma renin activity would be suppressed. Alternatively, the patient could be taking food or drugs on certain occasions. On further investigation, it was revealed that her symptoms and hypokalemia coincided with visits to her aunt who ran an old-fashioned sweet shop. The patient enjoyed chewing licorice root, which has a high concentration of glycyrrhizic acid, which causes sodium retention and potassium loss, mimicking mineralocorticoid hypertension. This is called pseudohyperaldosteronism since aldosterone is appropriately suppressed rather than raised. It is worthwhile noting that although hypokalemia is usually always present in such cases, the plasma sodium is usually normal or slightly raised.

Shift of potassium from extracellular fluid to intracellular fluid

An apparent hypokalemia can occur even when patients do not have a total potassium deficit, because potassium from extracellular fluid has been increasingly pumped into cells. Causes include:

- Insulin administration
- β-adrenoceptor agonists (for example, epinephrine or salbutamol)
- Alkalosis
- Familial periodic paralysis

In the case of insulin, the increased uptake of glucose provides a substrate for the generation of ATP to drive the cell-membrane ion pumps. Furthermore, as discussed in Chapter 7, in patients with alkalosis where there is a decreased plasma hydrogen-ion concentration, potassium replaces H^+ in exchange pumps. The result of this in the kidney can result in the loss of potassium to the urine.

Familial periodic paralysis is a rare autosomal dominant disorder, more common in males, in which potassium shifts from ECF to ICF via an as-yet unknown mechanism. Paralysis arises when potassium flows into muscle cells making them electrically inexcitable. Periods of paralysis last up to 24 hours and are brought on by physical and mental stress, alcohol, sleep, and carbohydrate- and sodium-enriched diets.

In many patients, potassium depletion is due to both inadequate intake and excessive output.

Clinical features of hypokalemia

In contrast to hyponatremia, where the major clinical changes are neurological, the major features of hypokalemia are predominantly muscular (**Table 6.4**). This is particularly true for cardiac muscle in which generalized muscle weakness can lead to cardiac arrest with decreasing plasma potassium. Classical ECG changes associated with hypokalemia (**Figure 6.4**) are a depressed ST wave, a flattened T wave, the appearance of a U wave, and a prolonged QT interval. An enhanced response to digoxin is also a feature of hypokalemia and should

TABLE 6.4 Clinical features of hypokalemia

Site or system	Clinical features
Cardiovascular system	ECG changes, for example prolonged QT interval
	Enhanced digoxin effect
	Cardiac arrest
Muscles	General muscle weakness
	Atony of the bowel
Kidney	Polyuria (ADH resistance)
	Alkalemia with acid urine (except with renal tubular acidosis)

ADH, antidiuretic hormone.

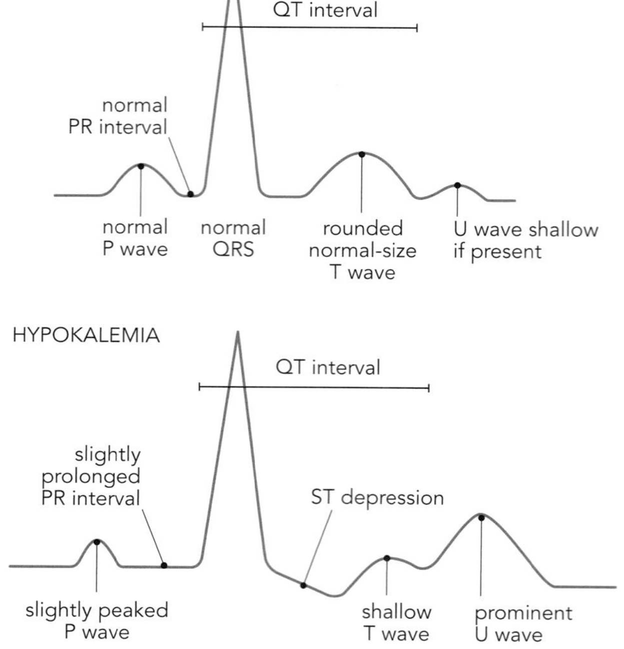

Figure 6.4 ECG changes in hypokalemia.
Typical changes that can be seen in a patient with severe hypokalemia, namely depression of the ST segment, a flattened T wave, and a raised U wave. In some cases, the appearance of the prominent U wave can merge with the T wave giving the impression of a more prolonged QT wave.

be borne in mind when treating patients who are being treated with this drug. The effects of digoxin may become toxic as plasma potassium falls and regular monitoring of plasma potassium is essential. Further manifestations of generalized weakness in skeletal muscle are atony of the bowel and myopathy.

In the kidney, hypokalemia induces resistance to antidiuretic hormone (ADH), giving rise to polyuria. Alkalemia with accompanying aciduria may result from increased urinary excretion of hydrogen ions, which replace potassium as the counterion in the renal reabsorption of sodium as plasma potassium concentration falls (see Chapter 4).

Clinical practice point 6.1

The greatest risk of cardiac arrhythmias is in older people and those with heart disease. However, young people are also at risk of sudden death from cardiac arrest, particularly when hypokalemia is a feature of an eating disorder with covert vomiting and/or laxative misuse.

Treatment of hypokalemia

In general, the underlying cause of hypokalemia should be addressed and potassium replaced. This should be done slowly and oral replacement is the safest method to prevent hyperkalemia, which may result from giving too much potassium too fast. If potassium is to be administered intravenously, the maximum rate of delivery should not exceed 20 mmol/h from pre-mixed potassium solutions of 20, 27, or 40 mmol/L (0.15, 0.2, and 0.3%, respectively). If patients are not in renal failure, approximately 50% of the potassium administered will be lost in the urine while most of the potassium retained by the body will be taken up into cells. In many cases where the amount of potassium to be replenished in the body may be several hundred mmol, treatment may take several days and plasma potassium should be monitored routinely at least once a day.

Summertime pseudohypokalemia

Although artifactual causes of hyperkalemia are well known, for example storage of samples in the cold, spurious causes of hypokalemia are not. The ideal temperature to store whole blood samples for potassium analysis is about 20°C, where the sodium/potassium pump can maintain the intracellular and extracellular potassium concentrations. In cold temperatures the pump does not work as effectively, leading to artifactually raised plasma potassium concentrations. This can be achieved also by mistakenly refrigerating samples, or transporting samples to the laboratory in a cold van. However, when samples are stored in ambient temperatures of >30°C, the sodium/potassium pump works harder and, since the blood sample is an isolated system, more potassium is pumped into the cells than leaks out. This gives rise to an artifactual hypokalemia, a temporary effect, which will be reversed after approximately five hours as the endogenous glucose is used up in the sample; at this time, an artifactual hyperkalemia will ensue.

This so-called summertime pseudohypokalemia is a very strong effect, and can be seen when comparing potassium results from the community with those from hospital patients (**Figure 6.5**). There is such a strong correlation with

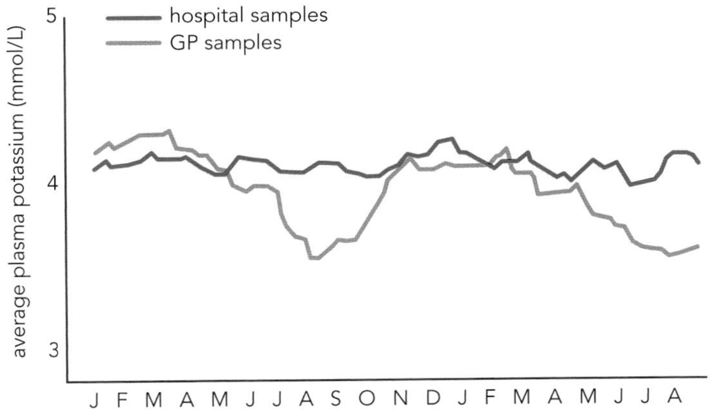

Figure 6.5 Measured potassium concentration in plasma derived from patients in the community (GP samples) and hospital settings. These figures were collected over an 18-month period, and clearly show the average community plasma potassium falling in the hot summer months compared with the rest of the year. Hospital samples did not show this trend, which was caused by the delay in community samples arriving in the laboratory before centrifugation, which in hot weather can cause artifactual lowered plasma potassium, or pseudohypokalemia.

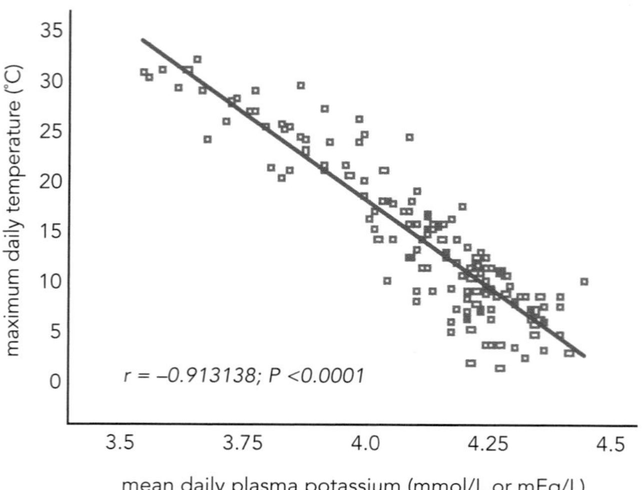

Figure 6.6 The effect of ambient temperature on plasma potassium concentration in blood samples taken in the community. The correlation between the maximum ambient daily temperature and the mean daily plasma potassium concentration is so significant that it is possible to predict the daily maximum temperature from the mean plasma potassium and vice versa.

maximum daily temperature that the peak temperature for any day can be calculated with reasonable accuracy from the mean potassium concentration of blood samples collected from the community (**Figure 6.6**).

6.3 HYPERKALEMIA

A condition of hyperkalemia occurs when plasma potassium concentration rises above 5.5 mmol/L (mEq/L). Because of the high concentration of intracellular potassium, hyperkalemia may arise with little change in the amount of total body potassium when there is movement of potassium from intracellular to extracellular fluid. The major causes of hyperkalemia can be divided into three main types:

- Artifactual
- Impaired excretion of potassium
- Increased input of potassium into the ECF

Artifactual hyperkalemia

Artifactual causes of hyperkalemia are very common, especially in blood samples taken in the community, away from the hospital laboratory, where they are probably the commonest reason for reporting hyperkalemia. Causes of artifactual hyperkalemia include:

- EDTA contamination
- Delayed separation of blood
- Storage of samples in the cold
- Hemolysis *in vitro*
- Drip-arm artifact
- Lysis of white blood cells and platelets *in vitro*

An obvious instance of artifactual hyperkalemia is the use of an inappropriate blood collection tube: potassium-EDTA tubes used primarily for determining blood hematology parameters. The potassium within the tube will artifactually raise the measured potassium value. The effect is so large that it can be seen even if the needle used in taking the blood sample is contaminated with a few drops of blood that have been in the EDTA tube. However, with a small degree of contamination, the resultant artifactually raised potassium result can be within a clinically possible range; therefore such artifacts may not be spotted and

Analytical practice point 6.1

Hyperkalemia may be fatal, but the commonest causes of high potassium results are artifactual. This is a particular problem for samples taken some distance from the laboratory or where transit times are over 4 hours. The clinical dilemma is whether to admit a patient to hospital unnecessarily or overlook a genuinely high result in the belief that it is an artifact. This is one of the biggest day-to-day problems for centralized laboratories.

inappropriate further testing or treatment may be undertaken. This artifact can be readily unearthed by simply measuring the plasma calcium, which binds very tightly to EDTA, giving rise to physiologically impossible results (<0.2 mmol/L or mEq/L).

Potassium is measured in plasma and any deterioration in sample quality before assay gives rise to high values. The high ICF potassium concentration in red blood cells is maintained by the membrane Na^+/K^+-ATPase pump, which requires ATP derived from glycolysis. If a sample is left at an ambient temperature of 20°C for 6–8 hours prior to assay, the glucose inside the red cell will become exhausted; the sodium/potassium pump in red blood cells then no longer works and potassium leaks out of the cells. It is thus important to separate plasma from cells at the earliest opportunity.

Temperature is also an important factor when storing blood samples prior to separation. As discussed earlier, artifactual hypokalemia can be found when storing samples at high temperatures. Artifactual hyperkalemia from storing samples at cold temperatures is a far more common finding. The colder the temperature, the more marked the effect. As samples are stored in the cold for prolonged periods, metabolism is slowed in the cells and the provision of ATP from glycolysis is insufficient to satisfy the sodium/potassium pump and potassium leakage will occur. The sodium/potassium pump decreases in activity by approximately 10% per degree fall in temperature from 37°C. In effect, at temperatures below 8°C, the sodium/potassium pump is switched off, leading to artifactual hyperkalemia. This effect is seen in wintertime and also from inappropriately refrigerating samples before they have been separated. Never cool blood samples intended for potassium analysis before they have been separated, and obviously do not freeze whole blood samples.

Any situation in which red cells are damaged (*in vitro* hemolysis) after sampling will yield a high potassium measurement, as intracellular potassium is released into the plasma. In this case, the plasma appears bright red on centrifugation due to the simultaneous release of hemoglobin from the cells. This should be distinguished from cases of real *in vivo* hemolysis, as seen in patients with parasitic infections such as malaria, who present with hyperkalemia. Here, the plasma has the red-brown appearance of cold tea due to partially processed hemoglobin (**Figure 6.7**).

Blood samples taken from a drip arm of a patient who is receiving parenteral potassium will obviously produce an artifactually raised potassium.

Finally, besides red blood cells, a patient's blood sample will contain white cells and platelets, and conditions in which these cells are damaged, such as the fragile white cells of early leukemia, or the presence of excessive platelets, will also produce spurious hyperkalemia. Artifactual hyperkalemia resulting from platelets and white cells breaking during processing is more likely to be encountered with clotted samples, especially when a barrier gel is used. To avoid this, whole blood potassium should be measured on a direct ion-selective electrode using a lithium heparin sample.

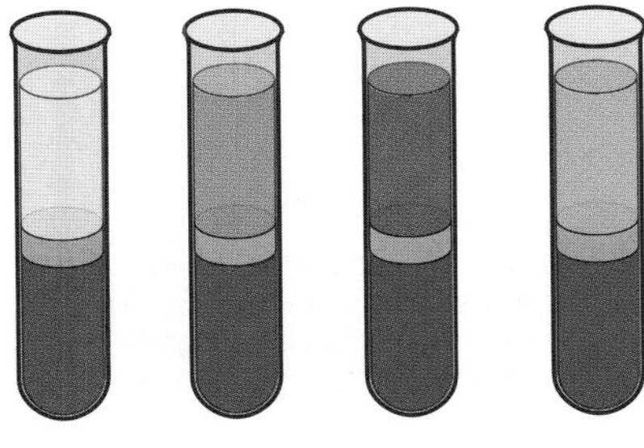

Figure 6.7 Hemolysis *in vivo* and *in vitro*.
Samples hemolyzed *in vitro* are red, samples hemolyzed *in vivo* are reddish brown, and jaundiced samples are brown.

CASE 6.3

A 37-year-old man is admitted to the hospital with a fever and some routine tests are performed on the patient. The clinical biochemistry laboratory reports the following results:

	SI units	Reference range	Conventional units	Reference range
Plasma				
Sodium	140 mmol/L	136–142	140 mEq/L	136–142
Potassium	Sample hemolyzed and unsuitable for analysis			

- **Why are some samples hemolyzed?**
- **Is hemolysis always due to poor venipuncture technique?**
- **Is it ever appropriate to report potassium results on hemolyzed samples?**

In hemolyzed samples, potassium is artifactually raised due to leakage of the cation from the damaged red blood cells and this is not uncommon in routine laboratory samples. Damage to red blood cells can occur during the blood-taking process (for example, by using a needle of too-fine bore) or afterward (for example, by shaking the sample or excessive centrifugation speed). In these situations, the plasma supernatant will look red after centrifugation as free hemoglobin does not spin down with the cells.

The requesting physician was by now aware that the patient had malarial parasites, which can cause intravascular hemolysis; that is, hemolysis *in vivo*. This contrasts with artifactual hemolysis *in vitro*. The laboratory reported the following results:

	SI units	Reference range	Conventional units	Reference range
Plasma				
Sodium	141 mmol/L	136–142	141 mEq/L	136–142
Potassium	Sample hemolyzed and unsuitable for analysis			

During intravascular hemolysis, potassium leaks out of the red blood cells in conjunction with hemoglobin, leading to a true raised potassium as opposed to the falsely raised value associated with hemolysis *in vitro*. In the current case, the patient may indeed have true hyperkalemia.

At first glance, plasma samples from patients with intravascular hemolysis may appear red, but on closer inspection they are more of a brown-red color, classically described as looking like cold tea, due to the presence of methemalbumin. Care must be taken when reporting such results so as not to miss a clinically important diagnosis, and true potential hyperkalemia. The laboratory re-reported the results, showing the patient did have hyperkalemia.

	SI units	Reference range	Conventional units	Reference range
Plasma				
Sodium	141 mmol/L	136–142	141 mEq/L	136–142
Potassium	6.1 mmol/L	3.5–5.0	6.1 mEq/L	3.5–5.0

Impaired excretion of potassium

Clinical situations that result in decreased excretion of potassium via the kidney include:

- Renal failure, especially acute
- Drugs, for example potassium-sparing diuretics or angiotensin-converting enzyme (ACE) inhibitors
- Adrenal failure (Addison's disease)
- Syndrome of hyporeninemic hypoaldosteronism (SHH)

Acute renal failure (acute kidney injury) is a common cause of hyperkalemia and plasma potassium rises as kidney function worsens. Hyperkalemia is not apparent in chronic renal failure until the glomerular filtration rate falls to less than 15 mL/min (chronic kidney disease stage 5, CKD 5); patients at this stage are often severely hyperkalemic prior to dialysis. Potassium-sparing diuretics such as spironolactone also raise plasma potassium by blocking potassium secretion in the kidney.

As described in Chapter 4, aldosterone promotes sodium reabsorption in the renal tubules in exchange for secretion of potassium. Thus, conditions in which aldosterone production is compromised give rise to hyperkalemia. These include administration of ACE inhibitors, which inhibit aldosterone production by blocking the conversion of angiotensin I to angiotensin II (see Chapter 4); adrenal failure (Addison's disease), where the hormone is not synthesized; and hyporeninemic hypoaldosteronism. An overview of drugs that may cause hyperkalemia is given in **Table 6.5**.

TABLE 6.5 Drugs giving rise to hyperkalemia

Action	Drug	Mechanism
Inhibit aldosterone production or action	β-Blockers	Inhibit renin secretion and inhibit β-adrenergic stimulation of cellular potassium uptake
	ACE inhibitors	Prevent angiotensin II formation
	Angiotensin receptor blockers	Prevent angiotensin II action
	Heparin	Inhibits aldosterone synthase
	Aldosterone receptor antagonists (spironolactone)	Inhibits aldosterone action
Inhibit prostaglandin formation	Nonsteroidal anti-inflammatory agents	Prostaglandins required for increased renal potassium secretion
Inhibit potassium transporters	Potassium-sparing diuretics (amiloride, triamterene)	Block potassium exit from collecting duct
	Certain antibiotics (trimethoprim, pentamidine)	Act as potassium-sparing diuretics

Increased input into ECF

The third group of causes of hyperkalemia includes those conditions in which the pumping of potassium from ECF back into ICF is compromised or massive amounts of potassium have been released from cells due to tissue breakdown. Causes include:

- Acidemia
- Crush injury/rhabdomyolysis

- Tumor lysis syndrome
- Gross hemolysis *in vivo*
- Iatrogenic (intravenous fluid)
- Familial hyperkalemic periodic paralysis

Thus in acidemia, hydrogen ions compete with potassium ions for the sodium pump and potassium, which leaks naturally from cells, is not returned to the ICF. The plasma potassium concentration rises approximately 0.6 mmol/L (mEq/L) for a fall in plasma pH of 0.1. Large quantities of potassium from ICF may also be released into plasma as a result of major burns or the trauma of crush injuries, and from muscle-cell damage throughout the body in rhabdomyolysis. Clinical treatment of large tumors with cytotoxic drugs may also result in massive cell damage and the release of potassium into ECF (tumor lysis syndrome).

Gross hemolysis *in vivo* as a result of parasitic infection has been described above. Iatrogenic causes of hyperkalemia include:

- Overzealous intravenous administration of potassium (too-rapid infusion of KCl)
- Administration of KCl to patients being treated with potassium-sparing diuretics
- Transfusions using old blood that has been stored for a long time
- Prolonged treatment of infections with potassium salts of antibiotics

Hyperkalemia is also a feature of a rare autosomal dominant condition called familial hyperkalemic periodic paralysis. This is characterized by sudden increases in plasma potassium, leading to muscle paralysis, particularly after exercise or a potassium-rich meal.

CASE 6.4

A 37-year-old woman with a long history of diabetes is being treated with the antibiotic trimethoprim for recurrent urinary tract infections. Routine blood tests show the following:

	SI units	Reference range	Conventional units	Reference range
Plasma				
Sodium	136 mmol/L	136–142	136 mEq/L	136–142
Potassium	6.0 mmol/L	3.5–5.0	6.0 mEq/L	3.5–5.0
Urea (BUN)	4.1 mmol/L	2.9–8.2	11.5 mg/dL	8–23
Creatinine	87 μmol/L	53–106	1.14 mg/dL	0.6–1.2

- **What is the most likely cause for the hyperkalemia?**

The patient has a mildly raised potassium result, which in many cases turns out to be artifactual (especially from samples taken in the community) for various reasons, including hemolysis *in vitro*, potassium-EDTA contamination, delays in separation, inappropriate refrigeration, and transport at low ambient temperature. In this case the result was a true result, and the most likely cause is trimethoprim, which acts like a potassium-sparing diuretic, leading to hyperkalemia in some individuals. This is more likely to happen if there is preexisting kidney disease, such as diabetic nephropathy, or the patient is already on drugs that tend to raise potassium, such as angiotensin-converting enzyme (ACE) inhibitors.

Clinical features and treatment of hyperkalemia

The classical clinical features of hyperkalemia are cardiac arrhythmias and characteristic changes on ECG, with peaked T waves when plasma potassium is greater than 6.5 mmol/L (mEq/L) due to accelerated repolarization of the cardiac action potential. This is followed by a prolonged PR interval, eventual loss of the P wave, and widening of the QRS complex as potassium rises above 7–8 mmol/L (mEq/L) (**Figure 6.8**). As the potassium concentration rises even further, the ECG pattern becomes sinusoidal and cardiac arrest may occur. Physically, hyperkalemia manifests initially as muscle weakness and if untreated may lead to cardiac arrest and death. In treating hyperkalemia it is essential to distinguish between true and iatrogenic causes and this is aided by examining the ECG. In true hyperkalemia, the immediate aim of treatment is to protect the myocardium by stabilizing the cell membranes against the effect of the hyperkalemia using intravenous calcium ions, and this will occur within a couple of minutes of administration of the infusion. Drugs are then used to promote potassium movement from ECF into ICF and to remove excess potassium by dietary intervention or, in some cases, using binding resins. The treatment of hyperkalemia is summarized in **Table 6.6**. Thus immediate protection of the myocardium by lowering plasma potassium is achieved by (i) driving potassium back into cells, (ii) removal of excess potassium by limiting dietary intake, and (iii) the use of potassium binders such as polystyrene sulfonate resins.

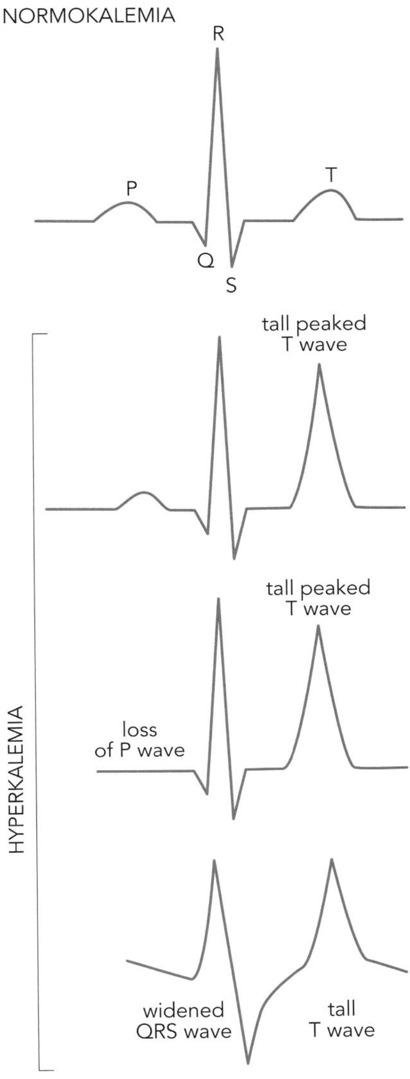

Figure 6.8 ECG changes in hyperkalemia.
Typical abnormalities that can be seen on a hyperkalemic patient's ECG.

TABLE 6.6 Clinical treatment of hyperkalemia

Mechanism	Therapy	Onset	Duration
Antagonize membrane effects	Intravenous calcium	1–2 minutes	30–60 minutes
Drive potassium into cells	Insulin (usually given with glucose)	30 minutes	4–6 hours
	β-adrenoceptor agonists	30 minutes	2–4 hours
Remove potassium from the body	Polystyrene sulfonate resins (potassium ions replaced by sodium ions)	1–2 hours	–
	Dialysis	Immediate but may take 4–6 hours to institute	–
	Diuretics	Hours (may require high doses with chronic hyperkalemia)	–

ACID–BASE DISORDERS

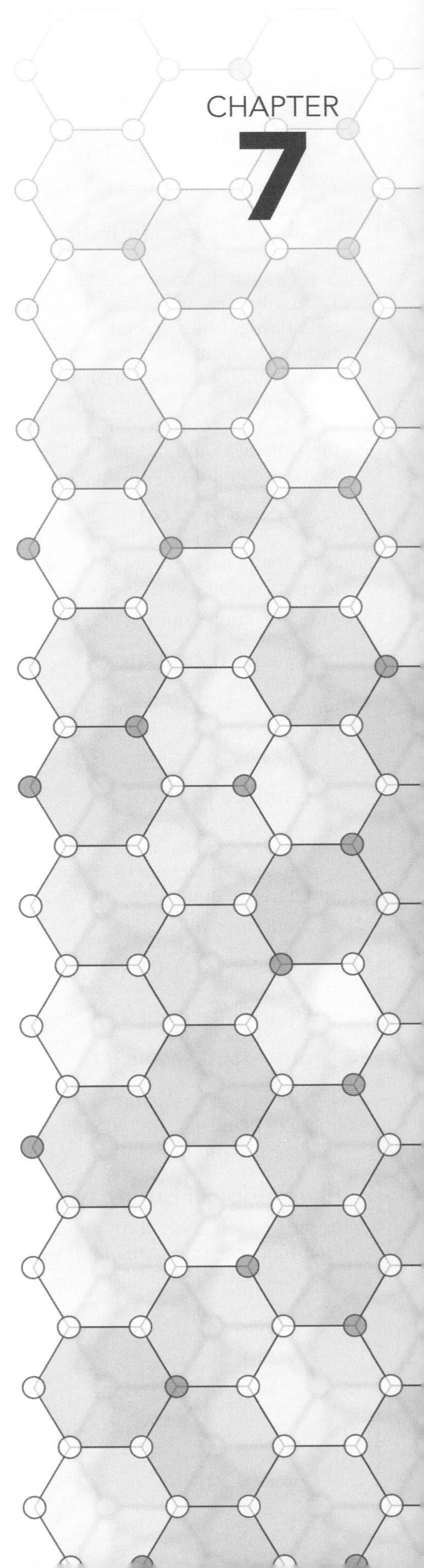

IN THIS CHAPTER

ACIDS AND BASES

HYDROGEN ION HOMEOSTASIS

DISORDERS OF ACID–BASE BALANCE: ACIDOSIS AND ALKALOSIS

Maintaining the concentration of hydrogen ions within narrow limits is of crucial importance to cellular function and health and presents a major challenge to the human body. Metabolism itself generates acid and large amounts must be excreted each day to maintain equilibrium. Both the lungs and the kidneys play important roles in regulating hydrogen ion concentration. Overproduction or underexcretion of acid or base lead to the clinical disorders of acidosis and alkalosis. These are classified as being of metabolic or respiratory origin and can be diagnosed by the measurement of a few simple parameters on a modern blood gas analyzer.

Acid–base disorders may be seen as either primary conditions or as a consequence of organ failure, particularly the lungs and kidneys.

7.1 ACIDS AND BASES

In its most fundamental terms, an acid is defined as a proton (H^+) donor and a base as a proton acceptor, such that any acid–base reaction involves a conjugate acid–base pair. The term alkali applies to a base that is soluble in water and produces hydroxyl ions, OH^-. In the context of biological fluids the two terms can be used interchangeably, although strictly they have slightly different chemical definitions. Any acid has a characteristic affinity for protons, determined by its chemical structure. Those with high affinity are poorly dissociated in aqueous solution and are termed weak acids. Conversely, those with low affinity are highly dissociated in aqueous solution and are strong acids. This tendency of an acid, HA, to dissociate to H^+ and A^- may be written as

$$K = [H^+][A^-]/[HA]$$

where K is the apparent dissociation constant. (For reasons beyond the scope of this text it is known as the apparent dissociation constant to distinguish it from the true thermodynamic dissociation constant.) This rearranges to

$$[H^+] = K[HA]/[A^-]$$

Taking logarithms to base 10 on each side of this equation,

$$-\log_{10}[H^+] = -\log_{10}K - \log_{10}[HA]/[A^-]$$

Since $-\log_{10}[H^+] = pH$, and $-\log_{10}K = pK$, this equation becomes

$$pH = pK + \log_{10}[A^-]/[HA]$$

a relationship known as the Henderson–Hasselbalch equation. Strong acids have a low p*K* while weak acids have a high p*K*. The Henderson–Hasselbalch equation is fundamental to the understanding of acid–base physiology in both homeostasis and pathological states and allows the calculation of the pH, p*K*, or molar ratio of the proton donor and proton acceptor when two of these factors are known.

It is worthwhile to derive pH from first principles to illustrate how it is used in physiological and medical settings. Pure water dissociates only weakly at 25°C and the equilibrium for the equation

$$H_2O \leftrightarrow H^+ + OH^-$$

is far to the left. In practice, H^+ is hydrated to become a hydronium ion, H_3O^+, but for the current discussion H^+ will suffice.

Thus the equilibrium process may be written

$$K_e = [H^+][OH^-]/[H_2O]$$

where K_e is the equilibrium constant. Since water dissociates little at 25°C, $[H_2O]$ is essentially 1000/18 = 55.5 M (the number of moles of H_2O, molecular weight 18, in 1 liter of water); thus, rearranging the above equation

$$55.5K_e = [H^+][OH^-]$$

This figure ($55.5K_e$) is the ionic product of pure water, K_w, and is equal to 10^{-14}, where $[H^+] = [OH^-] = 10^{-7}$. It is the basis of the pH scale which defines the $[H^+]$ and $[OH^-]$ of aqueous solutions.

As mentioned above,

$$pH = \log_{10}1/[H^+] = -\log_{10}[H^+]$$

and thus for a neutral solution, where $[H^+] = 10^{-7}$,

$$pH = -\log_{10}1/10^{-7} = 7$$

The use of a logarithmic scale overemphasizes the narrowness of the pH limits in health. A tenfold change in proton concentration is equivalent to only a single pH unit.

Physiological buffer systems

Because even short-term changes in $[H^+]$ may have a disastrous effect on cellular function, there are buffer systems in place which minimize the effect on pH of adding or removing protons. Buffer solutions (**Figure 7.1** and **Figure 7.2**)

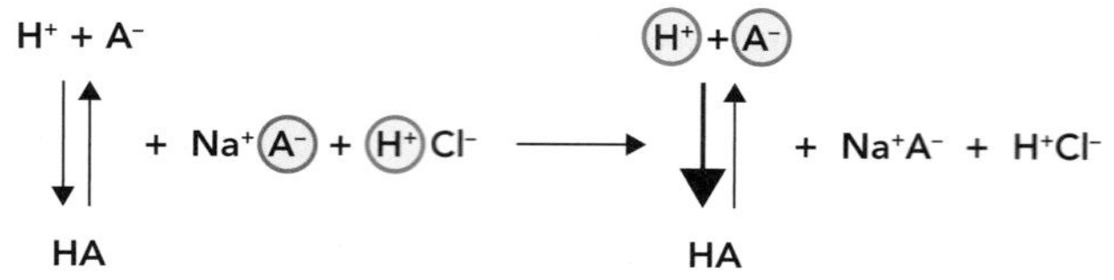

$$\begin{matrix} H^+ + A^- \\ \downarrow\uparrow \\ HA \end{matrix} + Na^+A^- + H^+Cl^- \longrightarrow \begin{matrix} H^+ + A^- \\ \Downarrow\uparrow \\ HA \end{matrix} + Na^+A^- + H^+Cl^-$$

Figure 7.1 Buffering of acid. A buffer solution consists of a weak acid (HA), which is in equilibrium with its dissociated ions, plus a salt of its base (in this case NaA), which dissociates fully. On addition of a strong acid (here, HCl), the additional H^+ (shown in blue) combines with the buffer base (shown in pink) and pushes the equilibrium toward the undissociated acid. There is a minimal fall in pH at the expense of consumption of buffer base. Further addition of acid will eventually deplete the buffer base to the point that pH will fall, unless the solution is replenished with base. In addition, the pH of the solution may be maintained by removal of undissociated buffer acid on the right-hand side of the equation, although this will cause even more rapid depletion of the buffer base.

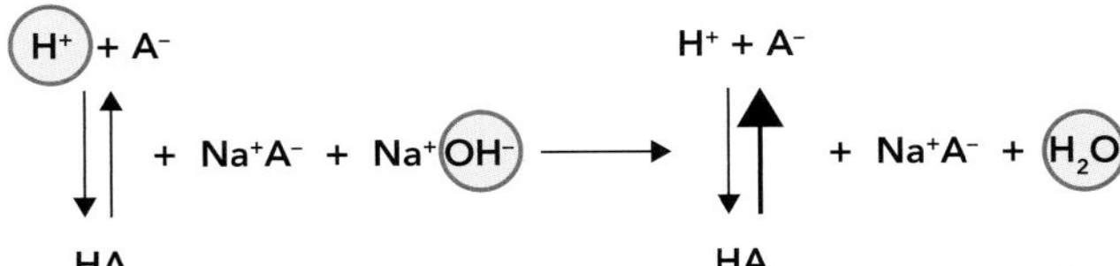

Figure 7.2 Buffering of base. The buffer solution is the same as described in Figure 7.1. Addition of base (NaOH) results in undissociated OH^- combining with H^+ from the buffer base to form water (shown in pink). Thus, a rise in pH is resisted at the expense of the depletion of buffer acid. Further addition of OH^- will eventually exhaust the buffer acid, unless it is replenished. The pH may also be maintained by addition of undissociated acid on the right-hand side of the equation.

are able to accept or donate H^+ and thereby resist changes in H^+ concentration and pH. Such systems consist of a weak acid and its conjugate base and are most effective at the pK_a of the acid; that is, the pH at which the acid is 50% ionized. In general, the optimal buffering range for any conjugate pair is within 1 pH unit of its pK_a, so that an ideal physiological buffer would have a pK_a close to the physiological pH of 7.4. In practice, of the three major buffer systems in the body (**Table 7.1**), only inorganic phosphate and the side chains of histidine and cysteine have a pK_a close to 7.4 and, even so, the two amino acid side chains are not major contributors to whole-body buffering. The principal body compartments in which these systems are effective are also shown in Table 7.1 and within each compartment the buffer systems are in equilibrium.

TABLE 7.1 Physiological buffering systems

Buffer	Acid–base pair	Principal effective body compartment
Phosphate	$H_2PO_4^-$ HPO_4^{2-}	Intracellular fluid, urine, plasma
Bicarbonate	H_2CO_3 HCO_3^-	Extracellular fluid
Protein	H.protein protein$^-$	Intracellular fluid

It is important to realize that buffer solutions are depleted by the addition or removal of protons, and will not resist changes in pH indefinitely. Addition of protons will consume the buffer base and increase the concentration of undissociated buffer acid. Removal of protons will consume undissociated acid and increase the concentration of buffer base. As the pH for any solution is determined by the ratio of base to acid, even a buffer system will eventually change its pH. There must, therefore, be other mechanisms in place to continually replenish the buffer system.

7.2 HYDROGEN ION HOMEOSTASIS

As with several other fundamental and tightly regulated aspects of the internal environment, pH affects the folding, shape, and therefore function of proteins. These include enzymes, structural proteins, and membrane transporters and channels, all of which are essential to cellular function. The intracellular fluid (ICF) hydrogen ion concentration [H^+] is maintained within narrow limits but is difficult to measure and varies between cells and tissues. Since the ICF [H^+] is linked to that of the extracellular fluid (ECF), this must also be regulated and is maintained at a pH range of 7.35–7.45, equivalent to [H^+] of 44–36 nmol/L.

The generation of hydrogen ions is a continuous process and is an inevitable result of cellular respiration. The human body produces, and must remove, between 12,000 and 22,000 mmol of CO_2 per day.

Carbon dioxide is known as respiratory (or volatile) acid since it can be cleared by the lungs. It is formed from complete oxidation of the major nutrients (carbohydrate, fat, and protein) and reacts with water to form carbonic acid, which dissociates to H^+ and bicarbonate (HCO_3^-) in a reaction catalyzed by carbonic anhydrase.

$$CO_2 + H_2O \rightarrow H_2CO_3 \rightarrow H^+ + HCO_3^-$$

Carbonic anhydrase is present in many cells and is particularly abundant in erythrocytes.

In addition to CO_2, a much smaller amount (70 to 100 mmol per day) of hydrogen ions is produced, mainly from incomplete oxidation of nutrients, which gives rise to metabolic or fixed acids. These cannot be excreted by the lungs and must be dealt with by the kidneys, although short-term changes in pH are prevented by buffering. Fixed acids become important in metabolic diseases that lead to their overproduction or underexcretion (kidney failure). The major fixed acids are lactic and keto acids. In addition, phosphoric and sulfuric acids are products of protein metabolism.

The pH of blood is maintained within a very narrow range; this is possible because the ECF is an open buffer system in which the PCO_2 can be modified rapidly by alteration of the rate of ventilation in the lungs. Thus an increase in the ventilation rate helps to remove CO_2 from the body and tends to counteract acidosis by lowering the concentration of carbonic acid. Although this does not restore the buffer base (HCO_3^-), it does increase the ratio of base to acid and therefore increases the pH back toward normal.

Since the pK_a of carbonic acid is 6.1, the equilibrium for the carbonic anhydrase reaction shown above lies far to the right, such that at physiological pH little, if any, H_2CO_3 is present in solution. The dissociation constant for H_2CO_3 may be defined as

$$K_a = [H^+][HCO_3^-]/[H_2CO_3]$$

which rearranges to

$$[H^+][HCO_3^-] = K_a[H_2CO_3]$$

Carbonic acid cannot be measured directly (and some even question its existence as other than a theoretical concept), but it can be considered as dissolved CO_2 for which the solubility constant (S) is known in terms of mmol/kPa or mmol/mmHg.

By accounting for the solubility of CO_2 in plasma and correcting for the necessary changes in units, this equation can be used to describe the relationship between H^+, HCO_3^-, and PCO_2 when body temperature and blood hemoglobin levels are within the normal physiological range; thus

$$H^+ \times HCO_3^- = K_a \times S \times PCO_2$$

Elimination of H^+ produced during oxidative metabolism

Protons are released in oxidation reactions involving NAD^+ as the electron acceptor; for example, the oxidation of glyceraldehyde 3-phosphate to 1,3-bisphosphoglycerate by glyceraldehyde 3-phosphate dehydrogenase (**Figure 7.3**) in the glycolytic pathway (as described in Chapter 2).

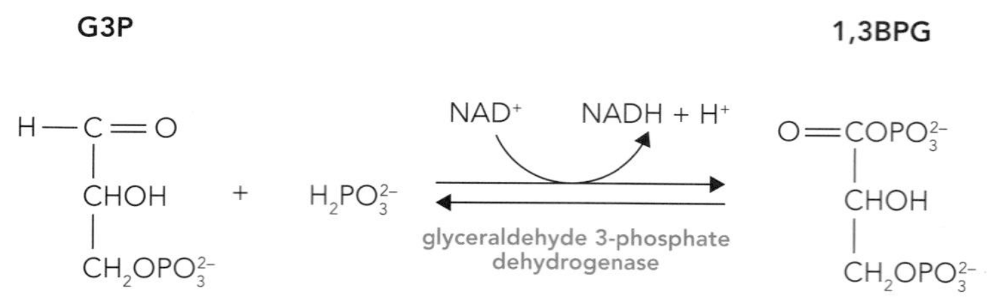

G3P: H—C=O | CHOH | $CH_2OPO_3^{2-}$ + $H_2PO_3^{2-}$ ⇌ (NAD⁺ → NADH + H⁺; glyceraldehyde 3-phosphate dehydrogenase) 1,3BPG: O=$COPO_3^{2-}$ | CHOH | $CH_2OPO_3^{2-}$

Figure 7.3 Oxidation of glyceraldehyde 3-phosphate (G3P) to 1,3-bisphosphoglycerate (1,3BPG). This reaction, catalyzed by glyceraldehyde 3-phosphate dehydrogenase, is an example of H^+ production in oxidative metabolism.

NAD^+ and ATP are formed on the inner mitochondrial membrane during electron transport and oxidative phosphorylation with consumption of H^+:

$$O_2 + NADH + 2H^+ + ADP^{3-} + P_i \rightarrow 2H_2O + 2NAD^+ + ATP^{4-}$$

In the oxidation of organic acids, for example of lactate to CO_2, there is consumption of one proton but generation of three molecules of CO_2, which in turn generate carbonic acid.

$$CH_3CHOHCOO^- + H^+ + 3O_2 \rightarrow 3CO_2 + 3H_2O$$

Elimination of H^+ during CO_2 transport

Carbon dioxide generated during the oxidation of fat and carbohydrate in the mitochondria of peripheral tissues such as heart and skeletal muscle is converted to H^+ and HCO_3^-, via carbonic acid, by carbonic anhydrase (**Figure 7.4**). Protons are exchanged for oxygen, which is delivered to the tissue by hemoglobin in red blood cells. The red cells circulate and deliver this hemoglobin.$2H^+$ to the lungs, where the reverse exchange occurs; protons are released to participate in the carbonic anhydrase reaction, re-forming CO_2 and water. The CO_2 is then exhaled.

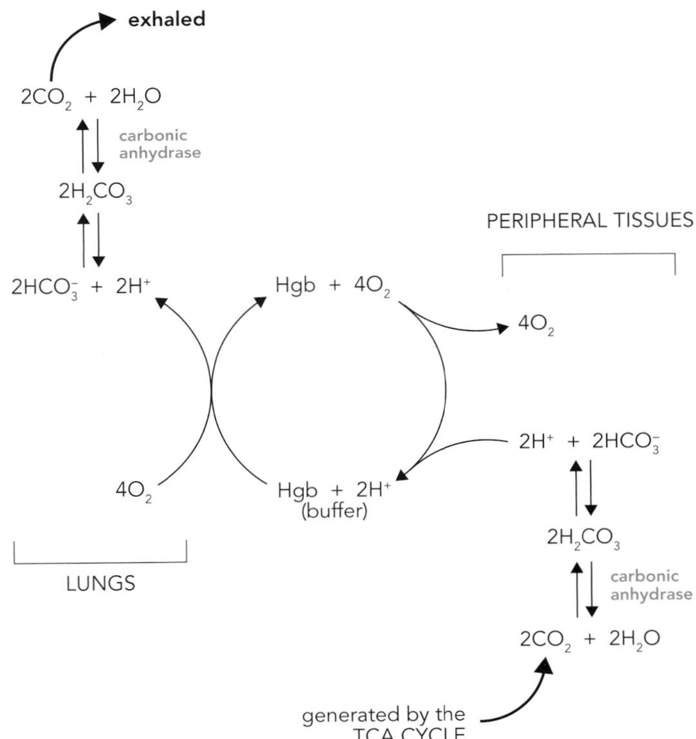

Figure 7.4 Elimination of H^+ during CO_2 transport within the extracellular fluid. Within the peripheral tissues, CO_2 generated by the aerobic metabolism of carbohydrate, fat, and protein in the tricarboxylic acid (TCA) cycle is dissolved in the extracellular fluid, where it is converted by the action of carbonic anhydrase into protons and bicarbonate ions. Some protons are buffered by deoxygenated hemoglobin (Hgb), are transported to the lungs, and are exchanged for oxygen, which is returned to the tissues. The protons released are recombined with bicarbonate, again under the action of carbonic anhydrase, and the resulting CO_2 is exhaled. Protons can be seen as a transport form of CO_2 from the tissues to the lungs. Any process that interrupts this transport, including decreased lung function, will result in accumulation of protons in the ECF (acidosis). Increased exhalation of CO_2 will reduce ECF protons (alkalosis).

Turnover of H^+ in the gastrointestinal tract

The secretions into the gastrointestinal (GI) tract in a healthy adult include approximately 0.3 mol/L of H^+ and this is balanced in health by recycling of H^+ as illustrated in **Figure 7.5**. Following a meal, the hydrogen ions generated by the gastric parietal cells and secreted into the stomach lumen result in bicarbonate ions being secreted into the extracellular fluid. This is known as the alkaline tide and lasts until the hydrogen ions meet bicarbonate-rich fluid that is secreted

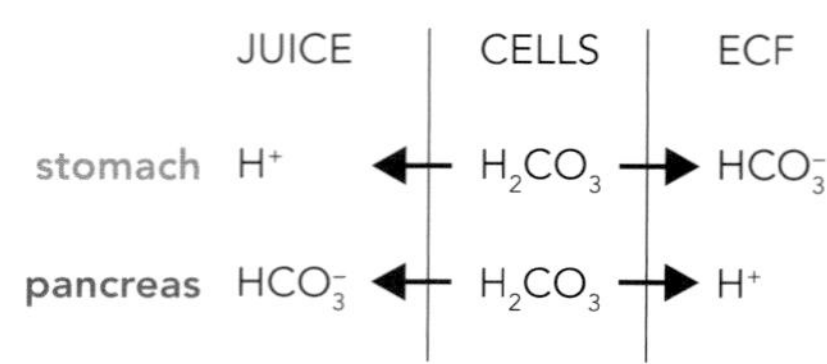

Figure 7.5 Proton turnover in the gastrointestinal tract. In stomach and pancreatic acinar cells, the protons and bicarbonate ions generated from the dissociation of carbonic acid are secreted in opposite directions for their respective juices. Stomach cells secrete acid into the stomach lumen, whilst pancreatic cells secrete bicarbonate into the duodenum. The bicarbonate secretion neutralizes the low pH of stomach contents to allow pancreatic enzymes to work effectively. The counterions, which are secreted into the ECF, give rise to a transient alkaline tide after a meal. It is important to note that there is no net loss or gain of protons, unless either stomach or pancreatic fluid alone escapes from the body, for example due to pre-pyloric vomiting or intestinal fistulas.

by the pancreas into the duodenum. Disturbances in H^+ homeostasis result from unbalanced losses, such as in vomiting or loss of pancreatic juice and the contents of the small intestine.

Elimination of H^+ by the kidneys

The role of the kidneys in acid–base balance is described in detail in Chapter 4. The importance of the kidneys is indicated by the amount of H^+ eliminated daily by these organs (**Table 7.2**). However, it is important to be aware that there are two separate mechanisms in the proximal and distal tubules, and they can be affected by different disease processes. Bicarbonate ions are freely filtered at the glomerulus and are reclaimed in the proximal tubule. The distal tubule, however, is able to regulate the acidity of urine by excreting hydrogen ions and adding new bicarbonate to the extracellular fluid. Disorders of either mechanism cause a systemic acidosis termed renal tubular acidosis. These are described in more detail later in this chapter. See also Case 7.3.

TABLE 7.2 Daily elimination of H^+ by kidneys in a healthy adult

Form of H^+	Amount (mmol/day)
Phosphate buffered	20–40
As NH_4^+	30–60
Free H^+	20–40

Relationship between pH and the concentration of components in the bicarbonate buffer system

The bicarbonate buffer system is of particular importance in clinical chemistry since it is quantitative, with components being easily measured and their concentration subject to physiological adjustment in various states. Furthermore, other buffers are regenerated from bicarbonate produced in the renal tubules (see Chapter 4).

The Henderson–Hasselbalch equation can be applied to the bicarbonate buffer system as follows:

$$\mathbf{pH = p}K_\mathbf{a} \mathbf{+ log_{10}[base]/[acid]}$$

The [acid] component can be replaced by PCO_2, multiplied by its solubility constant (S) in water, to give:

$$\mathbf{pH = 6.1 + log_{10}[HCO_3^-]/S \times PCO_2}$$

which can be simplified to:

$$\mathbf{pH \propto [HCO_3^-]/PCO_2}$$

When considering disorders of acid–base balance and interpreting blood results, HCO_3^- is considered to be the metabolic component and PCO_2 the respiratory component. These can be affected by different pathological processes and compensated for by altering the opposing component. For example, the respiratory center in the brain is sensitive to the pH of ECF and will compensate for abnormal concentrations of bicarbonate by changes in respiration rate. Such respiratory compensation is a short-term measure in acid–base disturbances, being particularly important in metabolic acidemia. In this case, respiratory rate increases in order to lower PCO_2 and hence compensate for the fall in $[HCO_3^-]$.

7.3 DISORDERS OF ACID–BASE BALANCE: ACIDOSIS AND ALKALOSIS

Definitions

The terms acidosis and acidemia (and the corresponding terms alkalosis and alkalemia) are often used somewhat interchangeably. They do, however, have slightly different and specific meanings as defined below.

Acidosis
A pathological condition which tends to cause a decrease in arterial blood pH; that is, an increase in $[H^+]$

Acidemia
A state of decreased arterial blood pH

Alkalosis
A pathological condition which tends to cause an increase in arterial blood pH; that is, a decrease in $[H^+]$

Alkalemia
A state of increased blood pH

Metabolic acidosis
A pathological condition where there is accumulation of acid other than carbonic acid, for example lactic acid; a condition in which a primary event is a reduction in plasma $[HCO_3^-]$

Metabolic acidosis with compensation
A pathological condition where arterial blood pH may be normal, but only because compensatory mechanisms operate, by decreasing PCO_2 in response to an initial fall in arterial blood pH.

Respiratory acidosis
A pathological condition in which the primary event is retention of CO_2, thereby increasing PCO_2 and plasma $[H^+]$

Respiratory acidosis with compensation
A pathological condition where arterial blood pH may be normal, but only because compensatory mechanisms operate, by the renal generation of bicarbonate and excretion of H^+, in response to an initial rise in arterial blood PCO_2.

Metabolic alkalosis
A pathological condition in which an increase in pH is caused by either the excessive loss of hydrogen ion or the gain of bicarbonate

Metabolic alkalosis with compensation
A pathological condition where arterial blood pH may be normal, but only because compensatory mechanisms operate. In response to an initial rise in

arterial blood pH, there is an increased loss of bicarbonate in urine (unless the primary defect is renal), decreased respiratory loss of CO_2, or both

Respiratory alkalosis
A pathological condition where the primary event is increased respiratory loss of CO_2, for example by hyperventilation, thereby reducing PCO_2 and $[H^+]$

Respiratory alkalosis with compensation
A pathological condition where arterial blood pH may be normal, but only because compensatory mechanisms operate, by the renal excretion of bicarbonate and retention of H^+, in response to an initial fall in arterial blood PCO_2

Metabolic and respiratory compensation

Changes in plasma bicarbonate and partial pressure of arterial CO_2 in response to acidosis and alkalosis are shown in **Figure 7.6** and **Figure 7.7**.

It follows from the equations shown earlier and Figures 7.6 and 7.7 that to maintain blood pH at 7.4, any changes in PCO_2 must be balanced by opposite changes in blood bicarbonate concentration $[HCO_3^-]$. Thus, a fall in PCO_2 as a result of hyperventilation causes a rise in pH; that is, alkalemia with blood pH $\geq$7.45. The kidneys respond by increasing the net excretion of HCO_3^- (see Chapter 4) in an attempt to return blood pH to normal (see Figure 7.7). Such a condition constitutes a respiratory alkalosis with metabolic compensation, because the initial cause of the change in pH was a change in ventilation rate.

In contrast, if large amounts of alkaline fluids are lost from the body, such as in intestinal secretions in diarrhea, blood pH will fall (acidemia when blood pH $\leq$7.35) and trigger an increased ventilation rate. To compensate for this fall in pH, or increase in $[H^+]$, there needs to be a reduction in PCO_2 by increasing the CO_2 expired by the lungs. In addition, the kidneys will excrete more H^+, and both of these compensatory responses will serve to raise blood pH toward the normal range. This is an example of metabolic acidosis, due to primary loss of bicarbonate, with mixed compensation, involving both the kidneys and the lungs.

In each situation the body responds by trying to compensate for the pH change arising from the metabolic (see Figure 7.6) or respiratory disorder (see Figure 7.7), but this compensation is seldom sufficient to restore pH to within the normal range and never overcompensates.

$$pH = 6.1 + \log \frac{[HCO_3^-]}{PCO_2 \times 0.225}$$

	metabolic disorder	
	metabolic acidosis	**metabolic alkalosis**
primary change	fall in bicarbonate fall in pH	loss of H^+ or rise in bicarbonate rise in pH
respiratory compensation	hyperventilation PCO_2 falls pH rises	hypoventilation PCO_2 rises pH falls

Figure 7.6 Metabolic disorders and their compensation.
The Henderson–Hasselbalch equation shows that pH is determined by the ratio of HCO_3^- to PCO_2. The two constants in the equation are 6.1, the dissociation constant (pK_a) for carbonic acid, and 0.225, the solubility constant for CO_2 in mmol/kPa (equivalent to 0.03 mmol/mmHg). In metabolic disorders, the primary change is a rise or fall in HCO_3^-, which must alter the pH in the same direction. Compensation consists of changes in the rate and depth of respiration by the lungs to alter PCO_2 and attempt to restore the ratio, and hence pH, toward normal. This is rarely fully effective and overcompensation never occurs.

$$pH = 6.1 + \log \frac{[HCO_3^-]}{PCO_2 \times 0.225}$$

	respiratory disorder	
	respiratory acidosis	**respiratory alkalosis**
primary change	increase in PCO_2 fall in pH	decrease in PCO_2 rise in pH
metabolic compensation	renal generation of bicarbonate rise in serum bicarbonate pH rises	renal loss of bicarbonate fall in serum bicarbonate pH falls

Figure 7.7 Respiratory disorders and their compensation. In respiratory disorders, the primary change is a rise or fall in PCO_2, which must alter the pH in the opposite direction. Compensation consists of renal alteration of bicarbonate to attempt to restore the ratio toward normal. This is rarely fully effective and overcompensation never occurs.

Assessment of acid–base status

Key parameters measured in the laboratory, and indices of metabolic status derived from them, which allow an assessment of acid–base status are listed in **Table 7.3**. Ion-selective electrodes in a blood gas analyzer are used to measure the pH, PO_2, and PCO_2 of arterial blood. These are the only parameters that are directly measured. While PO_2 does not directly affect blood pH, it is a useful indicator of the respiratory status of the patient. If it is not clinically necessary to measure PO_2, a venous blood sample is sufficient to measure pH and PCO_2 and is less painful to obtain than an arterial sample. Bicarbonate concentration can be calculated by inserting pH and PCO_2 into the Henderson–Hasselbalch equation. Other parameters can be calculated from the primary measurements using software built into the blood gas analyzer. One of these derived parameters is standard bicarbonate: an estimate of the expected bicarbonate after equilibration to a normal PCO_2 of 5.3 kPa (thereby removing the respiratory effect on bicarbonate of hyper- or hypoventilation). Another derived parameter is base excess. This is the theoretical amount of acid (in mmol) required to restore 1 litre of blood to normal pH at this PCO_2 of 5.3 kPa. Both derived parameters were invented as estimates of the metabolic component of an acid–base disorder in order to aid diagnosis where both respiratory and metabolic causes are present. A metabolic acidosis causes a negative base excess, as acid would need to be removed to restore the pH to normal. This is both a potentially confusing concept and has also been criticized on theoretical grounds, since in chronic respiratory disorders the kidney is able to adjust the secretion of HCO_3^- into the ECF. Fortunately, blood gas results can be interpreted without base excess and it is only included here for completeness, rather than because it is a useful concept. The normal ranges for arterial blood gases and related parameters are shown in **Table 7.4**.

Analytical practice point 7.1

Blood gas analysis may be performed on venous blood samples, rather than arterial, if there is no clinical reason to measure PO_2.

Analytical practice point 7.2

Blood gas analyzers measure pH, PCO_2, and PO_2. Other parameters may be calculated from these. However, base excess and standard bicarbonate are not useful and may be misleading.

TABLE 7.3 Parameters reported by blood gas analyzers

Method	Parameter
Measured by ion-selective electrode	pH
	PCO_2
	PO_2
	Lactate
Calculated from Henderson–Hasselbalch equation	Actual HCO_3^-
Calculated from other nomograms using onboard software	Standard HCO_3^-
	Base excess

TABLE 7.4 Reference ranges for arterial blood gases and related parameters

	SI units	Conventional units
pH	7.35–7.45	7.35–7.45
H^+	35–45 nmol/L	35–45 nEq/L
PCO_2	4.7–5.9 kPa	35–45 mmHg
PO_2	11–13 kPa	80–100 mmHg
HCO_3^-	22–26 mmol/L	22–26 mEq/L
Base excess	+2 to –2 mmol/L	+2 to –2 mEq/L
Lactate	0.5–1.6 mmol/L	4.5–14.4 mg/dL

Simple acid–base disturbances can be determined simply by using the Henderson–Hasselbalch equation. Alternatively, plotting a patient's pH (or hydrogen ion concentration) against their PCO_2 can deal with more complex, mixed disorders. The graph in **Figure 7.8** shows how these parameters are related in metabolic and respiratory alkalosis and acidosis. It is instructive to use the graph when considering the case scenarios in each section. The particular value of graphical interpretation is seen with mixed acid–base disorders, where plotted results will fall between the two bands.

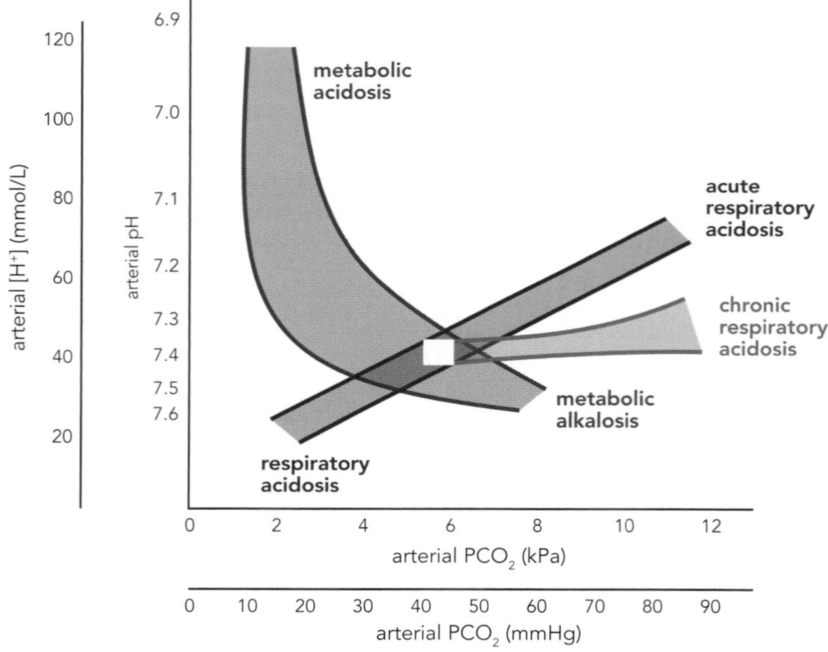

Figure 7.8 Relationship between arterial blood pH (H^+) and PCO_2 in different acid–base disorders. For single disorders, the results from a patient will fit inside one of the bands. For mixed disorders, the results will fall between two bands. Thus pH 7.2 and PCO_2 6 kPa (45 mmHg) can only occur if both a metabolic acidosis and respiratory acidosis are present. This method is the most effective way of diagnosing acid–base disorders and supersedes calculations of standard bicarbonate and base excess.

Acidosis

As shown in Figures 7.6 and 7.7, acidosis is caused by a primary reduction in the concentration of plasma HCO_3^- (metabolic) or an increase in PCO_2 (respiratory).

Metabolic acidosis

Metabolic acidosis may arise from:

- Consumption of HCO_3^- by accumulated endogenous or exogenous acid
- Loss of HCO_3^- from the gut
- Loss of HCO_3^- in urine
- Insufficient generation of HCO_3^- in the renal tubules

TABLE 7.5 Conditions which lead to consumption of bicarbonate

Mechanism	Example
Overproduction of acid	Lactic acidosis
	Ketoacidosis
	Poisons (for example methanol, ethanediol, or salicylate)
	Organic acidosis (inborn error of metabolism)
Impaired clearance of acid	Renal failure
Loss of bicarbonate	Diarrhea
	Intestinal fistulas

Clinical practice point 7.1

Diagnosis of acid–base disorders is most reliably achieved by plotting pH (or H^+) against PCO_2 on a clinical nomogram. This is particularly useful for mixed disorders.

Clinical conditions where these occur are listed in **Table 7.5**.

Loss of bicarbonate from the gut is particularly important in cases of fluids containing large volumes of pancreatic juices, such as fistulas and surgical drains, since the pancreas secretes large amounts of bicarbonate in order to neutralize the stomach acid as it enters the duodenum.

Renal disorders causing metabolic acidosis do so either by failure to reabsorb bicarbonate that is filtered by the glomerulus, or by an inability to excrete hydrogen ions (and at the same time generate bicarbonate for the extracellular fluid). This may occur as a generalized failure of renal function, or as a specific disorder, when it is termed renal tubular acidosis.

Anion gap

The plasma concentrations of sodium, potassium, bicarbonate, and chloride can be routinely measured, both in the clinical laboratory and by additional ion-specific electrodes in modern blood gas analyzers (although in this situation the bicarbonate is calculated). The combined concentrations of the two major negatively charged ions are less than the combined concentrations of the positively charged ions. This difference is known as the anion gap.

$$\textbf{Anion gap} = [Na^+ + K^+] - [HCO_3^- + Cl^-]$$

An alternative version of this equation does not include potassium, as it is a relatively small contributor to the cations (around 4 mmol/L, compared to 140 mmol/L for sodium).

Using the four ions in the above equation, the normal anion gap is 10–20 mmol/L. This gap consists of unmeasured negatively charged ions such as lactate and phosphate, plus a component from negatively charged proteins. The reference range may vary slightly between laboratories and over time, depending on the methods used to measure the component ions. However, for clinical purposes this matters little.

Metabolic acidosis can be classified as being associated with a high or normal anion gap. Normal anion gap metabolic acidosis can also be called hyperchloremic, because the chloride replaces the lowered bicarbonate when maintaining electroneutrality. When the anion gap is high, by inference there must be another negatively charged ion present. In some cases this can be easily measured, for example lactate, but it may not always be easily identified. Conditions affecting the anion gap are included in **Table 7.6**.

L-Lactic acidosis

As described in Chapter 2, lactic acid, produced by reduction of pyruvate, is the end product of anaerobic glycolysis. It is a normal metabolite, particularly of exercising muscle and red blood cells, and a healthy adult turns over approximately 1.2 mol (110 g) of lactate daily. This is either oxidized to CO_2 via the tricarboxylic acid (TCA) cycle:

$$CH_3CHOHCOO^- + H^+ + 3O_2 \rightarrow 3CO_2 + 3H_2O$$

TABLE 7.6 Conditions affecting the anion gap

Effect on anion gap	Condition
Raised (normochloremic) acidosis	Lactic acidosis
	Ketoacidosis
	Renal failure
	Poisoning (for example methanol, ethanediol, or salicylate)
	Organic acidosis (inborn error of metabolism)
Normal (hyperchloremic) acidosis	Hyperkalemia: renal failure, type 4 renal tubular acidosis
	Hypokalemia: diarrhea, types 1 and 2 renal tubular acidosis
	Ureterosigmoidostomy
Decreased or negative anion gap	Severe hypoalbuminemia
	Cationic paraprotein
	Bromide interference with chloride analysis

or converted to glucose via gluconeogenesis:

$$2CH_3CHOHCOO^- + H^+ \rightarrow C_6H_{12}O_6$$

Lactic acid is a strong acid ($pK_a = 3.9$) such that a metabolic acidemia occurs when it accumulates due to overproduction and/or impaired clearance. However, administration of lactate as a sodium or potassium salt causes a metabolic alkalosis, because the H^+ needed for its metabolism is derived from H_2CO_3 leaving an equimolar amount of HCO_3^-. An operational definition of lactic acidemia is a plasma lactate of >6 mmol/L in combination with an arterial blood pH of <7.3 (equivalent to $[H^+]$ >50 nmol/L). The causes of lactic acidosis are listed in **Table 7.7**. It can be broadly categorized as being due to inadequate tissue oxygen delivery (type A) or occurring despite adequate oxygen delivery (type B). Metformin, a widely used drug in type 2 diabetes, has been associated with type B lactic acidosis, as has the similar biguanide drug phenformin. Metformin-related lactic acidosis is rare, despite the popularity of the drug, but

TABLE 7.7 Causes of lactic acidosis

Categories	Causes
Type A: excessive anaerobic metabolism	Circulatory failure: central and peripheral
	Sepsis
	Strenuous muscular activity, including fits
	Very large tumor masses
Type B: interruption of electron-transport chain and oxidative phosphorylation	Severe liver failure
	Poisons including methanol, ethanediol and salicylate
	Mitochondrial myopathies
	Biguanide drugs: phenformin and metformin
	Severe acidemia, however caused
Inborn errors of metabolism involving enzymes of glycolytic pathway	Glucose 6-phosphatase deficiency
	Fructose 1,6-diphosphatase deficiency
	Pyruvate carboxylase deficiency
	Pyruvate dehydrogenase deficiency
D-Lactic acidosis	Overgrowth of gut flora

CASE 7.1

A 75-year-old woman with septicemia is moribund, hypotensive, and anuric.

	SI units	Reference range	Conventional units	Reference range
Arterial blood				
pH	6.88	7.35–7.45	6.88	7.35–7.45
H^+	132 nmol/L	35–45	132 nEq/L	35–45
PCO_2	2.7 kPa	4.7–5.9	20 mmHg	35–45
PO_2	17.2 kPa	11–13	152 mmHg	80–100
Plasma				
Sodium	140 mmol/L	136–142	140 mEq/L	136–142
Potassium	7.1 mmol/L	3.5–5.0	7.1 mEq/L	3.5–5.0
Bicarbonate	<5 mmol/L	21–28	<5 mEq/L	21–28
Chloride	100 mmol/L	96–106	100 mEq/L	96–106
Anion gap	>42 mmol/L	10–20	>42 mEq/L	10–20
Anion gap (exc. K)	>35 mmol/L	8–16	>35 mEq/L	8–16
Urea (BUN)	16.2 mmol/L	2.9–8.2	45 mg/dL	8–23
Creatinine	345 µmol/L	53–106	3.9 mg/dL	0.6–1.2
Lactate	21 mmol/L	0.6–1.7	189 mg/dL	5–15

- **What type of acid–base disorder is present?**
- **Why is the lactate high despite a raised PO_2?**

There is a profound metabolic acidosis as shown by the low pH and PCO_2. The PCO_2 has been lowered by increased ventilation rate and depth, driven by the respiratory center, in an attempt to compensate for the acidosis. The base excess is used as an indicator that the acidosis is metabolic, but is superfluous and only provided for completeness. The primary condition is lactic acidosis resulting from anaerobic tissue respiration, due to poor delivery of oxygenated blood, which occurs despite adequate oxygen exchange in the lungs. An important factor in this is low blood pressure, which has also reduced delivery of blood to the kidneys and caused acute kidney injury (shown by the raised urea and creatinine), which exacerbates the acidosis by reducing bicarbonate production. The potassium is raised due to kidney failure and the acidosis itself, which causes potassium to move from the intracellular to the extracellular space.

the risk increases as renal function falls. For that reason, metformin is discontinued when chronic kidney disease is indicated by plasma creatinine or estimated GFR, the exact value varying between expert guidelines.

D-Lactic acidosis

The D-isomer of lactic acid is formed by bacteria, particularly in the GI tract of ruminants, and D-lactic acidemia is well known in ruminant species. It is not normally found in humans but has been recognized more recently in patients with bacterial colonization of the small gut, usually after resection. Any D-lactic acid produced from glucose in the gut is absorbed but cannot be metabolized since humans lack the enzymes required to do this. D-Lactate thus accumulates giving rise to a metabolic acidosis. Patients hyperventilate and are stuporous, ataxic, and dysarthric. D-Lactic acid is not detected in plasma and urine by routine laboratory methods which use specific L-lactate dehydrogenase. However, D-lactate can be identified using chromatographic analysis.

If suspected, an organic acid screen of urine will readily detect the abnormality. Treatment involves administration of antibiotics by mouth to eliminate the offending gut bacteria.

Clinical practice point 7.2

Acidosis is usually accompanied by hyperkalemia, and alkalosis by hypokalemia.

Ketoacidosis

Ketoacidosis arising from excess production of acetoacetate from fatty acid oxidation in liver has been described in Chapter 2. Acetoacetate is produced even in healthy humans and is oxidized in many tissues, such as cardiac and skeletal muscle, but not in liver. Because of the reducing conditions present in liver in the catabolic state, much of the acetoacetate is further reduced to

CASE 7.2

A 16-year-old adolescent boy is admitted unconscious to the emergency department. His mother says he has been unwell for two days and started vomiting the previous day. He has been losing weight for a few weeks and been very thirsty recently. He is not taking any medication. On examination he is dehydrated.

	SI units	Reference range	Conventional units	Reference range
Arterial blood				
pH	7.18	7.35–7.45	7.18	7.35–7.45
H^+	66 nmol/L	35–45	66 nEq/L	35–45
PCO_2	2.5 kPa	4.7–5.9	19 mmHg	35–45
PO_2	11.5 kPa	11–13	86 mmHg	80–100
Plasma				
Sodium	137 mmol/L	136–142	137 mEq/L	136–142
Potassium	6.3 mmol/L	3.5–5.0	6.3 mEq/L	3.5–5.0
Bicarbonate	7.0 mmol/L	21–28	7.0 mEq/L	21–28
Chloride	105 mmol/L	96–106	195 mEq/L	96–106
Anion gap	31 mmol/L	10–20	31 mEq/L	10–20
Anion gap (exc. K)	25 mmol/L	8–16	25 mEq/L	8–16
Urea (BUN)	25 mmol/L	2.9–8.2	70 mg/dL	8–23
Creatinine	200 µmol/L	53–106	2.3 mg/dL	0.6–1.2
Glucose	36 mmol/L	3.9–6.1	649 mg/dL	70–110
Urine (dipstick)				
Glucose	+++	Negative	+++	Negative
Ketones	+++	Negative	+++	Negative

- What acid–base disorder is present?
- What is the underlying disease?

This patient has metabolic acidosis due to diabetic ketoacidosis. Hyperglycemia causes an osmotic diuresis leading to dehydration. Acidosis occurs as a consequence of ketone generation from free fatty acid catabolism, which occurs as glucose cannot be utilized as an energy source. In an attempt to compensate for the acidosis, the patient hyperventilates to reduce the PCO_2. This is not fully successful as there are limits to the degree of hyperventilation which can occur. If the patient becomes exhausted, the ventilation rate will drop and the PCO_2 will rise, worsening the acidosis. This patient will require rehydration with intravenous fluids plus insulin given according to the plasma glucose. Patients with ketoacidosis require lifelong insulin therapy (insulin-dependent or type 1 diabetes).

3-hydroxybutyrate such that the ratio of 3-hydroxybutyrate to acetoacetate in plasma is approximately 3:1. Both are usually undetectable in plasma by routine assays (<0.3 mmol/L). Acetoacetate (pK_a 3.6) and 3-hydroxybutyrate (pK_a 4.7) are strong acids. Under catabolic conditions, where the rate of lipolysis is high, overproduction and accumulation of the acids give rise to ketoacidosis, and in severe cases the plasma total keto acid concentration may rise above 25 mmol/L. Secondary lactic acidemia may also occur in severe ketoacidosis due to dehydration and poor perfusion of tissues with oxygenated blood. Causes of ketoacidosis are listed below.

- Uncontrolled type 1 (insulin-dependent) diabetes mellitus
- Alcoholic ketoacidosis
- Ethanol-induced ketotic hypoglycemia in children
- Some inborn errors of metabolism, for example glycogenosis type I, hereditary fructose intolerance, and organic acidemias

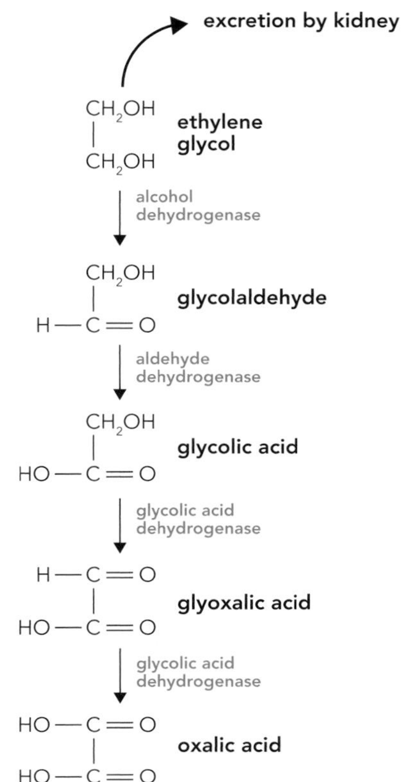

Figure 7.9 **Metabolism of ethanediol (ethylene glycol).** Ethylene glycol is itself relatively nontoxic, causing inebriation similar to ethanol. However, its sequential metabolism, initially by alcohol dehydrogenase in the liver, results in the production of glycolic, glyoxalic, and oxalic acids, which cause metabolic acidosis with a raised anion gap. Oxalic acid is also a toxin in its own right, causing precipitation of calcium throughout the body. Treatment involves competitive inhibition of alcohol dehydrogenase, either with ethanol or the drug fomepizole. Unmetabolized ethylene glycol is excreted by the kidney.

Poisoning with ethylene glycol (ethanediol)

Consumption of ethylene glycol, the major component of antifreeze, causes a severe metabolic acidosis which, unless diagnosed and treated immediately, causes death. Ethylene glycol is itself harmless, but is toxic when metabolized to glycolic and oxalic acids by the sequential action of the enzymes alcohol dehydrogenase and aldehyde dehydrogenase (**Figure 7.9**). Oxalic acid and the two intermediate products are strong acids which cause the severe metabolic acidosis.

Acute metabolic acidosis: principles of treatment

Treatment of acute metabolic acidosis may be directed to the immediate cause; for example:

- Restoration of circulation and administration of oxygen in lactic acidosis
- Administration of insulin in diabetic ketoacidosis
- Administration of ethanol or fomepizole in ethylene glycol (ethanediol) poisoning. These competitively inhibit alcohol dehydrogenase with an alternative substrate and decrease production of oxalic acid)

Administration of bicarbonate intravenously may appear to be an obvious choice of treatment and was used for many years, but it is controversial. There are a number of disadvantages including:

- Sodium overload
- Temporary increase in intracellular PCO_2 and further decrease in pH
- Shift of the oxygen dissociation curve of hemoglobin to the left
- Hypokalemia

Its use is therefore limited to severe metabolic acidosis not responding to other measures. Hemofiltration or hemodialysis may be used on intensive therapy and high dependency units to remove exogenous toxins such as salicylates or ethanediol.

Renal tubular acidosis

The previous examples of metabolic acidosis are all associated with an elevated anion gap, due to the accumulation of unmeasured anions (lactate, ketones, or oxalate). Renal tubular acidosis (RTA) occurs when the kidneys are unable to regulate bicarbonate excretion, resulting in metabolic acidosis with a normal anion gap. Type 1 RTA is the more severe form of acidosis and is due to a defect in the distal tubule which impairs the excretion of hydrogen ions and generation of bicarbonate for the extracellular fluid. In type 1 RTA the urine is inappropriately alkaline; that is, the pH cannot be lowered to less than 5.5. In type 2 RTA the defect lies in the proximal tubule, where bicarbonate, which has been filtered from the plasma, is not fully reabsorbed .This results in low plasma HCO_3^-, although it does not continue to fall below a threshold level.

In other words, there is a low renal threshold for bicarbonate. This is often accompanied by other proximal tubular defects causing glycosuria, again due to low renal threshold, as well as aminoaciduria and phosphaturia. Together this is called the Fanconi syndrome. Chronic but relatively mild acidosis results in mobilization of calcium from the skeleton, leading to a form of rickets and kidney stones.

Both types of RTA are accompanied by hypokalemia, which is an unusual finding in a systemic acidosis where high plasma potassium is both a cause and a consequence. A further type of RTA, type 4, also exists (there is no type 3) and is due to aldosterone deficiency. In contrast to the other two types, hyperkalemia is seen with type 4 RTA

Clinical practice point 7.3

The combination of acidosis and hypokalemia is unusual, and suggests renal tubular acidosis.

CASE 7.3

A 30-year-old woman presented with recent onset of muscle weakness. She had a history over several months of joint pains and dry eyes and mouth.

	SI units	Reference range	Conventional units	Reference range
Venous blood				
pH	7.14	7.35–7.45	7.14	7.35–7.45
H^+	72 nmol/L	35–45	72 nEq/L	35–45
PCO_2	2.7 kPa	4.7–5.9	20 mmHg	35–45
Plasma				
Sodium	138 mmol/L	136–142	138 mEq/L	136–142
Potassium	3.1 mmol/L	3.5–5.0	3.1 mEq/L	3.5–5.0
Bicarbonate	8 mmol/L	21–28	8 mEq/L	21–28
Chloride	119 mmol/L	96–106	119 mEq/L	96–106
Anion gap	11 mmol/L	10–20	11 mEq/L	10–20
Anion gap (exc. K)	8 mmol/L	8–16	8 mEq/L	8–16
Urea (BUN)	4.2 mmol/L	2.9–8.2	13 mg/dL	8–23
Creatinine	87 µmol/L	53–106	1.0 mg/dL	0.6–1.2
Urine				
pH	6.0	4.5–8.0	6.0	4.5–8.0

- What acid–base disorder is present?
- Why does she have muscle weakness?

There is a metabolic acidosis as shown by the low pH, PCO_2, and bicarbonate. There is a normal anion gap due to the high chloride. The urine pH is inappropriately alkaline (>5.5), indicating that there is a defect in the renal tubules; that is, renal tubular acidosis (RTA). A low plasma potassium is a characteristic of this condition, in contrast to other types of acidosis when potassium is raised. Hypokalemia is the cause of her muscle weakness. In this case, type 1RTA is present; that is, affecting the distal tubule. This is in turn due to the presence of Sjögren's syndrome, a type of autoimmune disease affecting the tear and salivary glands. It may occur as part of rheumatoid arthritis. There are increased levels of immunoglobulins in the serum, which affect distal tubular function, particularly the generation of bicarbonate.

Note that venous blood gases are measured here as there is no disorder of oxygenation suspected and arterial samples are invasive and unpleasant for the patient.

Respiratory acidosis

The causes of metabolic acidosis described above involve a decrease in $[HCO_3^-]$. With reference to the equation

$$pH \propto [HCO_3^-]/PCO_2$$

it can be seen that CO_2 retention giving rise to an increase in PCO_2 will also cause an acidosis—in this case, a respiratory acidosis. CO_2 retention may be due to impaired ventilation, lung perfusion, or gas exchange in the alveoli.

Common causes of CO_2 retention leading to respiratory acidosis are listed below.

- Obstructive airways disease (for example chronic bronchitis)
- Pneumonia
- Pulmonary edema
- Chest wall injury
- Myopathy
- Central nervous system depression

CASE 7.4

Blood taken from a 76-year-old man who has chronic bronchitis gave the following results on analysis.

	SI units	Reference range	Conventional units	Reference range
Arterial blood				
pH	7.35	7.35–7.45	7.35	7.35–7.45
H^+	45 nmol/L	35–45	45 nEq/L	35–45
PCO_2	7.8 kPa	4.7–5.9	59 mmHg	35–45
PO_2	7.6 kPa	11–13	57 mmHg	80–100

- **What type of acid–base disorder is present?**

He then becomes increasingly short of breath over two days and is coughing green sputum.

	SI units	Reference range	Conventional units	Reference range
Arterial blood				
pH	7.20	7.35–7.45	7.20	7.35–7.45
H^+	64 nmol/L	35–45	63 nEq/L	35–45
PCO_2	10.0 kPa	4.7–5.9	75 mmHg	35–45
PO_2	5.5 kPa	11–13	41 mmHg	80–100

- **What type of disorder is now present and why has it changed?**

He has chronic (compensated) respiratory acidosis due to poor gas exchange in the lungs. He has type 2 respiratory failure with low PO_2 and high PCO_2. Renal compensation returns the pH to normal by increasing plasma bicarbonate. In the second part of the case he has developed an acute chest infection in addition to his chronic disease. He now has acute respiratory acidosis. His gas exchange is even more impaired so that his CO_2 retention exceeds his renal compensation. Over the next few days the kidneys will attempt further compensation. Knowledge of bicarbonate is not actually required but can be calculated from the Henderson–Hasselbalch equation.

The compensation for a respiratory acidosis is renal. The distal tubule is able to increase net excretion of hydrogen ions into the urine, which simultaneously generates new bicarbonate ions in the plasma. The ratio of $[HCO^{3-}]$:PCO_2 is thus restored toward normal, albeit with higher concentrations of both. This renal compensation is slow compared to respiratory compensation for metabolic acidosis, taking hours to days.

Alkalosis

Alkalosis, a tendency to increased plasma pH, may also be metabolic—primarily as a result of increased bicarbonate—or respiratory, due to reduced PCO_2.

Metabolic alkalosis

Causes of metabolic alkalosis are:

- Loss of H^+, for example loss of stomach acid
- Redistribution of H^+ due to K^+ depletion
- Ingestion or infusion of alkali

Metabolic alkalosis is an uncommon acid–base disorder as the kidney can usually increase bicarbonate excretion, to produce an alkaline urine, before the plasma pH can increase. Metabolic alkalosis therefore requires both an initiating cause, for example loss of gastric acid due to vomiting, and a perpetuating factor. This is usually chloride deficiency, loss of plasma volume (hypovolemia), or both. Other factors that are important are potassium deficiency and mineralocorticoid secretion, but these are often coexistent and interdependent, so it may be difficult to determine their individual contributions, except for example in Conn's syndrome (see Chapter 8). Metabolic alkalosis can be subdivided clinically as saline responsive or non-saline responsive, depending on the response to an infusion of 0.9% saline, which will replace both extracellular fluid volume and chloride. A primary rise in bicarbonate, which occurs in metabolic alkalosis, is always accompanied by a fall in plasma chloride, in order to maintain electroneutrality. Hence, plasma chloride measurement is not helpful in predicting responsiveness to saline.

The classic example of saline-responsive metabolic alkalosis is vomiting due to pyloric stenosis. Vomiting usually results in loss of both gastric acid and duodenal fluid (which is high in bicarbonate) and so does not typically cause alkalosis. In pyloric stenosis (both as a congenital disorder in babies and in adults with cancer) only stomach contents are lost in the vomit. This results in loss of H^+, chloride, and extracellular fluid volume. Loss of H^+ causes an alkalosis and usually the kidney would respond by excretion of bicarbonate. However, this is prevented, as activation of the renin–angiotensin–aldosterone system by hypovolemia increases the reabsorption of sodium (in exchange for potassium) at the distal renal tubule. In order to maintain electroneutrality, a balancing anion must be reabsorbed. Usually, this would be chloride, but this ion is deficient due to its loss in vomit. Therefore, the balancing anion is bicarbonate, which perpetuates the alkalosis. The urine is inappropriately acidic, as the kidney is impaired in its ability to excrete bicarbonate. Infusion of sodium chloride replaces both extracellular fluid, hence reducing the stimulus to the renin–angiotensin–aldosterone system, and chloride, thus allowing renal excretion of bicarbonate.

Respiratory alkalosis

Respiratory alkalosis is a consequence of hyperventilation, causing reduced PCO_2. It occurs when the respiratory center in the brain is stimulated to increase ventilatory rate and depth. Usually a fall in PCO_2 acts to decrease ventilation, hence any stimulus must be sufficiently strong to overcome this.

Pulmonary causes for this include conditions where the stimulus to increase ventilation exceeds any impairment of gas exchange, such as mild asthma or a small pulmonary embolus. It may also occur in patients on mechanical ventilation when the rate is too high.

Hyperventilation due to anxiety or panic attacks may cause respiratory alkalosis and is usually self-limiting. Central causes for respiratory alkalosis, where the respiratory center is directly stimulated, include cerebral irritation, salicylate poisoning, and liver failure.

CASE 7.5

A 3-week-old newborn male has projectile vomiting after feeds for 3 days. On examination he is dehydrated.

	SI units	Reference range	Conventional units	Reference range
Venous blood				
pH	7.51	7.35–7.45	7.51	7.35–7.45
H^+	31 nmol/L	35–45	31 nEq/L	35–45
PCO_2	6.1 kPa	4.7–5.9	46 mmHg	35–45
Plasma				
Sodium	140 mmol/L	136–142	140 mEq/L	136–142
Potassium	2.9 mmol/L	3.5–5.0	2.9 mEq/L	3.5–5.0
Bicarbonate	40 mmol/L	21–28	40 mEq/L	21–28
Chloride	86 mmol/L	96–106	86 mEq/L	96–106
Anion gap	17 mmol/L	10–20	17 mEq/L	10–20
Anion gap (exc. K)	14 mmol/L	8–16	14 mEq/L	8–16
Urea (BUN)	20 mmol/L	2.9–8.2	56 mg/dL	8–23
Creatinine	112 μmol/L	53–106	1.27 mg/dL	0.6–1.2
Urine				
Sodium	6 mmol/l	Not applicable*	6 mEq/L	Not applicable*
Potassium	50 mmol/L	Not applicable*	50 mEq/L	Not applicable*
Urea	500 mmol/L	Not applicable*	1400 mg/dL	Not applicable*
Osmolality	650 mmol/kg	Not applicable*	650 mOsm/kg	Not applicable*
pH	4.3	4.5–8.0	4.3	4.5–8.0

*Interpret urine results as whether appropriate for clinical state and plasma results.

- **What type of acid–base disorder is present?**
- **What electrolyte losses initiate the disturbance?**
- **Should an acid or alkaline urine be produced?**

The newborn has pyloric stenosis. Vomiting stomach contents only causes loss of HCl and hence alkalosis. Dehydration (loss of ECF) causes maximal renin–aldosterone secretion with resulting renal reabsorption of Na^+ and loss of K^+ (hence hypokalemia). Since Cl^- has been lost by vomiting, HCO_3^- is reabsorbed with the Na^+ to maintain electroneutrality and the urine is inappropriately acid. **This perpetuates the alkalosis.** Note that there are no reference ranges for random urine, due to the ability of the kidney to adjust excretion over a wide range; results are interpreted as whether appropriate or not for the clinical context. In this case, urine sodium is low (that is, <10 mmol/L) and potassium is correspondingly high in response to aldosterone. The urea and osmolality are high in response to antidiuretic hormone, which is secreted in response to hypovolemia (see Chapter 5).

As with Case 7.3, blood gases have been measured on a venous sample as there is no clinical reason to measure PO_2. If it had been measured on arterial blood, it would tend to be slightly low, due to relative hypoventilation as an attempt to correct the alkalosis.

Respiratory alkalosis is often short-lived and resolves before renal compensation is able to take effect. Chronic respiratory alkalosis may occur during acclimatization to high altitude. Hyperventilation is stimulated by the low oxygen tension in the air, causing excessive loss of CO_2. The renal response is to increase excretion of bicarbonate, which falls in the plasma and restores the pH to normal.

Mixed acid–base disorders

The examples discussed above are those of single acid–base disturbances caused by one underlying disease. However, it is possible for two or more processes to be present at the same time, giving rise to a mixed acid–base disorder. This can be inferred if the measured PCO_2 and pH are not in agreement with each other as predicted by the graph in Figure 7.8.

The most common combination would be a dual acidosis—that is, with both a metabolic and respiratory component. In this situation, changes to both factors in the pH equation are affected; that is, there is simultaneously a fall in bicarbonate and a rise in PCO_2. The metabolic component is often lactic acidosis due to poor tissue oxygen delivery. This may be caused by reduced perfusion by oxygenated blood or by lung disease, which is also the origin of the respiratory component, either primarily or through an inability to increase ventilation to compensate for the metabolic acidosis. Examples of causes of mixed acidosis are:

- Cardiorespiratory arrest
- Drowning
- Smoke inhalation

Mixed disorders where there are opposite effects on pH are more difficult to identify, and may appear at first glance to be a compensated single disorder. However, as was stated previously, physiological overcompensation never occurs and if it appears to be present this would imply a mixed acid–base disorder. A metabolic acidosis plus a respiratory acidosis may occur with salicylate overdose and the pH would be higher than predicted from the low pCO_2 using Figure 7.8.

The only two disorders that cannot coexist are respiratory acidosis and respiratory alkalosis as they are due to hypo- and hyperventilation, respectively. Metabolic acidosis and alkalosis, however, can occur at the same time, albeit rarely. Whether an acidemia or alkalemia is the end result will depend on their relative contributions to the overall pH.

Theoretically, it is possible to even have a triple disorder consisting of metabolic alkalosis, metabolic acidosis, and a respiratory disorder (either hypo- or hyperventilation). Again, the final pH would depend upon the relative contributions of each disorder. This scenario is rarely encountered and a high anion gap in the presence of an alkalosis or normal pH may be the only clue to an additional metabolic acidosis.

The use of the graph of pH against PCO_2 (see Figure 7.8) is highly recommended for detecting mixed disorders.

Further clinical examples of mixed disorders are given in **Table 7.8**.

TABLE 7.8 Examples of mixed acid–base disorders

Disorders	Cause
Metabolic acidosis + respiratory acidosis	Respiratory failure, plus lactic acidosis from hypoxia
Metabolic alkalosis + respiratory alkalosis	Vomiting, plus hyperventilation
Metabolic alkalosis + respiratory acidosis	Diuretic treatment causing hypokalemia, plus chronic lung disease
Metabolic acidosis + respiratory alkalosis	Salicylate poisoning
Metabolic alkalosis + metabolic acidosis	Vomiting, plus ketoacidosis (increased anion gap, but HCO_3 unexpectedly high)

CASE 7.6

A 5-year-old boy is found at the bottom of a swimming pool. After cardiopulmonary resuscitation he had cardiac arrest on the way to hospital. On arrival he had blood gases measured.

	SI units	Reference range	Conventional units	Reference range
Arterial blood				
pH	6.87	7.35–7.45	6.87	7.35–7.45
H^+	135 nmol/L	35–45	135 nEq/L	35–45
PCO_2	14.0 kPa	4.7–5.9	105 mmHg	35–45
PO_2 (breathing 100% O_2)	9.8 kPa	11–13	74 mmHg	80–100
Lactate	11 mmol/L	0.6–1.7	99 mg/dL	5–15

- **What type of acid–base disorder is present?**

This is an extremely severe mixed acidosis, with results off the scale of the nomogram of pH versus PCO_2 (see Figure 7.8). The drowning has caused a respiratory acidosis due to lack of ventilation and CO_2 retention. Deprivation of tissues of oxygenated blood leads to anaerobic respiration and lactic acid generation, which causes metabolic acidosis. Thus, a "double acidosis" is present.

HYPERTENSION

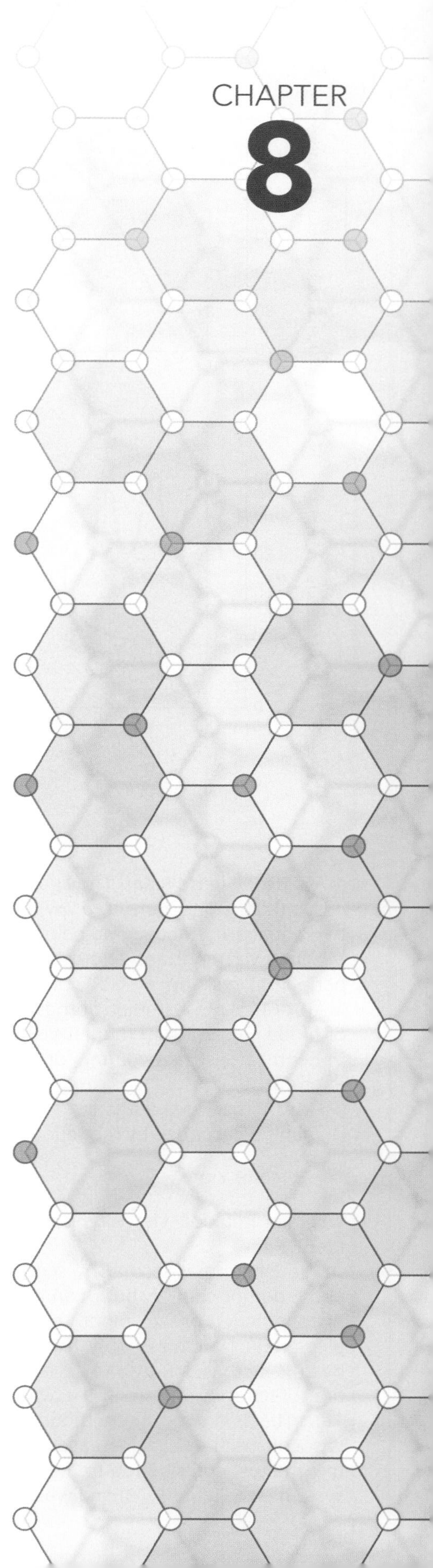

IN THIS CHAPTER

CARDIOVASCULAR PHYSIOLOGY

CLINICAL BIOCHEMISTRY OF HYPERTENSION

High blood pressure, or hypertension, is one of the key risk factors for cardiovascular disease. It is of huge public health significance, both in terms of its consequences and its treatment, with as many as 75 million adults having hypertension in the USA alone. In only a small minority of cases is the underlying cause understood: on a population basis, nearly all hypertension is labeled as essential, a euphemism for cause unknown. The physiology of blood pressure control is well understood, however, and many types of drug have been developed which target specific control mechanisms, with more drugs under development. It is important for the individual patient concerned not to miss the small minority of cases that do have a well-characterized cause in order to ensure that they receive the correct treatment.

8.1 CARDIOVASCULAR PHYSIOLOGY

Blood pressure

The blood in the cardiovascular system is pumped around the body in a series of closed vessels. As the left ventricle contracts (systole) and relaxes (diastole), the pressure that is exerted upon the walls of the arteries varies, normally rising to around 120 mmHg during systole and falling to around 80 mmHg during diastole (**Figure 8.1a**). Arterial blood pressure is therefore always quoted as two figures, such as 120/80 mmHg, where the first figure is the systolic blood pressure (SBP) and the second figure is the diastolic blood pressure (DBP). Such large variations in blood pressure over the cardiac cycle are only apparent in the arteries; by the time blood is flowing into the smallest arterioles and the capillaries, the blood pressure profile is smooth (**Figure 8.1b**). Tissue perfusion pressure is determined by the mean arterial blood pressure (MABP), which can be estimated using a simple equation:

$$\mathbf{MABP = DBP + [(SBP - DBP)/3]}$$

Inserting the normal blood pressure figures from above (120/80 mmHg), MABP is 93 mmHg. Mean arterial blood pressure is not simply the arithmetic mean of SBP and DBP since, at rest, the left ventricle spends twice as much time in diastole as it does in systole. Thus, DBP makes a greater contribution to

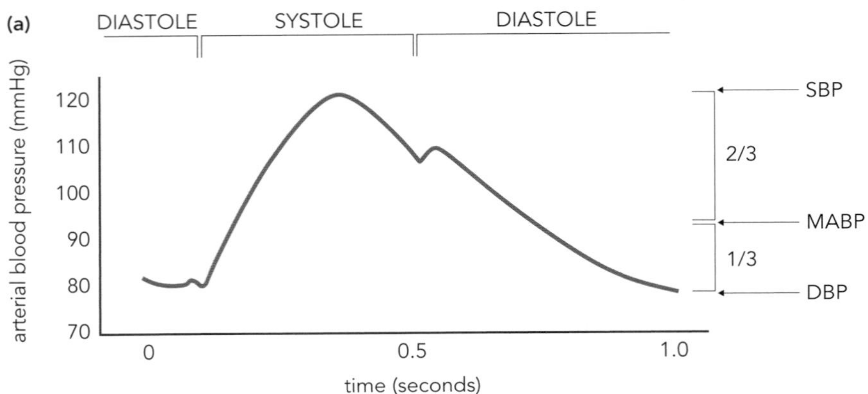

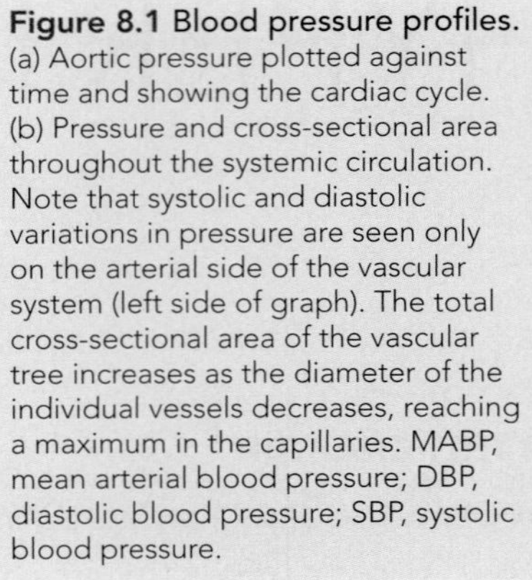
Figure 8.1 Blood pressure profiles. (a) Aortic pressure plotted against time and showing the cardiac cycle. (b) Pressure and cross-sectional area throughout the systemic circulation. Note that systolic and diastolic variations in pressure are seen only on the arterial side of the vascular system (left side of graph). The total cross-sectional area of the vascular tree increases as the diameter of the individual vessels decreases, reaching a maximum in the capillaries. MABP, mean arterial blood pressure; DBP, diastolic blood pressure; SBP, systolic blood pressure.

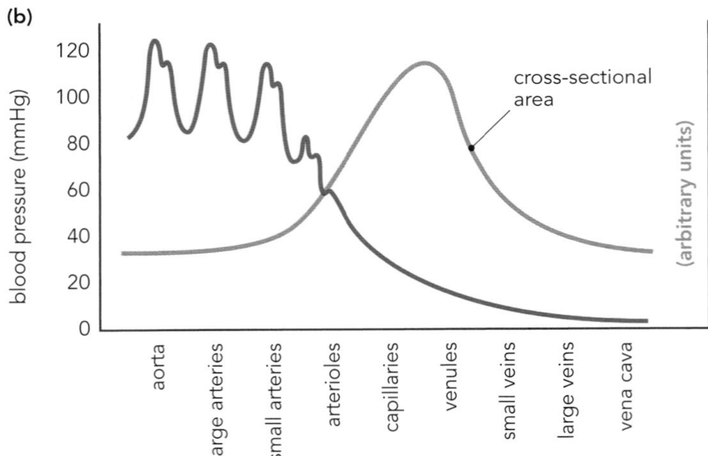

MABP (see Figure 8.1a). The figure for MABP is the point on the blood pressure curve that, if one were to draw a horizontal line across the curve, would give two equal areas above and below that line.

The two important determinants of MABP are cardiac output (CO) and total peripheral resistance (TPR), and increases in either or both tend to increase MABP. CO is the arithmetic product of heart rate and stroke volume (the volume of blood ejected from the left ventricle during each heart beat). Thus, changes in heart rate, stroke volume, or circulating blood volume have the potential to alter MABP.

MABP increases with increasing TPR. The factors affecting resistance to flow (R) can be described by the following equation:

$$\mathbf{R = 8\eta L/\pi r^4}$$

where η is blood viscosity, L is length of blood vessel, and r is blood vessel radius.

Under normal second-by-second circumstances in a given vascular bed, η and L do not change, but r can be influenced markedly by changing the degree of constriction of the smooth muscle in the wall of the blood vessel. For example, a reduction in r of only 10% results in a 52% increase in R, while reducing r by 20% increases R by 144%. Hence, vessel radius (and the degree of constriction of smooth muscle in the walls of blood vessels) has the largest effect on TPR. This is one of the methods that the body uses to alter MABP; generalized vasoconstriction can increase TPR and thereby raise MABP. Capillaries make the greatest contribution to the total cross-sectional area of the cardiovascular system (see Figure 8.1B) and while it might be thought that capillary constriction or dilatation might be important in altering TPR, capillary walls per se have no smooth muscle and thus little can be done to alter their cross-sectional area. However, flow to the capillary beds is controlled by the arterioles that feed them, and the walls of these arterioles are very muscular such that constriction or dilatation of these vessels results in large changes in TPR. For this reason, arterioles are commonly known as the resistance vessels of the cardiovascular system and are thought to make the greatest contribution to variations in TPR.

Neural mechanisms for control of blood pressure

Blood pressure is monitored by baroreceptors in the walls of blood vessels in various parts of the body. One of the major baroreceptor sites is in the carotid body, where distension of the wall of the carotid artery results in increased stimulation of specialized nerve endings and increased firing of those neurons. Other baroreceptors exist in the arch of the aorta. These baroreceptors send processes via the glossopharyngeal nerve (in the case of the carotid baroreceptors) and the vagus nerve (in the case of the aortic baroreceptors) to cardiovascular monitoring centers in the brain stem. Here, the neurons synapse with other neurons that control the firing rate of parasympathetic (vagal) and sympathetic (cardiac accelerator) nerve fibers to the heart, and sympathetic nerve fibers (vasomotor nerves) to the vasculature. In general, activation of baroreceptors by increased MABP leads to reduced sympathetic outflow from the cardiovascular center and increased vagal outflow to the heart. A reduction in sympathetic outflow by this mechanism would reduce heart rate and stroke volume and decrease vasomotor tone, thereby reducing CO, TPR, and MABP. An increase in parasympathetic outflow would reduce heart rate, thereby reducing CO and MABP. This negative-feedback loop functions to maintain MABP within relatively tight limits. One problem with baroreceptor control of blood pressure is that such receptors, if exposed to high levels of MABP for significant periods of time (1–2 days), reset their basal firing rate to suit the new levels of basal MABP. This can mean that sustained increases in MABP (as in hypertension) cannot be compensated for by normal mechanisms.

Endocrine mechanisms for control of blood pressure

Baroreceptor mechanisms in the body also serve to maintain blood pressure over the medium to long term. Such mechanisms are generally hormonal in nature and are therefore less rapid than the neural pathways described above. In view of the diverse nature of the potential hormonal responses to a change in blood pressure over the medium to long term, it would be simplistic to state that the consequences of a particular alteration in blood pressure would have a predictable effect on the blood concentration of a given set of circulating hormones. However, it is important to be able to predict what effect the various hormones individually might have on different parameters that have a role to play in determining blood pressure.

In general, the four major hormone systems that are at play are adrenal medullary norepinephrine/epinephrine, the renin–angiotensin–aldosterone system, adrenal cortical cortisol, and vasopressin (see **Table 8.1** for a summary of their actions).

The catecholamine hormone epinephrine is released from the adrenal medulla in response to a fall in MABP, as part of the sympathetic response. The adrenal medulla is a collection of innervated nerve cell bodies (chromaffin cells) which have not developed axonal processes but which normally discharge their transmitter (80% epinephrine, 20% norepinephrine) into the bloodstream. Thus, medullary epinephrine and norepinephrine are hormones but are released on activation of the sympathetic branch of the autonomic nervous system. These hormones act on α- and β-adrenoceptors on a wide range of body tissues to exert a physiological effect; in the case of the heart and blood vessels, they stimulate cardiac β-adrenoceptors and vascular α-adrenoceptors (which cause

TABLE 8.1 Major hormones affecting function of the cardiovascular system		
Hormone	**Specific effect**	**Effect on blood pressure**
(Nor)epinephrine	Peripheral vasoconstriction	Increased MABP, through increased TPR and increased CO
	Increased heart rate and force	
Angiotensin II	Peripheral vasoconstriction	Increased MABP, through increased TPR and increased CO
	Aldosterone secretion	
	Vasopressin secretion	
	Thirst	
Aldosterone	Increased salt and water reabsorption	Increased MABP, through increased CO
Cortisol	Permissive role—maintains/increases expression of adrenoceptors on peripheral tissues	Helps maintain or increase MABP
Vasopressin	Increased water reabsorption	Increased MABP, through increased CO and increased TPR
	Vasoconstriction	

CO, cardiac output; MABP, mean arterial blood pressure; TPR, total peripheral resistance.

vasoconstriction) to raise CO and TPR, respectively. This is also part of the general "fight or flight" response to a threatening stimulus. In conditions such as pheochromocytoma (where tumors develop from chromaffin cells), high circulating levels of these hormones can result in hypertension and a range of other effects.

The synthesis of epinephrine from norepinephrine in the adrenal medulla is influenced by cortisol, released from the adrenal cortex; exposure of chromaffin cells to cortisol results in the induction of the enzyme phenylethanolamine-*N*-methyl transferase (PNMT), which converts norepinephrine to epinephrine. Cortisol also maintains expression of adrenoceptors on the various target cells of the sympathetic nervous system such as the heart and blood vessels. Thus, when circulating levels of cortisol are abnormally low or high, this can result in low or high blood pressure, respectively.

The renin–angiotensin–aldosterone system is complex (**Figure 8.2**). Renin (released from the juxtaglomerular cells of the afferent arteriole of the renal

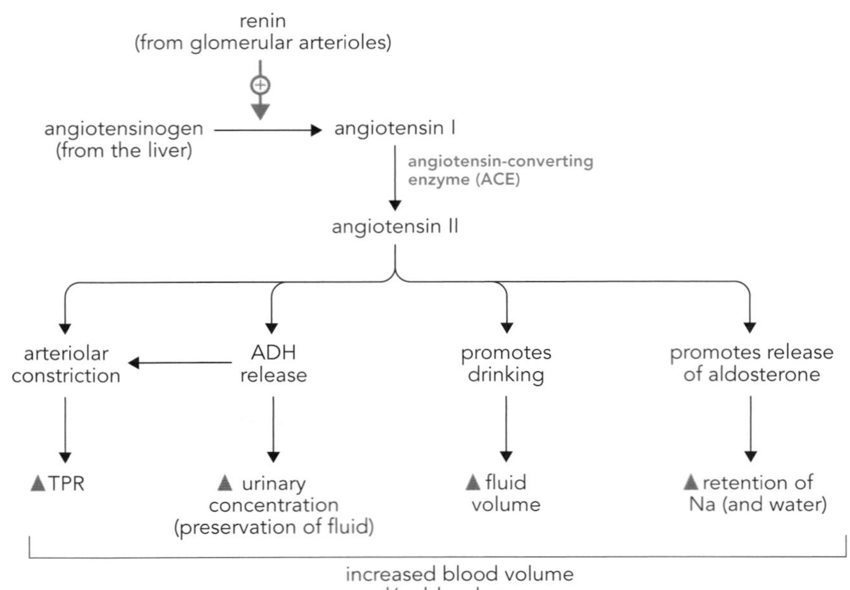

Figure 8.2 The role of the renin–angiotensin–aldosterone system in fluid balance and long-term regulation of blood pressure. Angiotensin II has multiple effects, the most important of which are release of aldosterone from the adrenal cortex and arteriolar constriction. ADH, antidiuretic hormone (vasopressin); TPR, total peripheral resistance.

corpuscle in response to a reduction in perfusion; see Chapter 4) circulates in the blood and converts angiotensinogen (a peptide produced by the liver) to angiotensin I. Angiotensin I is, in turn, converted to angiotensin II by the action of angiotensin-converting enzyme (ACE), which is present on vascular endothelial cells.

In situations where blood pressure is normal but there is significant renovascular disease (for example, renal artery stenosis), the resultant underperfusion of the kidney can result in renin release and trigger a cascade of events leading to elevated blood pressure. Diseases of the adrenal cortex, resulting in the under- or overproduction of adrenal cortical hormones such as aldosterone, can also have marked effects on fluid balance and blood pressure.

8.2 CLINICAL BIOCHEMISTRY OF HYPERTENSION

Elevated blood pressure is an important risk factor for cardiovascular disease, especially ischemic heart disease and stroke. Stroke may be due to either atherosclerotic disease or hemorrhage. Hypertension is involved in the pathogenesis of both types, as it is a risk factor for atheroma and increases the mechanical stress in areas of weakened arterial wall within the cerebral circulation. In common with other risk factors, such as blood lipids, blood pressure level is normally distributed within the population, with no threshold below which cardiovascular events never occur. Hence, the definition of hypertension is arbitrary but based on the observed risk of events in the population, particularly strokes. A 2 mmHg increase in average systolic blood pressure is associated with a 7% increase in death from heart disease and 10% from stroke. These are increments in relative risk, but modern cardiovascular risk assessment places much more emphasis on an individual's absolute risk, which is determined by all of their risk factors. This is expressed as a risk of that individual suffering a major cardiovascular event, such as a heart attack, in the next 10 years, and theoretically can be anywhere between 0 and 100%. The definition of high risk is arbitrary and varies over time, according to expert opinion in balancing the risks (and financial costs) against the benefits of treatment. The threshold has been lowered from 30% to 10% within the last two decades and may be reduced even further. Assessment of absolute risk can be done using charts or online calculators, which require input of factors such as age, gender, blood pressure, cholesterol, and smoking status. Definitions of hypertension and thresholds for treatment undergo periodic re-evaluation by expert panels as new evidence emerges. Currently, different stages of hypertension are recognized. Stage 1 is defined as office or clinic blood pressure (BP) of 140/90 mmHg or ambulatory BP of 135/85 mmHg. Stage 2 is 160/100 mmHg or 150/95 mmHg. Severe hypertension is BP 180/110 mmHg or above. The stage is determined by the highest BP; that is, the SBP or DBP.

Clinical practice point 8.1

The threshold blood pressure for the definition of hypertension is arbitrary.

Clinical practice point 8.2

An individual's absolute risk of cardiovascular events depends on all their risk factors.

Biochemical investigations are used to:

- Investigate the cause
- Monitor the effects of the disease and its treatment
- Assess other cardiovascular risk factors

The vast majority of hypertension is termed essential because its cause is not well understood. The pathogenesis appears to be a complex interplay of multiple genes involved in the regulation of neural and endocrine control of blood pressure, along with environmental factors such as salt and alcohol intake, obesity, and stress.

Clinical practice point 8.3

Most hypertension is essential: its cause is unknown.

A small proportion of hypertensive patients (around 5–10%) have an underlying disorder that raises blood pressure (termed secondary hypertension). This proportion is higher amongst patients who are younger, or who are resistant to treatment with standard drugs, or who have additional symptoms. All of these factors are indications to investigate further.

Clinical practice point 8.4

Secondary hypertension should be considered in young people, in resistance to drug therapy, and if additional symptoms or signs are present.

Pheochromocytoma

Pheochromocytoma is a tumor of catecholamine-secreting tissue in the sympathetic nervous system. Classically they arise from cells within the adrenal medulla, but extra-adrenal tumors make up about 10% of cases and may occur anywhere along the sympathetic chain. According to the 2004 World Health Organization classification, such tumors should be termed paraganglionomas rather than pheochromocytomas. Most commonly these tumors secrete norepinephrine, but epinephrine or very rarely dopamine may be the predominant hormone. Secretion is episodic and blood pressure may vary throughout the day, although sustained hypertension also occurs. Other symptoms associated with catecholamine release include palpitations, headache, pallor, abdominal pain, and severe anxiety attacks.

The majority of pheochromocytomas are benign in the sense that the cells do not metastasize but are not clinically benign in terms of their endocrine effects, which may eventually be fatal due to causing a hypertensive crisis. Around 10% are malignant tumors and a similar number are bilateral.

Biochemical tests to diagnose pheochromocytoma are based on the measurement of catecholamines or their metabolic products (metanephrines) and may be performed on urine or plasma. There are advantages and disadvantages to each of these.

Measurement of fractionated urinary catecholamines (epinephrine and norepinephrine) has been the most widely used test for the last two decades, having superseded the assay for vanillylmandelic acid (VMA), an end-stage metabolite of catecholamines. Measuring VMA is much less sensitive for detecting pheochromocytoma and is also prone to interference from foodstuffs and prescribed and over-the-counter medicines. The assay of catecholamines is relatively straightforward to perform and samples are stable as long as they are kept at acid pH. The 24-hour urine collection period allows an integrated measurement, which avoids the problem of intermittent secretion being missed. Plasma catecholamines can also be measured, but these are very unstable and the low concentrations are technically demanding to assay. Few laboratories therefore offer this test.

Recent studies have cast some doubt on the theoretical basis of using catecholamines to detect pheochromocytoma. This is because catecholamines may be metabolized to metanephrines (metanephrine and normetanephrine) within the tumor itself (**Figure 8.3**). Thus secretion of catecholamines into the blood, and ultimately the urine, will be lower than predicted. However, the diagnostic value of plasma or urine metanephrines will be higher. This has been confirmed in studies indicating that metanephrines have sensitivities of 97–99% compared to 85–90% for catecholamines.

Analytical practice point 8.1

24-hour urine metanephrines and catecholamines are the tests of choice for diagnosis of pheochromocytoma.

Sensitivity is the most important attribute for detecting a rare and potentially fatal tumor, but a high degree of specificity is also required to avoid false positive results requiring further testing. The degree of elevation above the

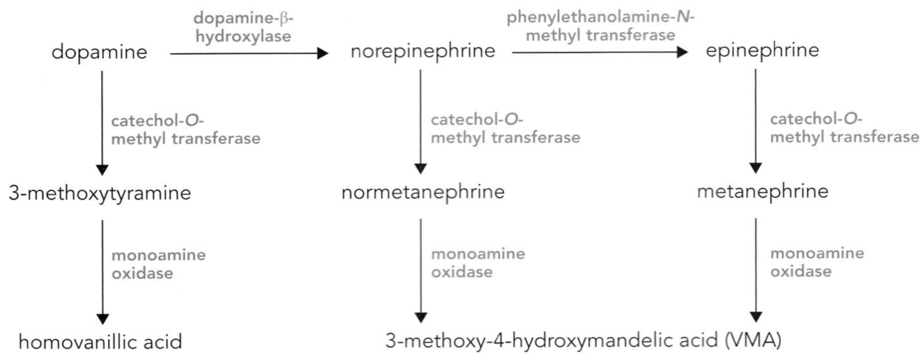

Figure 8.3 Metabolic steps in the interconversion of the catecholamines and breakdown of the catecholamines to their major metabolites. The three vasoactive hormones of the adrenal medulla—dopamine, norepinephrine, and epinephrine—are each converted by catechol-*O*-methyl transferase to metanephrines and then by monoamine oxidase to the end products homovanillic acid and 3-methoxy-4-hydroxymandelic acid (vanillylmandelic acid; VMA). All are excreted into the urine in increased amounts in the presence of a pheochromocytoma, an autonomously secreting tumor of the adrenal medulla.

reference range is important in determining the probability of a tumor, since minor rises are often seen with some drugs, including those used for treating hypertension. It is also recognized that hypertension itself is associated with modest increases in catecholamine secretion. It is therefore preferable to use a reference range that has been derived from a hypertensive population rather than a normotensive one.

Plasma metanephrines have a slightly higher sensitivity than 24-hour urine metanephrines and avoid the inherent difficulties in making a complete collection. They are, however, technically more difficult to measure at concentrations of less than 1 nmol/L, compared to levels of around 1 μmol/L in urine.

To improve the diagnostic value of plasma metanephrines, they may be measured before and after the administration of clonidine. In normal individuals, clonidine suppresses catecholamine secretion, but this does not occur in pheochromocytoma patients. An additional test that may help to confirm the presence of a pheochromocytoma is chromogranin A. This is a protein released from the vesicles containing catecholamines and levels are increased in pheochromocytoma, as well as in a number of other conditions where secretion from neuroendocrine cells is increased, for example carcinoid syndrome.

In a small number of pheochromocytomas, dopamine is co-secreted, and in an even smaller proportion of cases, dopamine alone is produced. These dopamine-only tumors do not cause hypertension or other specific clinical effects and so present late, once they have enlarged or spread. For this reason, dopamine-secreting tumors are believed to be more likely to be malignant. However, this is probably not an intrinsic biological property, just a reflection of late presentation. Dopamine is measured on the same chromatogram as epinephrine and norepinephrine when high performance liquid chromatography is used to measure urine catecholamines. Most high dopamine results are due to renal secretion, which increases in natriuresis. Occasionally, very high dopamine excretion is seen but this is more likely to be due to the therapeutic use of levodopa to treat Parkinson's disease rather than dopamine-secreting pheochromocytoma.

Primary hyperaldosteronism

Some aspects of hyperaldosteronism are discussed in Chapter 11. The underlying pathology may be an aldosterone-secreting tumor (usually benign [Conn's syndrome]), hyperplasia of the zona glomerulosa, or an interesting variant called glucocorticoid-suppressible hyperaldosteronism. This variant is due to a dominantly inherited hybrid gene which encodes an enzyme with properties of both 11β-hydroxylase and aldosterone synthase. This results in adrenocorticotropic hormone (ACTH)-dependent production of aldosterone in the zona fasciculata, which can be inhibited by giving glucocorticoids to suppress ACTH secretion by the anterior pituitary.

Primary hyperaldosteronism is suspected in patients with refractory hypertension, particularly when hypokalemia is present or easily provoked by low doses of thiazide diuretics. However, it is important to recognize that up to 40% of primary hyperaldosteronism cases may be associated with normokalemia. The prevalence of Conn's syndrome is often said to be up to 5% in hypertensive patients, but this is likely to be an overestimate due to selection bias. The initial investigation of suspected primary hyperaldosteronism is measurement of simultaneous renin and aldosterone, with the results expressed as a ratio. Classically, the aldosterone will be raised and the renin suppressed, giving a high ratio. Unfortunately, a number of drugs commonly used to treat hypertension interfere with the renin–angiotensin–aldosterone axis and may give rise to spuriously high or low ratios. If possible, patients should be taken off these drugs and managed with α-adrenoceptor antagonists (α-blockers), which have the least effect. This is often difficult to achieve, as the very reason for investigation is that blood pressure is poorly controlled, and clinicians may be unhappy to accept suboptimal treatment.

Analytical practice point 8.2

Aldosterone–renin ratio is the first-line test for primary hyperaldosteronism, but can be affected by antihypertensive drugs.

If a high aldosterone–renin ratio is found it requires further investigation, as it is not in itself diagnostic. Theoretically, adenomas are not sensitive to angiotensin II (release provoked by a change of posture to standing) but are sensitive to ACTH (which has diurnal variation), whilst the opposite is the case for hyperplasia. Unfortunately, postural tests to distinguish adenomas from hyperplasia are not reliable. Adrenal imaging has the disadvantage of detecting a high number of nonfunctioning adenomas, which are present incidentally (so-called incidentalomas). Sometimes, selective venous sampling from catheterization of the adrenal veins is required to determine if one side is secreting significantly more aldosterone than the other.

CASE 8.1

A 35-year-old man is found to have stage 2 hypertension (blood pressure 163/92 mmHg) at a routine health check. He is not overweight and does not drink alcohol. His BP does not improve with salt restriction and he is started on a thiazide. This does not improve his BP either and his plasma potassium is found to be low (2.9 mmol/L [2.9 mEq/L]). He is referred to a hospital hypertension clinic.

	SI units	Reference range	Conventional units	Reference range
Plasma				
Renin (activity)	0.1 pmol/L/h	1.1–2.7	0.05 ng/mL/h	0.6–4.3
Aldosterone	450 pmol/L	55–250	16 ng/dL	2–9
Aldosterone–renin ratio	4500	0–900	320	0–20

- **What is the significance of the hypokalemia and lack of response to treatment?**
- **What do the renin and aldosterone results indicate?**

A mass is seen in the right adrenal gland on computed tomography scanning. He undergoes selective venous sampling using cannulas placed in the adrenal veins.

	SI units	Reference range	Conventional units	Reference range
Aldosterone				
Right arm	450 pmol/L	55–250	16 ng/dL	2–9
Right adrenal vein	1200 pmol/L	55–250	43 ng/dL	2–9
Left adrenal vein	220 pmol/L	55–250	8 ng/dL	2–9

- **Why is aldosterone measured at three sites?**
- **What do the results indicate?**

Hypertension in a relatively young and fit individual raises the possibility of an underlying cause; that is, secondary hypertension. This is more likely when there is little response to treatment, especially if two or more drugs are required. Hypokalemia in the presence of hypertension suggests primary hyperaldosteronism and this is confirmed by suppressed renin, high aldosterone, and raised aldosterone–renin ratio.

The presence of an adrenal mass does not prove that this is the cause of the high aldosterone because incidental tumors are common. Selective venous sampling shows the right adrenal is the source of aldosterone, thus confirming the tumor is responsible. The left adrenal secretion is suppressed, due to low renin. The result from the right arm reflects dilution of aldosterone in the total blood volume.

The role of aldosterone in essential hypertension is a subject of significant interest and debate. Although classical primary hyperaldosteronism is relatively uncommon, a significant proportion of hypertensive patients respond to the aldosterone antagonist spironolactone. There is also growing evidence that aldosterone is directly toxic to the heart, kidneys, and blood vessels and contributes to the risk of cardiovascular disease. Spironolactone is a nonselective aldosterone antagonist with a number of side effects which limit its use, and more selective drugs are under development.

Other causes of hypertension

Hypertension may be a consequence of other diseases, although it may not be the most prominent feature. Of particular importance is kidney disease, where high blood pressure may be both a cause and a consequence of decreased renal function. Blood pressure control is an important aspect of slowing the progression of chronic kidney disease. Renal function can be monitored by both plasma markers (creatinine and estimated glomerular filtration rate, eGFR) and urine markers (protein–creatinine or albumin–creatinine ratio).

Clinical practice point 8.5

Hypertension is both a cause and a consequence of chronic kidney disease.

Endocrine disorders causing hypertension include Cushing's syndrome, thyroid disease, hyperparathyroidism, and acromegaly. Investigations for these conditions would usually only be considered if other clinical features were apparent, however. The probability of diagnosing one of these conditions in the presence of hypertension alone would be low.

Biochemical monitoring of the hypertensive patient

Monitoring of the effects of blood pressure on target organs and of drug treatment, together with assessment of other cardiovascular risk factors (particularly lipids), make up the majority of biochemical testing in hypertension. Mild to moderate hypertension may respond to lifestyle measures (weight loss, aerobic exercise, and restriction of dietary salt and alcohol), but in most cases drug therapy is also required. The large number of drug classes and mechanisms of action indicates the difficulty in treating hypertension adequately. Drug classes commonly used to treat high blood pressure are listed in **Table 8.2**.

Thiazides are currently considered to be the first-line treatment, with other drugs such as ACE inhibitors, angiotensin receptor blockers (ARBs), or calcium channel antagonists being used as alternatives if additional medical conditions are present. Second-line agents include aldosterone antagonists, such as spironolactone, and α-adrenoceptor antagonists. Two, three, or even four different agents may be needed to control hypertension adequately to the target blood pressure levels recommended by expert societies around the world.

TABLE 8.2 Drugs used to treat hypertension

Drug class	Example
Thiazide	Hydrochlorothiazide
Angiotensin-converting enzyme (ACE) inhibitor	Ramipril
Angiotensin II receptor blocker	Losartan
Calcium channel blocker	Amlodipine
β-Adrenoceptor antagonist (β-blocker)	Atenolol
Loop diuretic	Furosemide
Aldosterone antagonist	Spironolactone
α-Adrenoceptor antagonist (α-blocker)	Doxazosin
Vasodilator	Hydralazine

Depending on the doses and combinations, there may be metabolic side effects, which can be detected by routine monitoring of renal function (plasma creatinine and eGFR), electrolyte levels, and uric acid.

Thiazides are diuretics at high doses and can cause a number of metabolic side effects including hyponatremia, hypokalemia, hyperglycemia, and hyperuricemia leading to gout. However, these are rarely seen in current practice as the doses used are much lower than in previous decades. There is no additional antihypertensive effect from using higher doses. Hypokalemia occurring on low-dose thiazide raises the possibility of Conn's syndrome.

Drugs that block the renin–angiotensin–aldosterone system have a predictable effect on plasma potassium: they tend to raise it. In some patients this elevation may be sufficiently high that these drugs cannot be used at all. Another risk is worsening renal impairment due to decreased blood flow to the kidneys. This may occur when the renal artery is partially obstructed, perhaps by atheroma, and the hypoperfusion stimulates the release of renin, leading to increased angiotensin II and ultimately aldosterone. Renal perfusion is therefore maintained due to the high angiotensin II levels. When an ACE inhibitor or ARB is given, the angiotensin II level cannot be maintained and the kidney is hypoperfused once again. The result of this is a rise in plasma creatinine and a fall in eGFR. Thus renal function is carefully monitored after starting ACE inhibitors or ARBs and after dose increments. The risk is especially high in patients with peripheral vascular disease as this implies atheroma is also present in the renal arteries.

Hypertension as a cardiovascular risk factor

Hypertension is not a disease per se unless the blood pressure is particularly high (>180/120 mmHg) or enters an accelerated phase with end-organ damage (so-called malignant hypertension), when it becomes a medical emergency. In the vast majority of individuals labeled as hypertensive, the blood pressure is a risk factor for future cardiovascular events, such as myocardial infarction, heart failure, stroke, or kidney failure. However, elevated blood pressure is just one of a number of risk factors—albeit an important and treatable one—which include age, gender, smoking, diabetes, and plasma lipids. All of these should be assessed in a hypertensive patient, as multiple risk factor intervention is more effective for preventing cardiovascular events than treating just one factor. Plasma lipids should be measured as part of routine cardiovascular risk assessment. Increasingly, statin drugs (HMG-CoA reductase inhibitors) are given to individuals at high cardiovascular risk (for example, 20% risk of suffering a major cardiovascular event over 10 years), including those with hypertension, irrespective of the plasma cholesterol, because of the importance of reducing multiple risk factors.

CALCIUM, MAGNESIUM, AND PHOSPHATE

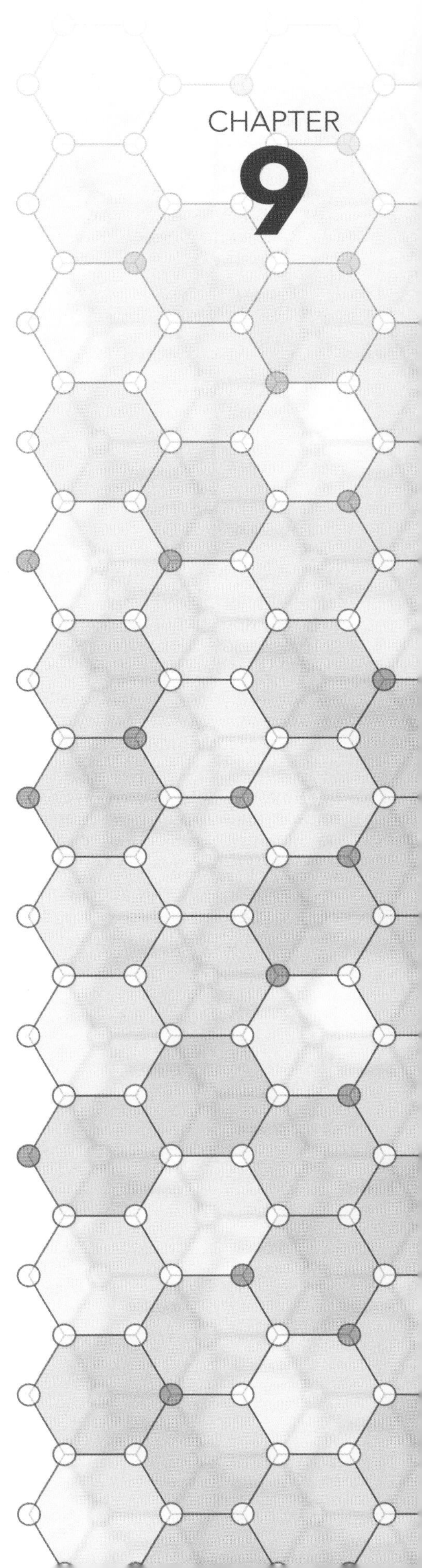

IN THIS CHAPTER

Calcium is the major inorganic element in the body, fulfilling a range of major physiological roles, both structural (for example, in bone) and functional (for example, in neuromuscular excitability and as an enzyme co-factor).

An average 70 kg adult human contains approximately 1.3 kg of calcium of which 99% is present as calcium hydroxyapatite [$Ca_{10}(PO_4)_6(OH)_2$] in bones and teeth, where it constitutes 70% of bone weight. The importance of the structural integrity of bone is illustrated by its functions, which include the following.

- Provides a rigid protective structure for soft tissues; for example, the skull protects the brain
- Serves as a store of minerals, especially calcium, magnesium, phosphorus, and sodium
- Structural role, providing the skeleton on which to hang other tissue
- Protects the bone marrow, the major site of synthesis of blood cells

It was originally thought that the skeleton was relatively inert but it is now apparent that, like other tissues, it undergoes turnover (remodeling), the skeleton being replaced every 3–10 years. In addition, approximately 500 mg of calcium move into and out of bone daily in response to fluctuations in plasma calcium concentration (**Figure 9.1**).

9.1 CALCIUM DISTRIBUTION IN THE BODY

About 1% of body calcium is present in the extracellular fluid (ECF) while less than 0.1% is in cells (intracellular fluid; ICF). However, it is the control of the ECF concentration in a narrow range of 2.2–2.6 mmol/L (8.8–10.4 mg/dL) that is

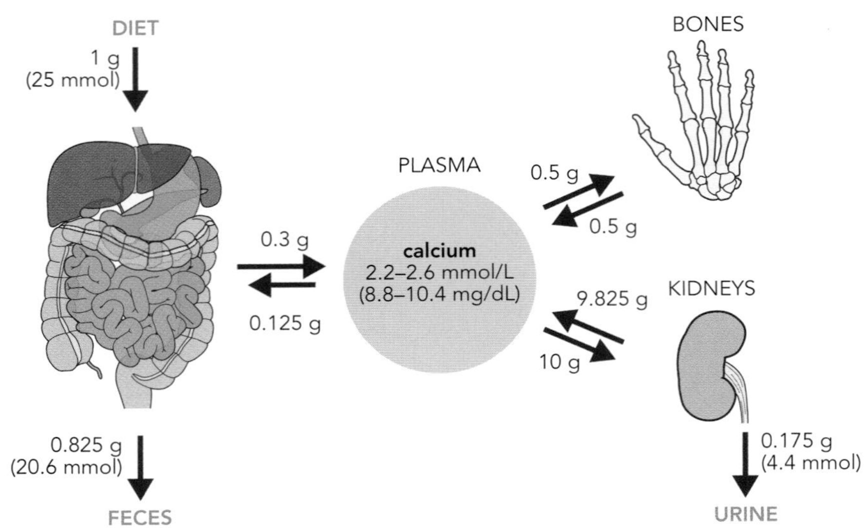

Figure 9.1 Calcium balance. From the daily 1 g of calcium found in the average diet, approximately 30% is absorbed via the gut into the body. However, the body excretes 0.125 g daily as part of gastrointestinal secretions; therefore the net absorption is less than 20% of the daily intake. The net absorption is partly dependent upon the body's calcium status; more can be absorbed when there is deficiency. The plasma concentration of calcium is kept within a very tight range, partly by varying the amount that can be absorbed via the gastrointestinal tract, partly via the flux between bone and plasma, and also by the excretion or reabsorption from the kidneys. When the system is in equilibrium there is no net loss or gain in calcium.

responsible for physiological well-being. About 50% of ECF calcium is present in the free ionized form, with approximately 40% bound to plasma proteins, mainly albumin and globulins, and 10% present in a diffusible form as calcium salts such as lactate and citrate (**Figure 9.2**). It is the free ionized form that is important physiologically and which is under homeostatic control. Reference ranges for calcium and other related analytes are given in **Table 9.1**.

Keeping the plasma calcium within this tight range is essential for the maintenance of membrane excitability in nervous tissue and muscle. Calcium in the ECF is also important as a co-factor for enzymes, including some proteases of the clotting cascade, and provides calcium for bone formation. The concentration of intracellular calcium is of three orders of magnitude lower than ECF calcium, in the submicromolar range. Cells that respond to stimuli by raising intracellular calcium levels have a store of calcium in the endoplasmic reticulum (compare with the sarcoplasmic reticulum in muscle cells, or myocytes) from which calcium is released via a signaling system involving inositol trisphosphate. The rise in intracellular calcium, which binds to high-affinity calcium-binding proteins such as calmodulin, elicits the intracellular response.

Ca^{2+}
1.25 mmol/L
Ca-albumin
0.8 mmol/L
complexed Ca
0.25 mmol/L
Ca-globulin
0.2 mmol/L

Figure 9.2 Calcium in the ECF. Calcium is found in the plasma in several different forms. The total calcium that is frequently measured is the sum of the different forms that can be found. In this illustration the total calcium is 2.50 mmol/L, and approximately half of this is the free or ionized calcium (Ca^{2+}), which is the form that the body responds to. The rest is either bound to proteins, mainly to albumin (32% of the total) and several globulins (8% of the total), or is complexed (10% of the total) with small molecules, for example citrate.

TABLE 9.1 Reference ranges in the analysis of plasma calcium, magnesium, phosphate, and related analytes

Analyte	Reference range	
	SI units	Conventional units
Calcium (adjusted)	2.20–2.60 mmol/L	8.8–10.4 mg/dL
Calcium (ionized)	1.15–1.30 mmol/L	4.6–5.2 mg/dL
Magnesium	0.70–1.00 mmol/L	1.7–2.3 mg/dL
Magnesium (ionized)	0.44–0.60 mmol/L	1.0–1.5 mg/dL
Phosphate	0.74–1.52 mmol/L (higher values are seen in children)	2.3–4.7 mg/dL
Alkaline phosphatase (ALP)	30–120 U/L (method dependent; higher values are seen in children)	30–120 U/L
Parathyroid hormone	1.06–6.90 pmol/L	10–65 pg/mL
25-Hydroxy vitamin D ($D_3 + D_2$)	37–104 nmol/L in winter 37–200 nmol/L in summer	15–42 ng/mL in winter 15–80 ng/mL in summer
1,25-Dihydroxy vitamin D	52–160 pmol/L	22–67 pg/mL

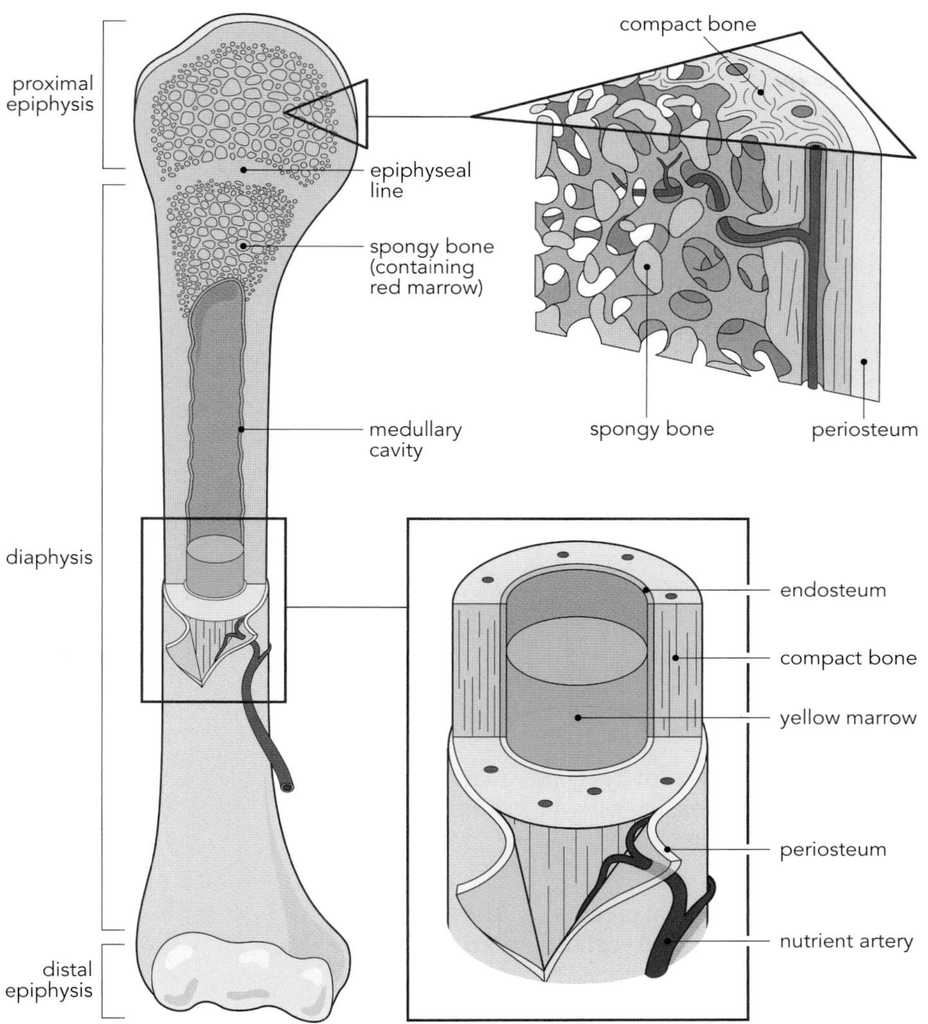

Figure 9.3 Structure of bone. The basic structure of bone, in which the calcium is found mineralized with phosphate onto a protein backbone structure. In addition, magnesium also forms part of the bone structure. The two main types of bone are spongy (trabecular) and compact (cortical).

Structure of bone

Bone consists of two general types, cortical and trabecular bone (**Figure 9.3**). Cortical bone gives strength, while trabecular bone is more metabolically active—although it only accounts for 20% of the bone mass, it is responsible for 80% of bone metabolic activity. The proportion of cortical to trabecular bone varies; a higher proportion of cortical bone is found in limbs, whilst a higher proportion of trabecular bone is found in the rib cage, backbone, and skull. Bones that have blood-forming marrow are limited mainly to the breastbone, spine, skull, hips, and ribs. The location of these types of bone becomes important when looking for changes associated with particular disease types.

Bone formation and resorption

Bone is a dynamic system that is being remodeled continuously in response to gravitational stress and the demands of calcium homeostasis (**Figure 9.4**). This remodeling process of bone resorption and formation is carried out through the actions of four principal cell types: osteoclasts, monocytes, osteoblasts, and osteocytes.

Osteoclasts

These are large multinucleated cells derived from hematopoietic stem cells of the monocyte/macrophage lineage which fuse under the action of 1,25-dihydroxy vitamin D [1,25(OH)$_2$D]. They resorb bone by secreting acid to dissolve the

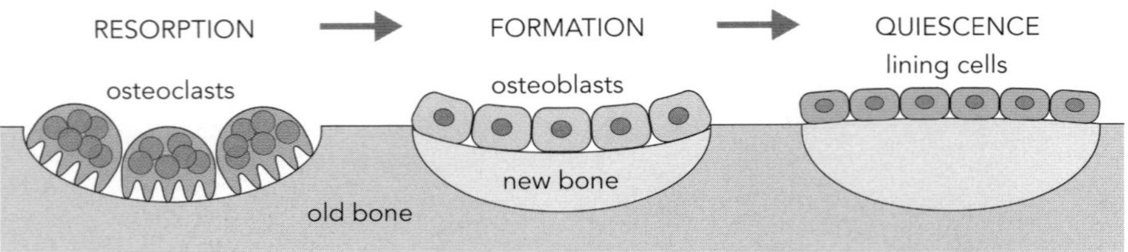

Figure 9.4 Resorption and formation of bone.
Bone is not inert, and different areas of bone will be turning over all the time. This process starts with resorption of old bone by osteoclasts, which break down the bone and release its contents, including calcium and phosphate, into the plasma. The next stage is formation, where new bone is then laid down by osteoblasts in the pits the osteoclasts have produced. In health, the system is in balance; that is, osteoclastic activity does not exceed osteoblastic activity and vice versa. In between these cycles of resorption and formation, the bone is said to be quiescent and is covered with lining cells.

inorganic mineral matrix, and using proteases and hyaluronidase to hydrolyze the exposed organic collagen matrix. They do not possess receptors for parathyroid hormone (PTH) but appear to respond indirectly through signals from osteoblasts.

Monocytes
These tidy up the resorption cavities formed by osteoclast activity by removing residual collagen and other proteins, and possibly by releasing growth factors which stimulate osteoblast proliferation and differentiation.

Osteoblasts
These major bone-forming cells are derived from mesenchymal precursors. They synthesize and secrete collagen to form the new organic extracellular matrix (osteoid) and control the mineralization of osteoid to form new bone. The major secreted protein is type I collagen. They also secrete a specific isoform of alkaline phosphatase and measurement of this isoenzyme in blood is indicative of osteoblast activity. Osteoblasts possess receptors for both PTH and $1{,}25(OH)_2D$.

Osteocytes
These are osteoblasts that have become entrapped in the bone matrix during bone growth. They lie in extracellular fluid within the canalicular spaces and communicate with each other via pseudopod extensions throughout the matrix. Their major function is to maintain the bone matrix and to facilitate the movement of calcium into and out of bone surfaces as part of the day-to-day maintenance of calcium homeostasis.

In homeostasis, the actions of these cells are tightly coupled by the process of remodeling such that there is no net change in bone mass (that is, amount of bone resorbed = amount of bone formed). When osteoclast activity exceeds osteoblast activity, the general bone mass will decrease, leading to osteoporosis, and since more metabolically active bone is restricted to certain sites in the skeleton, these sites will be where osteoporosis will present initially.

9.2 CONTROL OF CALCIUM HOMEOSTASIS

Three tissues—intestine, bone, and kidney—are intimately involved in the control of calcium homeostasis, which is regulated primarily through the actions of two hormones, PTH and $1{,}25(OH)_2D$ (**Figure 9.5**). A third hormone, calcitonin, is involved also in calcium homeostasis.

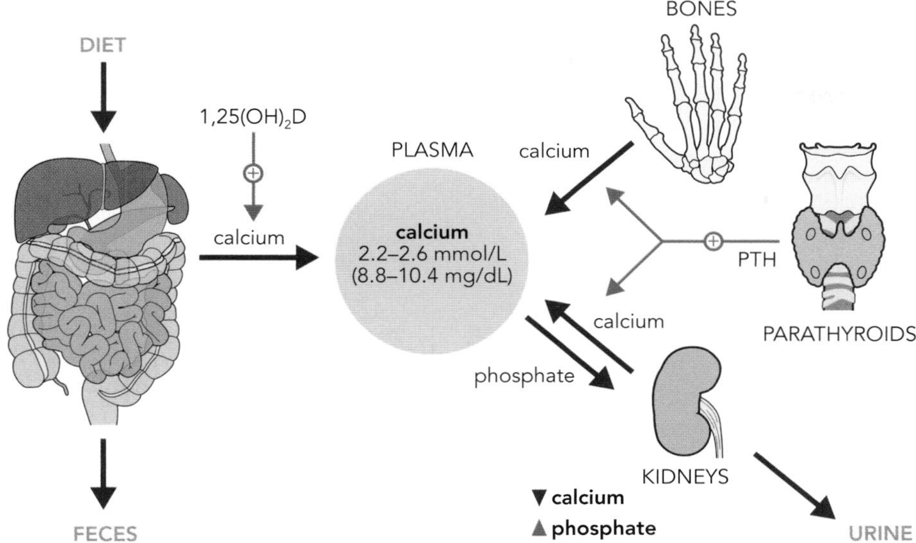

Figure 9.5 Control of calcium. Plasma calcium is kept tightly regulated in the main by the actions of two hormones, parathyroid hormone (PTH) and 1,25-dihydroxy vitamin D (1,25(OH)$_2$D). As the plasma concentration of calcium falls, the chief cells of the parathyroid gland sense the fall in the ionized fraction, which stimulates the release of PTH into the blood. PTH works to raise the plasma calcium back to normal in two ways: first, by increasing bone turnover in such a way as to increase the net flux of calcium from bone into plasma; and second, by decreasing the amount of calcium excreted into urine. This increased reabsorption of calcium by the kidneys is accompanied by an increased loss of phosphate into urine. When the plasma calcium rises back into the reference range, the parathyroid gland senses this rise in ionized calcium and switches off any more PTH release, thus keeping the system in balance. Although PTH does not directly increase calcium absorption from the gut, it does increase the enzymatic conversion of 1-hydroxy vitamin D to the much more active 1,25(OH)$_2$D, which will increase the absorption of calcium from the gut.

Acute (minute to minute) changes in plasma calcium are regulated by the actions of PTH on the mobilization of calcium from a labile pool on the surface of bone and on calcium excretion/resorption in the kidney. More chronic control (hours to days) depends on the action of PTH on the synthesis of 1,25(OH)$_2$D and the action of this metabolite on intestinal absorption of calcium.

Parathyroid hormone

PTH is synthesized in the four parathyroid glands located behind the lobes of the thyroid. Synthesis and secretion of PTH are exquisitely sensitive to deviations from an individual's set-point calcium, which are detected through a calcium sensor on the surface of the parathyroid cell. Through this sensor, the cells can respond to changes in serum ionized calcium as small as 25 μmol/L: a fall in plasma calcium results in an increase in PTH secretion, while a rise in plasma calcium results in a fall in PTH secretion. Although the secreted hormone is an 84-amino-acid peptide, the PTH gene encodes an initial translation product, prepro-PTH, of 115 amino acids (**Figure 9.6**). This peptide is synthesized on the rough endoplasmic reticulum and undergoes an initial proteolytic cleavage of a 25-amino-acid pro-sequence (signal or leader sequence) by a signal peptidase as it is translocated into the lumen of the endoplasmic reticulum. Further cleavage of a 6-amino-acid fragment occurs as it is packaged into secretory vesicles for storage prior to secretion. The biological activity of PTH resides in the N-terminal 1–34 amino acid residues; the circulating hormone undergoes further hydrolysis rapidly (half-life [T½] 3–5 minutes) in the hepatic Kupffer cells, yielding a biologically active N-terminal peptide (residues 1–36) and an inactive C-terminal peptide (residues 37–84). The N-terminal peptide is cleared by glomerular filtration and proteolysis by the kidney more rapidly (T½ 3–5 minutes) than the C-terminal peptide, and the serum concentration of the C-terminal peptide is greater than the N-terminal peptide.

Vitamin D

The most important members of the D family of vitamins are cholecalciferol (vitamin D_3), which is synthesized from the mammalian sterol cholesterol, and ergocalciferol (vitamin D_2) synthesized from the plant sterol ergosterol. Biologically they can be regarded as interchangeable in man. Vitamin D is obtained exogenously from the diet (D_2 and D_3) or endogenously from synthesis in skin (D_3). Oily fish, dairy products, and eggs are the major sources of dietary D_3, while fortification of margarine with D_2 from plant sources is quantitatively

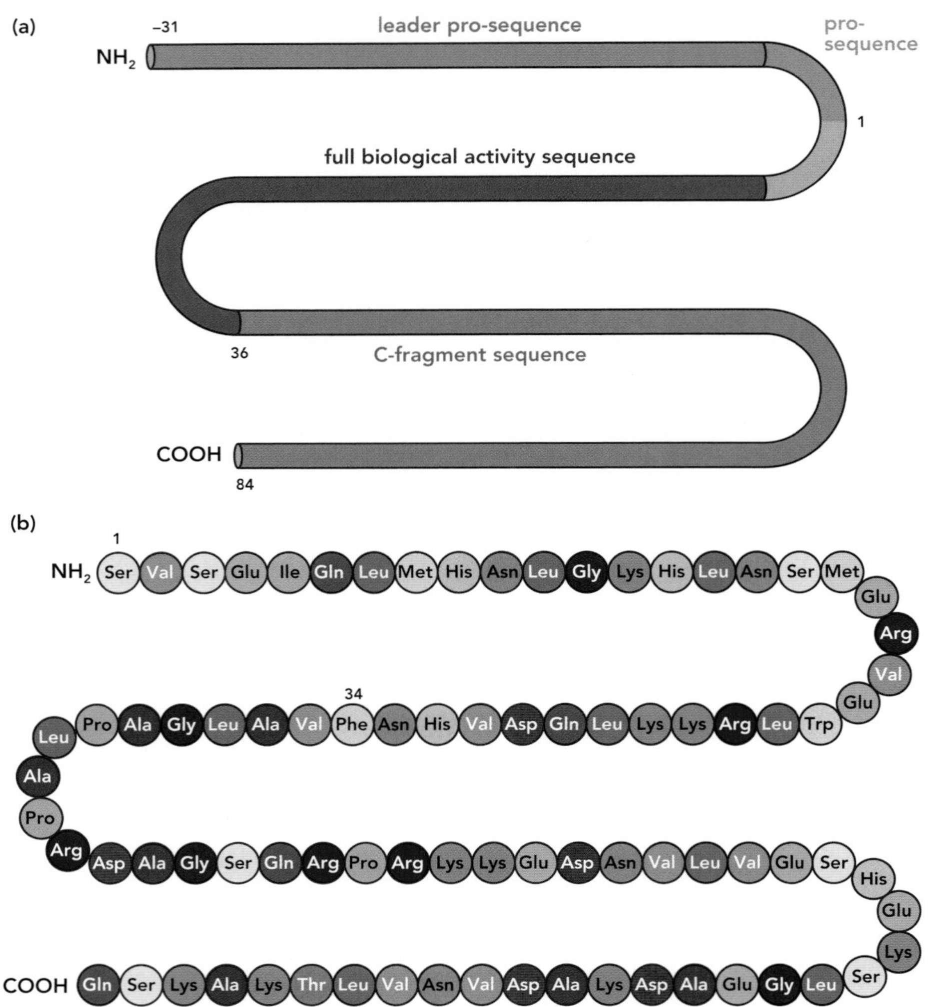

Figure 9.6 Parathyroid hormone. Parathyroid hormone (PTH) is synthesized in the parathyroid glands and stored in granules before it is released into the blood. It is stored as a prepro-hormone of 115 amino acids. (a) The prepro-hormone consists of the 1–84 amino acids associated with the intact PTH that is found in plasma, plus 31 amino acids attached to the N-terminal end. By convention these are numbered –1 to –31, and are referred to as the pro-sequence and leader pro-sequence. During the packaging and release process, 31 amino acids are cleaved from the N-terminal end, giving rise to (b) the intact PTH molecule of 84 amino acids found in plasma. It appears that only the first 34 amino acids of the N-terminal end of PTH are required for all of the functions related to calcium metabolism. The first six amino acids of the N-terminal are cleaved off during the degradation of the protein in blood, leaving the 7–84 fragment, which has no biological activity, but a long half-life and is found to accumulate in renal failure.

very important. In the absence of cutaneous synthesis, the dietary requirement of vitamin D to prevent clinical deficiency is about 2.5 μg/day for an adult and about 10 μg/day for a child.

Endogenous synthesis of vitamin D begins in the skin, where the action of sunlight (ultraviolet light) on 7-dehydrocholesterol in the epidermis opens the B-ring of cholesterol (nonenzymic photolysis) with the production of previtamin D, which then isomerizes to vitamin D. This endogenously synthesized vitamin D is transported to the liver by vitamin D-binding protein, a specific α-globulin, where it mixes with the exogenous vitamin D derived from absorption from the upper small intestine.

The parent vitamin D has little biological activity and has to undergo activation through two hydroxylations to give $1,25(OH)_2D$ (**Figure 9.7**). One of these reactions, 25-hydroxylation, occurs in the liver through the action of vitamin D 25-hydroxylase, a microsomal mixed function oxidase, which requires NADPH and molecular oxygen as co-substrates. Synthesis of 25-hydroxy vitamin D (25-OHD) is not product-inhibited and high circulating concentrations may be present, for example after vitamin D overdose. Although this is the major plasma form of vitamin D, it has little biological activity in terms of calcium absorption and bone resorption. It may be stored in fat and liver. The most important step in activation of vitamin D, 1α-hydroxylation, occurs in the proximal convoluted and straight tubules of the kidney, where $1,25(OH)_2D$ is produced by the action

7-dehydrocholesterol
UV LIGHT
DIET
VITAMIN D_3
(VITAMIN D_2 diet only)
LIVER
vitamin D 25-hydroxylase
25-OHD
$24,25(OH)_2D$
(inactive)
KIDNEYS
(25-OH)D
1α-hydroxylase
$1,25(OH)_2D$
(active hormone)

Figure 9.7 Vitamin D.
Vitamin D can be synthesized in human skin from 7-dehydrocholesterol with the aid of ultraviolet (UV) light (exposure of the skin to sunlight). This isomer is known as vitamin D_3, which can also be found in the diet, as can the other major isomer, vitamin D_2, which differs in an extra methyl group and double bond (shown in red). Vitamin D_2 is not synthesized in human skin. The D_2 and D_3 isomers have the same biological activity. The vitamin D has to go through two hydroxylation steps to make it fully biologically active. The first step occurs in the liver to give rise to 25-hydroxy vitamin D (25-OHD), followed by the second hydroxylation in the kidney to give rise to the fully active hormone, 1,25-dihydroxy vitamin D [$1,25(OH)_2D$]. Small amounts of 24,25-dihydroxy vitamin D [$24,25(OH)_2D$, made by a 24-hydroxylase enzyme in the kidney] can also be found, but this compound is believed to be biologically inactive.

of 25-hydroxy vitamin D_3 1α-hydroxylase [(25-OH)D 1α-hydroxylase], a mitochondrial mixed function oxidase which is up-regulated by PTH and decreased intracellular phosphate. The 1α-hydroxylase is strongly product-inhibited and, as the level of $1,25(OH)_2D$ rises, the 1α-hydroxylase activity falls while the renal 24-hydroxylase activity increases (see Figure 9.7). The precise order of the two hydroxylations of vitamin D is not critical for the production of the active metabolite, since the therapeutic administration of 1α-hydroxy vitamin D [1α-(OH) D; calcidiol] in vitamin D deficiency results in its hydroxylation to $1,25(OH)_2D$. Although the plasma concentration of $1,25(OH)_2D$ is of the order of a thousand times less than 25-OHD, it is the major biologically active metabolite of vitamin D, with important effects on calcium metabolism in bone and gut. All vitamin D metabolites are hydrophobic and are transported in the circulation by a hepatically synthesized vitamin D-binding protein.

Calcitonin

Calcitonin is a phylogenetically well-conserved protein, with only small differences in structure between such diverse species as salmon and man. It is secreted by the parafollicular C cells of the thyroid as a 32-amino-acid peptide (3.5 kDa). In man, the calcitonin gene encodes an initial translation product of 17 kDa and this undergoes co- and post-translational processing to yield the parent hormone (**Figure 9.8**). Although secretion of calcitonin is stimulated by a rise in plasma calcium and the gut hormone gastrin, its physiological role is unclear, since patients lacking measurable calcitonin appear to have no obvious skeletal abnormality or altered calcium homeostasis. However, calcitonin

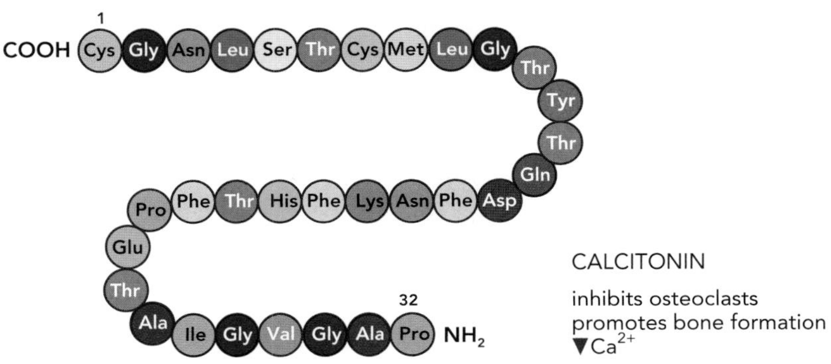

Figure 9.8 Calcitonin.
The amino acid sequence of the hormone, calcitonin, which has several functions including inhibiting osteoclasts and promoting bone formation. Interestingly, the calcitonin structure is phylogenetically well preserved; there are similar amino acids across wide ranges of species, so much so that calcitonin derived from salmon can be used to treat hypercalcemia in humans.

does have a transient inhibitory action on osteoclasts and may cause calcium to move from the plasma into bone. This latter action has led to the suggestion that the release of gastrin by the G cells of the gastric mucosa, following the arrival of food in the stomach, stimulates the secretion of calcitonin from the thyroid. A rise in plasma calcitonin leads to a transient hypocalcemia through promotion of calcium entry into bone. A transient hypocalcemia in turn stimulates PTH secretion from the parathyroid glands, and the overall effect will be a transient retention of dietary calcium absorbed by the intestine, as the raised PTH increases renal calcium retention while the body adjusts to calcium homeostasis. At pharmacological doses, calcitonin inhibits osteoclastic bone resorption and also inhibits calcium and phosphate reabsorption in the kidney and is used in the treatment of hypercalcemia.

The role of the intestine

An average diet provides approximately 1 g/day of calcium of which about 30% (300 mg) is actively absorbed via specific calcium transport proteins in the villi in the duodenum and proximal jejunum and by facilitated diffusion in the distal intestine. Although absorption is most active in the proximal intestine, the major portion of dietary calcium intake is absorbed in the distal segments because of their greater length and surface area. However, the net absorption (fractional absorption) is only about 50% of this (150 mg) since calcium is lost back to the gastrointestinal tract through pancreatic and biliary secretions and the sloughing off of intestinal epithelial cells. The fractional absorption is increased by the action of 1,25$(OH)_2$D on intestinal epithelial cells to stimulate the synthesis of calcium transport proteins involved in the absorptive process. The plasma levels of 1,25$(OH)_2$D are raised during hypocalcemia (for example, when on prolonged low-calcium diets) and the fractional absorption is increased. Furthermore, conditions that lead to an increase in plasma 1,25$(OH)_2$D can lead to a pathological increase in plasma calcium and vice versa. PTH has no direct actions on the intestine but regulates this process through its effect on the renal synthesis of 1,25$(OH)_2$D. The process of adaptation, which takes from hours to days to reach a new steady state, links changes in ECF calcium to compensatory changes in calcium absorption and the restoration of calcium homeostasis.

The role of the kidney

In contrast to calcium and phosphate co-transport in intestine and bone, the kidney regulates reabsorption of calcium and phosphate independently through the action of PTH. The kidneys filter about 10 g of non-protein-bound calcium each day of which >95% is subsequently reabsorbed by the renal tubules. The bulk of this reabsorption takes place in the proximal nephron and is hormone independent. The filtered load of calcium just exceeds the maximum capacity of tubular reabsorption (set by PTH) so that, as the filtered load increases, so does its excretion, and similarly conversely. In the proximal renal tubule, calcium and magnesium are absorbed passively, linked to sodium and chloride reabsorption down a concentration gradient. The remaining 5% of filtered calcium is subject to PTH-dependent reabsorption in the distal tubule. This process is exquisitely sensitive and is responsible for the immediate maintenance of plasma calcium homeostasis. An increase in plasma PTH, due for instance to hypocalcemia, leads to increased renal reabsorption of calcium, and urinary calcium excretion may fall to zero in some conditions such as primary hyperparathyroidism. Major causes of altered calcium excretion are shown in **Table 9.2**.

Under normal dietary conditions, however, humans are in zero calcium balance; approximately 0.175 g of calcium are lost in the urine daily and the renal output (0.175 g) plus fecal excretion (0.825 g) equals the dietary input (1 g). The other important action of PTH in the kidney is the stimulation of the 1α-hydroxylase responsible for the formation of 1,25$(OH)_2$D in the proximal tubule, as described above.

TABLE 9.2 Major causes of altered calcium excretion

Decreased calcium excretion	Increased calcium excretion
Reduced filtered load due to hypocalcemia	Increased filtered load due to hypercalcemia
Malabsorption of calcium due to intestinal disease	Increased intestinal absorption of calcium
Increased tubular reabsorption (PTH or PTHrP)	Renal leak of calcium (for example renal stones)
Increased proximal tubular sodium reabsorption	Reduced levels of PTH
Use of thiazide diuretics	

PTHrP, parathyroid hormone related peptide.

Further actions of PTH in the kidney are the inhibition of the reabsorption of phosphate (causing phosphaturia and hypophosphatemia) and bicarbonate (leading to proximal renal acidosis).

The role of bone

Calcium moves into and out of bone in response to hormonal stimuli and fluctuations in its concentration in ECF in order to maintain calcium homeostasis. Two separate processes occur in bone for the maintenance of calcium homeostasis. One, regulated by PTH and $1,25(OH)_2D$, is responsible for the day-to-day movement of calcium into and out of bone in response to small changes in plasma calcium as might occur after a meal rich in dairy products (rise), during an overnight fast (fall), or between meals (fall). This process involves the cells lining the bone surface (osteoblasts) and the osteocytes lying in the canalicular space. These cells have PTH receptors and are responsible for the movement of about 500 mg/day of calcium between bone and ECF (osteocytic osteolysis). The second process, bone remodeling, is essential for maintaining the structural integrity of bone, particularly in response to the normal occurrence of microfractures brought about by gravitational stress. This is the process of remodeling in which osteoclastic resorption, stimulated by PTH and $1,25(OH)_2D$, is tightly coupled to osteoblastic formation of bone (stimulated by PTH) such that under normal circumstances the amount of bone resorbed is balanced by that formed. In this case, although calcium is recycled through the ECF, there is no impact on calcium homeostasis. Mineralization of osteoid by osteoblasts requires the action of both PTH and $1,25(OH)_2D$, as demonstrated by the occurrence of rickets in vitamin D deficiency. A number of pathological conditions impose a stress such that net resorption arising from destruction of mineralized bone occurs in an effort to maintain ECF calcium homeostasis.

In summary, regulatory mechanisms are present within the body to maintain the plasma calcium concentration within a narrow range, even when dietary intake may vary throughout the day. Ingestion of calcium from, for instance, a meal rich in dairy products (milk, cheese, or yoghurt) produces a mild hypercalcemia which suppresses release of PTH from the parathyroid glands, with a consequent increase in renal calcium excretion and inhibition of bone resorption. These two actions combine to normalize plasma calcium. On the other hand, during periods of decreased or zero calcium intake (such as during an overnight fast), a slight reduction in plasma calcium leads to a transient increase in PTH and a subsequent increase in calcium reabsorption in the renal tubules, reducing renal calcium loss. The slight increase in PTH also stimulates the movement of calcium from labile bone stores, and again calcium concentration is normalized. Chronic hypocalcemia leads to raised PTH levels that stimulate the synthesis of $1,25(OH)_2D$ which, in turn, will lead to an increase in the fractional absorption of calcium in the intestine.

Other hormones can affect the rates of bone formation and resorption and should be taken into consideration when assessing bone pathology (**Table 9.3**).

TABLE 9.3 Summary of hormone effects on bone

Hormone	Action	Effect
Parathyroid hormone (PTH)	Increases turnover, stimulates osteoclasts	Net loss
1,25-dihydroxy vitamin D	Stimulates mineralization, osteoclast generation, and calcium absorption from gut	Net gain
Calcitonin	Retraction of osteoclast from bone surface	Net gain
Thyroid hormones	Vital for normal growth	Net loss in excess
Growth hormone	Vital for normal growth	Net gain
Glucocorticoids	Acute versus chronic	Long term, inhibits formation
Estrogens and androgens	Vital for growth, development, and maintenance; stimulate resorption	Net gain
Insulin	Stimulates bone matrix formation and mineralization	Net gain

9.3 MAGNESIUM BALANCE

Magnesium is the second most abundant intracellular metal ion and the fourth most abundant metal ion in the plasma. It is vital for most metabolic processes as ATP is only active when it is complexed with magnesium. Therefore energy-consuming or energy-producing processes (for example enzyme-catalyzed reactions or pumps) rely on a supply of magnesium. It is also required for the stability of membranes and for many other macromolecular structures.

Magnesium levels are under looser control than for calcium, with very little if any hormonal control over maintenance of the plasma concentration (**Figure 9.9**). The major site for control is the kidney, and the major influencing factor appears to be the plasma concentration of magnesium itself.

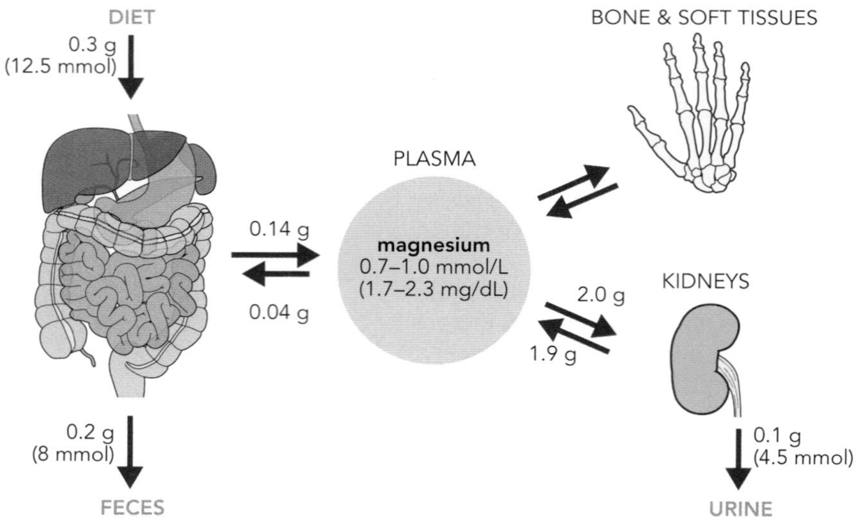

Figure 9.9 Magnesium balance. From the daily 0.3 g of magnesium found in the average diet, approximately 45% is absorbed via the gut into the body. However, the body excretes 0.04 g daily as part of gastrointestinal secretions, therefore the net absorption is about one-third of the daily intake. Similar to calcium, the net absorption is partly dependent upon the body's magnesium status; more can be absorbed when there is deficiency. Although there is flux between plasma and bone and soft tissues, the kidney is the main regulatory organ for maintaining magnesium levels. When the system is in equilibrium there is no net loss or gain in magnesium.

9.4 PHOSPHATE BALANCE

The inorganic element phosphorus, in the form of phosphate, is part of molecules that are essential for many aspects of both cell structure and function. Organic forms of phosphate (phosphomonoesters or phosphodiesters) are involved in the following processes:

- Structure (phospholipids, hydroxyapatite of bone)
- Replication (nucleotides, DNA, RNA)
- Signaling (cyclic adenosine monophosphate [cAMP], inositol trisphosphate, protein kinases)
- Cellular energy state (ATP, ADP, AMP)
- Cellular redox state (NAD/NADH, NADP/NADPH)
- Cell cycle (phosphorylated proteins)
- Metabolic intermediates (glycolysis, and so on)
- Modulation of oxygen binding to hemoglobin (action of 2,3-bisphosphoglycerate)

Inorganic phosphate is important in:

- Structure (calcium phosphate, hydroxyapatite)
- Buffering (as $H_2PO_4^-$, HPO_4^{2-})

Unlike calcium homeostasis, which is tightly regulated by the actions of $1,25(OH)_2D$ and PTH, the regulation of phosphate homeostasis is much looser and appears to have no specific hormonal control apart from the actions of PTH on the kidney. The normal range of ECF phosphate is 0.74–1.39 mmol/L (2.3–4.3 mg/dL).

In general, the absorption of phosphate via the intestine and its movement into and out of bone are linked to that of calcium, but the major tissue involved in maintaining serum phosphate within the normal range is the kidney. The mechanism of movement into and out of cells is, however, not yet fully understood; during metabolic acidosis, phosphate exits cells to produce a rise in ECF phosphate concentration, while, in alkalosis, phosphate enters cells and may lead to hypophosphatemia. Furthermore, hyperglycemia (such as may be seen during oral or parenteral infusions of glucose) may also lead to hypophosphatemia as phosphate follows glucose entry into cells.

Phosphate absorption in the intestine

An average Western diet contains 1–1.5 g of phosphate per day of which approximately two-thirds is absorbed (that is, a fractional absorption of ~67%). This remains relatively constant, although $1,25(OH)_2D$ may increase it to some extent. Like calcium, phosphate is absorbed by active processes in the duodenum and jejunum and by facilitated diffusion in the distal intestine. PTH has no direct effect on these processes.

Phosphate and bone

While bone is a major source of phosphate for ECF, phosphate movement into and out of bone is thought to be linked to that of calcium, which moves in response to the hormonal stimuli described earlier.

Phosphate and kidney

It is in the kidney that the movements of calcium and phosphate are independent, particularly in response to PTH. The filtered load of phosphate exceeds slightly the overall maximal tubular reabsorption (usually at or near the normal serum phosphate concentration) with the result that phosphate spills into urine. Phosphate is reabsorbed actively in the proximal convoluted tubule (PCT) linked to sodium. PTH lowers the renal tubular maximum for phosphate in the early part of the PCT thereby increasing phosphate excretion proximally. It also inhibits phosphate reabsorption in the later parts of the PCT and thus the combined effects of PTH increase urinary excretion of phosphate (phosphaturia). Between 5 and 15% of phosphate is also reabsorbed in the convoluted distal tubules.

Examples of the independent control of calcium and phosphate in the kidney include the following.

Hypocalcemia
Here, decreased ECF calcium stimulates PTH secretion, which in turn increases entry of both calcium and phosphate into ECF from bone and intestine via the effects of $1{,}25(OH)_2D$. Renal effects of PTH ensure absorption of calcium, while excess phosphate is excreted, resulting in increased ECF calcium with little change in ECF phosphate.

Hypophosphatemia
Decreased ECF phosphate stimulates renal $1{,}25(OH)_2D$ synthesis, resulting in increased calcium and phosphate absorption in the intestine and probably also resorption of bone. The consequent increase in ECF calcium then suppresses PTH secretion, thereby enhancing renal tubular reabsorption of phosphate and eliminating the increased filtered load of calcium.

Combined hypocalcemia and hypophosphatemia
Decreased ECF calcium leads to a PTH-driven increase in calcium through effects on the intestine (through the actions of $1{,}25(OH)_2D$) and bone. As the ECF calcium rises, so PTH secretion falls, with a concomitant progressive rise in the renal threshold for phosphate. Thus phosphate will now be retained and its concentration normalized.

Hypercalcemia
Here, the changes are the opposite of those seen in hypocalcemia.

Hyperphosphatemia
The homeostatic concentrations of calcium and phosphate are close to the solubility product of calcium phosphate, so that a rise in ECF phosphate leads to deposition of calcium phosphate in bone and soft tissues. The resulting fall in ECF calcium stimulates PTH secretion causing inhibition of renal phosphate reabsorption in the distal tubules, excretion of excess phosphate in the urine, and normalization of ECF phosphate.

9.5 WHAT IS MEASURED WHEN ASSESSING CALCIUM, MAGNESIUM, AND PHOSPHATE IN BONE DISORDERS?

The assessment of calcium, magnesium, and phosphate is common in the investigation of bone disorders; adult reference ranges for these parameters are given in Table 9.1. The laboratory measurement of calcium in routine practice is predominantly by one of two dye methods (Arsenazo III or ortho-cresolphthalein) or by an indirect ion-selective electrode (ISE). All these methods measure the total amount of calcium in plasma or serum. However, calcium can exist in blood in its free form, Ca^{2+} (free or ionized calcium), or bound to albumin and other proteins, or complexed with small anions such as citrate. Since it is the ionized form that the body responds to, efforts are made to reflect this in the routine assessment of calcium, and results can be reported as adjusted or corrected calcium, where it is reported back to a normal level of albumin as in the following example.

Plasma and serum calcium concentrations are most frequently measured as total calcium levels; that is, as the total measurement of all the different fractions. However, since up to 50–60% of the calcium can be bound to albumin,

$$\mathbf{Ca^{2+} + albumin \underset{acid}{\overset{alkali}{\longleftrightarrow}} Ca\text{-}albumin\ (\sim 60\%)}$$

and the fact that plasma albumin concentrations can vary considerably in health and disease, misleading results can be produced. For example, in many severe

Analytical practice point 9.1

Measurement of 24 hour urine calcium requires the urine to be collected into a bottle containing concentrated acid. This acidification stops calcium salt precipitation which occurs in neutral and alkaline conditions. However, most calcium methods were designed for the measurement of calcium in serum or plasma, and are very sensitive to the pH of the colorimetric reaction. Dependent upon the method employed the assay buffer maintains a pH between 6 to 8 for the reaction. This can cause problems with small urine collections (<500 mL of urine in 24 hours) which will have too much acid present, and the assay buffer will not be capable of achieving the correct assay pH. This gives rise to artifactually low results.

diseases albumin levels can be more than 50% lower than normal, which will in turn produce a total calcium measurement of less than the reference range and suggest hypocalcemia. Although in this case the total calcium will be low, the ionized fraction (which the body responds to) will not be, and the body will be biologically normocalcemic. In order to overcome this problem of potentially giving out misleading results, the total calcium result is adjusted (or corrected) back to a mean plasma albumin level, usually around 40 g/L (4 g/dL). Although correction factors will vary between laboratories, this approximately equates to a correction factor of 0.1 mmol/L (0.4 mg/dL) of calcium for every 5 g/L (0.5 g/dL) of albumin back to 40 g/L (4 g/dL). So, for a total calcium of 1.95 mmol/L (7.8 mg/dL, very low) where the patient's plasma albumin was 20 g/L (2 g/dL), the calcium would correct back to 2.35 mmol/L (9.4 mg/dL, normal) as follows:

For SI units:

Corrected calcium (mmol/L) = measured total Ca (mmol/L) + 0.1 × (40 – measured serum albumin [g/L])/5

(where 40 represents the normal albumin level in g/L)

$$= 1.95 + 0.1 \times (40 - 20)/5$$

$$= 2.35 \text{ mmol/L}$$

For conventional units:

Corrected calcium (mg/dL) = measured total Ca (mg/dL) + 0.4 × (4 – measured serum albumin [g/dL])/0.5

(where 4 represents the normal albumin level in g/dL)

$$= 7.8 + 0.4 \times (4 - 2)/0.5$$

$$= 9.4 \text{ mg/dL}$$

It should be noted that this correction or adjustment is only an approximation, and when the patient has a pH imbalance the correction will be invalid, as the amount of calcium that albumin binds is strongly dependent on pH. Alkaline conditions increase calcium binding whilst acid conditions decrease binding.

The ionized form of calcium also can be measured directly, usually by point-of-care equipment employing a direct ISE, which is found increasingly on acute wards. Although such measurements are unstable (they need to be analyzed within 60 minutes of collecting the blood sample), they are an ideal way of assessing the true ionized calcium and will be independent of the effects of small counterions and pH. The reference ranges are totally different, to total calcium results of course.

Magnesium can be measured by dye methods also (usually xylidyl blue) or by a direct ISE, while phosphate is usually measured by dye methods (phosphotungstic acid).

PTH is usually measured by immunoassay and reference values will vary between manufacturers, but most assays attempt to be specific to the whole molecule; however, there is a degree of cross-reaction with the 7–84 residue peptide, which may compete for PTH receptors with the intact molecule. Vitamin D can be measured by a variety of immunoassays and also by chromatographic methods, with high performance liquid chromatography (HPLC) tandem mass spectrometry giving the most accurate results. Most laboratories will measure the 25-OHD when a vitamin D level has been requested. Most immunoassays cross-react with both the D_2 and D_3 isomers and the results are a summation of the two isomers, whilst chromatographic methods can measure both the D_2 and D_3 isomers separately. The measurement of $1,25(OH)_2D$ is also used diagnostically, although it circulates at much lower concentrations and is consequently more difficult to measure. Calcitonin measurements are not required when assessing patients for disorders of bone or calcium; however, they are used when monitoring thyroid cell carcinoma.

9.6 HYPOCALCEMIA

Hypocalcemia by definition occurs when the adjusted calcium is below 2.20 mmol/L (<8.8 mg/dL) or the ionized calcium is below 1.15 mmol/L (<4.6 mg/dL). Since a low albumin is a common finding in many hospital patients, it is always advisable to adjust the calcium in order to take into account this problem. However, remember that the adjustment is only an approximation and does not take into account the acid–base status of the patient. Therefore, for example, in a patient with an acute respiratory alkalosis caused by hysterical overbreathing, the patient's high blood pH means that there is less ionized calcium so, although the adjusted calcium may be normal, the patient is functionally hypocalcemic and presents with symptoms such as paresthesia. Similarly, in patients with acidosis, there may be an apparent hypocalcemia with an adjusted calcium measurement, whereas the ionized calcium will be in the normal range due to the effect of the acidosis. The measurement of ionized calcium at the pH of the patient will overcome these problems.

Causes of hypocalcemia

Some of the causes of hypocalcemia are:

- Chronic kidney disease/acute kidney injury
- Vitamin D deficiency
- Defects in vitamin D synthesis or action
- Magnesium deficiency
- Acute pancreatitis
- Hypoparathyroidism
- Pseudohypoparathyroidism
- Infusion of phosphate, citrate, or albumin
- Calcium sensor abnormalities

In most cases hypocalcemia is essentially due to:

- Lack or functional lack of PTH
- Defect in vitamin D
    - Supply
    - Absorption
    - Activation
    - Action

As a rule of thumb when investigating hypocalcemic patients with normal renal function, if the phosphate is low, a lack or functional lack of vitamin D is probably the cause; if phosphate is high, PTH is the most likely defect. The hypocalcemia due to renal disease is linked to the reduced capacity of the kidney to convert 25-OHD to the $1{,}25(OH)_2D$ form (**Figure 9.10**). There is a rare disorder where the 1α-hydroxylase enzyme is defective and hypocalcemia can occur. This enzyme (also known as 25-hydroxy cholecalciferol 1-hydroxylase or calcidiol 1-monooxygenase) is encoded by the *CYP27B1* locus on human chromosome 12. A number of mutations in the gene are associated with vitamin D-dependent rickets type 1 (VDDR1), an autosomal recessive inherited disorder. The existence of a mutation hot spot in the gene is suggested by the observation that a 7-base-pair sequence duplication is shared by seven families

Analytical practice point 9.2

Blood collected into tubes containing EDTA as the anticoagulant (as used for full blood counts) will give artifactually low results for calcium and magnesium. These are usually undetectably low, but clinically credible results may be obtained in the case of contamination, for example by pouring blood between tubes or the wrong order of draw.

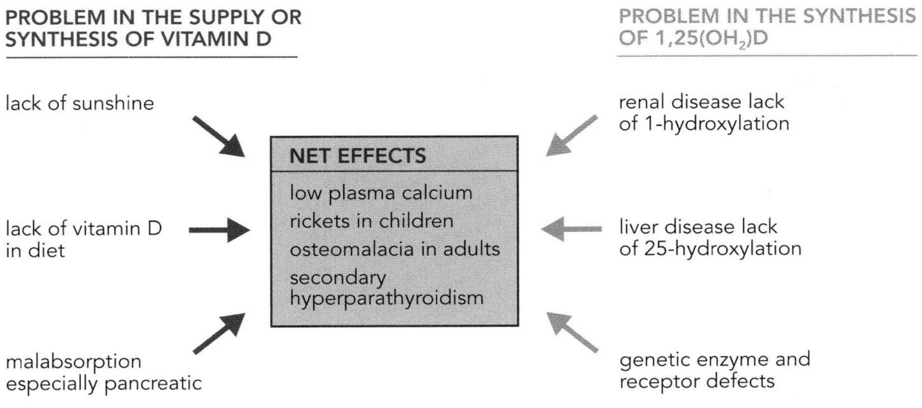

Figure 9.10 Causes of vitamin D deficiency.
There are several causes of functional vitamin D deficiency, and they can be often cumulative, with lack of exposure to sunlight (inhibition of endogenous vitamin D production) combined with dietary deficiency (exogenous source) being the commonest. Vitamin D deficiency will also occur in malabsorption, especially pancreatic related, where the general absorption of fats and hence fat-soluble vitamins like vitamin D is impaired. Functional vitamin D deficiency can also occur in liver and renal disease, where the capacity to hydroxylate vitamin D into its active hormone is reduced, and in cases of receptor deficiency. The net effect of vitamin D deficiency is decreased bone mineralization (rickets and osteomalacia) and in severe cases hypocalcemia will be observed. In addition, the falling calcium leads to enhanced PTH activity and osteoporosis.

of disparate ethnicity and is associated with four distinct short tandem repeat (STR) haplotypes.

Clearly vitamin D deficiency, where the measured plasma 25-OHD is less than 37 nmol/L (15 ng/mL), can be caused by a lack of exposure to sunlight, a lack of vitamin D in the diet, malabsorption, or a combination of these causes. As the levels of 25-OHD fall, eventually the concentration of the active hormone [1,25(OH)$_2$D] will fall and pathological effects will be seen. There will be reduced mineralization of bone and decreased absorption of calcium from the gut. The normal feedback action of 1,25(OH)$_2$D on PTH release will be reduced and this, combined with the hypocalcemia, leads to raised levels of PTH, secondary hyperparathyroidism, and an osteoporotic drive on the bone.

The effects of vitamin D deficiency on the mineralization of bone have been known for many years, it leading to rickets in children or osteomalacia (softening of bone) in adults (**Figure 9.11**). Although hypocalcemia can be a late symptom of vitamin D deficiency, most patients at the early stages may be normocalcemic, with the PTH drive enabling the blood levels to be kept within the normal range. Furthermore, many researchers are now stating that we should

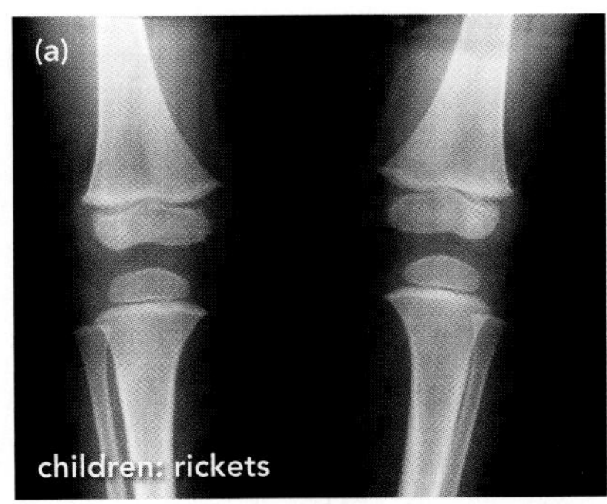

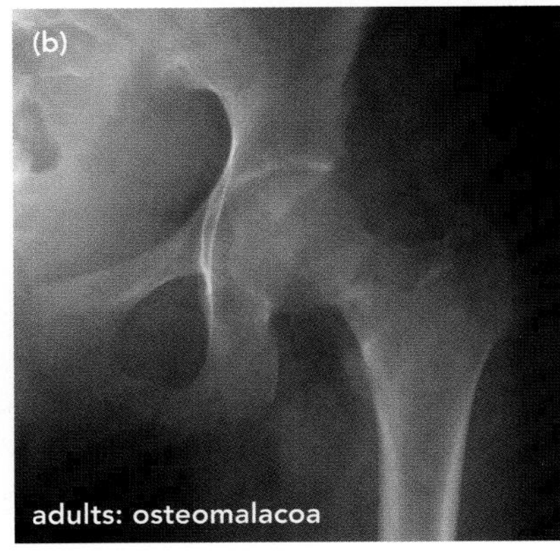

Figure 9.11 Bone effects of vitamin D deficiency.
Vitamin D deficiency reduces the mineral content of bone; that is, there is less calcium and phosphate relative to the protein matrix. This appears as less-dense areas on X-rays (a, b), and in children, where rickets can occur, larger gaps than normal appear at the joints (a). The lack of mineralization of the bone causes the bone to be softer (osteomalacia), leading to the characteristic bowing of the limbs seen in rickets. (a, courtesy of Nevit Dilmen under CC BY-SA 3.0.)

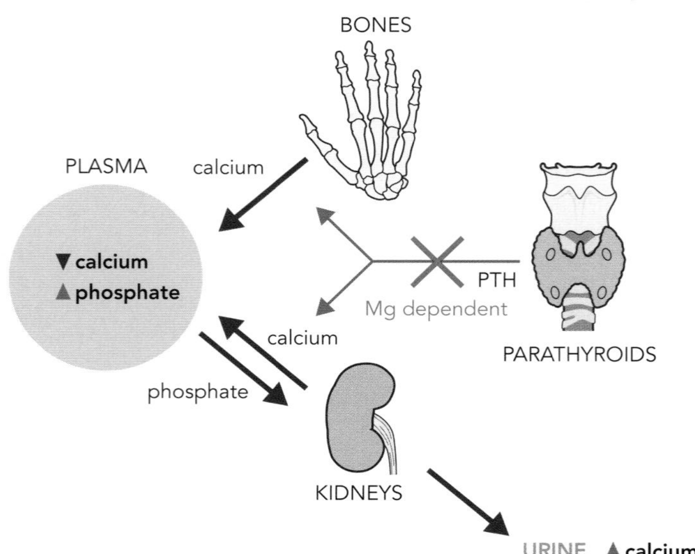

Figure 9.12 Lack of PTH. In primary hypoparathyroidism, the production and secretion of parathyroid hormone (PTH) from the parathyroid gland is impaired; therefore the normal rise in plasma PTH in response to falling calcium levels does not occur. Since there is no rise in PTH, the normal correction of plasma calcium by increased flux of calcium from the bone into the plasma, combined with increased reabsorption by the kidney, does not occur. This will lead to low levels of calcium and high levels of phosphate in the plasma. Since the release of PTH from the parathyroid gland is very dependent upon the plasma magnesium (Mg) level, severe hypomagnesemia can mimic primary hypoparathyroidism.

reassess what level of vitamin D we should consider as adequate, and have coined the phrase "vitamin D insufficient" to describe levels of 25-OHD of less than 75 nmol/L (30 ng/mL).

Hypocalcemia caused by a lack or functional lack of PTH is far less common than vitamin D deficiency. Autoimmune disease, tumor, surgery, and infarction have all been shown to cause damage to the parathyroid gland and lead to hypocalcemia (**Figure 9.12**). There can also be a functional lack of PTH caused by a genetically defective receptor for PTH in the kidney; therefore the patients cannot reabsorb calcium from the urine correctly. This disorder is called pseudohypoparathyroidism (**Figure 9.13**) and such patients present also with shortened metacarpals and metatarsals along with unusual facial features. Defects in the parathyroid hormone receptor 1 (*PTHR1*) locus on chromosome 3 have been observed in pseudohypoparathyroidism type 1b (PHP1b), a PTH-resistance

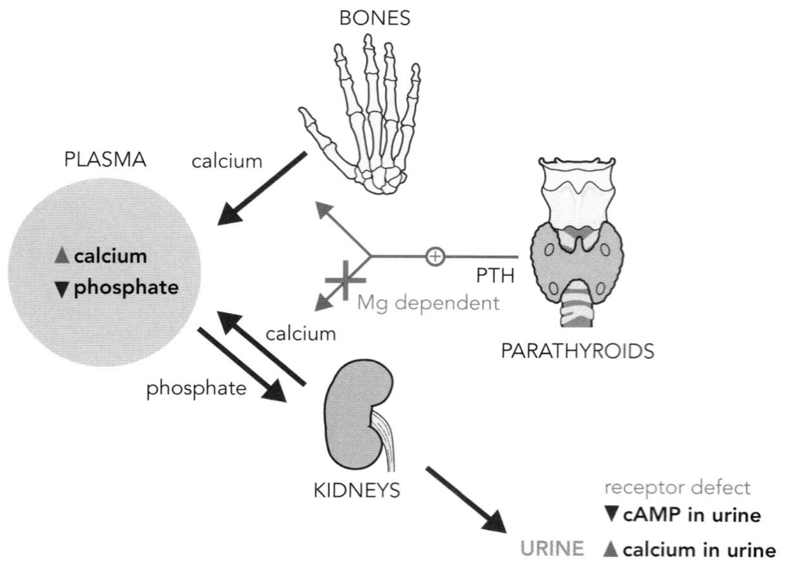

Figure 9.13 Pseudohypoparathyroidism. In patients with pseudohypoparathyroidism, the parathyroid gland can respond to falling calcium levels, and produces and secretes parathyroid hormone (PTH) normally; such patients have a normal or raised plasma PTH with hypocalcemia. The PTH stimulates bone turnover appropriately, to cause a net movement of calcium from the bone into plasma; however, the normal PTH effect on renal reabsorption of calcium is impaired due to a faulty receptor. In the most common mutation, this can be demonstrated by a lack of production of cAMP in the urine in response to exogenous PTH.

syndrome with impaired renal PTH-receptor-mediated signaling. An $(AAAG)_n$ tetranucleotide STR in the *PTHR1* gene has been associated with height and bone mineral density, but no consistent findings of sequence variation in the gene have been reported in PHP1b patients. The molecular lesions that cause PHP1b most commonly are sequence alterations and/or genomic imprinting changes affecting the *GNAS1* locus (chromosome 20), which encodes a G-protein α subunit that is crucial to PTHR1 receptor coupling in PTH-mediated signaling. Usually the genetic defect leads to an inability of the renal tubular cells to produce cAMP, and reduced levels can be seen in the kidney after a PTH challenge.

Magnesium deficiency can lead to hypocalcemia in two main ways. First, adequate magnesium concentration is required for the appropriate reabsorption of calcium by the kidneys; second, magnesium is also required for the synthesis and release of PTH from the parathyroid glands.

Transient falls in calcium can occur in several situations, including acute pancreatitis, where a possible reason could be the formation of calcium soaps by the free fatty acids released through the action of lipase that enters into the bloodstream. Infusions of citrate, phosphate, and gluconate can all lower the ionized calcium temporarily by complexing with ionized calcium. These effects are typically short-lived due to the usually rapid metabolism of these molecules by the liver. In some surgical procedures, for example liver transplant, where the liver may be bypassed for a long period of time and the patient is receiving a lot of blood transfusions and calcium salts, the measured total calcium can climb as high as 6.00 mmol/L (24 mg/dL), but the ionized calcium remains normal.

Effects of hypocalcemia

Patients with mild hypocalcemia may not have any specific symptoms relating to the low calcium itself, but their symptoms could be related to the underlying causes, for example osteomalacia in adults with vitamin D deficiency. The symptoms of hypocalcemia are predominantly neuromuscular and their frequency is dependent upon the severity of the hypocalcemia and the plasma pH of the patient. Symptoms include:

- Tetany (severe hypocalcemia only)
- Paresthesia (tingling in extremities)
- Cardiac arrhythmia
- Hypotension
- Cardiac failure

It may be possible also to demonstrate classical signs, such as the Trousseau or Chvostek signs, but these are not specific for hypocalcemia. Tetany is seen only usually in severe hypocalcemia.

Treatment of hypocalcemia

Mild hypocalcemia is not usually a medical emergency; however, urgent action may be required if the adjusted calcium falls rapidly below 2.00 mmol/L (8.00 mg/dL). If there are severe symptoms present, such as tetany, intravenous calcium gluconate may be required. In most cases oral calcium supplements will be required in conjunction with treating the underlying cause, for example magnesium or vitamin D supplementation. Although vitamin D can be given intramuscularly, it is usually given orally in conjunction with calcium as a cheap and effective treatment. With the increasing awareness of vitamin D insufficiency, there have been calls for the supplementation of food to combat the problem. In patients with renal disease or specific disorders of enzyme deficiencies, hydroxylated analogs of vitamin D are prescribed, such as 1α-hydroxy vitamin D (alfacalcidol) or $1{,}25(OH)_2D$ (calcitriol). These analogs would be used also in hypocalcemia caused by PTH deficiency.

CASE 9.1

A 35-year-old woman presents to her doctor with muscle aches and pains and an occasional tingling sensation. The following results were produced.

	SI units	Reference range	Conventional units	Reference range
Plasma				
Adjusted calcium	2.05 mmol/L	2.20–2.60	8.2 mg/dL	8.8–10.4
Phosphate	0.70 mmol/L	0.74–1.52	2.17 mg/dL	2.3–4.7
Alkaline phosphatase	356 U/L	30–120	356 U/L	30–120
PTH	13.3 pmol/L	1.06–6.90	125 pg/mL	10–65
25-Hydroxy vitamin D	<37 nmol/L	37–104 (winter)	<15 ng/mL	15–42 (winter)

- **What is the cause of her symptoms?**
- **Why is the PTH raised?**
- **What is the source of the alkaline phosphatase?**

The patient has presented with typical clinical symptoms of vitamin D deficiency. Her albumin-adjusted (corrected) calcium is unequivocally low and will be partly explaining her neuromuscular symptoms. The PTH is appropriately raised in the face of the hypocalcemia (secondary hyperparathyroidism), while the 25-hydroxy vitamin D is markedly reduced. These results are consistent with vitamin D deficiency.

The primary problem is the low vitamin D, often caused by a lack of exposure to sunshine plus some form of dietary deficiency. This deficiency reduces the ability to absorb calcium and phosphate from the gut, and will eventually lead to hypocalcemia and hypophosphatemia. It should be noted that many patients with vitamin D deficiency present with a plasma calcium within the reference range. In response to the deficiency, the parathyroid gland secretes more PTH, which in turn increases calcium reabsorption by the kidneys and increases phosphate excretion in the urine. In addition, the PTH increases bone turnover, hence the rise in alkaline phosphatase, which will help to maintain the plasma calcium within the reference range.

9.7 HYPERCALCEMIA

Hypercalcemia is defined by serum or plasma calcium levels greater than the upper limit of the reference range. For most laboratories, that is an adjusted calcium of greater than 2.60 mmol/L (10.4 mg/dL) or an ionized calcium greater than 1.30 mmol/L (5.2 mg/dL). As with hypocalcemia, pre-analytical factors (including acid–base status) can influence the result and its interpretation and thus should always be considered.

Causes of hypercalcemia

After erroneous causes of hypercalcemia have been eliminated, many pathological conditions can be considered as causes. More common causes (over 95% of cases found) include:

- Primary hyperparathyroidism
- Associated with renal disease
  - Tertiary hyperparathyroidism
  - Treatment with vitamin D analogs

- Malignancy
  - Solid tumors
  - Bone secondary growths
  - Hematological tumors, especially myeloma

Rarer causes of hypercalcemia include:

- Hyperthyroidism
- Granulomatous disease, for example sarcoid
- Drugs
  - Lithium
  - Thiazide
- Familial hypocalciuric hypercalcemia
- Milk alkali syndrome
- Vitamin D toxicity
- Immobilization in Paget's disease

However, the vast majority of cases of hypercalcemia are caused by primary hyperparathyroidism or malignancy, or are related to chronic kidney disease (CKD).

Primary hyperparathyroidism is probably the most common cause of hypercalcemia seen in the primary care setting. It is often caused by a benign, single adenoma in one of the parathyroid glands. Occasionally there may be multiple parathyroid glands affected and, more rarely, a malignant carcinoma may be the cause. The normal feedback mechanism, where rising calcium levels switch off further PTH secretion, appears to be blunted in primary hyperparathyroidism. The parathyroid gland requires a higher level of plasma Ca^{2+} to switch off further synthesis and secretion of PTH, leading to a new set point with characteristic typical blood results (**Figure 9.14**).

Many patients are asymptomatic and the condition can remain undiagnosed for many years. It is more common in women than men and seen far more frequently in the elderly. Familial hypocalciuric hypercalcemia (FHH) is a rare inherited disorder which can produce mild hypercalcemia with moderately elevated PTH. FHH type 1 (FHH1 or HHC1) is an autosomal dominant inherited

Clinical practice point 9.1

The two commonest causes of hypercalcemia are primary hyperparathyroidism and malignancy.

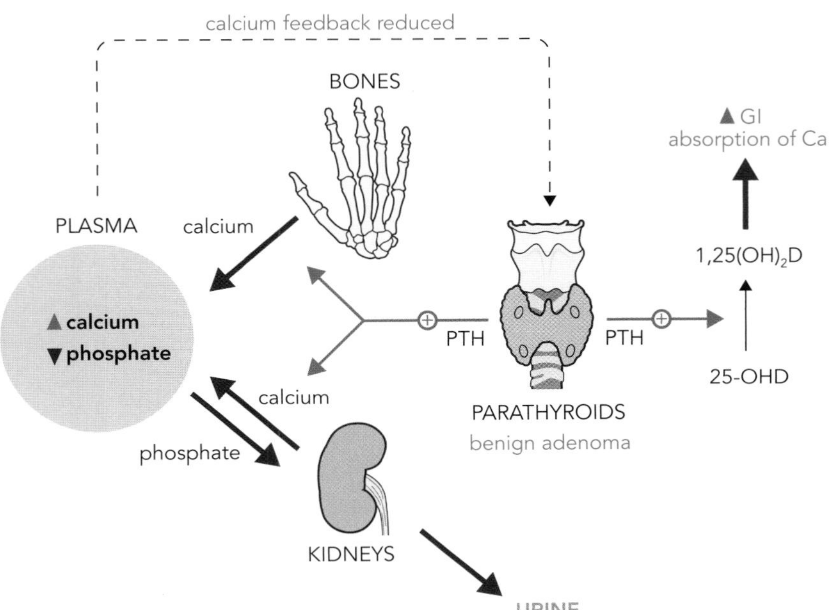

Figure 9.14 Primary hyperparathyroidism. Primary hyperparathyroidism is one of the commonest causes of hypercalcemia, especially in the elderly, and can remain asymptomatic for many years if the patient's plasma calcium has not been measured. It is usually due to a single, benign adenoma in one of the patient's four parathyroid glands, which has the effect of raising the level of plasma calcium required to switch off further production and secretion of parathyroid hormone (PTH). This usually leads initially to a slightly raised calcium with a raised plasma PTH. The raised PTH stimulates bone turnover, with resorption predominating, leading to a net flow of calcium from the bone and osteoporosis. In addition, the raised PTH also stimulates more reabsorption of calcium and loss of phosphate by the kidney, which combined with the bone effects leads to hypercalcemia and hypophosphatemia. The raised PTH also stimulates more calcium absorption from the gut, due to increasing the levels of $1,25(OH)_2D$.

disorder associated with mutations in the calcium sensing receptor (*CASR*) locus on chromosome 3. The CASR allows parathyroid cells to respond to extracellular calcium concentrations by altering phosphoinositide turnover and intracellular calcium signaling. CASR loss-of-function mutations cause FHH1 and neonatal severe hyperparathyroidism, while gain-of-function mutations cause autosomal dominant hypocalcemia. Although it is a rare cause of hypercalcemia, FHH should always be ruled out before surgical treatment for hyperparathyroidism is considered; this is because hypercalcemia will persist in FHH even after parathyroidectomy. The simplest way to establish a diagnosis of FHH in patients with raised PTH and calcium levels is to use the Calcium Excretion Index.

FHH can be distinguished from primary hyperparathyroidism in patients with high calcium and PTH levels by determining the Calcium Excretion Index from serum and a spot urine sample taken at the same time:

$$\frac{\textbf{serum creatinine (µmol/L)} \times \textbf{urine calcium (mmol/L)}}{\textbf{urine creatinine (µmol/L)}}$$

For FHH, the index should be <0.01; for primary hyperparathyroidism, it is >0.03.

When primary hyperparathyroidism is seen in the younger population, consideration should be given to rarer genetic causes, specifically multiple endocrine neoplasia (MEN; **Table 9.4**). MEN type 1 is caused by mutations in the *MEN1* gene, a tumor suppressor gene which maps to chromosome 11. MEN type 2a and type 2b are caused by mutations in the *RET* oncogene on chromosome 10. Diagnostic testing based upon mutational analysis of *MEN1* and *RET* is available widely, with some centers offering antenatal testing. The laboratory testing of the adrenal and pituitary abnormalities of MEN type 1 that occur in addition to the hyperparathyroidism is described in Chapter 11. With MEN type 2, the main finding is thyroid medullary cell carcinoma, which is generally diagnosed by analysis of the *RET* oncogene and monitored by serial measurements of calcitonin.

It is often suggested that malignancy is the most common cause of hypercalcemia in hospitalized patients. It is certainly the commonest cause of very high calcium results, but an increasing awareness of hypercalcemia in patients with CKD has revealed that, in hospitals with large renal units, CKD may be overall the commonest cause of hypercalcemia in these hospitals.

Clinical practice point 9.2

Familial hypocalciuric hypercalcemia should always be excluded before parathyroid surgery since such treatment is ineffective and unnecessary.

TABLE 9.4 Multiple endocrine neoplasia (MEN)

MEN type	Tumors
MEN type 1	Hyperparathyroidism
	Islet cell tumors, especially gastrinoma
	Pituitary tumors, especially prolactinoma
	Adrenal tumors
	Thymic and bronchial carcinoid
	Lipoma and angiofibroma
MEN type 2a (Sipple's syndrome)	Thyroid medullary cell carcinoma (all cases)
	Pheochromocytoma (50% cases)
	Hyperparathyroidism (20% cases)
MEN type 2b	Thyroid medullary cell carcinoma
	Pheochromocytoma
	Hyperparathyroidism does not occur

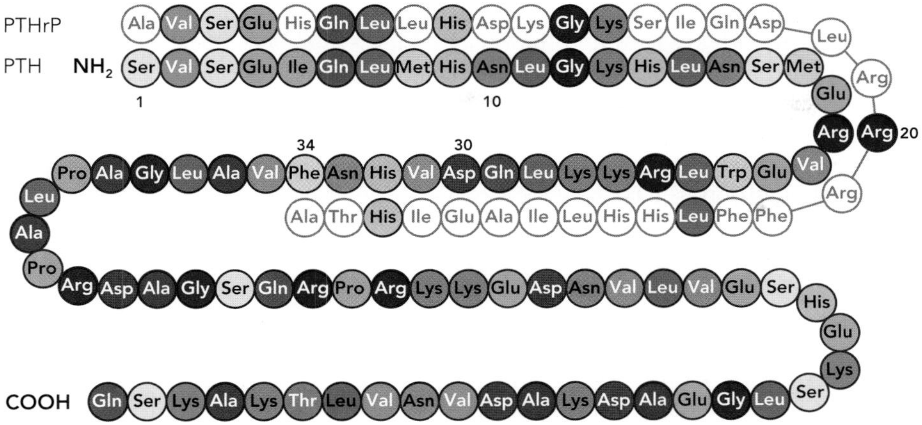

Figure 9.15 Similarities between parathyroid hormone and parathyroid hormone related peptide.
The first 34 N-terminal amino acids of parathyroid hormone related peptide (PTHrP, 144 amino acids) are very similar to those of parathyroid hormone (PTH, 84 amino acids). This means that PTHrP has similar effects to PTH.

Hypercalcemia due to malignancy can be separated into three main types:

1. Solid tumors producing parathyroid hormone related peptide (PTHrP)
2. Hematological tumors
3. Secondary growths in the bones

For many years it was believed that certain tumors, for example squamous cell carcinomas of the lung, caused hypercalcemia by the ectopic secretion of PTH. This idea arose because the PTH assays of the time cross-reacted with PTHrP, an onco-fetal protein which is involved in bone modeling during development and is produced also by certain tumors. Although ectopic PTH production has been shown, it is a very rare cause of hypercalcemia. PTH and PTHrP share similar molecular structures, especially the N-terminal amino acids (**Figure 9.15**), leading to similar clinical effects (**Figure 9.16**). PTH levels are suppressed in virtually all cases of hypercalcemia caused by malignancy due

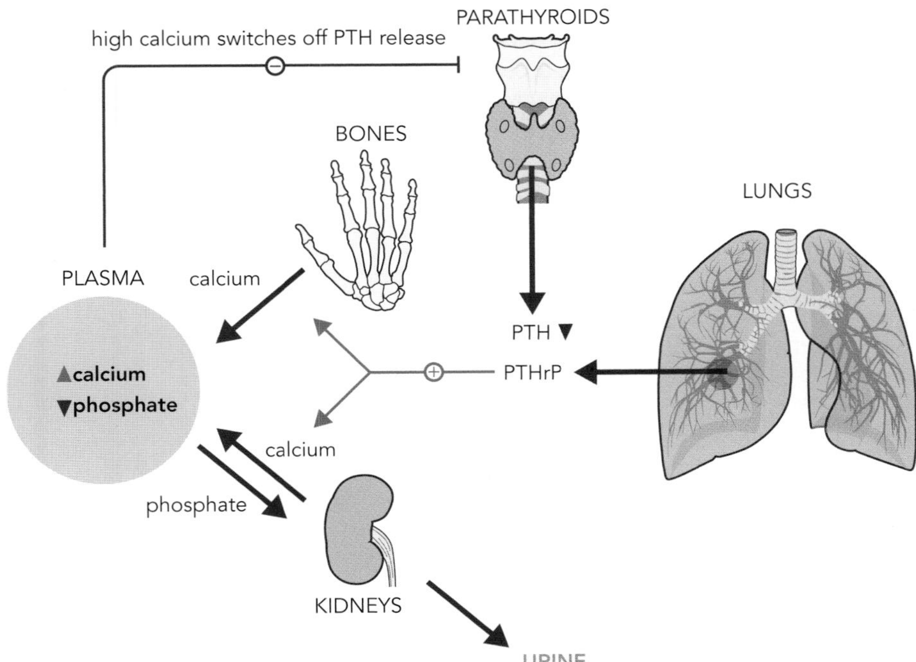

Figure 9.16 Hypercalcemia with parathyroid hormone related peptide.
Certain lung cancers will produce PTHrP, and since it has the same effects on calcium homeostasis as parathyroid hormone (PTH), calcium levels will be high due to increased flow of calcium from bone and increased calcium reabsorption by the kidney. The parathyroid gland is functionally intact, therefore its production of PTH is stopped by the high circulating calcium levels; thus, plasma PTH is appropriately suppressed in hypercalcemia. Patients with hypercalcemia due to PTHrP often present acutely, some with very high levels of calcium.

to feedback from the high calcium level. Higher calcium concentrations can be seen in malignancies compared to primary hyperparathyroidism and the onset is often far more rapid.

Hypercalcemia can be found in patients with hematological malignancies, especially myeloma. The mechanism causing the hypercalcemia is probably similar to that in patients with metastases in bone—that is, localized bone resorption hot spots leading to increased calcium levels—as opposed to the effects of a circulating factor. PTHrP may well be involved in the process at a localized level, along with several other mediating factors. Hypercalcemia is not a feature of primary bone cancers. The key factor in the differential diagnosis between primary hyperparathyroidism and malignancy is the measurement of PTH.

Hypercalcemia associated with CKD is sometimes caused by tertiary hyperparathyroidism. As CKD progresses, the ability of the kidney to perform the 1α-hydroxylation of 25-OHD falls, and there is a tendency toward hypocalcemia and secondary hyperparathyroidism. The normal feedback inhibition of PTH secretion from the parathyroid gland by $1,25(OH)_2D$ is also reduced and, combined with the reduced ability of the kidneys to reabsorb calcium and reduced clearance of PTH in renal failure, leads to very high PTH levels and renal osteodystrophy. In some patients the production of PTH by the parathyroid gland becomes autonomous and tertiary hyperparathyroidism develops. In a freshly diagnosed patient with hypercalcemia who has never had a blood test before, it would be impossible to distinguish a patient with primary hyperparathyroidism who has developed CKD from a patient with CKD who has developed tertiary hyperparathyroidism.

Mild hypercalcemia is quite common in CKD patients who are being treated with 1α-analogs of vitamin D, as one of the rate-limiting steps of normal synthesis (1α-hydroxylation) is not required. Transient mild hypercalcemia can also be seen in renal transplant patients after they regain their capacity for 1α-hydroxylation. Hypercalcemia due to patients in the normal population taking excessive quantities of vitamin D supplements is not common, as there is a requirement for conversion to the $1,25(OH)_2D$ form and such conversion should be reduced as the calcium level starts to rise. Where hypercalcemia has been described in these circumstances, patients were taking in excess of 50,000 units of vitamin D daily. As vitamin D is a fat-soluble vitamin, its effects are usually long term and the mild hypercalcemia can take several months to correct after the vitamin supplements have been removed from the diet.

Hyperthyroidism can produce a mild hypercalcemia, probably via a combination of increased resorption of calcium from bone and impaired excretion. Although not a common complication of hyperthyroidism, it is readily identifiable by the measurement of thyroid function.

The hypercalcemia associated with sarcoidosis is caused by the increased conversion of 25-OHD to $1,25(OH)_2D$ by the sarcoid tissue. Although serum angiotensin-converting enzyme levels are often raised in sarcoidosis, normal levels do not rule out the disease. The increased conversion of 25-OHD to $1,25(OH)_2D$ has also been put forward as the mechanism for hypercalcemia in several other disorders, such as lymphoma, tuberculosis, Wegener's granulomatosis, histiocytosis, and even beryllium poisoning.

Mild hypercalcemia due to thiazide diuretics is caused probably by an impaired secretion of calcium in the proximal tubule, in a similar way to the elevation of serum urate concentrations. Lithium is likely to cause hypercalcemia by a similar process. The hypercalcemia associated with immobile patients with Paget's disease is not a common finding, but may be due to increased focal release of calcium from bone. The so-called milk alkali syndrome associated with excessive use of calcium-containing antacids is also very rarely seen.

It is not unusual for patients to have multiple pathologies that could contribute to hypercalcemia. Since primary hyperparathyroidism is relatively common

Clinical practice point 9.3

Hypercalcemia of malignancy is usually a late-stage feature in a patient with a known tumor. Parathyroid hormone levels are suppressed and the patient is often unwell.

in older, postmenopausal women, it has been suggested that up to 10% of these patients may also have a malignancy present. In addition, since primary hyperparathyroidism can remain undetected for many years, other causes of hypercalcemia, for example thiazide diuretics, may unmask the underlying disorder. Finally, hypercalcemia can often be detected and exacerbated in patients with acute illnesses where there may be dehydration.

Effects of hypercalcemia

Many patients with mild hypercalcemia are said to be asymptomatic; however, the symptoms more commonly associated with hypercalcemia are listed in **Table 9.5**.

TABLE 9.5 Symptoms of hypercalcemia

System	Symptoms
Renal	Polydipsia and polyuria
	Dehydration
	Renal stones and damage
Bone	Osteoporosis
	Bone pain
	Fractures
Heart	Irregular rhythms
	Cardiac arrest
Gastrointestinal	Nausea and vomiting
	Constipation
Neuromuscular	Muscle weakness
	Disorientation
	Drowsiness and confusion
Other	Irritable eyes

Symptoms can be caused by the hypercalcemia or be related to the underlying disorder.

The symptoms of hypercalcemia are often said to be "stones, bones, and groans." Stones relates to the potential precipitation of calcium phosphate in the kidneys and urinary tract; bones to the fact that bone pain and/or damage is often associated with hypercalcemia; and groans refers to the gastrointestinal and neuromuscular problems. Several of the symptoms are due to the actual hypercalcemia. The polyuria and dehydration are caused by the direct effects of the high calcium levels on renal function. The low solubility of calcium phosphate salts will cause renal stones and precipitation in other tissues.

Cardiovascular and neuromuscular symptoms become more apparent as the levels of calcium become higher, and are not usually seen with adjusted calcium concentrations below 3.00 mmol/L (12 mg/dL). Most laboratories consider values above 3.00 mmol/L (12 mg/dL) to be potential medical emergencies, and an unexpectedly high value would usually be telephoned to the requesting physician.

Other symptoms are related to the underlying condition. For example, corneal calcification and osteoporosis, and the resulting fractures and bone pain, could be occurring because of the high resorptive drive associated with primary hyperparathyroidism. The investigation of hypercalcemia is described in **Figure 9.17**.

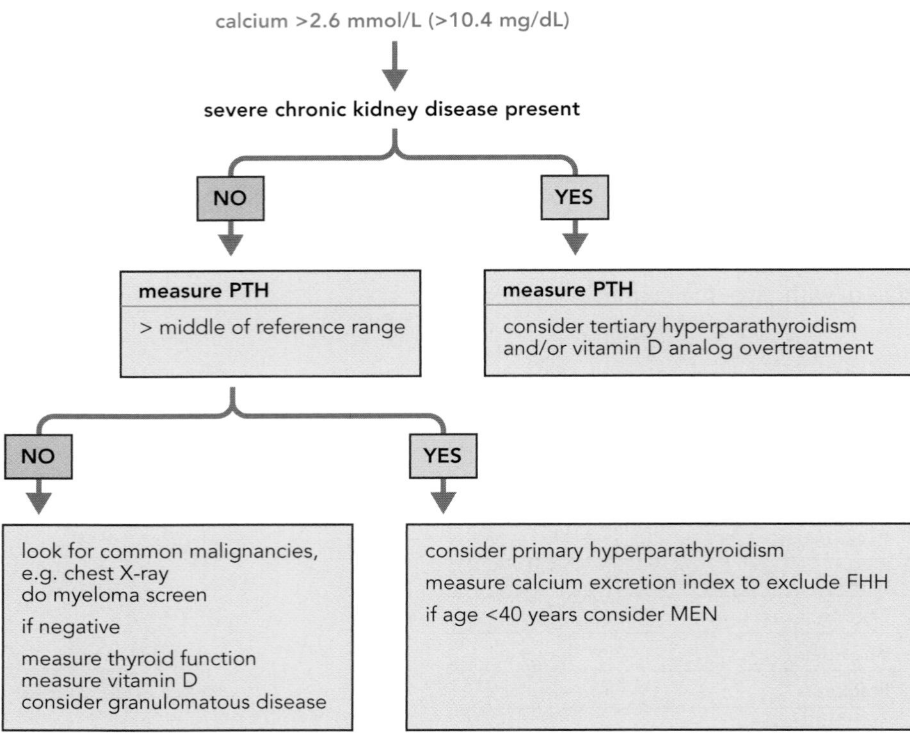

Figure 9.17 Summary of the investigation of hypercalcemia.

Treatment of hypercalcemia

The treatment of hypercalcemia depends on the presentation of the patient and the severity of the hypercalcemia. In many patients with mild (asymptomatic) hyperparathyroidism, no treatment may be required and regular monitoring of calcium levels and renal function will suffice. However, severe hypercalcemia (>3 mmol/L or >12 mg/dL) can be a medical emergency and requires immediate treatment. If the cause of hypercalcemia is unknown, a serum sample should always be taken for PTH assay before commencing treatment.

Since patients who present acutely with hypercalcemia are often severely dehydrated, this should be resolved first with intravenous saline. Bisphosphonates—drugs that inhibit osteoclastic resorption of calcium from bone—are often given intravenously to reduce calcium levels rapidly. Calcitonin can be used as an alternative but this is rare. A wide range of bisphosphonates are used also in the treatment of chronic hypercalcemia, especially in malignancies with bone involvement. Newer drugs, such as cinacalcet, inhibit PTH release and are used increasingly for the treatment of patients with primary hyperparathyroidism and for patients with secondary and tertiary hyperparathyroidism associated with chronic kidney disease.

Other treatment options will relate to the underlying disorder. Parathyroidectomy is often performed on younger patients with primary hyperparathyroidism or chronic kidney disease. Steroid therapy helps to resolve the hypercalcemia associated with sarcoid, while treating the hyperthyroidism resolves hypercalcemia in these patients. Obviously if the hypercalcemia is related to a drug or vitamin excess, the drug (for example a thiazide diuretic) should be stopped.

9.8 MAGNESIUM ABNORMALITIES

Although magnesium is the fourth most abundant cation in the body, there is less appreciation of abnormalities of magnesium compared to those of calcium.

CASE 9.2

A 65-year-old man presents to his doctor with tiredness and increased thirst. His blood glucose was within the reference range, excluding diabetes as a cause for the symptoms. His laboratory results were as follows.

	SI units	Reference range	Conventional units	Reference range
Plasma				
Adjusted calcium	2.95 mmol/L	2.20–2.60	11.8 mg/dL	8.8–10.4
Phosphate	0.56 mmol/L	0.74–1.52	1.73 mg/dL	2.3–4.7
Alkaline phosphatase	356 U/L	30–120	356 U/L	30–120
PTH	8.0 pmol/L	1.06–6.90	75 pg/mL	10–65

- **What is the likely cause for the patient's symptoms?**
- **Why is the PTH raised?**
- **What should be excluded in order for the diagnosis to be confirmed?**

The patient's plasma calcium is raised and may be high enough to cause the increased thirst and polyuria. His PTH is inappropriately high for the hypercalcemia, showing the patient has primary hyperparathyroidism. The high PTH will also explain the reduced phosphate and raised alkaline phosphatase seen in typical cases of primary hyperparathyroidism. Patients usually present with plasma calcium of less than 3.00 mmol/L (12 mg/dL), and are quite frequently asymptomatic. Note that the PTH does not have to be raised to make the diagnosis of primary hyperparathyroidism. A PTH in the top half of the reference range would be inappropriate for hypercalcemia, and consistent with primary hyperparathyroidism. In the majority of cases the underlying cause is a benign tumor (adenoma) of one of the four parathyroid glands. The only effective treatment is surgical removal (parathyroidectomy). However, before this is undertaken, urine calcium studies must be performed to exclude familial hypocalciuric hypercalcemia, which does not require treatment.

CASE 9.3

A 71-year-old man with terminal lung cancer presents to the emergency department with severe vomiting and confusion. His laboratory results are as follows.

	SI units	Reference range	Conventional units	Reference range
Plasma				
Adjusted calcium	3.75 mmol/L	2.20–2.60	15.0 mg/dL	8.8–10.4
Phosphate	0.78 mmol/L	0.74–1.52	2.41 mg/dL	2.3–4.7
Alkaline phosphatase	200 U/L	30–120	200 U/L	30–120
PTH	<1.00 pmol/L	1.06–6.90	<10 pg/mL	10–65

- **What does a calcium concentration this high usually indicate?**
- **What is the significance of the low PTH?**
- **Why is the calcium elevated?**

The patient has presented with a very high calcium concentration, which can cause nausea, vomiting, polyuria, dehydration, and mental changes. Calcium of this level is a strong indication that the patient has hypercalcemia due to malignancy, where calcium concentrations are often greater than 3.00 mmol/L (12 mg/dL). Furthermore, the rise in calcium can occur very rapidly, over days and weeks. The phosphate level can be low, and the alkaline phosphatase level can be very variable, sometimes driven by the increase in bone turnover driven by parathyroid hormone related peptide (PTHrP) or increased due to focal bone metastases (seen on bone scans as hot spots). Fundamental to the diagnosis is that the PTH is appropriately suppressed due to the hypercalcemia. In this case the hypercalcemia is probably due to the production of PTHrP by the lung cancer. With this type of hypercalcemia, the malignancy itself is usually straightforward to detect and the measurement of PTHrP, which is technically difficult and not generally available, is not usually required.

Hypomagnesemia is considerably more common than hypermagnesemia, with estimates suggesting that up to 7–11% of hospital patients have serum magnesium of less than 0.70 mmol/L (1.7 mg/dL). Since magnesium (like potassium) is predominantly an intracellular ion, patients may have imbalances in magnesium despite the serum level being well within the reference range, suggesting that deficiencies of magnesium may be underdiagnosed.

Clinical practice point 9.4

Magnesium deficiency often causes hypocalcemia due to functional hypoparathyroidism. Both PTH secretion and its end-organ effects are inhibited. In addition, hypokalemia may occur due to renal potassium wasting.

Hypomagnesemia

As described previously, hypomagnesemia is probably relatively common, with possibly up to 65% of critically ill patients and 90% of patients with hematological malignancies having low serum magnesium levels. There are numerous causes of hypomagnesemia, and often several interacting causes are present at once (**Table 9.6**).

The renal causes are linked to the fact that 70% of the reabsorption of magnesium occurs in the ascending loop of Henle, hence loop diuretics like furosemide can cause hypomagnesemia far more readily than thiazide diuretics. Many patients on cytotoxic drug therapy readily develop hypomagnesemia within 1–2 days of starting treatment. Poor diets and diarrhea are common causes of hypomagnesemia also, with the use of purgatives being associated especially with magnesium deficiency.

There is a very high incidence (>30%) of hypomagnesemia in alcoholics. The causes are multifactorial: poor intake, loss through diarrhea and vomiting, malabsorption, and chronic pancreatitis combined with enhanced renal loss due to the diuresis in these patients. It has been suggested that some of the effects of

TABLE 9.6 Causes of hypomagnesemia

System	Causes
Renal	Drug effect upon the kidney: Diuretics, especially loop diuretics Cytotoxic drugs, for example cisplatin Aminoglycosides, for example gentamicin Immunosuppressants, for example cyclosporine Other drugs, for example amphotericin B
	Renal damage, including post-obstructive nephropathy
	Post renal transplant
	Genetic abnormalities: Bartter's syndrome Gitelman's syndrome
Gastrointestinal	Reduced intake: Intravenous feeding, insufficient magnesium Dietary deficiency, for example stone diet
	Reduced absorption: Diseases causing malabsorption (for example celiac disease) Chronic diarrhea Purgative abuse Fistula Short bowel syndrome Inherited transport defect
Redistribution	Diabetic ketoacidosis correction
	Hungry bone syndrome
	Acidosis correction
	Catecholamine excess
Endocrine	Diabetes, osmotic loss
	Hypercalcemia

delirium tremens are directly caused by the hypomagnesemia, and correction of the magnesium deficiency can reduce some of the symptoms.

Many of the symptoms of hypomagnesemia are similar to those seen in hypocalcemia, with the additional finding that many patients appear to present with depression and have a lowered pain threshold (**Table 9.7**). Symptoms can become more marked as the magnesium concentration drops below 0.50 mmol/L (1.2 mg/dL).

TABLE 9.7 Symptoms in hypomagnesemia

Common symptoms	Rarer symptoms
Muscle weakness	Tetany
Twitching and tremor	Coma
Positive Chvostek sign and Trousseau sign	Seizures
Mild to moderate delirium	Vertigo
Paresthesias	Nystagmus
Cardiac disturbances	Ataxia
Premature ventricular beats	
Severe ventricular dysrhythmias	
Apathy	
Depression	

CASE 9.4

A 43-year-old woman undergoing radiotherapy for breast cancer presents in hospital with severe muscle spasms. She is on a number of medications, including a loop diuretic and a proton pump inhibitor (PPI). Her blood results were as follows.

	SI units	Reference range	Conventional units	Reference range
Plasma				
Adjusted calcium	1.85 mmol/L	2.20–2.60	7.35 mg/dL	8.8–10.4
Phosphate	1.96 mmol/L	0.74–1.52	6.07 mg/dL	2.3–4.7
Magnesium	0.30 mmol/L	0.70–1.00	0.7 mg/dL	1.7–2.3
Alkaline phosphatase	96 U/L	30–120	96 U/L	30–120
PTH	2.0 pmol/L	1.06–6.90	20 pg/mL	10–65

- What is the cause of her muscle symptoms?
- Is the PTH appropriate for her hypocalcemia?
- What is the significance of her low magnesium?

The patient has presented with a very low calcium, which explains her muscle spasms; in severe cases this can present as tetany. In contrast to vitamin D deficiency, her plasma phosphate is high, which in the absence of renal impairment suggests hypoparathyroidism as the cause of the hypocalcemia. This is confirmed with the plasma PTH being inappropriately low in the face of the hypocalcemia. Magnesium deficiency may cause reversible hypoparathyroidism and is caused by either reduced gut absorption or increased urine loss of magnesium, for example due to use of diuretics. Some patients develop hypomagnesemia (mechanism unknown) whilst on PPIs prescribed to reduce stomach acid. Unlike vitamin D deficiency, the plasma alkaline phosphatase is not raised, as there is no PTH drive. However, alkaline phosphatase can be raised by other factors (for example bone secondary growths from the breast cancer) so it is not always a useful pointer as to the cause of the hypocalcemia.

Hypermagnesemia

Hypermagnesemia is far less common than hypomagnesemia and is predominantly associated with renal failure.

Magnesium as a therapeutic agent

Magnesium, often given intravascularly, has been shown to be effective in treating some cases of asthma. It is also the treatment of choice for the prevention of eclampsia in pregnancy, and for treatment of torsades de pointes (a form of ventricular tachyarrhythmia). In these forms of therapy the magnesium levels are raised to above 2.00 mmol/L (4.9 mg/dL). As discussed earlier, it is also effective in treating and preventing seizures in alcohol withdrawal, possibly by decreasing the amount of acetylcholine liberated at the end plate by motor nerve impulses. This will block the neuromuscular transmission associated with seizure activity. Magnesium also has a depressant effect in the central nervous system.

9.9 PHOSPHATE ABNORMALITIES

Hypophosphatemia

Hypophosphatemia is defined by the majority of laboratories as a phosphate concentration of less than 0.74 mmol/L (2.29 mg/dL). It is overlooked quite frequently, yet it is a common problem in clinical practice, especially in hospital inpatients. Symptoms do not generally become apparent until the concentration falls below 0.40 mmol/L (1.24 mg/dL). Since phosphate is involved in so many metabolic functions, including ATP formation and usage, deficiencies can produce a wide range of symptoms and can become life threatening if the serum concentration falls below 0.15 mmol/L (0.46 mg/dL).

The causes of hypophosphatemia are many and varied but most frequently it can be attributed to intracellular shifts, increased loss in urine, lack of phosphate intake, or malabsorption (**Table 9.8**).

Since phosphate is found in most foodstuffs, hypophosphatemia due to deficient dietary intake is rare and found only in extremes such as starvation. Conversely, hypophosphatemia can be found in many types of malabsorption,

TABLE 9.8 Causes of hypophosphatemia

System	Causes
Gastrointestinal	Starvation
	Alcoholism
	Severe diarrhea
	Most forms of malabsorption
	Use of phosphate-binding antacids
Renal	Primary and secondary hyperparathyroidism
	Volume expansion treatment
	Diuretics
	Specific genetic defects (for example X-linked hypophosphatemia)
	Renal transplant
Redistribution	Respiratory alkalosis (for example hyperventilation)
	Refeeding syndrome
	Catecholamines
	Increased consumption by malignant cells

and is found with hypocalcemia in pancreatic malabsorption. Phosphate-binding antacids (for example aluminum hydroxide) will inhibit absorption and cause hypophosphatemia. Phosphate loss can become significant in severe diarrhea also.

In the absence of renal impairment, hypophosphatemia is seen frequently in hyperparathyroid states due to the increased excretion of phosphate driven by PTH. Stimulation of PTH release during volume replacement also causes hypophosphatemia, while the sudden increase in kidney function seen after renal transplant leads also to hypophosphatemia. Although genetic defects in renal phosphate transport are rare, they can lead to a profound hypophosphatemia. Several variants of vitamin D-resistant rickets, all genetic abnormalities, have been described. X-linked hypophosphatemia rickets is due to mutation in the phosphate-regulating endopeptidase (*PHEX*) gene which maps to Xp22.2-p22.1. Autosomal dominant hypophosphatemic rickets (ADHR) is caused by mutation in the *FGF23* gene, a member of the fibroblast growth factor gene family which maps to chromosome 12p13.3. The extremely rare autosomal recessive hypophosphatemic rickets (ARHR) is caused by private mutations in the dentin matrix acidic phosphoprotein 1 (*DMP1*) gene on chromosome 4q21. Generalized impairment of normal tubular function, such as in Fanconi syndrome or during the use of diuretics, will lead also to hypophosphatemia. Tumor-induced osteomalacia is a paraneoplastic disorder characterized by profound hypophosphatemia that is due to humoral overexpression of the *FGF23* gene, the same gene that is mutated in ADHR. Many disease states will lead to increased cellular uptake and utilization of phosphate and hence hypophosphatemia. This is seen commonly in respiratory alkalosis and should always be ruled out in the early stages of investigating a hypophosphatemic patient. Other conditions that lead to phosphate influx into cells, such as re-feeding or excess use of β-adrenoceptor agonists, will cause hypophosphatemia. Finally, rapidly metabolizing tumor cells can take up excess phosphate, causing hypophosphatemia.

Clinical practice point 9.5

Hypophosphatemia may occur transiently and without clinical effects in respiratory alkalosis due to hyperventilation. This can be caused by anxiety during venipuncture.

Symptoms of hypophosphatemia

The most common manifestation of hypophosphatemia is muscle weakness due to the impairment of normal high-energy phosphate metabolism. Other, less common symptoms increase in proportion to the degree of hypophosphatemia; these can include respiratory failure, rhabdomyolysis, myocardial depression, and impaired neurological function. Serum phosphate values of less than 0.20 mmol/L (0.62 mg/dL) should be considered to be medical emergencies and treated without delay.

Clinical practice point 9.6

Severe hypophosphatemia, when occurring on a background of total body phosphate depletion, can cause fatal multiorgan failure.

Investigation of hypophosphatemia

Most of the analyses used to investigate calcium disorders are employed also in the investigation of hypophosphatemia. In addition, investigation of the renal reabsorption of phosphate can be useful to identify tubular phosphate wastage.

Treatment of hypophosphatemia

In addition to dealing with the underlying cause, phosphate replacement therapy may be required. Oral preparations will suffice generally, though intravenous replacement may be necessary in severe hypophosphatemia.

Hyperphosphatemia

In adults, hyperphosphatemia is taken to mean a serum phosphate level above 1.40 mmol/L (4.33 mg/dL). The reference range for phosphate in children is higher, up to 1.80 mmol/L (5.57 mg/dL). As the phosphate concentration rises, phosphate starts to precipitate out of solution as calcium salts, leading to a range of problems and symptoms, not least a trend toward hypocalcemia.

The most common cause of hyperphosphatemia is renal impairment, due to the reduced excretion of phosphate by the kidneys, where serum phosphate can

TABLE 9.9 Causes of hyperphosphatemia

Primary system involvement	Causes
Renal	Acute or chronic renal impairment
Increased cell turnover and/or breakdown	Sepsis
	Rhabdomyolysis
	Intravascular hemolysis
	Severe trauma (for example crush injuries)
	Tumors (for example leukemia)
	Chemotherapy
Endocrine	Hypoparathyroidism
Redistribution	Acidosis
Excessive intake	Phosphate salts (oral, intravenous, enema)
Drug-induced	Bisphosphonates

be in excess of 3.00 mmol/L (9.29 mg/dL) in end-stage renal failure. Most of the other causes are related to cellular breakdown or to hormonal imbalances (**Table 9.9**).

Elevated phosphate levels are seen frequently in situations where there is massive cell breakdown, especially in rhabdomyolysis and in tumor cell lysis following chemotherapy. The decreased renal excretion of phosphate associated with any form of PTH deficiency will lead to raised serum phosphate concentrations also.

Symptoms of hyperphosphatemia

Since increased levels of phosphate can cause precipitation of calcium and therefore hypocalcemia, several of the symptoms of hyperphosphatemia are similar to those of hypocalcemia. Profound hypocalcemia can be seen in acute hyperphosphatemic states. Paresthesia and muscle cramps are seen frequently in patients at first presentation. The high levels of phosphate inhibit the action of PTH also, serving to exacerbate the hypocalcemic effect. With chronic hyperphosphatemia, the deposition of calcium phosphate salts can become widespread, involving many tissues and organs, and is known as metastatic calcification. Furthermore, the high plasma phosphate concentration causes enhanced phosphate uptake in vascular cells, leading to activation of the osteogenesis gene *CBFA1*, further calcification of blood vessels, and increased blood pressure. *CBFA1* (core-binding factor, Runt domain, alpha subunit 1) encodes a nuclear transcription factor that is up-regulated by low-level mechanical stress in human cultured osteoblasts and has been described as the primary osteogenic master switch.

Investigation of hyperphosphatemia

The investigation of hyperphosphatemia revolves around the routine investigation of the main causes outlined in Table 9.9.

Treatment of hyperphosphatemia

Apart from dealing with the underlying disorder, the mainstay of treatment for hyperphosphatemia is the use of phosphate-binding agents, which reduce the gastrointestinal absorption of phosphate, and dealing with the resulting hypocalcemia as it arises. In patients with chronic kidney disease, regular dialysis is used in conjunction with phosphate-binding agents to lower phosphate levels.

9.10 OTHER BONE DISORDERS

Patients can have bone pathology present without hypercalcemia or hypocalcemia. Indeed, primary bone cancers rarely exhibit abnormalities in serum calcium. However, although some disorders produce serum calcium abnormalities rarely, other serum and urine parameters can be used to monitor disease progression. A prime example is Paget's disease, where there is focal bone overactivity, and the disease can be monitored by measuring serum alkaline phosphatase (ALP). Very high levels of ALP (>1000 U/L) can be seen in Paget's disease but, apart from a few patients, hypercalcemia is not observed. Since the levels of ALP are so high, it is not usually necessary to measure the bone-specific isoenzyme of ALP in Paget's disease. However, with smaller rises in total ALP it is sometimes useful to undertake ALP isoenzyme analysis to try to determine the source of the raised ALP level.

Direct immunoassays for bone-specific ALP are available, but many laboratories undertake electrophoretic methods to determine the predominant isoenzyme present in a patient with raised ALP.

Many other markers of bone resorption and formation have been developed (**Table 9.10**) but few are used routinely in the management of metabolic bone disorders. They are of use in research and in monitoring the efficacy of new treatments for osteoporosis, bone secondary growths, and so on. They have limited, if any, use in the routine diagnosis of metabolic bone disease currently.

TABLE 9.10 Bone markers

Markers of bone formation	Markers of bone resorption
Total alkaline phosphatase	Tartrate-resistant acid phosphatase (TRAP)
Bone alkaline phosphatase	Hydroxyproline
Osteocalcin (bone GLA protein)	Hydroxylysine glycosides
C-terminal propeptide of type 1 procollagen (P1CP)	Pyridinoline
N-terminal propeptide of type 1 procollagen (P1NP)	Deoxypyridinoline
	N-terminal telopeptide of type 1 procollagen (NTX)
	C-terminal telopeptide of type 1 procollagen (CrossLaps®, CTx)

Bone GLA protein, bone gamma-carboxyglutamic acid protein.

IRON

CHAPTER 10

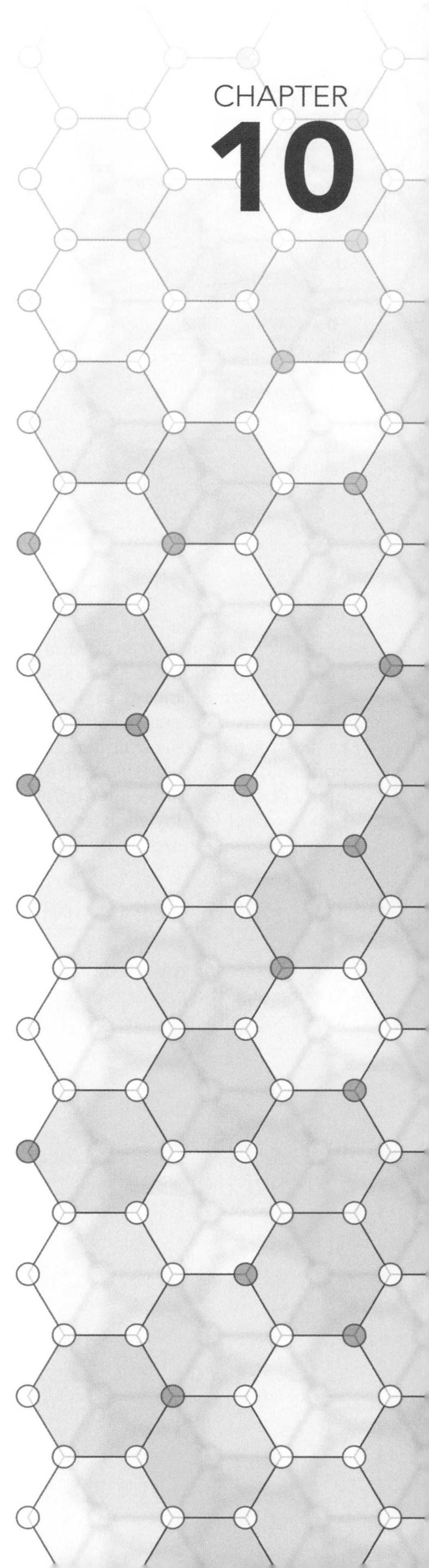

IN THIS CHAPTER

REFERENCE RANGES FOR IRON

HEME-CONTAINING PROTEINS

IRON HOMEOSTASIS AND TURNOVER

ASSESSMENT OF IRON STATUS

DISORDERS OF IRON HOMEOSTASIS

The biochemical importance of iron lies in its roles in oxygen transport and storage and in oxidative metabolism. It is an essential element in man, being a key component of heme proteins—including hemoglobin, myoglobin, and the cytochromes of both the mitochondrial and microsomal electron-transport systems—and a co-factor for a number of enzymes containing iron–sulfur complexes.

The total body iron content of a healthy neonate is approximately 300 mg and this increases during growth such that an adult male contains 3–4 g and an adult female 2–3 g of iron. Approximately a quarter of this is stored in tissues while the majority is present as functional iron.

10.1 REFERENCE RANGES FOR IRON

It is widely accepted that laboratory reference ranges for both hemoglobin and ferritin are lower for adult females than for males (**Table 10.1**). The explanation is simple enough—menstrual blood loss—but is this a physiological difference or a pathological one? Reference ranges are usually based on observed values in a defined population, but this per se need not necessarily imply that the range is ideal physiologically; for example, target concentrations of serum cholesterol for the minimization of cardiovascular risk are much lower than mean concentrations observed in Western societies. Furthermore, in the current chapter, the gender differences for iron and ferritin shown in Table 10.1 may be apparent rather than true and indicate a high prevalence of iron deficiency in adult females due to a combination of menstrual bleeding and inadequate iron intake. If males are viewed as non-menstruating females, a single range derived from the male population should be applicable to both sexes. In favor of this view are a number of observations. No gender differences are seen before or after the ages of menstruation in those primates which menstruate, and treating iron deficiency results in similar final hemoglobin concentrations in both sexes.

There are numerous possible effects of iron deficiency in the absence of overt anemia including hair loss, impaired response to infection, and reduced work capacity. Perhaps of greatest concern, however, is the effect on mental capacity

TABLE 10.1 Serum reference ranges in the analysis of iron and related metabolites			
Analyte	**Gender**	**Concentration (SI units)**	**Concentration (conventional units)**
Transferrin (increased in pregnancy; lower values in neonates)	M/F	25–50 μmol/L	200–400 mg/dL
>60 years	M/F	22–47 μmol/L	180–380 mg/dL
Iron	M	11.6–31.3 μmol/L	65–175 μg/dL
	F	9.0–30.4 μmol/L	50–170 μg/dL
Transferrin saturation	M/F	20–40%	20–40%
Total iron-binding capacity (TIBC)	M/F	44.8–80.6 μmol/L	250–450 μg/dL
Iron saturation	M	20–50	20–50
	F	15–50	15–50
Ferritin	M	45–562 pmol/L	20–250 ng/mL
	F	22–270 pmol/L	10–120 ng/mL
	F (postmenopausal)	45–562 pmol/L	20–250 ng/mL

and development. Iron is accumulated in the brain until early adulthood and iron deficiency in childhood is associated with reduced IQ scores. Failure to recognize and treat iron deficiency in women because reference ranges are used inappropriately may have serious effects on their health. A counterview is that using male ranges for females would result in a huge proportion (up to 50%) of the female population being classified as iron deficient, which seems scarcely credible to many observers. However, there has been no argument advanced to explain in biological terms why adult females should require lower ferritin and hemoglobin, so the conclusion that many are truly iron deficient is hard to escape.

10.2 HEME-CONTAINING PROTEINS

Hemoglobin and myoglobin are oxygen-binding proteins which enable the body to overcome the problem of the low solubility of oxygen in water. The principle iron-containing proteins are shown in **Table 10.2**.

TABLE 10.2 Principal iron-containing compounds	
Major types of iron-containing proteins	**Specific examples**
Heme proteins	Hemoglobin
	Myoglobin
	Cytochromes *b*, *c*
	Cytochrome oxidase
	Catalase
	Homogentisate oxidase
Flavoproteins	Cytochrome *c* reductase
	NADH dehydrogenase
	Xanthine oxidase
	Acyl-CoA dehydrogenase

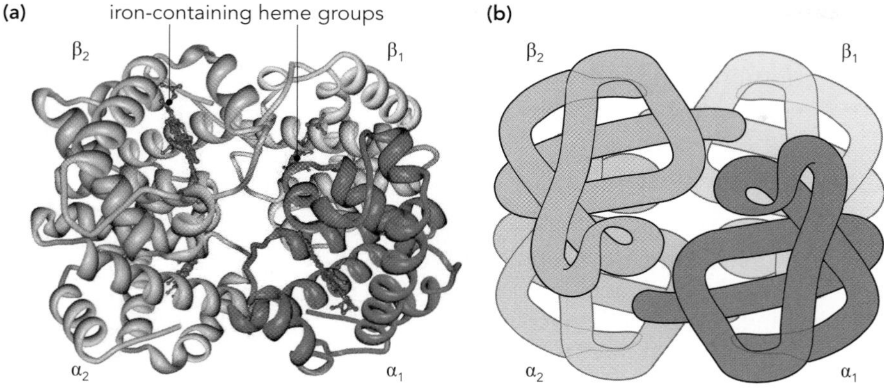

Figure 10.1 Structure of hemoglobin.
(a) The four polypeptide chains ($\alpha_2\beta_2$) are shown, each with its central iron atom (Fe^{2+}). (b) Simplified structure showing the helices.

Hemoglobin

Hemoglobin, found in red blood cells, is the carrier of oxygen in the circulation, delivering oxygen from the lungs to the tissues. It also transports CO_2 and H^+. It is a tetrameric protein consisting of two α globin and two β globin subunits arranged in a tetrahedral configuration which allows for multiple interactions between α and β subunits (**Figure 10.1**). A heme group is covalently attached to each of the four globin chains and consists of a protoporphyrin of four pyrrole rings to which ionic iron binds via the four nitrogens in the center of the protoporphyrin. When the iron is in the reduced, ferrous (Fe^{2+}) state it can bind a molecule of oxygen; thus each molecule of hemoglobin can bind up to four molecules of oxygen, one to each heme group. Initial oxygen binding exerts a cooperative effect such that it enhances the binding of subsequent molecules. Hemoglobin is also an allosteric protein and has binding sites for allosteric effectors, such as H^+, CO_2, and bisphosphoglycerate (BPG; also called diphosphoglycerate, DPG), which affect the affinity of heme for oxygen. Binding of oxygen does not alter the valence state of the iron in heme; indeed, oxidation of the iron from the ferrous to the ferric (Fe^{3+}) state (ferrihemoglobin or methemoglobin) leads to loss of affinity for oxygen.

Myoglobin

Myoglobin is present only in muscle; it serves to store oxygen in the muscle for aerobic respiration and also facilitates the movement of oxygen within the tissue. It is a compact, α-helix-rich monomeric globin polypeptide with a single associated heme group and binds just one molecule of oxygen.

Carbon monoxide (CO) competes with oxygen for binding to iron at the center of the heme prosthetic groups in both hemoglobin and myoglobin. The affinity of the heme group for CO is 200-fold greater than for oxygen, and thus CO will displace oxygen from hemoglobin, as seen in cases of carbon monoxide poisoning.

Cytochromes

Cytochromes *b*, *c*, c_1, and *a* are hemoproteins of the electron-transport chain on the inner membranes of mitochondria. In contrast to hemoglobin and myoglobin, the heme iron in these proteins undergoes reversible reduction and oxidation as electrons pass down the electron-transport chain to oxygen. Only the terminal cytochrome of the mitochondrial respiratory redox chain, cytochrome oxidase ($a + a_3$), has an oxygen-binding site for which CO will also compete and lead to inhibition of electron transport.

Haptoglobin

Haptoglobin is an acute-phase protein secreted by the liver and serves as a scavenger for heme released during breakdown of hemoglobin. It thereby avoids the toxic effects of free heme and also prevents major losses of iron from the body, particularly at times of excessive breakdown of red cells.

The structures and functions of transferrin, ferritin, and hemosiderin are described below.

10.3 IRON HOMEOSTASIS AND TURNOVER

Iron is unusual in that, unlike other cations such as sodium, potassium, and calcium, the body has no way of excreting an excess of the element apart from sweating, sloughing off of mucosal cells, and bleeding. The body therefore must have a reservoir of iron and an efficient iron recycling system to ensure provision of the element in times of low dietary iron, and also must have a mechanism to restrict excessive intake when the body is iron replete and dietary intake is high. The distribution of iron in the human body is shown in **Table 10.3**. The necessity to maintain a supply of iron for the synthesis of an optimal red cell mass is paramount. However, an excess of iron is toxic to the body and, in healthy individuals, iron homeostasis is normally maintained and iron overload prevented by regulation of absorption from the upper small intestine, although there are other routes by which iron may enter the body (**Table 10.4**).

TABLE 10.3 Approximate distribution of iron in adults

Iron store (50–70 mmol, 3–4 g)	Iron-containing proteins	%
Functional iron (75%)	Hemoglobin	70.0
	Myoglobin	4.9
	Other heme proteins (for example transferrin)	0.1
Stored iron (25%)	Bound to proteins (ferritin, hemosiderin)	25.0

TABLE 10.4 Routes of iron intake

Route	Specific example
Maternal transfer	Approximately 300 mg for a full-term infant
By mouth	Food: 10–20 mg/day for an adult on mixed diet
	Medicines
Parenteral	Iron preparations
	Blood transfusions (~300 mg/unit)

Absorption of iron

Iron is absorbed into the intestinal mucosal cells by a membrane component, divalent metal (ion) transporter 1 (DMT1), most efficiently in the reduced Fe^{2+} form from heme proteins of meat in the diet. Other dietary iron, usually present as ferric hydroxide [$Fe(OH)_3$], is not absorbed unless reduced to the ferrous state. The enzymes and acidic secretions of the stomach help to emulsify the ingested food mass and to release Fe^{2+} from heme, while a duodenal ferric reductase enzyme, duodenal cytochrome *b* (Dcyt*b*), reduces Fe^{3+} to Fe^{2+} during absorption. A number of factors can affect the absorption of iron including the nature of the iron-containing component(s) of the diet, other dietary components which might bind iron or impact on its redox state in the gut, the action of gastric juice on the dietary components, and the structural integrity of the mucosal surface across which absorption occurs. Also, the alkaline pancreatic secretion neutralizes the acidified food mass in the duodenum and this rise in pH helps to control

TABLE 10.5 Factors affecting the absorption of dietary iron

General factors	Specific examples
Intake	Food, especially meat (increase)
	Medicines (for example tetracycline) (decrease)
Dietary factors	Ascorbic acid (increase)
	Ethanol (decrease)
	Phytate (decrease)
	Tannin (decrease)
	Eggs (increase)
Gastric juice	Acid (increase)
	Pepsin (increase)
	Mucopolysaccharide (increase)
Pancreatic secretion	Alkali (decrease)
Mucosal integrity	Small bowel disease (decrease) (celiac disease; Crohn's disease)
Disorders associated with increased absorption	Iron deficiency not due to decreased absorption: Non-iron-deficiency anemia Hemochromatosis Porphyria cutanea tarda

excessive iron absorption. In addition, there are a number of clinical disorders which lead to increased absorption (**Table 10.5**).

Once the Fe^{2+} enters the mucosal cells, it is oxidized to the Fe^{3+} form in a reaction catalyzed by a transmembrane, copper-containing ferroxidase, hephaestin, a protein having sequence homology with ceruloplasmin. Hephaestin appears to mediate iron efflux from the enterocyte in conjunction with another transmembrane protein, ferroportin-1. Ferric ions are transported across the basolateral membrane of the enterocyte and bind to the iron-transport protein, transferrin (**Figure 10.2**), for transport in the circulation to various tissues. It is at this stage that the entry of iron into the circulation is regulated. The amount of iron moving into the blood and on to other body stores is controlled by the iron saturation of transferrin itself, which in turn reflects the iron status of the body. Any excess iron in the diet, above the amount required for homeostasis and daily turnover,

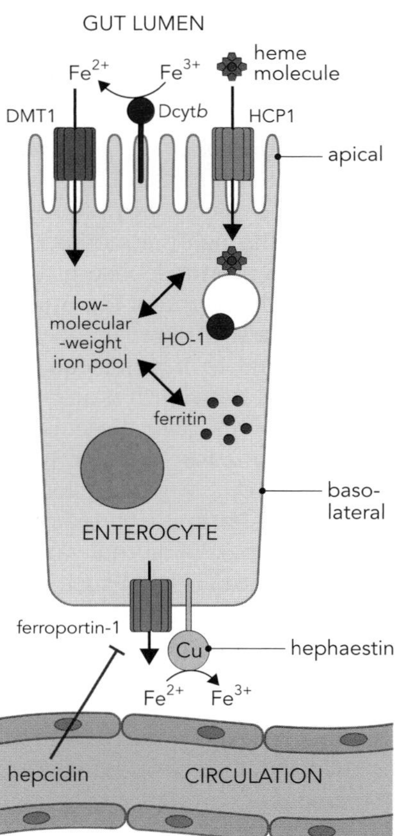

Figure 10.2 Regulation of iron transport through the enterocyte (intestinal epithelial cell).
Dietary iron is reduced to the ferrous state by a duodenal cytochrome *b* (Dcyt*b*) and absorbed into the enterocyte in this reduced state by the divalent metal (ion) transporter 1 (DMT1) in the cell membrane. It is oxidized to the Fe^{3+} form in a reaction catalyzed by a transmembrane, copper-containing ferroxidase, hephaestin. Hephaestin appears to mediate iron efflux from the enterocyte into the circulation in conjunction with another transmembrane protein, ferroportin-1, and delivers Fe^{3+} to the iron-transport protein, apotransferrin (see Figure 10.3), for transport in the circulation to various tissues. Excess iron is retained in the enterocyte bound to ferritin and is lost from the body during cell turnover. Hepcidin, a hepatically synthesized peptide hormone, is the principal regulator of systemic iron homeostasis. It controls plasma iron concentration and tissue distribution of iron by inhibiting intestinal iron absorption, iron recycling by macrophages, and iron mobilization from hepatic stores. HCP1, Heme Carrier Protein 1. Originally identified as Heme-Fe transporter, it was later identified as a folate transporter which also appears to act as a low affinity heme-Fe transport protein. HO-1, heme oxygenase 1, an enzyme which catalyses the breakdown of heme to biliverdin, iron, and carbon monoxide.

TABLE 10.6 Approximate iron losses in health

Route of loss	Approximate loss
Intestinal mucosa	1 mg/day
Menstruation	20 mg/month
Pregnancy	500 mg
Urine	<0.1 mg/day

remains in the enterocyte, probably stored as ferritin, and is lost from the body when the enterocytes slough off from the mucosal surface. This is important because, apart from bleeding, the body has no way of eliminating iron. The net absorption of iron in healthy adults is about 1 mg/day for men and postmenopausal women and 2 mg/day for pre-menopausal women. This replaces daily iron losses occurring mainly through shed intestinal mucosal cells (**Table 10.6**).

Larger amounts of iron are absorbed during pregnancy to provide (with some iron derived from stores) approximately 300 mg for a full-term baby and 200 mg for the placenta. Very little iron is excreted in urine. Iron absorption is also increased in iron-deficiency states, except in cases of malabsorption.

Hepcidin

The major controller of whole-body iron homeostasis is hepcidin, a peptide hormone synthesized in the liver. It acts by inhibiting the efflux of iron from cells by binding to the single iron transporter, ferroportin-1, thereby enhancing its degradation. Synthesis of hepcidin responds to the iron status of the body, being increased during times of iron overload and reduced in anemia and hypoxia. A deficiency in hepcidin is most likely responsible for most cases of hemochromatosis via mutations in the hepcidin gene or in one of the genes encoding regulators of hepcidin synthesis.

Iron transport

Free ferric iron (Fe^{3+}) is toxic and almost all of the iron in plasma is bound to the α-globulin transferrin, an 80 kDa glycoprotein synthesized in the liver. Plasma contains approximately 7 mg of iron and is the most active body compartment in terms of iron metabolism; transferrin undergoes a high rate of turnover as it supplies the needs of iron for red cell synthesis. The synthesis of transferrin is regulated through the modulation of mRNA stability. At low cellular iron concentrations, iron response proteins bind to iron response elements (sequence motifs) at the 3′ end of the transferrin mRNA and stabilize it, promoting transferrin synthesis. However, when iron concentrations are high, the iron response proteins bind to iron instead and the transferrin mRNA is degraded rapidly. Synthesis of transferrin is also stimulated by estrogens. Transferrin is one of the acute-phase response proteins whose synthesis is increased following infection and/or trauma (see Chapter 3).

Each molecule of transferrin has two Fe^{3+}-binding sites and, in healthy adults, transferrin is approximately 30% saturated. Plasma iron concentrations fluctuate as part of diurnal variation and, of course, following bleeding, such as during menstruation; in such situations, iron is shifted between transferrin and body iron stores. Most of the iron transported by transferrin is derived from the breakdown of hemoglobin from mature erythrocytes as they are cleared by the reticuloendothelial cells of the spleen at the end of their life span (~120 days). This recycling of iron in the body is essential to the provision of iron for hemoglobin synthesis for new red cells.

The absolute concentration of transferrin in plasma can be measured by immunoassay and is usually about 37.5 μmol/L (300 mg/dL). Clinically, however, it is important to know the proportion occupied of the total number of iron-binding sites on transferrin in plasma. This is known as transferrin saturation (TSAT) and in normal iron homeostasis is usually 30%. This can be calculated from the molecular weights of transferrin and iron and the fact that one molecule of transferrin will bind two molecules of iron. In the absence of a transferrin assay, total iron-binding capacity (TIBC) may be used as a measure of transferrin concentration in plasma. Plasma iron is first measured in plasma, and second in a similar sample of plasma to which excess inorganic ferric ions have been added and in which non-protein-bound ferric ions have subsequently been removed by ion-exchange resin chromatography. The iron concentration in the second sample, expressed in μg/dL in the USA, is the TIBC and reflects the iron bound to protein, usually transferrin. The degree of transferrin saturation may be calculated as the ratio of iron concentrations before and after the addition of saturating ferric ions. The only complicating situation is one whereby the plasma ferritin concentration is sufficient to bind ferric ions; here TIBC does not reflect the transferrin concentration. Plasma transferrin and iron concentrations are often low in inflammatory conditions due to increased degradation in the reticuloendothelial system.

Transferrin can donate iron to tissues via receptor-mediated endocytosis in which the protein enters the cell via the endocytotic vesicle. On fusion of this vesicle with the lysosome, the ferric ions dissociate from the protein due to the decrease in pH; transferrin is recycled to the plasma as iron-free apotransferrin while the ferric ions are retained by the cell. Apotransferrin can then pick up further ferric ions from the intestine for transport to cells (**Figure 10.3**).

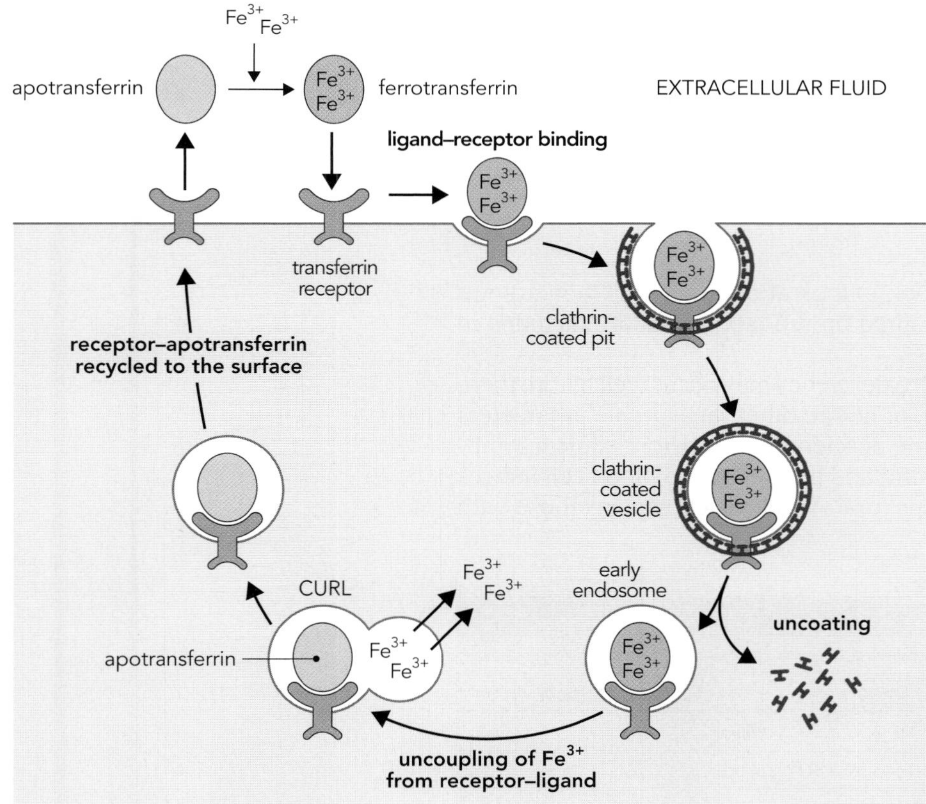

Figure 10.3 Regulation of intestinal iron transport.
Transferrin carrying ferric iron (Fe^{3+}) binds to its receptor on the target-cell membrane and the receptor–ligand complex is endocytosed in clathrin-coated vesicles. The internalized vesicles shed the clathrin coat and this early endosome fuses with lysosomes in the compartment for uncoupling of ligand from receptor (CURL). In the acidic milieu of the CURL, Fe^{3+} dissociates from the receptor-bound transferrin and the apotransferrin–receptor complex is recycled to the cell membrane. At the pH of the extracellular fluid, apotransferrin dissociates from the receptor and enters the circulation.

Iron storage

Healthy adults store appreciable amounts of iron as a complex, ferritin, which consists of ferric phosphate bound to apoferritin, a polymeric protein of 24 polypeptide subunits. Each molecule of ferritin can bind up to 2000 iron atoms. The total amount of iron stored bound to ferritin in adult humans is:

Men **~750 mg iron (range: 200–1200 mg)**

Women **~300 mg iron (range: 100–500 mg)**

Ferritin is widely distributed in tissues which express tissue-specific apoferritins, especially bone marrow and spleen, and ferritin in the circulation is usually in equilibrium with ferritin stores. It is worth noting that ferritin is an acute-phase protein and consequently its concentration in plasma may be increased in inflammatory conditions. In cases of iron overload, ferritin combines with denatured ferritin and other materials to form a poorly defined, insoluble aggregate known as hemosiderin. However, in contrast to the iron bound to ferritin, iron is not readily released from this complex mixture and is thus not freely available in times of general iron deficiency. Under overload conditions, aggregates of hemosiderin are found in the reticuloendothelial cells in bone marrow and spleen, particularly following bleeding and hemorrhage. Hemosiderin stains with ferricyanide and particles of hemosiderin are sufficiently large to be visible under light microscopy.

Lactoferrin is an iron-binding protein that is thought to have an antibacterial role. It is secreted by activated neutrophils and binds iron released from damaged cells, thereby depriving microorganisms of an essential growth factor.

10.4 ASSESSMENT OF IRON STATUS

The assessment of iron status may involve input from a number of diagnostic disciplines and the following parameters may be measured routinely:

- Hematology: hemoglobin, mean cell hemoglobin (MCH), mean cell volume (MCV)
- Clinical chemistry: serum ferritin, iron, transferrin
- Histology: stainable iron in tissues
- Medical physics: radio-iron kinetics

It is beyond the scope of this textbook to look at all aspects of hematology, but the basic parameters that are measured on red blood cells are included in **Table 10.7**.

As with any other vitamin or mineral, a deficiency may occur well before there are obvious clinical signs or symptoms, or before other measurable parameters become abnormal. A patient will become deficient in iron before clinical symptoms, such as tiredness, become apparent, and before the red blood cell indices are abnormal. In certain high-risk groups for iron deficiency, such as those with

TABLE 10.7 Red blood cell indices

Parameter	Sex	Adult reference ranges	
		SI Units	Conventional Units
Red blood cells (RBC)	F	$3.58–4.99 \times 10^{12}$ cells/L	
	M	$4.14–5.59 \times 10^{12}$ cells/L	
Hemoglobin (Hb)	F	115–160 g/L	11.5–16.0 g/dL
	M	135–175 g/L	13.5–17.5 g/dL
Mean cell volume (MCV)		80–100 fL	80–100 μm^3
Mean cell hemoglobin (MCH)		26–34 pg/cell	26–34 pg/cell
Mean cell hemoglobin concentration (MCHC)		30–35 g/dL	

celiac disease, it is probably worthwhile assessing iron status directly, as well as measuring the full blood count. However, the majority of patients have their iron status assessed after presenting to their doctor with clinical symptoms of anemia, and they have a full blood count undertaken afterward; that is, they are checked for anemia first.

In iron-deficient anemia, the classical finding in red blood cells is that the hemoglobin is below the reference range (anemia), the red blood cells are smaller than normal (microcytosis), and there is less pigmentation within each cell (hypochromia). The patient has a microcytic, hypochromic anemia and the most likely cause (but not the only cause) would be iron deficiency. This would be a straightforward pointer to then assess the patient's iron status. However, the full blood count may not be obviously pointing toward iron deficiency. For example, if the patient had other underlying illness that can make the red blood cell larger [for example vitamin B_{12} deficiency (see Chapter 24) or liver disease], this particular red blood cell index may be normal. Unless the cause of the anemia is obvious, most patients newly diagnosed will have their iron status assessed.

There are several parameters in common usage to assess iron status, and the main three—ferritin, iron, and transferrin—are shown in Table 10.1. Some situations in which serum iron and transferrin concentrations change are shown in **Table 10.8**.

TABLE 10.8 Changes in serum iron and transferrin in clinical situations

Change in serum iron concentration	Cause	Change in serum transferrin concentration
Decrease	Iron deficiency	Increase
	Chronic illness	Decrease
Increase	Anemia other than from iron deficiency	Not affected
	Chronic iron overload	Decrease
	Acute iron poisoning	Not affected
	Pregnancy	Increase
	Estrogen therapy	Increase

Serum ferritin is the first-line test to assess iron status and is widely available. It is measured by immunoassay, and there are variations between assays and differences in quoted reference ranges. In general, serum ferritin reflects body iron stores, low iron stores giving rise to low serum ferritin, and high iron stores to high serum ferritin. Unfortunately, since several unrelated diseases can also cause a raised ferritin, its measurement is limited to the diagnosis of iron deficiency and iron overload. Clinical conditions in which serum ferritin is raised include the following:

- Chronic iron overload, including hemochromatosis
- Malignancy, especially lymphomas
- Infection and inflammation
- Hepatocellular injury
- Acute leukemia
- Renal failure

Accepting these limitations, ferritin is still the best first-line test in the initial investigation of iron status; furthermore, since it can usually be assayed on an ethylenediaminetetraacetic acid (EDTA) sample used for the full blood count, the tests can be done on the same sample.

Clinical practice point 10.1

Low ferritin is a very specific marker of iron deficiency. However, ferritin lacks diagnostic sensitivity as it is increased in many inflammatory disorders. A normal, or even raised, ferritin does not exclude the presence of iron deficiency.

At first sight, it would have been thought that the measurement of serum iron would be the ideal test to assess a patient's iron status. However, as with many other parameters, the serum concentration of iron does not accurately reflect body stores of iron. Furthermore, serum iron is markedly affected by dietary status and the concentration of the transferrin carrying the iron in the serum. To overcome the latter problem, transferrin can be measured and the percentage saturation with iron can be determined. In iron deficiency the iron would be expected to be low, and the transferrin high, hence the TSAT would also be low. In iron overload the opposite would be expected to occur, and the TSAT would be high. For iron deficiency, ferritin is a more specific marker than TSAT, whilst for iron overload a TSAT is more specific than ferritin. Care must be taken when measuring serum iron, as the majority of methods use dye-binding to measure the metal. These methods are susceptible to interferences caused by anticoagulants. Blood samples collected into EDTA cannot be used, since EDTA interferes with virtually all of the dye methods, and several methods are also affected by lithium heparin. Furthermore, dye methods are unreliable in patients who are being treated for iron overload with desferrioxamine, as this can also interfere with such assays. Specialist centers often employ alternative technology (atomic absorption or plasma emission spectroscopy) to overcome these interferences.

Clinical practice point 10.2

Ferritin is a very sensitive marker of iron overload, but lacks diagnostic specificity, since many other disorders can cause a raised ferritin.

Analytical practice point 10.1

When assessing iron and percentage saturation it is important to take a fasting sample and to ensure that the patient has not taken iron within the previous 48 hours.

There are several other markers of iron status available, with differing claims of specificity and sensitivity. Some full-blood-count analyzers are equipped with a parameter called "percentage hypochromic red cells," which is supposed to be a specific and sensitive marker of iron deficiency, and has been advocated to be very useful in the assessment of iron status in renal failure. An alternative approach has been to measure the number of soluble transferrin receptors (STRs) in serum. In iron deficiency, more STRs would be expected, and the converse in iron overload. At the moment, these tests have had a limited uptake, but may be used more in the future.

10.5 DISORDERS OF IRON HOMEOSTASIS

Iron deficiency

Iron-deficiency states

Clinically, iron deficiency progresses from the pre-anemic to the anemic phase. It is commonly due to persistent bleeding from the gastrointestinal tract or uterus and less often to inadequate dietary intake or malabsorption (for example in Crohn's disease or celiac disease). Characteristics of iron deficiency determined in the laboratory are listed in **Table 10.9**.

Clinical practice point 10.3

Iron deficiency probably occurs before patients show clinical signs of anemia.

TABLE 10.9 Characteristics of iron deficiency

Parameter measured	Findings
Red cell indices	Decreased Hb, RBC, MCH, and MCV
Red cell	Hypochromia
	Microcytosis
	Anisocytosis
	Poikilocytosis
Red cell zinc	Zn-protoporphyrin increased
Serum	Iron decreased
	Transferrin increased
	Ferritin decreased
Bone marrow	Lack of stainable iron

CASE 10.1

A 68-year-old man complains of tiredness and shortness of breath on exertion. He looks pale on examination. His blood count indicated anemia. Further results were obtained on a fasting sample.

	SI units	Reference range (adult male)	Conventional units	Reference range (adult male)
Whole blood				
Hemoglobin	110 g/L	135–175	11.0 g/dL	13.5–17.5
Mean cell volume (MCV)	72 fL	80–100	72 μm3	80–100
Mean cell hemoglobin (MCH)	22 pg/cell	26–34	22 pg/cell	26–34
Serum				
Ferritin	11 pmol/L	45–562	4.9 ng/mL	20–250
Iron	7 μmol/L	11.6–31.3	39 μg/dL	65–175
Transferrin	5.2 μmol/L	25–50	422 mg/dL	200–400
Transferrin saturation	7%	20–40	7%	20–40

- **How do the MCV and MCH help classify the anemia?**
- **What is the significance of the ferritin and iron panel results?**
- **What is the cause of the anemia?**

Anemia is defined as low hemoglobin (Hb). It is usually classified by cell size as micro-, normo-, or macrocytic. In this case, the low mean cell volume (MCV) indicates microcytosis and the low mean cell Hb (MCH) indicates hypochromia. This is the usual pattern seen with an iron-deficiency anemia. Low ferritin is only caused by iron deficiency and is sufficient for diagnosis. The low serum iron and transferrin saturation are provided for illustration only and do not usually provide additional diagnostic information. It is not possible to diagnose the cause of the iron deficiency without further investigations. Often it is due to chronic low-level bleeding within the gastrointestinal tract. The most important test in an elderly person would be investigation of the lower bowel (for example by colonoscopy) to exclude cancer.

Iron-refractory iron-deficiency anemia

Iron-refractory iron-deficiency anemia (IRIDA) is a rare, autosomal recessive iron-malabsorption syndrome linked to mutations in the *TMPRSS6* gene on chromosome 22. Clinically, IRIDA presents as a hypochromic, microcytic anemia that does not respond to oral iron therapy and responds partially to parenteral iron therapy.

Chronic iron overload

There are multiple causes of iron overload, which may be due to excessive absorption of iron or may occur where intestinal absorption is bypassed by parenteral administration of iron (**Table 10.10**).

Chronic iron overload leads to accumulation of hemosiderin (aggregates of ferritin) in a number of tissues, with severe clinical consequences if untreated (**Table 10.11**). It presents with increased serum iron and ferritin and decreased transferrin. Hemosiderosis is a generic term describing chronic iron overload not due to hemochromatosis.

TABLE 10.10 Causes of chronic iron overload

General cause	Specific examples
Excessive iron absorption	Hemochromatosis
	Chronic anemia not due to iron deficiency
	High ethanol intake, especially with porphyria cutanea tarda
	Prolonged inappropriate iron treatment
	From food-preparation vessels (for example beer brewed in iron drums)
Bypass of intestinal absorption	Multiple blood transfusions except to replace external losses
	Excessive doses of parenteral iron preparations

TABLE 10.11 Clinical consequences of excessive hemosiderin deposition due to chronic iron overload

Tissue	Clinical consequence
Liver	Cirrhosis leading possibly to hepatoma
Heart	Cardiomyopathy leading to heart failure
Pancreas	Diabetes mellitus
Anterior pituitary	Panhypopituitarism
	Lethargy (decreased thyroid and adrenal hormones)
	Amenorrhea (decrease free water clearance)
Skin	Melanin pigmentation

Hereditary hemochromatosis

Hereditary hemochromatosis (HH) is an inherited single-gene disorder characterized by iron overload. It has an incidence of between 1 in 200 and 1 in 400 live births amongst Northern Europeans, and an observed carrier frequency of approximately 1 in 10. Affected individuals have up to an order of magnitude more iron than normal in their bodies (20–40 g) accumulated as hemosiderin, distributed in the heart, endocrine organs, and skin but particularly in the pancreas and liver. If untreated, this may give rise to serious clinical consequences. Characteristically, serum iron and ferritin are increased while transferrin is decreased. Though not completely elucidated, the disease mechanism involves substantially increased iron absorption via the gut, deposition of iron in the tissues, and eventual fibrosis (leading in the liver to cirrhosis). HH exhibits both locus and allelic heterogeneity; while the most common form of the disease is linked to the *HFE* gene (chromosome 6), juvenile hemochromatosis (JH) maps to the *Hepcidin* gene on chromosome 19, and HFE3 and HFE4 map to the *Transferrin Receptor 2* (chromosome 7) and *Ferroportin* (chromosome 2) genes, respectively. The predominant *HFE*-linked form of HH is associated with HLA-A3 and may be caused by at least two *HFE* gene mutations, the most common of which are C282Y and H63D. HH is diagnosed by means of the normal iron diagnostic tests and by polymerase chain reaction (PCR)-based mutation detection. Liver biopsy for stainable iron and evidence of cirrhosis is not performed routinely but can be useful in assessing the extent of liver damage. Treatment is by means of repeated venesection or, in a minority of cases, chelation therapy.

Hemolytic anemia

Hemolytic anemia results from increased destruction of red blood cells (erythrocytes) within the vasculature. The disease mechanism (hemolysis) can be

CASE 10.2

A 40-year-old man is seen in a hepatology clinic after he has been found to have abnormal liver function tests (LFTs) by his primary care doctor. He has no symptoms and his LFTs were tested for monitoring of statin treatment for high cholesterol.

	SI units	Reference range (adult male)	Conventional units	Reference range (adult male)
Serum				
Bilirubin	11 μmol/L	<21	0.6 mg/dL	<1.3
ALT	132 U/L	5–40	132 U/L	5–40
ALP	85 U/L	40–120	85 U/L	40–120
Albumin	35 g/L	32–52	3.5 g/dL	3.2–5.2
Ferritin	695 pmol/L	45–562	309 ng/mL	20–250
Iron	26.0 μmol/L	11.6–31.3	145 μg/dL	65–175
Transferrin	18 μmol/L	25–50	146 mg/dL	200–400
Transferrin saturation	72%	20–40	72%	20–40

- **What should be considered when interpreting the transferrin saturation result?**
- **What do the ferritin and transferrin saturation results suggest?**
- **What further test should be performed?**

The transferrin saturation is high, but this may be seen in non-fasting samples and in people taking iron supplements. If either is the case then the sample should be repeated. The raised ferritin may indicate an inflammatory disorder, liver disease, or an iron-overload state. The transferrin saturation (taken from a patient fasting and not on supplements) at this level is consistent with iron overload. There are several causes for this, but in an otherwise-well individual the most likely is hereditary hemochromatosis, which can be confirmed by genetic testing. There are only two common pathogenic mutations, which makes genetic testing more straightforward than for many other inherited conditions. If positive, family screening would be indicated and any affected homozygous relatives should have periodic monitoring of their ferritin and transferrin saturation. Treatment consists of venesection to remove iron from the body, there being no physiological iron-excretion mechanism.

triggered by multiple causes, giving rise to a wide range of types of hemolytic anemia. Broadly, these can be subdivided into hereditary and acquired hemolytic anemia. Causes in the hereditary group include inherited defects in components of the erythrocyte cytoskeleton and/or membrane (such as hereditary spherocytosis and hereditary elliptocytosis), inherited defects in hemoglobin (hemoglobinopathies and thalassemias), and metabolic defects that result in impaired erythrocyte stability (such as glucose 6-phosphate dehydrogenase deficiency). The acquired group encompasses a plethora of causes including immune mechanisms (for example transfusion reactions and hemolytic disease of the newborn) and non-immune mechanisms (hemolytic anemia secondary to trauma, systemic disease, infection, or adverse drug reactions).

Acute iron poisoning

Acute iron poisoning is rare nowadays, occurring usually in children, and is potentially fatal. Overdose using tablets may arise accidentally, when children mistake tablets for sweets, or intentionally when taken with suicidal intent in adults. Toxicity is due to free rather than protein-bound iron. Initial symptoms include irritation in the gastrointestinal tract presenting with vomiting (hematemesis), abdominal pain, and diarrhea. When the dose is sufficiently high and/or treatment is delayed, patients are at risk of liver and kidney injury. Treatment involves elimination of iron from the stomach by washout, and chelation with desferrioxamine of any iron remaining in the intestine. If the serum iron exceeds 80 μmol/L (447 mg/dL), desferrioxamine is given intravenously to chelate free iron, thereby rendering it nontoxic and able to be excreted into urine.

ENDOCRINE TISSUES

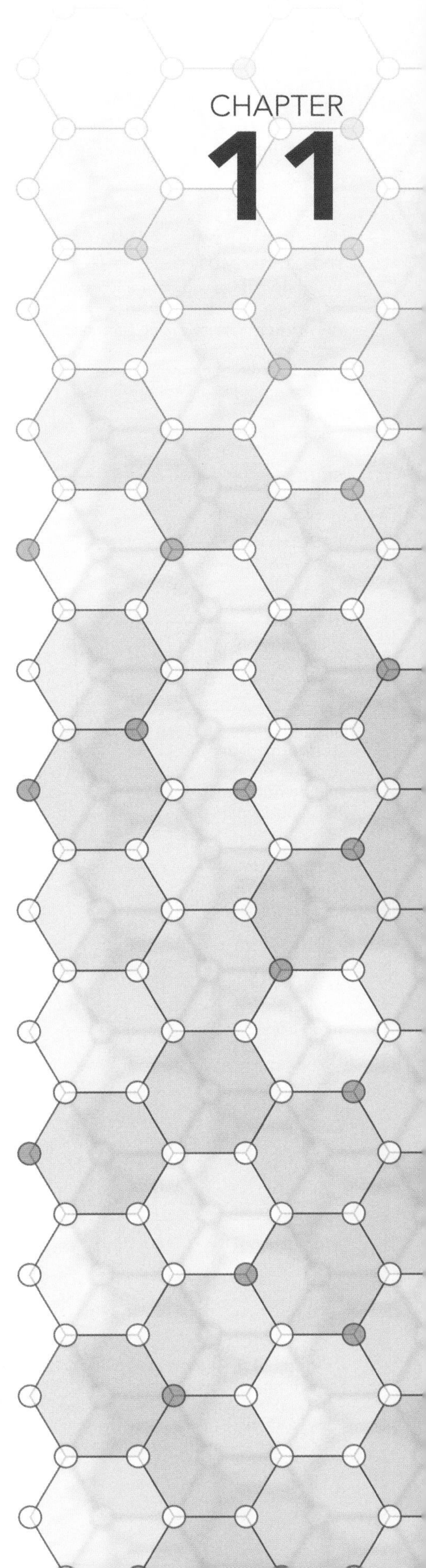

IN THIS CHAPTER

HYPOTHALAMUS

PITUITARY

THYROID

ADRENAL CORTEX

GONADAL ENDOCRINE FUNCTION

The ability of the body to maintain physiological and biochemical homeostasis, and to respond to induced changes in physiological status such as in pregnancy or trauma for example, is made possible by a network of endocrine systems. The anatomical distribution of the endocrine glands is shown in **Figure 11.1**. Inputs into the higher centers of the brain are focused on the hypothalamus, setting off a sequence of events which results ultimately in release of hormones from the end-organ endocrine tissue, allowing the body to respond metabolically to the initial stimulus. Under normal circumstances, a series of chemical feedback loops ensures that the response is limited temporally. An outline of the sequence of events is shown in **Figure 11.2**. In most cases, stimulation of the hypothalamus causes release of hormones, occasionally termed factors, which travel to the anterior pituitary where they bind to receptors on specific cells, stimulating these cells to release tropic hormones into the circulation. The pituitary hormones bind to specific receptors on target endocrine glands to elicit a physiological response, which usually includes the release of hormones. This cascade of events allows for amplification of the initial response, whereby the release of a small number of releasing factor molecules by the hypothalamus results eventually in the release of many thousands of molecules of hormone by the end-organ endocrine tissue. Control of the system is exerted by feedback inhibition of the release of hypothalamic releasing factors by hormones from both the anterior pituitary and the end-organ endocrine gland. Additionally, the release of the pituitary hormones is under feedback control by the end-organ hormones. The hypothalamus–pituitary system regulates five major endocrine axes controlling the biosynthesis and secretion of thyroid hormone, glucocorticoids, sex hormones, growth hormone, and prolactin. Three major endocrine axes will be considered in this chapter: hypothalamus–pituitary–thyroid; hypothalamus–pituitary–adrenal; and hypothalamus–pituitary–gonad. Reference to the gut axis is made in Chapter 14.

Reference ranges of analytes commonly measured in determining endocrine function of the hypothalamus–pituitary–thyroid and hypothalamus–pituitary–adrenal axes can be found in **Table 11.1**.

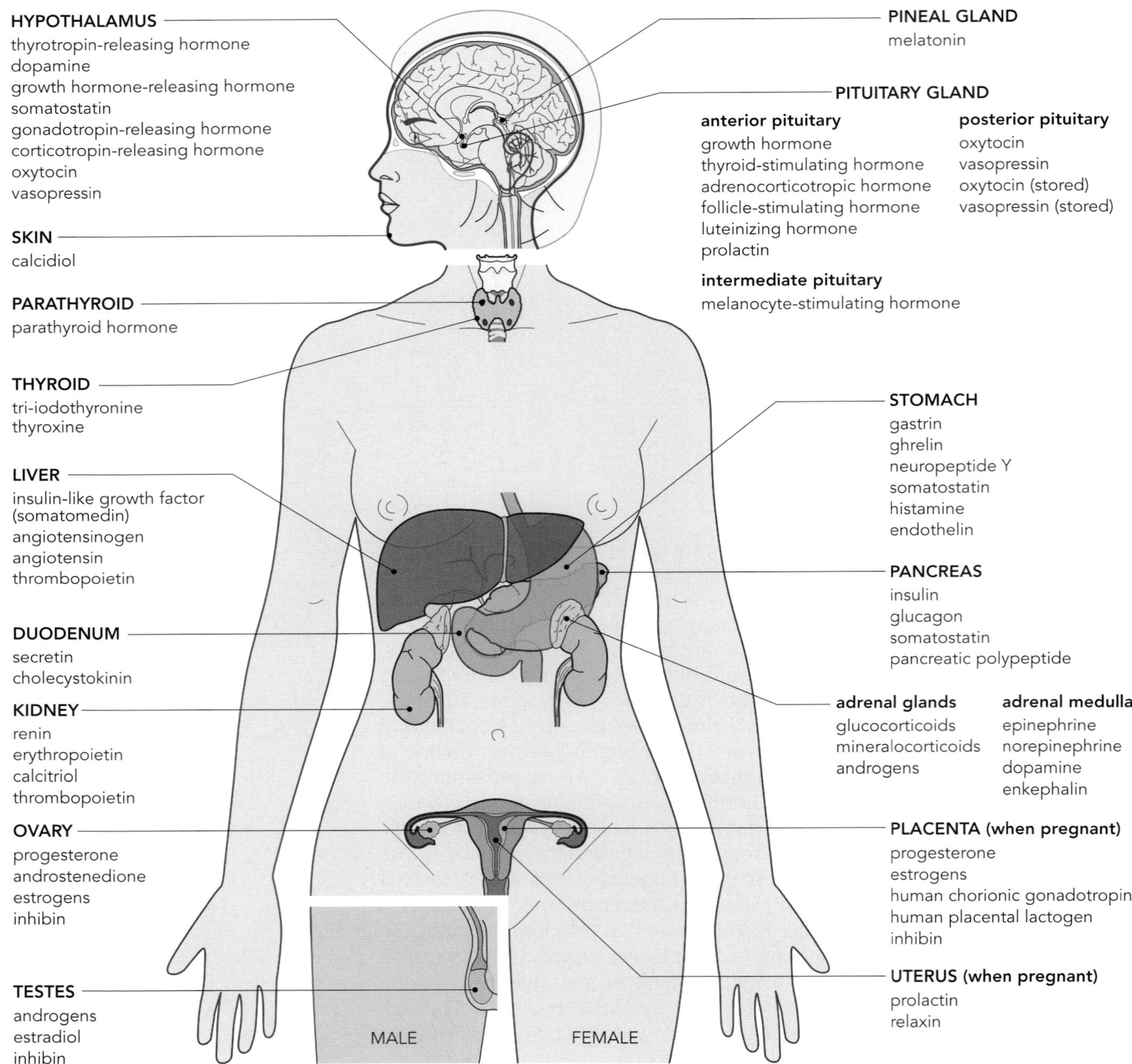

Figure 11.1 Anatomical relationship of tissues with endocrine function. The hormones synthesized by each tissue are also included. Discussions in the current chapter are restricted to the hypothalamus–pituitary–end organ axis systems, which include the thyroid, adrenal cortex, and sex glands. Other tissues are discussed elsewhere in the text.

Figure 11.2 Generic feedback control of hypothalamic–pituitary–end organ axes.
Releasing hormones from the hypothalamus activate the anterior pituitary to release specific tropic hormones, which in turn activate the adrenal cortex, thyroid, sex glands, or gut (end organs). Hormones secreted by the end organ inhibit the release of tropic hormones by the anterior pituitary (short feedback inhibition) and of releasing hormones from the hypothalamus (long-loop feedback inhibition).

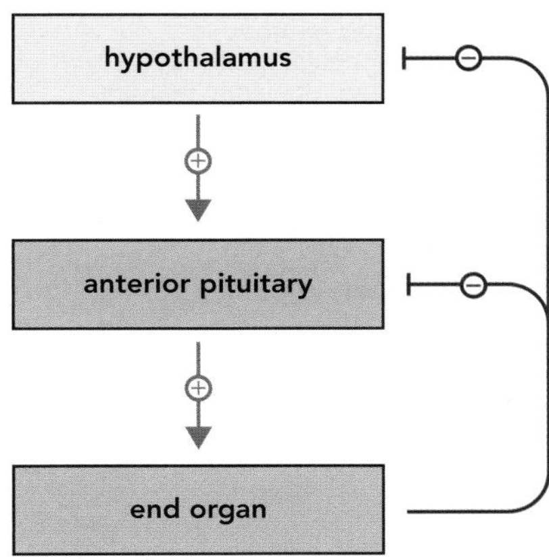

TABLE 11.1 Reference ranges of analytes commonly measured in determining endocrine function of the hypothalamus–pituitary–thyroid and hypothalamus–pituitary–adrenal axes

Analyte	Sample	Time and/or sex	Reference range	
			SI units	Conventional units
Adrenocorticotropic hormone (corticotropin or ACTH)	Plasma	08.00 h	<26 pmol/L	<120 pg/mL
		16.00–00.00 h	<19 pmol/L	<85 pg/mL
Cortisol	Serum or plasma	08.00 h	127–568 nmol/L	4.6–20.6 μg/dL
		16.00 h	82–413 nmol/L	3.0–15.0 μg/dL
		00.00 h	50% of value at 08.00 h	
	Urine	24-hour collection	55–248 nmol/day	2–9 μg/day
Corticosterone	Serum		1.5–45.0 nmol/L	53–1560 ng/dL
Dehydroepiandrosterone (DHEA)	Serum	Male	6.2–43.3 nmol/L	1.8–12.5 ng/mL
		Female	4.5–34.0 nmol/L	1.3–9.8 ng/mL
Dehydroepiandrosterone sulfate (DHEA-S)	Serum		1.6–12.2 μmol/L	50–450 μg/dL
11-Deoxycortisol	Serum		0.3–4.6 nmol/L	12–158 ng/dL
Growth hormone (hGH or somatotropin)	Serum	Adults	<18 μg/L	<18 ng/mL
17-Ketogenic steroids	Urine (24-hour collection)	Male	17–80 μmol/day	5–23 mg/day
		Female	10–52 μmol/day	3–15 mg/day
Prolactin	Serum	Male	174–617 pmol/L	4.0–14.2 ng/mL
		Female	165–1000 pmol/L	3.8–23.2 ng/mL
Testosterone	Serum	Male	10.4–41.6 nmol/L	300–1200 ng/dL
		Female	0.5–2.4 nmol/L	14–69 ng/dL
Thyroglobulin	Serum		3–42 μg/L	3–42 ng/mL
Thyroid-stimulating hormone (TSH; thyrotropin)	Serum		0.4–4.2 mIU/L	0.4–4.2 mIU/L
Thyroxine, free (fT_4)	Serum		10–31 pmol/L	0.8–2.4 ng/dL
Thyroxine, total (T_4)	Serum		59–142 nmol/L	4.6–11.0 μg/dL
Thyroid hormone-binding globulin (TBG)	Serum		11–21 mg/L	1.1–2.1 mg/dL
Tri-iodothyronine, free (fT_3)	Serum		4.0–7.4 pmol/L	260–480 pg/dL
Tri-iodothyronine, total (T_3)	Serum		1.54–3.08 nmol/L	100–200 ng/dL

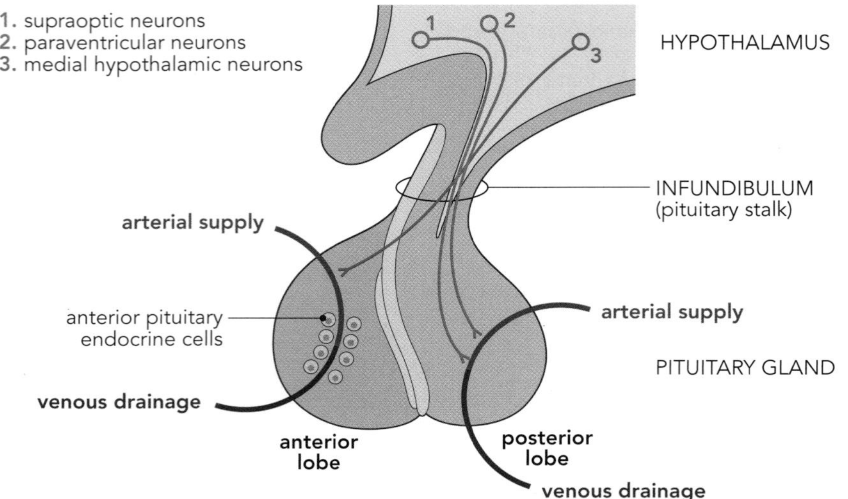

Figure 11.3 Anatomical relationship of the hypothalamus and pituitary. The hypothalamus forms the wall and floor of the third ventricle of the brain and lies above the pituitary stalk. Neurons with cell bodies in the supraoptic (1) and paraventricular nuclei (2) of the hypothalamus have projections into the posterior pituitary, where they release their hormones directly into the venous drainage of the gland. Neurons with cell bodies in the medial hypothalamic nuclei (3) project into the median eminence, where they release hormones into the pituitary portal veins. These hormones regulate the secretion of hormones from specific cells in the anterior lobe of the pituitary, which then enter into the venous drainage of the gland.

11.1 HYPOTHALAMUS

The hypothalamus, which contains the vital centers controlling appetite and thirst and also autonomic activity such as regulation of body temperature, lies in the wall and floor of the third ventricle of the brain, above the pituitary stalk (infundibulum). This stalk leads to the pituitary gland itself and carries the hypophyseal-pituitary blood supply (**Figure 11.3**). The hypothalamus communicates with the pituitary both directly (posterior pituitary) and indirectly (anterior pituitary) to regulate circadian rhythm, menstrual cycle activity, and the responses to stress, exercise, and emotional stimuli. Hormones are synthesized in different areas of the hypothalamus (**Table 11.2**). Thus, axons of neurons located in supraoptic and paraventricular nuclei of the hypothalamus release the hormones oxytocin and vasopressin directly into the venous drainage of the posterior pituitary. Neurons located in medial hypothalamic nuclei, however, secrete releasing factors into the pituitary portal veins in the median eminence of the hypothalamus and lower infundibulum, and stimulate secretion of hormones from cells of the anterior pituitary into the venous drainage from the gland. The hypothalamic releasing factors are synthesized by neural cell bodies and stored in the axonal terminals of neurons located in capillary networks within the median eminence, prior to secretion in response to a stimulus. Tight regulation of their synthesis and secretion is maintained via both long and short feedback loops involving target hormones at both the hypothalamic and pituitary levels. This tight regulation may be disturbed by a number of factors including stress, nutritional status, and systemic illness.

TABLE 11.2 Hormones synthesized and released by the hypothalamus

Region of synthesis of hormone	Hormone
Medial hypothalamic eminence	Thyrotropin-releasing hormone (TRH)
	Prolactin (PRL)-releasing hormone
	Growth hormone-releasing hormone (GHRH)
	Corticotropin-releasing hormone (CRH)
	Gonadotropin-releasing hormone (GnRH)
Supraoptic and paraventricular nuclei	Vasopressin (antidiuretic hormone; ADH)
	Oxytocin

Hypothalamic releasing factors stimulate the synthesis and secretion of tropic hormones by specific cells in the anterior pituitary. The releasing factors are peptides which bind to specific receptors on the target cell surface and regulate synthesis and secretion via transmembrane signal-transduction systems. Their mechanism of action may be complex, since a single hypothalamic factor can affect the secretion of more than one pituitary hormone.

11.2 PITUITARY

Anatomical relationships and morphology

The pituitary in an adult human is a small, oval-shaped tissue weighing 500–900 mg that lies immediately below the third ventricle of the brain. It is attached to the hypothalamus at the base of the brain by the pituitary stalk and is enclosed within a bony cavity, the pituitary fossa or sella turcica, with a membranous roof, the diaphragma sellae, through which the stalk passes. Direct communication, both neural and vascular, between the hypothalamus and pituitary is possible via the stalk. Part of the blood supply to the pituitary reaches it from the hypothalamus via portal vessels which run along the stalk. The optic chiasm and optic tracts are closely related to the stalk but do not form part of it. The pituitary gland consists of three lobes: the anterior lobe (adenohypophysis) and the posterior lobe (neurohypophysis), which have quite separate functions, and a vestigial intermediate lobe. The anterior lobe is derived embryologically from ectoderm and develops from the primitive pharynx, whereas the posterior lobe is a down-growth of the brain. In view of the migration of pharyngeal tissue to form the anterior pituitary, ectopic pituitary tissue can be left anywhere along the path of migration and this tissue may undergo hypertrophy or hyperplasia later in life to become a hormone-secreting tumor. Hormones released from the pituitary gland are secreted directly into the bloodstream, aided by the fact that the pituitary blood supply is outside of the normally tight blood–brain barrier.

Anterior pituitary

The anterior pituitary is responsible for the synthesis of six principal hormones—adrenocorticotropic hormone (ACTH; corticotropin), thyroid-stimulating hormone (TSH; thyrotropin), follicle-stimulating hormone (FSH), luteinizing hormone (LH), growth hormone (GH), and prolactin (PRL)—as well as locally acting paracrine growth factors (**Table 11.3**). Each hormone is synthesized in a specific cell type within the tissue and the secretion from these cells is controlled principally by the action of releasing hormones from the hypothalamus (**Table 11.4**). Five of the six major cell types in the anterior pituitary are components of the major endocrine axes involving the pituitary gland. The individual axes, including the actions of the pituitary hormones, are described in more detail in the

TABLE 11.3 Cell types and principal hormones synthesized in the anterior pituitary

Cell type (% of cells)	Tropic hormone	Target tissue
Corticotrophs (10–20%)	Adrenocorticotropic hormone (ACTH; corticotropin)	Adrenal cortex
Thyrotrophs (5%)	Thyroid-stimulating hormone (TSH; thyrotropin)	Thyroid
Gonadotrophs (10%)	Follicle-stimulating hormone (FSH)	Gonads
	Luteinizing hormone (LH)	Gonads
Somatotrophs (40–50%)	Growth hormone (GH)	Many
Lactotrophs (15–25%)	Prolactin (PRL)	Breast
Folliculostellate cells (<5%)	Paracrine growth factors	Neighboring cells

TABLE 11.4 Control of pituitary hormone secretion by hypothalamic hormones

Hypothalamic hormone	Pituitary hormone	Effect of increased secretion of hypothalamic hormone on secretion of pituitary hormone
CRH	ACTH	Increased
TRH	TSH	Increased
GnRH	FSH	Increased
GnRH	LH	Increased
GHRH	GH	Increased
Somatostatin	GH	Decreased
Dopamine	PRL	Decreased

appropriate sections below, but a feature common to all is that each is subject to feedback control mechanisms by hormones of the target endocrine gland (see Figure 11.2). Adrenocorticotropic hormone (ACTH), growth hormone (GH), and prolactin (PRL) are monomeric peptides, whereas the tropic hormones—thyroid-stimulating hormone (TSH), follicle-stimulating hormone (FSH), and luteinizing hormone (LH)—are dimeric glycoproteins.

Adrenocorticotropic hormone

ACTH is a 39-amino-acid polypeptide derived from a 241-amino-acid precursor, pro-opiomelanocortin (POMC) (**Figure 11.4**). Hydrolysis of the precursor peptide gives rise to equimolar amounts of several other peptides including γ-lipotropin, β-endorphin, and α-melanocyte-stimulating hormone (α-MSH). Initial cleavages of the POMC precursor are made by proteases at positively charged amino acid residues and the peptides are trimmed to produce the final products. Different corticotropic cells express different proteases to produce peptide hormones specific for each cell type. Those in the anterior lobe of the pituitary produce only ACTH and β-lipotropin, while γ-lipotropin, β-endorphin, and α-MSH are produced in cells of the intermediate lobe. These peptides are released from the pituitary in response to stimulation by corticotropin-releasing hormone (CRH) from the hypothalamus. Biological activity of ACTH appears to reside in amino acids 1–24 and the plasma half-life of the hormone is quite short (about 10 minutes). ACTH secretion exhibits a diurnal variation, with a peak at about 05.00 am, but is also increased by stress, both psychological and physical.

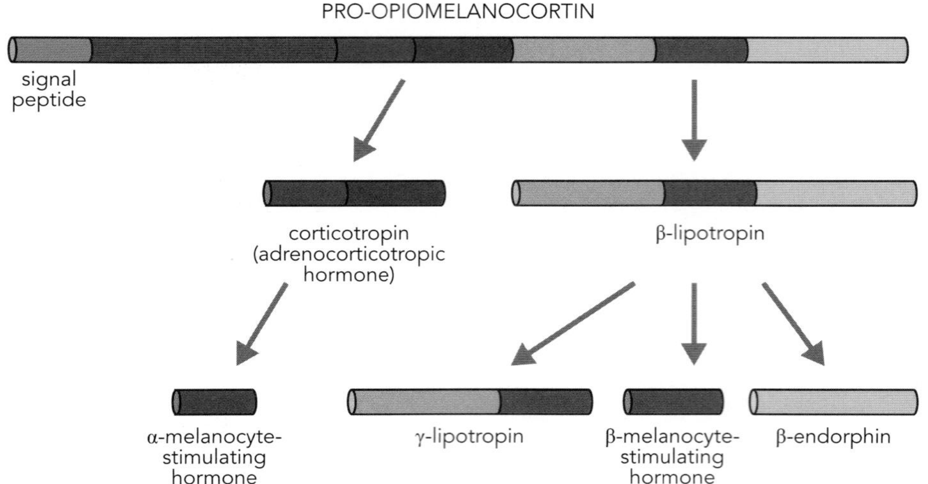

Figure 11.4 Processing of pro-opiomelanocortin. Pro-opiomelanocortin, synthesized in the anterior pituitary, is the precursor of many peptide hormones. The initial polypeptide loses its signal peptide sequence during packaging into vesicles in the Golgi prior to proteolysis at specific sites to produce the different hormones, including adrenocorticotropic hormone (ACTH), also called corticotropin.

Growth hormone

GH exists in blood in a number of variant forms, the predominant one being a 191-amino-acid (22 kDa) protein which circulates bound to a 29 kDa protein that has sequence homology with the extracellular domain of the growth hormone receptor. Binding to this protein helps to prolong the plasma half-life (20 minutes) of the hormone. The growth hormone peptide has two receptor-binding domains that allow it to bind to two separate receptor molecules on the target-cell surface, thereby inducing dimerization of the receptor. Dimerization of the receptors causes activation of an intracellular signaling cascade via tyrosine kinase activation. Release of GH is from somatotrophs and is pulsatile, with the highest plasma concentration occurring during sleep. Secretion is increased by hypoglycemia and decreased by administration of glucose. Its effect on growth is mediated by insulin-like growth factor 1 (IGF1, somatomedin C), a GH-dependent peptide synthesized in the liver; it is so named because of its weak insulin-like activity.

Prolactin

PRL is a 23 kDa protein which shares sequence homology with GH, suggesting that they are derived from a common ancestral gene. Secretion of PRL is pulsatile and is stimulated by dopamine antagonists (for example phenothiazines, metoclopramide, methyldopa, or reserpine) and inhibited by dopamine agonists (for example bromocriptine or pergolide). It has a plasma half-life of 20 minutes. Other physiological causes of increased secretion include hypoglycemia, exercise, stress, breast stimulation, convulsions, pituitary stalk section, primary hypothyroidism, and acromegaly.

Tropic hormones: TSH, FSH, and LH

The three tropic hormones—TSH, LH, and FSH—are dimeric glycoproteins which share a common, 92-amino-acid α subunit. Recognition of the three hormones by specific receptors on their respective target glands is via their similar-sized but unique β subunits (of size 118, 121, and 111 amino acids in TSH, LH, and FSH, respectively). It is likely that the α subunits also play a part in receptor recognition since the β subunits alone are not active. TSH and LH have similar plasma half-lives (50 minutes) while that of FSH is much longer (220 minutes).

Pituitary disorders

Disorders of the pituitary may be characterized by reduced (hypopituitarism) or increased (hyperpituitarism) hormone secretion (**Table 11.5**).

Hypopituitarism

Diseases of the pituitary gland can result in the reduced secretion of one or more hormones. When secretion of all of the pituitary hormones is reduced or absent this is termed pan-hypopituitarism. Pan-hypopituitarism arises as a result of destruction of the gland, usually by a benign tumor (nonfunctioning adenoma) but also as a result of trauma or infarction. In patients of all ages the condition

TABLE 11.5 Some disorders of the pituitary

Deficiency or excess	Disorder
Primary hormone deficiency	Generalized: pan-hypopituitarism
	Isolated: affecting a specific hormone (for example GH or ACTH)
Primary hormone excess	GH: gigantism; acromegaly
	PRL: prolactinoma (micro- and macroprolactinoma)
	ACTH: Cushing's syndrome

is characterized by tiredness, pallor, risk of mild hypoglycemia on fasting, and impaired resistance to physical and mental stress.

Although the gonadotropins are the first hormones to be lost, a deficiency of most or all of the pituitary and target-gland hormones occurs eventually. Consequently, hormone secretion from other endocrine glands (thyroid, adrenal cortex, and gonads) is reduced. As a consequence, growth failure and failure of puberty occurs in children with the disease. Amenorrhea and infertility are features of affected adult women. Adult men may suffer from impotence and decreased libido although this condition is most often associated with prolactinomas. There is also a loss of androgen-dependent hair growth.

Clinically, these consequent effects on the thyroid, adrenal cortex, and gonads are referred to as secondary hypothyroidism, hypoadrenalism (hypocortisolism), or hypogonadism, respectively. This is a distinction from the primary forms of endocrine failure, which are due to disorders affecting the target endocrine glands directly. In these cases of primary hypofunction, a lack of negative feedback leads to increased secretion of the pituitary tropic hormone. In hypopituitarism, however, there is a failure of negative feedback and the pituitary hormones are either undetectable or, more often, inappropriately low. For example, in primary hypothyroidism, the free thyroxine (that fraction of thyroxine that is not bound to plasma proteins) is low and the TSH is elevated, whereas secondary hypothyroidism is characterized by low free thyroxine together with TSH that is within or below the reference range.

A wide range of diseases can cause hypopituitarism, but the most common cause in adults is a tumor, the vast majority of which are benign adenomas (for example prolactinomas) (**Figure 11.5**). The effects of these vary according to the size and cell type involved, and so the presentation may be with one or more of the following:

- Hypersecretion from the tumor cells of one or more pituitary hormones, causing secondary hypersecretion from their target organs
- Loss of secretion of other pituitary hormones, causing secondary hyposecretion from their target organs
- Enlargement of the pituitary, causing space-occupying effects such as headache and direct pressure on adjacent structures. This often affects the optic chiasm and causes loss of the temporal visual fields (bitemporal hemianopia) (**Figure 11.6**)

Clinical practice point 11.1

Hypopituitarism results in low levels of the target hormone (for example cortisol, thyroxine) and inappropriately low or normal levels of tropic hormone (for example ACTH, TSH). One or more hormone axes may be affected. If all are involved this is called pan-hypopituitarism.

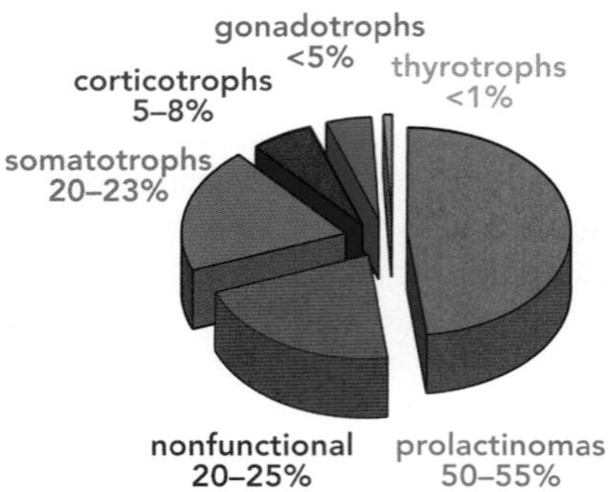

Figure 11.5 Cell-specific pituitary tumors.
The most common pituitary tumors are prolactinomas, constituting about 50% of the total. Tumors of growth hormone-producing cells (somatotrophs) and nonfunctional cells account for approximately 25% each, while tumors of other cell types are much less common.

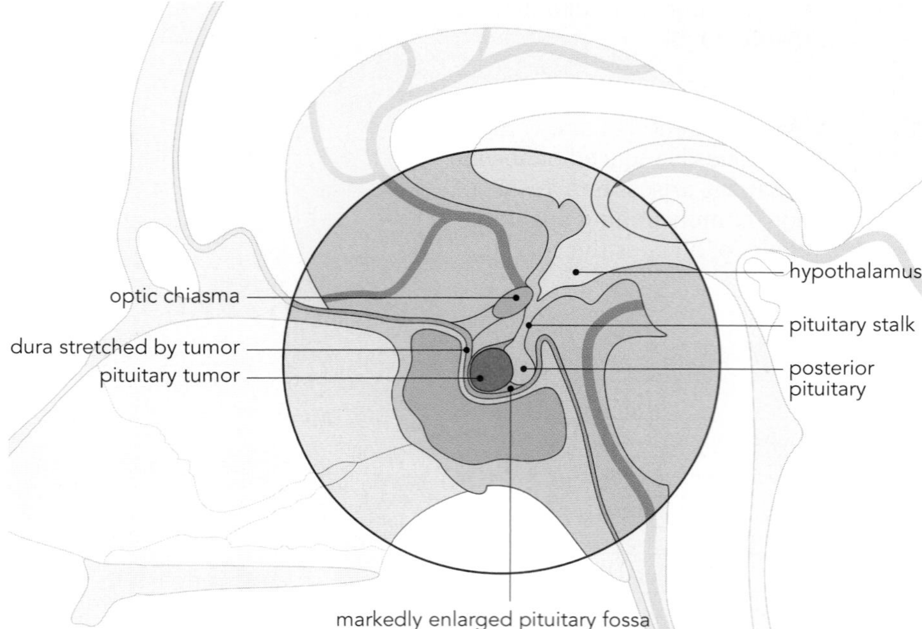

Figure 11.6 Pituitary tumors.
The close proximity of the pituitary gland to the optic chiasma means that, as the tumor grows, pressure is exerted on the optic chiasma, causing various visual problems including tunnel vision. Other effects include headache and loss of cerebrospinal fluid though the nasal cavity.

Diagnosis is made with the aid of computed tomography (CT) and magnetic resonance imaging (MRI) scans and pituitary function tests.

Pituitary tumors are not common and the cell types involved are not equally represented. Thus the most frequent types are prolactinomas and nonfunctioning tumors. In contrast, TSH-secreting tumors are extremely rare, comprising less than 1% of pituitary adenomas.

Isolated hormone defects

Isolated ACTH deficiency

This condition usually presents in children and is characterized by listlessness/tiredness and a mild fasting hypoglycemia due to compromised ability to perform gluconeogenesis (see Chapter 2). Plasma ACTH and cortisol concentrations are low and patients exhibit a slow response to prolonged stimulation with Synacthen® (a synthetic ACTH analog). There is no evidence of deficiency of any other hormones. Treatment involves administration of cortisol.

Growth hormone defects

Both excess and deficiency of GH can occur. GH excess causes the condition gigantism when it occurs before the closure of the epiphyseal growth plates in the long bones. Thus childhood GH excess causes increased height. In adults, inappropriately excessive secretion of GH, due usually to a benign tumor or rarely to ectopic growth hormone-releasing hormone (GHRH) secretion by a pancreatic tumor, leads to abnormal skeletal growth, a condition known as acromegaly (**Figure 11.7**). There is no additional gain in height, but patients with the condition exhibit characteristic enlargement of the face, hands, feet, and heart, plus coarse skin, sweating, and a husky voice. Multinodular thyroid goiter and arthropathy are also common. Amenorrhea or impotence and glycosuria may also present. Local effects of the condition are headaches and visual impairment due to raised intracranial pressure and impingement on the optic chiasm/optic tracts, respectively. Skull X-ray and CT/MRI scans are used for demonstration of the tumor.

GH deficiency may be suspected in children who are not growing taller at the expected rate, but this is a relatively unusual cause compared to genetic, nutritional, and other environmental factors. In rare cases, dwarfism may be caused by a gene deletion leading to a decreased or absent response to GHRH from the hypothalamus, but other possible causes include:

- Serious, chronic, non-endocrine illness, for example malabsorption or chronic renal failure
- Laron dwarfism due to the failure of GH to stimulate the hepatic production of IGF1
- Emotional deprivation giving rise to decreased GH secretion

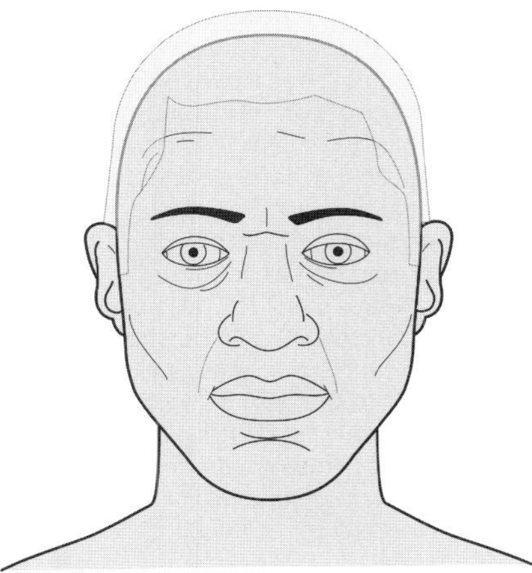

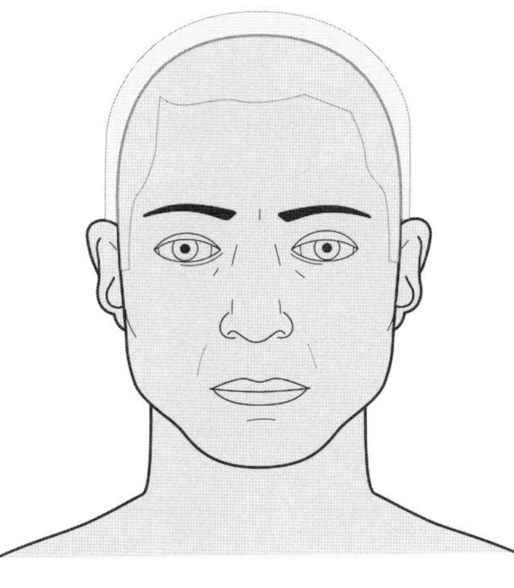

Figure 11.7 Growth hormone excess: acromegaly.
Acromegaly results from excessive growth hormone production by pituitary tumors. Suspicion of growth hormone excess is often raised by changes in facial appearance, including enlargement of the lower jaw, nose, and lips. Treatment may involve surgery to reduce tumor mass and/or drug therapy including cabergoline, a dopamine agonist and somatostatin agonist. Such treatments often produce a more normal physiognomy.

There is an age-related decline in GH in adulthood, which led to the view that it is an unnecessary hormone once growth is complete. GH deficiency in adults has become recognized as a genuine condition in recent years, following the increased availability of recombinant human GH as a therapeutic agent. However, pituitary disease rarely results in isolated GH deficiency and so it can be difficult to differentiate this from symptoms due to other hormone deficiencies.

Assessment of GH function is not straightforward as secretion occurs in a series of pulses throughout the day, with serum GH varying greatly within a short period of time. As with other endocrine axes, stimulation and suppression tests are often used, depending on whether excess or deficiency is suspected (**Figure 11.8**). In addition, the measurement of IGF1 is often a helpful marker of GH activity. Classically this has been considered to be secreted only by the liver as the sole mediator of GH action, but recent studies have also shown peripheral secretion of IGF1 acting in a paracrine or autocrine fashion. Units for reporting GH were standardized in 2007, in conjunction with the introduction of an international reference preparation with World Health Organization (WHO) approval (IS 98/574). This has removed the anomaly caused by reporting in IU/L in the UK and ng/mL in the USA. The agreed unit of measurement is now μg/L, with 1 μg/L equal to 3 mIU/L.

Suspected GH deficiency is investigated by a variety of stimulation tests. Historically, the test of choice was GH response to insulin-induced hypoglycemia, but this has obvious dangers and, following some well-publicized deaths in the UK (due to overtreatment of hypoglycemia with glucose), it fell out of favor. It remains the gold-standard test, however, and may still be performed in some specialist units with very close supervision. Alternatives to insulin include arginine, clonidine, and glucagon. Lack of response to one or more stimuli is usually defined as a peak of less than 5 μg/L, and growth hormone treatment may be commenced under specialist supervision.

GH excess may be suspected when an individual has symptoms and signs such as coarse facial features, enlarged hands, feet, and head, sweating, and hyperglycemia. Random GH measurements are of limited value, as mentioned above, due to its pulsatile secretion and its elevation by numerous factors such as sleep, malnutrition, stress, exercise, hypoglycemia, and liver disease. Acromegaly may occur in association with GH values which are not greatly elevated but are instead persistently detectable throughout the day due to more frequent secretion. Serial measurement of GH is therefore one method of determining excessive secretion. The gold-standard biochemical test, however, is the response to an oral glucose load (**Figure 11.9**). As in glucose loading for the diagnosis of diabetes, a 75 g dose is given. In a normal individual the GH falls to 0.2 μg/L or less, whilst in acromegaly there is a failure to suppress GH or even a paradoxical rise. The cutoff values used are being re-evaluated as more sensitive GH assays become available, which are able to measure lower concentrations.

Analytical practice point 11.1

Growth hormone is secreted in a pulsatile fashion and plasma levels are variable in normal individuals. Diagnosis of growth hormone excess or deficiency requires suppression or stimulation tests, respectively.

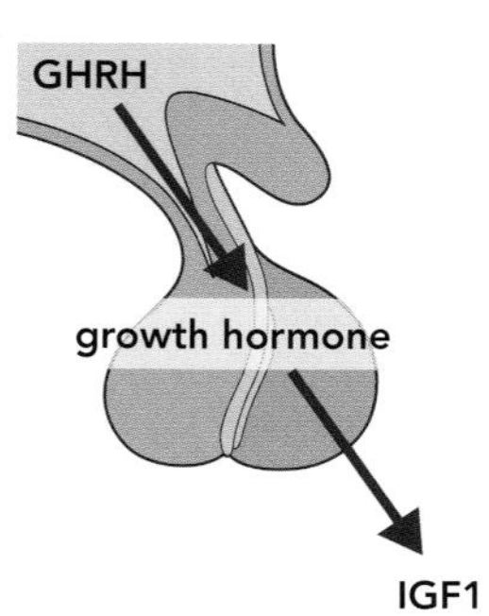

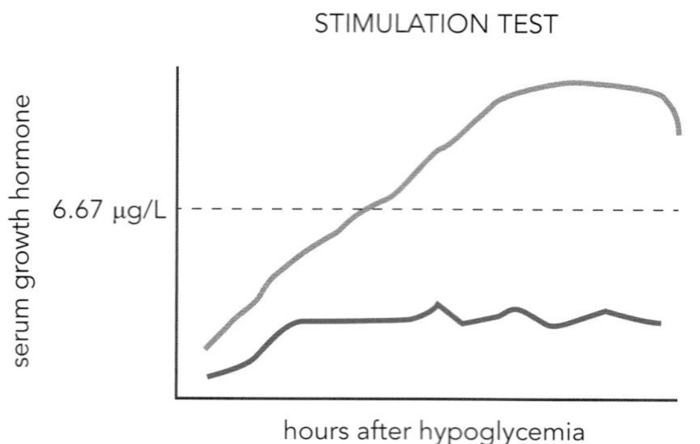

Figure 11.8 Dynamic function test for growth hormone deficiency. The measurement of serum growth hormone is very unreliable in the diagnosis of growth hormone deficiency. The initial diagnosis depends upon a dynamic function test, where the hypothalamic–pituitary axis (left-hand side of diagram) is stimulated by inducing hypoglycemia. Normally, the serum growth hormone concentration rises above 6.67 μg/L within a couple of hours (orange line), whilst it remains low in patients with growth hormone deficiency (blue line). Primary growth hormone deficiency presents as dwarfism in children and may arise through genetic abnormalities of structural proteins or enzymes involved in the synthesis or action of the hormone, including mutations in the genes encoding growth hormone-releasing hormone (GHRH) receptor, growth hormone, growth hormone receptor, and insulin-like growth factor 1 (IGF1).

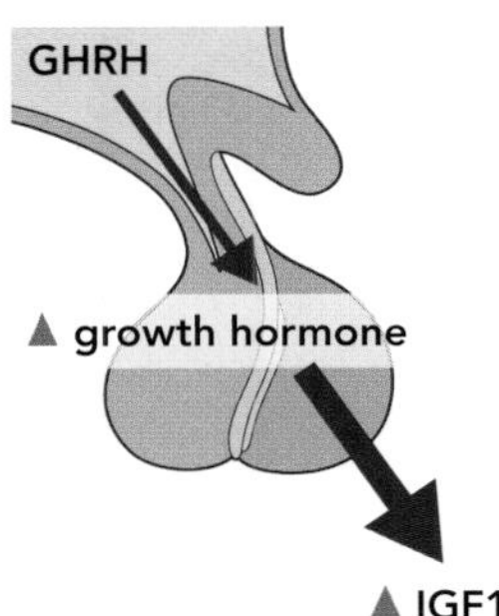

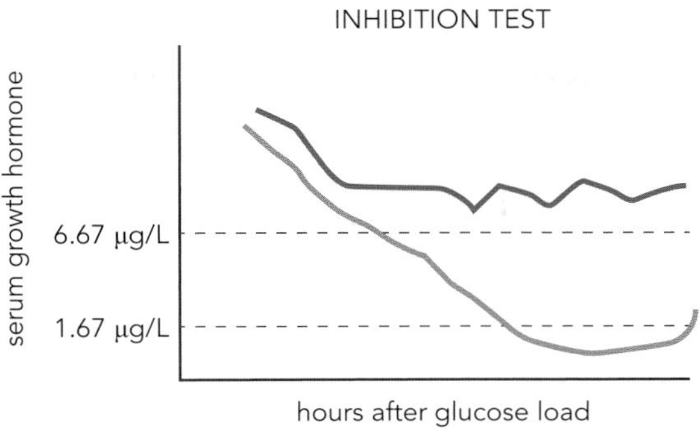

Figure 11.9 Dynamic function test for growth hormone excess. Growth hormone excess states are usually caused by pituitary tumors, rather than excessive production of growth hormone-releasing hormone (GHRH). Patients have high circulating concentrations of both growth hormone and insulin-like growth factor 1 (IGF1). However, baseline concentrations will overlap with those in the normal population. The response to a prolonged glucose load in normal subjects (orange line) is a fall in serum growth hormone to less than 1.67 μg/L, whereas serum growth hormone concentration remains above 6.67 μg/L in patients with growth hormone excess (blue line). Primary growth hormone excess presents as gigantism in children and acromegaly in adults. Growth hormone excess may also be seen in patients with tumors secreting forms of IGF1.

Prolactin

Prolactin is unique amongst the anterior pituitary hormones in being predominantly under inhibitory, rather than secretory, control. This occurs by the secretion of dopamine from the hypothalamus down the pituitary stalk. Although prolactin-secreting tumors (prolactinomas) are the most common type of pituitary tumor, elevated serum prolactin is more often due to another cause, in particular stress or medication. It may also occur if a non-prolactin-secreting tumor compresses the pituitary stalk and interrupts the flow of dopamine.

Prolactin-secreting tumors are usually small (<10 mm) and so rarely cause space-occupying effects. These are termed microprolactinomas. Their clinical effects are due to excess serum prolactin (**Figure 11.10**) which causes galactorrhea and hypogonadism manifest as oligo/amenorrhea in pre-menopausal women. There is, however, poor correlation between the degree of galactorrhea and the level of prolactin. Only about half of women with galactorrhea have elevated prolactin, and very high prolactin levels may be seen frequently in the absence of galactorrhea. On the other hand, in non-pregnant women with menstrual irregularities, prolactin should be measured since a number of such patients will have an occult prolactinoma. Microprolactinomas are usually associated with serum prolactin >8.70 nmol/L (>200 ng/mL). Similar levels may be seen with dopamine antagonist drugs such as antipsychotics and antinausea agents. Prolactin elevations of 870 pmol/L to 8.70 nmol/L (20–200 ng/mL) may occur with stress or non-prolactinoma pituitary tumors. It may, on occasion, be necessary to perform a series of measurements on serum taken from an indwelling venous catheter to eliminate the stress of venipuncture and its

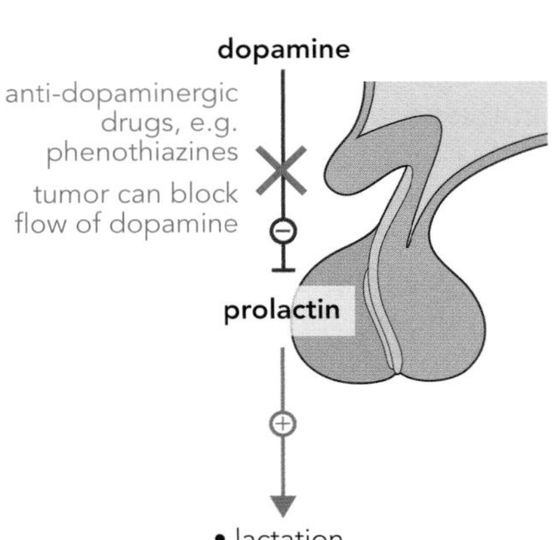

Figure 11.10 Hyperprolactinemia. Secretion of prolactin by the anterior pituitary is under negative dopaminergic control from the hypothalamus. Any blocking of this dopamine effect will lead to hyperprolactinemia. This can be due to tumor growth blocking the flow of dopamine down the hypothalamic stalk, or to the action of some anti-dopaminergic drugs, for example phenothiazines.

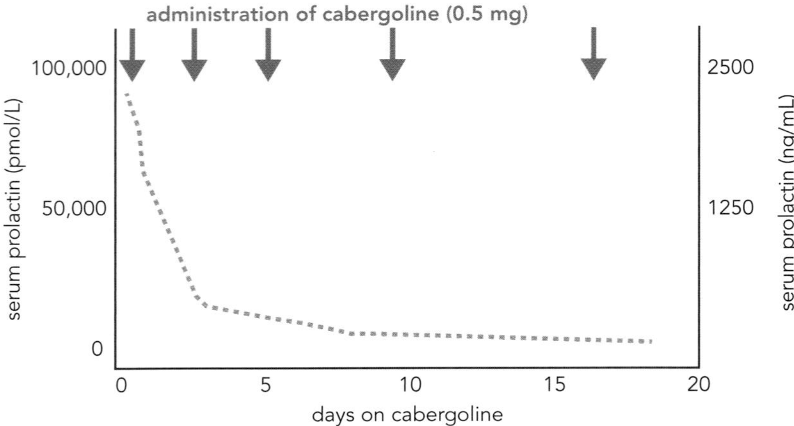

Figure 11.11 Monitoring the effect of a dopamine agonist (cabergoline) on serum prolactin in a patient with a prolactinoma. Many patients with a prolactinoma have very high concentrations of serum prolactin, up to 100–200 times the basal levels, and exhibit characteristic symptoms. Patients may be treated using dopamine agonists, usually cabergoline, which produce very rapid falls in circulating prolactin concentrations. In response to treatment, the tumor shrinks rapidly and the symptoms disappear equally rapidly.

potential confounding effect on prolactin release. Treatment with cabergoline reduces prolactin levels (**Figure 11.11**).

In men and postmenopausal women, prolactinomas often present much later in their natural history, by which time they have usually enlarged (macroprolactinomas, size >10 mm) to cause symptoms such as headache and visual field loss. Prolactin levels may be in excess of 500 ng/mL (22 μmol/L). A comparison between macro- and microprolactinomas is shown in **Table 11.6**.

Analytical practice point 11.2

Mild hyperprolactinemia is most commonly due to stress or medication.

TABLE 11.6 Comparison of macro- and microprolactinomas

Clinical findings	Present in macroprolactinomas	Present in microprolactinomas
Decreased secretion of other hormones	Yes	No
Visual impairment/headaches	Yes	No
Easily seen on imaging	Yes	No

In recent years it has become apparent that some individuals have high serum prolactin due to the presence of a complex of prolactin and an immunoglobulin. This is termed macroprolactinemia (not to be confused with macroprolactinoma described above). It has no adverse clinical effects, as the complex cannot interact with the usual receptor, but it may cause diagnostic confusion and unnecessary further investigation and treatment. It is now the practice of many clinical laboratories to look for macroprolactin by techniques such as polyethylene glycol precipitation or gel filtration in samples with elevated prolactin levels to avoid clinicians being misled.

Analytical practice point 11.3

Macroprolactin should be excluded as a spurious cause of hyperprolactinemia before any further investigations or treatment are initiated.

Gonadotropins: LH and FSH

The role of LH and FSH in the menstrual cycle is described later in this chapter in the section on the hypothalamus–pituitary–gonad (HPG) axis. Measurement of LH and FSH in women is usually requested as part of the investigation of menstrual disturbance: amenorrhea, oligomenorrhea, or infertility with normal cycles. Information that can be gained from this investigation is often useful in distinguishing primary ovarian failure (which eventually occurs in all women) from the much rarer hypothalamic–pituitary disorders. Because negative feedback from ovarian hormones is lost following the menopause, gonadotropins are higher in postmenopausal than in menstruating women (for example >30 U/L), and FSH is usually higher than LH. If the reverse pattern is seen it suggests a normal mid-cycle surge has occurred. In the transition to the menopause (the perimenopause), normal cycles may be interspersed with non-ovulatory cycles and long episodes of amenorrhea. During this time, levels of FSH and LH can vary greatly, even from one cycle to the next. It is therefore important not to diagnose menopause on the basis of biochemistry alone. The main concern is that a potentially fertile woman may stop contraception too early and have an unplanned pregnancy. It is advised that contraception is continued for two years following the last menstrual period in women under 50 years, and for one year in women over 50.

Clinical practice point 11.2

Menopause cannot be diagnosed by raised gonadotropins alone, although these are supportive. Diagnosis is clinical and based on amenorrhea for at least 12 months, hence can only be made retrospectively.

Low LH and FSH (<1 U/L) have a number of possible causes. Deliberate suppression of gonadotropins is the mechanism of action of the combined oral contraceptive pill and this hormone pattern is therefore expected in women using this. In pregnancy, the high levels of estrogen and progesterone also cause feedback suppression of LH and FSH (see Chapter 12). Thus a pregnancy test should be performed before considering the third possibility, which is hypothalamic–pituitary disease. It is not uncommon for stress or eating disorders to cause decreased gonadotropin secretion, and these should be considered before investigating further for structural pituitary disorders such as tumors.

Clinical practice point 11.3

Low gonadotropins in a woman of childbearing age raise the possibility of pregnancy if she is not on a combined oral contraceptive pill.

In men the measurement of gonadotropins is usually part of the investigation of erectile dysfunction, loss of libido, or infertility. As with women, high levels of gonadotropins are seen with primary gonadal failure. In men there is often a differential elevation of FSH with failure of spermatogenesis, and of LH with loss of testosterone production, although in many cases both arms of the HPG axis are affected. Low gonadotropins should lead to a consideration of exogenous androgen use before considering pituitary disease. Some men may use such drugs surreptitiously as anabolic steroids for bodybuilding or enhancement of athletic performance. Unfortunately, due to the effect of negative feedback, these drugs may also cause testicular atrophy.

Clinical practice point 11.4

Low gonadotropins in a man raise the possibility of anabolic steroid use.

Pituitary function tests

When hypopituitarism is suspected, the initial biochemical investigation is measurement of basal hormone levels at 09.00 am, when the levels of cortisol, ACTH, and testosterone in the blood should be at their highest. Thus TSH with free thyroxine, cortisol, and testosterone (males) or estradiol (females), with LH and FSH, GH with IGF1, and prolactin, are appropriate first-line tests. If these are normal, further testing may not be required. If partial deficiency is suspected because of equivocal basal hormone results, it may be necessary to assess pituitary reserve function by administering one or more pharmacological stimuli. The administration of insulin to induce hypoglycemia has already been referred to in the context of assessing GH reserve. In addition, hypoglycemia stimulates the secretion of ACTH and hence of cortisol. Thus a rise in cortisol is evidence that the pituitary and adrenal glands can respond adequately to stress, such as illness or surgery. Thyrotropin-releasing hormone (TRH) is rarely used as a stimulus for the TSH–thyroid axis since modern TSH assays are sufficiently sensitive to distinguish normal from low levels, which was not the case when the TRH test was first developed. Gonadotropin-releasing hormone (GnRH) may be administered to assess the pituitary–gonadal axis, but this gives no extra information in adults compared to the measurement of basal hormones.

Analytical practice point 11.4

Basal hormone measurements at 09.00 am are the first-line investigation for suspected pituitary disease.

Posterior pituitary

Tumors of the anterior pituitary rarely cause deficiency of the hormones released from the posterior pituitary. However, trauma or metastatic tumors may reduce posterior pituitary function.

The posterior pituitary secretes vasopressin and oxytocin. Oxytocin is involved in uterine smooth muscle contraction during childbirth and the ejection (not synthesis) of milk from the mammary glands in lactating women, but there are no known clinical disorders resulting from increased or decreased secretion and it is never measured in routine practice. Vasopressin, on the other hand, is extremely important in the homeostatic control of extracellular osmolality. It may be measured, but the assay is technically difficult and it is rarely required.

Analytical practice point 11.5

Vasopressin (antidiuretic hormone) function is assessed indirectly by measurement of urine osmolality.

It is assessed indirectly by measurement of plasma and urine osmolality, either in the basal state or in response to water deprivation. When plasma osmolality rises, such as when there is no access to water, osmoreceptors in the hypothalamus stimulate a vasopressin response.

The release of vasopressin from the posterior pituitary reduces free water excretion by increasing the permeability of the collecting ducts of the kidney to water; it increases the number of aquaporin-2 channels in the epithelial cells of this part of the nephron, thereby increasing water reabsorption and returning plasma osmolality to within normal levels. It is because of this effect of vasopressin on reabsorption of water in the kidney that it has the alternative name of antidiuretic hormone (ADH). A high urine osmolality (for example >750 mOsm/kg) is evidence that vasopressin is being released and that the kidney is able to respond to it. A urine osmolality less than this in the presence of raised plasma osmolality indicates that one or other of these mechanisms is impaired; this is termed diabetes insipidus (DI). Loss of vasopressin secretion is termed cranial or neurogenic DI, whilst an impaired renal response to vasopressin is called nephrogenic DI. They present clinically as polyuria (usually >5 L/day of abnormally dilute urine being produced) and an associated polydipsia. As long as access to water is unrestricted, such individuals may maintain their plasma osmolality at normal levels. However, they soon become severely dehydrated if water intake does not keep up with their urine output.

11.3 THYROID

Gross anatomy and microstructure of the thyroid

The thyroid is a butterfly-shaped gland situated in the neck in front of the trachea (**Figure 11.12**). It is an unusual endocrine tissue in that it stores hormones not in intracellular secretory vesicles, but extracellularly within colloid in a lumen surrounded by the cells which synthesize the hormones. The unit structure is the acinus or follicle consisting of a single layer of follicular cells which encloses the

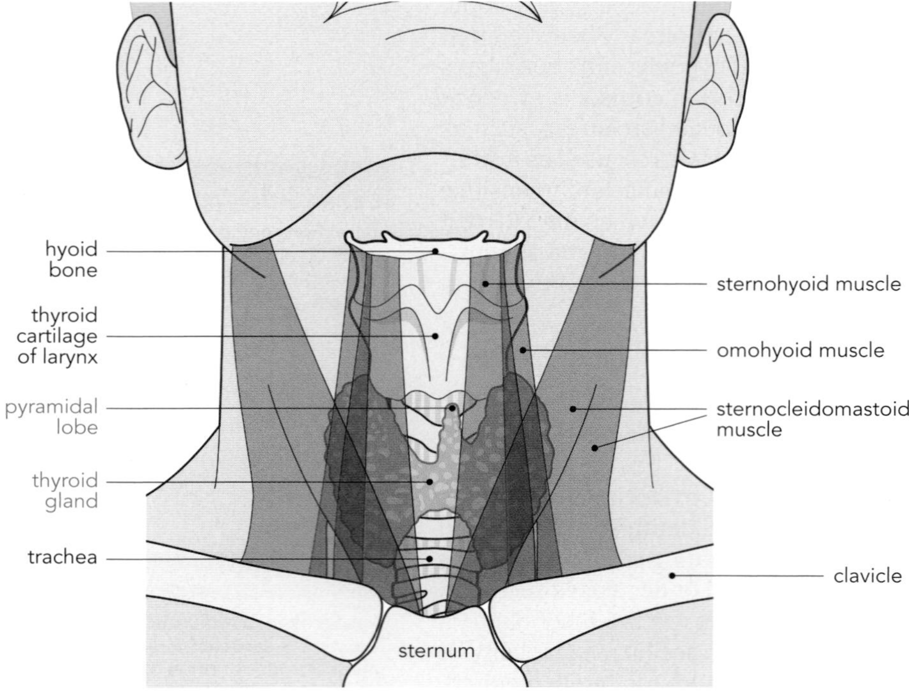

Figure 11.12 Anatomy of the thyroid.
The thyroid gland (shown in pink) is the largest endocrine gland (~20 g in weight in adults) and lies anterior to the thyroid cartilage and the trachea in the neck.

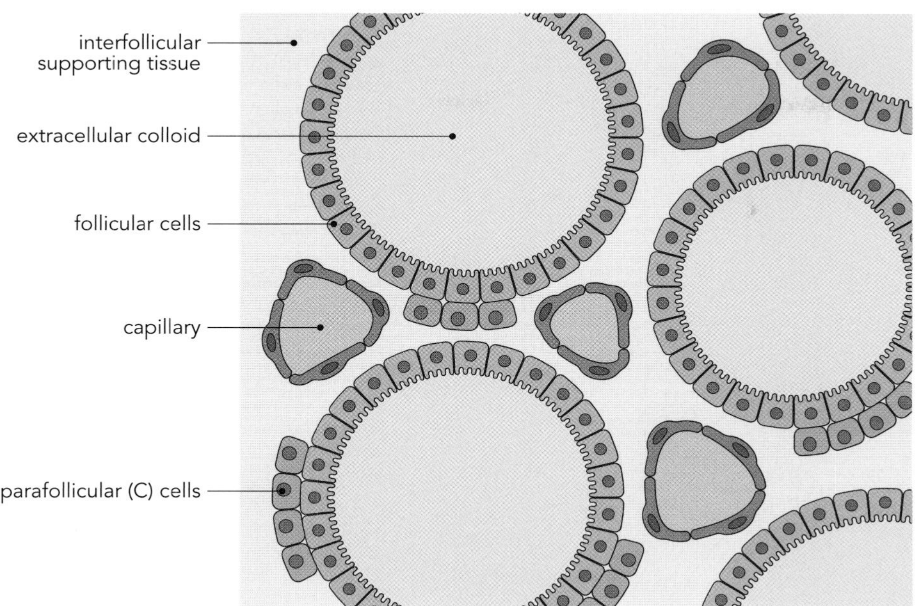

Figure 11.13 Microstructure of thyroid follicles.
Thyroid follicles are of variable size. Secretory epithelial cells of each follicle are flattened and cuboidal in shape, their height reflecting the rate of hormone production in the individual cells. The single layer of cells surround a lumen filled with proteinaceous colloid of which the major component (75%) is thyroglobulin. The sparse interfollicular supporting tissue is occupied mainly by capillaries (brown). The parafollicular C cells (green) located in connective tissue adjacent to the follicles are neuroendocrine cells and are the site of synthesis of calcitonin, a hormone involved in calcium metabolism.

follicular lumen (**Figure 11.13**). The follicles vary in size, and the height of the follicular cells reflects the rate of hormone synthesis; actively secreting cells are smaller than in less active tissue, where cells are more flattened and hormone is stored in colloid. Release of hormone from the tissue requires re-entry of the hormone precursor into the follicular cell from the follicular lumen, hydrolysis to release the parent hormones, and secretion into the interstitial space. The hormones then diffuse into the capillary network around each follicle and enter the circulation. A number of single cells are scattered between the follicles. These are the parafollicular or C cells which synthesize and secrete calcitonin as described in Chapter 9.

Thyroid hormones

Thyroid hormones exert a variety of actions in the body ranging from chronic effects, such as facilitation of skeletal growth and brain development and function, particularly post-partum, to acute effects, such as stimulation of basal metabolic rate, calorigenesis, and sensitization of tissues to the effects of catecholamines.

The thyroid hormones are the final products of the hypothalamus–pituitary–thyroid (HPT) axis (the hormones of which are shown in **Table 11.7**) and their concentration in plasma is subject to tight control through feedback loops (**Figure 11.14**) which are discussed in detail below. The structures of the major thyroid hormones, tetra-iodothyronine (T_4), commonly known as thyroxine, and tri-iodothyronine (T_3), are shown in **Figure 11.15**. The numerical subscripts indicate the number of iodine atoms in the respective molecules.

TABLE 11.7 Hormones of the hypothalamus–pituitary–thyroid axis

Tissue	Hormone
Thyroid	Thyroxine (T_4)
	Tri-iodothyronine (T_3)
Pituitary	Thyrotropin (thyroid-stimulating hormone; TSH)
Hypothalamus	Thyrotropin-releasing hormone (TRH; originally TSHRH)

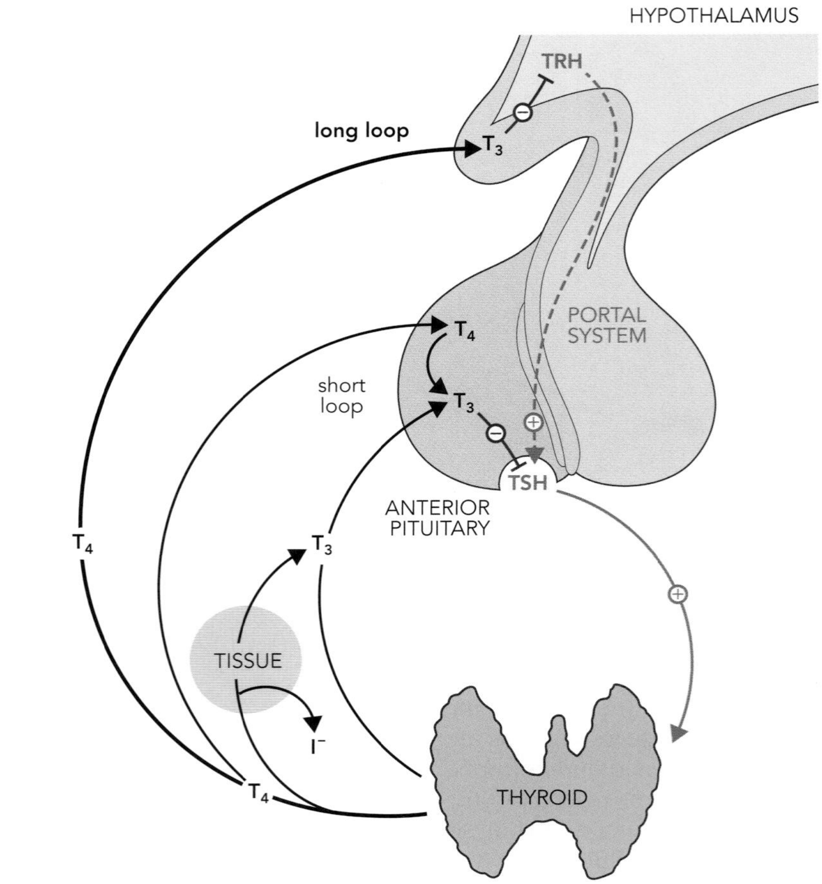

Figure 11.14 Feedback control of the hypothalamus–pituitary–thyroid axis.
The control of thyroid hormone synthesis and secretion is via feedback loops involving the hypothalamus, anterior pituitary, and the thyroid itself. Thyrotropin-releasing hormone (TRH) from the hypothalamus triggers the release of thyrotropin (thyroid-stimulating hormone, TSH) from the thyrotrophic cells in the anterior pituitary. TSH in turn stimulates the synthesis and release of thyroid hormones, T_4 and T_3, from the thyroid. Increased levels of plasma thyroid hormones inhibit the release of TSH from the pituitary (short loop) and of TRH from the hypothalamus (long loop) such that serum TSH concentration is inversely, log-linearly related to serum T_4 concentration. The major sites of deiodination of T_4 to the active hormone T_3, occurs in peripheral tissues with the release of iodide (I).

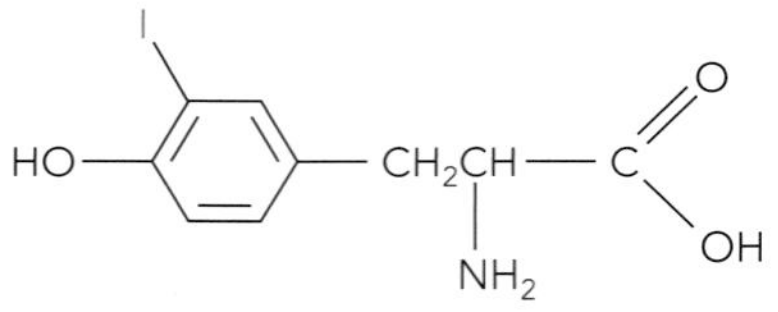

3-mono-iodotyrosine (MIT)

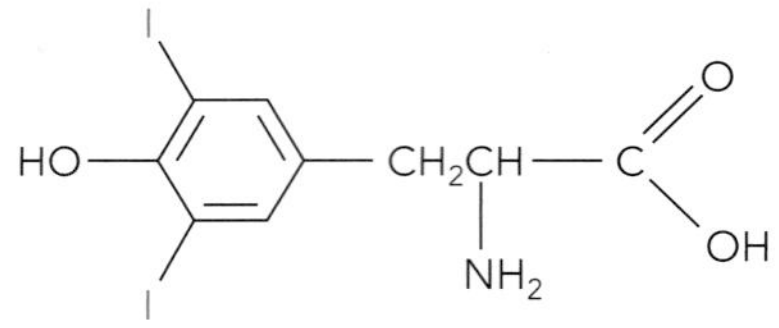

3,5-di-iodotyrosine (DIT)

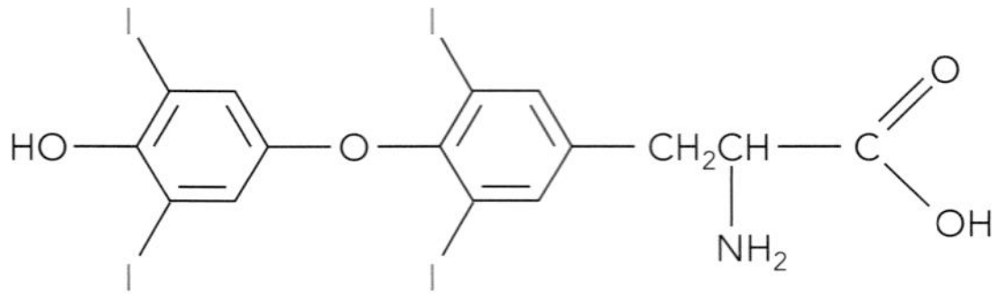

3,5,3',5'-tetra-iodothyronine (thyroxine, T_4)

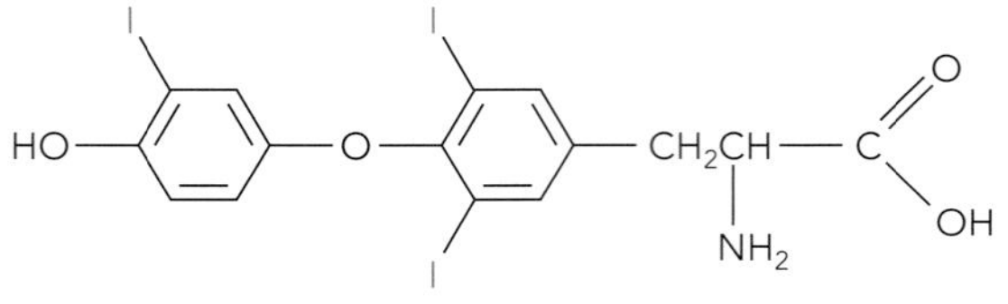

3,5,3'-tri-iodothyronine (T_3)

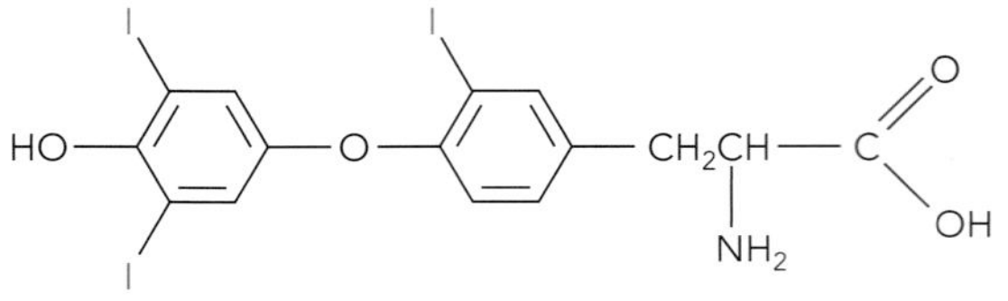

3,3',5'-tri-iodothyronine (reverse T_3, rT_3) (inactive)

Figure 11.15 Structures of thyroid hormones.
The major hormone secreted by the thyroid is thyroxine (T_4; 3,5,3′,5′-tetra-iodothyronine) which undergoes 5′-deiodination to the active hormone, 3,5,3′-tri-iodothyronine, in the peripheral circulation, primarily in liver and kidney. Thyroid hormones T_4 and T_3 are synthesized from two iodotyrosine residues in thyroglobulin, a high-molecular-weight protein synthesized and stored in the thyroid.

Biosynthesis of thyroid hormones

The biosynthesis of thyroid hormones involves three key steps:

1. Formation of a large (660 kDa) protein, thyroglobulin, in the follicular cells
2. Iodination of specific tyrosine residues on thyroglobulin
3. Coupling of iodinated tyrosine residues to form T_4 and T_3

Iodine metabolism

Iodine is a relatively scarce element but constitutes 65% by weight of T_4 and is essential for thyroid hormone synthesis. The thyroid concentrates and conserves iodine to help ensure its supply for this purpose. Optimal intake of iodine to maintain normal thyroid function is 150–300 μg/day, and intake of iodine worldwide varies widely between 20 and 700 μg/day. Inadequate intake leads to endemic goiter and, to ensure intake in the optimal range, iodine (as potassium iodide) is added to table salt in the UK and USA. Dietary iodine is reduced to iodide in the gut and is absorbed rapidly within 30 minutes. Iodide is cleared from the blood via the thyroid or into urine, with clearance into the thyroid being inversely proportional to its plasma concentration. An outline of whole-body iodide turnover is shown in **Figure 11.16**, which indicates that, under homeostasis, approximately 25% of dietary iodide is extracted by the thyroid. Iodide uptake into the thyroid is driven by an ATP-dependent transport system in the follicular cell membrane, which produces a concentration gradient for the anion of about 30:1 between the cell cytoplasm and plasma.

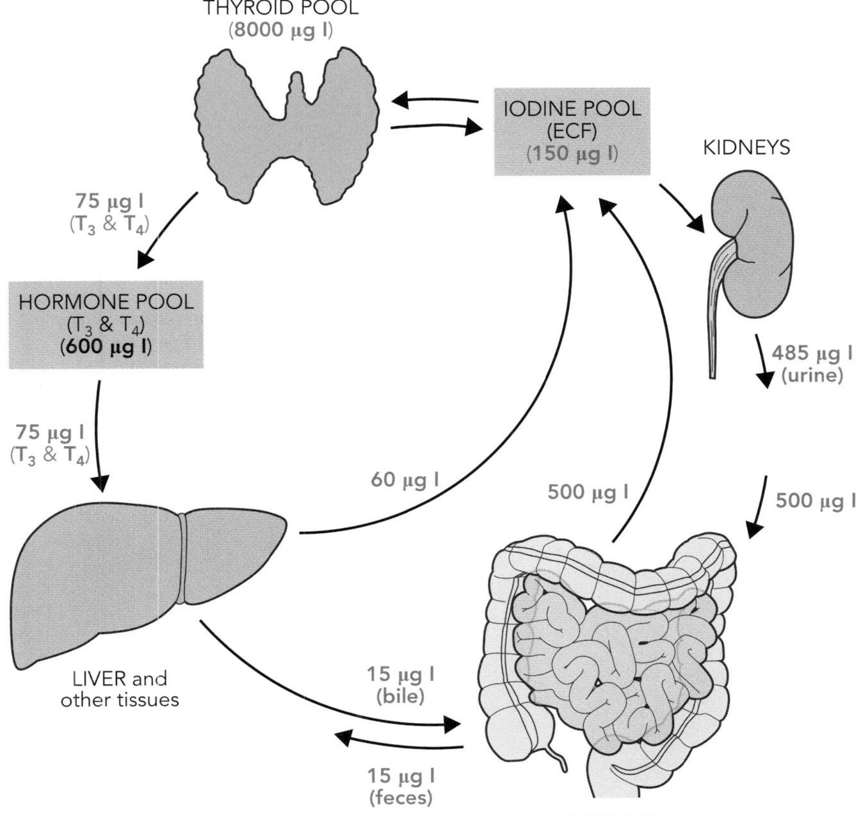

Figure 11.16 Whole-body iodide turnover. Approximately 8000 μg is stored in the thyroid gland linked covalently to tyrosine residues of thyroglobulin. Smaller amounts are present as free iodide anion in extracellular fluid (ECF; 150 μg) and circulating thyroid hormones (600 μg). Iodine metabolism is highly dynamic and the approximate daily flux of iodine (as iodide and thyroid hormones) between compartments is shown. Whole-body iodine homeostasis represents a balance between that entering the body via the diet [500 μg via the gastrointestinal (GI) tract] and that lost in urine (485 μg) and feces (15 μg).

Iodination of thyroglobulin and formation of thyroid hormones

Iodide concentrated in the follicular cells is rapidly incorporated into thyroglobulin, a 660 kDa dimeric glycoprotein synthesized by the cells. This organification process involves a peroxidase which catalyzes oxidation of iodide and its incorporation into tyrosine residues near the N-terminus of the protein. Iodination of thyroglobulin occurs at the apical border and results in the formation of mono-iodotyrosine and di-iodotyrosine residues (MIT and DIT, respectively) in the protein. Formation of T_4 and T_3 involves the coupling of the iodinated tyrosines in a process again catalyzed by the thyroid peroxidase; T_4 is formed from two molecules of DIT, and T_3 from one molecule of DIT and one molecule of MIT (**Figure 11.17**). At this stage, a molecule of thyroglobulin contains 6–7 MIT and 4–5 DIT molecules, and the iodinated protein is exocytosed into the colloid of the follicular lumen for storage. Under appropriate stimulation, thyroglobulin is endocytosed back into the follicular cell as a colloid droplet which, on fusing with lysosomes, undergoes proteolysis to release T_4, T_3, DIT, and MIT. Approximately 30–40 molecules of T_4 and 2–3 molecules of T_3 are formed per 10 molecules of thyroglobulin, and approximately 25 mg/day of thyroglobulin undergoes proteolysis, yielding about 110 nmol (90 μg) of thyroid hormones, and, in a healthy adult, the thyroid contains approximately 10 μg of T_4 and 1 μg of T_3. About 20% of T_4 is deiodinated in the thyroid gland itself, at the basal border prior to secretion, such that the ratio of T_4:T_3 secreted into plasma is 20:1 (approximately 100:5 nmol/day). In summary, the synthesis of T_4 involves the following steps (see Figure 11.17).

- Synthesis of thyroglobulin (TG)
- Iodide trapping in the thyroid by an ATP-dependent mechanism
- Iodide oxidation and incorporation into tyrosine residues in TG by peroxidase

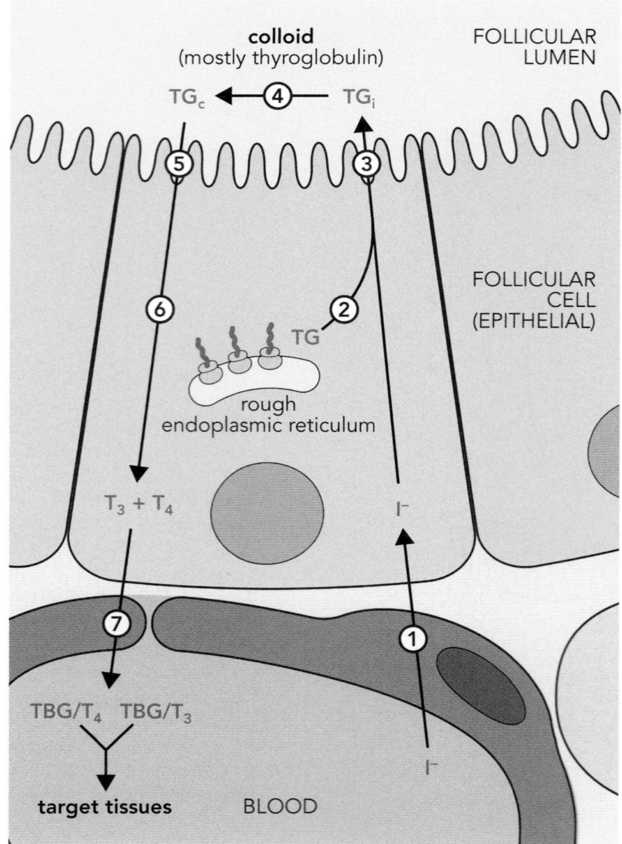

Figure 11.17 Formation of thyroid hormones from iodotyrosine residues of thyroglobulin. The formation of thyroid hormones by the thyroid follicular cells is shown in a series of numbered steps in the figure. (1) Concentration of iodide by the thyroid follicular cells; (2) synthesis of thyroglobulin (TG) on the rough endoplasmic reticulum; (3) iodination of tyrosine residues of thyroglobulin (TG_i) at the follicular border; (4) coupling of iodotyrosine residues in thyroglobulin and storage in the colloid of the follicular lumen (TG_c); (5) endocytosis of iodinated thyroglobulin from colloid; (6) lysosomal hydrolysis of thyroglobulin to release T_4 and T_3; (7) secretion of T_4 and T_3 into the circulation and binding to thyroid hormone-binding globulin (TBG).

- Storage of TG in extracellular colloid
- Endocytosis of TG into thyroid cells and coupling of DIT and MIT to form T_4 and T_3

Iodinated tyrosines, particularly MIT and DIT, liberated from thyroglobulin are deiodinated by a specific 5′-iodothyronine deiodinase to release iodide, which enters the plasma pool and can be salvaged by concentration in the thyroid. Thyroid hormones leave the gland via capillaries. Small amounts of intact thyroglobulin enter the blood via the lymphatics. Individual stages of thyroid hormone synthesis have been associated with separate inborn errors and have been targets for the action of different drugs such as thioureas. These stages include:

- Iodide trapping
- Organification (that is, formation of DIT and MIT)
- Coupling of DIT and MIT
- Dehalogenation of DIT and MIT

Release of thyroid hormones

Thyroid hormones are released when specific receptors on the thyroid cell surface are occupied by the normal physiological stimulator TSH, synthesized in the anterior pituitary, or by TSH receptor antibodies, as seen in the cases of Graves' disease. Overall control of thyroid hormone synthesis is via the HPT axis, which closely regulates secretion of T_4 and T_3 through long and short feedback loops as shown previously in Figure 11.14. In this system a tripeptide, TRH, formed from proteolysis of a 26 kDa pro-hormone, is initially released in a pulsatile, circadian manner from the hypothalamus and travels via the portal venous system to the anterior pituitary. TRH binds to specific receptors on the surface of thyrotrophic cells and stimulates the synthesis and secretion of TSH via an inositol trisphosphate-regulated mechanism. TSH is a 26 kDa glycoprotein consisting of two subunits, α and β; the α subunit is common to other pituitary glycoprotein hormones, whereas specificity of action is conferred by the unique β subunit. TSH is also secreted in a circadian rhythm and has a short plasma half-life (65 minutes). Its actions on the thyroid, which are mediated via intracellular production of cyclic adenosine monophosphate (cAMP) and protein kinases, serve to promote the synthesis and secretion of thyroid hormones. The important feedback control in the HPT axis lies in the inhibition of expression of TRH receptors in the anterior pituitary by both TSH and T_3. T_3 also feeds back to the hypothalamus to regulate secretion of TRH (see Figure 11.14).

Thyroid hormones in plasma

T_4 and T_3 secreted from the thyroid are relatively insoluble in aqueous solution and are solubilized in plasma by binding to carrier proteins. Both molecules bind to thyroid hormone-binding globulin (TBG) and albumin, while T_4 also binds to transthyretin (thyroid hormone-binding prealbumin; TBPA). The reference ranges for these proteins and relative affinity for T_4 are shown in **Table 11.8**. Thus in a healthy adult the ratio of bound to free hormone is 2000:1 for T_4 and 200:1 for T_3; binding to plasma proteins results in less than 1% of the total plasma thyroid hormone being present in the free form (~0.05% T_4 and 0.5% T_3). Despite

TABLE 11.8 Reference ranges for thyroid hormone-binding proteins

Protein	Reference range	Relative affinity for T_4
Thyroid hormone-binding globulin (TBG)	16–24 mg/L	++++
Thyroid hormone-binding prealbumin (transthyretin)	100–400 mg/L	++
Albumin	32–52 g/L	+

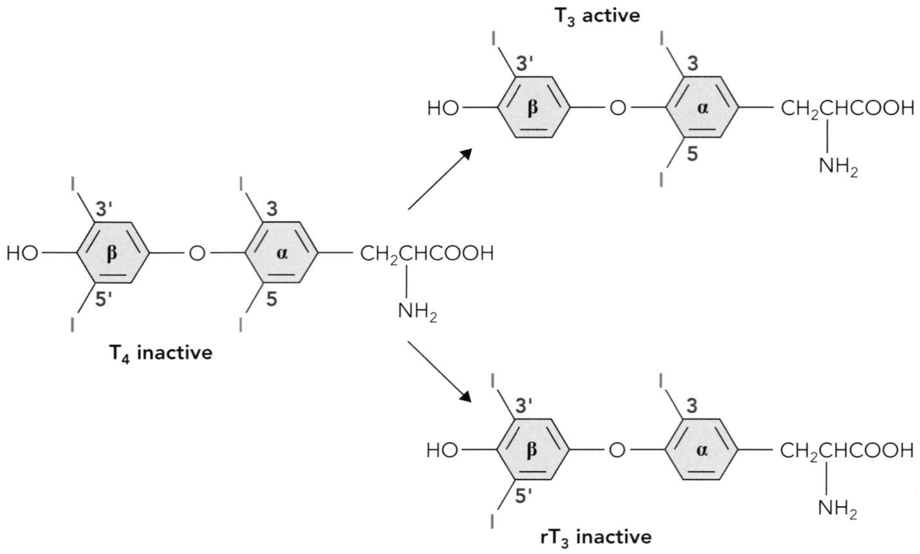

Figure 11.18 Review of the structure and actions of thyroid hormones.
Whilst thyroxine (T_4) is the most abundant thyroid hormone (>90%) secreted by the thyroid, it is biologically inactive and is activated only after loss of its 5′-iodine residue to become tri-iodothyronine (T_3). Loss of the 5-iodine residue from T_4 yields reverse T_3 (rT_3), which is also biologically inactive.

the high ratio of protein-bound to free hormone, it is thought that the unbound free fraction of T_3 is the biologically active fraction. The total hormone concentration—free and protein-bound in plasma—depends on the rate of production and secretion of hormone and the availability of binding sites on carrier proteins (that is, carrier protein concentration, especially of TBG) and competition for binding to carrier sites by other ligands, especially drugs.

T_3 alone has appreciable biological activity and T_4 is deiodinated peripherally to form the active hormone. In healthy individuals, approximately 80% of T_3 in the circulation is formed by mono-deiodination of T_4 by a specific deiodinase in the kidneys and liver. There are three forms of the enzyme responsible for the peripheral deiodination of T_4.The products of this process are active T_3 (40%) and biologically inactive reverse T_3 (rT_3; 60%), which arise from loss of iodine from the 5′- and 5-carbons of T_4, respectively (**Figure 11.18**). T_4 is also converted to T_3 in the brain and anterior pituitary, where it acts locally.

Further metabolism of T_4 and T_3

Further metabolism of thyroid hormones leads to their deactivation through deiodination, deamination, and conjugation. The metabolic products of these reactions are excreted into the intestine via the bile and any iodinated products undergo enterohepatic circulation to conserve iodine. The amount of T_4 or T_3 bound to protein affects the half-life of the hormone in plasma and this is reflected in the different half-lives for T_4 (6–7 days) and T_3 (<1 day).

Actions of thyroid hormone T_3

The whole-body actions of T_3 have been mentioned earlier. In general, these actions are of slow onset and chronic, resulting from stimulation of protein synthesis in the target cells. T_3 enters the target cell and forms a heterodimer with a retinoid before translocating to the nucleus. Here the heterodimer binds to specific receptor sequences on DNA adjacent to genes regulated by T_3 and induces transcription of a small number of specific genes, a process which takes about 30 minutes. The protein products of these genes promote a secondary response leading eventually to changes in transcription of tissue-specific genes. The particular response to T_3 in a given tissue resides in these cell-specific gene regulatory proteins. The biochemical responses to the actions of thyroid hormones, particularly T_3, include:

- Increased basal metabolic rate
- Stimulation of thermogenesis via increased mitochondrial oxidative metabolism to supply the ATP demand for sodium-pump activity

- Stimulation of lipolysis through activation of hormone-sensitive lipase to supply free fatty acids for thermogenesis
- Stimulation of hepatic glycogenolysis and gluconeogenesis to supply glucose for thermogenesis

Thyroid function tests—basic interpretation

TSH controls the secretion of T_4 and T_3 and is under negative-feedback control. Hence TSH rises as thyroid hormone production falls (hypothyroidism) and becomes suppressed when the thyroid becomes overactive (hyperthyroidism). When the source of the dysfunction is the thyroid gland itself it is termed primary, and primary hypothyroidism and hyperthyroidism are by far the most common types of thyroid disease. Less commonly a disorder may originate from excess or deficient secretion of TSH from the anterior pituitary. These are termed secondary hypo- and hyperthyroidism. The TSH in these conditions is inappropriate for the level of T_4; low in secondary hypothyroidism and high in secondary hyperthyroidism. In secondary hypothyroidism there are usually deficiencies of other pituitary hormones (LH, FSH, GH, ACTH), but very rarely an isolated TSH deficiency may occur. Also very rare is a TSH-secreting pituitary tumor (TSHoma), which is the cause of secondary hyperthyroidism. A common cause for the same pattern of thyroid function test (TFT) results as seen in secondary hypothyroidism is where patients are on treatment for primary hyperthyroidism. In this situation the pituitary thyrotroph cells, which secrete TSH, become suppressed by the high levels of T_4 and T_3. If antithyroid drugs or radioactive iodine are given, the secretion of thyroid hormones is reduced, which would normally stimulate increased TSH by negative feedback. However, the suppressed thyrotrophs take time—sometimes several months—to recover, hence this is a form of temporary secondary hypothyroidism. Unless this phenomenon is recognized, the TFT results may trigger unnecessary investigation for pituitary disease.

Choice of tests for assessing thyroid function

In the last two decades there has been an explosion in the demand for TFTs performed by clinical laboratories. In order to cope with this demand, technological developments have allowed automated and rapid testing of samples, which in turn has probably driven demand further. Prior to the advent of widespread automated immunoassays in the 1990s, TFTs consisted of total thyroxine and tri-iodothyronine, plus the assay of TSH. The sensitivity of early methods for the measurement of TSH, however, was insufficient to distinguish between normal and suppressed concentrations, thereby making diagnosis of hyperthyroidism difficult. Present-day TFTs now consist of free thyroid hormones (fT_4 and fT_3) plus more-sensitive TSH assays, which can distinguish low from normal concentrations. Measurement of the levels of the free hormones has obvious advantages, being unaffected by changes in thyroid hormone-binding globulin, which is commonly affected by pregnancy and by the oral contraceptive pill. In the early days of free hormone measurement there was considerable theoretical debate about its scientific validity. However, such assays have been in routine use now for many years and correlate extremely well with the clinical state of the patient in all but a small minority of cases. The scientific purists may still have their doubts but clinical pragmatism has won the argument. However, as with all laboratory tests, there are situations where free thyroid hormone measurement may be invalid, for example in the assessment of thyroid function in pre-term infants.

> **Analytical practice point 11.6**
>
> Standard thyroid function tests (TFTs) comprise TSH, free T_4, and free T_3. Some laboratories measure TSH alone as an initial test. The risk of missing cases of pituitary disease must be balanced against the cost of many unnecessary tests.

Many laboratories have developed strategies to optimize the use of resources in response to TFT requests. Most commonly this consists of a front-line TSH strategy, with no further tests being done if the TSH is within the reference range. A raised TSH leads to measurement of fT_4 as well, whilst a suppressed TSH triggers both fT_4 and fT_3 measurements. The drawback to this approach is that TSH may be normal in hypopituitarism (and when patients are on treatment for hyperthyroidism). For diagnosis of secondary hypothyroidism, the measurement of fT_4 is required in addition. Therefore, such a strategy requires good clinical information to be supplied to the laboratory. Even so, there is the risk of missing hypopituitarism, and for this reason some advocate measuring fT_4 in addition to TSH on all samples. Given the vast number of TFTs performed, this is an expensive strategy and not necessarily justified by the rarity of hypopituitarism. As health care budgets come under increasing pressure, such a strategy is unlikely to be viewed as cost effective.

Thyroid autoantibodies and thyroglobulin

Because thyroid disorders commonly have an autoimmune basis it is useful to quantitate the presence in serum of specific autoantibodies as an adjunct to thyroid function tests. In previous decades this was done by indirect immunofluorescence, whereby the patient's serum is incubated with a thin slice of thyroid tissue to allow antithyroid antibodies to bind to their target antigens. After washing excess serum away, the autoantibodies can be visualized under a microscope by adding a fluorescently labeled secondary antibody directed against them. Quantitation is performed by using serial dilutions of serum until the fluorescence is no longer visible. Using this technique two types of thyroid autoantibody were seen, with their antigens being the thyroid microsome and thyroglobulin. It was shown subsequently that the microsomal antigen was the enzyme peroxidase. Automated immunoassays for the quantitation of thyroid peroxidase (TPO) antibodies have replaced these cumbersome and slow manual techniques and have allowed testing to keep pace with the growth in demand for thyroid function tests.

It is also possible to measure antibodies against the TSH receptor, although immunoassays cannot distinguish between stimulatory and inhibitory antibodies. These are usually measured only in pregnancy because they can cross the placenta and cause neonatal thyrotoxicosis.

Thyroglobulin antibodies provide no additional information compared to TPO antibodies and are no longer routinely measured. The main exception to this is in the assay of thyroglobulin itself. This may be usefully measured in serum in patients who have undergone total thyroidectomy for thyroid carcinoma. In this situation it is valuable to have a marker of the presence of thyroid tissue, as this would indicate either incomplete removal, recurrence, or secondary spread of the tumor. Thyroglobulin is detectable in the serum of normal individuals and its measurement does not contribute to the diagnosis of thyroid carcinoma, or any other thyroid disease. Unfortunately, the presence of antithyroglobulin antibodies in the serum can interfere with the measurement of thyroglobulin. It is therefore usual practice to measure these antibodies at the same time to aid the interpretation of the results.

Thyroid disease

Thyroid disease is usually associated with deficient or excessive secretion of thyroid hormones, along with the expected clinical effects. However, some thyroid diseases do not affect hormone secretion significantly and are manifest only as local swelling of the gland (goiter). Hyperthyroidism and hypothyroidism are often detected as biochemical abnormalities at a stage before obvious clinical effects, due to the widespread use of thyroid function tests as part of a general screen in well-person checks or in response to nonspecific symptoms.

Hypothyroidism

The common causes of hypothyroidism are:

- Iodine deficiency (worldwide, but not in Western Europe or USA)
- Autoimmune destruction of thyroid (Hashimoto's thyroiditis)
- Post-ablative: surgery or treatment with radioactive iodine

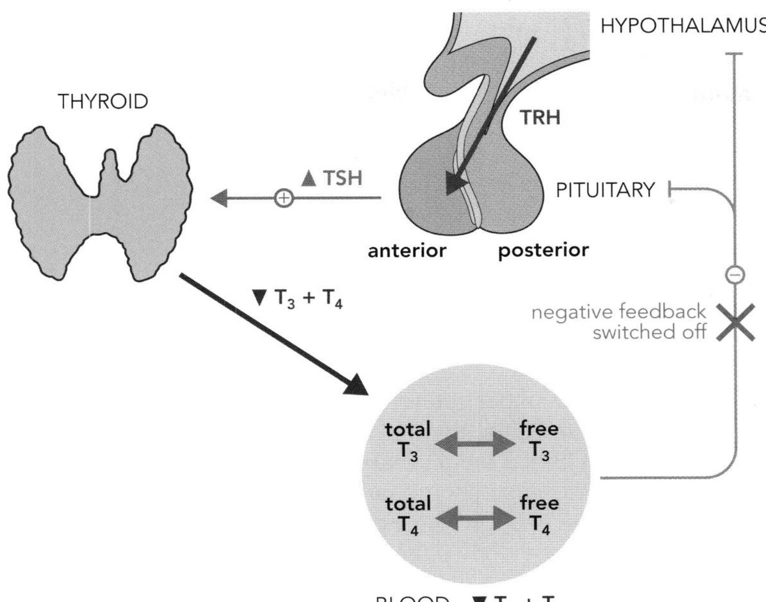

Figure 11.19 Primary hypothyroidism.
Primary hypothyroidism arises from damage to the thyroid, usually due to an autoimmune cause. Such damage reduces the ability of the gland to synthesize thyroid hormones.

A low plasma concentration of thyroid hormones and the consequent loss of feedback control on the pituitary leads to increased production of TSH by the tissue (**Figure 11.19**).

The symptoms of hypothyroidism (**Table 11.9**) are those of a decreased metabolic rate, as might be predicted from the cellular effects of thyroid hormones listed earlier. A suspicion of hypothyroidism, and a request for TFTs, is often triggered by weight gain or tiredness. A TSH result within the reference range excludes primary hypothyroidism as the cause of symptoms. It does not exclude secondary hypothyroidism due to pituitary disease, and this would require additional measurement of fT_4. However, this condition is much rarer and would only be considered if there was evidence of disturbance of other pituitary axes, such as decreased sex hormones or cortisol, or visual field defects suggestive of a pituitary tumor.

In primary hypothyroidism the expected pattern of results is elevated TSH in conjunction with low fT_4 (**Figure 11.20**). The measurement of fT_3 is not

TABLE 11.9 Clinical features of hypothyroidism

Physiological system	Symptoms	Signs
General	Lethargy	Weight gain
	Coarsening of features	Malar flush
	Dry hair	Loss of eyebrows
	Cold intolerance	
Cardiovascular	Angina	Bradycardia
		Hypercholesterolemia
Gastrointestinal	Constipation	Abdominal distension
Hematological	Fatigue, exertional dyspnea	Anemia
Neuromuscular	Muscle weakness	Myopathy
	Carpal tunnel syndrome	Slow tendon reflexes
	Depression	
Reproductive	Infertility	Menorrhagia

PRIMARY HYPOTHYROIDISM

free T_4 ▼ typically <10 pmol/L (<0.8 ng/dL)

TSH ▲ typically >10 mIU/L

free T_3 ▼ unreliable

Figure 11.20 Thyroid function test results in primary hypothyroidism. Whilst the serum concentration of thyroxine (T_4) is usually below the normal range, an increase in the serum concentration of thyroid-stimulating hormone is the most sensitive marker of primary hypothyroidism.

usually helpful in this situation. The aim of treatment is to return the patient to a euthyroid state and treatment consists of L-thyroxine administered orally once a day. The dose is titrated to achieve a TSH level within the reference range; this may often result in a fT_4 level that is slightly above the upper reference limit. The explanation for this is that since no T_3 is secreted from the thyroid, a higher fT_4 is required to achieve the same tissue level of fT_3. Since the pituitary thyrotroph cells take time to adjust their secretion of TSH, it is usually recommended that TFTs are not repeated at intervals of less than six weeks when assessing changes in L-thyroxine dose. Care is required to start elderly patients and those with ischemic heart disease on low doses of L-thyroxine with slow upward titration, as the sudden increase in metabolic rate increases demand on the heart and can trigger angina. It is important to avoid giving excessive doses of L-thyroxine in all patients as there are potential risks of atrial fibrillation and osteoporosis, especially in the elderly. The finding of TSH suppressed to below the detection limit (that is, <0.05 mIU/L) should prompt a reduction in

Analytical practice point 11.7

TFTs should not be repeated within 6 weeks of starting or changing the dose of L-thyroxine, as it takes at least this long for TSH to reach a new steady state.

CASE 11.1

A 40-year-old woman sees her primary care doctor with tiredness, poor concentration, and weight gain over several months. Her thyroid function tests show:

	SI units	Reference range	Conventional units	Reference range
Plasma				
TSH (thyrotropin)	43 mIU/L	0.4–4.2	43 mIU/L	0.4–4.2
Free T_4	5.2 pmol/L	10–31	0.4 ng/dL	0.8–2.4
Thyroid peroxidase (TPO) antibodies	300 kIU/L	<35	300 IU/mL	<35

- **What is the diagnosis?**
- **How is it treated?**
- **How is treatment monitored?**

The raised TSH and low free T_4 indicate primary hypothyroidism. The presence of TPO antibodies in high titer is consistent with autoimmune thyroid disease. This is the commonest cause of primary hypothyroidism and hyperthyroidism. Treatment is by replacement thyroxine given at a dose that will result in TSH within the lower part of the reference range. Excessive doses, as shown by TSH suppression to <0.1 mIU/L, are generally avoided due to the risk of atrial fibrillation and osteoporosis, particularly in older people. However, suppressive doses of thyroxine are usually given for hypothyroidism following surgical removal of the gland (total thyroidectomy) for thyroid cancer. This is to prevent TSH from stimulating any remaining malignant thyroid tissue.

L-thyroxine dose. The exceptions to this are where patients are on antithyroid drugs to treat hyperthyroidism and in secondary hypothyroidism, where TSH is not a valid indicator of hypothyroidism.

A common pattern of results is an elevated TSH without a decreased fT_4. Where the TSH is significantly raised (for example >10 mIU/L), this is considered to be primary hypothyroidism and treated with L-thyroxine. However, if the TSH is above the upper reference limit but below 10 mIU/L, this is called subclinical or borderline hypothyroidism. The term subclinical is misleading as it refers to a biochemical result without any knowledge of the patient's symptoms, or lack of them. The optimum treatment strategy is a source of debate; some favor early treatment, whilst others advocate waiting until TFTs become more obviously indicative of hypothyroidism. Sometimes the decision is helped by the measurement of TPO antibodies, since their presence at high titer is evidence of an underlying autoimmune process, which is more likely to progress. Long-term data have shown the difference to be quite small: the annual risk of hypothyroidism is less than 5% with TPO antibodies compared to 2.6% without them.

Secondary hypothyroidism arises from a fall in TSH production, as a consequence of pituitary damage (autoimmune) or a tumor, and the fall in plasma TSH results in low thyroid hormone production (**Figure 11.21**). Thyroid function test results typically seen in cases of secondary hypothyroidism are shown in **Figure 11.22**.

CASE 11.2

A 26-year-old woman attends her primary care physician feeling tired, but there are no other symptoms or physical signs. She has no known family history of thyroid disease. Thyroid function tests are performed.

	SI units	Reference range	Conventional units	Reference range
Plasma				
TSH (thyrotropin)	3.2 mIU/L	0.4–4.2	3.2 mIU/L	0.4–4.2
Free T_4	78 pmol/L	10–31	6.1 ng/dL	0.8–2.4
Free T_3	25 pmol/L	4.0–7.4	1623 pg/dL	260–480
Thyroid peroxidase (TPO) antibodies	25 kIU/L	<35	25 IU/mL	<35

- **What are the three most likely causes of these unusual results?**
- **How can they be distinguished?**

Raised free T_4 and free T_3 without a fully suppressed TSH is unusual and not the expected pattern for primary hyperthyroidism. The three most likely causes in descending order of probability are: antibody interference in the immunoassays; familial thyroid hormone resistance syndrome; and TSH-secreting pituitary tumor (TSHoma). Additional information to distinguish between these is from the clinical examination (with antibody interference, the patient is not clinically thyrotoxic), family history, levels of free α subunit of TSH (increased in TSHoma), pituitary imaging, and the TRH stimulation test. In the first instance, the sample was re-analyzed by another laboratory using a different method. Normal levels of free T_4 and free T_3 were obtained and the TSH was unchanged. Further testing proved the presence of interfering antibodies in the initial assays for free T_4 and T_3. These gave spuriously high results. Such antibodies in the patient's serum may be directed toward the hormones themselves or to the animal species in which the reagent antibodies were raised (for example rabbit or mouse).

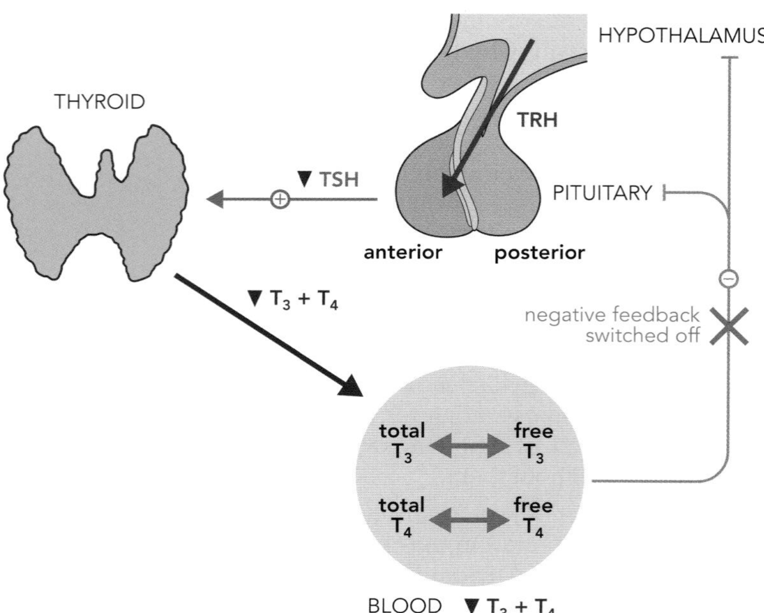

Figure 11.21 Secondary hypothyroidism.
Secondary hypothyroidism arises from damage to the pituitary gland via an autoimmune cause or a pituitary tumor. The primary defect here is a reduction in the synthesis of TSH by the pituitary with consequent falls in T_3/T_4 production in the thyroid. It is worthwhile noting that although the TSH is always low in this condition, it is not usually suppressed at diagnosis.

SECONDARY HYPOTHYROIDISM

free T_4	▼	typically <10 pmol/L	(<0.8 ng/dL)
TSH		typically 1–2 mIU/L	
free T_3	▼	typically <4 pmol/L	(<260 pg/dL)

Figure 11.22 Thyroid function test results in secondary hypothyroidism.
Secondary hypothyroidism is due to an inadequate production of TSH by the anterior pituitary. It may be expected that the main finding would be a very low serum TSH concentration but this is rarely found at initial diagnosis, as the pituitary gland has a reserve of TSH but this is inadequate to maintain normal circulating concentrations of thyroid hormones. Therefore, at presentation, the TSH is typically within the reference range but both T_3 and T_4 are low. All three hormones have to be measured to make a diagnosis.

Myxedema coma

On rare occasions hypothyroidism may be sufficiently severe to cause impaired mental function leading to confusion and coma. So-called myxedema coma usually occurs in the elderly and is accompanied by hypothermia.

Hyperthyroidism

Hyperthyroidism (thyrotoxicosis) is the clinical state caused by elevated levels of thyroid hormones in the blood (**Figure 11.23**). It may occur as a result of stimulation by excessive TSH (very rare) or by immunoglobulins with TSH-like activity causing overproduction of T_3/T_4. The latter is termed Graves'

PRIMARY HYPERTHYROIDISM

free T_4	▲	typically >30 pmol/L	(>2.3 ng/dL)	>98% of cases
TSH	▼	typically <0.1 mIU/L		
free T_3	▲	typically >10 pmol/L	(>650 pg/dL)	

Figure 11.23 Thyroid function test results in primary hyperthyroidism.
All patients with primary hyperthyroidism have suppressed TSH concentrations and raised thyroid hormones. A diagnosis of primary hyperthyroidism can only be made after measurement of TSH and T_3.

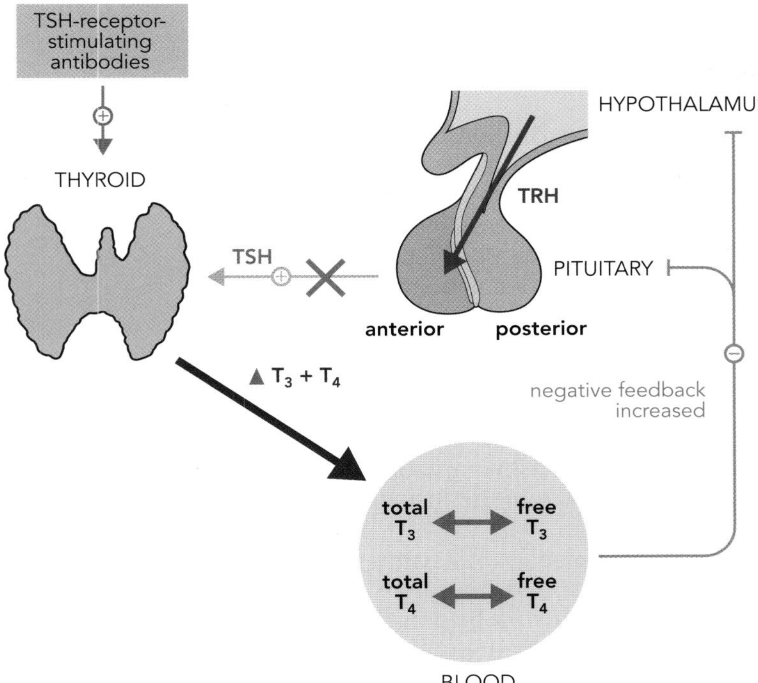

Figure 11.24 Primary hyperthyroidism: Graves' disease. Graves' disease is an autoimmune disease in which B lymphocytes synthesize immunoglobulins that bind to and activate the TSH receptor on the thyroid gland. Such binding stimulates tissue growth and excessive synthesis and secretion of thyroid hormones. Consequently, raised serum concentrations of thyroid hormones feed back to inhibit TRH and TSH synthesis and secretion by the hypothalamus and pituitary, respectively.

disease and is the most common cause of thyrotoxicosis (**Figure 11.24**). This is an autoimmune condition which is associated with other autoimmune endocrinopathies including hypothyroidism, Addison's disease, and type 1 diabetes mellitus. Graves' disease comprises hyperthyroidism plus specific signs in the eyes, and less commonly the skin, probably due to antibody cross-reactivity with antigens in connective tissue. Clinical features of hyperthyroidism are listed in **Table 11.10**. Although the TSH-receptor antibody responsible for

TABLE 11.10 Clinical features of hyperthyroidism

Physiological system	Symptoms	Signs
General	Fatigue Increased appetite Heat intolerance Goiter	Weight loss Hair loss
Cardiovascular	Palpitations	Vasodilation Tachycardia Atrial fibrillation
Neuromuscular	Myopathy	Agitation Psychosis
Gastrointestinal	Vomiting	Diarrhea
Reproductive	Infertility	Oligomenorrhea or amenorrhea
Skin	Pruritis	Pretibial myxedema*
Eyes	Visual disturbance	Exophthalmos/proptosis* Lid retraction; lid lag

*Only seen in Graves' disease; may be present also in patients who are hypothyroid or euthyroid.

Graves' disease can be detected in plasma, it is a technically demanding and time-consuming assay and so is rarely carried out. Instead, TPO antibodies, which are readily measurable on automated immunoassay analyzers, are used as evidence of an autoimmune process.

Other causes of thyrotoxicosis include autonomous overproduction of thyroid hormones due to multinodular goiter or a single adenoma, inflammation (thyroiditis), or release of preformed thyroid hormones due to destruction of tissue. Transient thyrotoxicosis often precedes hypothyroidism in Hashimoto's disease when thyroid hormones are liberated during autoimmune destruction of the follicular cells. Ingestion of exogenous thyroid hormones or compounds containing iodine may also cause thyrotoxicosis.

The terms hyperthyroidism and thyrotoxicosis are often used interchangeably but, strictly, hyperthyroidism only applies to the situation where excess thyroid hormone synthesis and secretion occur. Thus it applies to Graves' disease and toxic goiters, but does not apply to thyroiditis or exogenous thyroid hormone intake, although the clinical effects and pattern of TFTs are indistinguishable. The diagnosis of these conditions depends upon information from the history and examination, and sometimes thyroid imaging.

Hyperthyroidism is suspected clinically when patients present with symptoms of increased metabolic rate such as weight loss, anxiety, or tachycardia. The symptoms and signs specific to Graves' disease, particularly protrusion of the eyes (exophthalmos), may be present in addition. The principal biochemical findings are undetectably low TSH with elevated fT_4 and fT_3. Often the fT_3 rises before the fT_4, and at this stage the term T_3 toxicosis may be used. Tests

CASE 11.3

A 52-year-old man complains of unintended weight loss, despite good appetite, tremor, and pain in the throat. His symptoms began two weeks ago. On examination he had a tremor and tachycardia but no exophthalmos.

	SI units	Reference range	Conventional units	Reference range
Plasma				
TSH (thyrotropin)	<0.05 mIU/L	0.4–4.2	<0.05 mIU/L	0.4–4.2
Free T_4	45 pmol/L	10–31	3.5 ng/dL	0.8–2.4
Free T_3	10.0 pmol/L	4.0–7.4	650 pg/dL	260–480
Thyroid peroxidase (TPO) antibodies	75 kIU/L	<35	75 IU/mL	<35

- What is the diagnosis?
- What is likely to happen over the next few months?

The suppressed TSH and raised free T_4 and T_3 indicate primary hyperthyroidism. The commonest cause for this is Graves' disease due to immune stimulation of the thyroid by immunoglobulins that stimulate the TSH receptor. However, the short duration of his symptoms, localized pain, and only slightly abnormal TPO antibodies are more suggestive of thyroiditis due to viral infection of the thyroid and release of preformed thyroid hormones. The ratio of free T_4 to T_3 tends to be higher than in Graves' disease. Treatment with antithyroid drugs, radioactive iodine, or surgery is not indicated; in the short term, β-blockers and anti-inflammatory drugs may be used. After a few weeks the inflammation resolves and there may be a hypothyroid phase, which may become permanent and require thyroxine replacement.

to identify the cause include TPO antibodies, which point to Graves' disease if high, and thyroid imaging with ultrasound and isotope scans. Treatment options consist of antithyroid drugs (for example, carbimazole), surgery, and radioactive iodine (^{131}I). Each has advantages and disadvantages which are taken into account for the individual patient. In each case the patient is often rendered hypothyroid by treatment and requires replacement with L-thyroxine, titrated as in primary hypothyroidism. In the early stages of treatment the TSH may remain suppressed, sometimes for several months, thus making it useless for assessing adequacy of L-thyroxine doses. The fT_4 concentration is used for monitoring until the thyrotroph cells recover full feedback sensitivity. Antithyroid drugs may be difficult to titrate where the TFT results swing between hypo- and hyperthyroidism as doses are increased and reduced. A commonly used strategy to avoid this is the "block and replace" approach, where a sufficiently high dose of antithyroid drug is used to switch off all thyroid hormone synthesis, combined with a replacement dose of L-thyroxine. This allows more stable thyroid status to be maintained.

The counterpart to subclinical hypothyroidism, with suppressed TSH in the absence of elevated fT_3 and fT_4, may be seen in the elderly. This probably reflects the increasing incidence of toxic multinodular goiter with age. Subclinical hyperthyroidism is rarely treated, unless perhaps accompanied by atrial fibrillation, but is usually monitored at least annually due to the risk of progression to overt hyperthyroidism.

Secondary hyperthyroidism due to a rare TSH-secreting pituitary tumor (TSHoma) (**Figure 11.25**) would be suggested by the finding of elevated fT_4 and fT_3 in the presence of detectable (that is, nonsuppressed) TSH. However, this pattern of results may be seen in two other situations, both of which are more common than TSHomas. The first is analytical interference due to the presence of heterophilic antibodies directed against the animal species used in the immunoassay, or against thyroid hormones themselves. Further laboratory investigations using antibody-blocking tubes or assays with antibodies derived from different animals can usually identify these interferences. The second situation is thyroid hormone resistance syndrome, which is due to a genetic defect in the T_3 receptor, which segregates in families as either an autosomal dominant or recessive trait. The thyrotroph cells are not inhibited by T_3 and the lack

Analytical practice point 11.8

Following treatment of thyrotoxicosis the TSH may remain suppressed for several months, even if the free T_4 and free T_3 are low. It is not a reliable marker of thyroid function at this stage and can cause confusion with pituitary disease.

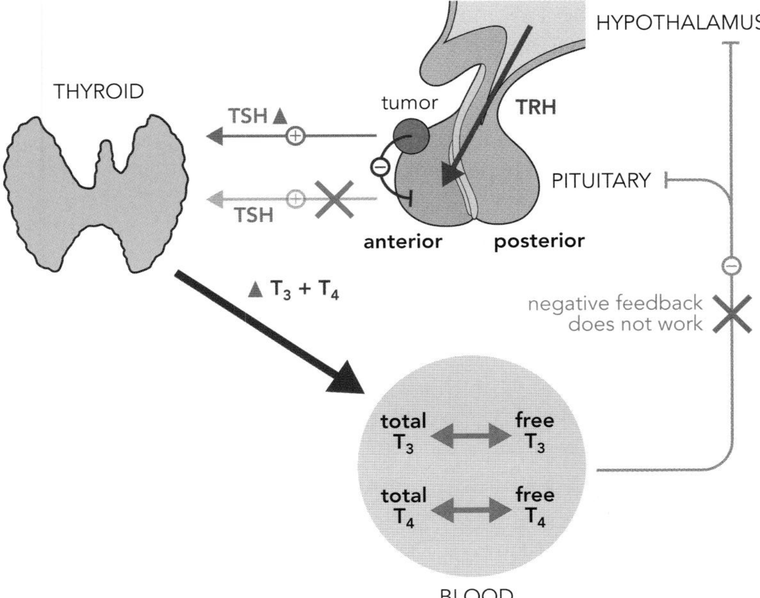

Figure 11.25 Secondary hyperthyroidism: TRH- and TSH-producing tumors. In these rare instances, increased ectopic production of either TRH or TSH by the tumor leads to raised serum TSH with consequent enlargement of the thyroid and increased synthesis and secretion of both T_4 and T_3. Serum concentrations of all thyroid hormones are thereby increased.

of negative feedback results in high fT_4 and fT_3. However, since the peripheral tissues are also lacking the functional receptor, the patient is usually clinically in a hypothyroid state. A variant of this condition affects only the thyrotrophs, thus the peripheral tissues respond to the high thyroid hormone levels and the patient appears hyperthyroid. Clinical treatment of hyperthyroidism may include the use of carbimazole to inhibit production of thyroid hormones, surgery to remove a part or the whole of the thyroid, or treatment with radioactive iodine (^{131}I) to destroy thyroid tissue. It is important to measure and monitor fT_3 during treatment; when fT_3 falls after treatment, TSH may remain suppressed for several weeks.

Analytical practice point 11.9

TSH-secreting tumors are extremely rare. Raised thyroid hormones without suppressed TSH are most likely to be due to interfering antibodies in the immunoassays.

Thyroid storm

The term thyroid storm is used when there is decompensation of one or more organ systems due to severe thyrotoxicosis. It is a life-threatening condition requiring aggressive treatment, but fortunately occurs rarely. Clinical features include fever, tachycardia with high-output heart failure, mental changes, diarrhea, vomiting, and jaundice. Hyperglycemia, hypercalcemia, and raised liver function tests are amongst the biochemical abnormalities that may occur. The pathogenesis is poorly understood. TFTs are no higher in thyroid storm than in many patients with uncomplicated thyrotoxicosis.

Sick euthyroid syndrome, non-thyroidal illness, and TFTs

Several changes in thyroid function occur during systemic illness, which affect the plasma levels of TSH and thyroid hormones. Sick euthyroid syndrome is a physiological abnormality resulting from changes in peripheral deiodination of T_4 and the rate of clearance of thyroid hormones. Exogenous causes of non-thyroidal illness include drugs that affect iodide transport, such as lithium, and others such as salicylate and phenytoin which displace thyroid hormones from proteins.

If these changes in TFTs are not understood, patients may be misdiagnosed with thyroid disease and treated inappropriately. There is no single pattern of TFT abnormalities which occurs in systemic illness, as test results are influenced by the severity and length of the condition, as well as some commonly used drugs, such as corticosteroids. The most sensitive indicator of illness is the free T_3, which falls readily, reflecting decreased activity of tissue deiodinases. In more severely unwell patients, such as those in the intensive therapy unit, plasma free T_4 decreases as well. TSH often falls in this situation, in contrast to the usual response of negative feedback. During recovery, the TSH may rise to supra-normal levels. Due to the difficulties with interpreting TFTs, many guidelines discourage their measurement in the acutely unwell inpatient unless there is a clear suspicion of thyroid disease. Even then, results may be misleading and should be confirmed in the weeks following recovery from the acute illness.

The interpretation of TFTs in primary and secondary hypo- and hyperthyroidism is summarized in **Table 11.11**.

Clinical practice point 11.5

TFTs are often affected by systemic illness and drugs and should not be measured in hospital patients without a good clinical reason.

TABLE 11.11 Thyroid function tests in primary and secondary hyper- and hypothyroidism

Clinical condition	TSH	Free T_4	Free T_3
Primary hypothyroidism	High	Low	Low
Secondary hypothyroidism	Normal or low	Low	Low
Primary hyperthyroidism	Undetectable	High	High
Secondary hyperthyroidism	Normal or high	High	High

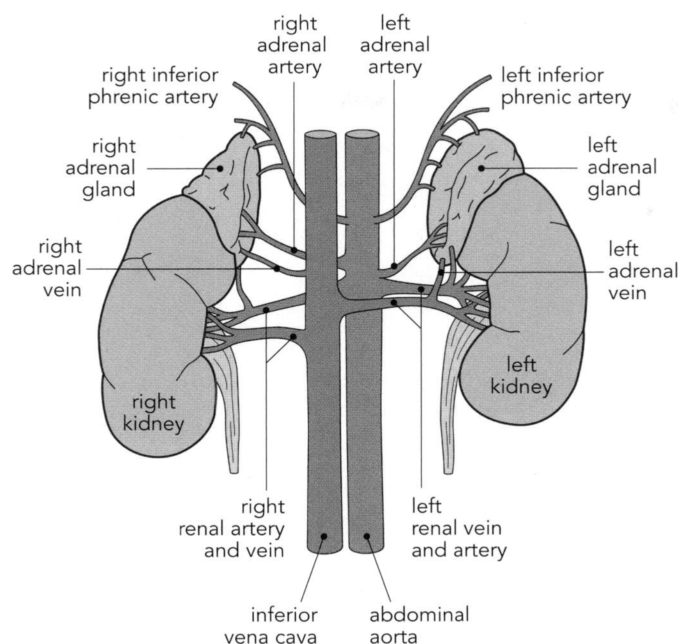

Figure 11.26 Gross anatomy of the adrenal glands, including blood supply.
The adrenal glands are small, flattened endocrine glands at the upper pole of each kidney. They consist of two types of endocrine tissue—the outer adrenal cortex and the inner adrenal medulla—which are quite distinct both developmentally and functionally, having different embryological origins.

11.4 ADRENAL CORTEX

Anatomical relationships and morphology

The adrenal glands are paired glands that, as their name suggests, are found near the superior poles of the kidneys on the posterior abdominal wall (**Figure 11.26**). The glands receive their blood supply from adrenal arteries that arise directly from the abdominal aorta and are drained by adrenal veins that enter the vena cava (in the case of the right adrenal gland) and the left renal vein (in the case of the left adrenal gland). Each adrenal gland is composed of two embryologically distinct tissues: the exterior cortical tissue, derived from mesoderm, and the interior medullary tissue, derived from the neural crest (**Figure 11.27**). The cells of the adrenal cortex synthesize and secrete steroid hormones, whereas those of the medulla (chromaffin cells) are part of the sympathetic nervous system and synthesize and secrete catecholamines. Blood flowing through the gland first perfuses the cortex, followed by the medulla, meaning that hormones released by the cortex have the capacity to act upon the chromaffin cells as these express steroid receptors. This is the principal reason why the chromaffin cells of the adrenal medulla, unlike the majority of sympathetic neurons, express phenylethanolamine-*N*-methyl transferase and convert norepinephrine to epinephrine, resulting in epinephrine being the principal catecholamine released from the adrenal medulla.

The adrenal cortex can be further subdivided into three zones named (from outside to inside) the zona glomerulosa, zona fasciculata, and zona reticularis. Each of these zones has the capacity to produce all classes of steroid hormones but they produce predominantly mineralocorticoids, glucocorticoids, and androgens, respectively (**Figure 11.28**). In view of the arrangement of the vascular drainage of the adrenal cortex, it will be apparent that steroid hormones released from the zona glomerulosa can enter the cells of the zona fasciculata and the zona reticularis, and that steroids produced by the cells of the zona fasciculata can enter the cells of the zona reticularis. This has implications for the function of these various zones, even if hyper- or hyposecretion of hormones is being caused by disease of only one layer of the cortex.

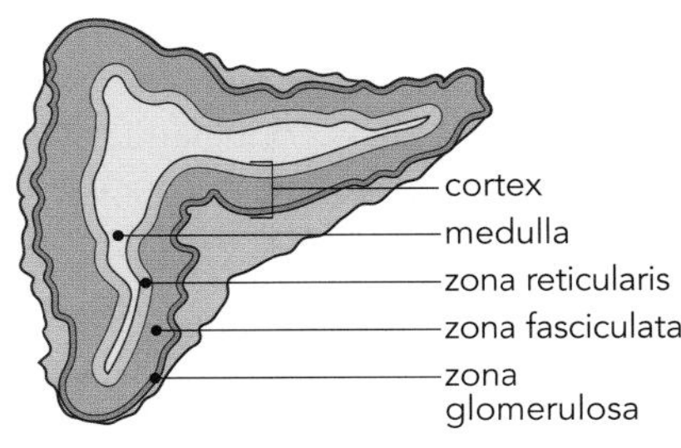

Figure 11.27 Relationship between the adrenal cortex and medulla.
The adrenal gland has two components—the medulla (center) and cortex (outer layer)—which serve quite separate functions. The cortex arises from mesoderm and, in response to stimulation by adrenocorticotropic hormone from the pituitary, synthesizes and secretes steroid hormones.
The medulla is a neuroendocrine tissue arising from embryological development of neural tissue and is the major site for catecholamine synthesis.

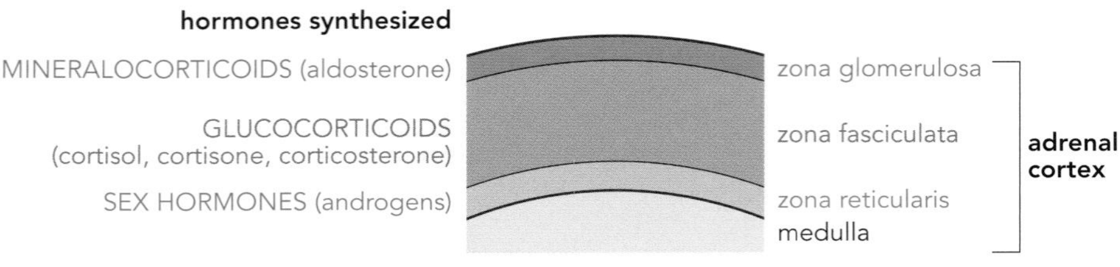

Figure 11.28 Morphology of the adrenal cortex. The adrenal cortex consists of three zones: the outer zona glomerulosa is the site of synthesis and secretion of mineralocorticoids, principally aldosterone; the intermediate and largest zone, the zona fasciculata, is the major site of synthesis and secretion of corticosteroids, principally cortisol; the inner zona reticularis is the site of synthesis and secretion of androgens, principally dehydroepiandrosterone and androstenedione.

Hormones of the adrenal cortex

The major hormones synthesized in the adrenal cortex are cortisol, corticosterone (both glucocorticoids), and aldosterone (a mineralocorticoid) (**Table 11.12**) but minor amounts of the sex hormones are also synthesized including androgens [androstenedione, testosterone, dehydroepiandrosterone (DHEA), and dehydroepiandrosterone sulfate (DHEA-S)] and estrogens (estradiol and estrone).

TABLE 11.12 Major hormones of the adrenal cortex

Hormone	Site of synthesis	Mode of action	
		Glucocorticoid	Mineralocorticoid
Cortisol	Zona fasciculata	+++	+
Corticosterone	Zona fasciculata	+++	+
Aldosterone	Zona glomerulosa	+	+++

The number of + symbols reflects the relative potency of the glucocorticoid or mineralocorticoid action of the hormone.

The control of synthesis of the glucocorticoids is via the hypothalamus–pituitary–adrenal axis, while mineralocorticoid synthesis is controlled by the renin–angiotensin system involving the kidneys, liver, and vascular endothelial cells. The immediate regulator of cortisol secretion is pituitary-derived ACTH (also known as corticotropin), the secretion of which is in turn controlled by CRH. CRH, a 41-amino-acid peptide, is released from the paraventricular nucleus of the hypothalamus in response to physiological or psychological stress. In homeostasis, the axis is controlled by both long and short feedback loops.

ACTH is derived from processing of a larger, precursor, 241-amino-acid peptide, pro-opiomelanocortin, in the pituitary gland (see Figure 11.4).

Hormone biosynthesis

The structures of the major hormones of the adrenal cortex and their biosynthetic pathways are shown in **Figure 11.29**. The initial substrate for the synthesis of steroid hormones is cholesterol, which may be synthesized endogenously and stored as cholesterol ester in the adrenal cortex, or donated to the tissue by circulating low-density lipoprotein. The first step of the pathway provides cholesterol from hydrolysis of the ester by a cholesterol esterase. The rate-limiting step of steroid synthesis involves cleavage of a six-carbon side chain at C21 by desmolase, a cytochrome P450-linked mixed function oxidase, to produce pregnenolone. Pregnenolone is converted to progesterone, the common precursor of the major steroid hormones, by the action of 3β-hydroxysteroid dehydrogenase-Δ4-5 isomerase, a reaction in which the double bond in the original cholesterol precursor moves from the B ring to the A ring and the 3β-hydroxyl group is oxidized to a keto group. Many of the steps of steroid synthesis require further action by cytochrome P450-linked hydroxylases; synthesis

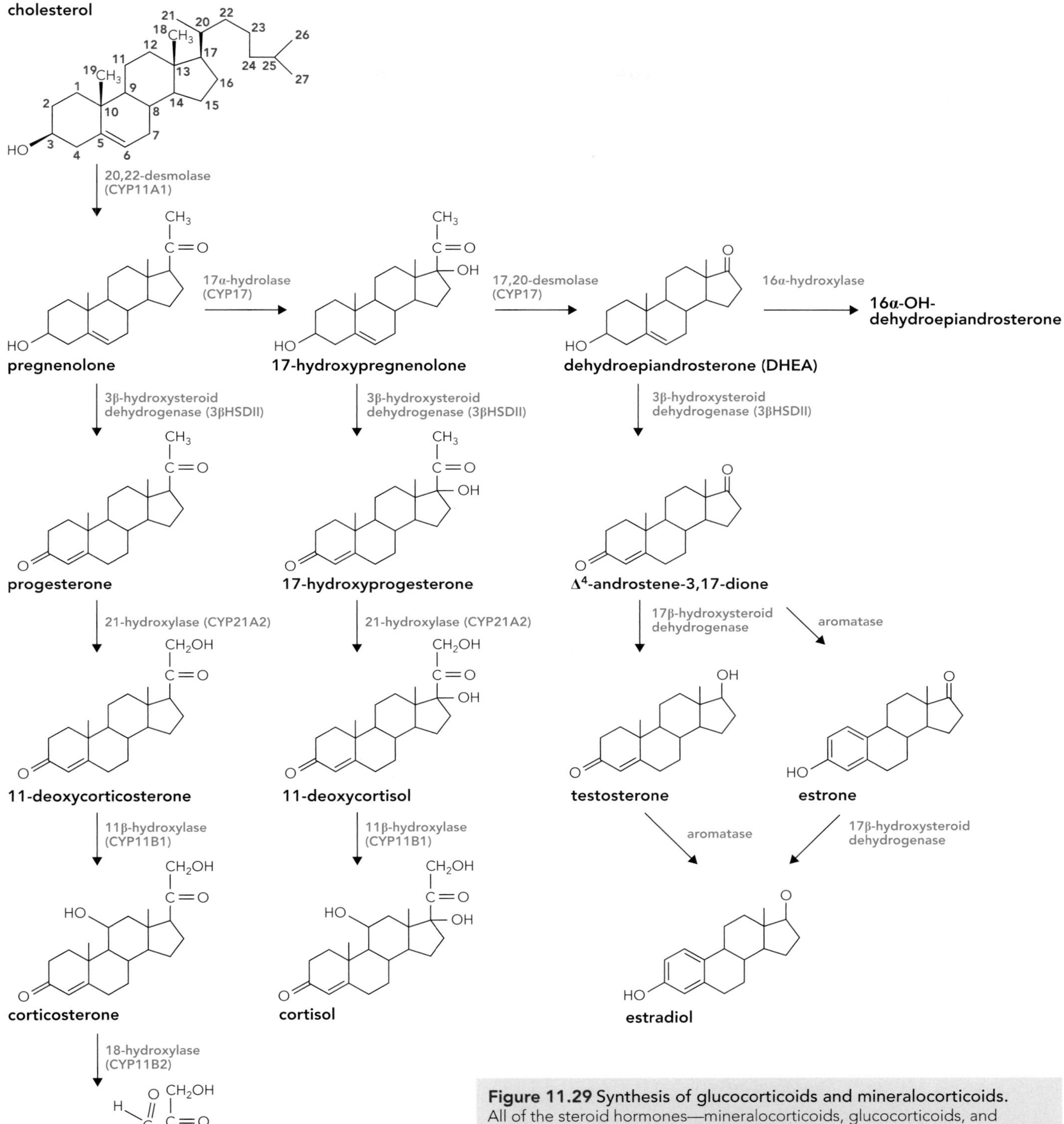

Figure 11.29 Synthesis of glucocorticoids and mineralocorticoids. All of the steroid hormones—mineralocorticoids, glucocorticoids, and androgens and estrogens—are synthesized from cholesterol delivered to the cell by low-density lipoprotein or from cholesterol synthesized *de novo* in the cell. Many of the enzymes involved in the pathways of synthesis are cytochrome P450-linked and are thus inducible. The 3-βHSDII isoform is synthesised in the adrenal and testes. A second isoform, 3-βHSDI, encoded by a different gene, is synthesised in the liver and placenta.

of cortisol, for example, requires sequential hydroxylations at C17, C21, and C11. Enzymes of the steroid hormone biosynthetic pathway from cholesterol include:

- Cholesterol side-chain cleavage enzyme (desmolase); CYP11A1
- 3β-Hydroxysteroid dehydrogenase-Δ4-5 isomerase (3βHSDH)
- 21α-Hydroxylase; CYP21A2
- 11β-Hydroxylase; CYP11B1
- Aldosterone synthase (see below); CYP11B2
- 17α-hydroxylase/17,20-lyase (C17,C20 lyase); CYP17
- 17β-hydroxysteroid dehydrogenase
- Aromatase CYP19

Aldosterone synthase catalyzes three reactions:

1. The formation of corticosterone from 11-deoxycorticosterone
2. Conversion of corticosterone to 18-hydroxycorticosterone
3. Formation of aldosterone from 18-hydroxycorticosterone

The localized production of specific hormones is made possible by the differential expression of genes leading to differing distribution of key hydroxylases in various parts of the cortex. For example:

- 3β-Hydroxysteroid dehydrogenase-Δ4-5 isomerase is absent from fetal cortex, which is able to synthesize only DHEA. Other hormones for fetal use must be transferred from the maternal circulation
- 17α-Hydroxylase is present in the zona fasciculata but not in the zona glomerulosa and thus the latter is unable to synthesize cortisol
- 18-hydroxylase and 18-steroid dehydrogenase are located only in the zona glomerulosa, the sole site of synthesis of aldosterone

Furthermore, synthesis of cortisol, for example, involves the movement of intermediates into and out of the mitochondrion: the 21-hydroxylase is located in the smooth endoplasmic reticulum whereas the final hydroxylase, 11β-hydroxylase, is mitochondrial (**Figure 11.30**). Approximate rates of secretion of corticosteroids in healthy adults are shown in **Table 11.13**.

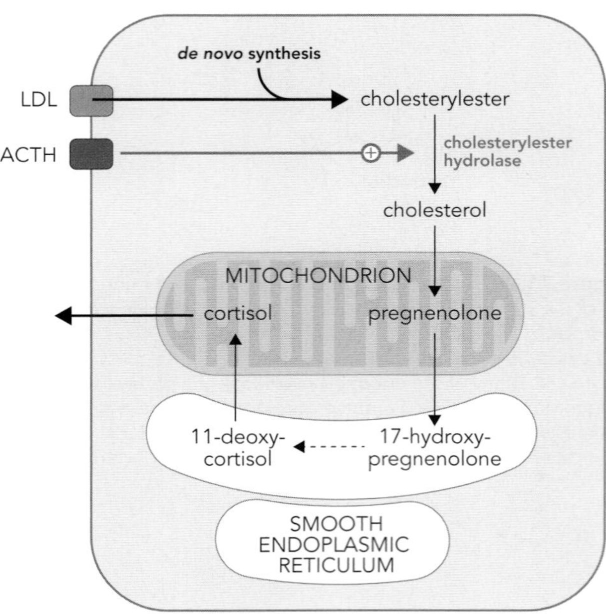

Figure 11.30 Subcellular location of hydroxylases involved in the synthesis of steroid hormone. Low-density lipoprotein (LDL) transports cholesterylester to the cell. Cholesterylester is hydrolyzed by a cholesterylester hydrolase which is activated by ACTH via a cAMP signaling pathway. Free cholesterol can then undergo metabolism to the steroid hormones. Conversion of cholesterol to pregnenolone by desmolase (CYP11A1) occurs in the mitochondrion. Pregnenolone then moves from the mitochondrion to the smooth endoplasmic reticulum for further metabolism by various hydroxylases. For example, 21-hydroxylase (CYP21A2), which converts 17-hydroxypregnenolone to 11-deoxycortisol, is located in the smooth endoplasmic reticulum, whereas the final hydroxylase of cortisol synthesis, 11β-hydroxylase (CYP11B1), is mitochondrial.

TABLE 11.13 Approximate rates of secretion of major corticosteroids in healthy adults

Hormone	Rate (mg/day)
Cortisol	20.0
Corticosterone	5.0
Aldosterone	0.1

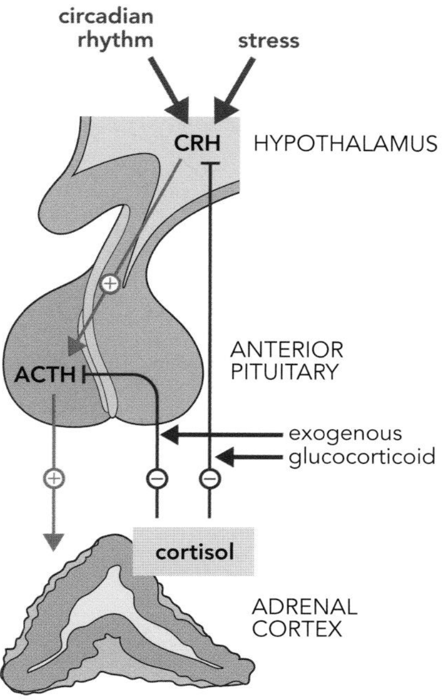

Figure 11.31 Feedback control in the hypothalamus–pituitary–adrenal axis.
As part of the circadian rhythm and also during stress, both physical and psychological, corticotropin-releasing hormone (CRH) is released by the hypothalamus and CRH stimulates the synthesis and secretion of corticotropin (adrenocorticotropic hormone; ACTH) by the anterior pituitary. ACTH in turn stimulates the synthesis and release of cortisol by the adrenal cortex. A rise in plasma cortisol inhibits the release of ACTH by the anterior pituitary (short loop) and of CRH by the hypothalamus (long loop). Administration of exogenous glucocorticoids (for example hydrocortisone) has similar inhibitory effects on ACTH and CRH release.

Control of cortisol secretion

An overview of the control of cortisol secretion through the hypothalamus–pituitary–adrenal axis is shown in **Figure 11.31** and the rates of secretion in Table 11.13. Synthesis and release of cortisol from the adrenal cortex are increased by stimulation of the hypothalamus by stress or diurnal rhythm via the sequential actions of CRH and ACTH. Cortisol (free fraction in plasma) also regulates its own synthesis via negative feedback at both the pituitary (on ACTH release) and hypothalamus (on CRH release). Exogenously administered glucocorticoids will also exert negative feedback in a similar manner. The action of CRH in stimulating ACTH production has been described earlier. ACTH is a 39-amino-acid peptide with biological activity residing in residues 1–24. While in healthy adults ACTH is formed from hydrolysis of POMC, it may also be produced ectopically in certain malignancies. ACTH is secreted from the pituitary in response to stress imposed on a diurnal rhythm, which peaks at around 05.00 am (that is, early morning), and secretion is subject to feedback control. It is unstable in plasma and has a short half-life of about 10 minutes. Binding of ACTH to specific receptors on the surface of adrenal cortical cells, particularly in the zona fasciculata (less so in the zona reticularis), stimulates the synthesis of cAMP, leading acutely, in less than 3 minutes, to activation of cholesterol esterase to provide cholesterol for cortisol synthesis. Chronic stimulation of the cells by ACTH induces transcription of genes encoding the enzymes of steroidogenesis and even greater rates of cortisol synthesis. Cortisol acts via short and long feedback loops to inhibit ACTH and CRH secretion, respectively, and thereby indirectly controls its own synthesis. ACTH also feedback-inhibits CRH secretion. Other factors regulating CRH and ACTH secretion are shown in **Table 11.14**. The concentration of cortisol in plasma exhibits a marked diurnal variation, peaking between 05.00 am and 09.00 am in response to the diurnal variation in ACTH, and reaching a nadir at around midnight. The normal response may be overridden in response to stress and in Cushing's syndrome; in both of these instances, high levels are sustained throughout the day and night.

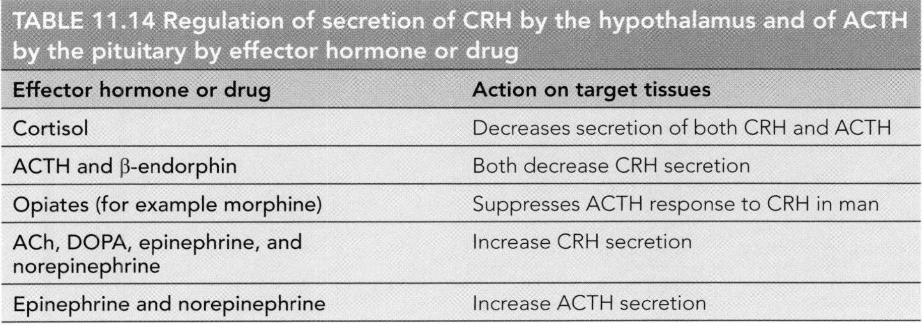

TABLE 11.14 Regulation of secretion of CRH by the hypothalamus and of ACTH by the pituitary by effector hormone or drug

Effector hormone or drug	Action on target tissues
Cortisol	Decreases secretion of both CRH and ACTH
ACTH and β-endorphin	Both decrease CRH secretion
Opiates (for example morphine)	Suppresses ACTH response to CRH in man
ACh, DOPA, epinephrine, and norepinephrine	Increase CRH secretion
Epinephrine and norepinephrine	Increase ACTH secretion

CRH, corticotropin-releasing hormone; ACTH, adrenocorticotropin; ACh, acetylcholine; DOPA, dihydroxyphenylalanine.

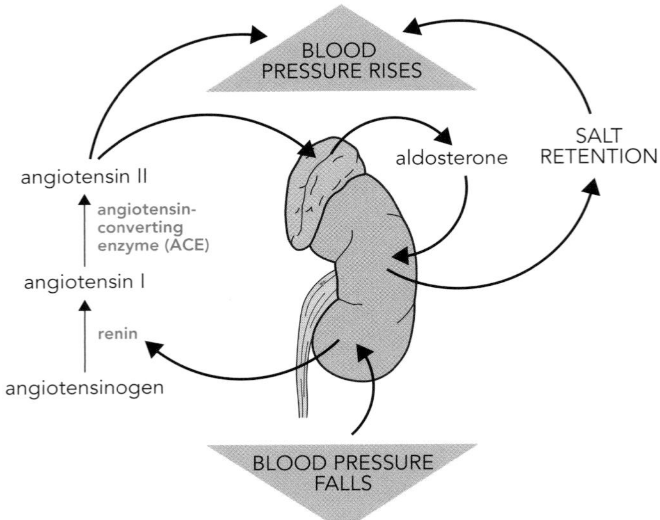

Figure 11.32 Control of aldosterone synthesis. The renal juxtaglomerular cells respond to a fall in renal artery blood pressure (decreased renal blood flow)—which stimulates the baroreceptors in the afferent arterioles—by releasing renin, a protease which hydrolyzes hepatically synthesized angiotensinogen (452 amino acids) to angiotensin I (10 amino acids) in the circulation. Angiotensin I undergoes further hydrolysis to angiotensin II (8 amino acids) by angiotensin-converting enzyme (ACE) during passage through the lungs. Angiotensin II binds to its receptor on the adrenal cortex and activates the synthesis and secretion of aldosterone. Aldosterone stimulates sodium reabsorption by the kidney. In addition, angiotensin II causes vasoconstriction of blood vessels, which with the retention of sodium, raises blood pressure. Other factors will also affect this process, including the plasma concentrations of sodium and potassium.

Control of aldosterone secretion

In contrast to the glucocorticoids, aldosterone (a mineralocorticoid) is not controlled via the hypothalamus–pituitary–adrenal axis but by signals emanating from the kidney in response to a fall in renal artery blood pressure (kidney perfusion pressure) (**Figure 11.32**). A decrease in volume of extracellular fluid, sodium depletion, or potassium overload stimulates the release of a peptide, renin, from the juxtaglomerular apparatus into plasma. Renin hydrolyzes angiotensinogen, a pro-hormone produced in the liver, to angiotensin I; this in turn undergoes hydrolysis by angiotensin-converting enzyme (ACE) on the surface of vascular endothelial cells to angiotensin II. Angiotensin II binds to angiotensin II receptors to cause contraction of smooth muscle cells and, more importantly in the current context, to stimulate the synthesis of aldosterone in the zona glomerulosa in the adrenal cortex. This is described in more detail in Chapter 4.

Adrenocortical hormones in plasma

Reference ranges for the individual hormones secreted by the adrenal cortex are given at the beginning of this chapter. The hormones are present in plasma in both free and protein-bound forms, which are in equilibrium (**Table 11.15**).

Only the free forms are active but the plasma half-life is determined by the extent of protein binding, particularly to albumin and cortisol-binding globulin, both of which are synthesized by the liver (half-life in healthy adults: cortisol, 60–100 minutes; aldosterone, approximately 20 minutes). The total hormone concentration is dependent on the rate of hormone turnover and the concentration of the carrier proteins.

TABLE 11.15 Approximate distribution of cortisol and aldosterone in healthy adults

	Free	Protein bound	
		Cortisol-binding globulin	Albumin
Cortisol	10%	75%	15%
Aldosterone	30%	20%	50%

TABLE 11.16 Reactions involved in metabolism (detoxification) of steroid hormones

Phase	Enzyme reaction
Phase I	Reduction of double bond in A ring
	Reduction of 3-keto to 3-hydroxyl (A ring)
	Cleavage between C20 and C21
	Hydroxylation at C6
	Oxidation of 11β-hydroxyl and 21-hydroxyl
Phase II	Conjugation with glucuronate and/or sulfate

Elimination of adrenocortical hormones

The hydrophobic nature of adrenocortical hormones militates against their direct excretion in urine and only small amounts of cortisol (50–270 nmol/day) and aldosterone (10–90 nmol/day) are eliminated unchanged by this route. Rather, the steroids undergo a series of phase I and phase II reactions (**Table 11.16**) to render the hormones relatively more polar before excretion in urine.

Mode of action of cortisol

In common with other steroid hormones, adrenocortical hormones are recognized by receptors in the target-cell cytoplasm and are translocated into the nucleus. Here they modulate the transcription of a number of genes encoding effector proteins.

Actions of cortisol

The physiological actions of cortisol are mainly permissive, with particular effects on glucose homeostasis, fat metabolism, and the excretion of free water (**Table 11.17**). More-specific actions of glucocorticoids are shown in **Table 11.18**.

TABLE 11.17 Major general actions of glucocorticoids

Target tissues	Tissue response
Liver	Increased gluconeogenesis; increased amino acid breakdown (anti-anabolic)
Peripheral tissues	Decreased insulin response; decreased protein synthesis (anti-anabolic)
Lymphoid tissues	Reduced inflammatory response; decreased prostaglandin synthesis
General	Reduced stress response

TABLE 11.18 Major specific actions of glucocorticoids

Increased/stimulated	Decreased/inhibited
Gluconeogenesis	Protein synthesis
Glycogen deposition	Host response to infection
Protein catabolism	Lymphocyte transformation
Fat deposition	Delayed hypersensitivity
Sodium retention	Circulating lymphocytes
Potassium loss	Circulating eosinophils
Free water clearance	Circulating neutrophils
Uric acid production	

Adrenal cortex—investigations

The most common direct, first-line test of adrenal cortical function is the measurement of plasma cortisol and, perhaps, 24-hour urinary cortisol. However, in the basal state, this gives only limited information since the cortisol level varies throughout the day, with highest levels in the morning, and rises with stress and illness. A single cortisol measurement that falls within the reference range cannot usually exclude adrenal hypofunction or hyperfunction and, if these are suspected clinically, some form of dynamic function test is required. ACTH may be measured as a second-line investigation, but usually only once it has been established that adrenal insufficiency or excess is present.

The measurement of ACTH (as a second-line investigation following assessment of cortisol status) is technically straightforward. Compared to other hormones, ACTH is relatively unstable and samples should be transported on ice and delivered immediately to the laboratory. A single measurement of ACTH can rarely be interpreted in isolation. Partly this is due to its circadian rhythm, whereby high levels are expected in the morning and low levels at midnight. ACTH deficiency cannot be diagnosed on the basis of an undetectable ACTH level, unless there is also evidence of cortisol deficiency. ACTH-secreting tumors cause Cushing's disease by stimulating excessive production of cortisol from the adrenal cortex. However, plasma ACTH levels may not be greatly elevated but are considered inappropriately high in the presence of cortisol excess. The use of ACTH in the investigation of adrenal disorders is considered below.

Clinical practice point 11.6

Random plasma cortisol is of limited value in the diagnosis or exclusion of adrenal hypo- or hyperfunction, unless results are very high or low. Results within the reference range do not exclude adrenal disease.

Adrenal insufficiency

Primary adrenal insufficiency, also termed Addison's disease, is due to pathology arising within the adrenal gland itself. In the developed world, autoimmune destruction has replaced tuberculosis as the most common cause. Clinical symptoms of Addison's disease include:

- Tiredness
- Weakness
- Anorexia and weight loss
- Dizziness and postural hypotension
- Hypotrophic glossitis
- Pigmentation
- Loss of body hair in females

Other causes of hypocorticism include infarction, surgery, and congenital hyperplasia. Secondary adrenal insufficiency occurs due to lack of ACTH secretion from the anterior pituitary. This in turn may be due to hypopituitarism or to prolonged use of corticosteroids (as anti-inflammatory drugs) which leads to suppression of ACTH release via negative feedback and adrenal atrophy. The different causes of primary and secondary hypocorticism are shown in **Figure 11.33**.

Adrenal insufficiency may be suspected in patients with fatigue, hypotension, or electrolyte abnormalities (hyponatremia and/or hyperkalemia) and would be confirmed by finding a random plasma cortisol level <100 nmol/L (3.6 μg/dL). It would be excluded if the cortisol was >500 nmol/L (>18.1 μg/dL).

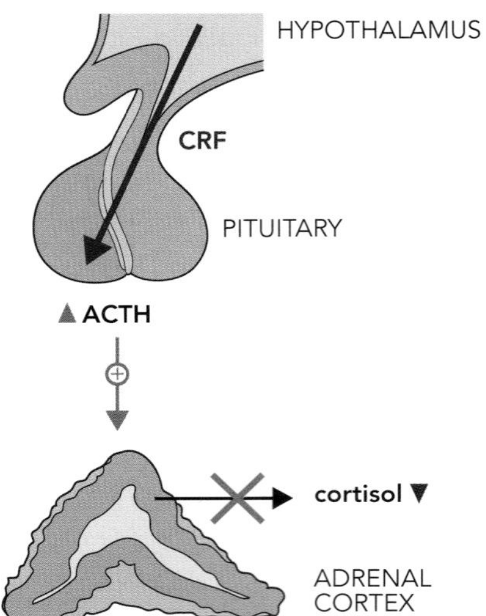

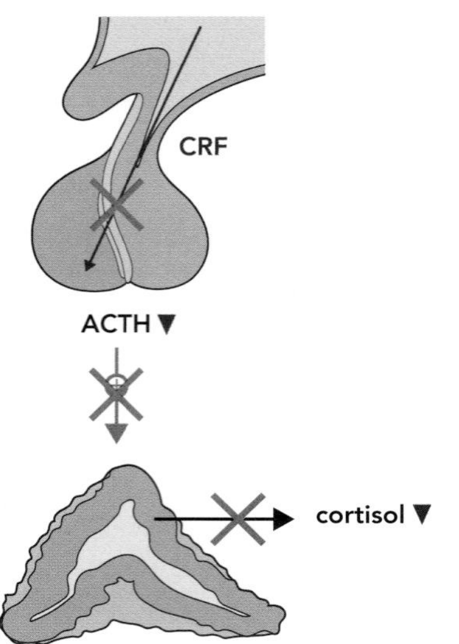

Figure 11.33 Adrenal hypofunction.
Adrenal hypofunction can be classified as primary or secondary. In primary hypofunction (Addison's disease), the circulating cortisol will be low, and since the normal feedback control is switched off, the plasma ACTH (corticotropin) level will be raised. In secondary hypofunction (pituitary failure leading to pan-hypopituitarism), both the circulating cortisol and ACTH will be low. Crosses indicate where the blocks in hormone synthesis occur.

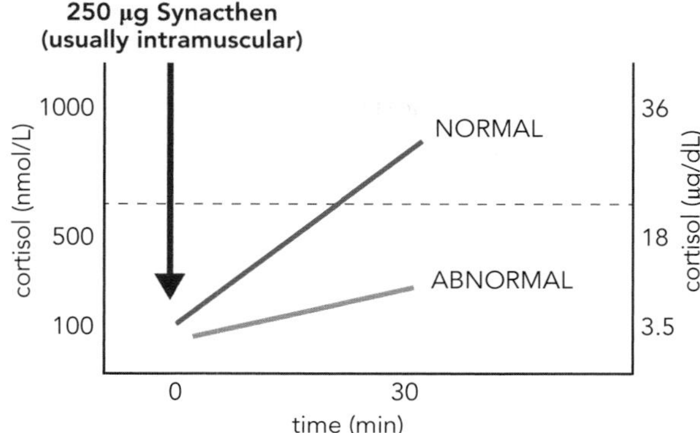

Figure 11.34 Synacthen test. The measurement of the basal level of plasma cortisol is unreliable to establish adrenal hypofunction. Patients with suspected hypofunction undergo the short Synacthen test (SST) in which plasma cortisol is measured before and 30 minutes after an injection of 250 μg of the ACTH analog Synacthen. In a control subject (normal response in blue), plasma cortisol rises from about 150 nmol/L to a value greater than 600 nmol/L. When the rise in plasma cortisol is blunted (<150 nmol/L pre-test), or the 30-minute plasma cortisol is less than 600 nmol/L (orange), the patient is considered to have impaired adrenal function and such a result warrants further investigation.

However, cortisol is often between these two levels and a stimulation test is required to assess adrenal reserve (**Figure 11.34**). Synthetic ACTH (tetracosactrin, also known as Synacthen) is injected and cortisol is measured immediately before and 30 minutes afterward. A rise in cortisol to at least 500 or 600 nmol/L (18–22 μg/dL) (depending on the assay used) excludes Addison's disease. However, it may not always exclude secondary adrenal insufficiency. A subnormal response to tetracosactrin may be further assessed by measurement

CASE 11.4

A 48-year-old Caucasian man complains of lack of energy, dizziness on standing, and weakness. On examination he has a low blood pressure when sitting, which falls further when he stands up (postural hypotension). He has dark pigmentation in the mouth. Routine blood tests show slightly low plasma sodium. He has a short tetracosactrin (Synacthen) test performed at 09.00 am. An injection of tetracosactrin is given, and blood samples are taken immediately before and 30 minutes afterward.

	SI units	Reference range	Conventional units	Reference range
Plasma				
Cortisol (0 min)	172 nmol/L	127–568 (am)	6.2 μg/dL	4.6–20.6 (am)
Cortisol (30 min)	223 nmol/L	>550	8.1 μg/dL	>20.0
Cortisol increment	51 nmol/L	>200	1.9 μg/dL	>7.3
ACTH (0 min)	250 pmol/L	<26	1137 pg/mL	<120

- **Why is a 09.00 am cortisol result not sufficient for the diagnosis?**
- **What do the results show?**
- **Why is ACTH measured at baseline?**

The doctor suspects adrenal insufficiency. A very low 09.00 am cortisol (<100 nmol/L) would confirm this and a very high result (>500–600 nmol/L) would exclude it. However, most results are within the reference range and a stimulation test is required to determine the ability of the adrenal cortex to respond to adrenocorticotropin hormone (ACTH). Tetracosactrin is a synthetic form of ACTH and a normal adrenal gland is able to respond to it by increasing cortisol secretion by at least 200 nmol/L to a 30-minute peak of >550 nmol/L. In this case, the patient fails to respond by both criteria, confirming adrenal insufficiency. The baseline ACTH is raised, indicating primary adrenal failure (Addison's disease) rather than pituitary disease.

of ACTH on a pre-dose sample. A very high level would indicate that the adrenal insufficiency was primary, whilst a low level would be expected if it was secondary.

Addison's disease may result in inadequate aldosterone production, as well as reduced cortisol. Both glucocorticoids and mineralocorticoids need to be replaced in this situation, by hydrocortisone and fludrocortisone, respectively. Since aldosterone is not ACTH dependent, but controlled by renin release and production of angiotensin II, hypopituitarism does not result in hypoaldosteronism and treatment with hydrocortisone alone is usually sufficient.

Isolated loss of aldosterone may occur when renin secretion from the kidney is reduced. This is sometimes seen in long-standing diabetes as a form of nephropathy, and manifests as hyperkalemia which is sometimes triggered by drugs such as ACE inhibitors, trimethoprim, or nonsteroidal anti-inflammatories.

Clinical practice point 11.7

The short tetracosactrin (Synacthen) test is indicated if a random cortisol measurement is not sufficiently high to exclude adrenal insufficiency.

Glucocorticoid excess—Cushing's syndrome

Cushing's syndrome refers to the clinical effects of excess glucocorticoids which may be due to increased endogenous production of cortisol (**Figure 11.35**) or, more commonly, due to pharmacological doses of synthetic corticosteroids such as prednisolone. Such drugs are extremely valuable in the management of inflammatory and autoimmune diseases, but cause long-term side effects identical to the symptoms and signs of excess cortisol secretion. Diagnosis is easily made from the history and biochemical investigation is not required. In contrast, diagnosis of Cushing's syndrome in patients not treated with steroids is one of the most challenging in endocrine practice.

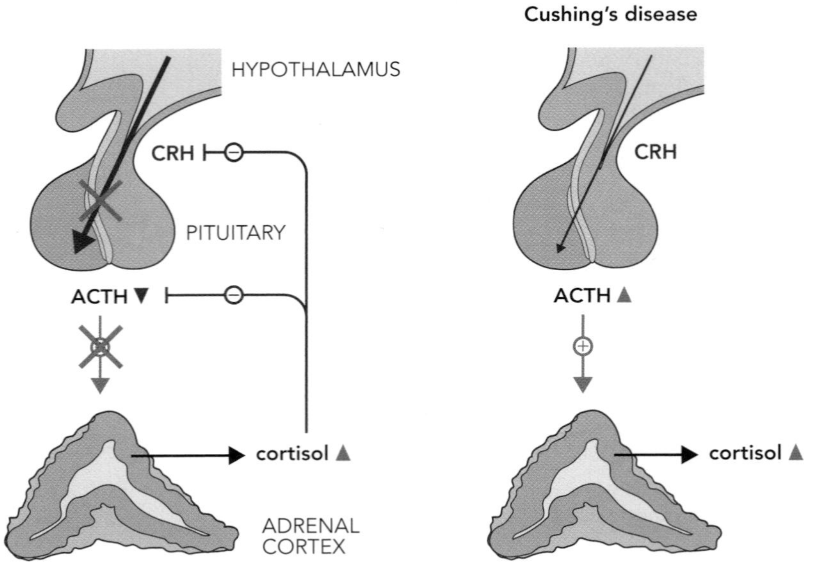

Figure 11.35 Cushing's syndrome.
Cushing's syndrome is the finding of raised plasma cortisol accompanied by the characteristic symptoms associated with cortisol excess (see the main text). When Cushing's syndrome is caused by a pituitary tumor producing excessive amounts of ACTH, the condition is known as Cushing's disease. In this instance, both plasma ACTH and cortisol are raised. Corticotropin-releasing hormone (CRH) production by the hypothalamus is reduced, but not usually totally switched off. In the other forms of Cushing's syndrome, the raised cortisol can be produced directly by the adrenal gland (usually due to an adrenal cancer) or by the ectopic production of ACTH (see Figure 11.36). When adrenal cancer is the cause of Cushing's syndrome, the primary defect is the excessive production of cortisol; therefore plasma cortisol is raised, whilst ACTH is often suppressed, and CRH production is switched off.

Symptoms of Cushing's syndrome

The clinical symptoms and signs of Cushing's syndrome are classically obesity of the trunk including the upper back (buffalo hump) with loss of muscle in the limbs, leading to the appearance of an orange on matchsticks. The face is rounded and reddened (moon face) and the skin is thin and fragile with purple stretch marks (striae) forming on the trunk, particularly the abdomen, buttocks, and breasts. Women may notice excessive growth of facial hair (hirsutism) with menstrual disturbance, and men may have reduced libido and erectile dysfunction. Other effects of exposure of tissues to long-term high levels of glucocorticoids include loss of bone mineral density (osteoporosis) leading to fractures, hyperglycemia causing polyuria and polydipsia, muscle weakness, and fatigue. Hypertension is due to the mineralocorticoid activity of cortisol which, although weak compared to aldosterone, is sufficient also to cause hypokalemia and metabolic alkalosis. Since obesity, hypertension, and diabetes are common conditions, both separately and in combination with each other, Cushing's syndrome is often suspected, although it is rare with an approximate incidence of 1 in 100,000 per year.

Clinical practice point 11.8

Endogenous Cushing's syndrome (that is, not due to prescribed steroids) is very rare.

Clinical causes of Cushing's syndrome

Cushing's syndrome arises from excess stimulation from ACTH or from autonomous (ACTH-independent) secretion from the adrenal gland. In all cases the underlying cause is a tumor secreting hormone inappropriately. In the case of ACTH this is most often due to a pituitary adenoma (when the term Cushing's disease is used), but on occasion there may be ectopic secretion of ACTH, particularly from small-cell carcinomas of the bronchus (**Figure 11.36**). Such tumors are of neuroendocrine origin and may secrete other hormones, such as antidiuretic hormone.

When Cushing's syndrome is suspected clinically, biochemical confirmation is required. Unfortunately, random plasma cortisol measurement is of almost no diagnostic value, as levels within the reference range do not exclude Cushing's syndrome and high levels may be due to stress or illness. The loss of diurnal variation in cortisol secretion is demonstrated by inappropriately high levels in a sample taken at midnight during sleep. However, this necessitates admission to a hospital bed and is often not a practical option. Cortisol can be measured in urine and a 24-hour collection allows integration of cortisol secretion over a day. Thus 24-hour urine free cortisol is a long-established test for Cushing's syndrome. There are, however, technical shortcomings with urine cortisol immunoassays which affect their reliability. The overnight dexamethasone suppression

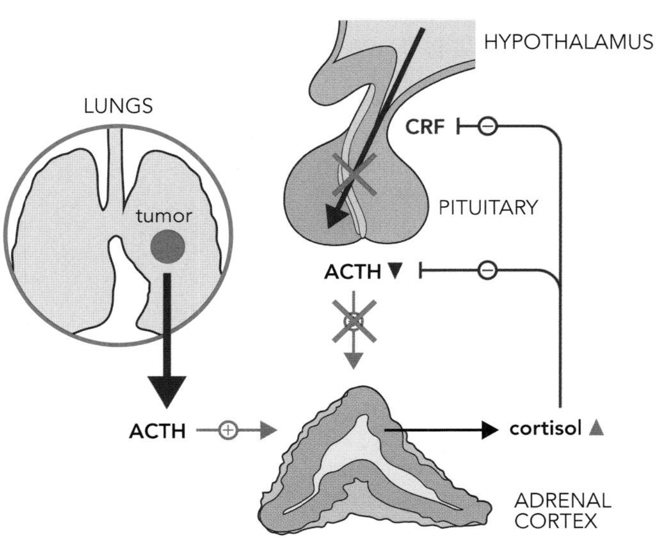

Figure 11.36 Ectopic production of ACTH.
Some tumors (for example lung cancer) will produce ACTH (adrenocorticotropin or corticotropin) in sites not associated with this peptide production. Such processes are termed ectopic hormone production. This leads to a very high circulating ACTH concentration and, as a consequence, very high cortisol concentration. A high cortisol concentration feeds back to inhibit ACTH production by the pituitary (red cross). However, there is no feedback inhibition on the ectopic ACTH production. Thus, although the measured plasma ACTH concentrations are characteristically very high, little or none of the hormone originates from the pituitary gland. Another characteristic of ectopic ACTH-dependent Cushing's syndrome is the presence of hypokalemia, with plasma potassium usually being less than 2.5 mmol/L. This is due to the high cortisol concentration overcoming the shuttle enzyme block in the renal cells, producing a mineralocorticoid effect.

test has become widely accepted as the most reliable first-line investigation for suspected Cushing's syndrome and has the advantage of being suitable for outpatient use. The patient takes a single 1 mg dose of dexamethasone at midnight and has plasma cortisol measured the following morning at 09.00 am. In a normal individual, the cortisol is usually undetectably low and certainly less than 50 nmol/L (1.8 μg/dL). A failure to suppress is not diagnostic of Cushing's syndrome and further investigations are required to confirm it. These include measurement of ACTH plus further dexamethasone suppression testing using low (0.5 mg 6-hourly) and high (2 mg 6-hourly) doses, each given for 48 hours. The low-dose test is to confirm the presence of Cushing's syndrome, whilst the latter is to determine its likely cause. In Cushing's disease, there is usually at least 50% suppression of plasma cortisol after 48 hours, with no such suppression expected with an adrenal tumor or ectopic ACTH secretion. In adrenal tumors, however, plasma ACTH is undetectably low, whilst in ectopic ACTH secretion it is usually very high. In Cushing's disease, the ACTH is usually detectable but inappropriately high.

Clinical practice point 11.9

Ectopic ACTH secretion may not be associated with the physical signs of Cushing's syndrome due to its short period of onset. However, the biochemical effects, particularly hypokalemia, may be more severe.

Analytical practice point 11.10

The overnight 1 mg dexamethasone suppression test is the most reliable first-line test for suspected Cushing's syndrome.

Corticotropin-releasing hormone test

Once a diagnosis of ACTH-dependent Cushing's syndrome has been made, it is often difficult to distinguish whether the origin is a tumor in the pituitary or an ectopic source. The respective tumors may be very small and not readily imaged, and the ACTH levels and response to dexamethasone often overlap. The CRH test may be used in this situation. In pituitary Cushing's disease, the infusion of CRH usually stimulates the release of ACTH and cortisol to a greater extent than in normal individuals, whereas this supranormal response is not typical of ectopic ACTH. The CRH test is not entirely reliable and selective sampling of venous blood from the petrosal sinuses, which drain from the pituitary, may be required. A clear difference in ACTH between the left and right sinuses (lateralization) is evidence of a pituitary tumor.

Hyperaldosteronism and Conn's syndrome

Excessive secretion of mineralocorticoids usually arises from an aldosterone-secreting adrenal tumor, which is termed Conn's syndrome. Hyperaldosteronism may also be due to adrenal hyperplasia. The effects of this are predictable from the known actions of aldosterone, which increases sodium reabsorption in exchange for potassium excretion in the distal nephron. The sodium retention (along with water) causes an expanded circulating volume and hence increases blood pressure. Thus Conn's syndrome may be suspected in individuals with hypertension in association with hypokalemia, particularly when the blood pressure is resistant to treatment. Initial investigation consists of measurement of both aldosterone and renin in plasma and expressing them as a ratio. In Conn's syndrome the raised aldosterone production leads to suppression of renin. The use of the ratio allows subtle alterations in this relationship to be detected. Ideally the test should be performed before treatment of hypertension as many of the drugs used for this affect the renin–angiotensin–aldosterone axis and make interpretation difficult. When a raised aldosterone–renin ratio has been found, it has been usual to perform further measurements at specified time points; that is, at 08.30 am before arising, at 09.00 am after ambulation, and at midday. In normal individuals, renin is stimulated by standing up and this in turn increases aldosterone. In Conn's syndrome there is no increase expected in renin or aldosterone. However, the reliability of the postural test has been shown to be poor, with many adenomas actually responding to standing upright. As an alternative, some centers measure the response of renin to the administration of the loop diuretic furosemide. Failure of renin to increase indicates primary hyperaldosteronism. Secondary hyperaldosteronism occurs in response to high renin secretion. This in turn may be due to decreased circulating volume (hypovolemia) or to renal artery stenosis. The former is associated with low blood pressure, whilst the latter is a cause of hypertension.

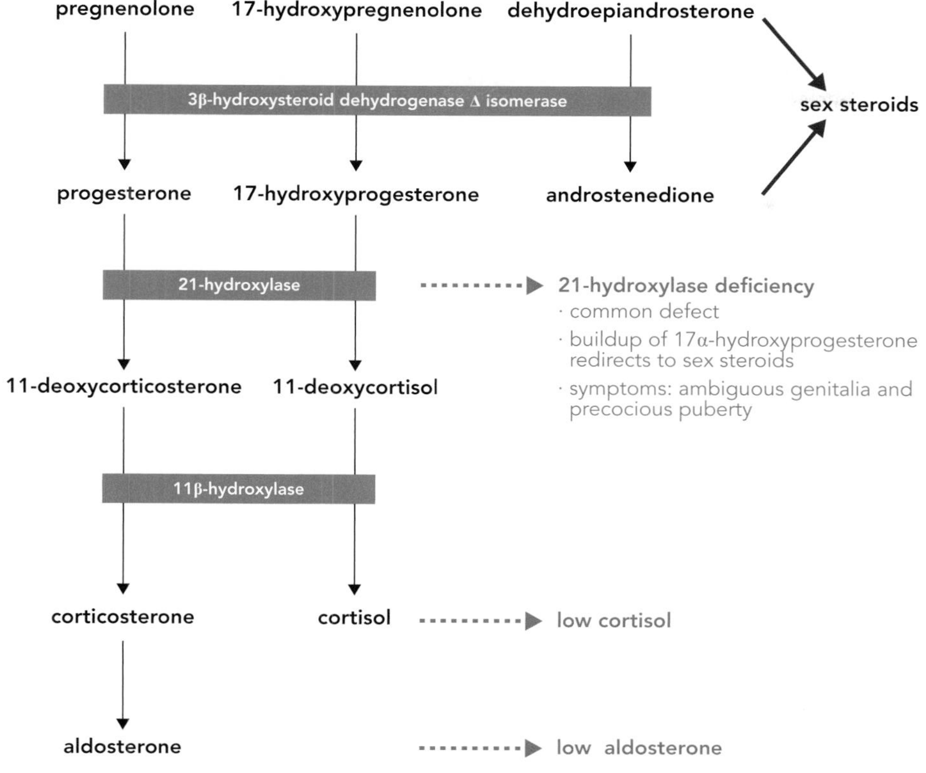

Figure 11.37 Diagnosis of congenital adrenal hyperplasia. Congenital adrenal hyperplasia (CAH) can be caused by mutations in several of the enzymes involved in the synthesis of cortisol in the adrenal cortex. The commonest causes of CAH (>90% of cases) are defects in 21-hydroxylase, which lead to impaired production of cortisol. In normal individuals, an increase in plasma cortisol concentration causes a feedback inhibition on the earlier steps of the synthetic pathway, thereby controlling its own synthesis. In patients with CAH, this feedback is no longer present and excessive amounts of precursors in the cortisol biosynthetic pathway (for example 17-hydroxyprogesterone) are produced; measurement of their concentrations in plasma help in the diagnosis of the disease. As a result of a block at the 21-hydroxylase step, the precursors in the pathway are metabolized to androstenedione and dehydroepiandrosterone (sex steroids), and increases in circulating levels of sex steroids can give rise to some classic symptoms such as ambiguous genitalia at birth and precocious puberty. All patients with 21-hydroxylase deficiency will have reduced plasma cortisol levels, and some will also have reduced aldosterone levels (a salt-losing condition). Treating patients with cortisol inhibits the production of cortisol precursors and the excessive production of sex steroids, thereby reversing the clinical symptoms. The salt-losing forms of CAH also need synthetic aldosterone treatment. Hormone structures can be found in Figure 11.29.

Congenital adrenal hyperplasia

Steroid synthesis in the adrenals requires a complex sequence of enzymatically regulated steps to produce glucocorticoids, mineralocorticoids, and androgens (**Figure 11.37**). Several inborn errors of metabolism are recognized in these pathways (**Table 11.19**) which cause impaired production of one or more of the hormone groups and an increase in their precursors, which themselves may have biological activity. The most common inborn enzyme defect, called classical congenital adrenal hyperplasia (CAH; autosomal recessive), affects 21-hydroxylase and impairs the synthesis of cortisol and aldosterone but not androgens. The lack of cortisol feedback to the anterior pituitary results in increased ACTH production, which stimulates adrenal production of androgens and growth (hyperplasia) of the gland. The condition usually presents in early childhood as a salt-losing condition with hyponatremia and hyperkalemia due to the lack of aldosterone, ambiguous genitalia in girls due to the androgen excess, or hypoglycemia due to cortisol deficiency. It is diagnosed by demonstrating high plasma levels of 17-hydroxyprogesterone (17-OHP) which is proximal to the enzyme block. High levels of 17-OHP are seen in the first few days of life in normal infants and repeat measurements may be needed for confirmation.

TABLE 11.19 Some inborn errors of the steroid hormone biosynthetic pathway

Enzyme deficiency	OMIM number
Congenital adrenal hyperplasia due to 21-hydroxylase deficiency	#201910
Congenital adrenal hyperplasia due to 11β-hydroxylase deficiency	#202010
Cytochrome P450 oxidoreductase deficiency	#201750
17β-hydroxysteroid dehydrogenase III deficiency	#264300

Other enzyme defects are much less common. If CAH is suspected and 17-OHP is not raised, a nonclassical form of the condition may be present and measurement of the urine steroid profile by a separation technique such as gas chromatography is required. By looking at the relative concentrations of different steroids and their precursors, the presence of a particular enzyme block can be deduced.

Analytical practice point 11.11

17-hydroxyprogesterone is often high in neonates, especially if unwell.

Adrenal medulla

There are a number of conditions that arise as a result of disease of the adrenal medulla, such as pheochromocytoma. Since these conditions often have hypertension as a presenting feature, they are covered in detail under hypertension in Chapter 8.

11.5 GONADAL ENDOCRINE FUNCTION

Hypothalamus–pituitary–gonad axis

The HPG axis is shown in **Figure 11.38** and reference ranges for its analytes in **Table 11.20**. Endocrine control of its target tissues—testes in males and ovaries in females—is mediated via the gonadotropins LH and FSH. The axis is controlled through the release of GnRH, a decapeptide synthesized from a 92-amino-acid precursor, from the hypothalamus. Release of GnRH by the hypothalamus is pulsatile, so that the number of GnRH receptors on the pituitary gonadotrophs is up-regulated continuously. The major action of GnRH is the stimulation of the synthesis and release of both LH and FSH by the gonadotropic cells of the anterior pituitary. Differential rates of synthesis and release of the two gonadotropins from these cells is made possible by differences in sensitivity to feedback control by a number of effectors including steroid and peptide hormones.

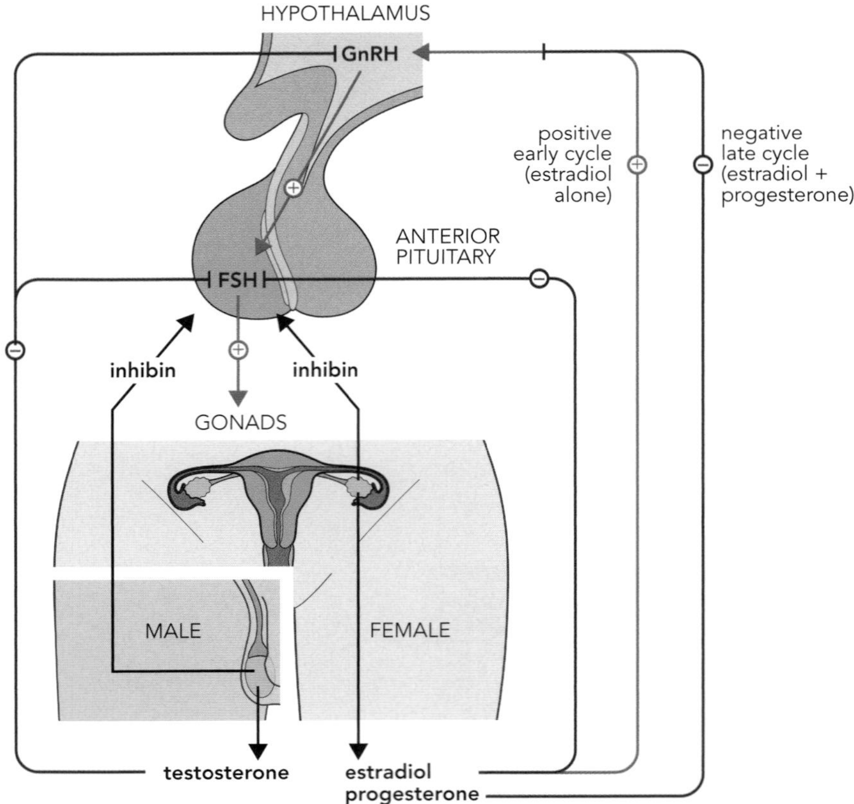

Figure 11.38 Control of the hypothalamus–pituitary–gonad axis. The synthesis and secretion of sex hormones—testosterone by males and estradiol in females—are controlled by follicle-stimulating hormone (FSH) and luteinizing hormone (LH) secreted by the anterior pituitary. Release of FSH is stimulated by gonadotropin-releasing hormone (GnRH) from the hypothalamus. The end-product sex hormones inhibit release of both FSH from the pituitary (short loop) and GnRH by the hypothalamus (long loop). Inhibins, synthesized by the testes and ovaries, also inhibit synthesis and secretion of FSH. Early in the menstrual cycle, estradiol (in the absence of progesterone) stimulates the release of GnRH by the hypothalamus, but later in the cycle, when the progesterone concentration rises, the overall effect of the two hormones is to inhibit GnRH secretion.

TABLE 11.20 Reference ranges for analytes of the hypothalamus–pituitary–gonad axis

Analyte	Reference Range*		
	Extra information	SI units	Conventional units
Estradiol	Male	36–147 nmol/L	10–40 pg/mL
	Female; pre-menopausal (values vary widely during cycle)	55–1300 pmol/L	15–350 pg/mL
	Female; postmenopausal	<36 pmol/L	<10 pg/mL
Follicle-stimulating hormone (FSH)	Male	1.0–18.0 IU/L	1.0–18.0 mIU/mL
	Female; follicular	3.9–8.8 IU/L	3.9–8.8 mIU/mL
	Female; mid-cycle	4.5–22.5 IU/L	4.5–22.5 mIU/mL
	Female; luteal	1.8–5.1 IU/L	1.8–5.1 mIU/mL
	Female; postmenopausal	16.7–113.6 IU/L	16.7–113.6 mIU/mL
17α-Hydroxyprogesterone (17-OHP)	Male	1.5–7.5 nmol/L	50–250 ng/dL
	Female; follicular	0.6–3.0 nmol/L	20–100 ng/dL
	Female; luteal	3.0–15.5 nmol/L	100–511 ng/dL
	Female; postmenopausal	<2.1 nmol/L	<70 ng/dL
Luteinizing hormone (LH)	Male	1.8–8.6 IU/L	1.8–8.6 mIU/mL
	Female; follicular	2.1–10.9 IU/L	2.1–10.9 mIU/mL
	Female; mid-cycle	20–100 IU/L	20–100 mIU/mL
	Female; luteal	1.2–12.9 IU/L	1.2–12.9 mIU/mL
	Female; postmenopausal	10–60 IU/L	10–60 mIU/mL
Progesterone	Male	0.6–4.5 nmol/L	0.20–1.40 ng/mL
	Female; follicular	0.6–4.8 nmol/L	0.20–1.50 ng/mL
	Female; mid-cycle	2.5–9.5 nmol/L	0.80–3.00 ng/mL
	Female; luteal	5.4–85.9 nmol/L	1.70–27.00 ng/mL
	Female; postmenopausal	<0.5–2.5 nmol/L	<0.15–0.80 ng/mL
Testosterone	Male	10.4–41.6 nmol/L	300–1200 ng/dL
	Female	0.5–2.4 nmol/L	14–69 ng/dL

*Values are for adults and for serum samples unless otherwise stated.

The feedback loops shown in Figure 11.38 are found primarily in adults; there is little activity in children since gonadotropin and sex steroid hormone concentrations are very low. Major changes occur in puberty when the levels of both gonadotropins and sex steroid hormones start to increase and trigger sexual development. The axis may also be perturbed after puberty when intense physical training or anorexia nervosa lead to decreased secretion of GnRH.

Hormones controlling the hypothalamus–pituitary–gonad axis

Luteinizing hormone and follicle-stimulating hormone

As mentioned earlier, the two gonadotropins are dimeric and bind to their respective receptors on target cells through their common α subunits, specificity for the receptor being manifest via the β subunit. The receptors are members of the G-protein-linked seven-transmembrane family and signal into the cell by stimulating the synthesis of cAMP. LH and FSH are involved in the control of sexual differentiation, steroid hormone synthesis, and gametogenesis and have quite different roles in males and females. In males, for example, LH and FSH induce spermatogenesis; FSH is involved in sperm maturation in Sertoli cells and seminiferous tubules and LH stimulates androgen production in Leydig cells in testes. In females, LH stimulates ovarian theca cells to synthesize androgen and other steroid hormone precursors, which are transported to granulosa cells where FSH stimulates the final steps of estrogen synthesis. The secretory pattern of the gonadotropins during menstruation is described later.

Inhibins

Inhibins are heterodimeric proteins consisting of α (20 kDa) and smaller β (14 kDa) subunits linked via disulfide bonds. Two inhibins, A and B, have been identified which have common α but different β subunits (**Figure 11.39**). In females, inhibins A and B are produced by the ovary but only inhibin A is produced by the placenta. In males, inhibin B is synthesized in the testes.

Inhibins exert negative feedback to the pituitary where they inhibit production and secretion of FSH. Increases in the concentration of inhibin A in the maternal circulation are detectable within 12 days of conception and readily apparent at 5 weeks of gestation, peaking at 8–10 weeks.

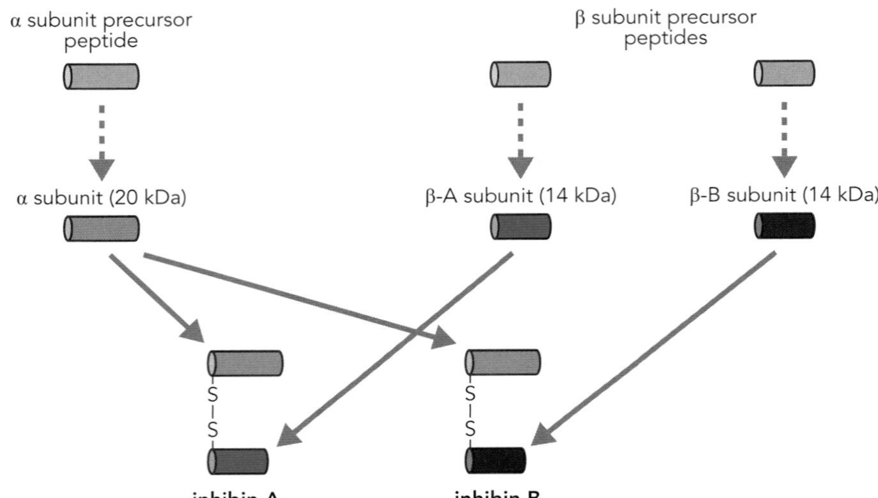

Figure 11.39 Structure of inhibins A and B.
Inhibins A and B are heterodimeric polypeptide hormones of the transforming growth factor-β (TGF-β) superfamily. They have common α subunits linked to either a β-A (inhibin A) or β-B subunit (inhibin B) via disulfide bridges. Subunits are differentially processed post-translationally, giving rise to various molecular forms. Free monomeric α subunits are found in the circulation but are inactive.

Estrogens—progesterone

The structure of progesterone is shown in **Figure 11.40**. It is synthesized from low-density lipoprotein (LDL)-derived cholesterol by the corpus luteum in response to FSH during the menstrual cycle (described below). Its plasma concentration in the luteal phase is 3.2–6.4 nmol/L (1–2 ng/mL) and this rises to a plateau value of 32–111 nmol/L (10–35 ng/mL) over the 7 days subsequent to the LH surge. In the absence of fertilization, the raised concentrations of both progesterone and inhibin feed back to the hypothalamus to inhibit GnRH secretion, and the consequent fall in FSH secretion by the pituitary causes degeneration of the corpus luteum and menstruation. If, however, conception occurs, the trophoblasts of the placenta secrete human chorionic gonadotropin (hCG) which helps to maintain the integrity of the corpus luteum so that it continues to secrete progesterone. An initial 6–8-fold rise in the concentration of progesterone in maternal blood at week 7 of pregnancy is followed by a further gradual increase from week 10 (from the last menstrual flow) through to term, when the concentration is 318–954 nmol/L (100–300 ng/mL).

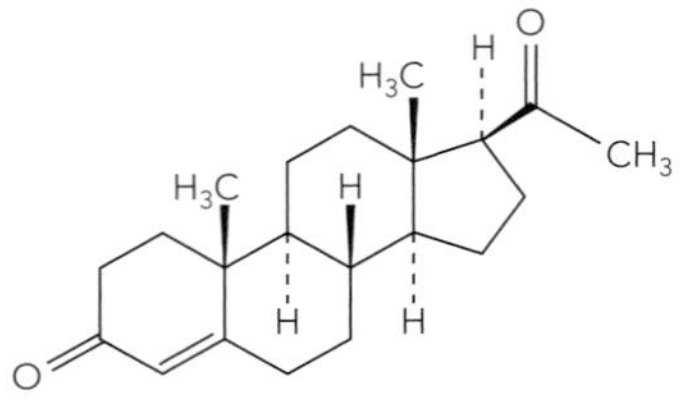

progesterone

Figure 11.40 Structure of progesterone.
Progesterone is a 21-carbon steroid hormone synthesized from LDL-derived cholesterol by the corpus luteum during menstruation under the control of FSH. During the early stages of pregnancy it is synthesized by the corpus luteum under the control of human chorionic gonadotropin synthesized by the placental trophoblasts.

Endocrine control of the testes

The testes, a pair of organs lying in the scrotal sac, control sexuality and fertility in males through their two major roles:

1. Synthesis (see Figure 11.29) and secretion of male sex hormones, mainly testosterone by the interstitial Leydig cells. The synthesis of testosterone in fetal Leydig cells is critical for the development of the male phenotype.
2. Production of spermatozoa by the Sertoli cells of the seminiferous tubules.

Leydig cells are able to synthesize testosterone from cholesterol derived either from LDL or *de novo* synthesis. The rate-limiting step of this pathway is the cleavage of the aliphatic side chain of cholesterol, catalyzed by the enzyme desmolase. Testosterone is released into the blood and transported bound to two plasma proteins, sex hormone-binding globulin (SHBG; testosterone-binding globulin; 44%) and albumin (54%). Approximately 2–3% of testosterone is present in blood as free, unbound hormone. The SHBG-bound fraction is tightly bound and acts as a storage form of the hormone, while the free and albumin-bound testosterone represents the bioavailable fraction. Besides having general anabolic actions, testosterone acts systemically to produce male secondary sexual characteristics, such as enlargement of the larynx at puberty and facial hair, and to maintain libido throughout life. Locally, it stimulates spermatogenesis in Sertoli cells. FSH stimulates the maturation of spermatozoa in these cells. Inhibin B is also synthesized by the Sertoli cells and regulates gametogenesis by negative feedback on FSH production in the pituitary. It may also have some local paracrine actions in the testes.

Endocrine control of the ovaries

The ovaries, a pair of organs lying in the peritoneal cavity, are responsible for the synthesis of eggs (oogenesis) early in fetal development and the release of a single ovum (secondary oocyte) from the ovaries (ovulation) into the Fallopian tube midway through the monthly menstrual cycle, which occurs during the 40 or so years between puberty (age 12–13 years) and the menopause. Oocyte development is arrested at the prophase of meiosis (primary oocytes) in the fetus and these oocytes, surrounded by a layer of epithelial cells, are stored as primordial follicles in the broad outer cortex of the ovary. Approximately 400,000 primordial follicles are present in the human ovaries at birth; the consequences of primary oocytes being present from birth are discussed later. At puberty, FSH induces the development of these primordial follicles into primary follicles; subsequently, each month, one of these primary follicles will progress to become a Graafian follicle, ready for ovulation, while the other primary and secondary follicles regress. The synthesis and secretion of estradiol by the developing follicle is induced by FSH and involves stimulation of the aromatase enzyme which converts androgens to estrogens (see Figure 11.29). Estradiol exerts negative feedback on further FSH release from the anterior pituitary such that only the most developed follicles ever become dominant. Furthermore, under the influence of estrogens and LH, the Graafian follicle releases a secondary oocyte into the Fallopian tube, where fertilization takes place and the fertilized egg is directed into the uterus for implantation. After ovulation, the granulosa cells that surrounded the ovum in the follicle remain in the ovary and become the corpus luteum (yellow body), which produces progesterone, a hormone responsible for the maintenance of the thick, secretory, highly vascularized endometrial lining of the uterus required for implantation of the fetus should fertilization occur. The two major functions of the ovary are thus the synthesis and secretion of steroid hormones and the monthly release of a mature ovum. If the ovum remains unfertilized, it is lost with the next menstrual bleed about 14 days after ovulation. Fertilization, on the other hand, results in pregnancy and retention of the fertilized ovum (zygote).

The female reproductive cycle (menstrual cycle)

The female reproductive or menstrual cycle, which is under strict endocrine control, involves coordinated responses in both the ovaries and uterus to ensure, respectively, maturation of the oocyte and preparation of the endometrium for the reception of a fertilized ovum. In the absence of fertilization, the surface of the endometrium sloughs off and is lost with the unfertilized ovum. Two distinct but interdependent cycles (ovarian and uterine) contribute to the menstrual cycle.

The menstrual cycle is controlled via positive and negative feedback signals of the HPG axis (see Figure 11.38). The cyclical release of GnRH from the hypothalamus induces cyclical release of LH and FSH from the pituitary, which in turn induce the cyclical production of estrogens and progesterone in the ovary (ovarian cycle). Cyclical changes in the release of steroid hormones lead to cyclical activity in the uterus (uterine cycle) and monthly bleeding in the absence of fertilization. The hormonal changes occurring during the menstrual cycle are shown in **Figure 11.41**. The cycle begins on the first day of bleeding (day 1) and has a median length of 28 days but may vary between 21 and 40 days.

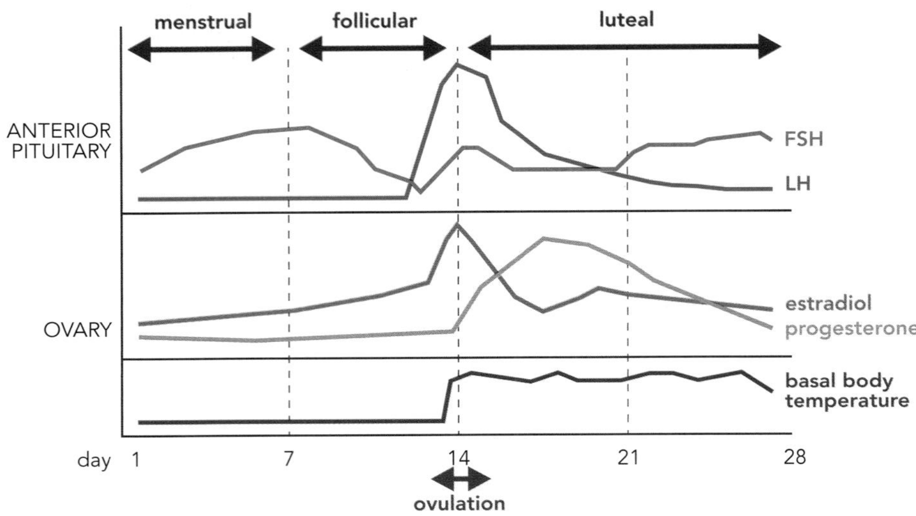

Figure 11.41 Endocrine changes during the menstrual cycle. The female reproductive cycle is composed of the ovarian and uterine cycles; hormonal changes in these cycles drive the menstrual cycle, resulting in menstruation approximately every 28 days. The ovarian cycle includes the follicular phase and the luteal phase, which are separated by ovulation. Ovulation is accompanied by a slight increase in basal body temperature. The uterine cycle begins on day 1 of menstruation.

The ovarian cycle

The ovarian cycle consists of two phases, a follicular phase and a luteal phase, separated by ovulation midway through the menstrual cycle.

Follicular phase

Up to twenty secondary follicles begin to mature during each ovarian cycle. Initiation of follicular growth occurs at about day 25 of the previous cycle in response to a rise in secretion of FSH by the anterior pituitary. This rise in FSH level continues into the follicular phase when, as its name implies, it stimulates the growth and development of the group of follicles. A further surge in FSH concentration occurs at ovulation. The result of the actions of FSH in the follicular phase is the selection of a single pre-ovulatory follicle from the original group, and only this follicle reaches maturity and undergoes ovulation; how this selection takes place is unknown at present. However, the dominant follicle suppresses the growth of the nonselected follicles through local actions and these probably act as an independent endocrine gland. As the selected follicle matures, the cells surrounding this follicle synthesize and secrete estrogen and inhibin. The level of inhibin B is maximal approximately midway through the follicular phase. Later, in the luteal phase of the cycle, as the corpus luteum, they also produce progesterone. Further maturation of the follicle induces changes in the oocyte, follicular cells, and the surrounding stromal tissue. For example, during follicular growth, the layer of granulosa cells surrounding the follicle produces estrogens (estradiol and estrone) in response to FSH, and plasma estrogen levels rise from about 7–8 days prior to ovulation, reaching a peak on about day 13 of the cycle. This increase in estrogen concentration, in the absence of progesterone, has a positive-feedback effect on GnRH secretion by the hypothalamus and thereby indirectly increases gonadotropin secretion by pituitary gonadotrophs. At this time, the effect is specifically on those gonadotrophs secreting LH and it initiates the LH surge that marks the end of the follicular phase. There is a concomitant increase in FSH secretion but this is of a much lower magnitude than for LH.

Plasma androgens also increase slightly during the follicular phase, with a peak concentration on the day of the LH surge. However, progesterone levels are not increased until immediately prior to the LH surge, from which time they rise markedly and reach a peak late in the luteal phase.

A number of morphological changes take place in the follicle during the follicular phase of the cycle. The growing follicle matures from a single layer of cells (theca interna) encircling the oocyte to a multilayered state and acquires a cavity filled with liquid (liquor folliculi) secreted by the granulosa cells. Also, the primary oocyte moves to an eccentric position in the follicle. The LH surge at about day 13 of the menstrual cycle causes the developing oocyte to complete its first meiotic division (meiosis I), which had stopped at prophase during fetal development, and become a secondary oocyte. Meiosis II now starts and halts in metaphase. These cells do not complete meiosis to become haploid cells until just prior to fertilization. Eventually, the follicle achieves a maximum size (Graafian follicle) and bulges on the ovarian surface.

In a regular 28 day cycle, about fourteen days after the start of the menstrual cycle and about seven days into the follicular phase, the Graafian follicle ruptures and release its secondary oocyte, still surrounded by its zona pellucida and corona radiata, into the peritoneal cavity, from where it travels to the Fallopian tube. This process is known as ovulation and is stimulated by the LH surge mentioned above. At this stage, all other growing follicles cease to grow and begin to involute in a process known as atresia.

Luteal phase

The luteal phase embraces the following fourteen days of the ovarian cycle, during which time the remaining tissue of the follicle, having shed the secondary oocyte and its surrounding cells, rearranges itself, acquires a yellowish tinge, and becomes the corpus luteum. Under the influence of LH, the corpus luteum secretes progesterone, estrogens, relaxin, and inhibin A. Plasma progesterone increases from a concentration of 3.2–6.4 nmol/L (1–2 ng/mL) to a maximum of 32–111 nmol/L (10–35 ng/mL) 6–8 days after the LH surge of the follicular phase. This rise in progesterone inhibits the development of new follicles and, in concert with inhibin A, inhibits contraction of the uterus thereby helping to prepare the endometrium for possible pregnancy. An intermediate peak in inhibin A is seen at the time of the luteal surge and is maximal in the mid-luteal phase. If fertilization does not occur and there is no source of hCG derived from the chorion of the conceptus, the corpus luteum is programmed to die by apoptosis after about two weeks.

The feedback control of the HPG axis is evident at this stage as the raised concentrations of both progesterone and estrogen feed back to the hypothalamus to inhibit the release of GnRH. This in turn inhibits further LH secretion from the pituitary and eventually the synthesis of progesterone in the corpus luteum. As the corpus luteum degenerates, plasma progesterone falls and the endometrium starts to break down. The decreased plasma concentrations of progesterone and estrogen toward the end of the luteal phase release the inhibition of GnRH synthesis and secretion and thus plasma GnRH begins to rise again, promoting FSH secretion by the pituitary and consequently stimulating follicular growth to start a new ovarian cycle. The inhibition of uterine contraction is also removed and menstruation begins, marked by detachment and shedding of the endometrial surface and loss of blood. However, if fertilization occurs at the ovulatory stage, the secondary oocyte starts to divide and the life span of the corpus luteum is prolonged, supported by hCG from the chorion of the conceptus from about eight days post-fertilization.

Uterine cycle

The uterine cycle begins on day 1 of the menstrual cycle with the loss of around 50–150 mL of blood, tissue fluid, and epithelial cells sloughed off from the endometrium; the loss of blood is usually 30–50 mL. The fall in progesterone secretion by the corpus luteum as it degenerates toward the end of the luteal phase of the ovarian cycle is accompanied by the secretion of prostaglandins, which cause constriction of the uterine spiral arterioles, leading to the death of endometrial epithelial cells through oxygen starvation.

Two phases of the uterine cycle can be distinguished—the proliferative and secretory phases—relating, respectively, to the follicular and luteal phases of the ovarian cycle. The pre-ovulatory, proliferative phase defines the period of proliferation of the endometrium and is driven by estrogens. Estrogens secreted by the growing ovarian follicles stimulate both the repair of the endometrium and the division of cells from the basal layer to produce a new stratum functionalis. During this time, estrogens from the theca interna around the growing follicle cause the endometrium to double in thickness to between 4 and 10 mm and proliferation of the simple tubular glands. Vascularization of the new stratum functionalis arises through coiling and stretching of arterioles into this layer from the stratum basalis.

In the secretory (post-ovulatory) phase of the uterine cycle, progesterone and estrogens from the corpus luteum promote the new endometrial glands to assume an irregular corkscrew shape and, one week after ovulation, to secrete fluid rich in glycogen to support the survival of the fertilized ovum in the uterus. The endometrium approaches its maximum thickness at this time. If fertilization does not occur, plasma progesterone levels fall as described earlier, the endometrial blood vessels contract, and the stratum functionalis is shed, together with fresh blood. This marks the beginning of the next cycle.

Clinical chemistry aspects of the HPG axis

There are a number of clinical situations where it is necessary to assess the endocrine aspects of gonadal function. In both sexes these include delayed or early puberty, loss of libido, infertility, the investigation of suspected hypopituitarism, and, in osteoporosis, where loss of gonadal steroids is an important cause of bone loss. The symptoms specific to females are menstrual disturbance (amenorrhea, oligomenorrhea) and hirsutism, whilst for males erectile dysfunction and gynecomastia may have an endocrine cause.

Measurement of plasma levels of male (testosterone) and female (estradiol and progesterone) gonadal steroids together with the gonadotropins LH and FSH are the usual first-line investigations. The interpretation of these tests is affected in females by the stage of the menstrual cycle, and by the time of day in the case of males, so it is important that this information is provided to the laboratory. Free sex hormone concentrations can be measured by equilibrium dialysis in specialized centers but this is rarely used in routine clinical practice.

Further investigations include the measurement of SHBG to allow the calculation of the free androgen index. The free androgen index is a simple ratio of testosterone to SHBG and gives an estimate of the unbound (and presumed physiologically active) testosterone. In some conditions SHBG, which is synthesized in the liver, may be increased or decreased making the total testosterone misleadingly high or low. Other androgens which may be measured routinely in plasma are androstenedione and DHEA-S. The former is the immediate precursor of testosterone and hence an increased level is confirmatory evidence of

Analytical practice point 11.12

Interpretation of gonadotropins, estradiol, and progesterone in pre-menopausal women requires knowledge of the phase of the menstrual cycle.

androgen excess. DHEA-S is predominantly of adrenal origin and is therefore used as a marker of increased adrenal androgen production in women with elevated testosterone.

Male hypogonadism

The two functions of the testes are spermatogenesis and testosterone production. Hypogonadism therefore results in infertility and symptoms of androgen deficiency such as loss of libido, erectile dysfunction, reduced energy, mental depression, and muscle wasting in association with low plasma testosterone. As testosterone is highest in the morning, this is the optimum time for sampling. A low plasma testosterone should always be confirmed on a sample taken at 09.00 am. Where the testosterone is borderline, the measurement of SHBG and calculation of the free androgen index may be helpful. It is accepted by most authorities that confirmed testosterone <8 nmol/L (<230 ng/dL) is an indication for testosterone replacement therapy, whilst some symptomatic patients may benefit from treatment between 8 and 12 nmol/L (230–345 ng/dL).

Testicular failure may be due to disease of the testes themselves or due to loss of gonadotropin drive from the anterior pituitary. In the latter case other endocrine deficiencies may be present, such as hypothyroidism or hypoadrenalism. Primary and secondary failure may be distinguished by the levels of gonadotropins. In primary hypogonadism the lack of negative feedback results in elevated LH and FSH, whilst in secondary hypogonadism the gonadotropins are low or within the normal range and considered to be inappropriately low.

Clinical practice point 11.10

The level of plasma testosterone at which hypogonadism is diagnosed and treated is controversial.

Erectile dysfunction

Most cases are due not to testosterone deficiency but are due to psychological, vascular, or neurological disorders, often in combination. Nevertheless, it is usual to exclude hypogonadism as it is a treatable cause.

Gynecomastia

Breast development in males is affected by the same hormones as in females, in particular estrogens. However, a complex interplay of hormones is required, including gonadotropins, prolactin, and growth hormone. Physiological breast enlargement occurs at birth and again around puberty due to the rises in hormone secretion at these times. When occurring in later life, gynecomastia may be caused by an increase in estrogen levels, either due to direct secretion from a testicular tumor or, more commonly, from increased conversion from testosterone by aromatase in the peripheral tissues. Aromatase activity is high in adipose tissue, hence higher estrogen levels occur in obesity. Paradoxically, high doses of androgens, which may be used as anabolic steroids by bodybuilders, may be converted to estradiol by aromatase giving rise to breast enlargement. This is in addition to the negative feedback on gonadotropin secretion, which leads to testicular atrophy. Some drugs may cause gynecomastia by displacing estradiol from SHBG, which binds estradiol with lower affinity than testosterone.

Analytical practice point 11.13

Routine assays for estradiol are not optimized to measure the low levels expected in men and subtle elevations may be missed. More specialized assays may be required from a reference center.

Amenorrhea

Failure to commence menstruation is called primary amenorrhea and has a number of possible causes, such as anatomical malformation or chromosomal disorders, in addition to endocrine disease. Complete cessation of menstrual periods in a previously menstruating female is termed secondary amenorrhea, and is usually due to ovarian failure. The reader should note that primary and secondary when used to describe amenorrhea should not be confused with the same terms when they refer to disease in the pituitary or the target endocrine gland. It is possible, for example, to have secondary amenorrhea due to primary ovarian failure, and vice versa. Age-related ovarian failure (menopause) is inevitable, although the age at which it occurs is variable. For most women

CASE 11.5

A 50-year-old man complains of tiredness, erectile dysfunction, and visual disturbance. The initial test requested by his physician was testosterone, which was low (09.00 am sample). Further tests were added by the laboratory's clinical scientist. The full results are shown below.

	SI units	Reference range	Conventional units	Reference range
Plasma				
Testosterone	2.1 nmol/L	10.4–41.6	60 ng/dL	300–1200
Luteinizing hormone (LH)	<0.1 IU/L	1.8–8.6	<0.1 mIU/mL	1.8–8.6
Follicle-stimulating hormone (FSH)	<0.1 IU/L	1.0–18.0	<0.1 mIU/mL	1.0–18.0
Prolactin	9500 pmol/L	174–617	218 ng/mL	4.0–14.2
TSH (thyrotropin)	1.2 mIU/L	0.4–4.2	1.2 mIU/L	0.4–4.2
Free T_4	6.3 pmol/L	10–31	0.5 ng/dL	0.8–2.4
Cortisol	235 nmol/L	127–568 (am)	8.5 μg/dL	4.6–20.6 (am)

- Why did the laboratory add on the further tests?
- What do they show?
- What further investigations are needed?

The initial finding of very low testosterone may be due to either testicular or pituitary disease. LH and FSH are raised in the former, which is more common, but are inappropriately low in the latter; hence these tests were added first. The finding of very low LH and FSH strongly suggests pituitary disease and further tests of the thyroid and adrenal axes were added, as well as prolactin. Low free T_3 and free T_4 with inappropriately low TSH indicate secondary hypothyroidism. Cortisol of this level does not exclude adrenal insufficiency due to low ACTH. (Basal ACTH is not usually helpful and requires a specially collected sample on ice, so was not added.) The very high prolactin is consistent with a prolactin-secreting tumor (prolactinoma). In men these are usually large tumors and may cause headaches and visual disturbance. The most important investigation is pituitary magnetic resonance imaging (MRI) to confirm the presence and size of the tumor. A tetracosactrin (Synacthen) test is also required to determine if glucocorticoid replacement is needed.

the menopause occurs in the fifth decade. When it occurs below the age of 45 years it is considered premature; in this case primary and secondary amenorrhea may be defined as indicated in **Table 11.21**. Biochemical evidence for the menopause is demonstration of elevated levels of FSH (for example to >30 IU/L) in the presence of amenorrhoea (**Figure 11.42**). Similar levels of FSH may be seen during the mid-cycle surge at the time of ovulation, but usually LH

TABLE 11.21 Presentation of primary and secondary amenorrhea

Cause	Presentation
Primary	Never menstruated; for example congenital adrenal hyperplasia
Secondary	Cessation of menstruation; for example ovarian failure

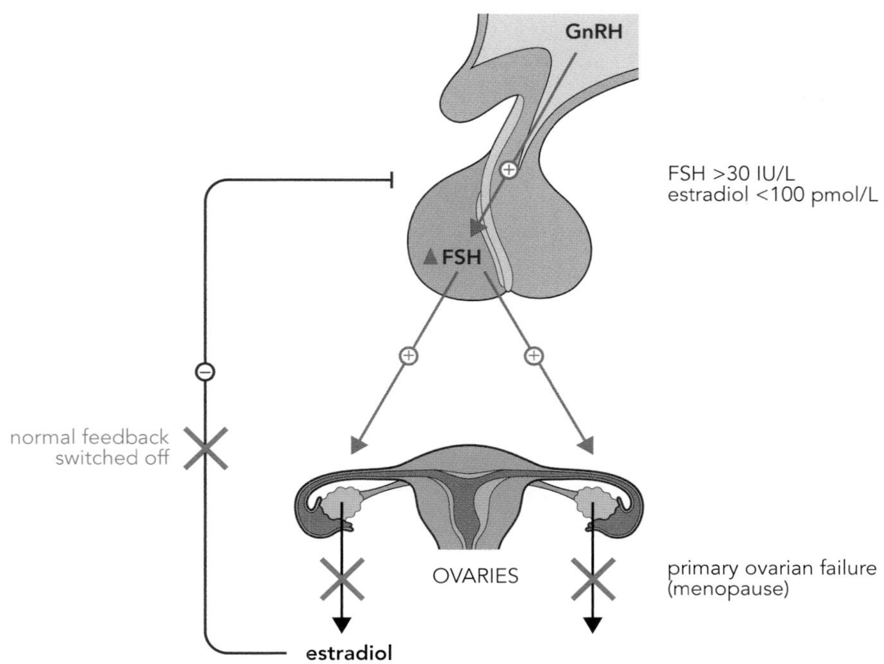

Figure 11.42 Biochemical changes in the menopause. The menopause is essentially primary ovarian failure; that is, a condition in which the ovaries can no longer produce sufficient estradiol (<100 pmol/L) to exert the normal feedback control over follicle-stimulating hormone (FSH) synthesis in the hypothalamus. As a consequence, circulating concentrations of FSH are raised (>30 IU/L), usually well above those normally seen in non-menopausal women. The red crosses show that the primary defect is the lack of estradiol production and the switching off of the normal feedback inhibition.

is higher than FSH in this situation and so it is helpful to measure both gonadotropins. It is important to emphasize that the menopause is not a single event, but a process which may take place over months, even years. During this variable transition phase, periods may become more irregular and gonadotropins may vary from low to high. Thus it is not possible to diagnose menopause by FSH alone, even if it is very high (for example 100 IU/L), and contraception should not be discontinued until amenorrhea has been present for at least 12 months in women over 50 years old, and for 24 months in younger women. The investigation of amenorrhea is summarized in **Table 11.22**. Ovarian failure due to pituitary disease is uncommon but would be suggested by the presence of low FSH and LH in association with decreased estradiol. A similar pattern is seen with hypothalamic amenorrhea, which occurs in response to emotional stress or severe weight loss. A third cause for this combination of results is the use of the combined oral contraceptive pill. In this situation a high dose of estrogens is used to deliberately suppress LH and FSH by negative feedback and hence prevent ovulation. The estrogens used are usually synthetic (for example ethinyloestadiol) and hence not detected in routine assays for estradiol. Low gonadotropins in combination with elevated estradiol are seen in pregnancy—a cause for amenorrhea in a pre-menopausal woman which should never be overlooked—which is confirmed by the finding of elevated hCG in urine or plasma.

TABLE 11.22 Initial investigation of amenorrhea

Investigation	Actions taken
Is the patient pregnant?	Urinary hCG (qualitative)
	Serum hCG (quantitative)
Has the patient reached the menopause?	Measure FSH (and estradiol)

CASE 11.6

A 28-year-old woman attends her gynecologist because of infertility. On questioning, she has not had a period for a year following her stopping the combined oral contraceptive and she has been having hot flushes.

	SI units	Reference range	Conventional units	Reference range
Plasma				
Luteinizing hormone (LH)	85 IU/L	Follicular: 2.1–10.9 Mid-cycle: 20–100 Luteal: 1.2–12.9 Postmenopausal: 10–60	85 mIU/mL	Follicular: 2.1–10.9 Mid-cycle: 20–100 Luteal: 1.2–12.9 Postmenopausal: 10–60
Follicle-stimulating hormone (FSH)	87 IU/L	Follicular: 3.9–8.8 Mid-cycle: 4.5–22.5 Luteal: 1.8–5.1 Postmenopausal: 16.7–113.6	87 mIU/mL	Follicular: 3.9–8.8 Mid-cycle: 4.5–22.5 Luteal: 1.8–5.1 Postmenopausal: 16.7–113.6
Estradiol	26 pmol/L	Pre-menopausal: 55–1300 Postmenopausal: <36	7 pg/mL	Pre-menopausal: 15–350 Postmenopausal: <10

- What do the results indicate?
- Can she ever conceive naturally?
- What other conditions should be considered?

Very high gonadotropins and low estradiol indicate primary ovarian failure. This is termed the menopause in older women, but is called premature ovarian failure (POF) if it occurs under the age of 40. Conception occurring without medical assistance is unlikely but still possible, as spontaneous ovulatory cycles may occur very infrequently. This is also true for the first few years after the menopause. High FSH and LH alone do not entirely rule out the possibility of pregnancy. POF is often autoimmune in origin and other endocrine glands may be affected, particularly thyroid and adrenals. The low estradiol increases the risk of osteoporosis and bone density measurement is usually advised.

Hirsutism

The unwanted appearance of male-distribution facial hair in a female is termed hirsutism. There are personal, cultural, and social thresholds for the degree of hirsutism that is acceptable before medical advice is sought. In principle, hirsutism is caused by the increased exposure of the hair follicle to androgens or increased sensitivity to their effects. Thus plasma testosterone concentration may not be greatly raised, and even the free androgen index may not be different to that seen in the non-hirsute female population. The majority of hirsutism which has an identifiable cause is due to polycystic ovary syndrome (PCOS), which is described in more detail below. Overinvestigation of hirsutism is discouraged, but it is important not to overlook a potentially more serious or treatable cause. In rare cases, hirsutism may be due to androgen-secreting tumors of either the ovary or adrenal cortex. Often the hirsutism is severe, of rapid onset, and associated with amenorrhea. Other evidence of virilization, such as clitoromegaly, may be present. Plasma testosterone is usually in excess of 5–6 nmol/L

(144–173 ng/dL) and may be into the male range. A second uncommon cause of hirsutism is late-onset congenital adrenal hyperplasia, which is characterized by partial deficiency of the enzyme 21-hydroxlase in the adrenal cortex and increased adrenal androgen secretion. This is diagnosed by increased secretion of 17-hydroxyprogesterone following tetracosactrin (Synacthen) stimulation. Some women may have a subnormal cortisol response, requiring glucocorticoid replacement, but this does not always improve the hirsutism.

Clinical practice point 11.11

Hirsutism is usually caused by hair follicles being overly sensitive to androgens, rather than androgen excess.

Infertility

Inability to conceive has many possible causes which may affect either the male or female partner. Failure of ovulation is one of the most common causes and initial investigation of infertility usually consists of endocrine investigations. Measurement of gonadotropins to exclude premature menopause is an important initial investigation. Fertility declines with age due to a decrease in ovarian reserve, even before the menopause, and this is reflected in gradually rising FSH with age. Thus, cycles may continue for years after ovulatory cycles have effectively finished. If menstrual cycles are irregular, they are usually anovulatory. The opposite is also true; almost all regular cycles result in ovulation. Nevertheless, to confirm ovulation and adequate function of the corpus luteum, it is usual to measure plasma progesterone in the mid-luteal phase of the cycle. In regular 28-day cycles, this would be on day 21. A progesterone level >30 nmol/L (>9.4 ng/mL) is regarded as indicating ovulation, whilst progesterone <10 nmol/L (<3.1 ng/mL) indicates an anovulatory cycle.

Clinical practice point 11.12

High testosterone in a female raises the possibility of an adrenal or ovarian tumor, especially if the history is short and there is other evidence of virilization. Pregnancy may also raise testosterone.

A rare cause of female infertility with an endocrine basis is androgen insensitivity syndrome (formerly testicular feminization syndrome). The patient is phenotypically female but has a normal male chromosome constitution (46, XY); the defect is typically a cytogenetic rearrangement affecting the X-linked *Androgen Receptor* (*AR*) locus. The underlying defect is failure to respond to androgens due to lack of functioning receptors. Lack of androgens but not of estrogens *in utero* results in failure to develop male sexual characteristics, and the diagnosis may come to medical attention only when treatment of infertility is sought. There is no negative feedback control of testosterone production from the testes (which are undescended and remain within the abdomen), but aromatization allows estrogens to be produced. Thus high levels of both testosterone and gonadotropins are seen in an apparently female patient.

Polycystic ovary syndrome

Although it is probably the most common female endocrine disorder in the Western world, PCOS has no single defining feature. International groups have placed emphasis on different aspects of PCOS but agree that it is a combination of menstrual disturbance (oligomenorrhea or amenorrhea), hyperandrogenism (either biochemical or clinical), and the presence of multiple ovarian follicles on ultrasound. Women may present with any or all of the following: irregular periods, infertility, hirsutism, and acne. In recent years there has been an increased understanding of the role of insulin resistance, which may be a precursor of type 2 diabetes mellitus and increased cardiovascular risk. This explains the clear relationship between obesity and PCOS and the rising prevalence in the population over the last two decades. However, as with type 2 diabetes, not all patients with PCOS are obese. As there are no definite endocrine criteria for PCOS, its diagnosis is based on combining clinical and ultrasound findings with hormone results and by excluding other pathology, such as virilizing tumors. Testosterone is usually mildly elevated [up to 5 nmol/L (144 ng/dL)] and SHBG reduced, giving a raised free androgen index. Often LH is high, although this is not a reliable diagnostic criterion, and the historical use of a ratio of LH to FSH greater than 4 was based on assays which are now obsolete.

Clinical practice point 11.13

Polycystic ovary syndrome is common and linked to insulin resistance, but there is no universal definition.

Precocious puberty

The definition of precocious puberty is the development of secondary sexual characteristics at an age more than 2.5 standard deviations below the mean for the population; typically 9 years of age in males. Because of the secular trend toward earlier female puberty, the previous age for considering it premature has been revised down from 8 years to 7 (Caucasians) and 6 (African-Americans). The incidence is 5–10 times higher in girls than boys. The most common cause is excessive production of gonadotropins secondary to GnRH secretion from the hypothalamus. This is termed central, complete, or true precocious puberty and is usually idiopathic, although it can be caused by some intracranial lesions. Less commonly, excessive estrogen or androgen production may occur peripherally, independently of gonadotropins. This is termed precocious pseudopuberty. Estrogen may be secreted from ovarian or adrenal tumors, whilst the origin of excess androgens is most usually congenital adrenal hyperplasia. In a girl, excess androgens cause virilization; that is, contrasexual precocious puberty. Measurement of plasma levels of gonadotropins following GnRH stimulation may be required. In central precocious puberty, the LH response exceeds that of FSH, which is the opposite of the normal pattern seen pre-pubertally.

PREGNANCY

CHAPTER 12

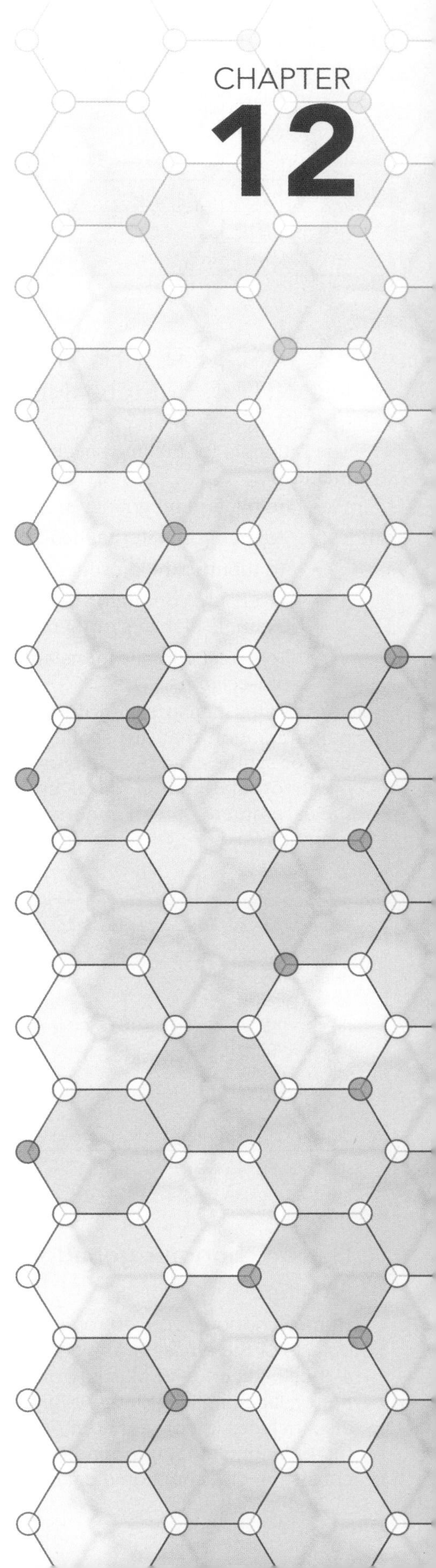

IN THIS CHAPTER

The cell that is formed by fusion of the sperm cell and ovum at fertilization is termed the zygote, and this cell divides many times over to form a ball of cells that can implant in the wall of the uterus and develop into an embryo, placenta, and embryonic membranes (together termed the conceptus). Pregnancy is confirmed clinically when a woman misses her menstrual period and the hormone human chorionic gonadotropin (hCG) is detected in her blood or urine. hCG is a glycoprotein hormone produced by the developing conceptus that serves to maintain the corpus luteum in the ovary of the pregnant woman until the placenta becomes established and can begin to produce estrogen and progesterone. Following a positive pregnancy test, the pregnancy is deemed, by convention, to have started on the first day of the woman's last menstrual period (LMP). A normal pregnancy lasts 40 weeks and is split into three trimesters, each lasting 3 months. The stages of pregnancy are outlined in **Table 12.1**.

It is important to note that ultrasonography is being used increasingly in clinical practice as an adjunct to clinical chemistry measurements. In this technique, high-frequency sound waves scan the abdomen and pelvic cavity of the mother and provide pictures of both the fetus and placenta. There are different sorts of scan that may be made throughout pregnancy to monitor the health of the fetus and, later in pregnancy, to diagnose congenital abnormalities.

Routine clinical chemical testing of pregnant mothers is performed to monitor maternal health and to exclude possible conditions that might harm the mother and/or fetus. Reference ranges for a number of analytes measured during pregnancy vary with gestational age and a few comprehensive studies have been published.

TABLE 12.1 Stages of pregnancy

Weeks since LMP	Trimester	Stages
2–13	First	Pre-embryonic development weeks 2–3; embryonic development weeks 4–10; fetal stage begins week 11
14–27	Second	Fetal development (fetus viable around week 23)
28–40	Third	Fetal maturation
41–42		Pregnancy complete (at term)

12.1 ANALYTES THAT MAY BE MEASURED TO MONITOR PREGNANCY

The rationale for making specific biochemical measurements during pregnancy includes:

- To confirm pregnancy
- To monitor maternal health
- To monitor the presence and continuing health of the fetus
- To ensure continuing health of mothers with preexisting conditions, for example diabetes or thyroid disorders
- To detect complications arising from pregnancy, for example pre-eclampsia

Analytes measured (**Table 12.2**) include molecules appearing in the maternal circulation that are synthesized by the mother as a result of pregnancy and/or synthesized by the fetus or placenta. For most pregnancies, measurements on maternal blood or urine suffice, but in some cases, particularly in the assessment of fetal abnormalities, further measurements may be made on amniotic fluid.

TABLE 12.2 Analytes that may be measured in pregnancy

Routine	Less common (more specialized)
Human chorionic gonadotropin (hCG) Alpha-fetoprotein Pregnancy-associated plasma protein A Unconjugated estriol Progesterone Inhibin A Glucose Urinary albumin/urinary total protein	Human placental lactogen Placental growth factor Soluble vascular endothelial growth factor receptor 1 (sVEGFR-1; sFlt-1)

sFlt-1, soluble fms-like tyrosine kinase 1.

Human chorionic gonadotropin

Structure

Human chorionic gonadotropin (hCG) is a glycoprotein synthesized by the trophoblasts of the placenta immediately after conception (it can be detected at the 4–8-cell stage), with plasma concentrations peaking at about week 9 of pregnancy. It belongs to a family of glycoprotein hormones of similar structure which also includes luteinizing hormone (LH), follicle-stimulating hormone (FSH), and thyroid-stimulating hormone (TSH) (see Chapter 11). Each hormone is dimeric, consisting of a common α chain of 92 amino acids and a unique β chain which

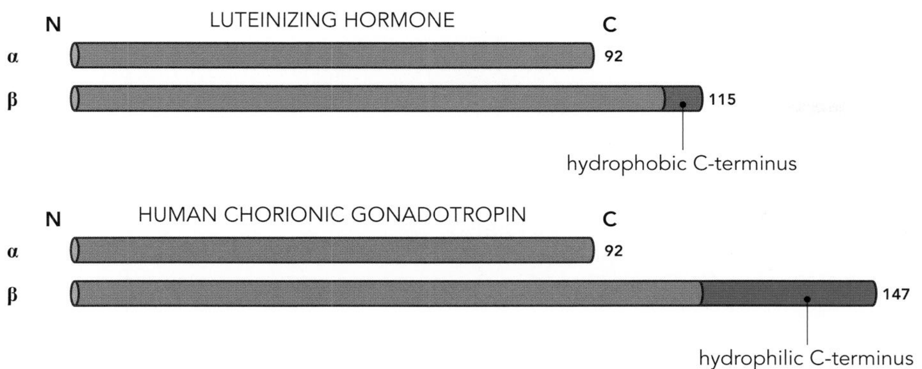

Figure 12.1 Similarity in structures of luteinizing hormone (LH) and human chorionic gonadotropin (hCG). LH and hCG are dimeric glycoprotein hormones with common α subunits (92 amino acids). The β subunits show >82% sequence homology, differing at their C-termini where LH (115 amino acids) has a short 7-amino-acid, hydrophobic sequence whereas hCG (147 amino acids) has a longer, more hydrophilic C-terminus. Glycosylation of both α and β subunits is required for biological activity and both subunits are involved in receptor binding. N and C indicate the N- and C-termini of the polypeptide chains. Numbers indicate the number of amino acids in each polypeptide chain.

defines its endocrine function. The β subunits of LH and hCG share 82% amino acid homology; major differences occur at the C-terminus where LH-β (115 amino acids) has a short hydrophobic sequence (7 amino acids) while hCG-β (147 amino acids) has a longer, more hydrophilic extension (**Figure 12.1**).

Function

The maintenance of the fetus in the first trimester of pregnancy requires a functioning endometrium, the viability of which is in turn maintained by the action of progesterone synthesized by the corpus luteum of the ovary. hCG is released by the trophoblast cells of the placenta within days of conception and stimulates the production of progesterone by the corpus luteum. The gonadotropin enters both the fetal and maternal circulations and, in the latter, its concentration doubles every 2–3 days, reaching a peak at between 8 and 13 weeks of gestation. Thereafter, its concentration falls rapidly to about 10% of the peak value.

Alpha-fetoprotein

Structure

Alpha-fetoprotein (AFP) is synthesized sequentially by the yolk sac, gastrointestinal tract, and fetal liver and is able to pass across the placenta into the maternal circulation. It is a glycoprotein with similar physical properties to albumin and is coded for by the same gene on chromosome 4 (4q11-4q13). It is the major protein in fetal plasma but its concentration in adult plasma is very low except in cases of hepatoma or teratoma.

Function

While AFP appears to function similarly to albumin, acting as an osmoregulator in fetal plasma, it has no known function in adults. It is able to cross the fetal kidney and enter the fetal urine so that it is readily detectable in amniotic fluid. Elevated concentrations of AFP in the maternal circulation are associated with a number of clinical conditions including:

- Neural tube defects, for example spina bifida
- Abdominal wall defects, for example gastroschisis, omphalocele
- Twin pregnancy
- Invasive procedures, for example amniocentesis
- Congenital nephritic syndrome
- Placental tumors
- Maternal liver tumor
- Intrauterine death

Decreased concentrations are found in cases where the fetus has Down syndrome (trisomy 21) or Edwards' syndrome (trisomy 18).

Pregnancy-associated plasma protein A

Structure

Pregnancy-associated plasma protein A (PAPPA), the largest of the pregnancy-associated glycoproteins, is a zinc metalloproteinase of mass approximately 400 kDa consisting of two identical subunits linked via disulfide bonds. It is synthesized by the placenta and is present in maternal plasma mainly as a complex with the precursor form of major basic protein (proMBP) to which it is also linked via disulfide bonds. High concentrations of proMBP are present in the placenta and serum of pregnant women. PAPPA may also be used as a marker of unstable angina and acute myocardial infarction.

Function

Analysis of the pattern of expression of PAPPA indicated its presence in antral follicles, being restricted to the granulosa cells, and in a subset of large luteal cells in the corpus luteum. It has been suggested that it might be a functional marker for the dominant follicle which becomes the corpus luteum. PAPPA is detectable in the first trimester and its concentration increases exponentially initially and then more slowly through pregnancy to term.

Unconjugated estriol

Although estriol is secreted by the placenta, its biosynthesis requires the participation of enzymes in the fetal adrenal gland and liver as well as the placenta, because the placenta lacks the ability to convert pregnenolone to 16-hydroxy-dehydroepiandrosterone sulfate (16-OH DHEAS), the precursor of estriol (**Figure 12.2**). Because of the involvement of fetal organs in its biosynthesis, the measurement of estriol has been used on occasion to monitor fetal health. However, the concentration of estriol is also affected by a number of pathological conditions and thus it is not measured routinely in clinical practice.

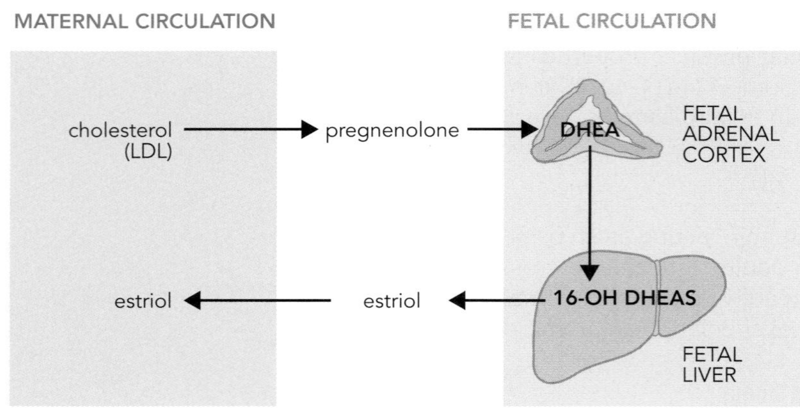

Figure 12.2 Biosynthesis of estriol involves the participation of maternal and fetal organs. The placenta synthesizes pregnenolone from circulating cholesterol but lacks the enzymes to convert pregnenolone to 16-hydroxy-dehydroepiandrosterone sulfate (16-OH DHEAS), the precursor for estriol synthesis. Pregnenolone from the placenta enters the fetal circulation and is metabolized to DHEA in the fetal adrenal cortex before sulfation to 16-OH DHEAS in the fetal liver. 16-OH DHEAS then circulates to the placenta where it can be converted to estriol. Estriol is synthesized in significant amounts only during pregnancy.

Progesterone

The structure and synthesis of progesterone is described in Chapter 11. During the luteal phase of the menstrual cycle, the plasma concentration of progesterone, synthesized by the corpus luteum, increases from 5.4 nmol/L (1.7 ng/mL) to 85.9 nmol/L (27 ng/mL). If fertilization occurs, progesterone is synthesized from LDL-derived cholesterol by the placenta from about week 7 of pregnancy, with an initial 6–8-fold increase in concentration in maternal blood. This is followed by a further gradual increase from week 10 from the last menstrual flow through to term, when the maternal progesterone plasma concentration reaches 318–954 nmol/L (100–300 ng/mL).

Inhibins

Structure

Inhibins are heterodimeric proteins consisting of α (20 kDa) and β (14 kDa) subunits linked via disulfide bonds. Two inhibins, A and B, have been identified which have common α but different β subunits (see Chapter 11 and Figure 11.39). Both inhibins A and B are produced by the ovary but only inhibin A is produced by the placenta.

Function

As described in Chapter 11, inhibins exert negative feedback on the pituitary where they inhibit production and secretion of FSH. Increases in the concentration of inhibin A in the maternal circulation are detectable within 12 days of conception, are readily apparent at 5 weeks of gestation, and peak at 8–10 weeks.

Human placental lactogen

Structure

Human placental lactogen (HPL), also known as human chorionic somatomammotropin (HCM), is a monomeric 22 kDa peptide (190 amino acids) with two intramolecular disulfide bridges. It has high amino acid sequence homology (>80%) with prolactin and growth hormone and has similar biological properties to both.

Function

HPL is secreted by trophoblasts of the placenta and is found mainly in the maternal circulation where, in contrast to hCG, its concentration rises with increasing gestational age. It is detectable during the fifth week of gestation and rises throughout pregnancy, reaching its highest concentration (5–7 ng/mL) in the third trimester. The concentration during pregnancy reflects placental size and, at term, is the highest of all placenta-derived protein hormones. Physiologically, it probably acts via insulin-like growth factor 1 (IGF1) as an insulin antagonist to maintain high concentrations of glucose, fatty acids, and amino acids in the maternal circulation to provide nutrition for the fetus. It may be one of the factors responsible for insulin resistance in the mother (see below). The concentration of HPL is further increased in multiple pregnancies, diabetes, and rhesus incompatibility and decreased in toxemia and placental insufficiency. Although it may give an indication of fetal well-being and growth, its measurement is of limited use clinically and not routine.

12.2 ENDOCRINE EFFECTS IN PREGNANCY

Effects of pregnancy on maternal physiology

Pregnancy involves a number of physiological changes in the mother which allow growth and development of the fetus and which prepare the mother for nurture of the fetus and delivery. There are changes to the function of the maternal hematological, cardiovascular, respiratory, gastrointestinal, and renal systems. These changes in maternal physiology are often brought about by production and secretion of a panoply of hormones by the fetus, the placenta, and the mother herself. The relative changes in plasma concentration of the major hormones of pregnancy—human chorionic gonadotropin, progesterone, estrogens, prolactin and HPL—are shown diagrammatically in **Figure 12.3**. The patency of the corpus luteum, which in the absence of fertilization degenerates within a

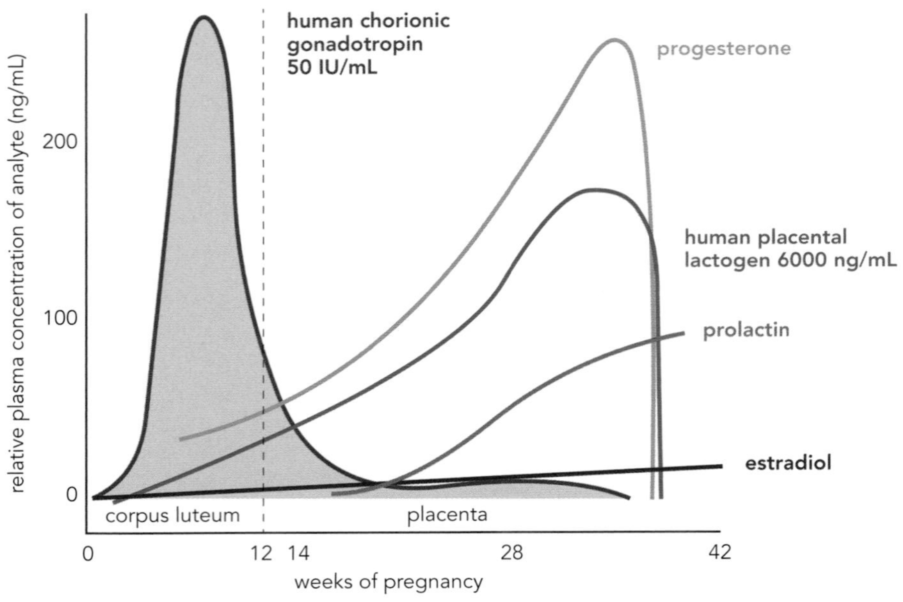

Figure 12.3 Relative changes in the plasma concentrations of major hormones in pregnancy. During the first two months of pregnancy, human chorionic gonadotropin (hCG), synthesized by trophoblasts, maintains the patency of the corpus luteum in the ovary. The peak secretion of hCG occurs at week 10 of gestation and falls as the corpus luteum degenerates. The corpus luteum is also the major source of both progesterone and estradiol in the first two months of pregnancy. The increase in concentrations of progesterone and estrogen from around week 12 of pregnancy is due to synthesis by the placenta. Prolactin and human placental lactogen (HPL) stimulate breast development in preparation for lactation after birth. Plasma concentrations of progesterone, estrogen, and HPL fall precipitously after birth as the placenta is delivered. Plasma concentrations of prolactin, synthesized by the pituitary, remain elevated after birth.

couple of weeks, is maintained in the early stages of pregnancy by the actions of hCG secreted by the trophoblasts as they merge with and burrow into the endometrium. Thus the signal to maintain the corpus luteum now comes from the conceptus rather than the mother, and hCG reaches a peak concentration at about ten weeks of gestation. The plasma concentration of hCG then falls rapidly to a low concentration by week 12, a concentration which is maintained throughout the rest of pregnancy. Besides maintaining a healthy corpus luteum, hCG, like LH, promotes the synthesis of steroid hormones, particularly progesterone and estradiol. During the first two months of pregnancy virtually all of the estrogens and progesterone are synthesized and secreted by the corpus luteum. Estrogens have major effects in preparing the mother for the development, delivery, and post-partum maintenance of the fetus by stimulating the growth of uterine muscle and of connective tissue forming the cervix. They promote also the development of ductal and alveolar tissue in the breast. Progesterone stimulates the formation of the maternal layers of the placenta (decidua) from the endometrium and has a general relaxant effect on uterine smooth muscle to inhibit uterine contractility during pregnancy. Its concentration in the maternal circulation rises from week 7 of pregnancy with a further gradual increase from week 10 through to term, when the concentration is 100–300 ng/mL. The rapid fall in hCG concentration between weeks 10 and 12 of pregnancy results in the degeneration of the corpus luteum and consequent loss of the tissue as a source of steroid hormones. From this time (week 12 of pregnancy) the trophoblast cells of the placenta become the major source of estrogens and progesterone. These cells have the enzymes required to synthesize progesterone from cholesterol and also the aromatase activity required for the conversion of androgens to estrogens. However, they lack the ability to metabolize progesterone to androgens, and thus the placenta can synthesize estrogens only from androgens derived from the maternal ovaries and adrenals or from the fetal adrenals. The raised concentrations of progesterone and estrogens in the maternal circulation during pregnancy, arising initially from the corpus luteum and latterly from the placenta, feed back to the hypothalamus and inhibit the secretion of GnRH with consequent inhibition of both LH and FSH by the anterior pituitary. The ovarian and uterine cycles of menstruation are thus inhibited during pregnancy. The actions of HPL contribute to fetal nourishment both during pregnancy and post-partum. Its growth hormone-like actions mobilize maternal fat and glucose production and inhibit maternal glucose utilization, thereby providing an energy supply for the growing fetus, while its prolactin-like activity stimulates breast development. Inhibin A, which exerts negative feedback on the pituitary to inhibit production and secretion of FSH, is measured occasionally in the monitoring of pregnancy. An increase in the concentration of inhibin A is detectable within 12 days of conception, is readily apparent at 5 weeks of gestation, and peaks at 8–10 weeks.

Other physiological aspects of pregnancy

Fetal nutrition

Cells of the part of the conceptus known as the trophoblast deeply invade the endometrial lining of the uterus around 6 days after fertilization, gaining access to the nutrients in the blood supply of the mother. Initially, blood sinuses known as trophoblastic lacunae form when the trophoblast disrupts maternal blood capillaries. Maternal blood flows through these vessels, and this blood supplies the embryo with adequate nutrients and oxygen to allow the early stages of development to take place. Later, fingerlike projections (chorionic villi) push their way into the trophoblast from the embryo, eventually forming blood vessels. The embryo develops into what is known as the fetus from the eleventh week of pregnancy. The placenta develops and acts as the interface between the fetus and the mother; fetal blood and maternal blood come into close apposition in the placenta, but do not mix. Development of the placenta is complete by around the fourteenth week of pregnancy and the fetus is surrounded by amniotic fluid within fetal membranes. Once the placenta has developed, the fetus is supplied with oxygen and nutrients entirely from the bloodstream of the mother. Several mechanisms ensure that the fetus gains as much from this relationship as is necessary for development. Fetal hemoglobin (HbF), for example, is different from adult hemoglobin (HbA); HbA is composed of two α-globin and two β-globin chains, but in HbF the β-globin chains are substituted by two γ-globin chains. This substitution results in HbF having higher affinity for oxygen than does HbA. Fetal blood also contains around 20% more hemoglobin than adult blood and so fetal blood can carry enough oxygen to meet the needs of the developing fetus, despite the fact that the partial pressure of oxygen (PO_2) of blood in the umbilical vein supplying the fetus may be as low as 30 mmHg (4 kPa), which is even lower than that of venous blood in an adult. The ratio of HbF:HbA in the later stages of pregnancy is about 4:1. As the baby starts to breathe after birth, gene switching causes a decrease in HbF and an increase in HbA (the genes encoding the β and γ chains of hemoglobin are both located on chromosome 11), so that very little HbF is present in normal human blood. This necessitates the synthesis of new red blood cells containing HbA and the destruction of older red blood cells containing HbF, with the heme component of the HbF being metabolized to bilirubin. Bilirubin is normally excreted into bile conjugated to glucuronic acid by the action of a glucuronyl transferase (see Chapter 3). The glucuronyl transferase is expressed toward the end of the third trimester and a delay in the expression of this enzyme, or premature birth, may lead to neonatal jaundice, a relatively common condition.

The ability of the fetus to obtain sufficient nutrients from the mother's bloodstream is shown by the incidence of gestational diabetes in pregnant women. Hormones produced by the placenta cause some degree of insulin resistance in all pregnancies but approximately 3% of pregnant women develop diabetes mellitus (high plasma glucose), usually manifested by tiredness and extreme thirst in the second trimester. This gestational diabetes results when the mother is unable to produce enough insulin to counteract the insulin resistance of the pregnant state. The induction of insulin resistance in the mother benefits the fetus by increasing the amount of glucose available for transfer across the placenta for use by the fetus. This increases the size of the fetus such that babies born to mothers with gestational diabetes become so large that delivery is often difficult. Thankfully, in the majority of cases, normal control by insulin of maternal plasma glucose concentration returns on delivery of the placenta, although gestational diabetes has been shown to be associated with the development of non-gestational diabetes mellitus later in life.

12.3 CHANGES TO MATERNAL HEMATOLOGICAL AND CARDIOVASCULAR PARAMETERS DURING PREGNANCY

The growing fetus presents an increased demand on the mother in terms of metabolism and blood supply to the placenta. Pregnancy is associated with an approximately 40% increase in blood volume (an extra 1–2 L) and a 20% increase in erythrocyte mass (total volume of red cells in the circulation) in an attempt to satisfy the increased oxygen use (15%) by the fetus as well as an increase in maternal basal metabolic rate. However, because plasma volume increases more than the red cell mass, there is a fall in plasma hemoglobin concentration, hematocrit, and red cell count during pregnancy. Maternal hemoglobin may also be reduced as iron is transferred to the fetus. The heart becomes slightly enlarged in mid-pregnancy due to wall thickening and, particularly, increased venous filling. There is also a 30% rise in cardiac output in the first trimester (heart rate increases from 70 to 85 beats per minutes and stroke volume from around 65 to 70 mL). Such increases have major implications for pregnancy in women with preexisting cardiac disease. Despite these alterations in cardiac output, it is normal for pregnant women to have slightly lower blood pressure than their non-pregnant counterparts, particularly in the first and second trimesters. This is thought to be caused by the relaxant effect of progesterone on smooth muscle, which causes generalized peripheral vasodilation to the extent that the influence of increased cardiac output on blood pressure is canceled out by reduced resistance to blood flow. Peripheral edema is also common in pregnancy, particularly in dependent areas like the ankles. This is caused by decreased oncotic pressure in the blood; as plasma volume increases to a greater extent than plasma protein production, less fluid is attracted back into the capillaries from the interstitial space, resulting in fluid accumulation in the interstitial space.

12.4 CHANGES TO MATERNAL RESPIRATORY AND RENAL PARAMETERS DURING PREGNANCY

Although, as mentioned above, there is a 15% rise in oxygen consumption during pregnancy, normally this does not constitute any major problems for the mother and her respiratory rate does not change. However, tidal volume increases by approximately 40% and thus the minute ventilation (breaths per minute × tidal volume) is also increased by the same percentage. Inspiratory volume increases progressively during pregnancy whilst various measures of expiratory flow rate remain unchanged. However, as the growing fetus pushes up on the maternal diaphragm, particularly in the later stages of pregnancy, it is not uncommon for pregnant women to become slightly breathless on exertion at this time. The kidneys also become enlarged during pregnancy due to an increase (by 70% in the third trimester) in renal parenchymal volume. This is accompanied by a 30–55% rise in renal blood flow, which is apparent as early as the first trimester. There is a simultaneous increase, by 50% after four months of pregnancy, in glomerular filtration rate (GFR) although this falls somewhat toward term. The resulting increased creatinine clearance, however, is not accompanied by increased creatinine or urea production at this time and so the plasma concentrations of these metabolites fall. A fall in serum uric acid concentration during pregnancy also reflects increased filtration without an increase in production or reabsorption of this substance. The filtered loads of sodium and potassium rise as a consequence of the increase in GFR and a parallel increase in tubular reabsorption leads to retention of both cations in maternal stores. Increased glucose excretion, possibly due to the increased GFR and/or decreased tubular reabsorption, is also a feature of pregnancy. This occurs independently of the plasma concentration of glucose and may be another cause of glycosuria.

12.5 CLINICAL CHEMISTRY OF PREGNANCY

Human chorionic gonadotropin and diagnosis of pregnancy

Assays to measure hCG in both serum and urine have improved greatly in their sensitivity in recent decades. It is possible to detect pregnancy as soon as a few days after implantation, as serum hCG rises above the non-pregnant range in an exponential fashion. Urinary hCG testing kits for home use by untrained individuals are inexpensive but very reliable. They have detection limits which vary from 20 to 100 IU/L, depending on the brand. There is a trade-off between the ability to detect very early pregnancy and the incidence of false positive results. More sensitive tests may detect implantation, which does not always result in a viable pregnancy and does not delay menstruation, so that without the test the woman would not have suspected she had conceived. False negative urine pregnancy testing may occur when the urine is very dilute and the hCG is close to the detection threshold. For this reason, testing is recommended on a first morning urine sample. Because urine testing is qualitative, with a positive or negative result, it cannot determine the week of gestation. Serum hCG does correlate with gestation, but there is a wide variation in the concentrations seen at different weeks, leading to a large degree of overlap with consecutive weeks (**Table 12.3**).

Analytical practice point 12.1

Urine assays for hCG may give false negative results in the first 2–3 weeks after conception, especially if the urine sample is very dilute.

When a conception outside the womb (ectopic pregnancy) or blighted ovum occurs (an ovum which does not grow, often leading to miscarriage in the first trimester), hCG does not rise as rapidly as normal and repeating the test at intervals of 24–48 hours can be helpful in diagnosis. Following miscarriage or termination of pregnancy, the hCG decays according to its half-life of 24–48 hours, and may remain detectable for several weeks. In twin and multiple pregnancies and trophoblastic disease, serum hCG is higher at any given week of gestation,

TABLE 12.3 Approximate expected ranges for hCG by week of gestation

Weeks after LMP	hCG (IU/L)
3	5–50
4	5–400
5	20–7500
6	1000–56,000
7–8	7000–230,000
9–12	25,000–290,000
13–16	13,000–250,000
17–24	4000–165,000
25–40	3640–117,000
Non-pregnant females	<2.0
Postmenopausal females	<10.0

whilst in nonviable and ectopic pregnancies it is lower. Although hCG is usually considered in pregnancy diagnosis, it may also be elevated in other conditions, particularly in germ-cell ovarian and testicular tumors, when it may be used as a tumor marker. Other tumors which occasionally secrete hCG include lung, breast, and gastrointestinal. Non-neoplastic disorders associated with raised hCG may include liver cirrhosis and inflammatory bowel disease. In postmenopausal women, hCG is higher than before the menopause. It is usually less than 10 IU/L, but may be present at severalfold higher concentrations, and is of pituitary origin. Before this finding was widely recognized it sometimes caused diagnostic confusion and unnecessary investigations for its source, with particular concern about hCG-secreting tumors. There have been several cases reported of surgery being performed on the basis of raised hCG, on the false assumption that the patient had a malignancy. It is important to be aware that routine assays for hCG may detect a number of closely related molecules with different glycosylation, as well as fragments. These may be differentially increased depending on their origin, and it may be necessary to use specialized assays at reference laboratories to characterize these related molecules when the clinical picture does not explain raised hCG results.

Clinical practice point 12.1

Serial measurements of hCG made 48 hours or more apart are often useful in distinguishing a viable pregnancy from ectopic pregnancy and miscarriage.

Analytical practice point 12.2

Postmenopausal women may have detectable, but nonrising, hCG of pituitary origin. This may lead to an incorrect diagnosis of pregnancy or hCG-secreting tumor if the clinician is not made aware of the phenomenon.

Serum hormone concentrations in pregnancy

Serum concentrations of hormones may change greatly during pregnancy either because they are directly involved in the establishment and maintenance of the pregnancy itself, or because of changes in hormone secretion or in the production of binding proteins (see Figure 12.3).

Progesterone, estrogens, and androgens

Progesterone is required for the maintenance of pregnancy and serum concentrations increase steadily throughout gestation by five- to tenfold from conception to a peak at week 40. For the first 8 weeks, the source of progesterone is predominantly the corpus luteum of the ovary in response to hCG, but the placenta takes over from the latter part of the first trimester. Measurement of progesterone in early pregnancy can be useful in suspected ectopic pregnancy or following assisted conception as a marker of viability. Estradiol and androstenedione rise greatly also as pregnancy progresses. As with progesterone, the source is initially the corpus luteum with the placenta taking over at around week 8 of gestation. Serum testosterone rises both directly, due to increased secretion, and also because of higher sex hormone-binding globulin (SHBG) levels. The synthesis of SHBG by the liver increases and the total serum pool of testosterone rises, although the free testosterone does not change.

Gonadotropins

LH and FSH secretion are switched off rapidly by negative feedback from rising estradiol concentrations. This inhibits further ovulation during pregnancy.

Prolactin

There is an approximate doubling in size of the pituitary during pregnancy. This reflects the increasing size and number of lactotroph cells, which secrete prolactin, under the stimulatory effect of estrogens. Serum prolactin concentration increases about tenfold or more as a result. Other pituitary cells are either unaffected in pregnancy (corticotrophs and thyrotrophs) or are reduced in number (gonadotrophs and somatotrophs).

Hypothalamus–pituitary–adrenal axis

The effect of pregnancy on the hypothalamus–pituitary–adrenal axis is extremely complex and not yet fully understood. Part of this is due to the placenta itself being an endocrine organ secreting identical hormones to the hypothalamus and anterior pituitary. Corticotropin-releasing hormone (CRH) is secreted by the placenta at a rate much greater than that from the hypothalamus. Plasma concentrations increase from late in the first trimester to reach a peak of up to 1000-fold higher than normal. However, there is not a correspondingly large increase in adrenocorticotropic hormone (ACTH) or cortisol due to most CRH being bound to a specific carrier protein which prevents interaction between CRH and its receptor. There is also down-regulation of the receptor on the corticotroph cells, which limits their secretion of ACTH. Nevertheless, there is a steady increase in ACTH of approximately threefold through pregnancy, some of which may be placental in origin, with a further dramatic rise just before parturition. Plasma cortisol rises due to a combination of increased secretion and an estrogen-stimulated rise in SHBG. However, progesterone competes with cortisol for protein binding sites and displaces some of it, resulting in a rise in urinary free cortisol. The maternal adrenal zona fasciculata expands during pregnancy, reflecting increased cortisol secretion, but the feto-placental unit also contributes to steroidogenesis. The changes in adrenal function in pregnancy overlap those seen in Cushing's syndrome in non-pregnant individuals. Interpretation of standard investigations is therefore complicated and diagnosis of true adrenal insufficiency and excess is difficult. Although dynamic function tests may be used, these have not been validated generally in pregnancy and diagnostic cut-offs are not defined, particularly with modern specific assays. Fortunately, both Cushing's syndrome and adrenal insufficiency rarely develop during pregnancy. Both conditions reduce fertility and newly occurring cases have the same causes as in non-pregnant individuals.

Growth hormone

The placenta secretes growth hormone which, along with IGF1, rises in the plasma from the first trimester. By around week 20, pituitary growth hormone is fully suppressed and all detectable growth hormone is placental. Women who are growth-hormone deficient before conception have no greater risk of maternal or fetal complications.

Thyroid hormones and thyroid-stimulating hormone

The effect of pregnancy on thyroid function tests (TFTs) is complex and varies according to the stage of gestation. There is structural similarity between TSH and hCG, such that the latter can activate the TSH receptor when present at very high concentration (>400,000 IU/L). In the first trimester the very large increase in hCG secretion may induce a state of mild biochemical (but rarely clinical) hyperthyroidism, with low (or even fully suppressed) TSH and raised thyroxine (T_4) and tri-iodothyronine (T_3). When hCG progressively falls from its peak at around 13 weeks, there is a reciprocal rise in TSH back to its pre-pregnancy concentration. Total T_4 and T_3 rise through pregnancy due to the estrogen-stimulated rise in thyroid hormone-binding globulin (TBG) from the liver. However, free T_4 and T_3 fall slightly, perhaps due to increased breakdown by the placenta (**Table 12.4**).

For some years, the true changes in free hormone concentrations were controversial due to debate around the reliability of early assay techniques. However, modern free thyroid hormone assays usually give consistent results as long as appropriate reference ranges are used for the stage of pregnancy.

Analytical practice point 12.3

Thyroid function tests change through pregnancy and trimester-appropriate reference ranges should be used.

TABLE 12.4 Thyroid function tests: normal values in pregnancy

Hormone	Units (SI)	Units (conventional)	First trimester		Second trimester		Third trimester	
			SI	Conventional	SI	Conventional	SI	Conventional
TSH	mIU/L	μIU/mL	<0.05–3.0	<0.05–3.0	<0.05–3.7	<0.05–3.7	0.1–3.8	0.1–3.8
Free T_4	pmol/L	ng/dL	11–23	0.84–1.79	8–17	0.62–1.32	8–14	0.62–1.09
Free T_3	pmol/L	pg/dL	3.0–6.2	194.8–402.6	2.5– 5.0	162.3–324.7	2.5–4.5	162.3–292.2

Pregnancy-associated plasma protein A

Low concentrations of this protein during the first trimester have been associated with early pregnancy failure, chromosomally abnormal fetuses, and with low-birth-weight babies. It has been used recently as a marker during the first trimester for Down syndrome (along with hCG and nuchal translucency) but was found to be no better than the triple test in the second trimester (hCG, unconjugated estriol, and AFP). Physiologically, PAPPA is a protease which rapidly inhibits IGF and may be a regulator of IGF activity; this gives rise to its alternative name of insulin-like growth factor binding protein 4 (IGFBP4).

12.6 PREGNANCY AND COMMON BIOCHEMICAL TESTS

The physiological changes of normal pregnancy affect the interpretation of several common biochemical investigations and, unless these changes are recognized, disease states may be incorrectly inferred from apparently abnormal results. There is also a risk of pathological changes being masked and true disease being missed.

Renal function and electrolytes

There is a significant increase in effective circulating volume and hence renal plasma flow, which peaks at the end of the first trimester. As a result of this, GFR increases by up to 50%. Markers of GFR, including plasma urea and creatinine, fall and creatinine clearance rises. Significant renal disease may be present with renal function tests that are normal for non-pregnant individuals. Similarly, there is an expected fall in blood pressure due to the relaxation of arterial smooth muscle induced by progesterone. Thus, hypertension is indicated by blood pressure that is normal compared to pre-conception levels. The rise in GFR explains, at least partially, a decrease in the renal threshold for glucose, which results in glycosuria in many pregnant women. This reduces the value of glycosuria as a screening test for diabetes mellitus. Other physiological changes include reduced serum urate, due to increased clearance, and lower serum sodium (by approximately 5 mmol/L), due to resetting of the threshold for secretion of vasopressin (also called arginine vasopressin, AVP, or antidiuretic hormone). Increased ventilation causes a chronic respiratory alkalosis. The renal compensation for this results in lowering of serum bicarbonate. This may be misinterpreted as metabolic acidosis.

Liver function tests

The major change in standard liver function tests is a progressive rise in alkaline phosphatase to a concentration of up to about three times the non-pregnant range. The enzyme is not, however, produced by the liver or biliary tract, but is instead a product of the placenta. The different isoforms (liver, bone, and placental) can be determined by electrophoretic techniques and differential heat stability, but in the context of pregnancy such investigations are rarely necessary. The increased circulating volume of pregnancy has a dilutional effect on proteins, such that albumin and total protein fall by up to about 10 g/L. The same mechanism explains the mild anemia which occurs also. However, this is not the full explanation as homeostasis would tend to correct the reduced oncotic pressure. There are changes also to protein synthesis in the liver (and red cell formation in the bone marrow) which blunt the homeostatic response. Other liver function tests are unaltered by pregnancy per se, such that rises in bilirubin or transaminases indicate pathology and require further investigation. Liver disease in pregnancy is not uncommon and can be due either to specific disorders of the pregnancy itself or to nonobstetric disease.

Analytical practice point 12.4

Raised alkaline phosphatase (up to 3 or more times the non-pregnant upper limit) and low albumin are normal findings in pregnancy. Raised bilirubin and transaminases are not expected and are likely to indicate liver disease.

Calcium metabolism

The growing fetus requires a large supply of calcium to be delivered across the placenta in order to mineralize its growing skeleton. To supply this there are a number of changes in maternal calcium metabolism which occur. Unfortunately, there has been a great deal of confusion over the last three decades about the exact sequence of events, due largely to limitations of older assay methods. For example, parathyroid hormone (PTH) was previously measured by assays that cross-reacted with inactive C-terminal fragments, giving spuriously high results in some situations, including pregnancy. Modern two-site intact PTH assays have shown that PTH tends to fall rather than rise during the first trimester. This occurs because there is an increase in 1,25-dihydroxy vitamin D, due to activation of 1α-hydroxylase in both the kidney and the placenta, which in turn increases calcium absorption from the gut. Plasma calcium appears to fall but this is due to the decrease in albumin, explained above, and if the albumin-adjusted calcium is calculated there is, if anything, a slight rise.

Lipids

There is an increase in both total cholesterol and high-density lipoprotein–cholesterol by around 15 to 20% during normal pregnancy. There is also a mild increase in serum triglycerides to approximately 3 mmol/L. Some women show much more dramatic increases, which usually represent the unmasking of a genetic hyperlipidemia, for example type III hyperlipoproteinemia. Such individuals often develop persistent hyperlipidemia later in life.

12.7 PRENATAL DIAGNOSIS OF CHROMOSOMAL AND GENETIC DISORDERS

The triple test (multiple marker screening), which may be performed during weeks 15–20 of pregnancy and preferably during weeks 16–18, includes measurements of AFP, hCG, and unconjugated estriol on maternal blood. AFP is synthesized by the fetus, hCG by the placenta, and estriol by both fetus and placenta. These measurements are used to assess the potential for genetic disorders in the developing fetus. Raised AFP is associated with open neural tube defects such as spina bifida and anencephaly, while low AFP, raised hCG, and low unconjugated estriol is associated with trisomy 21. All three markers are low with trisomy 18.

More definitive prenatal diagnoses of chromosomal and genetic disorders are made possible by the invasive techniques of amniocentesis and chorion villus sampling (CVS). Both of these procedures are associated with an increased risk of miscarriage, 0.5–1% for amniocentesis and 2% for CVS. In the UK, approximately 5 in 100 pregnant women are offered such prenatal diagnostic tests. Samples of amniotic fluid are withdrawn from the amniotic sac by insertion of a needle through the anterior abdominal wall, usually after 15 completed weeks of pregnancy, and are used mainly for karyotyping. It is important to verify that a sample obtained by amniocentesis is fetal and not maternal by testing the sample for HbF. Chorion villus sampling has largely replaced amniocentesis for detection of genetic disorders since chorionic samples may be taken in the first trimester: transcervically, between 8 and 12 weeks' gestation, and transabdominally from week 8 of pregnancy. The biopsies from the placenta attached to the uterus placental cells contain the same genetic material as cells of the developing fetus. The major indications for prenatal screening are shown in **Table 12.5**.

Screening for Down syndrome (trisomy 21)

The risk of giving birth to a baby with Down syndrome increases dramatically with age. At the age of 20 the risk is about 1 in 1500, rising to 1 in 100 by age 40 years. For this reason it used to be routine to offer amniocentesis to older

TABLE 12.5 Indications for prenatal screening

Stage of pregnancy	Reason for screening
Early pregnancy	Risk of chromosomal disorder due to maternal age
	Genetic or chromosomal disorder (for example neural tube defect) in a previous pregnancy
	Mother or partner is a carrier of a genetic disorder, for example: Sickle cell anemia Thalassemia Cystic fibrosis Duchenne muscular dystrophy
	Family history of genetic disorders; including rare inborn errors of metabolism such as: Glycogen storage disorders (for example Gaucher disease) Mucopolysaccharidoses Lipid metabolism (for example Niemann–Pick disease)
Later pregnancy	Fetal maturity, particularly lung (lecithin–sphingomyelin ratio)
	Hemolytic disease (Rhesus incompatibility)
	Chromosomal problems

women. However, the majority of pregnancies occur in younger women, and the smaller risk multiplied by the much greater number of individuals results in the majority of Down syndrome babies being born to mothers under 40. This may be changing as the number of older mothers increases in Western societies due to sociological and economic pressures, but the principle still holds.

It was discovered in the 1980s that there were differences in the average maternal serum concentrations of hCG (higher), AFP (lower), and unconjugated estriol (lower) in women carrying Down syndrome fetuses compared to normal pregnancies. From this observation, screening strategies were devised by using maternal serum and comparing results to median concentrations at the same week of gestation in order to calculate multiples of the median (MoM). These are then integrated with maternal age, weight, and clinical factors, such as smoking status, to calculate an overall risk of Down syndrome in the fetus. The so-called triple test is not diagnostic, however, and the definitive diagnosis requires fetal cells for chromosomal analysis obtained by amniocentesis or chorionic villus sampling. Being invasive tests, these carry a small but not negligible (0.5–1%) risk of fetal loss. A Down syndrome risk of around 1 in 200 to 250 is used as the threshold for offering diagnostic testing. This gives the optimal balance of true positives, false positives, and miscarriage caused by amniocentesis.

Developments in Down syndrome screening have included the discovery of other biochemical and radiological markers. Inhibin A can be added to the triple test (the four-marker quadruple test) to improve accuracy. High-resolution ultrasound can show an abnormality in the soft tissue of the neck in Down syndrome fetuses, termed increased nuchal translucency. This finding can be combined with biochemical markers to allow screening in the first rather than second trimester; this earlier screening is desirable, as the therapeutic option of termination of pregnancy is more difficult the further the pregnancy has advanced. Understandably, not all pregnant women wish to consider this possibility and may opt out of screening; others, whilst not prepared to terminate the pregnancy, may still wish to know that their baby will have Down syndrome.

12.8 NONINVASIVE PRENATAL TESTING

The increased efficiency, specificity, and decreasing relative cost of genomic sequencing techniques is beginning to make an impact on the detection of genetic abnormalities in the fetus. Noninvasive prenatal testing uses cell-free fetal DNA present in the plasma of pregnant women to screen for fetal aneuploidies—chromosomal mutations in which the chromosome number is abnormal. Currently, the technique can detect only trisomy13 (Patau syndrome), trisomy 18 (Edwards syndrome), and trisomy 21 (Down syndrome) with any degree of certainty and results are used as confirmatory or supporting data. Cell-free fetal DNA, thought to be derived primarily from the placenta, constitutes 3–13% of the cell-free maternal DNA in peripheral plasma. It is lost from the maternal circulation within hours post-partum.

Maternally derived, cell-free fetal DNA has been used in the determination of fetal blood types in screening for Rhesus incompatibility, as described below. It is likely to play an increasingly important role in the development of more wide-ranging programs that screen for genetic abnormalities in the fetus.

Analytical practice point 12.5

Advances in the analysis of fetal cell-free DNA in maternal plasma may supersede biochemical and ultrasound screening for chromosomal disorders such as trisomy 21.

12.9 DISORDERS OF PREGNANCY

Although pregnancy is a physiological state, there are a number of disorders which are unique to pregnancy. In some of these, biochemical measurements are used in diagnosis or management. Some of the more common examples are discussed here.

Hyperemesis gravidarum

Mild nausea and vomiting in early pregnancy (morning sickness) is almost universal. However, in a small proportion of women (around 1–2%), the symptoms are severe and intractable, leading in some cases to a need for intravenous fluids and nutritional support. This is termed hyperemesis gravidarum, meaning excessive vomiting of pregnancy. Hormonal causes are implicated, particularly hCG. Symptom severity correlates with the serum hCG concentration usually, which explains the improvement after the peak hCG around week 15. In up to a half of these cases, there are abnormalities of TFTs which suggest hyperthyroidism. However, these changes overlap with those due to pregnancy itself, and most authorities would recommend antithyroid drug treatment only where abnormal TFTs persist beyond around 18 weeks' gestation.

Hemolytic disease of the fetus and newborn

This condition, also known as Rhesus disease, may occur when a pregnant mother with Rhesus D negative (RhD–ve) blood carries a fetus with Rhesus D positive (RhD+ve) blood. In addition, the mother must have been sensitized previously to RhD+ve blood from an older live child, miscarriage, or, infrequently, a blood transfusion. The mother's immune response to RhD+ve fetal red blood cells is to produce RhD+ve antibodies (anti-D antibodies) which cross the placenta, enter the fetal circulation, and lyse fetal red blood cells. This attack on the baby's red blood cells may continue for a few months post-partum. While the mother is unaffected by the Rhesus incompatibility, lysis of the baby's red blood cells leads to anemia and jaundice which *in extremis* may lead to intrauterine death. The symptoms of hemolytic anemia may take up to three months to develop in mild cases and may include an increased rate of breathing and poor feeding.

While about one-half of all cases of Rhesus disease are mild and require little by way of treatment, monitoring of the baby is important to detect potential problems. However, since the severity of presentation may be acute, it is essential to determine the anti-D antibody status of the mother by at least week 28 of pregnancy to ensure appropriate clinical care of both mother and baby. Ideally, blood tests carried out early in pregnancy test for anemia, rubella, human immunodeficiency virus (HIV), hepatitis B, and blood group, including Rhesus D status, positive or negative. Anti-D antibodies should be measured in Rhesus D negative blood samples.

The introduction of molecular biological techniques, including polymerase chain reaction (PCR)-based amplification assays, to blood typing has made possible the determination of fetal blood group status using fetal DNA from amniotic fluid or chorionic villus samples. More recently, the finding of cell-free fetal DNA in maternal peripheral blood obviates the necessity to perform invasive techniques to obtain fetal DNA samples, and this maternally derived cell-free fetal DNA can be used to determine the blood group of the fetus.

Rhesus disease can be prevented by injection of the mother with anti-D immunoglobulins (antibodies) to neutralize any RhD+ve antigens which may have entered the maternal circulation. Anti-D immunoglobulins are administered routinely to nonsensitized RhD–ve mothers during the third trimester of pregnancy (routine antenatal anti-D prophylaxis; RAADP).

Ectopic pregnancy

In around 1% of pregnancies the fertilized ovum implants at a site other than the uterine cavity. This is almost always one of the Fallopian tubes, but other sites of implantation (cervix, peritoneal cavity) occur occasionally. A tubal pregnancy is potentially life threatening to the mother, as eventual rupture and catastrophic bleeding will occur if the embryo continues to grow. When an ectopic pregnancy is suspected clinically, measurement of hCG can be helpful. When serum hCG exceeds around 1500 IU/L, a normal uterine pregnancy is usually visible on ultrasound scanning. If none is seen, an ectopic site of implantation is inferred. A further application of hCG measurement is serial measurements at 24–48 hour

CASE 12.1

A 36-year-old woman presents to the emergency department with lower abdominal pain. She usually has regular menstrual cycles, but her period is 2 weeks late. Her urine pregnancy test (hCG) is positive, but a transvaginal ultrasound shows no embryo in either the uterine cavity or the Fallopian tubes. Her serum hCG is measured and repeated after 24 and 48 hours.

	SI units	Reference range	Conventional units	Reference range
Serum				
Chorionic gonadotropin (hCG) baseline	1011 IU/L	0–5	1011 mIU/mL	0–5
hCG 24 hour	1107 IU/L	0–5	1107 mIU/mL	0–5
hCG 48 hour	1325 IU/L	0–5	1325 mIU/mL	0–5

Reference range for non-pregnant females.

- **Why are serial hCG measurements taken?**
- **What would be the expected change?**
- **What do the results suggest?**

The history and positive urine test indicate pregnancy. However, in early pregnancy the embryo may not be visible. In a normal viable pregnancy there is an exponential rise in hCG in the first trimester, with an approximate doubling of the serum hCG every 48 hours, whilst following a miscarriage the hCG will fall in line with its half-life (24–36 hours). In this case, the hCG increases but at a lower than usual rate. The most likely cause of this is ectopic pregnancy, usually in one of the Fallopian tubes. If the pregnancy progresses there is a risk of tube rupture and severe intra-abdominal bleeding. Ectopic pregnancy has traditionally been treated by surgical removal of the affected tube, but some cases may be treated with methotrexate (a cytotoxic drug) and falling serial hCG measurements can be used to show it has been effective.

intervals. A uterine pregnancy usually causes hCG to double every 48 hours in the early first trimester. A slower rise is seen in ectopic pregnancies (or no rise if the embryo has spontaneously aborted). Serum progesterone may sometimes be a helpful marker of pregnancy viability: high concentrations implying a normal pregnancy and low concentrations indicating nonviability. However, intermediate concentrations are often obtained, which are not contributory.

Molar pregnancy

A molar pregnancy is known also as hydatidiform mole or trophoblastic disease. The disorder occurs from an aberrant fertilization event, due to chromosomal abnormality in the sperm cell, which leads to overgrowth of the trophoblastic cells that normally make up the placenta. The fetus either does not develop at all, or is abnormal and incapable of life outside the uterus. Because it consists entirely of trophoblastic cells, a molar pregnancy secretes more hCG at any stage of gestation and, more often than not, it causes hyperemesis gravidarum. Molar pregnancy is detected on ultrasound scanning. Because cells can persist in the mother following a molar pregnancy, and these can spread to other organs, hCG is measured periodically for several months. A rising concentration (in the absence of a new pregnancy) indicates residual trophoblastic disease needing chemotherapy.

Pre-eclampsia and HELLP syndrome

Pre-eclampsia is a common disorder of later pregnancy, occurring after week 20 in about 7% of women. It is characterized by hypertension and proteinuria. Tests that help with diagnosis include urinary albumin, placental growth factor (PlGF), and soluble vascular endothelial growth factor receptor 1 (sVEGFR-1; also known as sFlt-1); the latter two are, respectively, pro- and antiangiogenic factors released by the placenta into the maternal circulation. The rationale for the measurement of PlGF and its inhibitor, sVEGFR-1, arises from the suggestion that pre-eclampsia is due to impaired angiogenesis and endothelial dysfunction.

Severe edema, abdominal pain, and vomiting may occur in pre-eclampsia and a small proportion of those affected develop eclampsia, or seizures, which can be fatal to both mother and baby. A variant of pre-eclampsia is the HELLP (hemolysis, elevated liver enzymes, and low platelets) syndrome. Although supportive and antihypertensive treatment is given, the only cure for pre-eclampsia and its variants is delivery of the baby and the placenta.

Cholestasis of pregnancy

Cholestasis of pregnancy is also known as obstetric cholestasis. It affects about 1% of pregnancies and occurs in the third trimester. The cause is thought to be a combination of high hormone concentrations, particularly estrogens, with a genetic predisposition. The flow of bile in the intrahepatic ducts is reduced leading to increased serum bile acid concentration. These bile acids accumulate in the skin and are responsible for the cardinal symptom of pruritus, or itching. Measurement of serum bile acids is useful therefore in confirming a diagnosis. In more severe cases, liver transaminases and bilirubin may be elevated. The evidence for increased risk of obstetric cholestasis to the mother and baby is not clear, and it is argued by some that the common practice of early delivery causes more harm than good.

Neural tube defects

Neural tube defects occur during the first month of pregnancy and may affect the brain, spine, and spinal cord. The most common are:

- Spina bifida: the fetal spinal column does not close properly and leads to nerve damage and paralysis of the legs

- Anencephaly: much of the skull and brain does not develop and babies are usually stillborn or die soon after birth
- Chiari malformation: the brain extends into the spinal cord, eventually compromising the brain stem

The cause(s) of the defects are unknown but maternal folate deficiency has been associated with spina bifida. Prenatal diagnosis involves both laboratory tests and imaging. A raised AFP in the triple test blood screen (AFP, hCG, and estriol) performed in the second trimester is indicative of a problem, as is raised AFP and high acetylcholinesterase activity in amniotic fluid. Scanning techniques including X-ray, magnetic resonance imaging, computed tomography, and ultrasound may all be employed to determine spinal defects.

MUSCLE

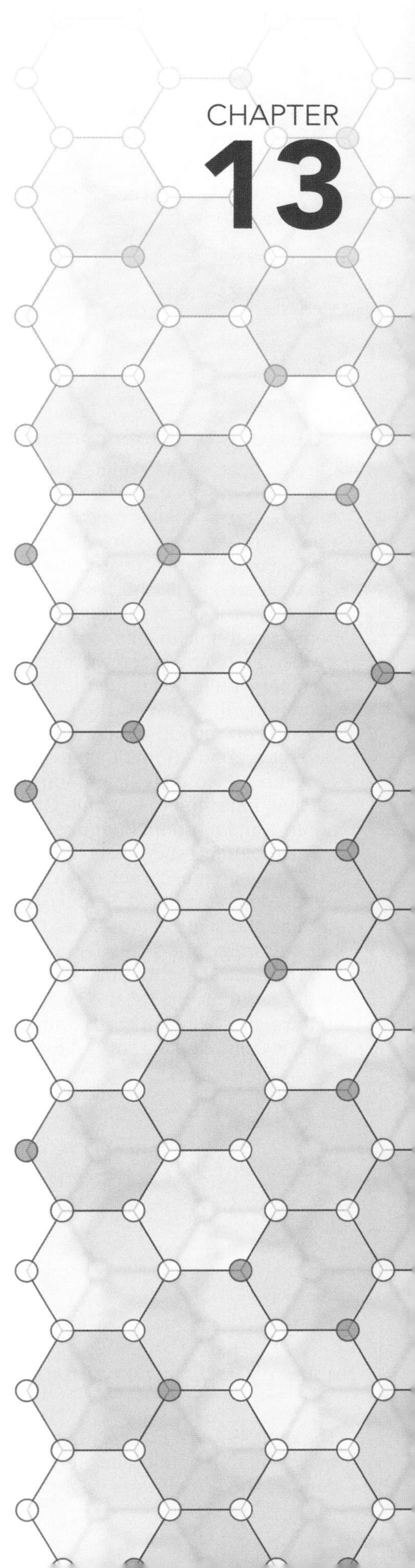

IN THIS CHAPTER

ENERGETICS OF MUSCLE CONTRACTION

CARDIAC MUSCLE BLOOD SUPPLY

CARDIAC MUSCLE METABOLISM

MARKERS OF CARDIAC DAMAGE

MEASUREMENTS MADE DURING MANAGEMENT OF MI

SKELETAL MUSCLE STRUCTURE AND METABOLISM

MARKERS OF SKELETAL MUSCLE DAMAGE

SPECIFIC DISORDERS OF SKELETAL MUSCLE

Muscle cells are characterized by their unique ability to undergo contraction and subsequent relaxation, the former process requiring the expenditure of metabolic energy through hydrolysis of adenosine triphosphate (ATP). There are three types of muscle in humans—cardiac, skeletal, and smooth—but in terms of significance to clinical chemistry, assessments of cardiac muscle function are by far the most relevant. Measurements on skeletal muscle are also useful in certain circumstances but no tests on smooth muscle are made routinely.

Cardiac, skeletal, and smooth muscle cells

Cardiac muscle consists of fully differentiated cells (cardiac myocytes) which can no longer divide. This is important because once the cells rupture, such as in myocardial infarction when the myocardium is damaged, they are unable to regenerate and repair the damaged tissue. Skeletal muscle, which accounts for approximately 40% of body weight in a healthy adult, is a major site of energy production both as chemical-bond energy in ATP, for intrinsic expenditure in contraction, and as heat, for maintenance of body temperature. Like cardiac myocytes, most cells in skeletal muscle are terminally differentiated and cannot divide. Skeletal muscle uniquely, however, also contains a few nondifferentiated muscle precursor cells (satellite cells) which, under an appropriate stimulus, can proliferate and divide into mature myocytes and repair damaged skeletal tissue. A layer of smooth muscle underlies the vascular endothelium and consists of cells (smooth muscle cells) which are not fully differentiated. Thus, these cells may divide in response to appropriate stimuli, such as hypertension, and repair damaged endothelium. They also participate in angiogenesis during recovery from major trauma. Smooth muscle is not discussed further in the current text.

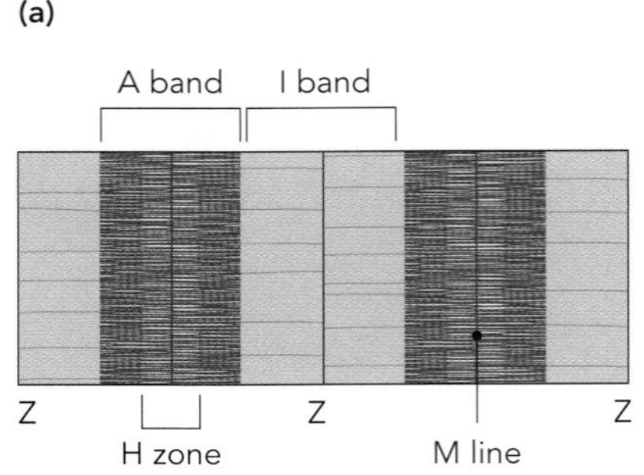

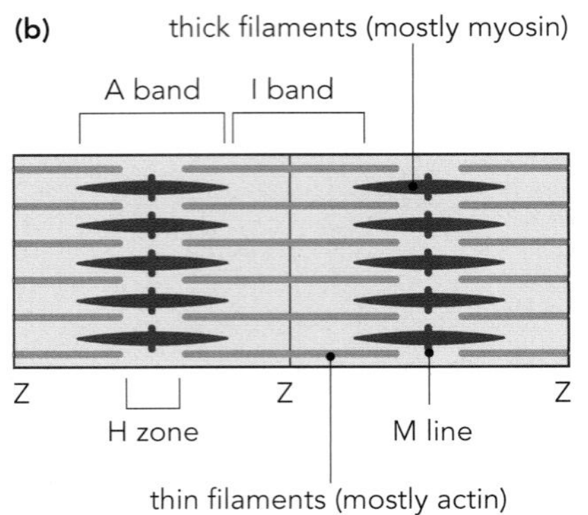

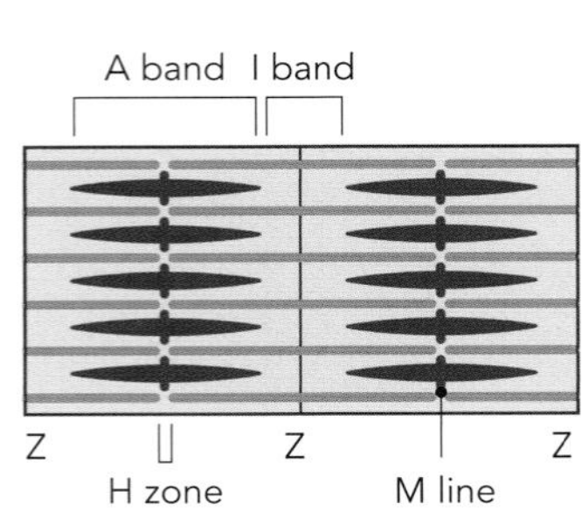

Figure 13.1 Schematic diagrams of (a) the arrangement of dark (A) and light (I) bands in striated muscle and the relationship between thick and thin filaments in (b) relaxed and (c) contracted muscle.
Note that, during contraction, the width of the A band remains constant but the width of the I band and the H zone decreases because of increasing overlap between the thick and thin filaments, with the thin filaments moving toward the M line. Z indicates the Z lines, which mark the boundaries of each sarcomere (contractile unit).

Structural features common to cardiac and skeletal muscle

Cardiac muscle and skeletal muscle are the two forms of striated muscle in the body. This means that when cells from these muscle types are viewed under the microscope, a regular repeating pattern of dark and light bands, termed striations, is observed (**Figure 13.1a**). Each striation consists of a highly organized arrangement of muscle proteins that are involved in contraction of the cell when the skeletal muscle shortens or the heart beats (**Figure 13.1b**). The dark bands (A bands) consist of thick filaments (myosin), whereas the light bands (I bands) consist of thin filaments (mostly actin, but with associated troponin and tropomyosin proteins). When the muscle cell contracts (**Figure 13.1c**), the thin filaments are pulled over the thick filaments, increasing the degree of overlap and shortening the muscle. The troponin proteins and tropomyosin help to control muscle contraction. When the sarcoplasmic calcium ion concentration is low (at rest around 10 nmol/L), tropomyosin obscures the binding site for myosin on actin. When the calcium ion concentration increases (to 1–10 μmol/L), the calcium ions bind to troponin C of the troponin complex, also consisting of troponin T and troponin I (**Figure 13.2**), which moves tropomyosin away from the myosin-binding site on actin, allowing myosin to bind to actin in the presence of ATP. When muscle cells are damaged, by ischemia for example, troponins are released into the blood. Slightly different isoforms of the troponins are present in cardiac and skeletal muscle and this is relevant in differentiating between damage to cardiac and skeletal muscle using laboratory assays.

The contractile proteins in muscle are anchored to the sarcolemma by a number of proteins, the most clinically significant of which is dystrophin. Mutations in the gene coding for dystrophin (*DMD*) result in the absence of the protein from muscle cells, causing several forms of muscular dystrophy.

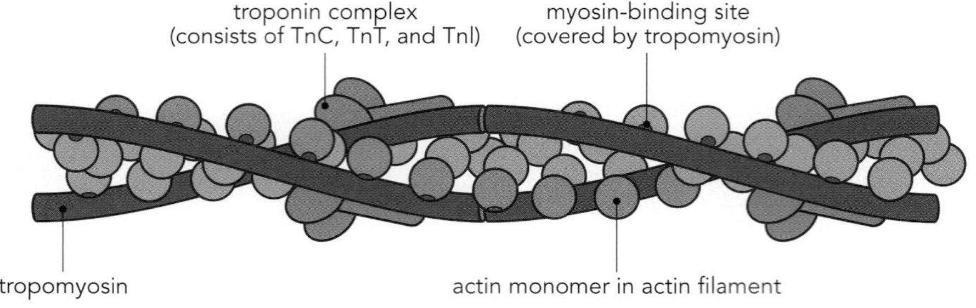

Figure 13.2 Diagram showing the relationship between tropomyosin, the troponin complex, and the actin polymer in the thin filament of striated muscle cells.
Tropomyosin sits in a groove formed by the spiral arrangement of the actin filament and, when calcium ion concentrations are low, covers the myosin-binding site on actin. When calcium ions are released from the sarcoplasmic reticulum, a calcium-induced conformational change in the troponin complex moves the tropomyosin out of the way of the myosin-binding site on actin, allowing muscle contraction to take place.
TnC, troponin C; TnT, troponin T; TnI, troponin I.

13.1 ENERGETICS OF MUSCLE CONTRACTION

The energetics of muscle contraction are similar for both cardiac and skeletal muscle in that both tissues rely on a continued supply of ATP generated by oxidation of body fuel stores of glucose and fatty acids. Under anaerobic conditions,

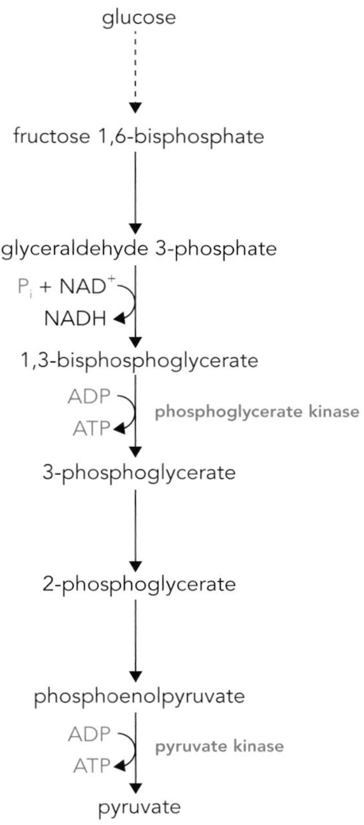

Figure 13.3 Anaerobic formation of adenosine triphosphate (ATP) by substrate-level phosphorylation in the glycolytic pathway.
For each molecule of glucose that enters the pathway, two molecules of glyceraldehyde 3-phosphate are formed, metabolism of each resulting in one molecule of NAD^+ reduced to NADH and two molecules of ATP formed from ADP, in reactions catalyzed by several enzymes. The names of the enzymes involved in the steps in which ATP is formed are given in blue. The dashed arrow indicates conversions that take at least two steps and solid arrows indicate single-step conversions. P_i, inorganic phosphate.

such as in rapid contractions in skeletal muscle during acute exercise (for example sprinting), energy for ATP synthesis is derived from glycolysis (**Figure 13.3**) and the end product of glycolysis in anaerobic conditions is lactate. With an adequate supply of oxygen, however, muscle cells are able to oxidize pyruvate to acetyl-CoA (see Chapter 2) and acetyl-CoA can enter the tricarboxylic acid (TCA) cycle for further oxidation to CO_2, with the subsequent oxidative phosphorylation generating greatly increased amounts of ATP. In this aerobic situation, fatty acids may also act as an energy substrate through the provision of acetyl-CoA via β-oxidation in mitochondria (**Figure 13.4**).

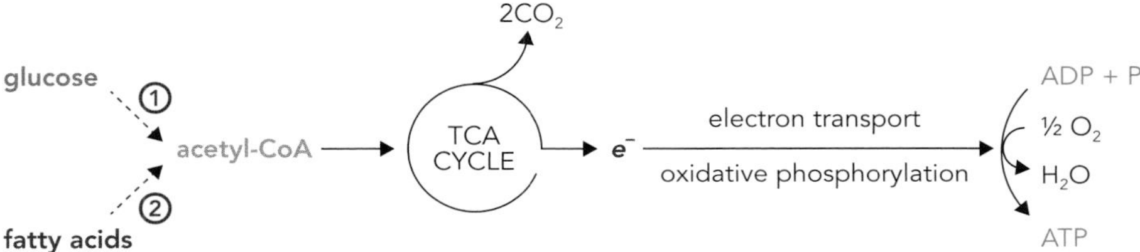

Figure 13.4 Aerobic formation of ATP by oxidative phosphorylation, where oxygen (O_2) is the final electron acceptor.
Acetyl-coenzyme A (acetyl-CoA) can be formed by glycolysis (1) or β-oxidation (2) and enters the tricarboxylic acid (TCA) cycle. The TCA cycle produces two molecules of carbon dioxide (CO_2) per molecule of acetyl-CoA and releases electrons that can be passed along the electron-transport chain in the mitochondrion. ATP is formed from ADP and inorganic phosphate (P_i) by the action of ATP synthase. Dashed arrows indicate conversions that take multiple steps.

Role of phosphocreatine

The total amount of adenine nucleotides (ATP + ADP + AMP) within the muscle cell, as in all cells, is limited and thus the store of ATP in muscle is theoretically maximal when all adenine nucleotides are converted to ATP. Muscle also contains another energy-rich molecule, phosphocreatine, which can donate phosphate to ADP, rapidly re-forming ATP during times of acute muscular activity. Creatine is formed in liver by methylation of guanidinoacetate, itself formed from arginine and glycine in the kidneys (**Figure 13.5**). Creatine enters the muscle cells via a specific transporter and is phosphorylated to phosphocreatine by the enzyme creatine kinase (formerly known as creatine phosphokinase, CPK) using

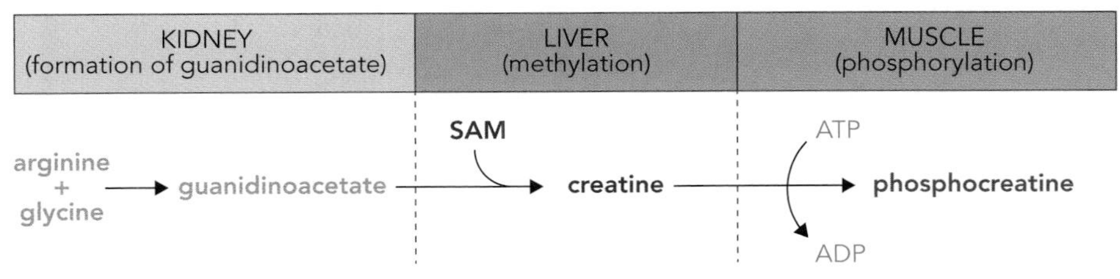

Figure 13.5 Synthesis pathway for phosphocreatine.
De novo synthesis of phosphocreatine involves three body tissues (kidney, liver, and muscle). SAM, *S*-adenosylmethionine, is the methyl donor for the formation of creatine from guanidinoacetate.

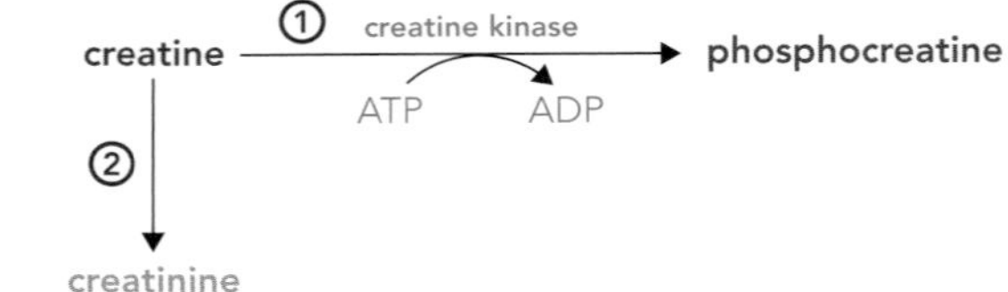

Figure 13.6 Metabolism of creatine.
1: Phosphorylation to phosphocreatine, catalyzed by creatine kinase.
2: Spontaneous degradation to the metabolic end-product, creatinine.

ATP as phosphate donor. Thus, phosphocreatine is formed when the muscle is at rest and ATP and oxygen are plentiful (**Figure 13.6**). Creatine kinase (CK) is a heterodimeric enzyme and tissue-specific differential expression of the two genes encoding the individual monomeric peptides (M and B) leads to three predominant CK isoforms (isoenzymes): MB in heart, MM in skeletal muscle, and BB in brain. There are high concentrations of CK in the sarcoplasmic reticulum of both cardiac and skeletal muscle. The CK isoform synthesized in cells of the myocardium, CKMB2, undergoes post-translational hydrolysis by a carboxypeptidase of its C-terminal lysine residue, yielding CKMB1 (**Figure 13.7**). This form is secreted and is normally present in the circulation. Damage to the myocardium releases nonprocessed CKMB2 and increases the CKMB2–CKMB1 ratio in the circulation.

As described earlier, ATP formed from normal oxidative metabolism in muscle at rest is used for synthesis of phosphocreatine and during bouts of muscular exercise, when ATP hydrolysis provides energy for contraction. ATP is regenerated by transfer of high-energy phosphate from phosphocreatine to ADP. Creatine degrades spontaneously to an inert molecule, creatinine, which diffuses out of the muscle and is excreted in the urine. Under normal physiological conditions, the concentration of creatine in muscle cells remains relatively constant and turns over regularly such that the daily urinary output of creatinine reflects the total muscle mass in the body.

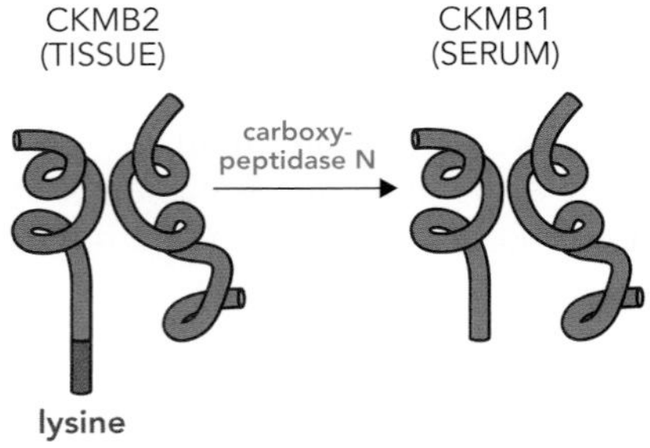

CKMB2 (TISSUE)
CKMB1 (SERUM)
carboxy-peptidase N
lysine

Figure 13.7 Creation of the serum-specific isoform of muscle/brain creatine kinase (CKMB1) by the action of carboxypeptidase N.

13.2 CARDIAC MUSCLE BLOOD SUPPLY

The heart is supplied with blood via two main coronary arteries (**Figure 13.8**), which originate from sinuses in the wall of the aorta just above two of the cusps of the aortic valve. The left coronary artery begins above the left posterior cusp

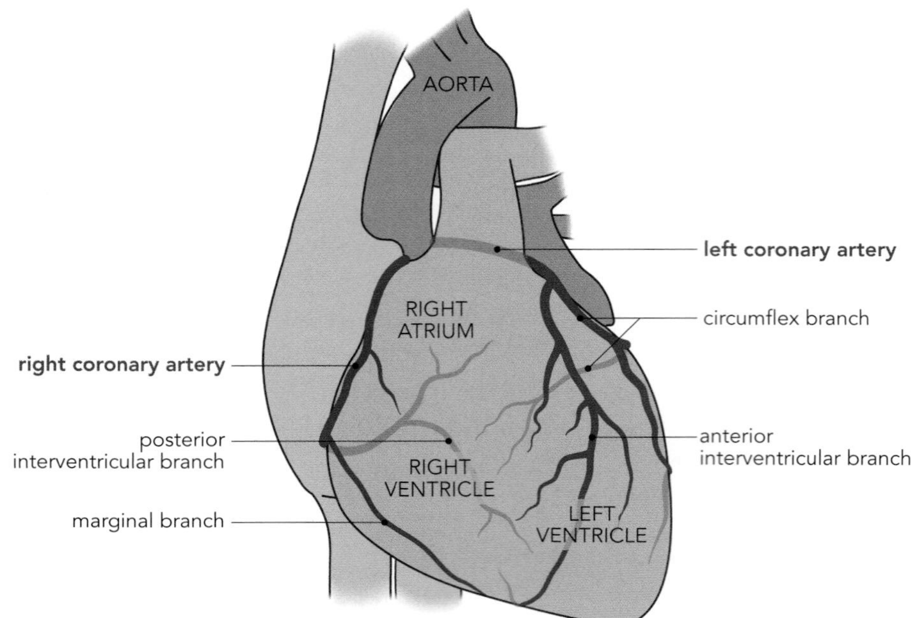

Figure 13.8 The coronary arteries and their main branches.
Note that the posterior parts of the heart are supplied by branches of both the right and the left coronary arteries.

of the aortic valve; the right coronary artery begins above the anterior cusp of the aortic valve. The left coronary artery and its branches (the anterior descending, or interventricular, branch and the circumflex branch) predominantly supply the left ventricle and left atrium. The right coronary artery and its branches (for example the marginal artery) predominantly supply the right ventricle and right atrium. However, there are some overlapping territories for both arteries, and small branches of both coronary arteries supply the heart's pacemaker, the sinoatrial node.

Coronary blood flow is matched to the metabolic demands of the tissue by a process known as autoregulation, whereby the by-products of cellular metabolism such as H^+ ions cause vasodilation. Thus, as activity of the heart increases, a vasodilator response allows blood flow (and oxygen availability) to increase.

If narrowing of the branches of the coronary arteries occurs due to the development of atherosclerosis, thromboembolism, or coronary vasospasm, the ability of the blood vessels to supply adequate amounts of blood and oxygen to meet the demands of the heart is compromised. The degree of occlusion of the arteries determines the clinical picture in the patient. With a relatively small reduction in the lumen of the artery, inadequate blood flow to meet demand may only occur during exercise; the chest pain that is characteristic of this is known as stable angina pectoris (stable because the chest discomfort or pain occurs only when the patient is active or exercising). When the degree of occlusion is more marked, inadequate blood flow to meet demand may occur at rest; the chest pain that is characteristic of this is known as unstable angina pectoris (unstable because it may occur at any time). In angina, chest pain can be relieved by nitrate tablets such as glyceryl trinitrate, which increase the degree of vasodilation of the arteries. Complete occlusion of a branch of one of the coronary arteries is known as myocardial infarction, and is characterized by central, "crushing" chest pain that is not relieved by rest or by nitrates. Myocardial infarction results in the death of myocardial cells and can cause long-term problems with ventricular function.

13.3 CARDIAC MUSCLE METABOLISM

Cardiac muscle metabolism under aerobic conditions

Unlike skeletal muscle, the activity of cardiac muscle is constant and rhythmic, the heart beating more than 2×10^9 times over a life span of 70 years. Fatigue of heart muscle leads to death and this is prevented in the tissue by high levels of aerobic metabolism. Cardiac muscle cells use oxidative metabolism to generate high amounts of ATP to provide the energy for contraction and also, because the heart is an excitable tissue, to maintain polarization of the plasma membrane by active pumping of various ions. The high level of oxidative metabolism is made possible by two factors: (i) the high number of mitochondria in cardiac myocytes; and (ii) by the heart being well vascularized, thereby being provided with a plentiful supply of oxygen and respiratory substrates such as glucose and fatty acids. Metabolic fuels are thus oxidized to CO_2 and water. In addition to ATP, a significant amount of energy is stored in cardiac muscle as phosphocreatine (creatine phosphate). Cardiac muscle cells also contain the oxygen-binding pigment myoglobin (see Chapter 10) which acts as an oxygen store in the tissue and helps to maintain a high partial pressure difference between coronary arterial blood and the inside of the cardiac muscle cell, thereby increasing diffusion of oxygen into the cell. Coronary blood flow is closely matched to cardiac activity thereby ensuring (a) that cardiac myocytes, again unlike skeletal muscle cells, do not build up an oxygen debt and (b) the rapid removal of carbon dioxide from the tissue.

In contrast to skeletal muscle, intrinsic fuel reserves of heart muscle are in short supply, being limited to small amounts of glycogen and phosphocreatine. The heart is thus dependent on fuels from blood to meet its metabolic demands.

In the fed state following a meal, glucose is the preferred, abundant substrate and is metabolized via the glycolytic pathway to pyruvate, which enters the mitochondrion and is oxidized to acetyl-CoA by pyruvate dehydrogenase (see Chapter 2). Acetyl-CoA is in turn oxidized in the TCA cycle to CO_2 and water with the release of energy to drive the synthesis of ATP. In this situation the respiratory quotient of the heart (the ratio of CO_2 produced to O_2 utilized) is 1.0.

As blood glucose levels fall to fasting levels (4–5 mmol/L or 70–90 mg/dL) between meals, albumin-borne fatty acids released from adipose-tissue triglyceride are preferentially extracted from the coronary circulation and become the major source of fuel for the heart. Following activation to their CoA esters in the cytosol, fatty acids are transported into mitochondria and undergo β-oxidation to acetyl-CoA. With a plentiful supply of oxygen, acetyl-CoA is oxidized via the TCA cycle, again yielding energy to drive ATP synthesis (see Figure 13.4). In this situation, the respiratory quotient for the heart is 0.75. This switch from glucose to fatty acids, even during short-term fasting between meals, conserves glucose for glucose-dependent tissues such as brain and nervous tissue.

As the body enters a more prolonged fasting state, ketone bodies (acetoacetate and 3-hydroxybutyrate) synthesized in liver from acetyl-CoA, become a preferred metabolic fuel for the heart. Acetoacetate is converted to acetoacetyl-CoA by the action of β-ketoacyl transferase, and β-ketothiolase hydrolyzes acetoacetyl-CoA to two molecules of acetyl-CoA (**Figure 13.9**). Here again, a plentiful supply of oxygen ensures oxidation in the TCA cycle to CO_2 and water and a maximum yield of ATP. 3-Hydroxybutyrate is oxidized to acetoacetate before following the same catabolic route.

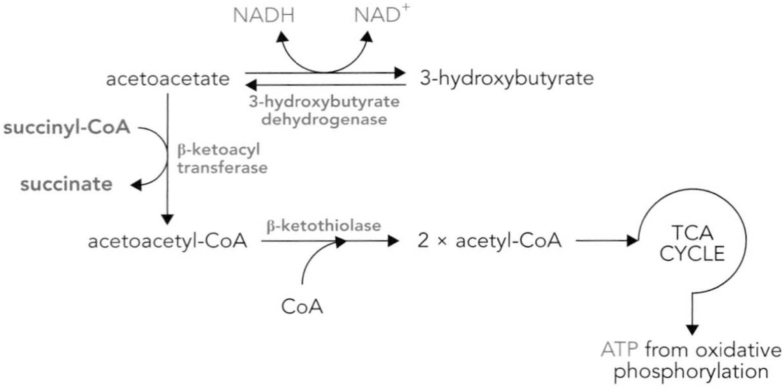

Figure 13.9 Metabolism of ketone bodies in muscle.
Important enzymes in ketone body metabolism are 3-hydroxybutyrate dehydrogenase, β-ketoacyl transferase, and β-ketothiolase. The coenzyme A (CoA) donor in green is succinyl-CoA, derived from the TCA cycle. The enzyme used for this part of the reaction is succinyl-CoA:acetoacetate-CoA transferase.

Cardiac muscle metabolism when oxygen supply is compromised

Despite these mechanisms to support the energy requirements of cardiac muscle, the cells cannot sustain activity if oxygen supply to the myocardium does not meet demand. Under such circumstances, oxidation of acetyl-CoA via the TCA cycle is compromised with a consequent reduction in ATP production, and the buildup of acetyl-CoA in the mitochondrion inhibits pyruvate oxidation by pyruvate dehydrogenase and β-oxidation of fatty acids. In order to continue the generation of ATP by anaerobic glycolysis, pyruvate is now reduced to lactate to generate the NAD^+ required for the continued activity of the glycolytic glyceraldehyde 3-phosphate dehydrogenase reaction (**Figure 13.10** and see Chapter 2).

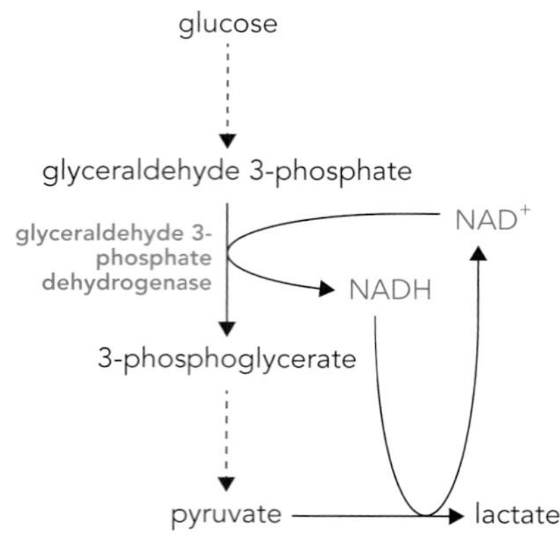

Figure 13.10 Regeneration of nicotinamide adenine dinucleotide (NAD^+) from reduced NAD (NADH).
The formation of lactate from pyruvate is an important reaction in maintaining a supply of NAD^+ for use in glycolysis. Dashed arrows indicate conversions requiring multiple steps.

The lactate dehydrogenase which catalyzes the redox reaction between pyruvate and lactate in cardiac muscle is a different isoform from that found in the liver and can be measured in plasma following myocardial damage.

Lactate is a strong acid and causes cell damage. Furthermore, a fall in ATP production results in defective pumping at the cell membrane with a consequent decrease in intracellular potassium concentration, and an inability to regenerate adenine nucleotides raises intracellular adenosine levels. Thus when coronary blood flow is compromised in any way, cardiac activity may not be able to meet the demands placed on the heart. Such a situation exists in stable angina pectoris, where a narrowing of the coronary arteries results in chest pain during exercise as the patient increases cardiac work above that made possible by the reduced cardiac blood supply. In this situation, the reduced blood supply is sufficient to maintain cardiac metabolism at rest. In unstable angina, however, where the cardiac blood flow is more restricted, pain may also occur at rest as damaged cardiac myocytes release lactate, potassium ions, and adenosine, all of which activate nociceptors and cause pain.

When a coronary artery is completely occluded, chest pain becomes acute and is associated with a myocardial infarction or heart attack. The pathological events leading to infarction may be summarized as follows:

- ATP synthesis starts to fall within 2 minutes and falls by 50% in 10 minutes
- Reduced ATP leads to membrane channel (Na^+/K^+-ATPase) disruption and an increase in cell membrane permeability
- Increased permeability leads to myocyte swelling
- Calcium influx activates hydrolases to disrupt cell function; cell death in 15–20 minutes

Without an adequate supply of oxygen, the cardiac myocytes become damaged and eventually necrose, releasing their contents into the circulation. This includes enzymes and other proteins present in the cytosol of these cells. Hence the cardiac-specific troponins and CKMB can be used in clinical assays as markers of myocardial cell damage. Whilst myoglobin is also released from the myocardium during damage, it is also present in skeletal muscle cells so is not per se a specific marker of myocardial cell damage.

13.4 MARKERS OF CARDIAC DAMAGE

As the understanding of the processes involved during a myocardial infarction (MI) has progressed and the treatment has become more successful, so the diagnostic tools to detect an MI have become more specific and sensitive. An MI occurs when a coronary artery becomes blocked and the resultant lack of oxygenated blood leads to acute tissue necrosis. This in turn leads to the classic symptoms (**Figure 13.11**), changes to the electrocardiogram (ECG; **Figure 13.12**), and the release of the contents of the heart cells into the bloodstream. Initially, cardiac enzymes released into the blood were employed as markers of cardiac damage; hence, the term "cardiac enzymes" was and is still used to request cardiac markers. Ideally, markers of cardiac damage should satisfy the following parameters:

- Be specific for heart muscle and not released by other damaged tissues (non-cardiac events)
- Be not normally present in serum
- Be always raised following myocardial infarction
- Show a significant rise post-damage
- Be detectable rapidly post-damage (within 6 hours of cardiac event)
- Be detectable early after clinical symptoms present

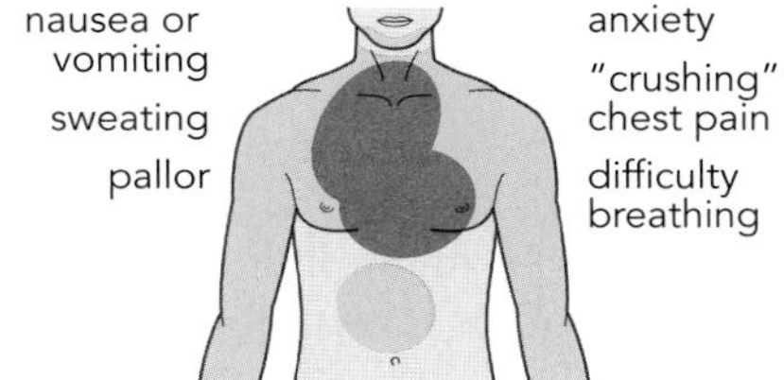

Figure 13.11 Common signs and symptoms of myocardial infarction. Patients undergoing myocardial infarction may exhibit a range of different signs and symptoms. The most common symptom is "crushing" chest pain, most likely in the area shaded red, but occasionally extending to the areas shaded pink. Note that all signs and symptoms will not be present in all patients and so a patient may present without the classical symptom of chest pain.

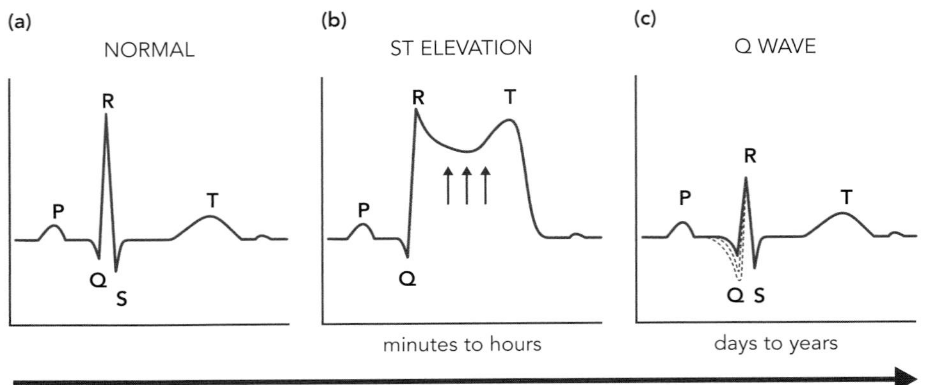

Figure 13.12 Characteristic changes to the electrocardiogram (ECG) caused by full-thickness myocardial infarction. (a) The normal ECG has a P wave representing atrial depolarization, a QRS complex representing ventricular depolarization, and a T wave representing ventricular repolarization. In the period between the QRS complex and the T wave (the ST segment), the ECG is normally isoelectric, running along the baseline of the trace. (b) Around the time of the myocardial infarction, the ST segment of the ECG becomes elevated. (c) The characteristic change in the ECG that is seen in patients who have suffered a previous myocardial infarction is the presence of pathological Q waves, where the Q wave becomes progressively broader and deeper than normal.

Candidate enzymes that were used in the past included lactate dehydrogenase (LDH), CK, and aspartate transaminase (AST), all of which had varying degrees of sensitivity to detect an MI. However, none was specific, since each enzyme was also present in tissues other than cardiac muscle, and they did not show early increases post-infarction. All three enzymes have measurable background activity in the plasma due to normal cell turnover (background noise; see Chapter 1) such that, when an acute event happens, there has to be a significant rise in the marker concentration above this background noise before a cardiac source can be confirmed with any degree of certainty. Clearly, the higher this background noise, the greater the rise in cardiac marker, and hence degree of tissue damage, that has to occur before a significant change can be measured in the blood.

The earliest detectable increase in the cardiac marker will be dependent upon the background noise and also upon the type and site of the cardiac damage. With a partial or total occlusion of the artery, the blood flow downstream of the blockage will either slow down or stop completely. The appearance of some markers in the peripheral bloodstream, however, may arise from alternative drainage from the damage site (for example via the thoracic system) and become detectable many hours after the acute event. The relative size of the increase in a cardiac marker will depend on the amount of damage that has occurred, the background noise, and the rate of clearance from the bloodstream, and raised concentrations may only be detectable many days after the acute event.

Damage to the myocardium is readily seen as acute and chronic changes in the ECG; an ST elevation develops within a few minutes to hours, and pathological Q waves develop over hours to days post-injury. The search for biochemical markers that might indicate the severity of injury and allow monitoring of the recovery of the patient following injury has been ongoing for a number of years and a list of candidates is given below. The figures in parentheses represent approximate molecular mass of tthe molecular marker.

- Aspartate transaminase (54.5 kDa)
- Creatine kinase MB mass (82 kDa)
- Creatine kinase MB1 and MB2 isoforms
- Creatine kinase total
- Glycogen phosphorylase b BB isoenzyme (195 kDa)
- Heart fatty acid-binding protein (15 kDa)
- Lactate dehydrogenase (140 [4 x 35] kDa)
- Myoglobin (16.7 kDa)
- Thrombus precursor protein (96.9 kDa)

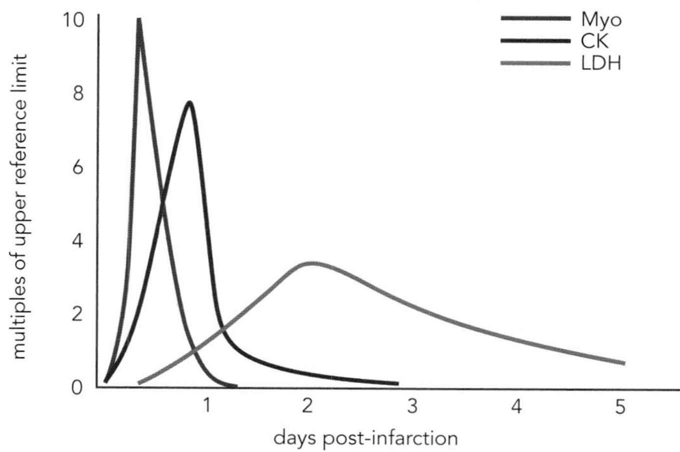

Figure 13.13 Time course of the appearance of protein markers of myocardial infarction in the blood.
Myoglobin (Myo) is released early, followed by creatine kinase (CK). Lactate dehydrogenase (LDH) is released late following myocardial infarction, reaching peak concentrations by 48 hours post-infarction. Note that none of these protein markers of myocardial infarction are ideal for its assessment, based either on tissue specificity and/or time of release.

- Troponin I (cTnI) (18 kDa)
- Troponin T (cTnT) (39 kDa)
- Ventricular myosin light/heavy chain (51/230 kDa)

Although all have been used as markers of cardiac damage at some stage, only two—cardiac troponins and CK—are in routine use currently in clinical practice. Differences in the size (mass) and intracellular location of these proteins impact on their relative efficacy as markers; the appearance of myoglobin, CK, and LDH is shown in **Figure 13.13** as an example.

Using the criteria listed above, myoglobin and LDH have proved to be poor markers. Myoglobin, the oxygen-binding protein in the myocyte cytosol (see Chapter 10), is rapidly released from damaged cells and also rapidly cleared from the circulation. However, as it is indistinguishable from skeletal muscle-derived myoglobin, it lacks specificity. LDH, also present in the cytosol, is a tetrameric enzyme with distinct isoforms. Although the liver and heart isoenzymes are readily distinguishable, this is not so for the cardiac and skeletal muscle enzymes; as such, LDH measurement is of low specificity and sensitivity. The same lack of specificity is obtained for AST and myosin light chain.

Fatty acid-binding protein, which is involved in myocardial lipid homeostasis, is abundant in the cytoplasm of cardiac myocytes and is released rapidly during cardiac injury. In addition, it is cleared rapidly from the circulation, which suggests a useful role as a marker of cardiac damage. While its measurement in plasma has been used on occasion to estimate infarct size, the heart is not the only source of the protein and it thus lacks specificity. In addition, no viable commercial assay is available due to problems with the assay calibration, stability, and matrix. Similar problems were experienced with the development of an assay for glycogen phosphorylase b BB isoenzyme. Because of such difficulties in interpreting results using these proteins, only two markers (cardiac CK and troponins) are in routine use clinically.

Creatine kinase

Measurements of plasma CK and LDH activities were routinely made in patients suspected of having cardiac damage but, as shown in **Figure 13.14**, neither total CK nor LDH activities rise early post-injury and are not specific to cardiac events.

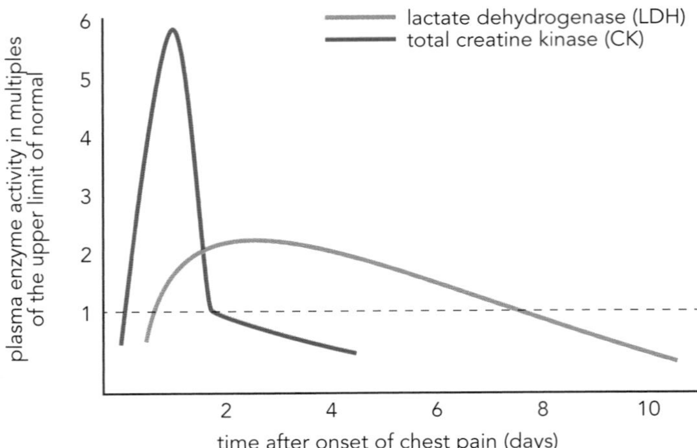

Figure 13.14 Time course of release of muscle enzymes following myocardial infarction, as assessed by enzyme activity assays. Creatine kinase is released early, followed by lactate dehydrogenase. However, note that because of a lack of tissue specificity and/or time of release, neither of these enzyme markers is ideal for the assessment of myocardial infarction.

Differential tissue expression of the genes encoding the isoforms of CK—CKMM, CKMB, and CKBB (described earlier)—allowed some improvement in specificity by measuring the CKMB isoform activity post-injury. However, although the CKMB isoform is more specific for cardiac damage it is not uniquely so, since CKMB contributes to 1–2% of skeletal muscle CK activity; thus, here too, measurement of CKMB enzyme activity is of limited use due to low specificity. Some improvement in specificity is gained by measuring changes in CKMB activity relative to total CK activity, but rises in CKMB activity occur not much earlier than total CK, and because of a general lack of specificity and difficulties in interpretation of results, this approach often did not justify the cost of measurement. As CKMB is the product of a different gene and hence has a unique protein structure, measurement of isoform mass by immunoassay greatly improves specificity. Even so, the rise in plasma CKMB mass does not occur much earlier than total CK activity (**Figure 13.15**). As described earlier, both CKMM and CKMB are themselves isoforms, the initial gene products undergoing proteolytic cleavage before appearing in plasma. Measurement of plasma CKMB1 is a promising candidate as a marker for early detection of cardiac injury.

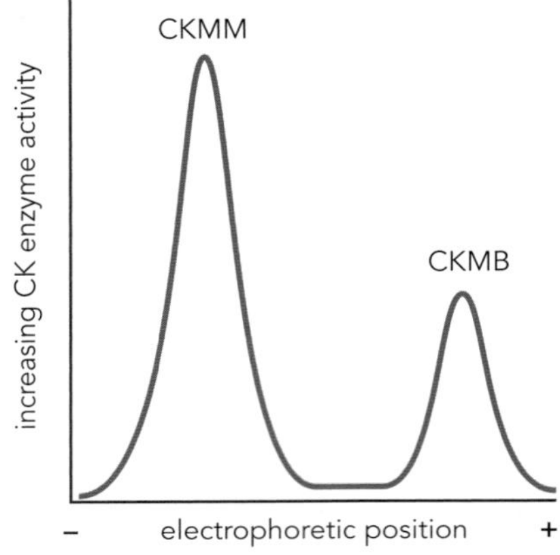

Figure 13.15 Relative abundance of creatine kinase (CK) isoenzymes in myocardial tissue. The cytosolic proteins of ventricular muscle were subjected to gel electrophoresis and the gel was subsequently stained for CK enzyme activity and scanned to quantify the CK activity. Two major CK fractions were identified: one (the muscle–muscle isoform, CKMM) with 72% of total CK activity, and the other (the muscle–brain isoform, CKMB) with 28% of the total CK activity. Cardiac muscle is the most abundant source of CKMB. Hence, the presence of CKMB in the patient's blood most likely indicates cardiac damage, although there is a small chance that it reflects damage to skeletal muscle (in which 1–2% of the CK exists as the CKMB isoform).

Troponins

Unlike CKMB, the troponins of cardiac muscle are tissue specific (**Figure 13.16**). The two cardiac troponins, cTnI and cTnT, are proteins present in the sarcomere and, having primary sequences quite distinct from their skeletal muscle counterparts, are readily identified. These proteins, which are not normally present in blood, are released early from damaged myocytes (**Figure 13.17**) and are easily detectable. A summary of the kinetics of their release from the time of initial cardiac damage and the range of clinical cutoff points is shown in **Table 13.1** and **Table 13.2**. While measurement of both cTnI and cTnT is very sensitive, particularly at 12 hours post-event, and also specific for cardiac damage, most studies in the UK have measured cTnT because of the problems of establishing a reliable cutoff for cTnI.

The differences in time over which the CKMB2 and CKMB1 isoforms are released compared to the troponins provide valuable clinical information with respect to the nature of the myocardial damage. It should be remembered that total serum CK amount/activity will include a contribution from skeletal muscle CK and this may increase with even minor trauma to this tissue. The specificity of the troponins as markers of cardiac damage can be seen in cases of severe skeletal muscle damage, such as rhabdomyolysis, where CK levels can rise to greater than 100,000 U/L while troponin levels remain within normal reference ranges. This specificity also contributes to the lack of background noise in the

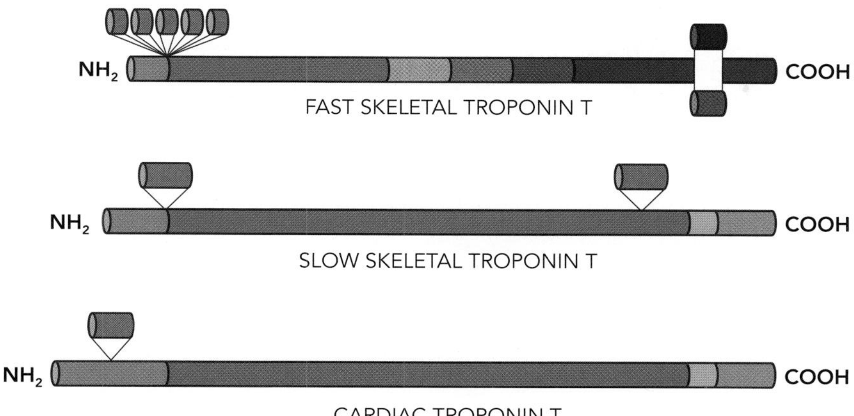

Figure 13.16 Specificity of various forms of troponin T for various types of striated muscle. All members of the troponin T superfamily of genes have a reasonably high degree of homology, with specific genes expressed in different types of striated muscle. Further variation in the isoform expressed in a given striated muscle cell type is also possible due to the potential for alternative splicing of the gene products, especially near the sections of the genes that encode for the N- and C-terminal portions of the proteins. Differences in the primary sequence of the various forms of troponin T have allowed development of blood tests that are specific for cardiac troponin T, which would be released from cardiac muscle during a myocardial infarction but not released by damage to other muscle types.

troponin assay, so, for example, in a particular MI where there may be a rise of two- to threefold in CK, there might well be a 100-fold rise in troponin T or troponin I. Thus, troponin assays detect much smaller degrees of damage than the older cardiac markers (CK or LDH). Currently, troponin T and I assays will detect virtually all MI cases 12 hours after the acute event. Equally as useful

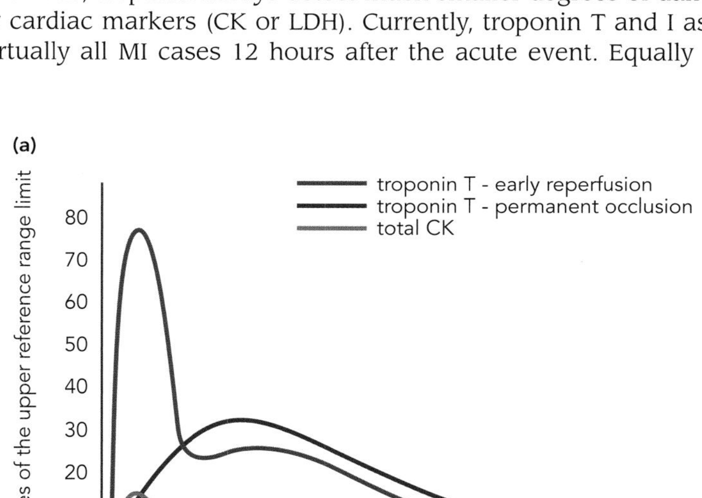

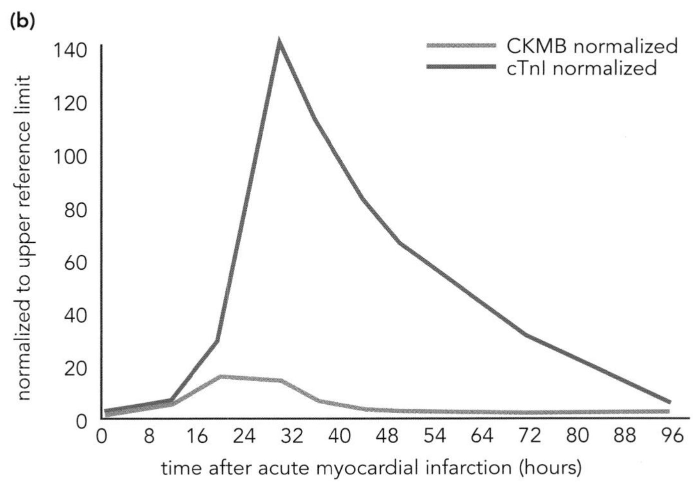

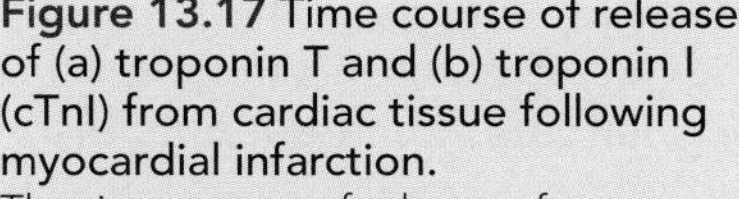

Figure 13.17 Time course of release of (a) troponin T and (b) troponin I (cTnI) from cardiac tissue following myocardial infarction. The time course of release of troponin T differs in situations where early reperfusion of the tissue is facilitated by thrombolysis or coronary angioplasty compared to situations where the occlusion is permanent. Note that whilst the time course of release of the troponins is not appreciably different from creatine kinase [total CK in (a) and CKMB in (b)], the specificity of the troponins for cardiac tissue and the amount released relative to normal blood levels make these the current gold-standard cardiac markers.

TABLE 13.1 Release kinetics for troponins post-myocardial infarction

Troponin form	Elevated	Peak	Duration
Troponin T	1–6 hours	10–24 hours	14 days
Troponin I	3–6 hours	18–20 hours	5–7 days

TABLE 13.2 Clinical cutoff points for troponins post-myocardial infarction

Troponin form	Clinical cutoff points
Troponin T	100–200 ng/L (0.1–0.2 ng/mL) – commonly reported
Troponin I	Different values are reported: 10 μg/L (Cummins et al.)[a] 6 μg/L (Hunt et al.)[b] 3.1 μg/L (Adams et al.)[c] 2.1 μg/L (Hafner et al.)[d] 0.1 μg/L (Trinquier et al.)[e]

a, Cummins B, Auckland ML & Cummins P (1987) *Am Heart J* 113:1333–1344; b, Hunt AC, Chow SL, Shiu MF, Chilton DC, Cummins B & Cummins P (1991) *Eur Heart J* 12:690–693; c, Adams JE 3rd, Bodor GS, Davila-Roman VG, Delmez JA, Apple FS, Ladenson JH & Jaffe AS (1993) *Circulation* 88:101–106; d, Hafner G, Thome-Kromer B, Schaube J, Kupferwasser I, Ehrenthal W, Cummins P, Prellwitz W & Michel G (1994) *Clin Chem* 40:1790–1791; e, Trinquier S, Flécheux O, Bullenger M & Castex F (1995) *Clin Chem* 41:1675–1676.

clinically, a troponin T or I result within the reference range rules out the diagnosis of an MI if the sample had been taken 12 hours after the supposed acute event. This vastly increased sensitivity has resulted in new terminology when assessing patients with MI. Since detectable rises in troponins can be seen without classical ECG changes, the term acute coronary syndrome (ACS) is used to cover a spectrum of different degrees of acute cardiac damage. Newer assays for troponin T can measure as little as 3 ng/L (0.003 ng/mL) with a diagnostic cutoff of 14 ng/L (0.014 ng/mL). Therefore very small degrees of acute myocardial damage can now be detected. However, this greatly improved sensitivity has raised new problems in cases where small increases in troponins, which were not due to classical acute myocardial damage, have been detected. Apart from a few artifacts, such as "macro-troponin" and heterophilic antibodies, most cases of minimally raised troponins are due to spillage of the protein into the bloodstream or the slowing down of its removal from the bloodstream. Inflammatory conditions that can affect heart tissue may also lead to chronically, slightly raised troponin levels of up to 100 ng/L (0.1 ng/mL) for troponin T. In addition, some patients with chronic renal failure have slightly raised troponin T levels—usually <100 ng/L (<0.1 ng/mL)—due to reduced clearance from the bloodstream. Such patients also have a greater risk of developing acute myocardial damage, due possibly to the minor chronic damage of inflammation. It is thought that troponins could be slightly raised in many chronic conditions and, in order to establish unequivocally the presence or absence of acute myocardial damage in patients without other definitive evidence of this (for example ECG), a second blood sample should be taken 6 hours after the first sample; a change of >20% in troponin is usually indicative of acute myocardial damage.

Clinical practice point 13.1

Cardiac troponins are extremely sensitive and specific markers of myocardial cell damage. However, they do not indicate the cause of the damage and may be raised in a number of non-cardiac systemic illnesses as well as acute coronary syndromes. The magnitude and time course of the rise and fall in troponin, plus the clinical history, are essential for interpretation.

Myoglobin

As for all the other factors discussed earlier, the molecular nature of the marker itself (size and three-dimensional shape) also probably influences the rate at which it is raised significantly in blood. It is likely that globular and low-molecular-weight proteins will appear earlier in the blood than larger proteins. Myoglobin for example is often detectable in plasma within a couple of hours after the acute event and its measurement is used in conjunction with CKMB and troponin I in a point-of-care device for the early detection of MI. Myoglobin is cleared rapidly from the circulation by the kidneys, and plasma concentrations may be within the reference range 12–24 hours post-event; however, while not specific for cardiac muscle, it can be used in the right clinical setting as an early test to exclude MI. Many other markers have been put forward but none are used routinely to any great extent.

An alternative approach being considered currently is to search for proteins of the hemostatic system which may be present in plasma during or immediately before the process of the clot formation in the coronary artery. At this early stage of injury where there is still blood flow, such markers might be measurable immediately in the bloodstream, making them ideal for detection of MI. Apart from a couple of speculative candidates (for example thrombus precursor protein) no such assays are available at present but they may represent the future in this field.

13.5 MEASUREMENTS MADE DURING MANAGEMENT OF MI

The mainstay of modern treatment of an MI is early intervention to minimize the extent of the damage to the cardiac cells, which continues for many hours and even days after the initial occlusion. The revascularization of the cardiac tissue

CASE 13.1

A 49-year-old man is seen by his primary care physician. Three days previously he had awoken with central chest pain radiating down his left arm. The pain only lasted 30 minutes and as he was away on business he did not seek medical attention the same day. He had previously experienced similar but less severe pain when jogging. An ECG showed no significant changes, but cardiac enzymes were requested and showed the following.

	SI units	Reference range	Conventional units	Reference range
Plasma				
Creatine kinase	123 U/L	40–150	123 U/L	40–150
Cardiac troponin T	97 ng/L	0–10	0.097 ng/mL	0.0–0.010

- **What does the troponin T signify?**
- **What is the most likely diagnosis?**
- **What two explanations are there for the normal CK?**

An elevated level of cardiac troponin T (or troponin I) signifies cardiac muscle injury. Although this is often presumed to indicate myocardial infarction (MI) there are other conditions which may cause low-level myocardial injury, and the test result must always be interpreted in conjunction with the clinical presentation. In this case there is a classic history of a heart attack and this is the likeliest cause, despite the nondiagnostic ECG. He also gave a history of exertional angina, which is an indicator of underlying coronary artery disease. The CK is often raised in MI and its peak level is proportional to the amount of myocardial necrosis that has occurred. A normal CK may be seen if there is only minimal necrosis or if the peak (at 24–48 hours) is missed. Troponin remains elevated for several days, and up to 2 weeks in some cases, depending on its peak level and rate of clearance.

greatly improves mortality and morbidity. Clot-buster drugs like streptokinase are given routinely to patients to hydrolyze the clot and so minimize the cardiac damage. In addition, the emergency use of cardiac artery stents is also used widely. Whichever therapy is chosen, the earlier it is used the better. It is recommended that streptokinase is given within 12 hours of the MI, and ideally within one hour. Clearly, the earlier a cardiac marker rises, the more useful it will be. Due to the nature of the damage occurring during an MI in some patients, several hours may elapse before the marker becomes significantly raised. Similarly, the ECG can remain normal for several hours after the event. Many patients will have significant rises in troponins within a few hours, but they can only be used to rule out an MI, with any degree of certainty, 6–12 hours after the acute event. This has led to several other markers being put forward as potential very early markers of myocardial damage.

Cardiac markers may also be used to screen patients for cardiac failure. Levels of cardiac peptides such as atrial natriuretic peptide (ANP) and the inappropriately named brain natriuretic peptide (BNP) are found to be higher in the circulation of patients with cardiac failure. BNP and its more stable precursor protein pro-BNP are used routinely to screen patients being referred for echocardiography, and may also be of use in the emergency setting to triage patients with shortness of breath.

13.6 SKELETAL MUSCLE STRUCTURE AND METABOLISM

Skeletal muscles are organized into bundles of muscle fibers called fascicles. Within each fascicle there exists a mixture of different types of muscle fibers. Some will be fast muscle fibers and some slow, reflecting the kinetics of ATP hydrolysis by the myosin in the different types. The fast fibers usually have a greater diameter than the slow fibers and are especially prevalent in muscles where short bursts of powerful contractions are required; the majority of fibers in the thigh muscles of a sprinter would be of this type. In view of their pattern of use, the fast fibers do not rely on oxidative metabolism; they are categorized as fast glycolytic and fast oxidative-glycolytic. In comparison, the slow fibers have a smaller diameter and higher amounts of myoglobin and are especially prevalent in muscles used for maintenance of posture, which have to resist fatigue. The difference in diameter of the two types is related to the ability of oxygen to diffuse to the center of the muscle fiber; since diffusion is inversely proportional to the distance over which the oxygen has to diffuse (Fick's law), there is an upper limit imposed on the diameter of the slow fibers by the need for oxidative metabolism. As in cardiac muscle, skeletal muscle fibers also contain CK, this time mainly the CKMM isoform.

Skeletal muscle metabolism

This discussion will not be concerned with metabolic differences between slow- and fast-twitch skeletal muscle fibers. Metabolism is described only to illustrate material relevant to clinical chemistry investigations into muscle damage.

Skeletal muscle is a major site of storage of glycogen, which provides an immediate source of glucose for energy production in the tissue. When oxygen delivery to skeletal muscle is adequate, as in aerobic exercise, glucose is metabolized to CO_2 and water via glycolysis and the TCA cycle to provide ATP for muscle contraction. Similarly, between meals and during prolonged low-intensity exercise, fatty acids may also serve as a substrate for energy production, provided that the oxygen supply to the tissue is sufficient. Skeletal muscle regenerates its store of phosphocreatine, using ATP and its tissue-specific isoform of CK (CKMM), between periods of physical activity.

CASE 13.2

A 78-year-old woman is found collapsed on the floor of her home. Although now conscious, she had been unable to get up and had spent 18 hours lying in the same position. On arrival in hospital, her biochemistry results showed the following.

	SI units	Reference range	Conventional units	Reference range
Plasma				
Sodium	137 mmol/L	136–142	137 mEq/L	136–142
Potassium	6.1 mmol/L	3.5–5.0	6.1 mEq/L	3.5–5.0
Urea (blood urea nitrogen, BUN)	23 mmol/L	2.9–8.2	64 mg/dL	8–23
Creatinine	352 μmol/L	53–106	4.0 mg/dL	0.6–1.2
Creatine kinase	30,000 U/L	40–150	30,000 U/L	40–150
Cardiac troponin T	253 ng/L	0–10	0.253 ng/mL	0.0–0.010

- **What does the creatine kinase (CK) of this level indicate?**
- **What complication has occurred?**
- **What is the significance of the troponin T result?**

CK is released from both cardiac and skeletal muscle following cell necrosis. However, because skeletal muscle mass is so much larger than cardiac muscle mass, extremely high levels of CK (>5000 U/L) indicate primarily skeletal muscle damage. Generalized muscle injury is called rhabdomyolysis and may have ischemic, toxic (for example drugs or snake venom), or physical (for example electrocution) causes. In this case, lying in one position for many hours caused pressure necrosis and ischemic injury to muscle. The myoglobin released is toxic to the renal tubules and can cause acute kidney injury, as is seen here. This may also be contributed to by dehydration from lack of access to fluids. Troponin T is highly cardiac specific and an elevated level indicates myocardial injury. This is most likely to be a myocardial infarction (MI) as the underlying cause for her collapse. In the elderly, the classic symptom of chest pain may be absent: so-called "silent MI."

However, when physical exercise becomes more intense, such as in sprinting, and oxygen supply to the muscle does not meet the demand for oxidative phosphorylation, pyruvate metabolism switches from oxidation to acetyl-CoA to reduction to lactate as the cells build up an oxygen debt. As mentioned earlier, this production of lactate is necessary to oxidize NADH generated during glycolysis to NAD^+, to allow glycolysis to continue. This is essential, since in the absence of oxidative phosphorylation, glycolysis is the only source of ATP for continuing muscle contraction. Lactate is a toxic, dead-end product in skeletal muscle and it diffuses out into the blood and circulates to the liver. Here it is oxidized back to pyruvate, which may undergo gluconeogenesis to glucose (see Chapter 2 for the Cori cycle) or be oxidized to CO_2 and water.

Although it is relatively rare for skeletal muscles to become ischemic, they can of course be damaged by trauma, such as fractures or crush injuries, or in compartment syndrome where pressure in a given fascial compartment rises due to inflammation or injury. Death and destruction of skeletal muscle cells is termed rhabdomyolysis. In these circumstances the contents of the cells can enter the circulation and may potentially be used as markers of skeletal muscle damage. However, the most important consequence of increasing concentrations of myoglobin, potassium ions, and calcium ions is the potential damage that these substances can do to other body systems.

Myoglobin is nephrotoxic and, when present in the blood, can cause acute renal injury and renal failure. Potassium ions leaking from damaged muscle cells can lead to hyperkalemia and this has the potential to cause fatal cardiac arrhythmias due to its effects on membrane potential in cardiac muscle. Hyperkalemia may potentially be exacerbated by acute renal injury, caused by myoglobin, because excess potassium is normally excreted via the kidneys.

Muscle cells can also be damaged by degeneration and necrosis in the various forms of muscular dystrophy. Levels of total CK can be highly elevated in these patients.

Analytical practice point 13.1

CK may be 100 or more times the upper reference limit in rhabdomyolysis, compared to 10 times in a large myocardial infarction. This is due to the relative masses of skeletal and cardiac muscle.

13.7 MARKERS OF SKELETAL MUSCLE DAMAGE

Disorders of skeletal muscle are relatively uncommon conditions but include both genetically determined and acquired disorders. Examples of such genetically determined disorders include:

- Dystrophies
- Metabolic myopathies
- Periodic paralysis
- Myotonias

whilst examples of acquired disorders include:

- Myasthenic disorders
- Polymyositis and other inflammatory disorders
- Hyperthyroidism, Cushing's syndrome, or high-dose steroid
- Osteomalacia
- Drug- and ethanol-induced myopathies
- Tumors

Investigations include serum CK activity and electromyography. Skeletal muscle biopsies and the techniques of molecular biology are invaluable for investigation into specific enzyme deficiencies associated with metabolic and mitochondrial myopathies and other genetic abnormalities. In practice, clinical chemistry investigations of damaged skeletal muscle are generally restricted to measurement of the serum CK amount/activity. In this case, the CKMM isoform is in high concentration in the cell cytosol and is released into the circulation on damage to the tissue. CKMM is raised in blood in many diseases of skeletal muscle and also in a number of other situations not associated per se with disease. For example, serum CK remains high for several days in normal adults following strenuous exercise, such as lifting heavy weights, and also in the following instances:

- Minor muscle trauma
- Adults with large muscle mass
- Some individuals of African origin
- Involuntary prolonged muscle contraction (tetany)
- Motor neuron disease and other neuromuscular disorders

Measurement of other enzymes including LDH and the transaminases (AST and alanine transaminase, ALT) has been made on occasion in the past, but these enzymes are also present in high concentration in the liver and are released into the circulation during liver disease. Their measurement thus lacks both sensitivity and specificity for skeletal muscle damage and is no longer considered to be of use diagnostically.

13.8 SPECIFIC DISORDERS OF SKELETAL MUSCLE

Duchenne muscular dystrophy

Duchenne muscular dystrophy, an X-linked recessive disorder, is the most common muscular dystrophy with an incidence of 1 in 3000 in boys. The defect presents as a lack of expression of functional dystrophin and mutations in the gene are not uncommon. It presents by the age of 3–4 years and the progressive destruction of skeletal and cardiac muscle usually leads to death by the age of 25 years. Clinical diagnosis is confirmed by a very high serum CK activity. This activity declines progressively with the gradual loss of muscle, as does the urinary creatinine. Carriers can be detected by measurement of serum CK activity and the use of a genetic probe; the latter is available for prenatal diagnosis of the disorder.

Myasthenia gravis

Myasthenia gravis, which presents with muscle fatigue, is manifest mainly in middle-aged women. Patients synthesize autoantibodies to the acetylcholine receptor, leading to its destruction. The disease is associated with other autoimmune disorders and disorders of the thymus (70% of myasthenia gravis patients have thymus hyperplasia, and 10% have a thymus tumor). It is important to differentiate the disease from Eaton–Lambert myasthenia and small-cell carcinoma of the bronchus.

Malignant hyperthermia

Malignant hyperthermia is a rare autosomal dominant disorder characterized by hyperthermia, lactic acidemia, and myoglobinuria during and following anesthesia. It is due to the release of excess calcium from the sarcoplasmic reticulum, leading to uncoupling of oxidative phosphorylation and release of oxidative energy as heat. Serum CK activity is raised by approximately 7% in at-risk individuals.

CASE 13.3

A 5-year-old boy is seen by a pediatrician because of difficulty walking and muscle weakness. On examination he has muscle wasting affecting his thighs and shoulders but enlargement of the calf muscles. An initial blood test shows the following.

	SI units	Reference range	Conventional units	Reference range
Plasma				
Creatine kinase	10,500 U/L	40–150	10,500 U/L	40–150

- **What is the source of the high CK?**
- **Why is it significant that the patient is male?**
- **What is the prognosis of his condition?**

The CK is extremely high, indicating severe skeletal muscle pathology. The boy has Duchenne muscular dystrophy, an X-linked genetic disease. The underlying disorder is a mutation in the gene for dystrophin (an important structural protein in muscle) which is carried on the X-chromosome. Females may be carriers but only males are clinically affected. There is progressive loss of skeletal muscle function and, by early adulthood, nearly all patients are paralyzed below the neck. Life expectancy is around 25 years, although some people may survive longer. Cardiac muscle may become involved in later stages of the disease.

GASTROINTESTINAL TRACT

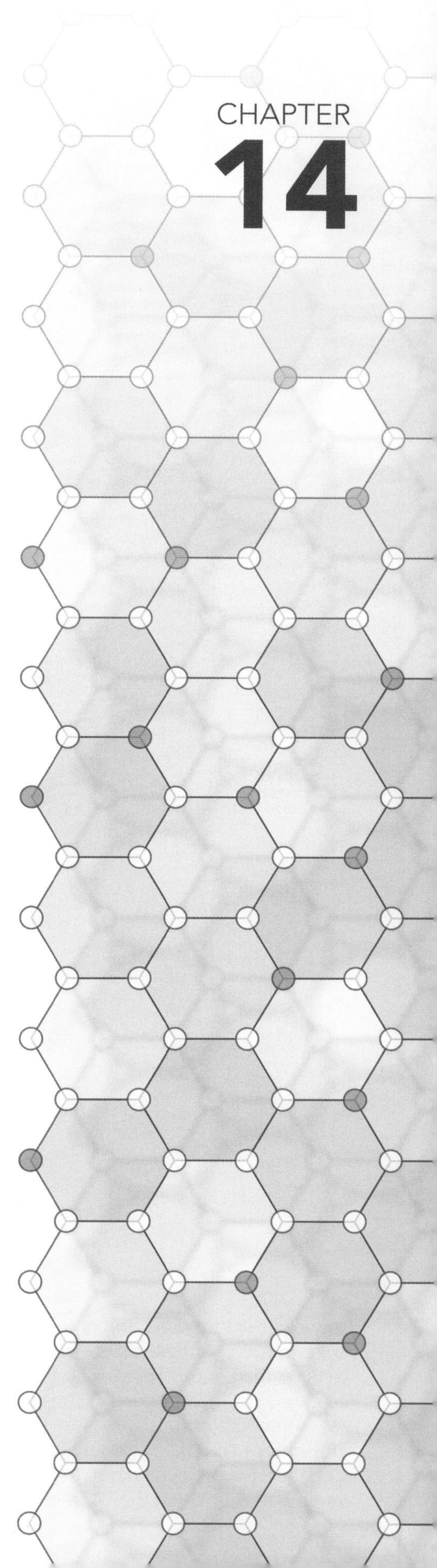

IN THIS CHAPTER

The gastrointestinal (GI) tract is a set of organs whose main functions are the digestion and absorption of food and the excretion of solid waste matter in the form of feces. From oral to anal end, the parts of the GI tract are: mouth and oropharynx, esophagus, stomach, small intestine (duodenum, jejunum, and ileum), large intestine (cecum, ascending colon, transverse colon, descending colon, sigmoid colon, and rectum), anal canal, and anus (**Figure 14.1**). Each of these regions has a specific role to play in digestion and absorption of food, as shown in **Table 14.1**.

14.1 MAJOR FUNCTIONS OF THE GI TRACT

The GI tract is the means by which we obtain nutrients from our environment. Food is consumed and is broken down mechanically by chewing and by churning movements of the GI tract, during which it is mixed with digestive juices that help to break it down into its chemical constituents. The digestive juices are produced by glands in the wall of the GI tract or by accessory structures such as the pancreas and liver. The chemical constituents of food, comprising sugars, fats, amino acids, vitamins, and minerals, are absorbed across the wall of the GI tract into the body before being used in metabolic processes or else stored for future use in the form of complex carbohydrates (for example glycogen) or as adipose tissue.

The GI tract has other functions in addition to dealing with the food that we consume. It also has endocrine functions and produces a number of hormones that are released into the bloodstream. Some of these hormones are designed to alter the activity of other parts of the GI tract or its accessory glands whilst others have more widespread roles in the body. For example, secretin is a hormone produced by the small intestine that alters stomach and small-intestinal function as well as increasing the alkalinity of the secretions being produced by the pancreas. Its role is therefore relatively local and serves to create the correct

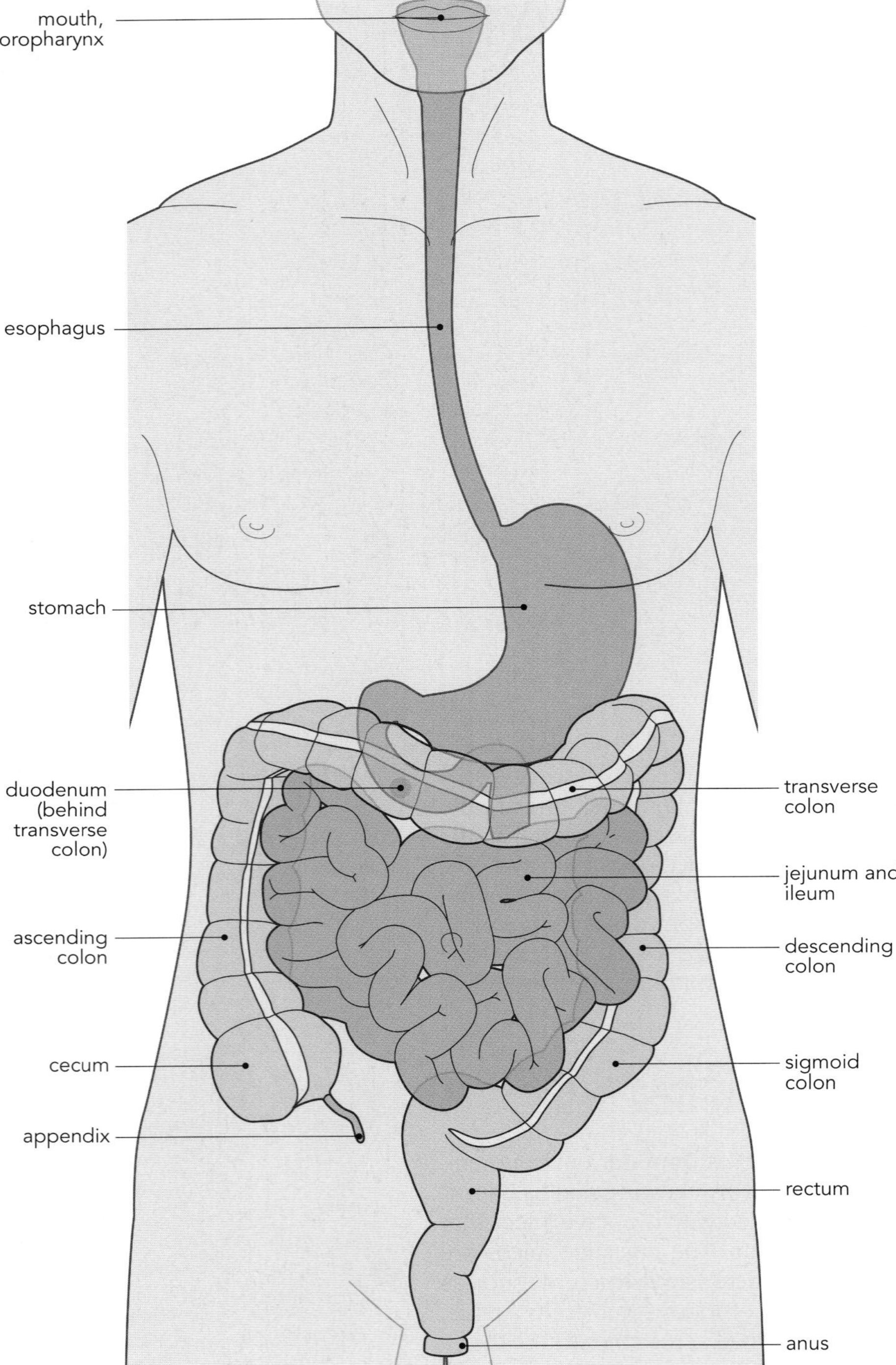

Figure 14.1 Schematic diagram of the parts of the gastrointestinal tract.

environment for digestion of food in the small intestine. The hormone cholecystokinin (CCK), on the other hand, as well as having local effects on the GI tract, acts on the central nervous system (certain brainstem and hypothalamic nuclei) to cause feelings of satiety. Thus its effects are both local and distant.

A third major function of the GI tract is that it acts as a barrier to infection and poisoning. The GI tract is lined with epithelial cells and the lumen of the GI tract can therefore be considered to be an outside surface of the body. There are three main ways in which the GI tract performs this barrier function. First, once food

TABLE 14.1 Broad functions of the parts of the gastrointestinal tract

Part of GI tract	Functions
Mouth and oropharynx	Lubrication, chewing, swallowing
Esophagus	Coordinated propulsion of food to stomach
Stomach	Mechanical digestion, chemical digestion, defense against infection (acid environment), regulation of appetite
Small intestine	Chemical digestion and absorption of nutrients
Large intestine	Final absorption of water and electrolytes, storage, compaction
Rectum	Storage of feces
Anal canal and anus	Continence/defecation

has passed down the esophagus and entered the stomach, it is immersed in the highly acidic environment of the stomach (see below). The pH of the stomach contents after a meal is usually just above 3 (that is, the H^+ ion concentration is somewhere between 0.5 and 1 mmol/L), which is said to be sufficiently acidic to kill many bacteria. Second, a large amount of mucosa-associated lymphoid tissue (MALT) exists in the wall of the GI tract, with the greatest amount in the small intestine. Like lymphoid tissue elsewhere, the function of these patches of white blood cells is to act as sentinels and facilitate the destruction of foreign organisms that may have broken through the epithelial cell barrier in the GI tract. Third, the large number of chemoreceptive cells in the gut can signal the presence of noxious substances (such as alcohol or bacterial toxins) and induce vomiting.

It should be mentioned that not all microorganisms that find their way into the GI tract are necessarily harmful to us. A large number of commensal bacteria live in the GI tract. The purpose of these bacteria is to help us to digest food and excrete substances such as bilirubin (see Chapter 3). One only has to take a course of oral antibiotics, which can kill some of these bacteria, in order to discover how important they are for gut function; diarrhea is one of the most common side effects of oral antibiotic therapy.

The accessory organs of digestion, such as the liver and pancreas, have ducts that connect them to the gut tube. Secretions produced by these organs (bile in the case of the liver and a bicarbonate- and enzyme-rich secretion in the case of the pancreas) pass down these ducts and enter the lumen of the gut, where they are able to exert their effects.

14.2 SALIVARY SECRETION AND ESOPHAGUS

Salivary glands and secretion

Food that has been broken down into smaller pieces by chewing can then begin to be digested by enzymes present in saliva. Saliva is a hypo-osmotic secretion that contains proteins and mucus and has a neutral pH. It is produced by three sets of ducted glands present in the oral cavity; the sublingual, submandibular, and parotid glands. The sublingual and submandibular glands produce a secretion that is rich in mucins (large, heavily glycosylated proteins), which help to lubricate the mouth. The sublingual glands have more mucus-producing cells than the submandibular glands. The parotid glands produce a secretion that is rich in enzymes, such as amylase, and other proteins that have an antimicrobial action. Saliva also contains significant amounts of lysozyme, immunoglobulin A (IgA), and lactoferrin, all of which improve defense against infection.

The purpose of the enzymes secreted by the parotid glands has been the focus of recent debate. In a test tube and in a pH-neutral environment, it is true that salivary amylase can catalyze the breakdown of starch into disaccharides such as maltose and other longer-chain oligosaccharides. However, it is not clear that this is the role of amylase *in vivo*; whether enough amylase penetrates the relatively large bolus of food in the mouth or whether there is enough time for the amylase to have its effect is uncertain. An alternative theory suggests that the role of amylase is to help keep the mouth clean by limiting any buildup of starchy foodstuffs, thereby limiting the likelihood that the growth of microorganisms in the mouth may become problematic. The role of bicarbonate ions may be to help neutralize acids that are present in foods and produced by bacteria. If the function of salivary glands is impaired, this can lead to tooth decay, yeast infections, and inflammation in the mouth. A small amount of lipase (lingual lipase) is produced in the mouth, but this enzyme only acts on the small amount of dietary triglycerides containing short-chain fatty acids.

Swallowing and peristalsis

The act of swallowing is a complex response to the introduction of food or saliva to the posterior surface of the tongue. Muscles supplied by the glossopharyngeal and vagus nerves (cranial nerves IX and X) propel the food backward, the cricopharyngeal sphincter (upper esophageal sphincter, UES) opens, and the nasopharynx closes, allowing food to pass into the esophagus. The majority of these actions are mediated (in both sensory and motor terms) by the vagus nerve and, to a lesser extent, the glossopharyngeal nerve. Thus, brain injuries (for example a stroke) or neurodegeneration in the medulla can lead to potentially life-threatening deficits in this process.

Once the bolus of food has been introduced into the esophagus, it is propelled toward the stomach by a process known as peristalsis, whereby coordinated waves of smooth muscle relaxation and contraction help propel food along the gut tube. Peristalsis is an intrinsic mechanism of the gut tube; it persists even when parts of the gut are removed and thereby isolated from neural and hormonal influences. The trigger for this coordinated sequence of events is stimulation of the intrinsic nerve plexuses of the GI tract. In its simplest terms, peristalsis involves receptive relaxation of the segment ahead of the bolus of food and then contraction of the segment behind the bolus (propulsion) in order to move the food along (**Figure 14.2**). Once the bolus of food has been moved on, the same process starts over again in the next segment of GI tract, although when viewed in real time, this can look like a smooth, coordinated squeezing motion. It is only by experimental analysis of neurotransmission and electrophysiology that it has been possible to distinguish the basic components of this intrinsic reflex.

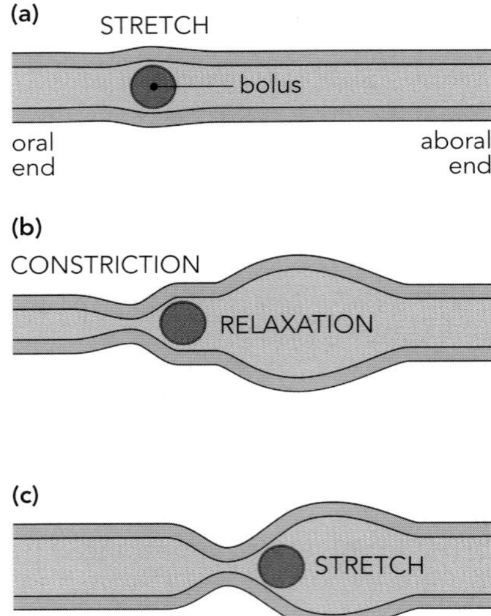

Figure 14.2 The process of peristalsis.
(a) Stretch of the gut wall aboral to the bolus of food induces (b) aboral receptive relaxation and constriction on the oral side of the bolus. (c) This causes the bolus to be propelled from the oral toward the aboral end of the gut tube, inducing stretch in the next aboral segment.

Gastro-esophageal junction

The esophagus is lined with stratified squamous epithelium and contains many mucus-producing glands in the mucosa and submucosa. The mucus produced by these glands serves to lubricate the passage of the bolus of food down the esophagus and the stratified squamous epithelium is designed to be friction resisting. At the gastro-esophageal junction, the epithelium changes form to become simple columnar and noticeably more glandular in the stomach, in order to protect the wall of the stomach from the hydrochloric acid that is produced to activate stomach enzymes (see below). Reflux of stomach contents into the esophagus (for example in gastro-esophageal reflux disease, GERD) can lead to the development of Barrett's esophagus, a form of metaplasia where the esophageal epithelium becomes simple and glandular in an effort to protect the esophagus from the acidic stomach contents. When this change is seen in patients with GERD, it can be indicative of future cancerous change in the esophageal epithelium. However, in most people, gastro-esophageal reflux is prevented by the functionally defined lower esophageal sphincter (LES). No anatomical sphincter can be seen here (unlike at the UES); the LES is formed in part by the folded arrangement of the gastro-esophageal junction and in part by the constricting action of the diaphragm where the esophagus passes between the thoracic and abdominal cavities.

14.3 STOMACH AND GASTRIC SECRETION

Parts of the stomach

The stomach is important in the first phases of digestion of food. Glands in the stomach produce many different secretions that facilitate the process of digestion. However, the stomach also has an important role in the storage of food and regulates the amount of food that enters the small intestine in a given period of time so that small-intestinal digestive and absorptive processes are not overwhelmed. Gastric emptying is under the control of neural and hormonal influences from the small intestine.

The stomach consists of four main anatomical parts: the cardia, the fundus, the body, and the antrum (**Figure 14.3**). These various parts of the stomach are said to be analogous to the hopper and mill components of the process of grinding cereals like wheat and corn; the cardia, fundus, and upper parts of the body act as storage areas (the hopper), whereas the lower parts of the body and the antrum participate more in the process of mechanical and chemical digestion of food (the mill). Of the various parts, the cardia and fundus regions are important in the process of receptive relaxation; an active relaxatory response that allows more food to enter the stomach after a meal. The fundus and body are important in producing gastric secretions that are designed to chemically digest food (particularly protein), and the antrum plays an important role in regulation of gastric secretion, via gastrin release, and the retention of substances in the stomach or the allowing of food to enter the duodenum.

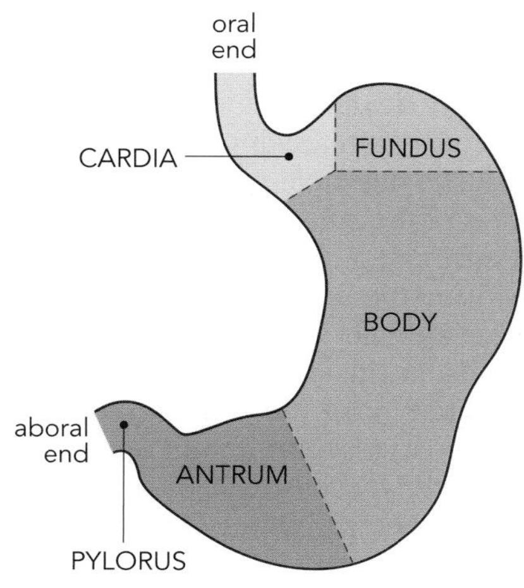

Figure 14.3 Parts of the stomach.

Cardiac and gastric glands

As with other parts of the GI tract, the stomach is lined with epithelium. In the cardia (**Figure 14.4a**), some of the epithelial cells are arranged into coiled

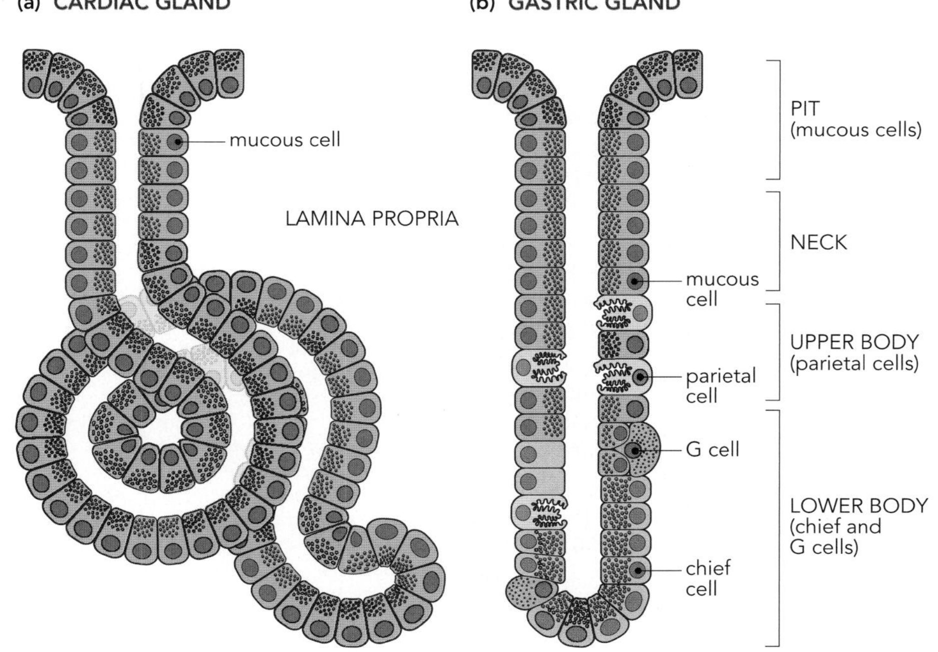

Figure 14.4 Types of secretory glands in the stomach.
(a) Cardiac glands; (b) gastric glands. In gastric glands, the various parts of the gland include different types of specialized epithelial cells. G cells, gastrin-producing cells.

glands and secrete large amounts of mucus; this is important not only in lubricating the bolus of food but also in protecting the epithelium from damage that might be caused by the acid secretion of other stomach glands.

In the fundus and body, the epithelium is arranged into gastric glands (**Figure 14.4b**), tubular structures that are composed of epithelial cells with a number of different phenotypes. Four main epithelial cell types exist in gastric glands: mucous cells in the neck region of the gastric gland; parietal cells (producing hydrochloric acid and intrinsic factor); and, in the lower region of the gastric glands, chief cells (producing the enzyme precursor pepsinogen) and a few G cells (enteroendocrine cells that release the hormone gastrin into the bloodstream). A few other enteroendocrine cells, known as D cells, are also present and release somatostatin, which acts in a paracrine fashion to inhibit gastric acid secretion. In the antrum, the glands are similar in appearance, but they contain fewer parietal cells and more chief cells and G cells.

The digestive properties of gastric secretion are determined by the composition of the substance being produced, which is a mixture of all these secretions. Pepsinogen, released from chief cells, is converted to an active peptidase (pepsin) by the action of hydrochloric acid released from parietal cells. The pepsin can then act at low pH to break proteins down into small peptides and amino acids. Intrinsic factor, released from parietal cells, facilitates the later absorption of vitamin B_{12} in the ileum. Vitamin B_{12} is essential for the production of mature erythrocytes and vitamin B_{12} deficiency can cause pernicious anemia. The mucus secreted by cells near the surface of the stomach lumen is also slightly alkaline, which helps in protecting the stomach lining from damage by hydrochloric acid produced by parietal cells. The G cells, which are endocrine and therefore release their secretions into the bloodstream, produce the hormone gastrin, which is an important factor in stimulating the parietal cells. Other influences on gastric secretion include acetylcholine released from parasympathetic nerves and histamine released from resident enterochromaffin-like cells in the stomach.

Gastric acid secretion

The secretion of hydrochloric acid (HCl) by parietal cells (also known as oxyntic cells) does not, at first glance, seem compatible with life. After a meal, the pH of stomach contents is around 3, meaning that the H^+ ion concentration is around 1 mmol/L. This low pH is a requirement for both the activation of pepsinogen and the action of pepsin itself, with the added advantage of being bactericidal. The bacterium *Helicobacter pylori*, a major cause of gastric ulcers, has evolved a mechanism to avoid destruction in the acid environment of the stomach. This bacterium uses urease to metabolize urea to form ammonia (NH_3) and carbon dioxide. The ammonia can then help to buffer H^+ ions in the immediate vicinity of the bacterium, allowing it to survive in the stomach. This specific action of bacterial urease is utilized in the diagnosis of *H. pylori* infection in the urea breath test.

The formation of HCl by parietal cells is an elegant example of many of the processes involved in exocrine secretion, involving active pumping of ions, enzymatic formation of intermediates, and electrically neutral exchange of ions by facilitated diffusion. Whilst actively producing HCl, parietal cells require large amounts of ATP to run ion pumps, and the production of ATP by these metabolically active cells results in the production of CO_2. ATP and a ready source of CO_2 in the aqueous (water-rich) environment inside the parietal cell are all that the parietal cell needs to produce HCl (**Figure 14.5**). Like all cells, parietal cells contain ion transport proteins, ion channels, and ion pumps. Parietal cells have basolateral and apical surfaces, and trafficking of the various ion pumps, ion channels, and ion transporters to either the basolateral or apical membranes is centrally important in the process of HCl secretion.

Parietal cells have H^+/K^+-ATPase pumps (so-called proton pumps) on their apical surfaces. This pump removes H^+ ions from the cell and pumps K^+ ions into

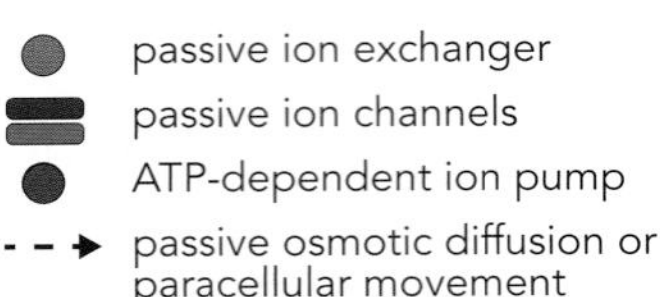

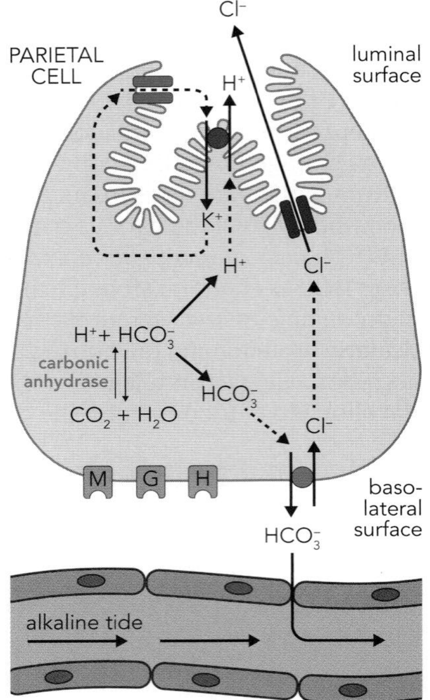

Figure 14.5 Cell model of hydrochloric acid (HCl) secretion by parietal cells in the stomach. HCl is produced via a complex mechanism that involves the action of carbonic anhydrase and a number of different energy-dependent and energy-independent transport systems and channels for various ions. The major mechanism by which H^+ ions are pumped into the gastric lumen is via an ATP-dependent H^+/K^+ pump, often termed the proton pump. ACh, acetylcholine.

the cell. The K^+ ions are able to leak back out of the cell through K^+ channels (down their concentration gradient) and together with the availability of ATP, this keeps the pump working. However, if the H^+ concentration inside the cell became low, this could lead to pump failure. The parietal cells therefore have a useful mechanism for generating H^+ ions. H_2O is combined with the plentiful supply of CO_2 to form carbonic acid (H_2CO_3), a reaction catalyzed by carbonic anhydrase. At physiological pH, H_2CO_3 rapidly dissociates into HCO_3^- ions and H^+ ions. The H^+ ions are the source of H^+ to keep the apical H^+/K^+-ATPases running. However, a buildup of HCO_3^- ions would result in a significant degree of alkalinization of the parietal cell intracellular fluid; so, HCO_3^-/Cl^- exchangers, placed on the basolateral membranes, allow HCO_3^- ions to leave the cell and Cl^- ions to enter the cell down their concentration gradients (facilitated diffusion). The HCO_3^- ions are removed from the extracellular fluid by the high levels of blood flow during gastric acid secretion (this results in an alkaline tide in blood leaving the stomach after a meal); the Cl^- ions leave the cell through Cl^- ion channels on the apical surface of the parietal cells. These Cl^- ions combine with the H^+ ions that have been pumped out by the apical H^+/K^+-ATPases to form HCl.

Parietal cells are stimulated to produce gastric acid by three main molecules: acetylcholine (ACh), gastrin, and histamine. ACh released from nerves and gastrin released from G cells in the stomach act on muscarinic ACh receptors and CCK_2 receptors (the form of CCK receptor that responds to both CCK and gastrin), respectively, to cause an increase in intracellular Ca^{2+} ions. In both cases, this is sufficient to cause an increase in HCl secretion. Histamine released from enterochromaffin-like cells in the stomach activates histamine H_2 receptors, which switches on adenylyl cyclase and leads to an increase in cyclic AMP (cAMP) levels in parietal cells. When ACh and histamine, or gastrin and histamine, or all three substances are released, the combination of high levels of cAMP and high levels of intracellular Ca^{2+} ions produces an increase in HCl secretion that is significantly greater than the additive effect of the compounds when applied individually. This seems to be related to an increased ability of the parietal cells to pump H^+ and K^+ ions at the apical surface, but the mechanism is not well understood. However, this helps us understand the reasons why the two most effective pharmacological treatments for excess gastric acid secretion are used. H_2 histamine receptor antagonists like cimetidine were the mainstays of treatment for such conditions for many years. In recent years, drugs such as omeprazole, a proton pump inhibitor that blocks the action of the H^+/K^+-ATPases on parietal cells, have become important.

Phases of gastric secretion

Gastric secretions increase when we eat. However, the pattern of gastric secretion in response to a meal has several phases. These are known as the cephalic phase, the gastric phase, and the intestinal phase of gastric secretion. During the cephalic phase, the mere sight and smell of food is enough to increase gastric secretions, but tasting food, chewing, and swallowing also have a large part to play. The increase in gastric secretions that is observed during the cephalic phase is mediated by neural pathways, the final output of which is parasympathetic output in the vagus nerve. ACh released from vagal fibers activates both the parietal cells and the G cells, and the gastrin release from G cells further activates the parietal cells. The gastric phase of gastric secretion is produced by distension of the stomach when food enters and, later, by the presence of digested peptides. Local neural reflexes and longer loop reflexes, with afferent and efferent arms in the vagus nerve, produce ACh release and gastrin release, as in the cephalic phase. Once some digestion of proteins has taken place, the presence of peptides in the stomach also increases gastrin release and increases parietal cell secretion. In the final phase of gastric secretion, the presence of the first (usually liquid) component of the meal in the duodenum provides a final top-up (10%) of parietal cell secretion, caused by gastrin release from the duodenum, in an effort to digest the remaining solid matter that is still in the stomach. Toward the end of the intestinal phase, the presence of chyme in the small intestine shuts down gastric secretion by releasing the hormone secretin; eventually, the presence of fatty acids and salts in the duodenum releases gastric inhibitory peptide (and other hormones), which switches off parietal cell secretion.

Control of gastric emptying

The pattern of smooth muscle motility in the stomach is complex. In simple terms, the contractile activity of the hopper part of the stomach (cardia, fundus, and upper body) consists mainly of tonic contractions that move the food around a little, and the activity of the mill parts of the stomach (lower body and antrum) are more phasic, rapid, and propulsive. This antral pump serves two purposes: it churns the food and gastric secretions around and propels the food toward the pyloric sphincter between the stomach and duodenum. The activity pattern that leads to the antral pump is myogenic, meaning that the basic pattern of activity is independent of neural and hormonal influences. However, vagal input and circulating hormones can modulate this basic pattern of activity.

This pattern of activity (**Figure 14.6**) only allows a small amount of food to exit the stomach on each contraction. This is because it takes longer for the contraction on the greater curvature side of the stomach (trailing contraction) to reach the pylorus than the contraction on the lesser curvature side (leading contraction). By the time that the trailing contraction reaches the pylorus, the pyloric sphincter has been closed by the leading contraction. The majority of food is propelled back toward the antrum on each contraction, and only a small amount of food (liquid components of a meal first, followed by more solid matter) is released into the duodenum. The rate of gastric emptying is proportional to the frequency and strength of the contractions in the gastric antrum.

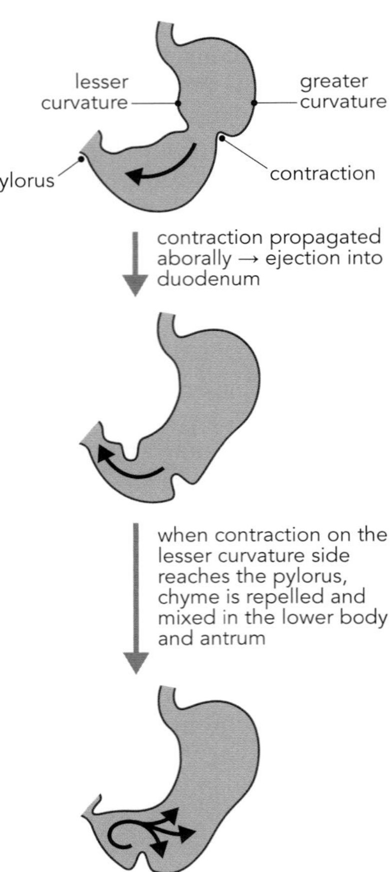

Figure 14.6 Mechanism of controlled gastric emptying and churning of chyme in the antrum.

14.4 CHEMICAL DIGESTION OF FOOD

One of the major factors that determine gastric emptying is the composition of the liquid food (chyme) in the first part of the small intestine, the duodenum. When the calorific value of the chyme in the duodenum increases (fatty acids are the most calorific form of food that the duodenum encounters), this results in a reduction in gastric emptying by reducing gastric motility. The most important hormone mediator of this effect is CCK, which also causes contraction of the gallbladder to release bile (the name cholecystokinin suggests that it moves the gall bladder) and stimulation of pancreatic secretion. Secretin, as mentioned above, inhibits gastric secretion but also encourages the production of a bicarbonate-rich serous secretion from the pancreas, helping to neutralize stomach acid and create an appropriate alkaline environment for the action of many pancreatic enzymes. Both of these effects exemplify the high degree of coordination of gastric emptying and digestion of food in the small intestine.

Digestion of carbohydrates

Nutritional experts (for example British Nutrition Foundation) recommend that around 50% of the calories consumed in the diet should come from carbohydrates. This includes sugars and complex carbohydrates like starch. Some dietary carbohydrate is not absorbed; this is known as fiber, which helps to keep the gastrointestinal tract healthy. The dietary energy requirements of a healthy 70 kg man who is moderately active is around 2500 kiloCalories (kCal) and therefore 1250 kCal should ideally come from carbohydrates. Carbohydrates have a calorific value around 4 kCal/g and so a healthy 70 kg man should consume just over 300 g carbohydrate per day.

Carbohydrates are absorbed in the small intestine as monosaccharide sugars, and so complex carbohydrates like starch must be chemically digested by

TABLE 14.2 Major enzymes involved in carbohydrate digestion in the gut

Enzyme	Source	Reactions catalyzed
Salivary amylase	Secretions from parotid gland	Starch to maltose, maltotriose, and α-limit dextrins
Pancreatic amylase	Pancreatic secretion	Starch to maltose, maltotriose, and α-limit dextrins
Brush-border enzymes	Part of luminal plasma membrane of epithelial cells in the small intestine involved in absorption of carbohydrates	Maltose, maltotriose, α-limit dextrins, sucrose, and lactose to monosaccharides

the action of several enzymes; in the case of carbohydrate digestion, there are three main digestive enzymes (**Table 14.2**). The first of these is salivary amylase, which is secreted by the cells of the parotid gland. Salivary amylase is mixed with food entering the mouth during chewing and its action is facilitated by the neutral pH of salivary secretions. Salivary amylase catalyzes the conversion of polysaccharides like starch to maltose (a disaccharide), maltotriose (a trisaccharide), and α-limit dextrins (pentasaccharides). There is debate about whether a significant amount of digestion of carbohydrates is produced by salivary amylase, because ingested food is rapidly swallowed and enters the acidic environment of the stomach. However, it is possible that the central portion of pieces of food that have entered the stomach are protected from this acidic environment, giving the salivary amylase more time to act. The second enzyme involved in the digestion of carbohydrates is pancreatic amylase. This is secreted as an integral part of the alkaline pancreatic secretion, which helps to neutralize the acidic chyme from the stomach. Pancreatic amylase is therefore provided with a pH-neutral environment in which to act and produces the same digestion products as salivary amylase. The third set of enzymes involved in the digestion of carbohydrates to monosaccharides are brush-border disaccharidases. These are proteins associated with the highly folded plasma membrane of the cells in the small intestine that actually absorb monosaccharides. The products of carbohydrate digestion by amylases, together with sucrose and lactose, are further digested to the monosaccharides glucose, fructose, and galactose before being absorbed by cells in the small intestine.

Digestion of proteins

Around 15% of the calories consumed in the diet should ordinarily come from proteins. This is low compared to carbohydrates because amino acids (the monomers that make up proteins) are used primarily for rebuilding muscle in the body rather than for energy production. Proteins have a calorific value of around 4 kCal/g and so a healthy 70 kg man should consume just over 90 g protein per day.

Proteins are absorbed in the small intestine as short peptides or amino acids and so, as with carbohydrates, they must be chemically digested by the action of several enzymes. In the case of protein digestion, the digestive enzymes come from three main sources (**Table 14.3**). The first of these is pepsin, produced in the stomach. Chief cells in the stomach epithelium produce an inactive precursor of pepsin known as pepsinogen. Pepsin digests proteins into smaller polypeptides by acting on specific bonds within protein molecules and it is thus known as an endopeptidase. Although it is important, a lack of pepsin activity in the stomach can be compensated for by the action of another major group of peptidases, which are produced by the pancreas. These are produced as precursor molecules (pro-enzymes) that are activated to become trypsin, chymotrypsin,

TABLE 14.3 Major enzymes involved in protein digestion in the gut

Enzyme	Source	Reactions catalyzed
Pepsin	Chief cells in stomach epithelium; pepsinogen hydrolyzed to pepsin	Protein to polypeptides
Pancreatic peptidases	Pancreatic secretion	Protein and polypeptides to peptides and amino acids
Brush-border peptidases	Part of luminal plasma membrane of epithelial cells in the small intestine involved in absorption of tripeptides, dipeptides, and amino acids	Larger peptides to tripeptides, dipeptides, and amino acids

elastase, and carboxypeptidase, each of which have a specific role in the digestion of proteins and polypeptides to peptides and amino acids. Trypsin is formed from trypsinogen by the action of enterokinase, which is an enzyme produced by small-intestinal cells. Trypsin then activates the various other pancreatically secreted pro-enzymes. Pancreatic peptidases are very important and diseases that affect the function of the pancreas (such as cystic fibrosis) can result in malabsorption and malnutrition. The third set of enzymes involved in the digestion of proteins are brush-border peptidases. Amino acids and proteins are absorbed by specific small-intestinal transport processes.

Pancreatic secretion

The small intestine represents the major site of digestion and absorption of food and the microvillous surface of the epithelial cells lining the tissue presents a huge surface area for absorption. The epithelium is rapidly turned over as it is replaced by cells emanating from the crypts. Loss of these surface cells releases enzymes into the lumen of the small intestine; one of these enzymes, enterokinase, activates trypsinogen to trypsin. This initiates a hydrolytic cascade in which trypsin activates further molecules of trypsinogen to trypsin (autocatalysis), which activates the other inactive precursor hydrolases secreted by the exocrine pancreas. The overall activity of these hydrolases is to digest the macromolecules of food to much smaller molecules which can be absorbed by specific transfer mechanisms across the brush border and into the enterocytes. Thus, proteins are hydrolyzed to tripeptides, dipeptides, and amino acids; carbohydrate, principally starch, to the disaccharide maltose; and triglyceride to monoglyceride and fatty acids.

Digestion of lipids

Around 35% of the calories consumed in the diet should ordinarily come from fats. Fat and carbohydrates are the major sources of fuel for energy production in the body. However, fat has a calorific content of 9 kCal/g, so a healthy 70 kg man should consume only 90–100 g fat per day. Under circumstances where we consume more calories than required for our activities, fat is stored in the body as adipose tissue. Each kilogram of fat tissue has a calorific value of around 9000 kCal; this means that if a man who wishes to lose weight cuts his calorie intake from the recommended 2500 kCal/day to 2250 kCal/day, it would theoretically take at least 36 days for him to lose 1 kg of fat.

Triglycerides, phospholipids, and cholesterol (the major forms of fat intake in the diet) are absorbed in the small intestine as monoglycerides, fatty acids, lysophosphatidylcholine, and free cholesterol. All of these components are absorbed by a process known as micellar solubilization, which depends on bile secretion (see below). Three main pancreatic enzyme groups are involved in fat digestion: pancreatic lipases, phospholipase A_2, and cholesterol esterase. Triglycerides (an assembly of glycerol and fatty acids) are digested to

2-monoglyceride and free fatty acid by the combined action of pancreatic lipase and co-lipase. Phosphatidylcholine (also known as lecithin, a combination of glycerol, choline, and two fatty acids) is digested by phospholipase A_2 to form free fatty acid and lysophosphatidylcholine. Cholesterol esterase converts the major dietary form of cholesterol, cholesterol ester (a combination of cholesterol and a fatty acid), to free cholesterol, so that it can be absorbed in the small intestine.

Bile secretion and release

The major secretory product of the liver is bile. Bile contains bicarbonate ions, sodium ions, bile salts, phospholipids, cholesterol, and bilirubin, the end product of the metabolism of heme groups in hemoglobin. Bile salts are derived from cholesterol and act as detergent molecules since they have both hydrophilic and hydrophobic domains. Bile is stored in the gall bladder until required and it is also concentrated there by secondary active transport of sodium ions and thence osmotic movement of water. One of the triggers for gallbladder contraction is release of CCK from the intestinal epithelium into the bloodstream in response to the introduction of calorifically rich food (including fats) into the duodenum. CCK also causes relaxation of the sphincter of Oddi (between the common bile duct and the duodenum). Once introduced into the intestinal tract, the bile salts are able to emulsify 2-monoglycerides and fatty acids prior to absorption.

14.5 ABSORPTION OF NUTRIENTS

Absorption of carbohydrates

The major dietary carbohydrate is starch, a polymer of glucose derived mainly from plants. The disaccharide sucrose from cane sugar (containing glucose and fructose) and lactose from milk (containing galactose and glucose) may also be significant components of the diet. Other monosaccharides including glucose and fructose in fruit may also be present in the diet. Sugars are absorbed as monosaccharides and the terminal step in the hydrolysis of disaccharides occurs on the brush-border membrane of the small intestine. The villi of the small intestine are lined by epithelial cells, each of which has a highly folded luminal membrane. These folds are known as microvilli and are apparent as the brush border of the intestinal epithelium; their purpose is to maximize the surface area available for absorption of nutrients (**Figure 14.7a**). Final hydrolysis of carbohydrates to monosaccharides is followed by transport of the monosaccharide products into the enterocyte (**Figure 14.7b**). Intestinal uptake of glucose and

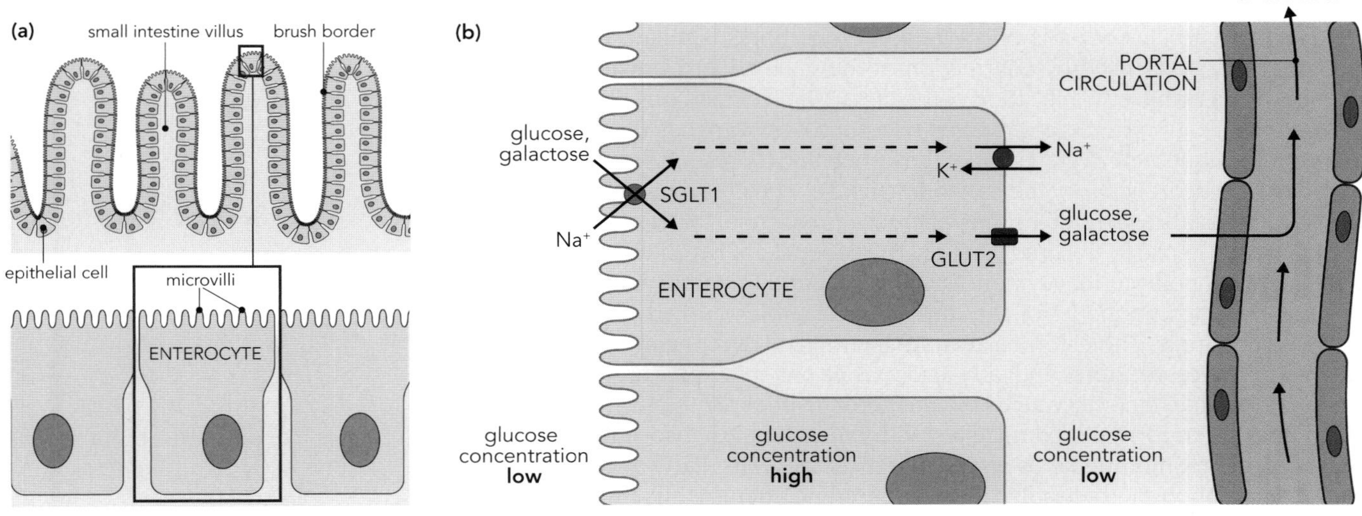

Figure 14.7 Absorption of glucose and galactose in the small intestine, showing (a) the arrangement of enterocytes into villi and (b) a cell model for absorption. Na^+-linked glucose absorption by enterocytes is driven by the action of Na^+/K^+ pumps on the basolateral membranes; these create a low concentration of Na^+ ions inside the cell, thereby driving the movement of Na^+ and glucose into the cell from the lumen of the small intestine via SGLT1. Glucose can then leave the basolateral side of the cell via GLUT2 in order to access the interstitial space and the portal circulation. GLUT2, sodium-independent glucose transporter 2; SGLT1, sodium–glucose linked transporter 1.

galactose requires the co-transport of sodium and water by the sodium–glucose linked transporter (SGLT1). The sodium gradient across the cell membrane is maintained through the action of an energy-dependent Na^+/K^+-ATPase (sodium pump) and so the luminal transport of glucose and galactose takes place via secondary active transport. This permits the facilitated diffusion of glucose and galactose from a region of low concentration (the intestinal lumen) to a region of high concentration (the cytoplasm of the enterocyte). Galactose and glucose then leave the enterocyte via the sodium-independent glucose transporter GLUT2 (down their concentration gradients) and enter the portal circulation.

Fructose derived mainly from the hydrolysis of sucrose by sucrase on the brush-border membrane is transported by another sodium-independent facilitative glucose transporter (GLUT5) that is present on both the apical and basement membranes. GLUT5 is energy-independent and has a low affinity for glucose. Also, like glucose and galactose, fructose can be transported across the basement membrane into the blood via GLUT2. These sugars are then transported to the liver via the hepatic portal vein.

Absorption of amino acids and peptides

There are several different sodium-dependent amino acid transporters, some specific but others more general, with overlapping substrate specificities. For example, there exist separate transporters for acidic amino acids, basic amino acids, aromatic amino acids, neutral aliphatic amino acids, and imino acids (proline and hydroxyproline) and these co-transport sodium ions and the amino acids into the villus cell.

Some oligopeptides derived from hydrolysis of proteins in the lumen of the small intestine by pancreatic peptidases undergo further hydrolysis to amino acids and di- and tripeptides by the action of endopeptidases on the brush-border surface. Free amino acids produced in this way are transported on the sodium-dependent transporters described above, while the di- and tripeptides are absorbed across the enterocyte membrane by a proton-coupled oligopeptide transporter (PepT1). Once inside the cell, these di- and tripeptides are hydrolyzed to their constituent amino acids by intracellular peptidases.

Amino acids are then transported across the basement membrane into the portal blood, where they are transported to the liver.

Absorption of lipids

The major lipid component of the diet is triglyceride (triacylglycerol). The extremely hydrophobic nature of the molecule presents problems both in terms of its hydrolysis and absorption. Apart from a little hydrolysis of triglycerides containing short-chain fatty acids by the lingual lipase, most of the dietary lipid arrives in the duodenum intact. Peristalsis helps in breaking down the lipid mass into smaller particles but it is only when the food bolus is mixed with bile salts (see Chapter 3), released by contraction of the gallbladder in response to CCK, that any appreciable emulsification of the dietary lipid occurs (micellar solubilization). Emulsification of the lipid serves two purposes: it makes the lipid compatible with the aqueous environment of the lumen and it presents the lipid in a form which can be hydrolyzed by the pancreatic lipase. The presence of a smaller protein, co-lipase (also secreted by the pancreas), is essential for this latter action, as it anchors the lipase to the lipid surface. The products of hydrolysis of the triglyceride are 2-monoglyceride (2-monoacylglycerol) and free fatty acids.

Other lipid components of the diet are free cholesterol, cholesterol esters, and phospholipids. Cholesterol esters are hydrolyzed by a pancreatic cholesterol esterase to free cholesterol and fatty acids, while phospholipids are hydrolyzed to lysophospholipids and fatty acids by a pancreatic phospholipase A_2. The hydrolyzed lipids then form mixed micelles with bile salts and absorption across the brush-border membrane takes place from these micelles. Fatty acids and monoglyceride are thought to dissolve in the membrane and are thus transported into the cell down a concentration gradient. A specific transporter which regulates the movement of cholesterol into the enterocyte is thought to exist, and this transporter can be inhibited by ezetimibe, a drug used to decrease the absorption of dietary cholesterol in hypercholesterolemic patients. Bile salts undergo enterohepatic circulation and are reabsorbed not from the mixed micelle in the jejunum but in the terminal ileum.

Once inside the enterocyte, the absorbed lipid is reconverted to the form in which it occurred in the diet. Thus triglyceride is synthesized from monoglyceride and fatty acids. Cholesterol ester and phospholipids are also resynthesized and packaged with the triglyceride and some newly synthesized hepatic protein into chylomicrons. These lipoprotein particles are transported not into the portal circulation but into the lymphatic system, from which they exit into the peripheral blood via the thoracic duct.

Absorption of vitamins

The fat-soluble vitamins (A, D, E, and K) in the diet are incorporated into the mixed micelles described above and absorbed in the jejunum. Inside the enterocyte, they are incorporated into the chylomicrons and eventually reach the liver as part of the chylomicron remnant, the product of chylomicron hydrolysis in the periphery.

Most of the water-soluble vitamins are absorbed by facilitated transport on specific transporters in the jejunum. The one exception is vitamin B_{12} (cobalamin). Vitamin B_{12} binds to intrinsic factor in the jejunum and the complex passes into the ileum where it binds to a specific receptor. The vitamin is transported into the enterocyte and from there into the blood where it circulates bound to a specific transport protein, transcobalamin II.

14.6 FLUID AND ELECTROLYTE REABSORPTION

Absorption of fluid and electrolytes

Pumping of ions to maintain the correct balance of various ions on either side of the cell membrane is central to all cellular activities. It is therefore crucially important to maintain normal fluid and electrolyte balance in the face of variable intake of fluid and electrolytes on a daily basis. **Table 14.4** shows the average amounts of fluid consumed, secreted, absorbed, and lost in the gastrointestinal tract during a normal day in a healthy 70 kg male subject. It is evident that the fluid and electrolytes absorbed from the large intestine, although relatively

TABLE 14.4 Sources and fate of fluid in the GI tract

Part of gut	Amount ingested	Amount secreted (net)	Amount reabsorbed (net)	Net loss to feces
Mouth	2.0 L			
Saliva		1.4 L		
Stomach		2.3 L		
Biliary tree		0.6 L		
Pancreas		1.6 L		
Duodenum		1.1 L		
Jejunum			5.4 L	
Ileum			2.2 L	
Colon			1.2 L	
TOTAL	2.0 L	7.0 L	8.8 L	0.2 L

The small intestine secretes as well as absorbs fluid but the net effect is absorption.

small, are important. If the large intestine did not absorb these amounts of fluid and electrolytes and diarrhea resulted, it would not take long before bodily functions could be compromised.

As in the small intestine, large-intestinal fluid and electrolyte absorption is carried out by epithelial cells lining the wall of the gut. In view of the fact that the amount of fluid and electrolytes to be absorbed is relatively small, and because fecal material is present in the large intestine for a long period of time, the lining of the large intestine is folded but is not thrown into villi in the same way as the small intestine. **Figure 14.8** shows a schematic view of the processes involved in electrolyte absorption by large-intestinal epithelial cells. Na^+/K^+-ATPase pumps are placed on the basolateral membranes of colonic epithelial cells, creating a concentration gradient for Na^+ ions between the luminal fluid and the interior of the epithelial cell. Na^+ channels are placed at the apical membranes of the epithelial cells, facilitating the diffusion of Na^+ ions into the cell from the lumen of the colon. K^+ ions pass through channels on both the basolateral and apical membranes of colonic epithelial cells and are also secreted as a component of colonic mucus, but this secretion of K^+ can be limited by the action of a H^+/K^+-ATPase on the luminal membrane that actively transports both H^+ ions and K^+ ions against their concentration gradients and helps keep K^+ ions in the cell. However, long-lasting diarrhea can be problematic because significant numbers of K^+ ions can still be lost in the feces, resulting in reduced plasma K^+ (hypokalemia). This can have potentially fatal effects on the excitability of the

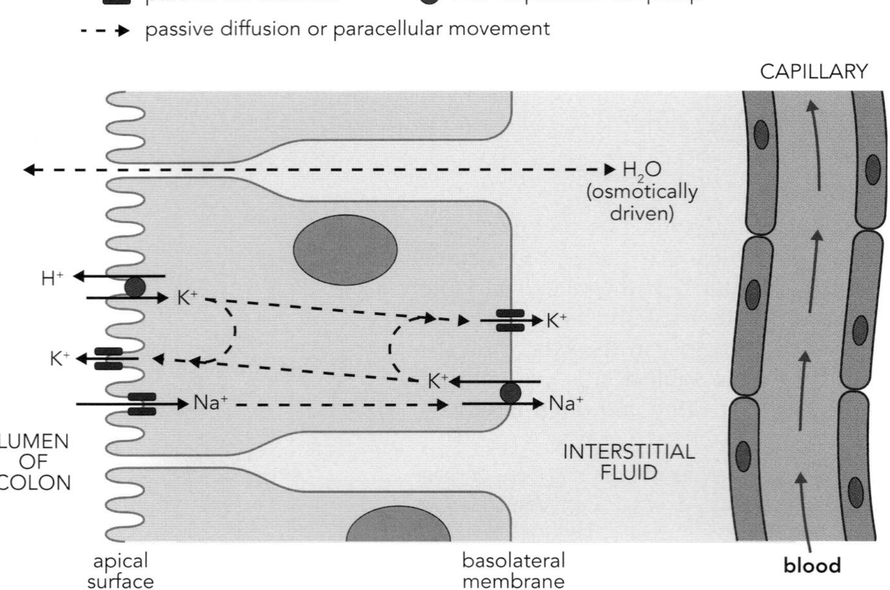

Figure 14.8 Cell model for electrolyte and water reabsorption in the colon.
The example shown here is for the movement of Na^+ and K^+ ions. Basolateral and apical ATP-dependent pumps such as the Na^+/K^+-ATPase (often called the sodium pump) and H^+/K^+-ATPase (often called the proton pump) allow for the movement of electrolytes across the epithelial surface of the colon. Passive ion channels such as Na^+ and K^+ channels are also involved, in this case allowing Na^+ ions to enter the cell and feed the basolateral Na^+/K^+-ATPase, and allowing K^+ ions to leave the cell to provide sufficient extracellular K^+ ions to maintain the action of the ATPases on both surfaces. Water molecules follow the net absorption of solutes, driven by osmotic force. This means that if, for example, additional Na^+ is absorbed, water reabsorption will increase; if less Na^+ is absorbed, water reabsorption will decrease.

heart and conduction of action potentials in heart tissue (see Chapter 6). Water moves from the lumen of the gut to the interstitial fluid of the mucosa or vice versa, driven by osmotic forces; when osmotic pressure on either side of the intestinal epithelial cell is balanced, there should be no net movement of water.

14.7 CLINICAL CHEMISTRY OF THE GI TRACT

Patients with disorders of the GI tract present with the symptoms and signs expected from loss of digestive or absorptive function, namely weight loss, diarrhea, or specific nutrient deficiencies (**Table 14.5**). Abdominal pain, which can be acute or chronic, may arise from perforation or obstruction of the gut or from inflammation of the gut or pancreas. Clinical chemistry investigations play an important role in diagnosis of these conditions and their complications, such as fluid and electrolyte disturbances.

TABLE 14.5 Major consequences of malabsorption

Clinical condition	Clinical consequences
Steatorrhea	Consequent decreased energy intake
	Growth failure (in part) when prolonged
	Weight loss
Deficiency of fat-soluble vitamins	Night blindness (vitamin A)
	Rickets/osteomalacia (vitamin D)
	Bleeding disorder (vitamin K)
Osmotic diarrhea due to decreased absorption of fats and disaccharides, which are osmotically active	Dehydration
Anemia due to decreased absorption of iron, folate, and vitamin B_{12}	Tiredness/lethargy
	Shortness of breath
	Pallor
Osteomalacia due to calcium and/or magnesium deficiency	Looser's zone fractures
	Ostealgia
	Muscle weakness

Diarrhea

Diarrhea is a common complaint and is generally defined as three or more loose or watery stools in a day. It can vary from a mild, self-limiting disorder up to a severe, life-threatening condition with massive loss of fluids and electrolytes. In developing countries, diarrhea is a major cause of death, particularly in young children. There are many causes of diarrhea (**Table 14.6**), with infection being the commonest and most serious worldwide. In terms of mechanisms, diarrhea can be classified as being secretory, osmotic, or inflammatory (or exudative) or due to increased gut motility.

Secretory diarrhea occurs when the function of ion channels in the gut epithelium is impaired, usually by a microbial toxin. This leads to excessive secretion or reduced absorption of ions and water but without significant damage to the cells themselves. Diarrhea associated with cholera is a good example of this.

Osmotic diarrhea is the result of failure to absorb small, osmotically active molecules from the gut lumen. This results in an osmotic pull of water into the stools. This may be seen with disaccharidase deficiency (for example lactose intolerance) and with celiac disease. Both are described below. In some cases, excessive use of osmotic laxatives such as magnesium salts may cause osmotic diarrhea.

Exudative diarrhea occurs as a result of an inflammatory process which causes structural damage to the gut epithelium, with subsequent passive leakage of plasma contents, including protein. There may also be blood and pus in the stools as a result of the inflammation. Diarrhea of this type may be seen with inflammatory bowel disease (Crohn's disease and ulcerative colitis), some types of infection due to bacteria or parasites, and sometimes due to bowel cancer.

If gut motility is increased, undigested food may transit through the bowel too rapidly for the effective absorption of nutrients, electrolytes, and water by

TABLE 14.6 Causes of diarrhea

Causes	Causative agents
Infections (including food poisoning)	Viruses: Rotavirus; Norovirus
	Bacteria: *Campylobacter jejuni*; *Salmonella*; *Escherichia coli*; *Shigella*; *Yersinia enterocolitica*; *Clostridium difficile*; *Vibrio cholerae*
	Parasites: *Cryptosporidium*; *Giardia lamblia*
Inflammatory bowel disease	Crohn's disease
	Ulcerative colitis
Autonomic neuropathy	Loss of nervous control of bowel
	Diabetes mellitus
Drug induced	Antibiotics
	Laxatives
	Magnesium-containing antacids
	Opiate withdrawal
	Chemotherapy
Food allergies or intolerances	Cow's milk protein allergy
	Soy protein allergy
	Multiple food allergies
	Methylxanthines (caffeine, theobromine, theophylline)
Malabsorption	Celiac disease
	Glucose–galactose malabsorption
	Sucrase–isomaltase deficiency
	Late-onset (adult-type) lactose intolerance
	Chronic pancreatitis
	Cystic fibrosis
Endocrine	Thyrotoxicosis
	Carcinoid syndrome
	VIPoma
Vitamin deficiencies	Niacin
	Folate
Metals	Excess of copper, tin, zinc
Toxins	Plants and fungi (for example hyacinths, daffodils, azalea, mistletoe, *Amanita* species mushrooms)

VIP, vasoactive intestinal polypeptide.

the brush-border epithelium. This may occur with some types of neuropathy affecting the autonomic nervous system, most commonly as a complication of diabetes mellitus. Thyrotoxicosis may also cause diarrhea by increasing gut motility and reducing stool transit time.

Investigation of diarrhea

A number of modes of investigation for diarrhea may be employed, guided by the history and physical examination. These include imaging the interior of the bowel by endoscopy, biopsy, and histology of the luminal tissue, imaging (barium contrast radiology, ultrasound, computerized tomography, and magnetic resonance imaging), microbiology of the stools, and investigations directed at specific causes. Some of these are described in further detail below. Biochemical analysis of diarrhea fluid is rarely helpful and not usually performed. Unless the stool is watery, analysis of its electrolyte composition and osmolality is technically difficult and does not give a specific diagnosis. However, analysis of stool magnesium may occasionally be useful to confirm excess ingestion of magnesium salts. If abuse of other types of laxatives is suspected, these may be detected in urine as they are systemically absorbed and renally excreted.

Fecal calprotectin

Calprotectin is a calcium-binding protein found in the cytosol of leukocytes. It is not degraded by digestive enzymes and so can be detected in increased amounts in the stool when gut inflammation occurs. It has applications as a marker of the inflammatory bowel diseases ulcerative colitis and Crohn's disease. These are relapsing and remitting diseases characterized by intermittent flares, which are not predictable. Although histology is required for diagnosis, the degree of disease activity and the response to therapy can be followed by calprotectin measurement. The test has also been validated in some patient groups for the differentiation of irritable bowel syndrome from inflammatory bowel disease without the need for invasive investigations such as colonoscopy.

Malabsorption

As referred to above, the clinical syndrome of malabsorption refers to the inability to digest and/or absorb nutrients from the gastrointestinal tract. It may be generalized or specific to a small number of nutrients, depending on the mechanism. Generalized malabsorption may be suspected when protein and energy malnutrition occur with fat loss in the stools (steatorrhea). This is often noticed by the patient as offensive floating stools which are hard to flush away. The causes can be broadly divided into pancreatobiliary disease due to loss of digestive enzymes and biliary secretion, and small-intestinal disease due to lack of absorptive function. Selective malabsorption will result in deficiency diseases specific to particular vitamins or trace elements. An example would be lack of intrinsic factor causing vitamin B_{12} malabsorption.

Biochemical investigations for malabsorption have changed significantly over the last two decades, and tests which were previously considered important have been replaced by better alternatives. A good example is the analysis of fecal fat content of stool samples collected over 3 or 5 days. This was the mainstay for diagnosis for many years. As well as being unpleasant for both the patient and laboratory, it was realized that fat excretion was increased in many causes of diarrhea other than malabsorption, due to shortened transit time, and it is possible to have malabsorption without elevated fecal fats. The test lacks both sensitivity and specificity and furthermore does not give a diagnosis of the cause of malabsorption.

Celiac disease

Previously called nontropical sprue, to distinguish it from infective conditions endemic in the Tropics, celiac disease is characterized in the most serious cases by villous atrophy on biopsy samples taken from the small intestine. As a consequence of atrophied villi reducing the surface area for nutrient absorption, the patient can develop malabsorption syndromes. The underlying cause of celiac disease is an immune-mediated reaction to dietary gluten, which causes inflammation in the gut. Gluten is a protein found predominantly in wheat, barley, and rye. An alternative name is thus gluten-sensitive enteropathy. Celiac disease was previously thought to be a childhood condition, with only rare cases in adults. However, the emergence of reliable serological tests in the last two decades has led to more widespread testing and the recognition that it is common in the population (prevalence approximately 1 in 200). The traditional presentation with diarrhea, weight loss, and iron deficiency has been supplemented with other features such as abnormal liver function and even neurological disorders. Treatment is to follow a lifelong gluten-free diet.

Clinical practice point 14.1

Celiac disease is common and can be diagnosed at any age. It is increasingly recognized as having manifestations outside of the GI tract, for example neuropathy.

Diagnosis of celiac disease can be made with a high degree of certainty by detection of antibodies in the serum. Initially, these tests were directed at antibodies against the antigen in gluten (gliadin), but these have been superseded by tests for antibodies against the gut—endomysial or tissue transglutaminase antibodies. The latter test can be automated and reported as a numerical value, whereas endomysial antibodies are detected by indirect immunofluorescence under a microscope. Most of these tests detect IgA antibodies and false negative results will occur in IgA-deficient individuals. Immunoglobulin G (IgG)-based tests can be used in their place. As serological testing is not 100% sensitive or specific, and a gluten-free diet is lifelong, most authorities still recommend duodenal biopsy as the gold-standard test.

Pancreatic function tests

Assessment of the functional integrity of the exocrine pancreas was initially done by giving a test meal through a nasogastric tube and collecting duodenal fluid for analysis (the Lundh meal test). Such tests were difficult and unpleasant and were superseded by tubeless tests based on absorption of specific chemicals following cleavage from parent molecules by pancreatic enzymes. An example was the fluorescein dilaurate (pancreolauryl) test, which measured the urinary excretion of fluorescein following the ingestion of capsules containing fluorescein coupled to a fatty acid. Free fluorescein is released by the action of pancreatic lipase, absorbed in the gut, and excreted into the urine, which is collected over several hours. On a second day, the procedure is repeated using fluorescein alone as a control. These tests have also been superseded.

Current tests to investigate pancreatic function rely on detection of stable enzymes in stool samples. Fecal elastase, which is undegraded by passage along the gut, is measured in untimed stool samples and can detect severe impairment of pancreatic function. It is, however, relatively insensitive for the detection of mild to moderate pancreatic insufficiency.

Analytical practice point 14.1

Normal pancreatic exocrine function may be retained even with extensive (up to 90%) destruction of the pancreas.

Isolated specific absorption defects

Specific mutations in the genes encoding enzymes and transport proteins involved in absorption in the intestine have been described for many of the dietary molecules required by the body and include:

- Deficiency of disaccharidases
- Glucose–galactose malabsorption
- Cystinuria; Hartnup disease
- Metal malabsorption: Cu, Zn, Mg
- Folate- and cobalamin-deficiency anemias
- Triglyceride malabsorption; abetalipoproteinemia
- Familial chloride diarrhea

CASE 14.1

A 46-year-old woman presents with chronic tiredness, weight loss, and episodes of bloating and diarrhea. These do not appear to be related to any particular food and she is eating a varied diet with no avoidances. She is on thyroxine for hypothyroidism, which has always been well controlled, and her sister has type 1 diabetes from childhood.

	SI units	Reference range	Conventional units	Reference range
Plasma				
TSH	1.2 mIU/L	0.4–4.2	1.2 mIU/L	0.4–4.2
Ferritin	11 pmol/L	45–562	4.9 ng/mL	20–250
Tissue transglutaminase antibodies	25 u/mL	0–5	25 u/mL	0–5
Endomysial antibodies	Positive	Negative	Positive	Negative
IgA	2900 mg/L	400–3500	290 mg/dL	40–350

- **What do the TSH and ferritin show?**
- **What do the antibody tests indicate and why has IgA been measured?**
- **What further test is required and what precautions are necessary?**

The normal TSH excludes overtreatment with thyroxine as the cause of diarrhea and the ferritin indicates iron deficiency. The raised tissue transglutaminase (TTG) and positive endomysial antibodies (EMA) are highly specific for celiac disease. This is an immunologically mediated response to the protein gluten in wheat, rye, and barley. There is an inflammatory response within the small intestine which results in malabsorption and gastrointestinal symptoms of variable severity. Although it is relatively common, it is more likely to occur in people with a personal or family history of autoimmune endocrine disorders, such as thyroid disease and type 1 diabetes. IgA deficiency is also associated with celiac disease and since TTG and EMA are IgA antibodies, false negative results may be seen. Thus, low TTG and EMA results in the presence of IgA deficiency should be followed up by corresponding IgG-based tests. Serology alone is not considered entirely diagnostic (although some guidelines are changing position on this) and definitive diagnosis is by endoscopic biopsy of the duodenum demonstrating villous atrophy. It is important that intake of gluten is not restricted before the diagnostic testing is completed, to avoid false negative results. However, lifelong adherence to a strict gluten-free diet is the only effective treatment.

Disaccharidase deficiencies

Apart from acquired lactase deficiency, which is relatively common (see below), the disaccharidase deficiencies are rare conditions usually affecting a single enzyme of the cohort of disaccharidase activities of the epithelial brush border. The deficiency may be due to an inborn error, where the particular enzyme is not expressed because of a mutation in its gene, such as for:

- Maltase
- Isomaltase/sucrase
- Lactase
- Trehalase

or the deficiency may be acquired after damage to the absorbing epithelial surface, by enteritis, for example, or surgical resection of the small intestine. In the case of enteritis, loss of activity of all enzymes may occur, but loss of lactase activity is particularly relevant. Absence of brush-border enzyme activity causes intolerance of dietary disaccharide(s), and as the disaccharides are not digested, they accumulate in the small bowel where they are osmotically active and draw water and ions into the lumen of the small intestine. When this watery mass reaches the colon, the disaccharide is digested by intestinal bacteria, leading to watery diarrhea with bloating, colic, and flatus. Diagnosis of the absence of disaccharide deficiency is by disaccharide challenge and observation of symptoms, plus measurement of breath hydrogen and of disaccharides in feces and urine. Patients are treated by avoidance of the corresponding dietary sugar. Acquired lactase deficiency, usually due to decreased enzyme synthesis, is common (up to 90% or even 100%) in many populations, especially Asian, African, and Native American. In these ethnic groups, lactose intolerance may appear as early as 2 years of age, with accompanying inability to digest lactose from milk and dairy products.

Absorption of amino acids

As described earlier, amino acids are absorbed into the intestinal epithelial cells via active transport mechanisms linked to sodium. The principal amino acid transporters are shown in **Table 14.7**. Although inborn errors affecting individual amino acid transporters have been described, patients do not become amino acid deficient, due to the activity of the other transport systems in the brush border for di-and tripeptides derived from hydrolysis of dietary proteins. However, the absence of these transporters in the kidneys has major clinical consequences (see Chapter 4).

TABLE 14.7 Active transport systems for amino acids (small intestine and renal tubules)

Generic amino acid family	Amino acids
Neutral I	Pro, hydroxyproline, Gly
Neutral II	Ala, Ser, Val, Leu, Phe, Tyr, Trp, His, Thr
Dibasic	Cys, ornithine, Lys, Arg
Dicarboxylic	Glu, Asp
β-Amino acids	β-Ala, β-AIB, taurine

AIB, amino-isobutyrate.

Breath tests

Breath testing is a noninvasive form of investigation of gastrointestinal disorders. The general principle is that gut absorption of specific molecules can be measured by their presence in expired air. Incorporation of an isotopically labeled carbon atom into a substrate which is metabolized to CO_2 is a related technique. A number of breath tests have been developed over the years, but some are little used today and only two are described in this section.

Hydrogen breath test for lactose intolerance

An inability to digest the milk sugar lactose to its component monosaccharides glucose and galactose results in unabsorbed lactose reaching the distal gut, where bacteria are able to ferment it. This produces symptoms such as bloating, abdominal pain, and diarrhea. In many parts of the world, the enzyme lactase is not expressed following weaning and there is little consumption of dairy products. Northern European populations have supplemented their food intake with cows' milk for many centuries and so usually retain lactase into adulthood. Withdrawal of dairy products from the diet or gastrointestinal infections may trigger the development of lactose intolerance. Objective diagnosis of this condition can be made by breath testing. A dose of lactose is given by mouth. In the absence of lactase, the gut bacteria produce hydrogen and methane gases, which can be measured in expired air.

Breath hydrogen testing may also be used to diagnose bacterial overgrowth in the small intestine as a cause of gastrointestinal symptoms, which may overlap with those of lactose intolerance, irritable bowel syndrome, or, in some cases, nutrient malabsorption. In this case, glucose or lactulose is given as the substrate and increases in expired hydrogen or methane are used as markers of bacterial overgrowth. Humans cannot produce hydrogen from food, but bacteria can; if bacterial overgrowth into the small intestine from the colon occurs, then sugars and carbohydrates in the small intestine can be metabolized to hydrogen, which passes into the patient's bloodstream and is exhaled at the lungs.

Urea breath test for *H. pylori* infection

Until the 1980s, it was accepted wisdom that peptic ulcer disease was primarily caused by excessive acid secretion, and therapies—both pharmacological and surgical—were developed to target this mechanism. The discovery of *H. pylori* as the causative agent by Marshall and Warren (including them deliberately ingesting the organism) revolutionized the understanding of peptic ulcer disease and led to their Nobel Prize in 2005. The mainstay of treatment is now antibacterial therapy.

In order to survive within the acid environment of the stomach, the organism uses the enzyme urease to produce ammonia and carbon dioxide. This allows it to create a microenvironment at a higher pH. The presence of urease in the stomach is exploited as a diagnostic test for *H. pylori*. The patient ingests a dose of urea containing an isotopically labeled carbon atom, either ^{14}C or ^{13}C, which is detectable in the CO_2 of expired air. The urea breath test can be used to diagnose *H. pylori* infection and also to confirm eradication after antibiotic treatment. Other tests for the organism are detection of urease activity in gastric biopsy samples, antigens in the stool, and antibodies in the blood. The first two tests are less acceptable for some patients, whilst the latter has the drawback of remaining positive after treatment; that is, it does not distinguish current from previous infection.

Disorders of gut hormones: neuroendocrine tumors

The endocrine cells of the gut and pancreas can, in rare cases, undergo neoplastic transformation and produce unregulated amounts of their specific hormone. This causes particular clinical syndromes which can be predicted from the end-organ effects of the hormones. Neuroendocrine tumors (NETs) of the gut are classified as pancreatic NETs (PNETs) or as carcinoid tumors, the latter being subclassified as foregut (stomach to first part of duodenum), midgut (second part of duodenum to ascending colon), or hind gut (transverse colon to rectum). Other NETs may occur in the ovary, lung, heart, paraganglia, and adrenal medulla, the last of these being called pheochromocytomas, which are a cause of hypertension (see Chapter 8). A characteristic of NETs is their overexpression of somatostatin receptors (of which there are five subtypes). This provides a target for both imaging NETs and their treatment. Somatostatin analogs, such as octreotide, are used to control hormonal secretion and can be labeled with radioactive emitters, which are detected using nuclear medicine techniques to show very small tumors. Approximately one-third of NETs are nonfunctional; that is, they do not secrete excess hormone, and cause symptoms only by their anatomical presence due to their size, position, or metastatic spread. As many as 90% of PNETs (both functional and nonfunctional types) are malignant. The exception to this is insulinoma, where the majority of tumors are benign. Insulinomas are discussed in Chapter 2.

Zollinger–Ellison syndrome

This is due to a gastrin-secreting PNET (gastrinoma) causing gross overproduction of stomach acid. This results in severe, intractable peptic ulceration and sometimes malabsorption and diarrhea, due to inactivation of digestive hormones in the proximal small intestine. High levels of gastrin are detectable in the serum, but interpretation of these data requires knowledge of the underlying gastric acid output. A high gastrin plus a high acid secretion is an inappropriate combination and diagnostic of Zollinger–Ellison syndrome. However, high gastrin is a physiological response to low acid output, such as occurs with atrophic gastritis or, more commonly, the use of acid-suppressant drugs such as proton pump inhibitors or histamine H_2 receptor antagonists. A high gastrin cannot therefore be interpreted in a patient on a proton pump inhibitor and the sample should be taken before treatment is started, as it may be difficult to discontinue treatment later.

VIPoma

Vasoactive intestinal polypeptide (VIP) stimulates intestinal water and electrolyte secretion. Excess secretion causes severe watery diarrhea, leading to dehydration and hypovolemia. High VIP levels in the plasma may be used for diagnosis.

Glucagonoma

Excessive secretion of glucagon causes hyperglycemia and a characteristic rash called migratory necrolytic erythema. Plasma glucagon levels are elevated and can be used to confirm a diagnosis.

Carcinoid syndrome

Carcinoid tumors arise from the argentaffin cells of the gut, and predominantly occur in the appendix, terminal ileum, and rectum. They can secrete serotonin (5-hydroxytryptamine) plus other vasoactive molecules such as kinins and histamine. Since venous blood from the gut drains to the liver via the portal vein, these products are metabolized before they can enter the systemic circulation. However, if the tumor metastasizes to the liver itself, the secretory products are able to enter the circulation via the hepatic vein. Symptoms include flushing episodes, wheezing, and diarrhea, and over time the hormones can induce fibrosis of the heart valves. This clinical picture is termed the carcinoid syndrome.

Biochemical tests to confirm and monitor the progress of the carcinoid syndrome include the quantitation of serotonin in blood or its product 5-hydroxyindole acetic acid (5-HIAA) in urine. The level of serum chromogranin A may also be a useful tumor marker for this and other neuroendocrine tumors, including the ones described above.

Pancreatitis

Inflammation of the exocrine pancreas can be acute or chronic. Both conditions are serious, and acute pancreatitis is often fatal. The conditions are relatively common and becoming more prevalent in developed countries, since a major etiological factor is alcohol consumption.

Acute pancreatitis

Acute pancreatitis is the result of premature activation of digestive enzymes within the pancreas, rather than in the gut. As would be predicted, this results in the digestion of pancreatic and surrounding tissue, causing severe abdominal pain, vomiting, hypovolemia, shock, and, in around 10% of cases, death. The two commonest causes of acute pancreatitis are gallstones and excess alcohol consumption. Other causes include hypertriglyceridemia, mumps, and hypercalcemia.

The biochemistry laboratory plays an important role in the diagnosis and management of acute pancreatitis. When the pancreas becomes acutely inflamed, enzymes are released into the circulation. The two most often measured are amylase and lipase. There is little to choose between them in most cases, but lipase has a longer serum half-life and so is more useful if the patient presents late to hospital. Amylase is rapidly cleared by renal excretion and so may have disappeared after 2 or 3 days. It may still be detectable in urine after the serum levels have fallen back to normal. Neither enzyme is specific for pancreatitis and levels may be elevated in other causes of acute abdominal pain. However, the higher the levels, the higher the probability of acute pancreatitis, such that levels of three or four times the upper reference limit or higher are usually considered diagnostic.

One recognized diagnostic pitfall with measurement of amylase is the presence in some individuals of a form complexed with immunoglobulin to form macroamylase. This has a longer serum half-life and is too large to be renally excreted. The consequence is that measured serum amylase is persistently high and may lead to false diagnosis of pancreatic (or salivary) disease. Another pitfall occurs when falsely low levels of amylase are measured in the presence of hypertriglyceridemia. This may be due to the presence of an inhibitor substance. Severe hypertriglyceridemia is a recognized cause of acute pancreatitis, and falsely low amylase levels can result in an important misdiagnosis and inappropriate surgery.

The prognosis of acute pancreatitis can be predicted by scoring systems which include biochemical results. These enable high-risk patients to be moved rapidly to intensive care units for aggressive treatment. A number of scoring systems exist, but one which has long-standing acceptance is the Glasgow score.

Analytical practice point 14.2

Amylase is unusual in being a diagnostic enzyme that is detectable in both plasma and urine. This can be useful in delayed presentation of acute pancreatitis and when lipemia is present.

Modified Glasgow score for the severity of acute pancreatitis

- **P**O_2 (partial pressure of oxygen in arterial blood) <8 kPa (60 mmHg)
- **A**ge >55 years
- **N**eutrophils: white blood cell count $>15 \times 10^9$/L
- **C**alcium <2 mmol/L (8 mg/dL)
- **R**enal function: urea >16 mmol/L (45 mg/dL)
- **E**nzymes: lactate dehydrogenase (LDH) >600 IU/L or aspartate transaminase (AST) >200 IU/L
- **A**lbumin <32 g/L (3.2 g/dL)
- **S**ugar: glucose >10 mmol/L (180 mg/dL)

Score 1 point for each feature: a score ≥3 indicates severe pancreatitis.

Chronic pancreatitis

Chronic inflammation of the pancreas, most often due to long-term excess alcohol intake, causes malabsorption due to decreased enzyme secretion, together with ongoing pain. Serum amylase and lipase are not usually significantly elevated, except if an episode of acute pancreatitis also occurs. Treatment consists of enzyme supplements that can be taken with food to replace the missing protease, lipase, and amylase. Chronic pancreatitis carries an increased risk of malignant change in the acinar cells giving rise to pancreatic cancer, which is difficult to diagnose and treat.

CASE 14.2

A 47-year-old man is admitted as an emergency with severe upper abdominal pain, radiating into his back, with vomiting. He does not drink alcohol and has no history of gallstones. On examination he has abdominal tenderness with guarding (reflex spasm of the abdominal muscles).

	SI units	Reference range	Conventional units	Reference range
Arterial blood				
PO_2	8.1 kPa	11–13	61 mmHg	80–100
Plasma				
Sodium	135 mmol/L	136–142	135 mEq/L	136–142
Potassium	4.5 mmol/L	3.5–5.0	4.5 mEq/L	3.5–5.0
Urea (blood urea nitrogen, BUN)	20.0 mmol/L	2.9–8.2	56 mg/dL	8–23
Creatinine	187 μmol/L	53–106	2.1 mg/dL	0.6–1.2
Glucose	12.3 mmol/L	3.9–6.1 (fasting)	222 mg/dL	70–110
Amylase	775 IU/L	27–131	775 IU/L	27–131
Calcium (total)	1.90 mmol/L	2.05–2.55	7.6 mg/dL	8.2–10.2
Albumin	28 g/L	32–52	2.8 g/dL	3.2–5.2
Corrected calcium	2.20 mmolL	2.20–2.60	8.8 mg/dL	8.8–10.4
LDH	305 IU/L	100–200	305 IU/L	100–200

- What is the most likely cause of the acute abdominal pain?
- How can the biochemical abnormalities be explained?
- How can the severity of this condition be graded from the results given?

There are many causes of acute abdominal pain, but very high plasma amylase is highly suggestive of acute pancreatitis. Although gallstones and alcohol excess are the commonest causes, a significant proportion of cases have no identifiable cause. Acute pancreatitis results in extracellular fluid loss due to vomiting and fluid accumulation in the abdominal cavity. This causes dehydration and may lead to acute kidney injury: the urea and creatinine reflect this. Low albumin is due to leakage from the vascular space, with capillary walls becoming more permeable both in the damaged pancreas and generally as part of the systemic response to injury. Raised glucose is due to a combination of a stress response (release of catecholamines and cortisol) and impaired insulin release. Low PO_2 is a result of impaired gas exchange due to fluid accumulation in the lungs plus restricted movement of the diaphragm from pain and abdominal swelling. The formation of soaps from the action of lipase on fatty acids reduces calcium due to chelation in the abdominal cavity. Markers of severity here are urea >16 mmol/L, glucose >10 mmol/L, calcium (uncorrected) <2.00 mmol/L, and albumin <32 g/L. Since there are three or more positive markers, this is classified as severe pancreatitis which should be managed in an intensive care or high-dependency unit.

CEREBROSPINAL FLUID AND THE NERVOUS SYSTEM

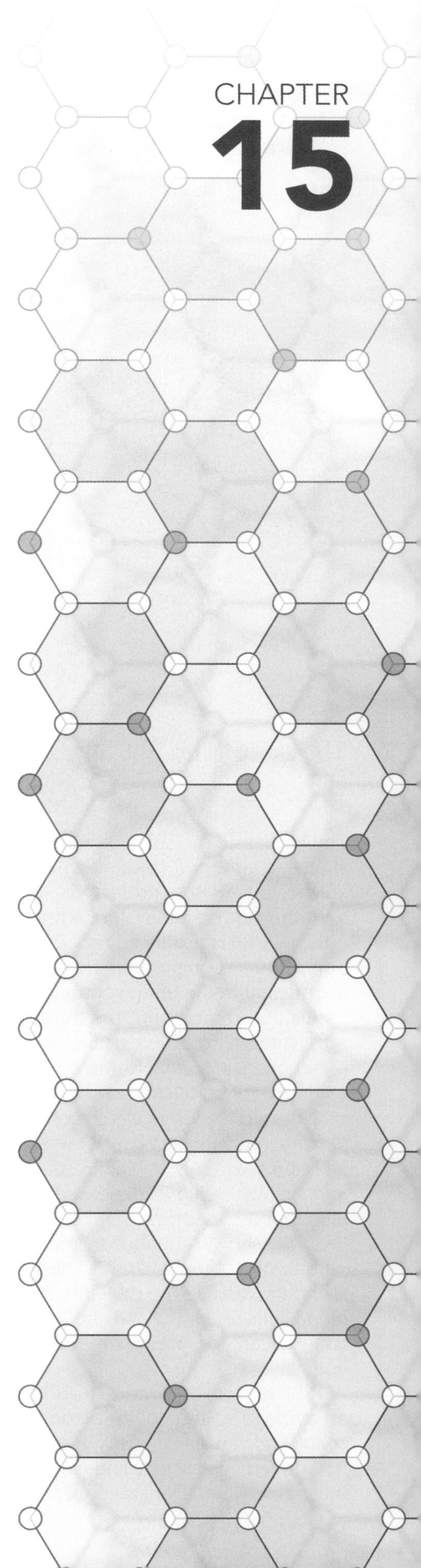

IN THIS CHAPTER

ANATOMY OF THE MENINGES

PRODUCTION, COMPOSITION, AND FLOW OF CSF

CSF SAMPLING

CLINICAL CHEMISTRY INVESTIGATION OF CSF

Changes in brain chemistry can provide important insights into neurological conditions. Direct chemical measurements of brain in patients are difficult, although not impossible (for example by using nuclear magnetic resonance or microdialysis). Therefore, cerebrospinal fluid (CSF), which is more easily sampled, is often used as a surrogate to give insight into changes in brain parenchyma, although it can also give important information on diseases that directly affect the CSF system (for example meningitis).

15.1 ANATOMY OF THE MENINGES

The brain and spinal cord are located within the bony compartments of the skull and the vertebral column, respectively. These organs are covered in three layers of meningeal membranes called the pia mater, arachnoid mater, and the dura mater (**Figure 15.1**).

The pia mater is a single cell layer that adheres tightly to the surface of the brain and spinal cord. It is pierced by smaller blood vessels as they pass into the brain and spinal cord tissue. The arachnoid mater is also a single-cell layer membrane which exists outside the pia mater. A space exists between the arachnoid mater and the pia mater and is called the subarachnoid space. This space is filled with CSF and also contains the arteries supplying the brain and spinal cord. The dura mater is the outermost, toughest membrane, which is closely associated with the arachnoid mater. The arachnoid has projections through the fluid-filled subarachnoid space connected to the pia. This gives rise to its name, as it appears like a spider's web. In the head, the dura mater actually consists of two layers: an inner meningeal layer and an outer periosteal layer, which is closely adherent to the skull. In certain places these two layers are separated to form (dural) venous sinuses. The space between the arachnoid mater and the meningeal layer of the dura (the subdural space) is only a very small space occupied by bridging veins that help drain the venous blood from the brain into the venous sinuses, which then empty into the internal jugular veins.

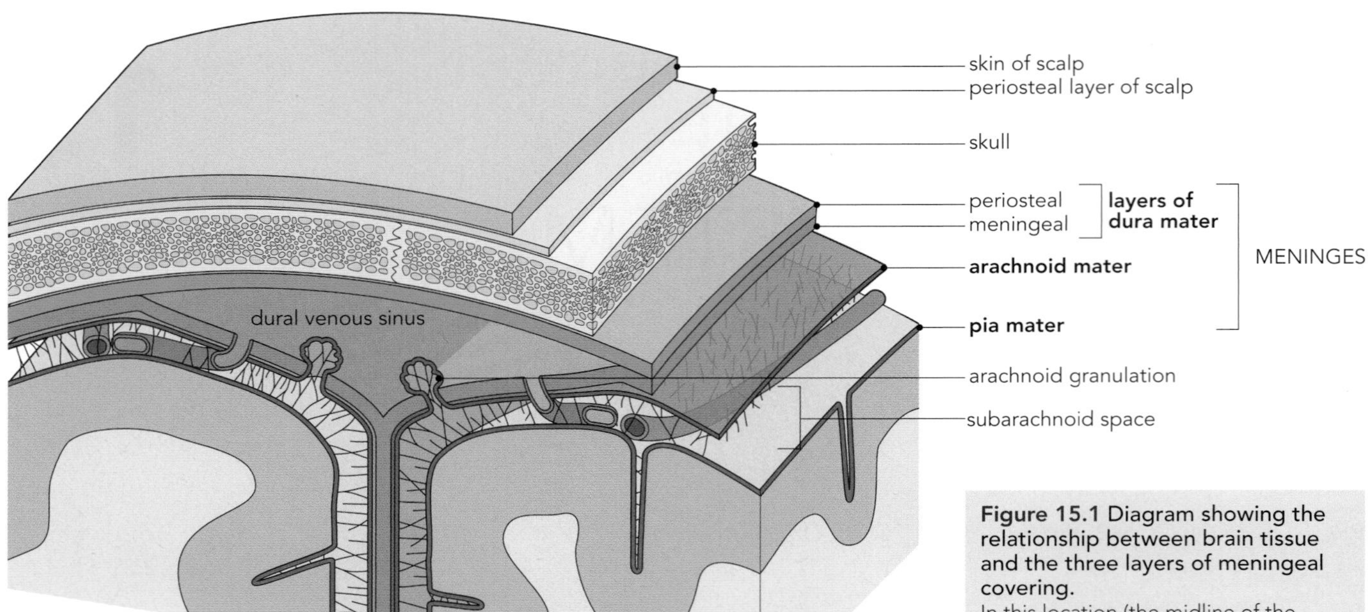

Figure 15.1 Diagram showing the relationship between brain tissue and the three layers of meningeal covering.
In this location (the midline of the skull, viewed here in coronal section), the dura is separated into two layers, one of which (the periosteal layer) is tightly adherent to the skull. The space between the dura and the skull is known as the extradural (or epidural) space. This is, in fact, only a potential space because of the tight adherence of the dura to the skull. If a blood vessel is damaged by a skull fracture, the blood would commonly collect in this space (an extradural hematoma). Between the dura (orange) and the arachnoid mater (green), another potential space exists, the subdural space. Subdural hematoma commonly occurs when the small bridging veins in this space become torn. Between the arachnoid and the pia mater (pink), an actual space known as the subarachnoid space exists; this space is filled with blood vessels and cerebrospinal fluid (CSF). Blood coming from an aneurysm of one of the medium-sized arteries supplying the brain would commonly collect in this space (a subarachnoid hemorrhage). CSF can be sampled if one can safely introduce a needle into this subarachnoid space. The pia mater is tightly adherent to the surface of the brain tissue. These meningeal layers continue down to cover the spinal cord within the vertebral column as well as the brain within the skull.

15.2 PRODUCTION, COMPOSITION, AND FLOW OF CSF

The brain and the spinal cord are relatively soft and are fairly mobile within the skull and vertebral column, which means that they can be prone to injury if they are not adequately protected. Some protection is provided by CSF, which fills both the ventricular system within the brain and the subarachnoid space to the outside. CSF is produced by a combination of filtration of the plasma and secretion of other components by the epithelial cells of the choroid plexuses, which are present in the brain ventricles. There are choroid plexuses in the lateral ventricles in the forebrain, the third ventricle in the region around the hypothalamus, and the fourth ventricle in the brain stem. About 12% of CSF comes from metabolic water generation by the brain parenchyma when glucose undergoes oxidative metabolism.

Normal CSF is a clear, colorless liquid that is similar in composition to plasma, but with very much lower concentrations of protein (including immunoglobulins), slightly lower levels of glucose, calcium ions, and potassium ions, and slightly higher levels of sodium ions and chloride ions (**Table 15.1**). Protein electrophoresis of CSF shows more than 1000 proteins; compared to serum the CSF shows enhanced prealbumin, two transferrin bands, and decreased γ-immunoglobulins. One of the transferrin bands is a variant, Tau protein, not found in plasma. The CSF should contain very few white blood cells and no red blood cells under normal circumstances. Changes in the composition of the CSF can indicate the presence of infectious, inflammatory, hematological, or metabolic disease.

CSF is produced and reabsorbed at a rate of approximately 500 mL/day with an average volume of approximately 120 mL existing within the subarachnoid space and ventricles at any given point in time. The CSF flows in a pulsatile manner, driven by vascular and respiratory pulses, through the ventricular system of the brain. A significant amount finds its way into the subarachnoid space via three

TABLE 15.1 Normal values for protein, glucose, and lactate and sodium, calcium, potassium, chloride, and magnesium ions in plasma and CSF

Constituent	Plasma concentration		CSF concentration	
	SI units	Conventional units	SI units	Conventional units
Protein	65–80 g/L	6.5–8.0 g/dL	0.15–0.45 g/L	15–45 mg/dL
Glucose	3.6–6.0 mmol/L	70–110 mg/dL	>67% of plasma concentration	
Na^+	136–142 mmol/L	136–142 mEq/L	147–151 mmol/L	147–151 mEq/L
Cl^-	98–106 mmol/L	98–106 mEq/L	118–132 mmol/L	118–132 mEq/L
Ca^{2+} (free/ionized)	1.15–1.3 mmol/L	4.6–5.2 mg/dL	1.00–1.35 mmol/L	4.0–5.4 mg/dL
K^+	3.5–5.0 mmol/L	3.5–5.0 mEq/L	2.8–3.2 mmol/L	2.8–3.2 mEq/L
Mg^{2+} (free/ionized)	0.44–0.6 mmol/L	1.0–1.5 mg/dL	0.78–1.26 mmol/L	1.77–2.1 mg/dL
Lactate	0.6–1.7 mmol/L	5.0–15.0 mg/dL	0.85–-2.0 mmol/L	7.7–-18.0 mg/dL

foramina in the region of the brain stem—the (medial) foramen of Magendie and the (lateral) foramina of Luschka. At any one time, the ventricular system and central canal contain approximately 20 mL of CSF, with approximately 100 mL in the subarachnoid space. Given that 500 mL is produced each day, a significant amount is reabsorbed each day; this is done via arachnoid granulations (also known as villi), which allow CSF to drain from the subarachnoid space into the dural venous sinuses and, from there, into the circulatory system (**Figure 15.2**), although there is evidence of other drainage sites.

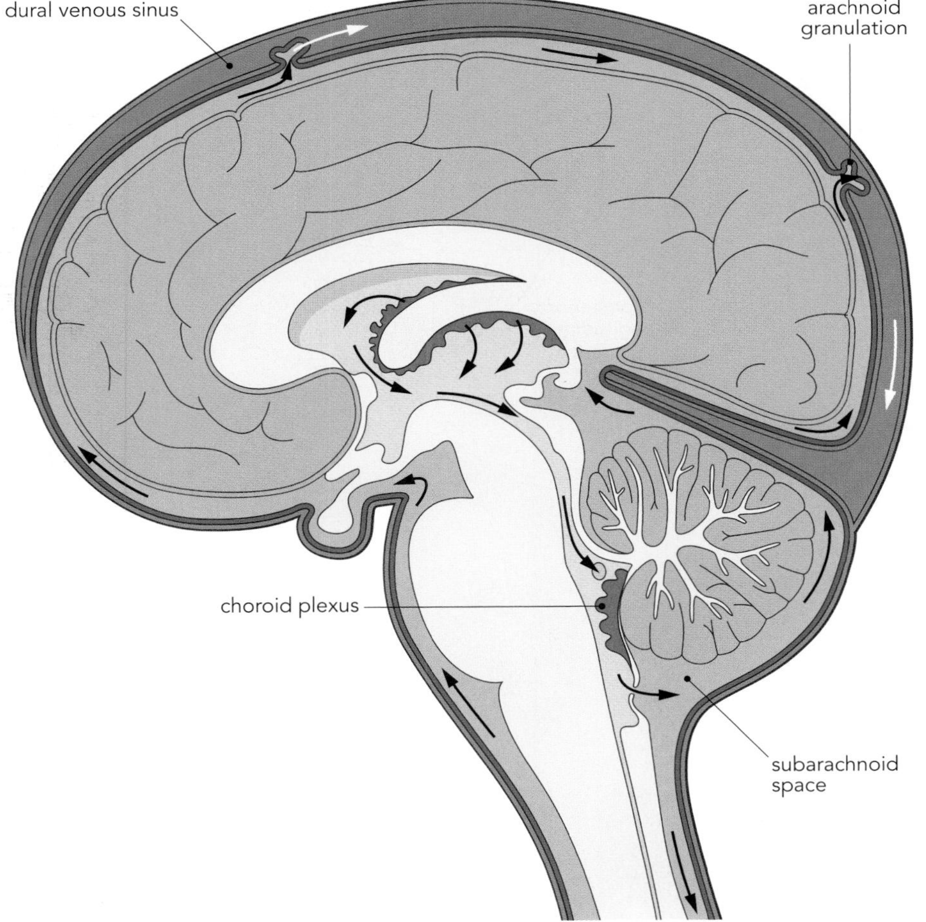

Figure 15.2 Mid-sagittal diagram of the brain, brain stem, and upper spinal cord showing the route of flow of cerebrospinal fluid (CSF). CSF is produced by choroid plexus (bright pink), which exists in all ventricles of the brain. The fluid fills the ventricular system of the brain and finds its way into the subarachnoid space (light blue) via foramina in the fourth ventricle of the medulla. From the subarachnoid space, the fluid is able to find its way into the circulatory system at the dural venous sinuses (darker blue) via arachnoid granulations/villi (white arrows).

If the flow of CSF is obstructed (for example by a tumor such as a pineal gland tumor that is pressing on the third ventricle, bleeding, infection, inflammation, or developmental defects), accumulation of fluid can result in swelling of the ventricular system, which can cause obstructive hydrocephalus and raised intracranial pressure. Hydrocephalus can also result from blockage of the CSF drainage pathway by intraventricular or subarachnoid blood, or by congenital malformations leading to obstruction of the cerebral aqueduct, the narrow passage between the third and fourth ventricles. Such obstruction is termed aqueductal stenosis.

15.3 CSF SAMPLING

It is important to note that there are several situations in which a lumbar puncture is contraindicated. One particular example is when there are signs of raised intracranial pressure due to a space-occupying lesion such as a tumor; to perform a lumbar puncture in this situation risks the patient dying from tonsillar herniation, where the cerebellar tonsils and brain stem are forced down through the foramen magnum when the pressure in the subarachnoid space is reduced because of CSF sampling. Bleeding diathesis and intake of certain blood thinners are also contraindications for a lumbar puncture due to the risk for bleeding and compression of the spinal cord or nerve roots.

CSF sampling is normally done by lumbar puncture (spinal tap). This is performed in the lower lumbar region below the level of termination of the spinal cord, usually between the third and fourth lumbar vertebrae (L3 and L4) in an adult and one level lower in children (**Figure 15.3**).

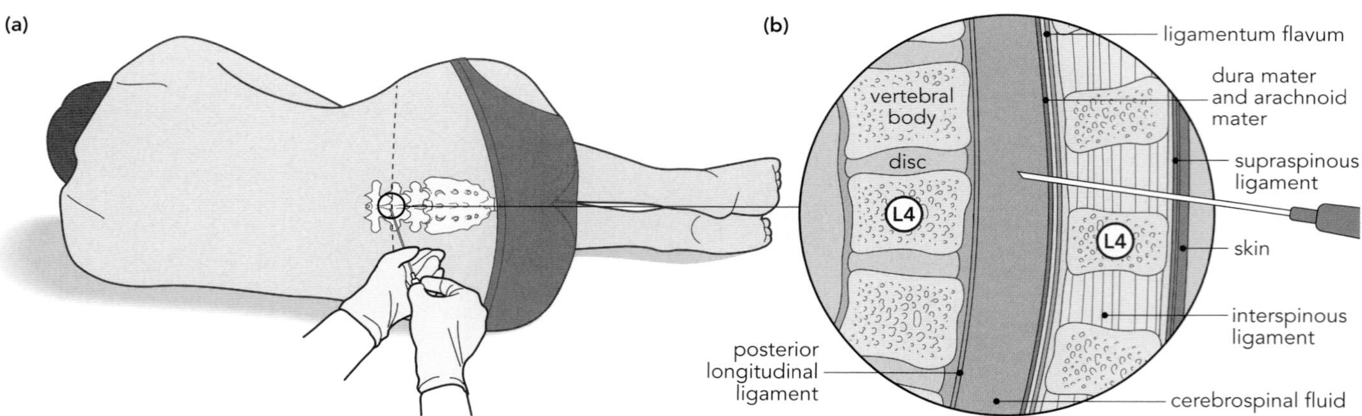

Figure 15.3 Diagrams showing (a) the usual position adopted by the patient during sampling of cerebrospinal fluid (CSF) from the lumbar region (lumbar puncture) and (b) the layers through which the doctor must pass the needle in order to remove CSF from the subarachnoid space.
(a) During lumbar puncture, the patient is commonly asked to lie on their side, to put their chin on their chest, and to draw their knees up into a fetal position. This serves to reduce the degree of lumbar lordosis and open up the space between the spinous processes of the vertebrae into which the sampling needle will be introduced (b). The position for sampling of the CSF is identified by drawing a line between the posterior iliac crests; this marks the position of the L4 spinous process and so is below the level of termination of the spinal cord in adults at approximately L1/L2. Lumbar puncture can therefore safely be performed in adults in either the L3/L4 or L4/L5 interspaces (in children, the cord ends at L3 so lumbar puncture must be performed at L4/L5). (b) Following preparation of the site and local anesthesia, CSF can be sampled by passing the needle through the skin, the interspinous ligament, and the dura mater. The dura and arachnoid mater are tightly adherent to each other and so piercing the dura will usually result in the needle entering the subarachnoid space, allowing some CSF to be removed.

In both cases, use of this site reduces the risk of damaging the spinal cord, which ends at L1/L2 in adults and in L3/L4 in children. A needle and cannula are introduced, under local anesthetic, into the subarachnoid space and a small amount of CSF is drained, which can then be analyzed with respect to color, turbidity, and its cellular and chemical composition. The normal CSF opening pressure is 7–18 cm H_2O. The amount of CSF drained during a lumbar puncture depends on the indication and the studies needed. Up to 40–50 mL of CSF can be withdrawn in adults, although 5–15 mL is sufficient for the analyses required in most cases. In conditions such as communicating hydrocephalus and pseudotumor cerebri, a significant amount of CSF may need to be drained to relieve pressure.

A sample for glucose measurement should be collected into fluoride tubes and analyzed immediately. Great care should be exercised in handling CSF samples because of their potentially infectious nature.

15.4 CLINICAL CHEMISTRY INVESTIGATION OF CSF

Biochemical analysis of CSF can be useful in a number of neurological conditions. However, obtaining a sample is painful for the patient, and is more difficult and inherently more risky than for blood, so lumbar puncture is usually only performed on clinical suspicion of a limited number of dangerous conditions. Most commonly, these are meningitis or subarachnoid hemorrhage (SAH); in the former instance, a lumbar puncture is performed even if there is a low suspicion for meningitis, since CSF analysis is the only way to make the diagnosis. Clinically useful investigations that can be performed include cell counts and differential (important for making the diagnosis of meningitis), smear and culture (important in identifying the organism in meningitis), glucose, proteins, lactate, and bilirubin (xanthochromia). Occasionally, measurement of CSF lactate may be useful in the diagnosis of defects of the respiratory transport chain in children. CSF analysis also has a role in the diagnosis of demyelinating conditions such as multiple sclerosis (MS) and research is ongoing in other disease processes.

It is important to remember that any change in CSF may simply reflect a change in plasma and the concentration of any particular analyte in CSF should always be compared with that in plasma. This is particularly true of glucose, where CSF concentration should be no lower than two-thirds of the plasma value. Diseases in which changes in CSF are manifest are listed in **Table 15.2**.

TABLE 15.2 Clinical changes in CSF and their causes

Altered appearance	Causes
Blood-stained	Traumatic lumbar puncture
	CNS hemorrhage
Xanthochromia	Old hemorrhage
	Spinal block
Turbidity	Pyogenic meningitis
Clots (fibrin)	Meningeal malignancy
	Tuberculous meningitis
	Spinal block
Change in glucose concentration	
Decreased	Pyogenic meningitis
	Prolonged hypoglycemia
Increased	Prolonged hyperglycemia

Identification of CSF

Sometimes following trauma or surgery, a breach in the meninges may allow CSF to escape and leak from the nose (CSF rhinorrhea) or ear (CSF otorrhea). It is very important to recognize this because such a breach may allow the entry of microorganisms into the meninges and lead to meningitis. In this case, surgical repair of the meningeal defect would be required. β-2 Transferrin is a carbohydrate-free (desialated) isoform of transferrin, which is found almost only in the CSF. It is not found in blood, mucus, or tears, thus making it a specific marker of cerebrospinal fluid. Its presence in nasal or ear discharge fluids indicates a CSF leak. The laboratory can identify the presence of β-2 transferrin using a form of protein electrophoresis called isoelectric focusing.

Meningitis

Meningitis describes an inflammation of the meninges and central nervous system, usually caused by bacterial, viral, or fungal infection. Bacterial meningitis is the more severe and is rapidly fatal in many cases, particularly when due to meningococcus. Treatment with antibiotics must be started as soon as the diagnosis is suspected, but diagnosis will be confirmed by microscopy of CSF, looking for organisms and leukocytes, followed by microbiological culture. Supporting biochemical evidence for the condition, which has the advantage of being rapidly available, would be the combination of raised protein and low glucose, reflecting inflammation and increased cellular respiration, respectively.

CASE 15.1

A 10-year-old boy is admitted as an emergency with recent onset of fever, headache, neck stiffness, and photophobia. He has a rash that does not blanch when pressed. He has a lumbar puncture (spinal tap) performed. Some of the CSF is sent to the biochemistry laboratory for analysis.

	SI units	Reference range	Conventional units	Reference range
CSF				
Protein	1.3 g/L	0.15–0.45	130 mg/dL	15–45
Glucose	1.9 mmol/L	2.5–4.4	34 mg/dL	40–80
Lactate	4.75 mmol/L	0.85–2.00	43 mg/dL	7.7–18.0

- What is the likely diagnosis?
- How do the biochemical results help?
- How is the diagnosis confirmed?

The patient is suspected to have meningitis. This may be due to a viral or bacterial infection, with the latter being more serious and life threatening. The presence of a nonblanching rash is highly suggestive of meningococcal infection. The biochemical results, which can be obtained rapidly, show a raised protein (due to inflammation) and low glucose (due to bacterial and white blood cell metabolism). Neither is specific for bacterial infection, but the probability increases with the degree of abnormality. CSF lactate is raised to a much greater extent in bacterial than in viral meningitis, but this marker is probably underused. Both glucose and lactate must be measured in samples collected into fluoride-oxalate tubes (as for blood glucose) to prevent *in vitro* cellular metabolism. The definitive diagnosis requires culture of the CSF in a microbiology laboratory to show the organism, but this may take 48 hours. Antibiotic treatment is started before culture results are available, but may be modified when the organism's sensitivity is known.

Increased respiration is due to the activity of the host leukocytes rather than the bacterial cells themselves. Where meningitis is due to *Mycobacterium tuberculosis* infection, the glucose and protein of CSF are often strikingly low and high, respectively, compared to infection with other bacteria such as meningococcus or streptococcus.

Subarachnoid hemorrhage

Bleeding from an intracranial aneurysm is often rapidly fatal, but if the patient survives a first bleed it is important to make the correct diagnosis and prevent further episodes. This requires surgery or endovascular coiling after identifying the anatomy of the aneurysm by vascular imaging techniques. These carry a risk of neurological damage, as contrast material must be injected. It is therefore necessary to be confident of the diagnosis so that only patients with a high probability of SAH are investigated. In the early presentation, the diagnosis is made predominantly by computerized axial tomographic (CAT) scanning of the head.

CASE 15.2

A 26-year-old man presents to the emergency department with severe occipital headache which began 48 hours previously. He has neck stiffness but no other neurological signs. A head CAT scan is reported as normal. He has a lumbar puncture (spinal tap) performed to obtain cerebrospinal fluid (CSF) and this is analyzed in the laboratory for the presence of xanthochromia.

	SI units	Reference range	Conventional units	Reference range
CSF				
Net oxyhemoglobin absorbance	0.03 AU	<0.02	0.03 AU	<0.02
Net bilirubin absorbance	0.015 AU	<0.007	0.015 AU	<0.007
Protein	0.75 g/L	0.15–0.45	75 mg/dL	15–45

AU, absorbance units.

- What is xanthochromia?
- What is the significance of a positive result?

Xanthochromia is the presence in the CSF of an abnormally high concentration of bilirubin. Most serum bilirubin is unconjugated and bound to albumin, which usually prevents it from crossing the blood–brain barrier into the CSF. Elevated levels of CSF bilirubin, as shown by net bilirubin absorbance (NBA) at 476 nm indicate the presence of blood in the CSF due to breakdown of heme. For a more detailed description of the method for calculating NBA, see Chalmers (2001)[a] and Chalmers and Kiley (1998)[b].

Oxyhemoglobin is detected at 415 nm but is not specific for bleeding within the CSF as it may be introduced into the sample by the lumbar puncture itself. Subarachnoid hemorrhage (SAH) may be detected on a CAT scan but this is progressively less sensitive after 24 hours. Xanthochromia may therefore be the only objective evidence of SAH for presentations after this time. Confirmation is required by further tests such as cerebral angiography to show the source of bleeding (usually an arterial aneurysm). Another cause of xanthochromia is meningitis, either from infection or another insult, which allows passive transfer of serum bilirubin into the CSF. This is accompanied by high CSF protein.

a, Chalmers AH (2001) *Clin Chem* 47:147–148; b, Chalmers AH & Kiley M (1998) *Clin Chem* 44: 1740–1742.

However, if CAT scanning is performed beyond 24 hours of symptom onset, the sensitivity of CAT scanning for the detection of SAH is lower than if the scan is performed earlier. In these patients, biochemical analysis of the CSF is essential if one wishes to exclude SAH.

It was recognized for many years that CSF following SAH was yellow tinged when looked at against a white background (xanthochromia). This is due to the presence of bilirubin at low concentration, which is itself formed from the degradation of hemoglobin. However, it is no longer acceptable to designate a sample as positive or negative for bilirubin on the basis of its subjective appearance to the human eye, especially as bilirubin may be present at abnormally high levels and still be invisible. A sensitive and quantitative method is required. The concentration of bilirubin in CSF is below the level that can be reliably measured by the methods used for serum. However, it can be assessed indirectly by use of a spectrophotometer, due to its absorbance at 476 nm. Bilirubin is detectable from 12 hours following SAH and may be present for up to 10 days. The presence of an oxyhemoglobin peak between 410 and 418 nm is expected soon after SAH but this diminishes with time as hemoglobin is degraded. Oxyhemoglobin alone, however, is not sufficient to diagnose SAH, since blood may be introduced into the sample by minor tissue trauma caused by the lumbar puncture needle. This is called a bloody tap. Interpretation of CSF bilirubin also requires knowledge of CSF protein and serum bilirubin, since inflammatory conditions which reduce the integrity of the blood–brain barrier may allow transfer of serum constituents.

Analytical practice point 15.1

The presence of bilirubin in CSF (xanthochromia) must be confirmed by spectrophotometry: visual inspection is insufficiently sensitive or specific.

Multiple sclerosis

MS is the most common disabling neurological disease affecting young adults in the third and fourth decades of life. In MS, an autoimmune process causes damage to the myelin nerve sheaths, leading to a variety of neurological symptoms affecting balance, vision, muscle function, and cognition to varying extents and with variable severity over time. Historically, CSF examination was a key investigation for suspected MS, but magnetic resonance imaging (MRI) has now superseded this in most cases, leaving biochemistry as an adjunct where the diagnosis is not clear. The principle of biochemical investigation is the demonstration of increased immunoglobulin synthesis within the central nervous system. This is shown by the presence of oligoclonal banding on protein electrophoresis (**Figure 15.4**). The pattern obtained must be compared to that from serum taken simultaneously to exclude the possibility of passive leakage of serum proteins across the blood–brain barrier. The ratios of serum and CSF albumin and immunoglobulin G (IgG) can be calculated to assess the permeability of the blood–brain barrier.

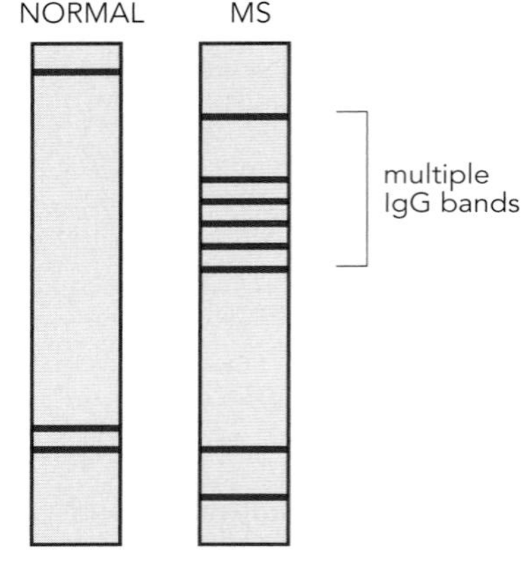

Figure 15.4 Oligoclonal banding pattern demonstrated by gel electrophoresis of a sample of cerebrospinal fluid (CSF) from a patient with multiple sclerosis (MS) compared to normal CSF.
In MS, the myelin in the central nervous system is attacked by the patient's own immune system and multiple clones of immunoglobulin G (IgG) can be detected in the patient's CSF.

Neurological disease

Clinical biochemistry investigations play a relatively small but important role in the diagnosis of neurological conditions. Due to the presence of the blood–brain barrier, potential biomarkers of neurological disease are unable to enter the plasma, where they might be detected. It is possible to sample the CSF but, as mentioned previously, this is invasive and only of proven value in the conditions described above. In addition, reference ranges for new CSF markers are difficult to establish due to the limited numbers of samples available, especially from healthy subjects. On the other hand, the brain is susceptible to many types of metabolic insult which can be investigated by tests done on plasma. However, new research has identified several CSF biomarkers that are specific for Alzheimer's disease and, once validated, these are likely to be available for clinical use.

Acute encephalopathy and coma

There are many causes of confusional states and loss of consciousness. Those causes with a metabolic basis are shown in **Table 15.3**. Liver and kidney failure

TABLE 15.3 Clinical conditions with a metabolic basis causing acute encephalopathy

Clinical condition	Biochemical feature
Liver failure Renal failure	Accumulation of endogenous toxins
Hyperglycemia Hypoglycemia	Change in plasma glucose concentration
Hypernatremia Hyponatremia	Change in plasma sodium concentration and osmolality
Alcohol Drugs	Poisoning and intoxication
Wernicke's encephalopathy	Vitamin deficiency

both result in impaired clearance of metabolic products, which can accumulate and interfere with neurological function if they are able to cross the blood–brain barrier. A well-recognized example of this is the accumulation of ammonia in liver failure. Other, less-well-recognized toxins undoubtedly also play a part by affecting neurotransmitter activities, and are responsible for encephalopathy of uremia which may occur in severe renal failure (an estimated glomerular filtration rate <15 mL/min).

A good supply of glucose is essential for neuronal function; hypoglycemia (glucose <3 mmol/L) causes impaired brain function resulting in confusion, abnormal behavior, and eventually coma. The commonest reason for hypoglycemia is overtreatment of diabetes, but other causes need to be considered and are explained in detail in Chapter 2.

Peripheral neuropathy

The peripheral nerves are the part of the nervous system outside of the brain and spinal cord, and comprise sensory, motor, and autonomic nerves. Symptoms and signs of neuropathy depend on which of these is predominantly affected. Sensory neuropathy can cause pain, paresthesia, or loss of reflexes, whilst motor neuropathy only affects motor function, causing weakness and loss of reflexes. Autonomic neuropathy may affect the gastrointestinal tract (causing vomiting and diarrhea), the bladder (causing urinary retention), or the cardiovascular system (causing hypotension).

There are some causes of peripheral neuropathy with a biochemical or metabolic basis and these include:

- Diabetes
- Alcohol excess
- Paraprotein/amyloidosis
- Renal failure
- Liver disease
- Endocrine disease
  - Hypothyroidism
  - Acromegaly
- Vitamin deficiencies
  - B_{12}
  - Thiamine
  - B_6
  - E
- Lead poisoning
- Porphyrias

The best recognized cause of peripheral neuropathy is diabetes mellitus, where neuropathy is one of the principal microvascular complications. This may affect any or all of the peripheral nerve systems, but most commonly it presents as a sensory neuropathy affecting the lower limbs and starting in the feet. Initial symptoms are usually pain and paresthesia but, later, complete loss of sensation may occur. This may then result in unrecognized injury, which can lead to ulceration, infection, and ultimately loss of the limb. CSF protein levels can sometimes be elevated in diabetic neuropathy.

Neuropathies may also be seen with excessive alcohol intake, vitamin deficiencies, and lead poisoning. They may also occur in renal, liver, and endocrine disease. Rarely, a monoclonal protein may cause peripheral neuropathy and this can be diagnosed by serum protein electrophoresis and serum or urine light-chain measurement. Other rare causes of peripheral neuropathies are the porphyrias, which are caused by deficiencies (acquired or genetic) in one of the heme biosynthesis enzymes (see Chapter 16). In most of these, there is an accumulation of porphobilinogen, which can be detected in urine. In cases of peripheral neuropathy of unknown cause, investigations for some or all of these conditions may be required, as directed by the clinical history and, if present, other physical signs.

HEME BIOSYNTHESIS AND PORPHYRIAS

CHAPTER 16

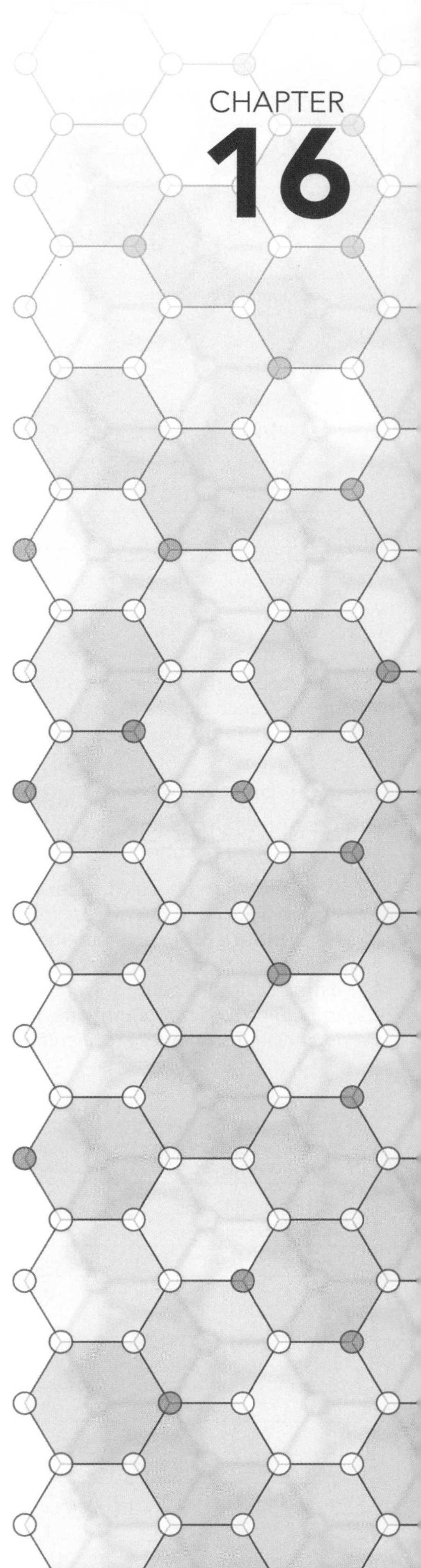

IN THIS CHAPTER

PORPHYRINS AND THE EFFECTS OF STRUCTURE ON ROUTE OF EXCRETION

BIOSYNTHESIS OF HEME

PORPHYRIAS

INVESTIGATIONS OF PATIENTS FOR PORPHYRIA

Heme is a macromolecule found in all mammalian cells and is essential for life. Its major role is in the transport and delivery of oxygen to cells and also in oxidative metabolism. Heme-containing molecules include but are not limited to:

- Hemoglobin, neuroglobin, cytoglobin
- Myoglobin
- Cytochromes a, $a + c_3$ (cytochrome oxidase), b, b_5, and c_1, other hemoproteins
- Cytochrome P450-linked enzymes
- Oxidative enzymes, for example catalase

Heme consists of a porphyrin ring chelated to iron using the nitrogen atoms of the four cyclic tetrapyrrole rings. The most common type of heme, heme B (ferroprotoporphyrin IX), is shown in **Figure 16.1**; the six β substituents are four methyl, two vinyl, and two propionyl groups. No function is known in man for porphyrins other than as part of the heme molecule and the biosynthetic route is unique to this molecule. Heme is synthesized in all cells for assembly into mitochondrial cytochromes, but the major sites of synthesis are liver (50 mg/day) and erythrocyte precursor cells (300 mg/day in bone marrow). Total heme production in erythropoietic cells reflects cell number. A deficiency in any of the

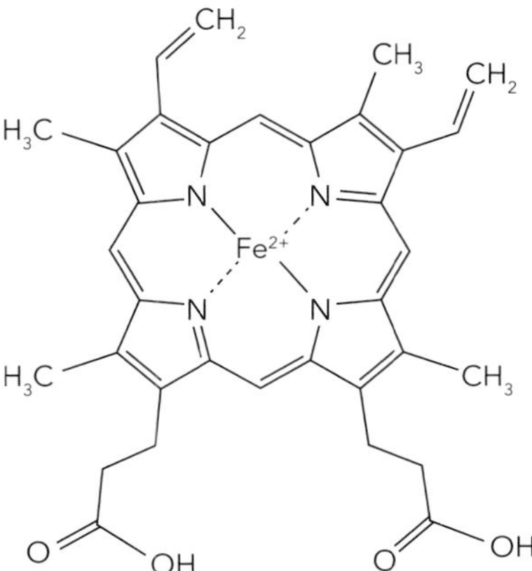

Figure 16.1 Structure of heme (ferroprotoporphyrin IX).
Heme is a conjugated porphyrin in which a single ferrous iron (Fe^{2+}) is chelated to each nitrogen atom of the four tetrapyrrole rings. Substituents on the individual pyrrole rings are methyl and either vinyl or propionyl.

TABLE 16.1 Reference ranges for some analytes measured in diagnoses of porphyrias

Analyte	Erythrocytes	Urine (24 hour)	Feces (24 hour)	Serum
Total porphyrin	16–60 µg/dL			
Uroporphyrin(U)	<2 µg/dL	Total ≤30 nmol/day (≤19.6 µg/day)	UI <183 nmol/day (120 µg/day) UIII <76 nmol/day (<50 µg/day)	
Coproporphyrin (C)	<2 µg/dL	Total male ≤230 nmol/day (≤130 µg/day) Total female ≤168 nmol/day (≤109 µg/day)	CI <760 nmol/day (<500 µg/day) CIII <611 nmol/day (<400 µg/day) IsoC <305 nmol/day (<200 µg/day)	
Protoporphyrin	16–60 µg/dL		<2.7 µmol/day (<1500 µg/day)	
PBG	ND	<2.2 µmol/day (<0.5mg/day)	ND	
ALA	ND	<15 µmol/L (<2 mg/L)	ND	1.1–1.8 µmol/L (15–23 µg/dL)

UI, Uroporhyrin I; UIII, Uroporphyrin III; CI, Coproporphyrin I; CIII, Coproporphyrin III; IsoC, Isocoproporphyrin; PBG, Porphibilinogen; ALA, Aminolevulinic Acid; ND, not determined.

enzymes of the heme biosynthetic pathway gives rise to the porphyrias, and the reference ranges for analytes measured in diagnosis of these disorders are shown in **Table 16.1**.

16.1 PORPHYRINS AND THE EFFECTS OF STRUCTURE ON ROUTE OF EXCRETION

Porphyrins are classified according to substituents at the angle carbon atoms (**Table 16.2**) and the arrangement of substituents around the porphyrin structure (**Table 16.3**). Small amounts of porphyrins with five, six, and seven COOH groups are found in body fluids and, in general, the pH at which porphyrins are extracted into organic solvents depends on the number of carboxyl groups in the molecule. Protoporphyrins, including protoporphyrin IX to which Fe^{2+} (Fe^{II}) is chelated in heme, are insufficiently polar to be excreted into urine, whilst

TABLE 16.2 Classification of porphyrins according to angle carbon substituents

Porphyrin type	Number of –COOH groups	Groups
Uroporphyrins	8	4 acetyl, 4 propionyl
Coproporphyrins	4	4 methyl, 4 propionyl
Protoporphyrins	2	4 methyl, 2 vinyl, 2 propionyl

TABLE 16.3 Classification of porphyrins according to arrangement of substituents

Type	Structure	Example	Order
I	Symmetrical and alternate	Uroporphyrin I	AP, AP, AP, AP
III	Asymmetrical and irregular	Uroporphyrin III	AP, AP, AP, PA

Note: Only types I and III occur in humans. A, acetate; P, propionate.

significant amounts of uroporphyrins can be found in urine. In general, intermediates such as 5-aminolevulinic acid (ALA) and porphobilinogen (PBG) at the beginning of the heme biosynthetic pathway are much more water soluble than those later in the pathway and are excreted almost entirely in the urine; those intermediates derived from later down the pathway, such as the protoporphyrins with two carboxyl groups, are excreted entirely in the feces. Thus, uroporphyrins, with eight carboxyl groups, are excreted almost entirely through the kidneys, whilst coproporphyrins, with four carboxyl groups, are, as the prefix copro implies, excreted predominantly in the feces (about 70%). The chemical structures of intermediates in the heme biosynthetic pathway and metabolites are shown in **Figure 16.2**.

ALA, the building block of porphyrin synthesis, and PBG are colorless and nonfluorescent whilst porphyrins are reddish and fluoresce on exposure to long-wave ultraviolet light. Deposition of porphyrins in teeth causes them to appear reddish brown (erythrodontia) while deposition in bone may lead to extensive

(i) porphobilinogen

(ii) hydroxymethylbilane

(iii) uroporphyrinogen III and uroporphyrinogen I

(iv) coproporphyrinogen III and coproporphyrinogen I

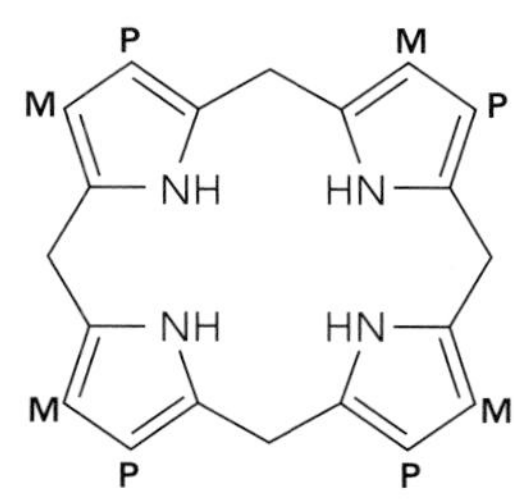

(v) protoporphyrinogen IX

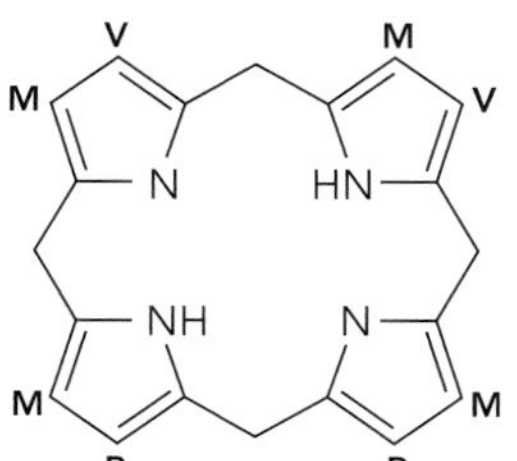

(vi) protoporphyrin X

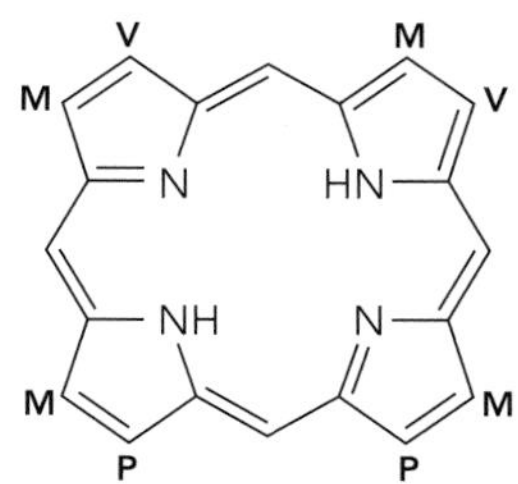

Figure 16.2 Structures of intermediates of the heme biosynthetic pathway.
(i) Porphobilinogen (PBG) is formed from condensation of two molecules of 5-aminolevulinic acid (ALA). (ii) Hydroxymethylbilane (HMB) is formed from the polymerization of four molecules of PBG. A and P indicate acetyl and propionyl residues, respectively. The linear molecule is shown here to illustrate the precursor prior to ring closure, which occurs in the subsequent reaction. (iii) Uroporphyrinogen III. Enzyme-catalyzed ring closure of HMB involving ring D inversion yields uroporphyrinogen III. Spontaneous, non-enzyme-catalyzed ring closure does not involve ring D inversion and yields uroporphyrinogen I. Consequently, A and P residues in uroporphyrinogen I are alternate around the molecule, while the A and P residues on one of the pyrrole rings are reversed in uroporphyrinogen III. (iv) Coproporphyrinogen III and coproporphyrinogen I are derived from decarboxylation of acetyl residues of uroporphyrinogens III and I, respectively. The acetyl residues are replaced by methyl residues (M) on the pyrrole rings. Coproporphyrinogen I is metabolically inactive. (v) Protoporphyrinogen IX is formed from oxidation of two of the propionyl residues of coproporphyrinogen III to vinyl residues (V). (vi) Protoporphyrin IX. Oxidation of the methylene bridges linking the four pyrrole rings of protoporphyrinogen IX yields protoporphyrin IX, the immediate precursor of heme.

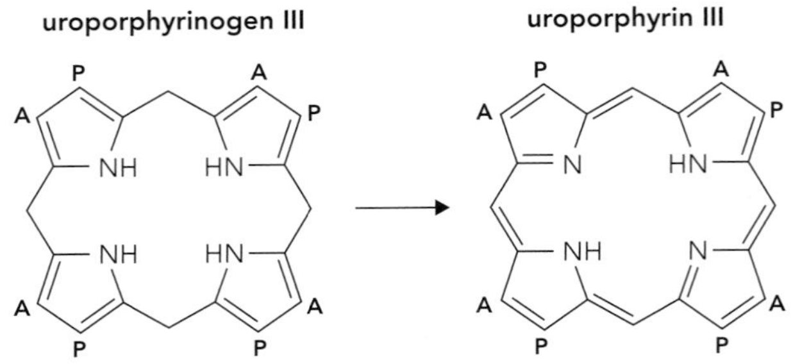

Figure 16.3 Spontaneous oxidation of uroporphyrinogen III to uroporphyrin III.
Oxidation of the methyl bridges in uroporphyrinogens and coproporphyrinogens in extracellular fluid yields uroporphyrins and coproporphyrins, respectively. These are metabolic dead-end molecules which may be measured in the clinical laboratory. The oxidation of uroporphyrinogen III to uroporphyrin III is shown here as an example.

demineralization. Under normal conditions, ALA and PBG are excreted in urine in significant amounts, but other intermediates of the heme biosynthetic pathway are conserved during heme synthesis and only small amounts are excreted. However, significant amounts of intermediates may be excreted in enzyme deficiencies that lead to porphyrias, and the route of excretion of these intermediates will depend upon the water solubility of the intermediate, as described above. Porphyrinogens undergo auto-oxidation outside of cells and are excreted mainly as porphyrins, the molecules measured in the clinical chemistry laboratory (**Figure 16.3**).

16.2 BIOSYNTHESIS OF HEME

The rate-limiting step of heme synthesis is the initial production of the key intermediate 5-aminolevulinic acid (ALA) from succinyl-CoA and glycine, catalyzed by the pyridoxal-dependent enzyme ALA synthase (**Figure 16.4**). This enzyme, in common with the final three enzymes of the heme biosynthetic pathway, is present in mitochondria. There appears to be no role for ALA other than for heme biosynthesis. The rate of ALA production is dependent on the amount of enzyme and, in the liver, this is increased by drugs and other compounds that induce cytochrome P450 enzymes (for example sulfonamides, anticonvulsants, and griseofulvin) and decreased by heme and glucose. Heme is not a direct physiological inhibitor of ALA synthase activity; rather, it inhibits its own synthesis via feedback inhibition of ALA synthase transcription by reducing the stability of ALA synthase mRNA. It also inhibits the import of the enzyme into mitochondria.

Subsequent steps in the biosynthesis of heme are shown in **Figure 16.5**. The next step is the condensation of two molecules of ALA under the action

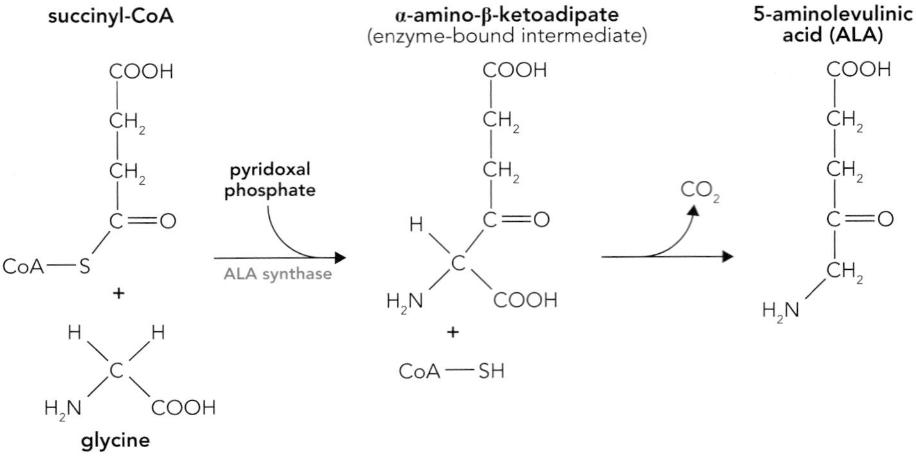

Figure 16.4 Synthesis of 5-aminolevulinic acid (ALA) by ALA synthase.
The first step in the synthesis of heme is the production of ALA from glycine and succinyl-CoA in the mitochondria, in a reaction catalyzed by the enzyme aminolevulinic acid synthase (ALA synthase). The reaction requires pyridoxal phosphate as a co-factor and the initial formation of α-amino-β-ketoadipate as an enzyme-bound intermediate, which undergoes the loss of CO_2 in forming ALA.

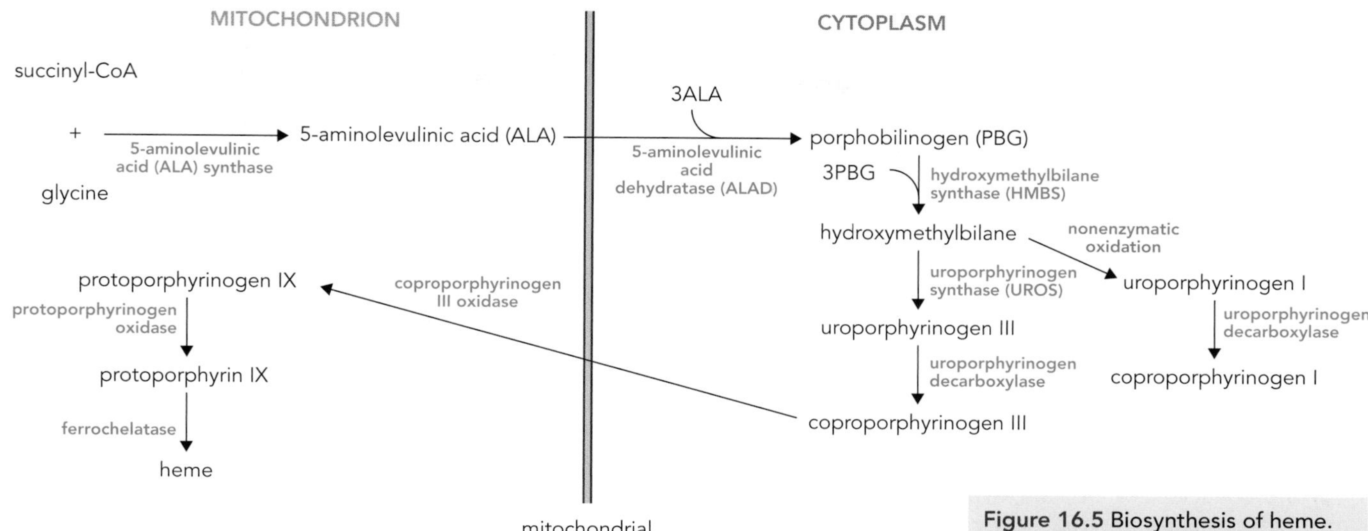

Figure 16.5 Biosynthesis of heme. Reactions indicated on the left are located in the mitochondrion whilst those on the right are located in the cytoplasm. The oxidation of hydroxymethylbilane to uroporphyrinogen I in the cytoplasm is nonenzymatic. The further formation of coproporphyrinogen I from uroporphyrinogen I represents a metabolic dead end.

of ALA dehydratase to form the pyrrole ring structure of PBG. Further condensation of four molecules of PBG yields a linear tetrapyrrole, hydroxymethylbilane (HMB), in a reaction catalyzed by hydroxymethylbilane synthase (HMBS; also known as PBG deaminase) with the release of an ammonium ion for every methylene bridge formed. Formation of the large asymmetric, macrocyclic molecule, uroporphyrinogen III, which involves cyclization of HMB by inversion of the final pyrrole ring (ring D isomerization) and linkage of this ring to the first pyrrole unit (A), is catalyzed by uroporphyrinogen III synthase (UROS). Ring closure, not involving inversion of ring D, can occur nonenzymatically at high concentrations of PBG, with the formation of uroporphyrinogen I, a symmetrical molecule. Uroporphyrinogen I is converted by uroporphyrinogen decarboxylase (UROD) to coproporphyrinogen I, a metabolically inactive molecule which does not participate in heme synthesis. The next reactions in heme synthesis modify the side chains and degree of saturation of the porphyrin ring. This involves sequential decarboxylation of four of the eight side chains of uroporphyrinogen III via hepta-, hexa-, penta-, to tetra-coproporphyrinogen III, catalyzed by UROD. Decarboxylation-dehydrogenation of the 2- and 4-carboxyethyl groups of coproporphyrinogen III to form protoporphyrinogen IX (which contains two vinyl substituents) is catalyzed by coproporphyrinogen oxidase. Further oxidation of protoporphyrinogen IX by protoporphyrinogen oxidase to protoporphyrin IX requires the loss of six hydrogens (four from the methylene bridges, two from ring nitrogens). Ferrochelatase completes the formation of heme (iron protoporphyrin IX) by catalyzing the addition of ferrous iron to protoporphyrin IX.

16.3 PORPHYRIAS

Porphyrias are a group of rare disorders in each of which there is an inherited or acquired defect that results in the decrease in activity of a specific enzyme of heme biosynthesis. This decreased enzyme activity does not usually result in significantly reduced concentrations of heme, but does lead to the accumulation of heme precursors and further metabolites related to the specific enzyme defect. These excess precursors and porphyrins derived from them are excreted into feces and/or urine or occasionally accumulate in liver, teeth, and bones. Clinically, it is important to differentiate between porphyrias and those acquired disorders in which abnormal porphyrin metabolism arises as a consequence of other conditions; examples of the latter are shown in **Table 16.4**. The eight

TABLE 16.4 Examples of abnormal porphyrin metabolism which are not caused by one of the known genetic porphyrias

Increased porphyrin or metabolite	Disease
Urine aminolevulinic acid (ALA)	Lead poisoning
Urine coproporphyrin I	Chronic liver disease
Urine coproporphyrin III	Lead poisoning
Red cell protoporphyrin	Lead poisoning, iron deficiency

main types of porphyria are due to specific defects in one of the seven main enzyme steps shown in Figure 16.2. The specific enzyme defect, pattern of inheritance, and clinical presentation of the individual porphyrias are summarized in **Table 16.5**.

While a deficiency of ALA synthase does not result in porphyria, it does cause sideroblastic anemia, since in this condition there is insufficient protoporphyrin to incorporate the iron delivered to mitochondria, with consequent iron overload of this organelle. The feedback inhibition by heme on iron uptake into maturing red blood cells is also compromised. Aminolevulinic acid dehydratase (ALAD) deficiency porphyria (ADP), congenital erythropoietic porphyria (CEP), erythropoietic porphyria (EPP), and hepatoerythropoietic porphyria (HEP) are inherited as autosomal recessive disorders, HEP being due to the concurrent presence of a low-activity, high-frequency polymorphism (low-expression allele and major mutation). The other porphyrias are autosomal dominant disorders with incomplete penetrance. As the name implies, in X-linked protoporphyria, the disorder-causing gene is located on the X chromosome and thus, while the disorder can appear in both males and females, males are usually more severely affected than females.

TABLE 16.5 Enzyme defects, patterns of inheritance, and clinical presentations of porphyria

Enzyme defect	Disorder	Pattern of inheritance	Presence of acute or neurological crises	Photosensitivity	Mixed symptomology
Aminolevulinic acid (ALA) dehydratase*	ALA dehydratase deficiency porphyria (ADP)	Autosomal recessive	+		
Hydroxymethylbilane synthase**	Acute intermittent porphyria (AIP)	Autosomal dominant	+		
Uroporphyrinogen III synthase	Congenital erythropoietic porphyria (CEP)	Autosomal recessive		+	
Uroporphyrinogen decarboxylase	Porphyria cutanea tarda (PCT)	Autosomal dominant		+	
Uroporphyrinogen decarboxylase	Hepatoerythropoietic porphyria (HEP)	Autosomal recessive		+	
Coproporphyrinogen oxidase	Hereditary coproporphyria (HCP)	Autosomal dominant			+
Protoporphyrinogen oxidase	Variegate porphyria (VP)	Autosomal dominant			+
Ferrochelatase	Erythropoietic protoporphyria (EPP)	Autosomal recessive		+	
ALA synthase 2	X-linked protoporphyria	X-linked		+	

*Also known as porphobilinogen synthase; **also known as porphobilinogen deaminase.

Whilst the classification of porphyrias by enzyme deficiency is precise, it is sometimes useful clinically to categorize the disorders by clinical presentation based on the major tissue involved in overproduction of the heme-pathway intermediate (hepatic or erythropoietic) or on the major symptoms presenting—neurovisceral or cutaneous, acute or photosensitive. As implied in the name of the disorders, EP and CEP are classified as erythropoietic because the bone marrow (erythroid cells) is the main site of overproduction of porphyrins. Four porphyrias—acute intermittent porphyria (AIP), porphyria cutanea tarda (PCT), hereditary coproporphyria (HCP), and variegate porphyria (VP)—are classified as hepatic, whilst HEP affects both hepatic and erythropoietic systems.

Patients may present with acute attacks, which may or may not be accompanied by skin lesions (never in AIP or ADP, sometimes in VP and HCP), or with skin lesions only (PCT, VP, CEP, EPP, and, rarely, HCP). There are two distinct types of skin presentation: (i) acute, painful photosensitivity without skin fragility or bullae (EPP); and (ii) with skin fragility and bullae (the bullous porphyrias—CEP, PCT, VP, and HCP).

Acute episodic

Acute episodes in porphyria (seen in ADP, AIP, HCP, and VP), which are often precipitated by drugs such as sulfonamides, barbiturates, and other anticonvulsants, are associated with excretion of excess ALA and PBG. Other important precipitants are endocrine factors; for example, premenstrual attacks, alcohol, fasting, infections, and stress. The episodes are characterized by the following clinical symptoms:

- Abdominal pain
- Nausea and vomiting
- Constipation
- Neuropsychiatric symptoms
- Neuropathy
- Hypertension
- Impaired water excretion

These symptoms can be a medical emergency and in most cases are partly relieved by an infusion of molecules that reduce the amount of ALA synthase and thereby the production of excess intermediates. Currently, hematin is the first-line treatment for severe acute attacks but partial relief is also obtained by infusion of glucose or heme arginate.

Photosensitive

Because of their limited water solubility and low rate of elimination from the body via urine, any excess porphyrins produced are deposited in the skin. Their highly resonating structure renders them susceptible to photolysis under the action of ultraviolet light. Absorbed light energy raises porphyrins to a triplet excited state which can return to the ground state through reaction with other molecules, most notably oxygen. The reactive oxygen species thus formed cause intracellular damage through reaction with various macromolecules including nucleic acids and membrane components (lipids and proteins). The type of skin damage is probably determined by the site of accumulation of porphyrin within the cell, which is dependent on its solubility. Such damage may present either as acute painful swelling or fragility of the area of exposed skin. Fragility presents as repeated tears on minimal trauma, easy bruising, or multiple small scars.

A similar situation may also occur secondarily to lysosomal injury where the normal catabolism of porphyrins is compromised. Both conditions can lead to a spectrum of physical effects including erythema, blistering, scarring, and pigmentation. The occurrence of photosensitivity in different porphyrias is shown in Table 16.5.

Mixed

Two types of porphyria (VP and HCP) present with symptoms both acute and neurovisceral. In some cases, as in PCT, patients are sensitive to sunlight but, in the majority, photosensitivity is present only during acute episodes.

Different types of porphyria

Aminolevulinic acid dehydratase (ALAD) deficiency porphyria

This is a very rare form of porphyria caused by a deficiency of aminolevulinic acid dehydratase and may present in children. It does not cause photosensitivity and symptoms are similar to acute intermittent porphyria but with very severe neuropathy. This disease will rarely be found in clinical practice (ten known cases worldwide) but is included here for completeness.

Acute intermittent porphyria

A deficiency of HMBS, previously known as porphobilinogen deaminase, leads to excess excretion of PBG and ALA, excretion of PBG always exceeding that of ALA. Urinary concentrations of PBG and ALA may occasionally also rise between episodes. The condition is manifest in only about 10% of affected individuals and rarely before puberty. Measurement of HMBS activity in white cells was originally used to identify the disorder, but this has now been replaced by mutational analysis of the HMBS gene as the test of choice. Family follow-up includes the offer of DNA analysis to identify affected relatives. Episodes may be precipitated by fasting, intercurrent illness, and drugs, especially those mentioned earlier in this chapter, and are characterized by abdominal pain, constipation, psychosis, neuropathy, hypertension, and impaired water excretion. Fresh urine samples are colorless but turn to a purple color on standing due to nonenzymatic polymerization of PBG (**Figure 16.6**). There is no excess porphyrin in either feces or erythrocytes. Patients are asymptomatic between episodes and do not exhibit photosensitivity. Day-to-day management of patients includes avoidance of drugs and periods of fasting, and treatment, including infusion of hematin or sometimes of heme arginate or glucose, during episodes.

It is recommended that all relatives of someone with AIP be tested to determine carriers of the trait. While most people who inherit the gene for AIP never develop symptoms, it is essential that all those who test positive be educated in measures to help avoid attacks—prevention is the key in such cases.

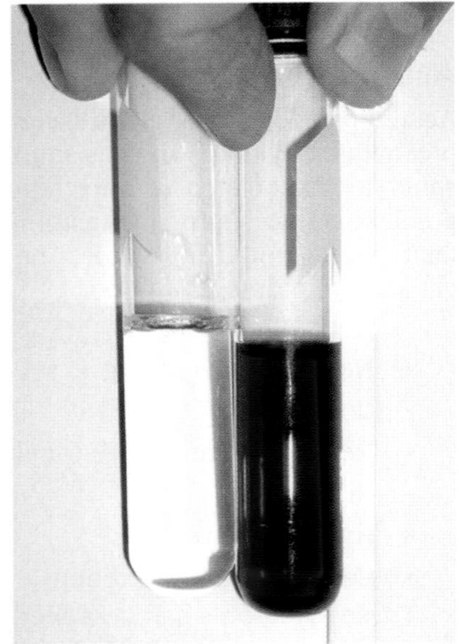

Figure 16.6 Urine from a patient with acute intermittent porphyria (AIP). Patients with AIP can have increased levels of porphobilinogen (PBG) in their urine, especially during attacks. PBG will polymerize on exposure to light, turning the urine a dark purple color. (From Lin CS-Y, Krishnan AV, Lee M-J et al (2008) *Brain* 13:2510–2519. With permission from Oxford University Press.)

Congenital erythropoietic porphyria

This very rare condition, sometimes called Gunther's disease, is characterized by severe photosensitivity, red teeth, hirsutism, and hemolytic anemia. Most patients present in infancy. Patients have a deficiency of uroporphyrinogen III synthase which leads to overproduction of type I porphyrins. Management involves rigorous protection from light.

Porphyria cutanea tarda

This is the commonest form of porphyria (approximately 1 in 25,000 people can develop this disorder) and is caused by a decrease in the activity of hepatic uroporphyrinogen decarboxylase. For symptoms to develop, the enzyme activity has to fall to about 20% of its normal activity. Most cases are not due to an inherited deficiency of the enzyme; they are caused by some intrahepatic inhibition of the enzyme activity and are referred to as sporadic or type I porphyria cutanea tarda. About one in five patients has an inherited deficiency resulting in a decrease in uroporphyrinogen decarboxylase activity in all tissues. However, these patients do not always develop symptoms, as they are heterozygous for the disorder and have 50% of the normal enzyme activity. There have to be other predisposing factors that reduce the enzyme activity down to 20% of normal before symptoms develop. Major risk factors for both forms of PCT are excess

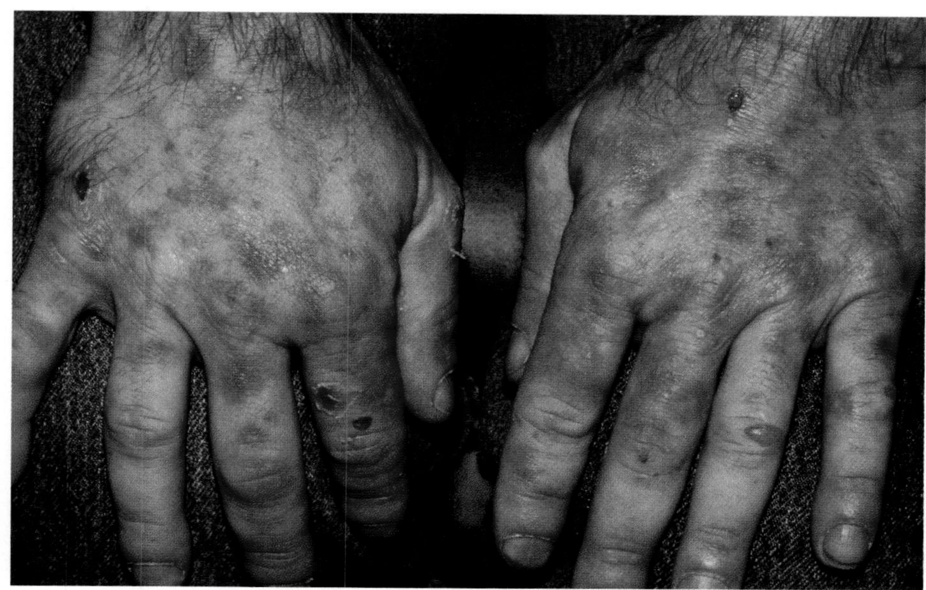

Figure 16.7 Hands of a patient with porphyria cutanea tarda (PCT). Typical blistering seen on the hands of a patient with PCT exposed to ultraviolet light. (Courtesy of Gary M. White, M.D., www.regionalderm.com)

ethanol intake, estrogens, hepatitis C, and hemochromatosis mutations. In past times, the major cause was excess alcohol intake, but currently the majority of cases in Western Europe are associated with hepatitis C and this should be tested for. Almost all patients have hepatic iron overload but usually not to the extent found in overt hemochromatosis. A particularly toxic presentation is manifest with intake of hexachlorobenzene or dioxin. Patients exhibit moderate photosensitivity and no acute episodes. Examples of the photosensitivity are shown in **Figure 16.7**.

The enzyme deficiency in porphyria cutanea tarda is usually of the order of 50%, but there is a more severe form of the disorder in which the enzyme activity is decreased by 90%, which is sometimes classified as hepatoerythropoietic porphyria. In this case, patients have two abnormal genes (not necessarily the same mutation) for uroporphyrinogen decarboxylase.

Management of patients includes venesection to remove excess iron (which also reduces uroporphyrin excretion) and protection from exposure to sunlight; low-dose chloroquine is a widely used and effective alternative to venesection.

Hepatoerythropoietic porphyria

Hepatoerythropoietic porphyria is a very rare type of porphyria due to mutations in both alleles of the uroporphyrinogen decarboxylase gene and may be considered as a homozygous form of PCT. The manifestations of this autosomal recessive disorder resemble those of CEP, with skin blistering, usually beginning in infancy. Porphyrins are increased in bone marrow and red blood cells, in contrast to PCT, as well as in liver, plasma, urine, and feces.

Hereditary coproporphyria

This condition, which is due to a deficiency of coproporphyrinogen oxidase, is characterized by episodes of abdominal pain, psychosis, and neuropathy, which may be precipitated by drugs or fasting; photosensitization is uncommon.

Variegate porphyria

A deficiency of protoporphyrinogen oxidase gives rise to photosensitive skin lesions which may be accompanied by episodes of abdominal pain, psychosis, and neuropathy. Management includes avoidance of drugs, fasting, and exposure to sunlight. Infusions of hematin and/or glucose are given during acute episodes, as in cases of acute intermittent porphyria.

Erythropoietic protoporphyria

This recessive disorder results from a deficiency of ferrochelatase, which leads to the accumulation in red cells of protoporphyrin in the free, non-zinc-chelated form. The clinical features of this condition, usually presenting in childhood, are burning and itching sensations, erythema, and swelling occurring within minutes in areas of skin exposed to sunlight. Erythropoiesis and iron metabolism are generally normal and hemolysis is uncommon. Liver function is also usually normal but a minority of patients with EPP are susceptible to the deposition of pigmented gallstones and also cirrhosis as a result of significant accumulation of protoporphyrins in the liver. Severe liver complications that are difficult to treat may result from the latter and some patients may require liver transplantation.

X-linked protoporphyria

Another disorder with a clinical presentation similar to erythropoietic protoporphyria is X-linked protoporphyria, sometimes referred to as variant EPP. In this disorder, however, the genetic defect is not in ferrochelatase but in ALA synthase 2. Gain-of-function mutations in the C-terminal region of the enzyme result in an increased activity of erythroid ALA synthase 2, leading to an increase in free and zinc-bound protoporphyrin in red cells. Clinical presentation of the disorder is very similar to that of EPP, with severe liver complications due to hepatic accumulation of protoporphyrins.

There are two ALA synthase genes, one located on the X chromosome for the erythroid enzyme (the absence of which leads to sideroblastic anemia) and the other, for all other tissues, on chromosome 3, which is not associated with any known disease.

16.4 INVESTIGATIONS OF PATIENTS FOR PORPHYRIA

The investigation of porphyria in patients will depend upon three primary indications:

1. Acute episodes of abdominal pain, psychosis, and neuropathy
2. Photosensitivity
3. Family history of porphyria

As indicated in Table 16.5, suspicion of a particular porphyria being present will be dependent upon whether acute attacks and/or photosensitivity are present. In addition, if there is a family history of porphyria, genetic testing is the preferred route of investigation. The majority of clinical laboratories usually provide only basic screening tests for the initial investigation of these patients, providing qualitative (positive/negative) or semiquantitative tests and a diagnosis based mainly on the identification by high performance liquid chromatography (HPLC) of characteristic patterns of excess porphyrin precursors and heme biosynthetic pathway intermediates in body fluids. Definitive diagnosis and further investigation of patients should be undertaken by a specialist center with experience in the investigation of these relatively rare disorders and the facilities to measure enzyme activities, to conduct porphyrin isomer quantitation and ALA and PBG quantitation, and to sequence heme-pathway genes. The key clinical chemistry findings for each of the porphyrias are listed in **Table 16.6**.

Investigation of acute intermittent porphyria, variegate porphyria, and hereditary coproporphyria

Patients may typically present as an emergency with acute symptoms, especially severe abdominal pain of unknown cause. A random urine sample is needed as the first line of investigation. In these three disorders, during attacks, urine PBG becomes raised (normally <1.0 μmol/mmol creatinine). PBG is usually at least

TABLE 16.6 Clinical chemistry of porphyrias

Porphyria	Tissue analyzed			
	Urine	Feces	Red blood cells	Plasma
ALA dehydratase deficiency porphyria (ADP)	Raised ALA Raised coproporphyrin III		Raised protoporphyrin IX	Raised ALA
Acute intermittent porphyria (AIP)	PBG and ALA markedly raised (>tenfold)		Erythrocyte HMBS activity reduced	
Congenital erythropoietic porphyria (CEP)	No excess of ALA or PBG Raised type I uroporphyrins and coproporphyrins	Raised type I coproporphyrins	Excess type I porphyrins	
Porphyria cutanea tarda (PCT)	Raised uroporphyrin. No excess of ALA or PBG	Raised coproporphyrins and protoporphyrins	No excess porphyrins	Increased iron and ferritin
Hepatoerythropoietic porphyria (HEP)	Raised uroporphyrins	Raised coproporphyrins and heptacarboxyporphyrins	Raised Zn-protoporphyrins	Raised uroporphyrins Characteristic fluorescence emission at 400–410nm
Hereditary coproporphyria (HCP)		High fecal coproporphyrin III Protoporphyrin normal		
Variegate porphyria (VP)	Raised uroporphyrins, ALA, and PBG	Raised coproporphyrins and protoporphyrins	No excess porphyrins	Characteristic fluorescence emission at 626 nm
Erythropoietic protoporphyria (EPP)	No excess of porphyrins, ALA, or PBG	Raised protoporphyrins	Excess free protoporphyrins	
X-linked protoporphyria		Raised protoporphyrins	Raised protoporphyrins Increased ratio of Free: Zn-protoporphyrin	

10 times the upper limit of normal (1.5 μmol/mmol creatinine) during an acute attack. Patients with AIP, HCP, or VP are often asymptomatic in between attacks. Furthermore, some of the metabolites, for example porphobilinogen, may not be raised in between attacks. In AIP, PBG levels will remain markedly elevated for many years following an acute attack, whereas levels may return to normal after a few months in VP and HCP. Therefore, it is important to take samples during attacks to maximize the chances of finding a positive result. It is also important to take the correct sample type; this is dependent upon which porphyria is suspected, and whole blood, feces, and urine may all be required. Since some of the porphyrins can be degraded by sunlight, samples should be protected from light. Often laboratories will undertake a rapid, simple qualitative screening test measuring urinary PBG. If the result is positive, the patient should be further investigated for one of the three main porphyrias that cause acute symptoms. Since PBG is unstable in urine, it is essential to obtain a fresh random urine sample for quantitative analysis, not a 24-hour collection. Negative results rule out these three porphyrias as the cause of the symptoms in this attack (but not any previous attack); however, further testing may have to be undertaken to rule out the very rare 5-aminolevulinic acid dehydratase deficiency porphyria.

As mentioned earlier, it is important to note that, dependent upon the severity and frequency of attacks, PBG can remain high in between episodes; thus, a positive result may not always explain the symptoms of that particular attack, but does indicate the presence of one of these three porphyrias.

Clinical practice point 16.1

The absence of elevated porphobilinogen (PBG) in urine, using a suitably sensitive method, excludes an acute porphyria as the cause of the symptoms. However, PBG excretion may be normal between attacks and the test cannot exclude porphyria in an asymptomatic relative of a known porphyric individual.

Acute intermittent porphyria

Although the absence of photosensitivity would indicate acute intermittent porphyria as the cause of the acute symptoms, further testing is essential before making the final diagnosis. Few specialist laboratories now measure HMBS to confirm the diagnosis of AIP because it has poor diagnostic accuracy, particularly in acutely ill patients. The diagnosis should be established by demonstrating increased urinary PBG excretion, with exclusion of VP and HCP by fecal and

CASE 16.1

A 23-year-old woman attends the emergency department with severe abdominal pain and constipation. She has been on a low-calorie diet to lose weight. There is no abdominal rigidity on examination. She has a tachycardia and her blood pressure is raised at 160/100 mmHg. No diagnosis is made but such is the severity of her pain she is given opiates. Her urine was initially noted to be normal, but on standing developed a definite red color. Urine dipstick testing for blood was negative. Laboratory results show the following.

	SI units	Reference range	Conventional units	Reference range
Plasma				
Sodium	124 mmol/L	136–142	124 mEq/L	136–142
Potassium	4.1 mmol/L	3.5–5.0	4.1 mEq/L	3.5–5.0
Urea (blood urea nitrogen, BUN)	2.1 mmol/L	2.9–8.2	5.9 mg/dL	8–23
Creatinine	50 µmol/L	53–106	0.57 mg/dL	0.6–1.2
Osmolality	264 mmol/kg	275–295	264 mOsm/kg	275–295
Urine				
Osmolality	540 mmol/kg	Not applicable*	540 mOsm/kg	Not applicable*
Porphobilinogen	Positive +++	Negative	Positive +++	Negative
Porphyrins	Negative	Negative	Negative	Negative

*Urine sodium and osmolality are interpreted as whenever appropriate for plasma results and clinical status.

- **What is the cause of the abdominal pain and red urine?**
- **What has precipitated this episode?**
- **What is it important to avoid in future?**

The patient has an acute porphyria, with the severe symptoms caused by the presence of porphobilinogen (PBG), which is toxic to neural tissue. On standing, the urine PBG polymerizes into a red-colored compound. Clinical effects include abdominal pain, peripheral and autonomic neuropathy, and neuropsychiatric conditions. In addition, the syndrome of inappropriate antidiuretic hormone (SIADH) can develop in acute porphyria, which may cause hyponatremia and low serum osmolality with a paradoxically higher urine osmolality; that is, inappropriately concentrated urine. There are four acute porphyrias: aminolevulinic acid dehydratase (ALAD) deficiency, also called porphobilinogen (PBG) synthase deficiency; acute intermittent porphyria (AIP); variegate porphyria (VP); and hereditary coproporphyria (HCP). ALAD deficiency is very rare and seldom encountered in routine practice. The initial laboratory results suggest the patient has AIP, as there is no excess of porphyrins. Distinction between the porphyrias is by analysis of urine, stool, and blood porphyrin patterns and genetic testing. The elevation in PBG is less with ALAD deficiency than the other three conditions and may not be detected. Episodes may be precipitated by fasting, alcohol intake, menstruation, infection, surgery, or other metabolic stress. Certain drugs may also cause attacks and only ones on a safe list should be used. Photosensitivity is not seen in AIP but can be seen in VP and HCP.

plasma porphyrin analyses. DNA testing of the proband is needed for family studies but is rarely required as a primary diagnostic test. For acute intermittent porphyria there should be no accumulation of porphyrins. DNA analysis to determine the mutation present is the gold-standard confirmatory test.

Hereditary coproporphyria and variegate porphyria

Raised porphyrins in feces are a feature of HCP and VP but the particular pattern will be different for the two disorders. Patients with HCP have diagnostically raised levels of fecal coproporphyrin III, whilst those with VP show increased levels of both fecal coproporphyrin III and protoporphyrin. In addition, patients with VP also show a distinctive fluorescence peak at 626 nm in their plasma. These abnormalities are present in these patients in between attacks.

Investigation of porphyria cutanea tarda, congenital erythropoietic porphyria, and erythropoietic protoporphyria

These patients always present due to their photosensitivity and are not usually a medical emergency.

Analytical practice point 16.1

Normal red blood cell and plasma total porphyrin results exclude cutaneous porphyrias as a cause of skin lesions and no further samples or testing are required.

Porphyria cutanea tarda

Patients with PCT often present with a combination of photosensitivity and abnormal liver function tests. Indeed, some patients may go on to develop severe liver disease due to increased iron stores. These patients demonstrate markedly raised porphyrins, derived from intermediates of the UROD reaction, in urine, plasma, and feces, although this in itself is not diagnostic; however, they have a distinctive pattern of fecal porphyrins which is diagnostic. In urine, ALA and PBG are within the normal range; there is increased uroporphyrin I (and, to a lesser extent, uroporphyrin III), increased porphyrins with seven COOH groups, and smaller increases in porphyrins with six, five, and four COOH groups. Feces contain porphyrins with seven, six, five, and four COOH groups. Raised uroporphyrin I and uroporphyrin III are found, as well as coproporphyrin I, but coproporphyrin III is not raised.

Although the measurement of uroporphyrinogen decarboxylase activity in red blood cells can distinguish the familial causes of the disease from the sporadic causes, because of the complicated genetics involved, molecular biological analysis is required to define the specific disorder. However, neither approach is indicated for the clinical management of most patients with PCT.

Congenital erythropoietic porphyria

Patients with CEP present with the most severe forms of photosensitivity and, due to the deficiency of uroporphyrinogen III synthase, they show a characteristic pattern of uroporphyrin and coproporphyrin isomers in feces. Coproporphyrin I is massively increased, whilst coproporphyrin III and uroporphyrin III are usually absent or present in very small amounts. Increased erythrocyte type I porphyrin is also an important feature of CEP and genetic analysis helps to confirm the diagnosis.

Erythropoietic protoporphyria and X-linked protoporphyria

The diagnosis of EPP and the related X-linked protoporphyria can be made by examining the total protoporphyrins and relative amounts of zinc-bound and free protoporphyrins in red blood cells:

- Raised free protoporphyrins relative to zinc-bound porphyrins indicates EPP
- Raised total protoporphyrins where zinc-bound and free protoporphyrins are equal indicates X-linked protoporphyria
- Raised zinc-bound protoporphyrins relative to free protoporphyrins indicates a non-porphyria cause

A useful protocol for investigation into porphyrias is shown in **Figure 16.8** (reprinted with thanks to and permission from the website of the American Porphyria Foundation). This website (www.porphyriafoundation.com) and that of the American Association of Clinical Chemistry (aacc.org) are recommended as excellent sources of information on porphyrias for both the lay reader and clinician alike.

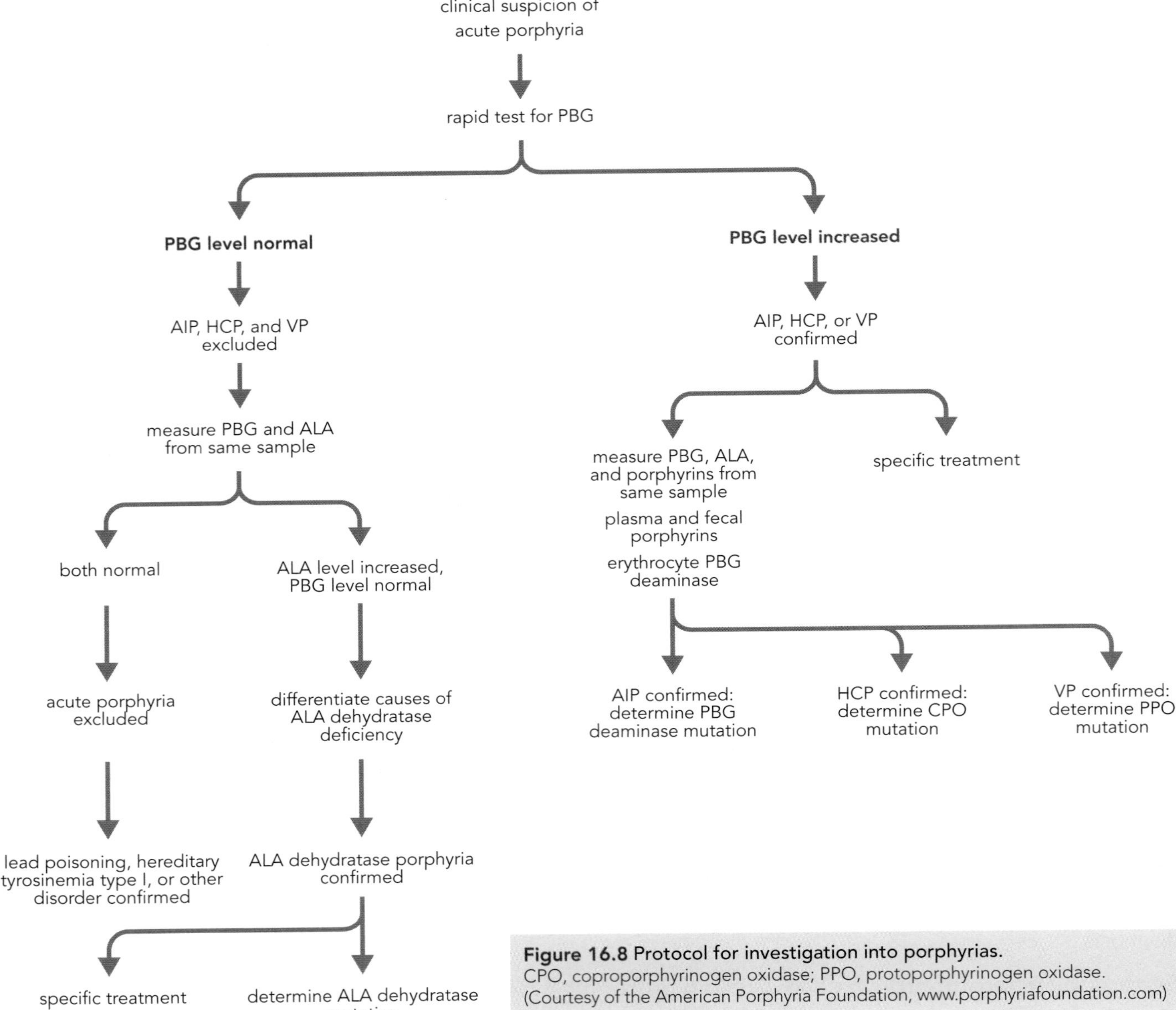

Figure 16.8 Protocol for investigation into porphyrias. CPO, coproporphyrinogen oxidase; PPO, protoporphyrinogen oxidase. (Courtesy of the American Porphyria Foundation, www.porphyriafoundation.com)

CASE 16.2

A 35-year-old man presents with blistering of his hands and face when working outside. He is otherwise well, but admits to drinking up to one bottle of wine each day (approximately 70 units per week). On further questioning he reports discolored urine on some occasions.

	Result	Reference range
Plasma		
Porphyrins	Increased	Not applicable
Red blood cell		
Porphyrins	Not increased	Not applicable
Urine		
Porphobilinogen	Negative	Negative
Porphyrins	Positive +++	Negative
Feces		
Porphyrins	Positive: increased coproporphyrins	Negative

- **What porphyria is present?**
- **What is the cause of the skin lesions?**

The patient has porphyria cutanea tarda (PCT) due to deficiency of uroporphyrinogen decarboxylase (UROD). Unlike the other porphyrias, it is usually acquired (80% of cases) rather than due to a genetic enzyme deficiency. Factors that increase iron in the liver (such as alcohol, hemochromatosis, and viral hepatitis) cause down-regulation of UROD activity and result in excess carboxylated porphyrins being released into the circulation. The greater the number of carboxyl groups present, the greater the water solubility. The porphyrins with eight carboxyl groups are excreted in urine (hence being called uroporphyrins) and those with four carboxyl groups are excreted in feces (copro- and isocoproporphyrins). These molecules are photoactive and absorb energy from violet light in the visible spectrum, subsequently causing oxidative damage to the surrounding skin tissue. Specialist centers can conduct detailed analysis of the fecal porphyrins to make a definitive diagnosis; in PCT, high fecal uroporphyrin III and coproporphyrin I levels would be expected with relatively low coproporphyrin III levels. In addition, genetic testing is usually recommended to check for inherited forms of the disease.

URIC ACID AND PURINE METABOLISM

CHAPTER 17

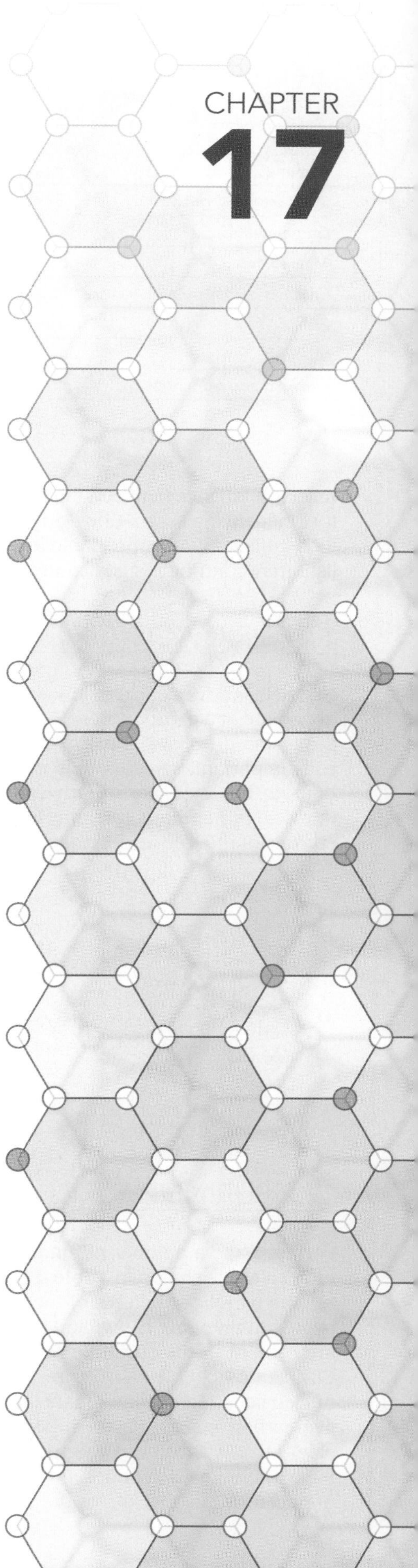

IN THIS CHAPTER

Uric acid is the major excretory product arising from the catabolism of purines. The differing reference ranges for the uric acid concentration in serum in adult males and females, as well as those for children, are given in **Table 17.1**. Since uric acid has limited solubility in water, it is important that the concentrations of uric acid do not go over these reference ranges, as whatever the cause of hyperuricemia, the net effect is the precipitation of urate and uric acid crystals. The solubility product of uric acid in physiological saline at room temperature is 570 μmol/L. However, this is increased in blood due to effects of protein and counterions on urate solubility, such that concentrations in excess of 570 μmol/L can be tolerated without precipitation occurring. These protective effects do not occur in other body fluids, such as synovial fluid and glomerular filtrate, and uric acid crystals deposit more readily in joints, kidneys, and the urinary tract when the urate concentration exceeds 570 μmol/L in these fluids.

The clinical consequences of hyperuricemia range from kidney stones, which can lead to renal failure, to joint inflammation and destruction. An operational definition of hyperuricemia is a serum urate concentration >420 μmol/L in adult men and >350 μmol/L in adult women. As with any metabolite, its concentration

TABLE 17.1 Reference ranges in the analysis of uric acid

		Concentration (SI units)	Concentration (conventional units)
Serum	Adult female	150–350 μmol/L	2.5–5.9 mg/dL
	Adult male	200–420 μmol/L	3.4–7.1 mg/dL
	Neonates (0–5 days)	120–470 μmol/L	2.0–8.0 mg/dL
	Children <12 years	110–320 μmol/L	1.9–5.4 mg/dL
Urine	Adults*	1475–4425 μmol/24h	250–750 mg/24h

*Reference ranges for neonates and children are unavailable.

TABLE 17.2 Foods with high purine content

Food	Example
Legumes	Peas, beans, broad beans
Meat extracts, gravies	Broths, bouillons, consommés
Game meats	Goose, duck, partridge
Minced meats	Ground beef, lamb, pork, venison
Offal	Brain, heart, kidneys, liver, sweetbreads
Fish	Herring, mackerel, sardines, anchovies, caviar, roe
Shellfish	Scallops, mussels
Drinks made from hops	Beer

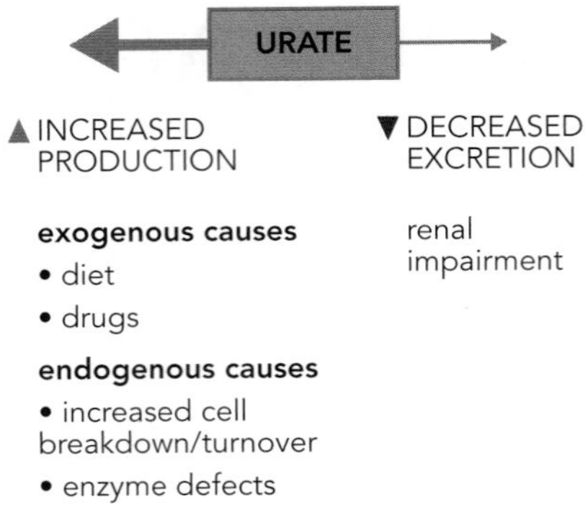

Figure 17.1 Major factors influencing serum uric acid (urate) concentration.
A rise in the concentration of serum urate may arise from increased production due to both exogenous (dietary nucleic acids, drug effects) and endogenous causes (increased cell breakdown or turnover, enzyme defects) or, more commonly, from decreased urinary excretion due to kidney disease.

reflects a balance between rates of production and excretion, and the major factors influencing serum uric acid concentration are shown in **Figure 17.1**. Many foodstuffs contain appreciable levels of purines (**Table 17.2**) and their catabolism may contribute significantly to uric acid production.

17.1 URIC ACID OR URATE?

Uric acid is a weak acid of low water solubility; in its enol tautomeric form it will dissociate at a pH above its pKa of 5.7 to urate and a proton (**Figure 17.2**). Thus at plasma pH of 7.4, most of the uric acid is dissociated to urate. This is critically important, since urate is more soluble than uric acid in aqueous solution. In urine, however, particularly when it is slightly acidic, appreciable amounts of uric acid rather than urate may be present and may precipitate out of solution as crystals of the free acid in the urinary tract.

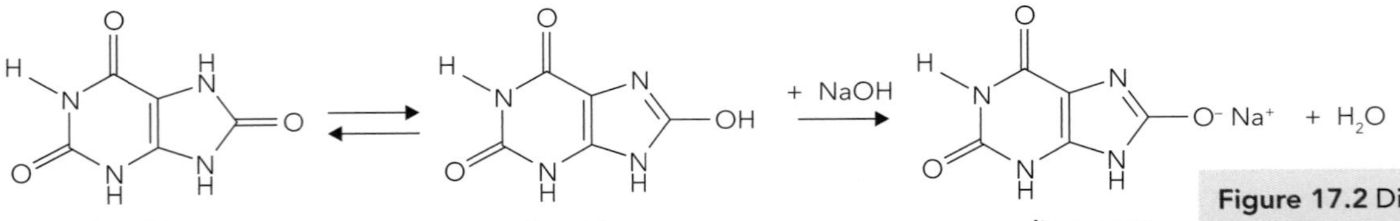

Figure 17.2 Dissociation of uric acid to urate.
Uric acid undergoes keto-enol tautomerism at physiological pH, yielding the urate cation and a proton.

17.2 PURINES

Purines have a number of fundamental roles in the cell as components of key molecules, as shown in **Table 17.3**.

The bicyclic structures of the common purine bases—adenine, guanine, and hypoxanthine—are shown in **Figure 17.3**; differences between these structures are evident in the substituents at C2 and C6, with further oxidation at carbon C8 in uric acid. The structure of caffeine is included for comparison. Purine nucleosides (adenosine, guanosine, and inosine) are formed by addition of a five-carbon sugar—ribose (ribonucleoside) or deoxyribose (deoxyribonucleoside)—to N9. Inosine is the trivial name given to the nucleoside derived from hypoxanthine. Nucleotides are formed by phosphorylation of the sugar residues of nucleosides.

TABLE 17.3 Biological importance of purines: molecules containing purines and pathways in which purines play a key role	
Biological roles of purines	**Examples**
Components of nucleic acids	DNA, RNA
Second messengers of hormone action	Cyclic AMP, cyclic GMP
High-energy phosphate compounds	ATP, GTP
Enzyme co-factors	NAD, NADP, FMN, FAD
Drugs	Azathioprine, 6-mercaptopurine*, allopurinol, theophylline, caffeine
Methyl-group transfer	*S*-adenosyl-methionine
Sulfate assimilation	3-Phosphoadenosine-5′-phosphosulfate

* 6-mercaptopurine is a metabolite of azathioprine. DNA, deoxyribonucleic acid; RNA, ribonucleic acid; AMP, adenosine monophosphate; GMP guanosine monophosphate; ATP, adenosine triphosphate; GTP, guanosine triphosphate; NAD, nicotinamide adenine dinucleotide; NADP, nicotinamide adenine dinucleotide phosphate; FMN, flavin mononucleotide; FAD, flavin adenine dinucleotide.

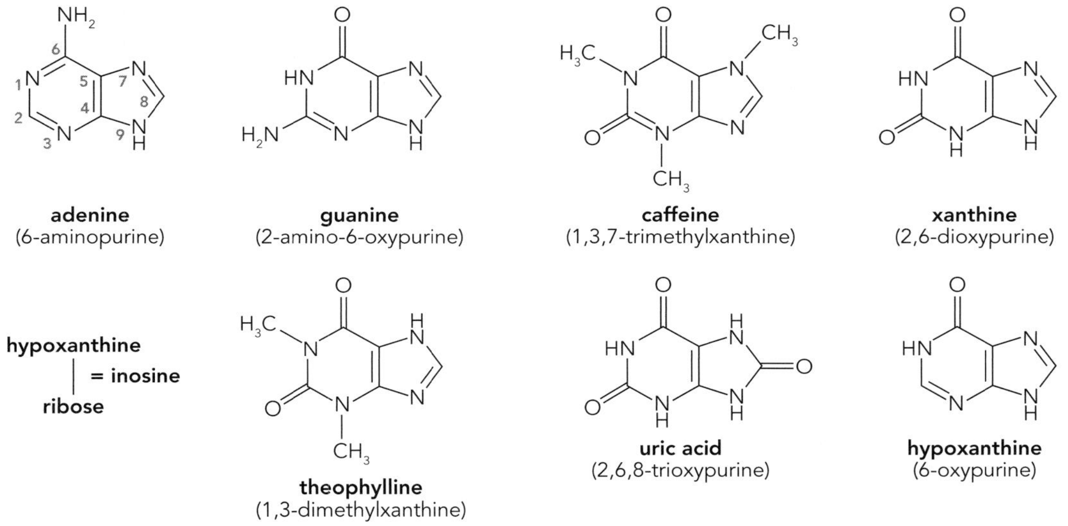

adenine
(6-aminopurine)

guanine
(2-amino-6-oxypurine)

caffeine
(1,3,7-trimethylxanthine)

xanthine
(2,6-dioxypurine)

hypoxanthine
| = inosine
ribose

theophylline
(1,3-dimethylxanthine)

uric acid
(2,6,8-trioxypurine)

hypoxanthine
(6-oxypurine)

Figure 17.3 Chemical structures of the major purines. Adenine and guanine are purines found in nucleic acids and are metabolized to uric acid via xanthine. Hypoxanthine is an intermediate in the formation of xanthine from adenine, while inosine is 9-ribosyl-hypoxanthine. The structures of the methylxanthines caffeine and theophylline, components of beverages such as coffee, are included for comparison.

Purine synthesis

The major purine nucleotides, adenosine monophosphate (AMP) and guanosine monophosphate (GMP), are formed from inosine monophosphate (IMP), which is synthesized in turn from simple building blocks: aspartate, CO_2, glycine, glutamine, and the one-carbon donor N^{10}-formyltetrahydrofolate. The biosynthesis of IMP is complex, involving a number of steps and considerable input of energy. The early steps in the biosynthetic pathway of purines are shown in **Figure 17.4**. A detailed description of the latter part of the pathway is beyond the scope of this text and the reader is referred to standard biochemistry texts for this. The rate-limiting step of the pathway is the formation of β-5-phosphoribosyl-1-amine catalyzed by phosphoribosylpyrophosphate amidotransferase (glutamine phosphoribosylpyrophosphate amidotransferase).

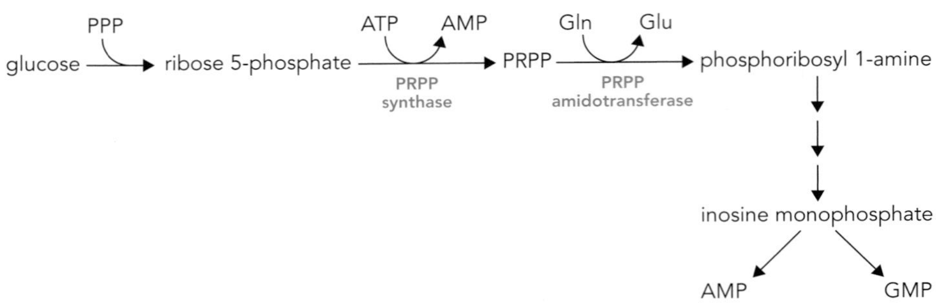

Figure 17.4 An outline of purine biosynthesis.
The pathway of biosynthesis of AMP and GMP begins with ribose 5-phosphate formed in the pentose phosphate pathway (PPP). Ribose 5-phosphate undergoes pyrophosphorylation by ATP to form 5-phosphoribosyl-1-pyrophosphate (PRPP), a reaction catalyzed by PRPP synthase. This product reacts with glutamine to form β-5-phosphoribosyl-1-amine in a reaction catalyzed by PRPP amidotransferase, where the amide nitrogen of glutamine displaces the pyrophosphate of PRPP with the release of glutamic acid and pyrophosphate. In the next step, a glycinamide ribonucleotide is formed from the reaction of the 1-amino group of β-5-phosphoribosyl-1-amine with the carboxyl group of glycine, a reaction requiring ATP. The purine ring is then built up around the glycinamide moiety in a complex series of nine steps yielding eventually AMP and GMP via inosine monophosphate (IMP). Gln, glutamine; Glu, glutamic acid. PRPP synthase is also known as ribose 5-phosphate pyrophosphokinase; PRPP amidotransferase is also glutamine PRPP amidotransferase.

The product of the pathway, IMP, does not accumulate within the cell and is converted to either AMP or GMP.

Purine catabolism and generation of uric acid

Uric acid is formed by the action of xanthine oxidase on xanthine (dioxypurine), a product common to the catabolism of the purine nucleosides adenosine and guanosine (**Figure 17.5**). Formation of uric acid from adenine requires oxidations at C2, C6, and C8$_8$, while formation from guanine requires oxidations at C66 and C8.

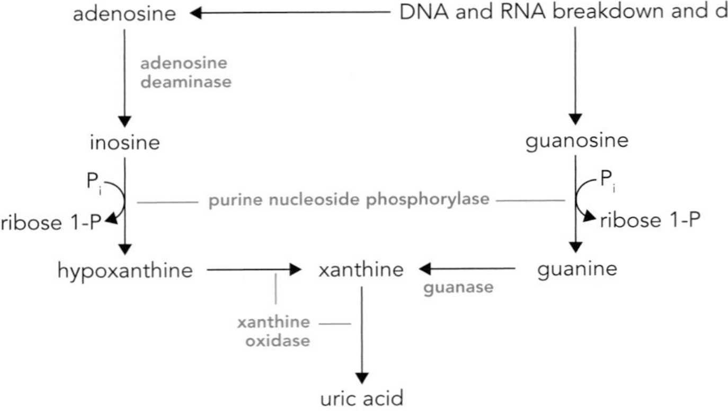

Figure 17.5 Generation of uric acid.
Purine nucleosides, adenosine and guanosine, are released by hydrolysis of nucleic acids derived from exogenous (dietary) or endogenous sources. Adenosine loses its amino group to form inosine in a reaction catalyzed by adenosine deaminase. Purine nucleoside phosphorylase hydrolyzes the ribose moiety from both inosine and guanosine, forming hypoxanthine and guanine, respectively. Both purines are then converted to xanthine; guanine by guanase and hypoxanthine by xanthine oxidase. Xanthine oxidase catalyzes the oxidation of xanthine to uric acid.

Nucleosides are derived principally from the breakdown of nucleic acids (DNA and RNA) as shown below:

$$\textbf{nucleic acids} \xrightarrow{\text{nucleases, phosphodiesterases}} \textbf{nucleoside monophosphates (NMP)}$$

$$\textbf{NMP} + \textbf{H}_2\textbf{O} \xrightarrow{\text{nucleotidases/phosphatases}} \textbf{nucleoside} + \textbf{inorganic phosphate}$$

$$\textbf{nucleoside} + \textbf{H}_2\textbf{O} \xrightarrow{\text{nucleosidases}} \textbf{base} + \textbf{ribose (deoxyribose)}$$

$$\textbf{nucleoside} + \textbf{inorganic phosphate} \xrightarrow{\text{purine nucleoside phosphorylase (PNP)}} \textbf{base} + \textbf{ribose-1-phosphate}$$

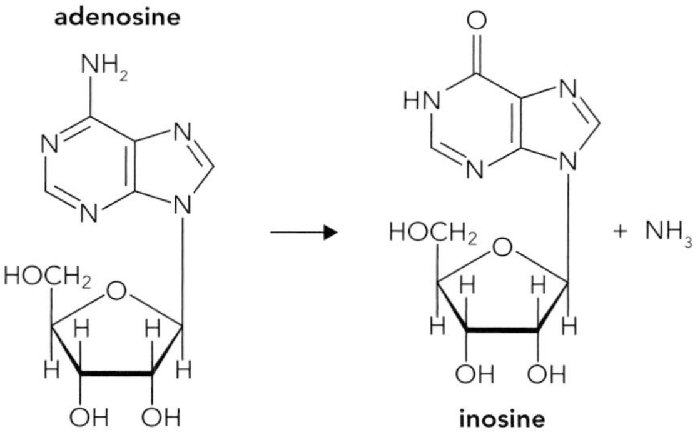

Figure 17.6 Adenosine deaminase.
The metabolism of adenosine to inosine is catalyzed by adenosine deaminase and involves the removal of the amino group at C6 (red) and formation of ammonia (NH_3).

For the purposes of the current chapter, the bases are the purines guanine and adenine. In general, nucleosides are hydrolyzed by the enzyme purine nucleoside phosphorylase (PNP), which releases ribose-1-phosphate and the purine base. However, adenosine is not a substrate for PNP and adenosine is converted first to inosine by adenosine deaminase (ADA) (**Figure 17.6**). Both inosine and guanosine are converted to their respective purines (hypoxanthine and guanine) by purine nucleoside phosphorylase (**Figure 17.7**). Two separate enzymes, xanthine oxidase and guanase (guanine deaminase), convert hypoxanthine and guanine, respectively, to xanthine (**Figure 17.8**). Xanthine oxidase then oxidizes xanthine to uric acid. Thus xanthine oxidase is involved in two steps in the oxidation of adenosine to uric acid, illustrating the dual nature of the enzyme. Most of the purines ingested in the diet are degraded immediately via these routes.

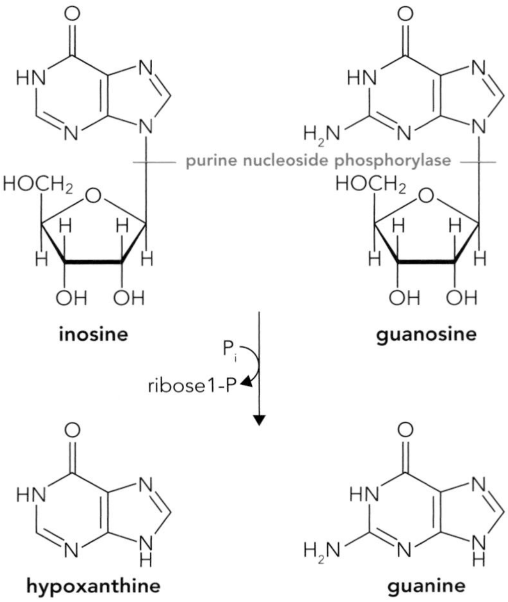

Figure 17.7 Purine nucleoside phosphorylase.
Purine nucleoside phosphorylase catalyzes the loss of ribose 1-phosphate from the purine nucleosides inosine and guanosine with the formation of hypoxanthine and guanine, respectively.

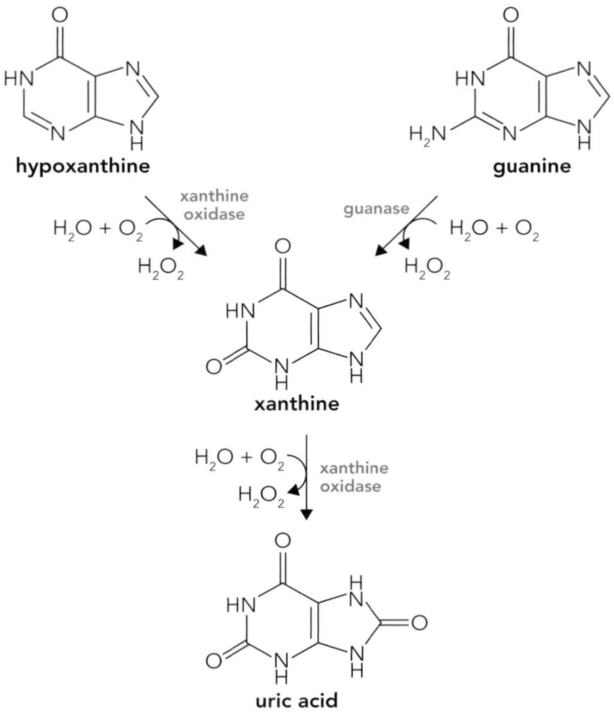

Figure 17.8 Formation of uric acid from hypoxanthine and guanine.
Xanthine oxidase, a molybdenum- and iron-containing flavoprotein, oxidizes hypoxanthine to xanthine, and xanthine to uric acid. Both reactions require molecular oxygen as the oxidant species and hydrogen peroxide is a reaction product. Hydrogen peroxide can be converted to oxygen and water by catalase. The conversion of guanine to xanthine involves replacement of the N2 amino group with a carbonyl group, a reaction catalyzed by guanase.

Purine salvage

Because of the vital role played by nucleotides in many cellular processes, it is essential that their concentration does not fall below a critical level. Two major factors contribute to maintenance of this level: (i) the nucleotidases, which hydrolyze nucleotides to nucleosides, are under tight metabolic control; and (ii) salvage pathways exist to rescue and recycle nucleosides. Salvage provides a means of recovering purines and resynthesizing nucleotides from them using purine phosphoribosyl transferases. It also ensures a continuing supply of nucleotides in cells at rest.

An outline of the metabolic interconversion of purine nucleotides and nucleosides and the rescue of purine bases from the catabolic pathway can be seen in **Figure 17.9**. There are two salvage enzymes of different specificity—adenine phosphoribosyl transferase (APRT) and hypoxanthine-guanine phosphoribosyl transferase (HGPRT)—which convert the respective, eponymous purine base to its corresponding ribonucleotide, adenylate (AMP) or guanylate (GMP), using 5-phosphoribosyl-1-pyrophosphate as the donor of ribose phosphate. This somewhat complex series of salvage and interconversion reactions enables the cell to maintain a critical concentration of nucleotides. The absence of key enzymes of this pathway leads to metabolic disturbances as discussed below.

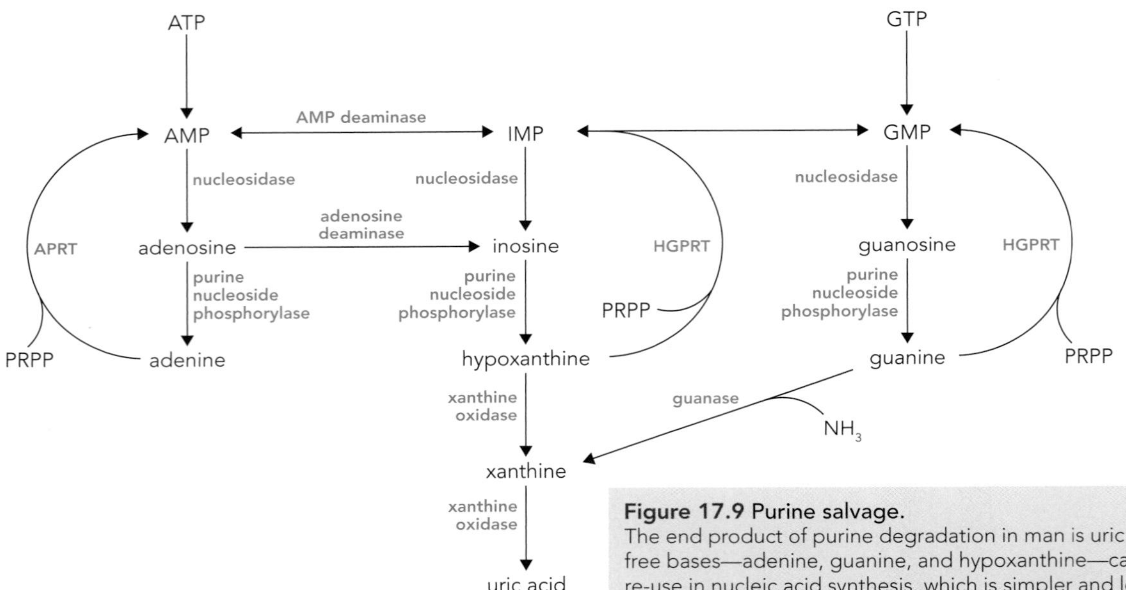

Figure 17.9 Purine salvage.
The end product of purine degradation in man is uric acid. However, the free bases—adenine, guanine, and hypoxanthine—can be salvaged for re-use in nucleic acid synthesis, which is simpler and less costly energetically than *de novo* synthesis of nucleotides. Phosphoribosyl transferases transfer ribose phosphate from phosphoribosylpyrophosphate (PRPP) to the free base to form the corresponding mononucleotide. Adenine phosphoribosyl transferase (APRT) catalyzes the formation of adenylate (AMP) while hypoxanthine-guanine phosphoribosyl transferase (HGPRT) catalyzes the formation of inosinate (IMP) and guanylate (GMP) from hypoxanthine and guanine, respectively. Inosine monophosphate is also the precursor for the synthesis of both adenosine and guanosine monophosphates.

Purine balance

The critical level of purine nucleotides within the cell reflects a balance between purines derived from the diet and from *de novo* synthesis, and those lost through catabolism to uric acid and excreted into urine. A quantitative idea of the daily throughput of purines and uric acid excretion is shown in **Figure 17.10**. On average, an adult male consumes about 600 mg of purines per day in the diet and this mixes with the purines synthesized *de novo* (about 300–600 mg). These

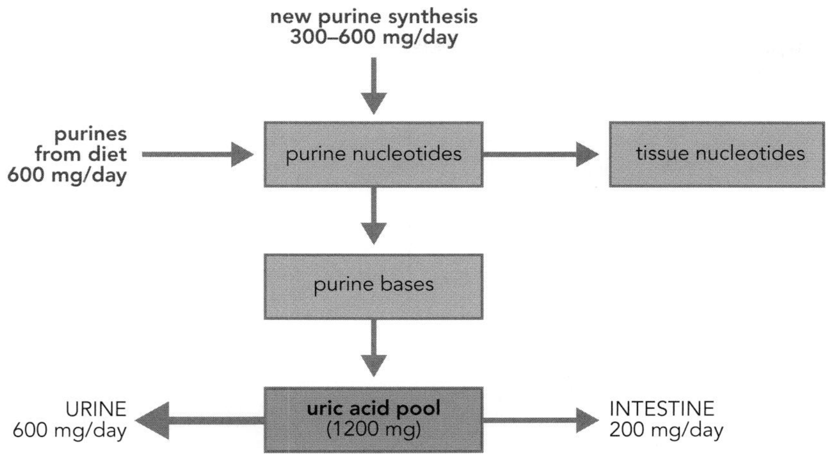

Figure 17.10 Daily throughput of purines and uric acid. The body pool of uric acid in serum is maintained by a balance between dietary and endogenously synthesized purines and excretion of uric acid via the kidneys and intestine. *De novo* synthesis will be determined by the amount derived from the diet.

equilibrate with the tissue purine pools, and excess purines are degraded to uric acid which is excreted in urine via the kidneys (about 600 mg) or the intestine (about 200 mg). The serum uric acid concentration reflects the rate of purine catabolism.

17.3 RENAL HANDLING OF URATE

In healthy humans, most (>95%) of the blood urate is filtered through the glomerulus and all of this is reabsorbed in the early proximal tubule (**Table 17.4**). Approximately 50% of this fraction is secreted in the proximal tubule and 80% of this secreted urate (40% of the filtered urate) is again reabsorbed, such that only about 10% of the filtered urate is finally excreted. The reabsorption rate rarely exceeds 15 mg/min.

TABLE 17.4 Renal handling of urate

Urate handling through the nephron	% Total urate
Urate filtered through the glomerulus	>95%
Filtrate reabsorbed in the proximal tubule	~100%
Reabsorbed urate secreted in the proximal tubule	~50%
Reabsorbed urate in the distal tubule	~40%

Overall, approximately 10% of the filtrate is excreted and the maximum reabsorption capacity (15 mg/min) is rarely exceeded.

The renal tubular excretion and reabsorptive processes are mediated by a number of membrane antiport transporter proteins (OATs, organic anion transporters) which co-transport sodium and anions. These transporters are present in the basolateral (for example OAT1, OAT3) and apical membranes (for example URAT1, MRP4) of the cells of the proximal tubule (**Figure 17.11**) and are responsible for the movement of anions and urate between plasma and urine. They transport in both directions, dependent on the relative concentrations of the ligands being transported. The compensating ligand for antiport transfer of urate is a dicarboxylic acid such as glutarate or α-ketoglutarate. Raised plasma urate enters the proximal tubular cell in exchange for a dicarboxylic acid via

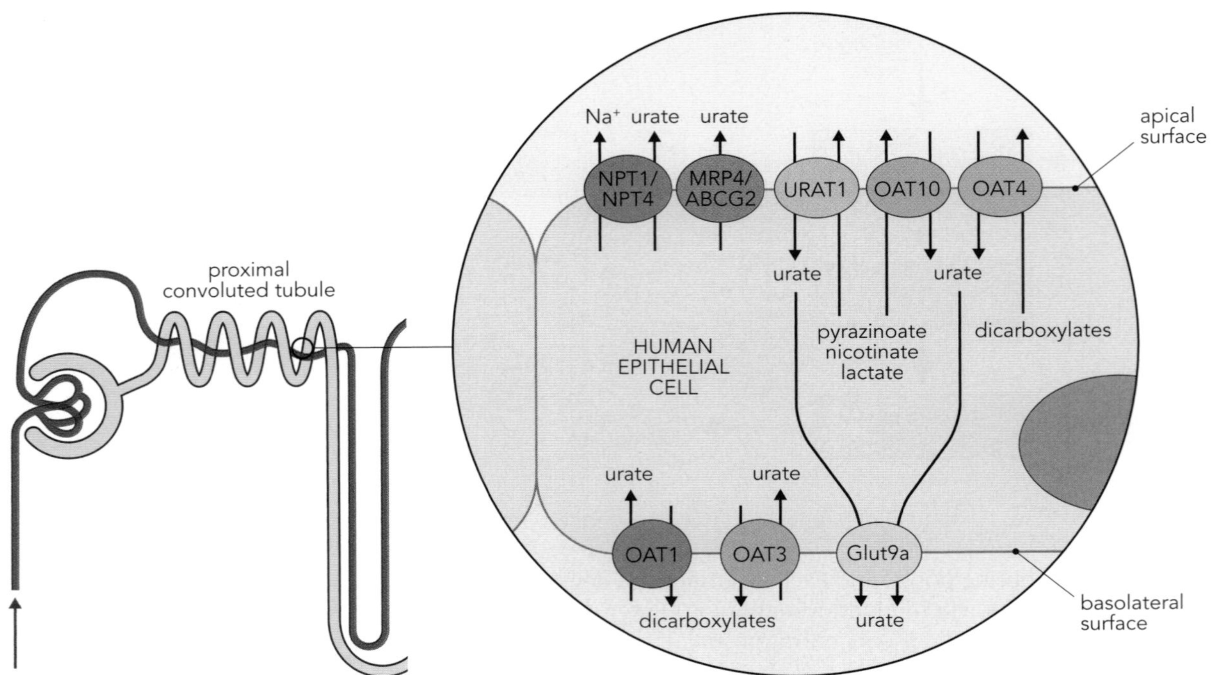

Figure 17.11 Renal handling of urate.
The serum concentration of urate is reflected in a balance between urate excretion and reabsorption by the epithelial cells of the proximal convoluted tubules in the kidney. The excretion of urate occurs in the cells of the proximal convoluted tubules. Urate transport into the cells from the circulation is affected by urate/dicarboxylate exchange proteins, OAT1 and OAT3, in the basolateral membrane, while urate is excreted into the tubular lumen via apical membrane transporter proteins, MRP4 and ABCG2, as well as sodium/phosphate co-transporters, NPT1 and NPT4. Urate reabsorption involves antiport proteins URAT1, OAT4, and OAT10 in the apical membrane which transport urate into the cell in exchange for organic anions: monocarboxylic anions such as lactate, nicotinate, and pyrazinoate for URAT1 and OAT10, and dicarboxylic anions for OAT4. Transport of intracellular urate back into the circulation is via the transport protein Glut9a in the basolateral membrane. Increased intracellular concentrations of monocarboxylic anions, such as lactate and drugs (including diuretics such as torasemide and hydrochlorothiazide), which are substrates of OAT4 and OAT10, increase urate reabsorption, leading eventually to increased serum urate. OAT, organic anion transporter; URAT, urate transporter; NPT, sodium/phosphate co-transporter; MRP, multidrug resistance protein; ABC, ATP-binding cassette protein.

OAT1, for example, and exits the cell via one of the apical membrane transporters. When the concentration of urate in the filtrate rises significantly, it is reabsorbed back into the cell via URAT1, for example, in exchange for a dicarboxylic acid anion. In conditions of excessive anion secretion, such as diabetic ketoacidosis or high alcohol intake, excretion of urate is impaired due to a lack of compensating antiport anion and cellular urate, and eventually plasma urate rises. Conversely, inhibition of URAT1 by drugs such as probenecid decreases reabsorption of urate, leading to a decrease in blood uric acid. A number of drugs can affect these transport systems, thereby compromising renal handling of urate and giving rise to hyperuricemia or, in some cases, to excessive urate excretion (**Table 17.5**).

TABLE 17.5 Examples of drugs that affect blood urate concentration

Drugs that increase blood urate	Drugs that decrease blood urate
Alcohol	Allopurinol
Ascorbic acid	High-dose aspirin
Aspirin	Azathioprine
Caffeine	Clofibrate
Cisplatin and other cytotoxics	Corticosteroids
Diazoxide	Estrogens
Diuretics	Glucose infusion
Epinephrine, levodopa, m-dopa	Guaifenesin
Ethambutol, pyrazinamide	Mannitol
Cyclosporine	Probenecid
Nicotinic acid	Warfarin
Phenothiazines	
Theophylline	

17.4 HYPERURICEMIA AND DISORDERS OF PURINE METABOLISM

Hyperuricemia may arise from overproduction of urate, reduced excretion of urate, or a combination of both. A number of physiological states and clinical conditions are associated with elevated urate production and these include increased cell breakdown or turnover, diet, drugs, and key enzyme defects.

Secondary hyperuricemia

Many clinical conditions present with hyperuricemia as a result of impaired renal function impinging on urate excretion and these are thus secondary causes of hyperuricemia; hyperuricemia is subsequent to some other disorder or treatment. Also, as mentioned above, many drugs affect the renal handling of urate and defects in renal urate transporter proteins may also lead to altered rates of urate excretion. When assessing hyperuricemia, it is important to measure serum creatinine to eliminate impaired renal function as a major cause.

Analytical practice point 17.1

Always measure creatinine to assess renal function when assessing a patient's urate status.

Primary hyperuricemia

Primary hyperuricemia, on the other hand, is rare and arises from specific defects in genes of purine metabolism. Some forms of gout are the best-known primary causes of hyperuricemia, but other primary disorders include Lesch–Nyhan syndrome and ADA deficiency.

Gout

The precipitation of urate crystals in the joints is termed gout. The exact definition follows later, but the precipitation of monosodium urate crystals in the joints causes the consequent inflammatory response presenting as acute and later as chronic arthritis. While the majority of cases of hyperuricemia are idiopathic, there is a significant genetic contribution to disease liability in some cases. Several quantitative trait loci (QTLs) have been identified which play a significant role in determining serum uric acid concentrations, including UAQTL1 and UAQTL2, both of which map to chromosome 4. In addition, Mendelian forms of gout have been identified including phosphoribosylpyrophosphate synthase I superactivity and partial hypoxanthine-guanine phosphoribosyl transferase (HGPRT) deficiency (Kelley–Seegmiller syndrome), both of which are X-linked traits.

Lesch–Nyhan Syndrome

Lesch–Nyhan syndrome (LNS) is caused by inherited defects in the HGPRT gene but, unlike Kelley–Seegmiller syndrome, the resultant enzyme defect is total. The defect eliminates the ability to salvage purines such that (i) there is excessive degradation of purine bases leading to increased production of uric acid, and (ii) the resulting deficiency of GMP and IMP causes increased *de novo* biosynthesis of purines which compounds uric acid overproduction. Clinical signs of LNS include, predictably, hyperuricemia with associated urate crystal deposition in the joints and the formation of kidney stones, but also a range of neurological complications—cognitive impairment, motor delay, hypotonia, spasticity, and self-injurious behavior. The rationale for the neurological complications is not known but may be related to abnormal dopamine function in the brains of patients with LNS. Studies measuring the binding of ligand to dopamine transporters to calculate the density of dopamine-containing neurons showed a 50–63% reduction in binding to dopamine transporters in caudate and a 64–75% reduction in putamen of brains from patients with LNS compared to control brains. This was supported by volume magnetic imaging which demonstrated a 30% reduction in caudate volume in LNS patients. Similar large dopaminergic deficits, decreased dopaminergic nerve terminals and cell bodies, have been shown using positron emission tomography in brains of patients with LNS, where all dopaminergic pathways, not just those in the basal ganglia, appear to be involved.

Adenosine deaminase deficiency

ADA deficiency, a rare genetic disorder due to mutation at the *ADA* locus on chromosome 20, is characterized by a severe combined immune deficiency (SCID). Deficient ADA activity results in the accumulation of adenine and its corresponding nucleosides and nucleotides, which are toxic to T lymphocytes in particular. Also, the defect depresses the production of deoxyribonucleotides generally through the inhibition of ribonucleotide reductase at elevated cytosolic deoxyadenosine triphosphate (dATP) concentrations. ADA deficiency is characterized by a range of clinical and laboratory abnormalities and, since immune system function is severely compromised, is fatal in early childhood. While it can be managed by enzyme replacement therapy or by bone marrow transplant, ADA deficiency has been the subject of a number of experimental gene-therapy trials in recent years.

Pseudohyperuricemia (adenine phosphoribosyl transferase deficiency)

Adenine phosphoribosyl transferase deficiency (APRTD) is a rare genetic condition characterized by the accumulation of the insoluble purine 2,8-dihydroxyadenine (2,8-DHA) in kidney and formation of urinary stones. As many as 50% of patients with this condition are asymptomatic but onset of symptoms may appear from the age of 5 months to late adulthood and present with renal colic, hematuria, dysuria, and, in some cases, renal failure. From a clinical chemistry perspective, it is essential to avoid interference by 2,8-DHA in uric acid assays by first separating the two molecules by selective capillary zone electrophoresis, for example.

Clinical significance of hyperuricemia and gout

As has been referred to earlier, the clinical importance of urate is due to its relatively low solubility in aqueous solution and, in particular, in body fluids and its consequent precipitation as sodium urate crystals in joints, kidney, and subcutaneous tissues. Unlike other mammals, humans and many other primates lack the enzyme urate oxidase which converts urate to a more water-soluble product, allantoin (**Figure 17.12**). The loss of this enzyme during evolution is somewhat enigmatic but may relate to the antioxidant properties of urate, which help to protect the body against tissue damage by reactive oxygen

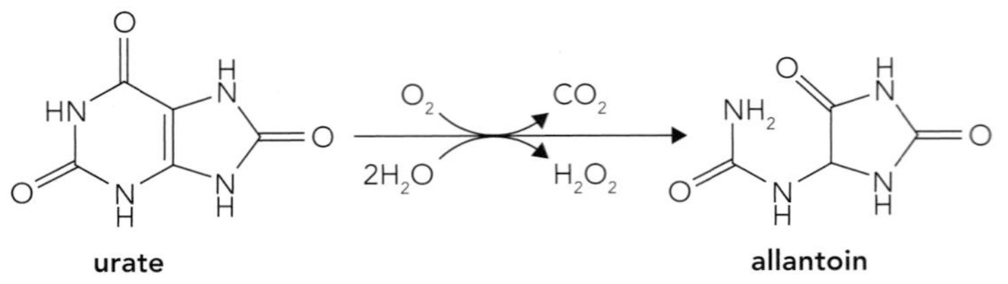

Figure 17.12 Conversion of urate to allantoin by urate oxidase. In humans and other primates, uric acid is the end product of purine catabolism. Other mammals possess the enzyme urate oxidase, which oxidizes urate to the more water-soluble product allantoin with the release of carbon dioxide (CO_2) and hydrogen peroxide (H_2O_2). Hydrogen peroxide is converted to oxygen and water by catalase.

species, with evolution choosing a longer, possibly gouty life rather than a life shortened by free-radical-initiated tissue damage. Clearly, urate only becomes a problem when it is in excess.

Gout is the clinical term used to describe precipitation of sodium urate in a joint. Precipitation, which may occur when serum urate changes rapidly, induces an acute local inflammatory response resulting in severe pain, swelling, redness, and loss of mobility of the joint. Gout is the most common type of crystal arthropathy and can be distinguished from pseudogout (caused by precipitation of calcium pyrophosphate dihydrate, which presents with a similar inflammatory picture) by microscopic examination of crystals aspirated from the affected joint.

Gout is a relatively common disorder affecting about 1% of adults, and the incidence in men is twice that in women. The incidence of hypercholesterolemia and hypertension in patients with gout is also higher than in the general population. In many instances, hyperuricemia may exist for many years (up to 20 years) before the development of obvious clinical symptoms resulting from crystal deposition which, if untreated, results in joint destruction and renal disease.

High concentrations of urate in the low-pH environment of the renal distal tubule favor the formation of urate crystals and may present as renal stones or calculi. These may be distinguished from more common renal stones of calcium oxalate by microscopic examination. Uric acid stones are radiolucent on plain X-ray but are seen as filling effects on an intravenous urogram. In the longer term, urate may also precipitate in soft tissues and present as tophi, particularly on the ears and fingers (**Figure 17.13**). Although tophi may appear unsightly, clinically they are not intrinsically problematic unless they ulcerate through the skin and form a site of infection. They are a sign of poorly controlled hyperuricemia and are far less common nowadays due to effective therapeutic regimes.

Urate precipitation and blood urate

Whereas the risk of developing renal stones, tophi, and urate deposition in joints increases with increasing concentration of blood urate (**Figure 17.14**), only a minority of individuals with elevated urate develop clinical disease. On the other hand, a small percentage of patients with blood urate within the reference range develop gout and/or renal stones. In these cases, protective factors

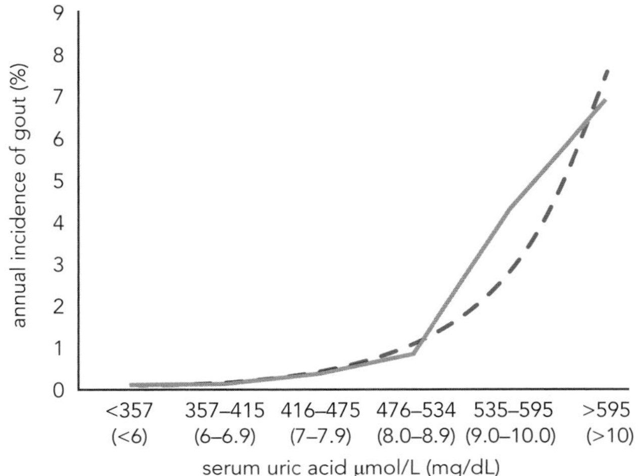

Figure 17.14 Hyperuricemia and the risk of gout in males.
The shape of the incidence plot against serum uric acid concentration is suggestive of a threshold effect with precipitation of urate occurring between 476 and 534 μmol/L (8.0–8.9 mg/dL). The solid line reflects the actual data, whilst the dotted line is the exponential projection.

Clinical practice point 17.1

The diagnosis of gout is not made by measuring serum urate levels, but is made by the aspiration and examination of synovial fluid.

Clinical practice point 17.2

Patients with hyperuricemia normally do not have gout, and some patients with normal levels of urate do have gout.

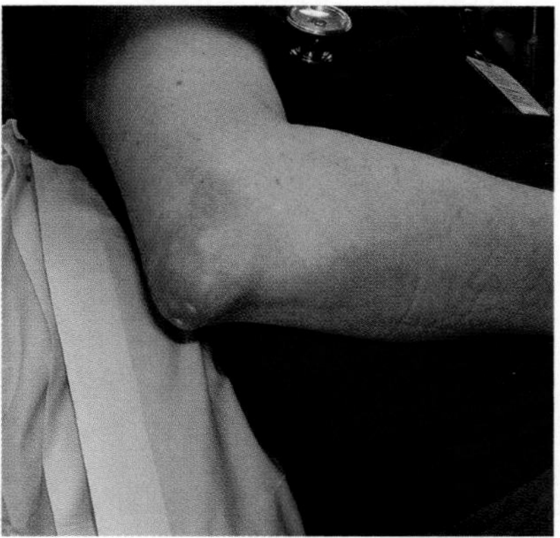

Figure 17.13 Tophi.
Tophi are often found in chronic gout as hard nodules in soft tissues, particularly on the ears and fingers, but may occur anywhere in the body including the tips of elbows, toe joints, vocal cords, and around the spinal cord. The white, chalky material of the tophi contains crystalline and amorphous monosodium urate dihydrate, often surrounded by mononuclear cells and fibroblasts.

which normally increase urate solubility, such as certain proteins, presumably are lacking.

Measurement of serum urate is essential in the assessment of patients with gout and in patients at risk of developing hyperuricemia. As described earlier in this chapter (see Figure 17.1), serum urate may be elevated as a consequence of increased intake of purine-rich foods, increased turnover of endogenous purines, or, more commonly, by decreased renal excretion of urate. Measurement of urinary excretion of urate over a 24-hour period in patients fed a low-purine diet may help to determine the cause of raised serum urate, but this is seldom done in practice.

In patients with established gout but who are pain-free, a sudden rise in serum urate may cause the exquisite pain classically associated with the disease. In the pain-free condition, urate crystals deposited in joints are frequently coated with apoproteins B and E which reduce immunoglobulin (Ig) binding to the crystals and the associated inflammatory effects. A sudden increase in blood urate causes urate in the supersaturated fluid around the joint to crystallize. These new crystals bind IgG and form a nidus for the accumulation of neutrophils. Phagocytosis by the neutrophils causes the release of local cytokines and initiation of inflammation resulting in localized pain. Certain alcoholic beverages, port being the oft-quoted classical example, are well

Analytical practice point 17.2

A 24-hour urine analysis can be used when discriminating between underexcretion and overproduction of urate.

CASE 17.1

At an occupational health screen, a 39-year-old man gives a history of three episodes of severe pain in the metatarsophalangeal joint of his big toe. The joint has become swollen, hot, red, and extremely tender. On each occasion he has taken anti-inflammatory drugs (bought over the counter) rather than seek medical advice. The pain has gradually subsided over a few days.

	SI units	Reference range	Conventional units	Reference range
Plasma				
Creatinine	85 μmol/L	53–106	0.96 mg/dL	0.6–1.2
Uric acid (urate)	660 μmol/L	200–420 (male range)	11.1 mg/dL	3.4–7.1 (male range)

- **What is the most likely cause of his joint pain?**
- **How can further attacks be prevented?**
- **Why is it important to measure creatinine?**

The site of the pain and its description are highly suggestive of gout. This is caused by crystals of uric acid forming in soft tissue and joints. It is thought that temperature plays a role, with gout most likely in the joints most exposed to cold. Uric acid crystals within a joint initiate an intense inflammatory reaction and extreme pain. Whilst acute attacks are self-limiting and may occur in individuals with normal uric acid levels, people with recurrent attacks and elevated plasma uric acid may benefit from uric acid-lowering drugs, such as allopurinol or febuxostat. Such treatment is usually lifelong. There is a risk of causing an acute attack of gout during the first few months due to the change in uric acid levels, and anti-inflammatory drugs are usually co-prescribed initially. Food and drink high in purines and drugs which can raise urate levels should also be avoided.
It is important to monitor kidney function in these patients for three reasons. Hyperuricemia may be due to impaired renal clearance, uric acid stones and nephropathy may be consequences of hyperuricemia, and allopurinol doses must be reduced in kidney disease.

known for their effects on gout sufferers. The effect of alcohol consumption on serum urate is shown in **Figure 17.15**.

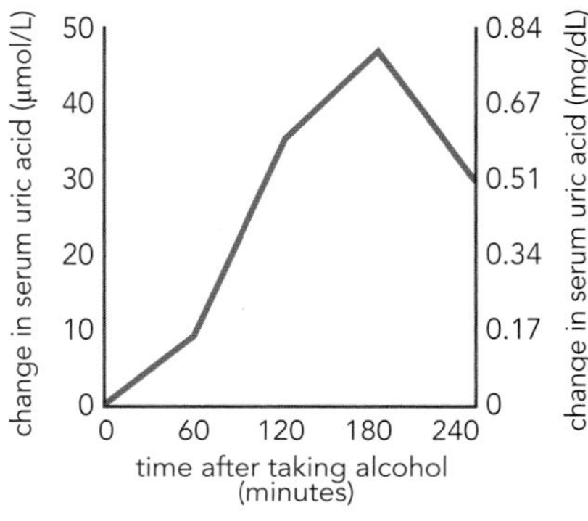

Figure 17.15 Acute effect of alcohol consumption on serum urate concentration. Ethanol accelerates the conversion of ATP to AMP and hence to urate and also increases the intracellular lactate–pyruvate ratio, thus decreasing urate excretion. These factors serve to increase serum urate concentration. Additionally, beer contains guanosine which is catabolized to urate.

Management of gout

Minimizing the exposure to increases in serum urate is the ultimate objective in the management of patients with gout and will involve the following:

- Treatment of acute attacks as soon as they arise
- Prevention of the incidence of acute attacks
- Lowering of serum urate concentration

Thus initially this will involve immediate treatment with nonsteroidal anti-inflammatory drugs (NSAIDS), such as diclofenac or indomethacin, to resolve the acute, painful, inflammatory attack. Other older remedies, including the use of colchicine which decreases the inflammatory response by inhibiting leukocyte migration, are also available.

Once this situation has been resolved, drug therapy may be initiated to reduce serum uric acid. Although isolated hyperuricemia is not normally treated, a reduction in serum urate is indicated when end-organ effects have occurred. Two therapeutic strategies are available for this:

1. Decrease urate production by inhibition of a key enzyme: for example, allopurinol inhibits xanthine oxidase (**Figure 17.16**); or
2. Increase renal excretion of urate: for example, probenecid and sulfinpyrazone inhibit urate reabsorption in the proximal tubule.

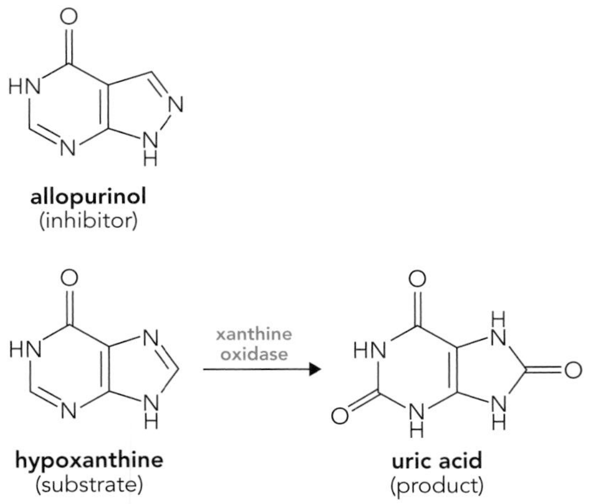

allopurinol (inhibitor)

hypoxanthine (substrate) —xanthine oxidase→ **uric acid** (product)

Figure 17.16 Allopurinol: an inhibitor of xanthine oxidase. Allopurinol inhibits the conversion of hypoxanthine to uric acid and has been used extensively in the treatment of gout to prevent a rise in serum urate and consequent urate stone formation. Oxypurinol, an active metabolite of allopurinol, is a stronger inhibitor of xanthine oxidase than allopurinol and has been used in the treatment of allopurinol-intolerant gout.

With relation to the latter therapy, it is important to consider the possible effects on the secretion of other drugs, such as antibiotics, which the patient might also be receiving. Paradoxically, and by mechanisms not fully understood, a sudden fall in serum urate may precipitate further attacks of pain in patients with gout. Since these drugs (probenecid and sulfinpyrazone) do not have an anti-inflammatory action, NSAIDs are usually prescribed prophylactically for three months when treatment for gout is started.

Longer-term management of patients with gout is designed to limit the recurrence of acute symptoms caused by deposition of crystals of monosodium urate and urolithiasis. Thus it is important to determine the underlying causes of gout, which include obesity, chronic alcohol abuse, hyperthyroidism, hypertension, hyperlipidemia, and chronic renal failure. In practice this often translates to a reduction in dietary intake of purine-rich foods (see Table 17.2), reduction in alcohol consumption (ideally to zero), and care in the use of drugs that raise serum urate (for example thiazide diuretics).

17.5 ACUTE MEDICINE AND URATE

Tumor lysis syndrome

It has been mentioned earlier that urate is produced during cell breakdown due to catabolism of nucleic acids. Increased cellular breakdown will increase urate production and such a situation can arise clinically when malignant cells undergo rapid lysis—so-called tumor lysis syndrome. This presents as an oncologic emergency where large amounts of cellular contents are released into the systemic circulation. Tumor lysis syndrome may occur spontaneously, but most commonly it occurs in hematological malignancies, such as lymphoma and

CASE 17.2

A 37-year-old woman with known acute leukemia is admitted to hospital for chemotherapy. After 48 hours she becomes acutely unwell with confusion and muscle weakness and spasms (tetany). Fluid charts show her urine output has fallen. She has urgent blood tests performed which show the following.

	SI units	Reference range	Conventional units	Reference range
Plasma				
Sodium	138 mmol/L	136–142	138 mEq/L	136–142
Potassium	6.9 mmol/L	3.5–5.0	6.9 mEq/L	3.5–5.0
Urea (blood urea nitrogen, BUN)	16.5 mmol/L	2.9–8.2	46 mg/dL	8–23
Creatinine	425 μmol/L	53–106	4.8 mg/dL	0.6–1.2
Calcium (albumin-adjusted)	1.86 mmol/L	2.20–2.60	7.4 mg/dL	8.8–10.4
Phosphate	2.4 mmol/L	0.74–1.39	7.4 mg/dL	2.3–4.3
Magnesium	1.2 mmol/L	0.70–1.00	2.8 mg/dL	1.7–2.3
Uric acid (urate)	20 μmol/L	150–350 (female range)	0.3 mg/dL	2.5–5.9 (female range)

- What condition has occurred and why?
- How are the electrolyte abnormalities explained?
- Why is the uric acid level so low?

The patient has developed tumor lysis syndrome. This occurs in certain malignant diseases where both the rate of cell division and sensitivity to treatment are very high. Acute leukemias and high-grade lymphomas are most commonly implicated. The simultaneous necrosis of large numbers of tumor cells releases their contents into the circulation. This includes potassium, magnesium, and phosphate. The latter can precipitate with calcium in the soft tissues, including the kidneys, causing hypocalcemia and impaired renal function. Breakdown of nucleic acids from tumor cells releases purines which are converted to uric acid. This, in turn, can cause acute uric acid nephropathy and acute kidney injury (AKI) due to the poor solubility of uric acid in the low-pH environment of the renal tubules. The AKI further raises potassium to a level that may cause cardiac arrest. The plasma uric acid is very low as the patient has been given rasburicase, an enzyme which breaks down uric acid *in vivo*. However, rasburicase continues to break down uric acid *in vitro* after the blood sample has been taken, giving rise to very low results, which are incorrect. To obtain an accurate uric acid measurement the sample must be collected on ice and analyzed without delay.

leukemia, often 2–3 days after the start of treatment. It is accompanied by hyperkalemia, hyperphosphatemia, and hypocalcemia. Without appropriate therapy, acute uric acid nephropathy may occur and it is thus usual to treat patients with allopurinol before and during chemotherapy.

Pre-eclampsia

A further use of urate measurement in acute medicine is in pre-eclampsia. Patients with pre-eclampsia and hyperuricemia are seven times more likely to give birth prematurely, and as such many centers measure serum urate in at-risk pregnant women before deciding upon appropriate clinical management. Interestingly, pregnant women with hyperuricemia but without hypertension and proteinuria are not at increased risk of premature delivery.

17.6 RASBURICASE AND THE CASE OF THE DISAPPEARING SERUM URATE

Release of intracellular contents including potassium, phosphate, and purines causes hyperkalemia, hypocalcemia, and hyperuricemia. The latter may result in acute urate nephropathy due to precipitation of urate in the renal tubules. In order to prevent this, patients are given allopurinol or, if they are at particularly high risk, rasburicase. The mechanisms of action of these two drugs are quite different. Allopurinol inhibits xanthine oxidase and hence urate production. However, xanthine is also insoluble (unlike hypoxanthine) and may itself precipitate in the kidney, particularly if the urine is alkaline. For this reason, urinary alkalinization, although making urate more soluble, should be avoided when allopurinol is given. In addition, allopurinol is not effective when high urate levels are already present. Rasburicase is recombinant urate oxidase (from *Saccharomyces cerevisiae*), which converts urate to the far more soluble allantoin (see Figure 17.12). Following an intravenous infusion of rasburicase, plasma urate concentrations fall rapidly and continue to fall *in vitro* following venipuncture. By the time blood reaches the laboratory for assay, the urate may be much lower than it was when collected from the patient, sometimes giving nonsensical below-zero values. In order to minimize this, samples are placed on ice and transported as quickly as possible.

Hypouricemia (xanthine oxidase deficiency)

A marked decrease in uric acid concentration in serum and urine is found in patients with the rare genetic disorder of xanthine oxidase deficiency, as patients with this condition are unable to metabolize xanthine to uric acid. The condition is characterized by excretion of large amounts of xanthine in urine and a tendency to form xanthine stones, which may deposit in the kidney.

PLASMA LIPIDS AND LIPOPROTEINS

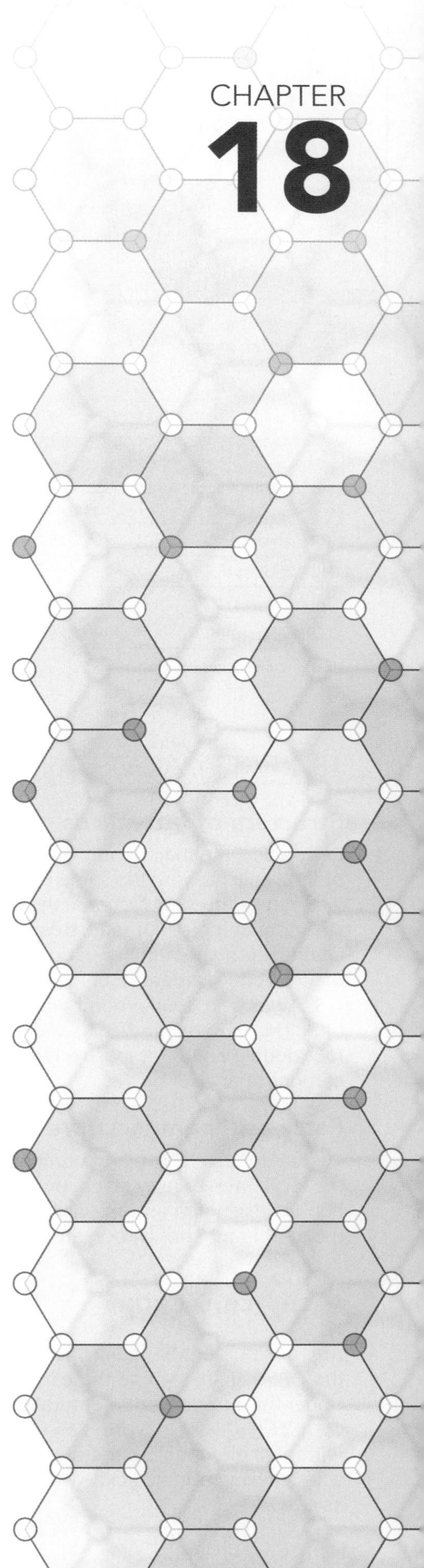

IN THIS CHAPTER

FATTY ACIDS AND TRIGLYCERIDES

CHOLESTEROL

LIPOPROTEINS

MEASUREMENT OF PLASMA LIPIDS

CLINICAL DISORDERS OF LIPIDS AND LIPOPROTEINS

There is no single, universally recognized definition of the term lipid, but it encompasses a range of molecules, including fatty acids and steroids, which are generally insoluble in water.

Lipids constitute approximately 20% of body weight in humans, mostly as an energy store (as triglyceride) in adipose tissue and as structural lipids (phospholipids) in cell membranes. The high energy yield from oxidation of fatty acids—twice that of carbohydrate—allows them to serve as the major respiratory substrate for most tissues apart from brain, nervous tissue, erythrocytes, and the kidney, and the movement of diet-derived triglyceride from the gut to adipose tissue, and of fatty acids from adipose tissue to other tissues, including liver, is essential for metabolic homeostasis. Increases in the concentration of blood lipids—measured in the clinical laboratory as cholesterol and triglycerides—may be features of pathological conditions, such as liver disease, hypothyroidism, and diabetes mellitus, and high concentrations of blood lipids are risk factors for the development of atherosclerosis.

The major lipid molecules found in blood are free fatty acids, phospholipids, triglycerides, and cholesteryl ester. The structures of the last three of these are shown in **Figure 18.1**. By definition, their hydrophobic nature precludes lipids from circulating free in the blood. Thus, fatty acids circulate bound to albumin whilst triglycerides, cholesterol, and phospholipids are constituents of lipoprotein particles. The plasma concentrations of clinically relevant lipids are not usually expressed as conventional reference ranges based on observed values in the population. This is because they vary across the world and in Western countries mean levels are higher than are associated with good health. Expert societies and government health departments therefore use target ranges for individuals and health professionals to aim for in order to minimize cardiovascular disease, particularly coronary heart disease (CHD). These ranges are periodically updated. The most recent National Cholesterol Education Program (NCEP) ranges are given in **Table 18.1**.

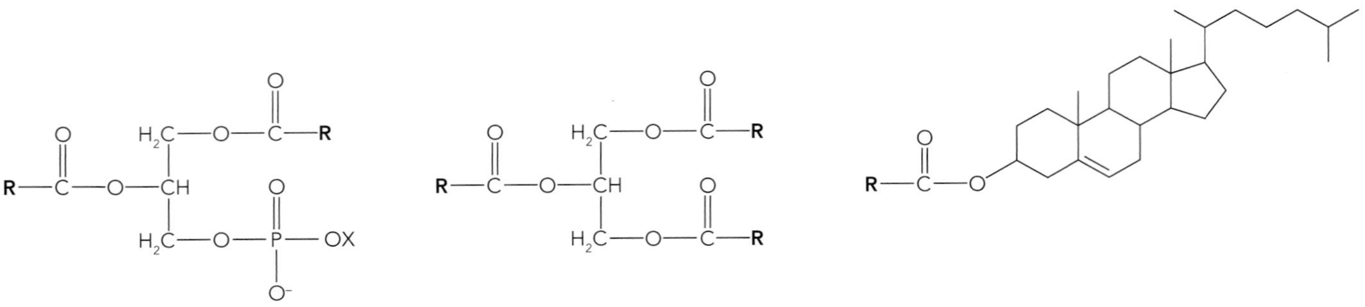

(a) phospholipid (b) triglyceride (c) cholesteryl ester

Figure 18.1 Structures of lipid molecules.
(a) Phospholipid. Phospholipids consist of a hydrophobic tail and a hydrophilic head. The tail comprises a diglyceride (two long-chain fatty acids), whilst the head has a phosphate group and an organic molecule such as choline. R, alkyl chain of fatty acid (may be the same or different in the phospholipid molecule); X, choline, ethanolamine, serine, or inositol. (b) Triglyceride (triacylglycerol). Triglycerides consist of a glycerol backbone (three carbons) to which three fatty acids are esterified. These may be the same or different from each other. (c) Cholesteryl ester. Cholesterol is a four-ring structure and an alcohol. The usual form in plasma is cholesteryl ester with a long-chain fatty acid in place of the –OH group. This is much more hydrophobic than free cholesterol.

TABLE 18.1 NCEP guidelines for plasma total and individual lipoprotein cholesterol concentrations

	Desirable	Borderline	Undesirable
Total cholesterol	<5.2 (200)	5.2–6.2 (200–240)	>6.2 (240)
HDL-cholesterol	>1.55 (60)	1.0–1.5 (40–59)	<1.0 (40)
Triglycerides	<1.69 (150)	1.69–5.62 (150–499)	>5.63 (500)
LDL-cholesterol	<2.6 (100)	3.4–4.1 (130–159)	4.1–4.9 (160–189)

Units are mmol/L (mg/dL).

18.1 FATTY ACIDS AND TRIGLYCERIDES

Structure of fatty acids

Fatty acids, the major component of body and dietary lipids, constitute a family of molecules having the same basic properties. All mammalian fatty acids have an even number of carbons, the common saturated fatty acids being myristic (14:0), palmitic (16:0), and stearic (18:0), where the numbers indicate carbon chain length and number of double bonds, respectively. Unsaturated fatty acids contain one or more double bonds in the alkyl chain; for example, oleic acid (18:1) has one double bond while the polyunsaturated fatty acids (PUFAs) linoleic (18:2), linolenic (18:3), and arachidonic (20:4) acids have two, three, and four double bonds, respectively. In these molecules the double bonds are three carbons apart.

Fatty acid nomenclature

Two alternative shorthand numbering nomenclatures have evolved to describe the structures of fatty acids. In the systematic nomenclature, the carboxyl carbon is designated as carbon 1, the number of CH_2 units in the chain as n, and the total number of carbon atoms (including the terminal methyl group) is n+2:

$$\overset{n+2}{CH_3}\text{-}(CH_2)_n\text{-}\overset{1}{C}OOH$$

Any unsaturated bonds in the molecule are indicated in superscript following the Greek delta sign as the carbon number at which the double bond occurs. Under this nomenclature, saturated 16-carbon palmitic acid (hexadecanoic acid) $CH_3(CH_2)_{14}COOH$ is written as C16:0; 18-carbon oleic acid (octadecanoic acid), with a single double bond at carbon 9, as C18:Δ^9; and 20-carbon arachidonic acid (eicosatetranoic acid) with four double bonds at carbons 5, 8, 11, and 14, as C20:$\Delta^{5,8,11,14}$.

In the alternative nomenclature, the carboxyl carbon is designated the α-carbon (alpha carbon) while the terminal methyl carbon is designated the ω-carbon (omega carbon):

$$\overset{\omega}{CH_3}\text{-}(CH_2)_n\text{-}CH_2\text{-}\overset{\alpha}{COOH}$$

Humans lack the ability to synthesize fatty acids with a double bond beyond carbon 9 and thus polyunsaturated fatty acids such as arachidonate, the precursor for prostaglandins, prostacyclins, thromboxanes, and leukotrienes, must be provided by the diet. Such fatty acids are known as essential fatty acids. The fish-oil fatty acids, eicosapentaenoic acid (EPA) and docosahexaenoic acid (DHA), which have received interest in relation to their possible role in the prevention of vascular disease, have a longer chain (C20 and C22) and are more unsaturated (five and six double bonds, respectively) than those of mammalian sources. The two sources of polyunsaturated fatty acids differ in the position of the terminal double bond in relation to the ω-carbon in the molecule and this gives rise to two series of PUFAs. If the last double bond in the fatty acid chain is six carbons from the ω-carbon (as in arachidonic acid, for example), the fatty acid is a member of the ω-6 series (C20:4ω-6) whereas if the last double bond is three carbons from the ω-carbon, as in EPA and DHA from fish oils, the fatty acid is a member of the ω-3 series (EPA is C20:5ω-3 and DHA C22:6ω-3).

Mobilization of fat

Excess fatty acids derived from the diet are stored as triglycerides (triacylglycerols) in adipose tissue. Triglycerides consist of three molecules of fatty acid esterified to the hydroxyl groups of glycerol. Fatty acids are mobilized from stored triglycerides by the action of a hormone-sensitive lipase activated via a cyclic adenosine monophosphate (cAMP) second messenger-mediated system by glucagon or epinephrine (**Figure 18.2**). Thus after the immediate postprandial period when blood insulin levels fall back to fasting levels, glucagon will activate the hormone-sensitive lipase to release fatty acids into the circulation for use as an energy source by most tissues. Albumin has a number of high-capacity binding sites for fatty acids and transports them through the circulation. Uptake of fatty acids into tissues such as skeletal muscle, heart, and liver is controlled by their concentration in blood. In catabolic states such as fasting, the bulk of the fatty acids are oxidized to CO_2 and H_2O in mitochondria in extrahepatic tissues. In overwhelmingly catabolic states, particularly where there is peripheral resistance to insulin, such as uncontrolled diabetes mellitus, trauma or acute alcohol intoxication, hepatic resynthesis of triglyceride becomes significant.

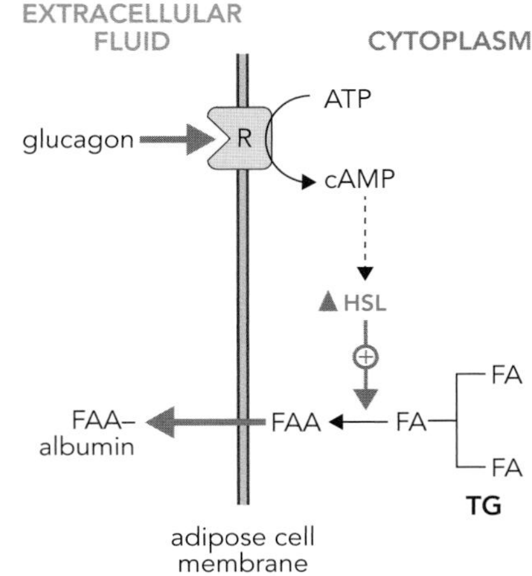

Figure 18.2 Mobilization of free fatty acids from adipose tissue. Fatty acids (FA) are stored in adipose cells in the form of triglycerides (TG), with three FAs per glycerol molecule. When glucagon, or epinephrine (not shown), binds to its receptor (R), it activates hormone-sensitive lipase (HSL) via cyclic AMP. HSL releases the FAs from triglyceride to produce free fatty acids (FFA), which are exported into the plasma and transported bound to albumin.

Fatty acid oxidation

Fatty acids are oxidized mainly in mitochondria by β-oxidation. On entry into the cell they are activated to their coenzyme A (CoA) derivative and transported into mitochondria via a carnitine shuttle (**Figure 18.3**). Fatty acyl coenzyme A in the mitochondrial matrix then enters the β-oxidation cycle being metabolized to acetyl-CoA. The four reactions which shorten the fatty acid chain length at each round of the process involve desaturation between carbons 2 and 3 (α and β), addition of water across the resultant double bond, and dehydrogenation (oxidation) of the hydroxyacyl intermediate to produce the shorter (by two carbons) fatty acyl CoA and the release of acetyl-CoA (**Figure 18.4**). Unsaturated fatty acids additionally require an isomerase and a reductase to allow β-oxidation to proceed to completion.

The process of β-oxidation yields reducing equivalents in the form of reduced nucleotides, NADH and $FADH_2$, which donate electrons to the electron-transport chain on the inner mitochondrial membrane and thereby generate ATP via

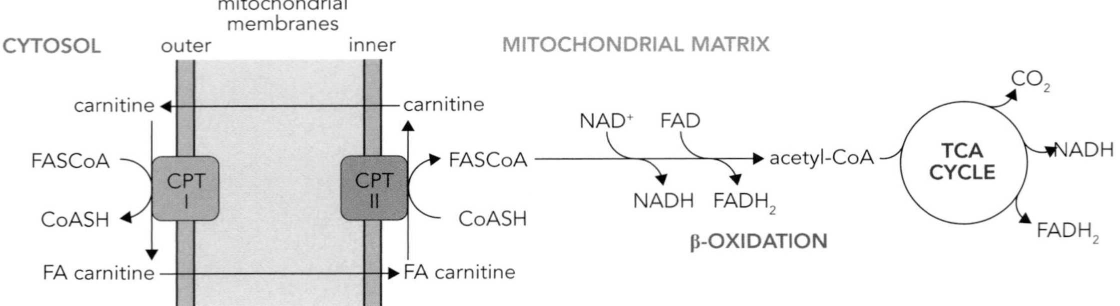

Figure 18.3 Energy production from fatty acids.
The process of generating cellular energy in the form of ATP from fatty acids is called β-oxidation and it occurs in the mitochondrial matrix. Coenzyme A (CoA), via its sulfhydryl group (SH), binds to a fatty acid (FA) in the cytosol. The FA enters the mitochondrion by a shuttle mechanism involving carnitine and the enzymes carnitine palmitoyl transferase (CPT) I and II on the outer and inner mitochondrial membranes. FAs with a chain length of less than 10 carbon atoms can diffuse into the mitochondrial matrix independently. Fatty acyl coenzyme A (FASCoA) within the mitochondrion undergoes β-oxidation, with sequential removal of two-carbon units and the generation of acetyl-CoA, which enters the TCA cycle. NADH and $FADH_2$ generated by β-oxidation and the TCA cycle donate electrons to the electron-transport chain, which generates ATP.

oxidative phosphorylation. Acetyl-CoA is oxidized in the tricarboxylic acid (TCA) cycle, generating CO_2 and further reduced nucleotides, which again provide electrons to drive ATP synthesis via oxidative phosphorylation (see Chapter 2).

Oxidation of fatty acids with even-numbered chain lengths yields two molecules of acetyl-CoA from the final four carbon atoms. However, odd-numbered chains (of plant or marine origin) yield a final product with three carbon atoms, propionyl-CoA. This is converted to succinyl-CoA (four carbons) via methylmalonyl-CoA in a step involving vitamin B_{12}, and succinic acid enters the TCA cycle. Genetic defects in the enzymes involved in this pathway result in the organic acidemias, propionic acidemia and methylmalonic acidemia (see Chapter 23).

Inherited defects in any of the key enzymes of fatty acid oxidation severely impair the body's ability to use fats for energy. Such disorders usually present in early childhood as severe, potentially fatal, hypoglycemia, particularly under conditions of metabolic stress, such as infection. These are described in more detail in Chapter 23.

R—CH_2(β)—CH_2(α)—C(=O)SCoA
acyl CoA dehydrogenase ① FAD → $FADH_2$
R—CH=CH—C(=O)SCoA
enoyl CoA hydratase ② H_2O
R—CH(OH)—CH_2—C(=O)SCoA
hydroxyacyl CoA dehydrogenase ③ NAD⁺ → NADH
R—C(=O)—CH_2—C(=O)SCoA
β ketothiolase ④ CoA
R—C(=O)SCoA + H_3C—C(=O)SCoA

Figure 18.4 β-Oxidation.
The process of releasing a two-carbon unit from a fatty acid (as fatty acyl CoA) consists of four steps: (1) Dehydrogenation by FAD to produce a double bond between carbon 2 and carbon 3 (also referred to as α and β, as shown), catalyzed by acyl CoA dehydrogenase. (2) Hydration of the double bond by enoyl CoA hydratase. (3) Oxidation by NAD of the hydroxyl group to a keto group by hydroxyacyl CoA dehydrogenase. (4) Thiolysis by another molecule of CoA of the bond between carbon 2 and carbon 3 to produce a new fatty acyl CoA, two carbons shorter than the first, and acetyl-CoA containing the original CoA molecule.

Fatty acid synthesis

The synthesis of fatty acids and their incorporation into triglycerides represents a mechanism for conserving the energy of dietary fat, carbohydrate, and the carbon skeletons of some amino acids which are in excess to the immediate requirements of the body. Under anabolic conditions, where the insulin–glucagon ratio is raised, the liver and adipose tissue are able to synthesize fatty acids from blood-derived glucose via acetyl-CoA (**Figure 18.5**) and to incorporate these fatty acids into triglyceride. Under catabolic conditions, the liver can also make acetyl-CoA from non-carbohydrate sources such as amino acids and lactate and this becomes relevant in stress situations. The rate-limiting enzyme of fatty acid synthesis is acetyl-CoA carboxylase, the product of which is malonyl-CoA (**Figure 18.6**). The enzymes of fatty acid synthesis are listed in **Table 18.2**. Acyl carrier protein (ACP) is an essential co-factor. Each round of the fatty acid synthetic cycle elongates the growing fatty acyl chain by two carbons donated by the three-carbon malonyl-CoA with the release of CO_2. Chemically, the four reactions involved in converting the ketoacyl-ACP to acyl-ACP are the reverse

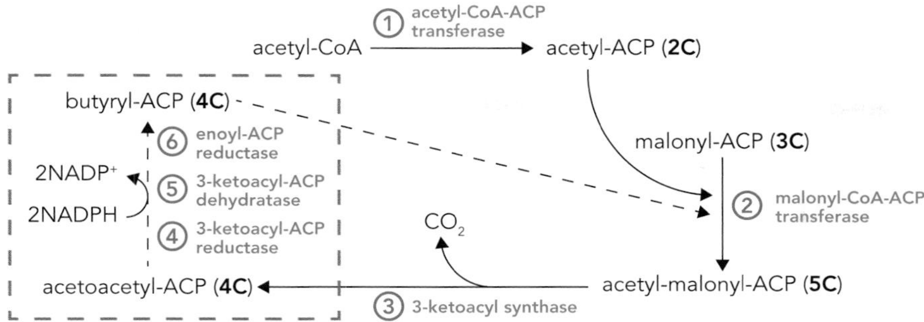

Figure 18.5 Fatty acid synthesis.
Saturated straight-chain fatty acid synthesis occurs by a series of six reactions, which add two-carbon units until palmitic acid is formed (16 carbons). Medium-chain fatty acids can be formed by early termination. Although the enzymatic steps are listed in sequence, they are all performed by different domains of the same multifunctional protein: fatty acid synthase. (1) Activation of acetyl-CoA by transfer to acyl carrier protein (ACP). (2) Activation of malonyl-CoA, also by transfer to ACP, to form a transient five-carbon (5C) unit. (3) Loss of CO_2 results in the four-carbon unit acetoacetyl-ACP; the free energy released drives the reaction. (4) Reduction of the ketone at carbon 3 to a hydroxyl group. (5) Removal of water and creation of a double bond. (6) Reduction of the double bond between carbons 2 and 3. Steps 4, 5, and 6 are shown in greater detail in Figure 18.7

$$^{-}HCO_3 + CH_3{-}C({=}O){-}SCoA \xrightarrow[\text{ATP} \rightarrow \text{ADP} + P_i + H^+]{\text{acetyl-CoA carboxylase}} {}^{-}O{-}C({=}O){-}CH_2{-}C({=}O){-}SCoA$$

bicarbonate **acetyl-CoA** **malonyl-CoA**

Figure 18.6 Synthesis of malonyl-CoA.
Malonyl-CoA, a three-carbon molecule, provides two-carbon units for fatty acid synthesis and is formed by the carboxylation of acetyl-CoA. This is an energy-dependent step requiring ATP and is catalyzed by acetyl-CoA carboxylase. During fatty acid synthesis, malonyl-CoA is able to regulate β-oxidation by inhibiting the association of free fatty acids with carnitine, thus preventing their entry into the mitochondria.

TABLE 18.2 Enzymes of fatty acid synthesis

Enzyme	Protein-bound product
1. Acetyl-CoA-ACP transferase	Acetyl-ACP
2. Malonyl-CoA-ACP transferase	Acetyl-malonyl-ACP
3. 3-Ketoacyl synthase	3-Ketoacyl-ACP (loss of CO_2)
4. 3-Ketoacyl-ACP reductase	Hydroxyacyl-ACP
5. 3-Ketoacyl-ACP dehydratase	Enoyl-ACP
6. Enoyl-ACP reductase	Acyl-ACP

Numbers refer to enzymes shown in Figures 18.5 and 18.7.

of those seen in β-oxidation (**Figure 18.7**) but the electron donor in this reductive biosynthesis is NADPH, produced by the pentose phosphate pathway (see Chapter 2). The overall reaction for the synthesis of palmitic acid (C16:0) from two-carbon acetyl-CoA is:

Acetyl-CoA + 7 malonyl-CoA + 14 NADPH + 14H⁺ → palmitate + 7 CO_2 + 8 CoA + 14 NADP⁺ + 6 H_2O

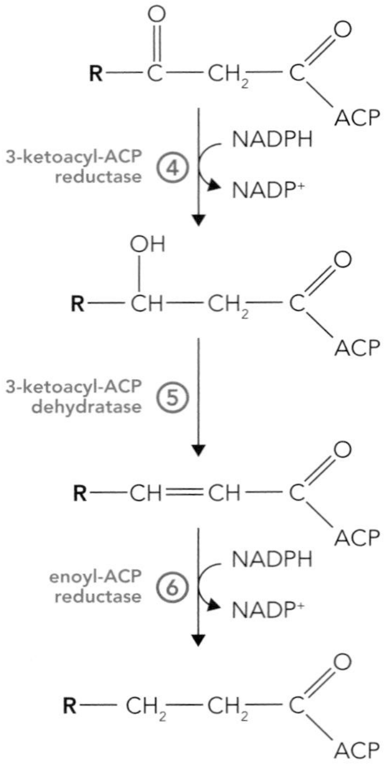

Figure 18.7 Later reactions involved in two-carbon addition in fatty acid synthesis.
An overview of fatty acid synthesis is shown in Figure 18.5. The final three steps, all catalyzed by domains of fatty acid synthase, are shown here in more detail. The final product is two carbon units longer until a maximum of 16 is reached. (4) Reduction of the ketone at carbon 3 to a hydroxyl group. (5) Removal of water and creation of a double bond. (6) Reduction of the double bond between carbons 2 and 3.

Malonyl-CoA itself has an inhibitory effect on β-oxidation and prevents this catabolic pathway from supplying acetyl-CoA to fatty acid synthesis, which is anabolic; it thereby prevents futile cycling of acetyl-CoA. The activities of acetyl-CoA carboxylase and fatty acid synthase are high in liver, adipose tissue, and mammary gland.

Triglyceride synthesis

Triglyceride synthesis in the liver and adipose tissue involves the esterification of two fatty acids to glycerol 3-phosphate to form the short-lived intermediate, phosphatidic acid, which in turn undergoes dephosphorylation. Triglyceride is formed by esterification of the resultant diglyceride with a third molecule of fatty acyl CoA (**Figure 18.8**). In addition to fatty acids formed from glucose metabolism, the liver uses dietary fatty acids derived from chylomicron remnants for triglyceride synthesis. The biosynthesis of triglycerides in the small intestine differs from that in liver and adipose tissue. In this tissue, triglyceride is formed by esterification of monoglyceride with two molecules of fatty acyl CoA (see Chapter 14).

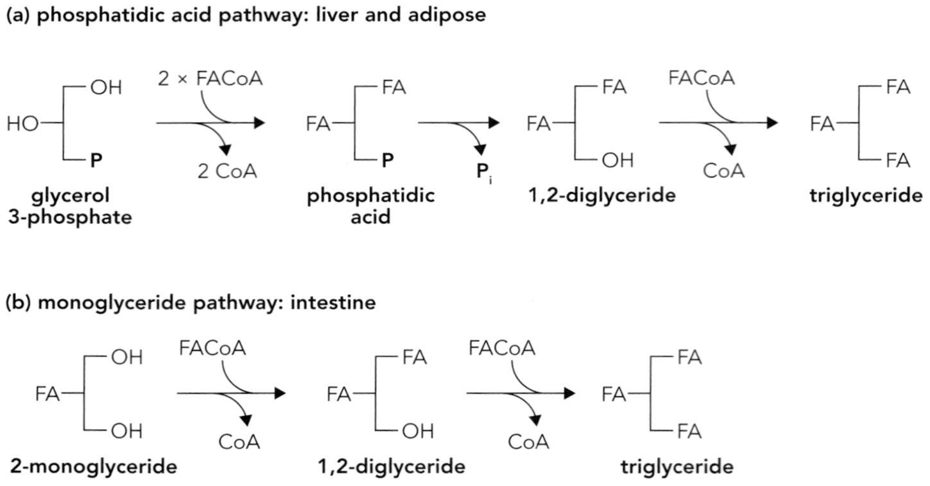

Figure 18.8 Synthesis of triglyceride.
Triglycerides consist of three fatty acid chains covalently bound to a single molecule of glycerol by esterification. The process of triglyceride synthesis differs between (a) adipose tissue and liver and (b) intestine. (a) Adipose cells lack the enzyme glycerol kinase, hence cannot re-utilize the glycerol produced by the action of lipoprotein lipase (see Figure 18.14). Instead, dihydroxyacetone phosphate is diverted from glycolysis to form glycerol 3-phosphate. This combines with two molecules of fatty acyl CoA (FACoA) to form phosphatidic acid, which loses phosphate to form 1,2-diglyceride. The third and final fatty acid is added by acyl transferase to form the triglyceride molecule. (b) Intestinal epithelial cells resynthesize triglyceride from digested dietary lipids. 2-Monoglyceride and free fatty acids are produced by the action of pancreatic lipase in the small intestine, taken up by intestinal cells, and undergo stepwise recombination by acyl transferases to form triglyceride.

18.2 CHOLESTEROL

Cholesterol has been described as "the most highly decorated small molecule in biology" by Michael Brown and Joseph Goldstein, recipients of a Nobel Prize in 1985 for their work in describing the low-density lipoprotein (LDL) receptor and the basis of familial hypercholesterolemia. It was originally isolated in the late eighteenth century from gallstones, which have a high cholesterol content. During the late twentieth century the association between the concentration of cholesterol in the blood and cardiovascular disease was established, and drugs that inhibit cholesterol synthesis have since become the most widely prescribed in the developed world.

Cholesterol synthesis

Cholesterol in the body may be derived exogenously from the diet or endogenously from *de novo* synthesis from acetate (outlined in **Figure 18.9**), predominantly in the liver and intestine. The initial reactions in this 26-step process involve the formation of two five-carbon molecules, isopentenyl pyrophosphate and its isomer dimethylallyl pyrophosphate, the building blocks of the cholesterol molecule. The first step is the formation of four-carbon acetoacetyl-CoA from two molecules of acetyl-CoA, which is catalyzed by thiolase. Acetoacetyl-CoA combines with a third molecule of acetyl-CoA to form six-carbon 3-hydroxy 3-methylglutaryl-CoA (HMG-CoA), a reaction catalyzed by HMG-CoA synthase. HMG-CoA is then converted to mevalonate by HMG-CoA reductase, which is the committed, rate-limiting step for cholesterol synthesis and a target for the class of cholesterol-lowering drugs known as statins. The formation of five-carbon isopentenyl pyrophosphate from mevalonate involves three enzymatic steps all requiring ATP and, in the last of these, 5-pyrophosphomevalonate undergoes decarboxylation with the loss of CO_2.

The second stage of cholesterol synthesis begins with the isomerization of isopentenyl pyrophosphate to dimethylallyl pyrophosphate and the condensation of the two five-carbon isomers to a ten-carbon product, geranyl pyrophosphate. A 15-carbon intermediate, farnesyl pyrophosphate, is then formed by condensation of geranyl pyrophosphate with another molecule of isopentenyl pyrophosphate. Reductive condensation of two molecules of farnesyl pyrophosphate yields 30-carbon squalene, the largest linear molecule of the cholesterol biosynthetic pathway.

The second stage of this pathway may be summarized as:

$$C_5 + C_5 \rightarrow C_{10};\quad C_{10} + C_5 \rightarrow C_{15};\quad C_{15} + C_{15} \rightarrow C_{30}$$

In the third stage of cholesterol biosynthesis, squalene is converted to squalene epoxide in a reaction requiring oxygen and NADPH and this epoxide then cyclizes to lanosterol, the first intermediate of the pathway with a structure containing the four rings of cholesterol. However, more than a dozen further enzymatic steps are required to complete the biosynthetic pathway to cholesterol.

Free cholesterol is toxic within cells, although it is a vital component of their membranes, and esterification, particularly in liver, represents a mechanism to lower the intracellular free cholesterol concentration. The enzyme, acylcoenzyme A:cholesterol acyl transferase (ACAT) catalyzes this reaction. The major form of cholesterol in blood is cholesteryl ester (see Figure 18.1), in which a fatty acid is esterified to carbon 3 in the A ring and this circulates as a component of lipoproteins.

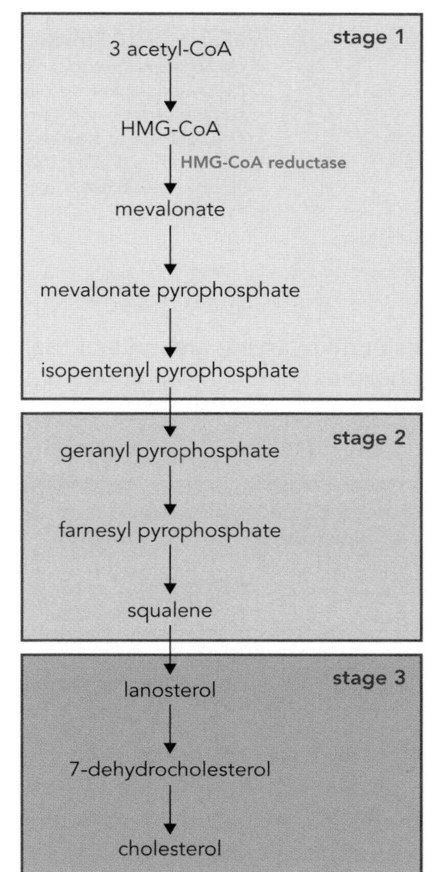

Figure 18.9 Cholesterol synthesis. Cholesterol is a cyclic hydrocarbon containing 27 carbon atoms, all originating from acetyl-CoA. Biosynthesis is complex but can be considered in three stages. (1) Formation of isopentenyl pyrophosphate from mevalonate and ATP. Mevalonate is formed from three molecules of acetyl-CoA, which combine to form 3-hydroxy-3 methylglutaryl-CoA (HMG-CoA). The conversion of this to mevalonate is catalyzed by HMG-CoA reductase, the rate-limiting step in cholesterol synthesis and the target of statin drugs. (2) Condensation of six molecules of isopentenyl pyrophosphate via geranyl and farnesyl pyrophosphate results in squalene. (3) Cyclization of squalene results in lanosterol which is finally converted to cholesterol.

18.3 LIPOPROTEINS

Since lipids by their very nature are hydrophobic molecules, an efficient means of transport of both exogenously and endogenously derived lipid through the aqueous environment of the circulation is required. Furthermore, a mechanism for targeting circulating lipids to specific tissues must also exist. These two requirements are fulfilled by lipoproteins, which provide a vehicle both for transport and recognition by specific receptors on tissues. The generic shape of all lipoproteins is spherical and they consist of an oily lipophilic core of cholesteryl

Figure 18.10 Structure of a lipoprotein.
Lipoproteins are macromolecular complexes that allow insoluble lipids to circulate in aqueous plasma. They have a hydrophobic core and hydrophilic surface. Apolipoproteins on the surface interact with receptors and enzymes. The four major classes of lipoproteins are chylomicrons, very-low-density lipoproteins (VLDL), low-density lipoproteins (LDL), and high-density lipoproteins (HDL).

surface monolayer of phospholipids and free cholesterol

apolipoproteins

hydrophobic core of triglyceride and cholesteryl esters

ester and varying amounts of triglyceride surrounded by a monolayer of phospholipids containing apoproteins and free cholesterol (**Figure 18.10**). The families of apoproteins (also called apolipoproteins) are listed in **Table 18.3**.

TABLE 18.3 Families of apolipoproteins

Apolipoprotein	Subtypes
A	AI, AII
B	B48, B100
C	CI, CII, CIII
E	E2, E3, E4

The lipoproteins have been named according to their density, which reflects the proportion of lipid in the core of the particle. The lipoproteins in order of increasing density are: chylomicrons (CM); very-low-density lipoproteins (VLDL); low-density lipoproteins (LDL); and high-density lipoproteins (HDL) (**Figure 18.11**). A further subgroup, the short-lived intermediate-density lipoproteins (IDL), is formed from metabolism of VLDL and may be further metabolized to LDL. The difference in density is used in the isolation of lipoproteins by sequential flotation in media of increasing density. The properties of the major lipoproteins are shown in **Table 18.4**. It should be appreciated, however, that such a classification into four major groups is something of an oversimplification and that in the circulation lipoproteins are distributed along a density continuum from a density of 1.00 g/mL (CMs) to 1.2 g/mL (high-density lipoproteins). The differences in apolipoprotein composition give the lipoproteins different charges and thus differing electrophilic mobilities (α-HDL, pre-β-VLDL, and β-LDL), which may be of diagnostic value in qualitative analysis in patients with defects in lipoprotein metabolism. The properties of the major apolipoproteins are shown in **Table 18.5**. Lipoproteins are highly dynamic structures whose particle size and composition are undergoing continual change, with both lipids and apolipoproteins moving within lipoprotein classes.

In the following discussion of lipoprotein metabolism the following abbreviations are used.

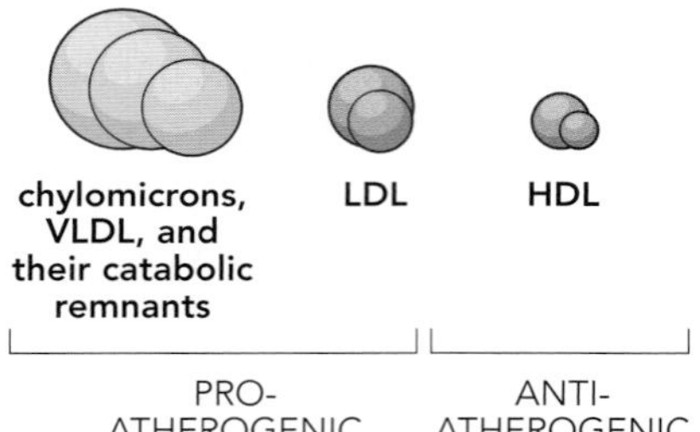

Figure 18.11 Lipoprotein classes.
Although all lipoproteins have the same basic structure as shown in Figure 18.10, they vary in their composition, size, density, apolipoproteins, and tendency to cause atheroma. The triglyceride-containing particles very-low-density lipoprotein (VLDL) and chylomicrons are large and tend to float on separated plasma. The cholesterol-containing particles low- and high-density lipoprotein (LDL and HDL) are smaller and denser. Alone amongst the lipoproteins, HDL is negatively correlated with disease caused by atheroma.

LPL

Lipoprotein lipase is an extracellular enzyme present on the surface of the capillary endothelium which hydrolyzes triglyceride borne by CMs and VLDL to free fatty acids and glycerol. Synthesized by muscle and adipose cells, it is released from its anchor site on the endothelium by heparin.

LCAT

Lecithin-cholesterol acyl transferase is an enzyme synthesized and secreted by the liver. It catalyzes the formation of cholesteryl ester from free cholesterol and phosphatidylcholine (lecithin) in HDL and is activated by apolipoprotein A1.

TABLE 18.4 Properties of the principal lipoproteins

Property		Chylomicron	VLDL	IDL	LDL	HDL
Density (g/mL)		<1.00	<1.006	1.006–1.019	1.019–1.063	1.063–1.21
Diameter (nm)		80–100	30–80	25–30	20–25	8–13
Electrophoretic mobility		Uncharged	Pre-β	β	β	α
Major components (% by weight)	TG	90–95	50–65	25–40	4–6	7
	CE	2–4	8–14	20–35	35–45	10–20
	C	1	4–7	7–11	6–15	5
	PL	2–6	12–16	16–24	22–26	25
	Protein	1–2	5–10	12–16	22–26	45
Major apolipoproteins (% total protein)		AI (31)	C (40–50)	B100 (60–80)	B100 (>95)	AI (65)
		C (32)	B100 (30–40)	C (10–20)	C (<1)	AII (10–23)
		E (10)	E (10–15)	E (10–15)	E (<1)	C (5–15)
		B48 (5–8)				E (1–3)

TG, triglyceride; CE, cholesteryl ester; C, cholesterol; PL, phospholipid.

TABLE 18.5 Properties of major apolipoproteins

Apolipoprotein	Molecular mass (Da)	Major location	Function
AI	28,000	HDL, CM	Binding to AI receptor
			Activator of LCAT
AII	17,000	HDL	Unknown, possibly redundant
B100	512,000	LDL, IDL, VLDL	Binding to LDL (apoB100) receptor
			Assembly of VLDL
B48	245,000	CM	Assembly of CM
CII	9000	CM, VLDL, IDL, HDL	Binding to and activation of LPL
E	34,000	IDL	Binding to apoE receptor
			Inhibitor of LPL

CETP

Cholesteryl ester transfer protein is synthesized mainly by liver and mediates the transfer of cholesteryl ester between lipoprotein species, in particular from HDL to remnant particles formed from metabolism of CMs and VLDL.

ACAT

Acylcoenzyme A:cholesterol acyl transferase is an enzyme of the endoplasmic reticulum which catalyzes the intracellular synthesis of cholesteryl ester from free cholesterol and acylcoenzyme A.

CEH

Cholesteryl ester hydrolase is an enzyme which hydrolyzes stored intracellular cholesteryl ester to free cholesterol and fatty acid. There are at least two species: an acidic lysosomal enzyme and an enzyme which is active at neutral pH.

Triglyceride-rich lipoproteins

The two largest and least dense lipoproteins are responsible for the transport of exogenous diet-derived triglyceride (CM) and endogenous hepatically synthesized triglyceride (VLDL) in the circulation. Their assembly, in the intestine

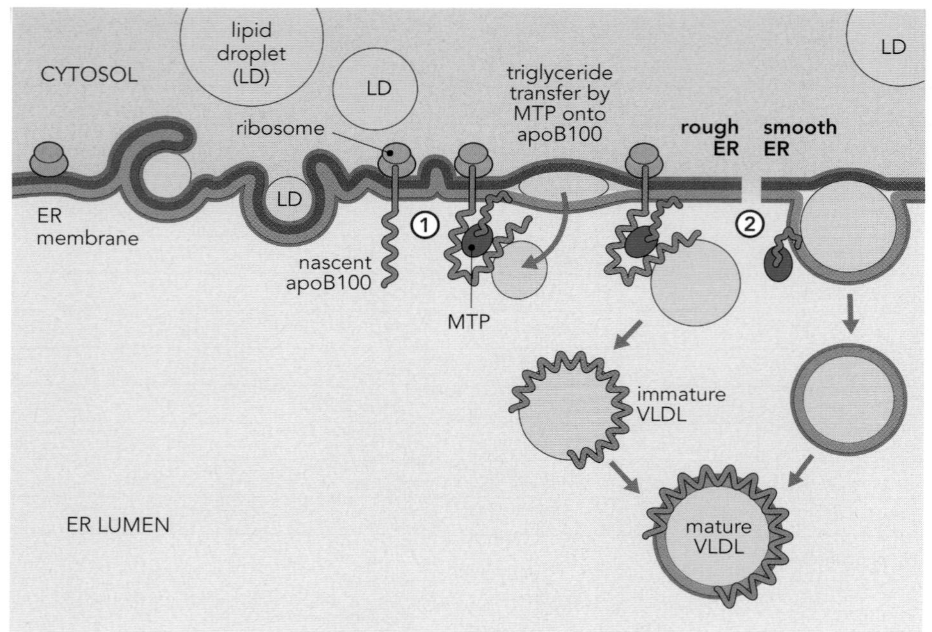

Figure 18.12 Microsomal triglyceride transfer protein (MTP) in VLDL and CM assembly. Nascent apoB100 within the rough endoplasmic reticulum (ER) must be rapidly combined with lipid to avoid proteolysis. There are two lipidation mechanisms. (1) MTP can bind apoB and remove lipid from the ER membrane until the VLDL is large enough to bud off. (2) MTP may also combine with a lipid droplet before presenting it to apoB. Additional lipid from the smooth ER is required for VLDL maturation before it is secreted into the blood.

and liver, respectively, involves the participation of a protein, microsomal triglyceride transport protein (MTP), which moves lipid from its site of synthesis on the smooth endoplasmic reticulum to nascent apolipoprotein B (apoB) as it folds during translation (**Figure 18.12**). Although MTP is not present in the final lipoproteins, lack of MTP activity during CM and VLDL assembly due to mutations in the MTP gene prevents synthesis of these triglyceride-rich lipoproteins. Nonassembly of CMs leads to accumulation of lipid in the enterocyte and lipid is lost as the enterocyte sloughs off, producing an effective fat malabsorption. Thus patients who are unable to synthesize apoB present with fat malabsorption (steatorrhea) and also fat-soluble vitamin deficiency, since vitamins A, D, E, and K are incorporated into CMs for transport to the liver. This results in very low concentrations of plasma lipids, the condition of abetalipoproteinemia, which is discussed later. The following list summarizes the properties of MTP:

Function

- Expressed in endoplasmic reticulum of liver and intestine where it facilitates assembly and secretion of apoB-containing lipoproteins
- Transfers lipids (triglyceride, cholesterol, and phospholipid) to nascent apoB

Structure

- Heterodimer of a 59 kDa protein disulfide isomerase and a unique larger subunit (97 kDa) which has extensive homology to amphibian lipovitellin

Expression

- Increased by dietary fats especially saturated fatty acids and cholesterol
- Decreased by insulin and consumption of ethanol

The major lipoproteins of both CM and VLDL are products of the apoB gene. In the liver, the translated apoB is a 512 kDa monomeric polypeptide of 4536 amino acids. This product has been given the trivial name of apoB100; the major lipid-binding domain is within the N-terminal half of the protein and the domain which binds to the apoB100 receptor (LDL receptor) is within the C-terminal half. Apo B100 is a component of VLDL. This contrasts with the situation in the intestine, where post-transcriptional editing of the apoB mRNA converts codon 2153 (CAA), which specifies glutamine, to UAA, a termination codon. Consequently, the translated apoB in intestine is only 48% of the length of the apoB100 synthesized in liver and is given the trivial name of apoB48. Critically, while apoB48 has the lipid-binding domain, it lacks the domain for binding to the apoB100 receptor (LDL receptor) and thus neither CMs nor CM remnant particles are cleared by the LDL receptor.

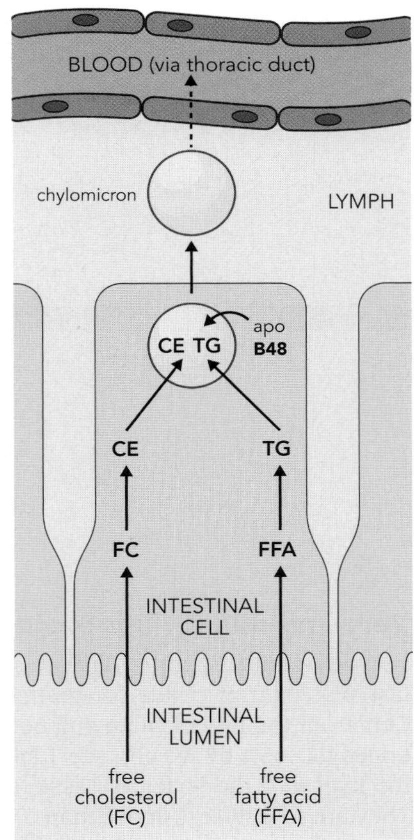

Figure 18.13 Formation of chylomicrons in enterocytes. Chylomicrons are lipoproteins produced in the intestinal wall that enable transport of triglycerides from a fat-containing meal to the circulation. Free fatty acids and cholesterol are esterified to triglyceride (TG) and cholesteryl ester (CE), packaged with apolipoprotein B48, and secreted into the lymphatic system. The small vessels, called lacteals, eventually drain into a single thoracic duct, which empties into the superior vena cava and the right side of the heart.

Chylomicrons

CMs, the largest and least dense of the lipoproteins, are synthesized in the wall of the small intestine, secreted into the lymphatic system, and drain into the blood via the thoracic duct, directly into the right side of the heart and avoiding the liver (**Figure 18.13**). They are secreted following a meal and the postprandial CM concentration reflects the fat content of the meal. Their lipid composition in general reflects that of the diet, including fat-soluble vitamins but excluding short- and medium-chain fatty acids, which can be absorbed directly into the hepatic portal blood without the need for packaging into lipoproteins (see Chapter 14). The extremely low density of the CM particles, due to 90% of their weight being triglyceride, causes them to float on the top of a plasma sample when left at 4°C overnight, identical to the way in which cream rises to the surface of whole milk when left in the refrigerator. The hydrophobic triglyceride-rich core is surrounded by a surface monolayer of phospholipid containing additionally cholesterol and apoproteins AI and AII. A single molecule of apoB48 forms a structural scaffold for the particle.

CMs are secreted from the enterocyte into the lymphatic system and enter the circulation through the thoracic duct. Nascent CM particles contain apoB48 and apoA1 but are deficient in apolipoprotein CII which is required for interaction with LPL in the capillary beds of extrahepatic tissues such as muscle and adipose tissue. They acquire apoCII and apoE during metabolism from HDL, which enables them to bind to LPL and results in extracellular hydrolysis of the core triglyceride to free fatty acids and glycerol. The glycerol released is returned to the liver whilst the free fatty acids are taken up into the extrahepatic tissue and stored as triglyceride (adipose tissue) or metabolized to CO_2 and water (muscle). Resynthesis of triglyceride in adipose tissue requires glycerol 3-phosphate from glycolysis. CMs are synthesized in the early postprandial phase, being present in the circulation when blood glucose and hence insulin concentrations are raised. Insulin, which activates LPL, also promotes uptake of glucose into adipose tissue and stimulates glycolysis to provide the glycerol 3-phosphate acceptor of fatty acids derived from the LPL hydrolysis of CM triglycerides, leading to the intracellular storage of triglyceride.

Approximately 80–90% of CM triglyceride is hydrolyzed by LPL and gradual loss of triglyceride from the core of the particle leads to a reduction in size with loss of apoCII, apoAI, and cholesteryl ester to HDL and decreased affinity for LPL. The particle is now known as a CM remnant (CMR) (**Figure 18.14**). As the affinity for LPL decreases, the previously latent apoE is expressed on the surface of the remnant particle, allowing it to bind to and be cleared by apoE receptors (LRP; LDL receptor-related protein) in the liver. Virtually all CMR are cleared as particles of density less than 1.019 g/mL and are not normally found in the LDL fraction.

Dietary fat-soluble vitamins incorporated in CMs during assembly in the intestine are delivered to the liver via CMR and it has been suggested that this transport of dietary fat-soluble vitamins to the liver is the primary role of the CM.

The clearance of CMs from the circulation is rapid and in healthy adults they are absent from plasma after an overnight fast. However, deficiency of lipoprotein lipase or apoCII results in accumulation of CMs in plasma (lipemia). They also accumulate when plasma VLDL concentrations are very high due to competition for LPL sites in the capillary endothelium.

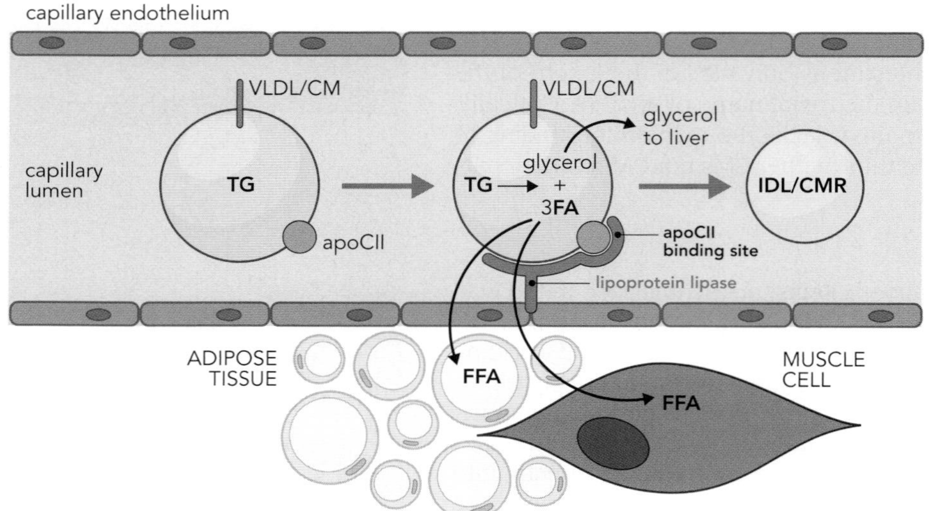

Figure 18.14 Action of lipoprotein lipase. Triglyceride (TG)-containing lipoproteins VLDL and chylomicrons (CM) interact with the endothelial-bound enzyme lipoprotein lipase through the binding of apolipoprotein CII. Triglycerides are broken down to their component free fatty acids, which are taken up by muscle (for β-oxidation to produce energy) or adipose cells (for storage). The triglyceride-depleted lipoprotein particle is called a chylomicron remnant (CMR) or intermediate-density lipoprotein (IDL) if derived from VLDL.

Very-low-density lipoproteins

Very-low-density lipoproteins are synthesized in the liver and, after an overnight fast, plasma triglyceride concentration primarily reflects only that in VLDL, with CMs from the last meal having been cleared. Like CMs, they are rich in triglycerides (50–65% by weight; see Table 18.4) but are assembled in and secreted by the liver into the space of Disse, entering the circulation via the hepatic vein. They are slightly less dense than CMs but are still light enough to float on plasma when left to stand at 4°C overnight. Up to 40% of the lipid content of the particles is endogenously synthesized triglyceride. The half-life of a VLDL particle is of the order of 30 minutes.

The source of fatty acids for VLDL triglyceride is from endogenous synthesis in the liver or from adipose-tissue fatty acids delivered to the liver bound to albumin. The cholesteryl ester, phospholipid, free cholesterol, and apolipoprotein content (apoB and apoC) is higher than in CMs; interestingly, each VLDL particle contains only a single molecule of apoB100. The size of VLDL particles reflects the triglyceride content and larger particles are secreted from the liver under conditions of increased triglyceride synthesis, such as excessive alcohol consumption, obesity, and type 2 diabetes mellitus.

After the nascent VLDL particles are secreted into the blood they acquire apoCII and apoE from HDL. The metabolism of VLDL is referred to as the endogenous lipid pathway (**Figure 18.15**) to distinguish it from the exogenous pathway concerned with chylomicrons derived from dietary fat.

The initial stages of metabolism are similar to those described for CMs. Thus, apoCII on the surface of the VLDL particles interacts with LPL on the surface endothelium of peripheral tissues, causing activation of the lipase and hydrolysis of VLDL triglyceride with concomitant transport of free fatty acids into the tissues, particularly cardiac and skeletal muscle. In the fully mature VLDL particle, the C-terminal portion of apoB100 is buried in the hydrophobic interior of the lipoprotein and is not available to bind to LDL receptors on the surface of peripheral tissues. As triglyceride hydrolysis continues and the lipid core is depleted, the VLDL particles increase in density and decrease in size, becoming intermediate-density lipoproteins (VLDL remnants, VLDLR). During this process the cholesteryl ester is transferred to VLDL such that IDL have increased ester content relative to VLDL and, importantly, apoE and the C-terminal of apoprotein B100 are expressed on the surface of the remnant particles.

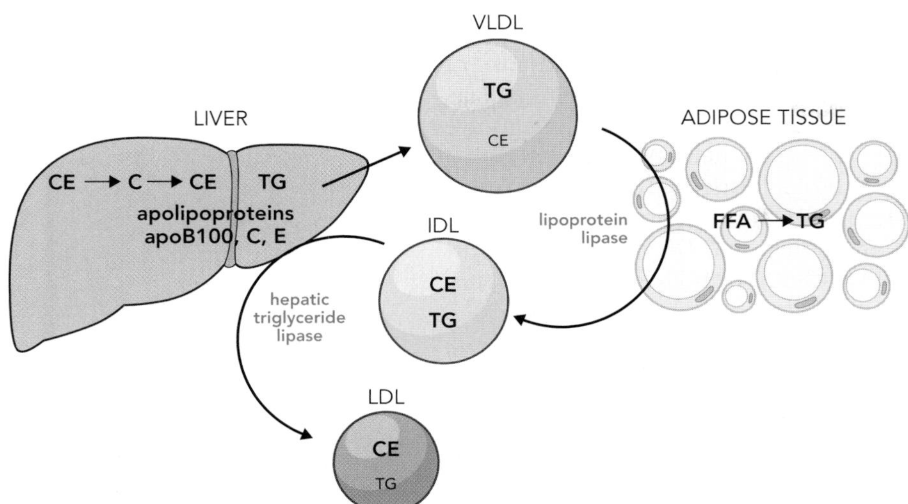

Figure 18.15 Very-low-density lipoprotein metabolism. VLDL is synthesized in the liver as a predominantly triglyceride (TG)-containing particle, with small amounts of cholesteryl ester (CE). It acquires apoCII from HDL (not shown) which allows it to interact with lipoprotein lipase in the same way as chylomicrons. The resulting remnant particle is called intermediate-density lipoprotein (IDL), rich in both TG and CE. Further TG is removed from IDL by hepatic triglyceride lipase to leave a particle enriched in CE called low-density lipoprotein (LDL). IDL is also cleared by interaction of apoE (also acquired from HDL) with remnant receptors (see Figure 18.16).

Intermediate-density lipoproteins

There are two possible fates of IDL, both of which occur very rapidly such that the plasma concentration of IDL is very low in normal individuals. The appearance of apoB100 and apoE on the surface of IDL allows the particle to bind to LDL receptors (apoB100 receptors) and LRP on hepatocytes and approximately 60% of IDL is cleared by the liver by receptor-mediated endocytosis, thereby returning cholesterol, phospholipid, and unhydrolyzed triglyceride to the tissue. The remaining 40% of IDL undergoes further hydrolysis by a hepatic triglyceride lipase (HTGL) bound to the hepatocyte membrane. This results in a change in protein and lipid configuration with transfer of the apoC and apoE, cholesterol, and phospholipid to HDL and the formation of a cholesteryl ester-rich, low-density lipoprotein (LDL) containing a single molecule of apoB100 as its only apolipoprotein (see Figure 18.15). It appears *in vivo* that larger VLDL particles enriched in apoE are rapidly metabolized and endocytosed as IDL, while smaller VLDL particles are metabolized more slowly through to LDL. Mutations in apoE which result in decreased binding to its receptors lead to increased plasma IDL with increased cholesterol and triglyceride. This is the basis of type III hyperlipoproteinemia, which is discussed later.

Cholesterol-rich lipoproteins

The two cholesterol-rich lipoproteins are low-density and high-density lipoproteins (LDL and HDL) which perform what might be simplistically described as opposing roles; LDL transports cholesterol to extrahepatic tissues, and HDL transports cholesterol from extrahepatic tissues to liver. This function of HDL is called reverse cholesterol transport and is extremely important since elimination of cholesterol from the body occurs mainly via hepatically produced bile (see Chapter 3). It is often stated that LDL is the "bad guy" of lipoprotein metabolism because of its association with cardiovascular disease, but it should be remembered that all tissues require cholesterol as a component of cell membranes and lack of LDL itself causes clinical problems. In addition, cholesterol is the precursor for steroid synthesis in the adrenal cortex and gonads and also for vitamin D in skin. Under normal circumstances, tissues obtain most, if not all, of their cholesterol from circulating LDL, despite having a full complement of the 28 enzymes required for *de novo* synthesis of the molecule.

Low-density lipoproteins

Low-density lipoproteins arise from metabolism of IDL following the loss of apoE and lipid to HDL. The physical and chemical properties of LDL and its metabolism are shown diagrammatically in **Figure 18.16**. These particles are the primary cholesterol transport particles in the circulation, with cholesteryl ester accounting for 35–45% and free cholesterol a further 6–15% of the particle weight. LDL serves as a donor of cholesterol to both peripheral tissues and the liver, with the liver being responsible for 50% of the total LDL uptake. LDL circulates with a half-life of 2–3 days. The removal of LDL by both liver and extrahepatic tissues occurs via receptor-mediated endocytosis involving the apoB100/E receptor (LDL receptor). This 839-amino-acid protein is anchored to the cell surface by a membrane-spanning region near its C-terminal. Part of the protein shows sequence homology with epidermal growth factor, and the LDL-binding region, consisting of eight repeat sequences enriched in negatively charged amino acids (especially cysteine), is located at the N-terminus on the extracellular surface. This negatively charged region binds the positively charged region of the C-terminus of apoB100. Each LDL particle has only a single molecule of apoB100 such that binding of LDL to its receptor is monovalent; that is, one particle binds to only one receptor.

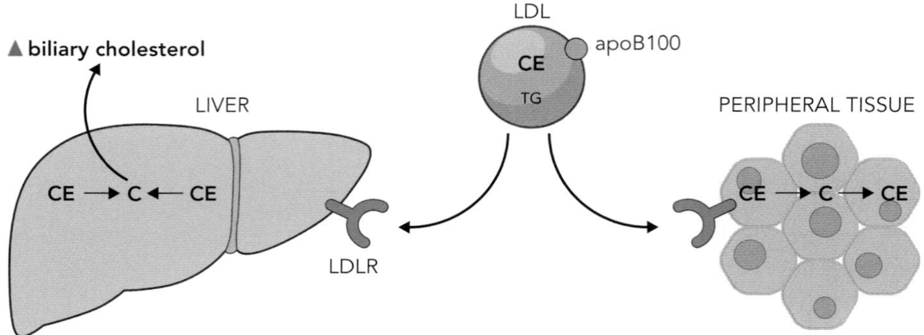

Figure 18.16 Low-density lipoprotein metabolism. LDL is formed in the circulation from IDL as shown in Figure 18.15. Each particle carries one molecule of apoB100, which binds to the LDL receptor (LDLR) present on the cell surface of the liver and most other tissues. The LDL is thereby internalized and the cholesteryl ester (CE) is released to be utilized within the cell as free cholesterol (C) or re-esterified. Most intracellular cholesterol is obtained from LDL rather than *de novo* synthesis. In the liver, excess cholesterol can be excreted into the bile which enters the gut lumen, resulting in a net loss from the body. Mutations in the LDLR or, less commonly, in apoB100 which result in reduced binding between the two cause the condition familial hypercholesterolemia with elevated levels of LDL in plasma.

Endocytosis of LDL results in the formation of an endosome in the tissue with the LDL particle bound to its receptor inside the vesicle. Acidification of the endosome causes dissociation of the lipoprotein from its receptor such that many receptors are recycled to the cell surface while the LDL is digested by lysosomal hydrolases. Amino acids derived from hydrolysis of apolipoprotein B100 enter the cellular amino acid pool and cholesteryl ester is hydrolyzed by a cholesteryl ester hydrolase to free cholesterol and fatty acid.

Under normal conditions, the free concentration of cholesterol in the cell represents a balance between endogenous synthesis, exogenous LDL-derived cholesterol, and esterification by ACAT (acylcoenzyme A:cholesterol acyl transferase). Raised concentrations of free cholesterol are toxic for the cell and the increase in free cholesterol concentration arising from endocytosis of LDL has important consequences for cellular cholesterol metabolism, each of which attempts to restore the homeostatic intracellular free cholesterol concentration:

- Inhibition of HMG-CoA reductase, thereby inhibiting endogenous cholesterol synthesis
- Activation of ACAT, which converts free cholesterol to cholesteryl ester
- Down-regulation of LDL receptor expression on the cell surface, limiting the entry into the cell of exogenous cholesterol from LDL

These effects are likely mediated by an oxy-metabolite of cholesterol.

Thus in the liver, a major cholesterol-synthesizing tissue, the actions of LDL-derived cholesterol have an important role in controlling endogenous cholesterol synthesis and removing LDL from the circulation via LDL receptors. This provides a target for therapeutic intervention to lower plasma LDL, as described later.

In addition to the LDL-mediated entry of cholesterol into the cell, there are also concentration-dependent mechanisms of entry of LDL into the cell not coupled to the control of cholesterol synthesis. These include uptake via a non-receptor-dependent process and uptake of oxidatively modified LDL via a scavenger receptor on liver cells and macrophages. It is thought that oxidation of LDL is a key step in making it atherogenic. Some particles of LDL are more easily oxidized, particularly those that are small and dense. These tend to be produced in greater proportion in some diseases, particularly insulin resistance states and type 2 diabetes mellitus.

High-density lipoproteins

High-density lipoprotein is considered the "good guy" in terms of blood lipids since there is strong epidemiological evidence of a negative association between HDL and rates of cardiovascular disease. Whereas VLDL and LDL transport cholesterol away from liver to extrahepatic tissue, HDL particles are thought to be responsible for the transfer of cholesterol from peripheral tissues to the liver for elimination from the body. However, there is considerable debate and uncertainty as to whether this entirely explains the cardioprotective effect of HDL. Other possibilities include its antioxidant effects and the presence in HDL of the enzyme paraoxonase (PON). This may reduce the oxidation of LDL particles and hence their propensity to cause atheroma formation. The reverse transport of cholesterol is of utmost importance since the liver is the principal route of excretion of cholesterol from the body; the absence of HDL, seen in patients with Tangier disease, leads to the accumulation of cholesterol in extrahepatic tissues.

HDL is synthesized and secreted mainly by the liver, but also in intestine, as a nascent particle consisting of a phospholipid bilayer disc containing free cholesterol and apoAI as its major apolipoprotein with trace amounts of apoE (**Figure 18.17**). There is evidence that similar discoidal particles containing both apoAI and apoAII are also secreted but only apoAI-containing particles appear to participate in reverse cholesterol transport. The disc-shaped,

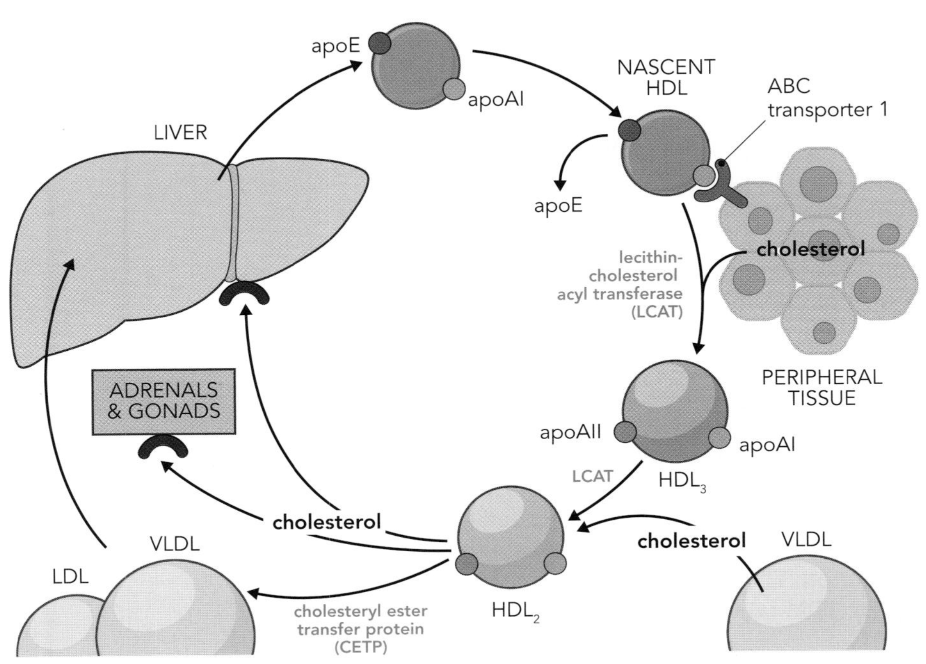

Figure 18.17 High-density lipoprotein metabolism. HDL is secreted by the liver as nascent, disc-shaped particles containing phospholipid, apoAI, and apoE. It collects free cholesterol from peripheral cells following binding of apoAI to ABC transporter 1 on cell membranes. The enzyme lecithin-cholesterol acyl transferase (LCAT) esterifies the cholesterol. ApoE is transferred to VLDL and apoAII is acquired to form HDL$_3$. Further accumulation of cholesterol from tissues and VLDL forms HDL$_2$. Cholesteryl ester transfer protein (CETP) facilitates exchange of cholesteryl ester for triglycerides between lipoprotein classes. HDL delivers cholesterol back to the liver and steroidogenic endocrine tissues (adrenals and gonads) directly by receptor-mediated uptake. It also returns it indirectly to the liver by transfer back to LDL and VLDL particles using CETP.

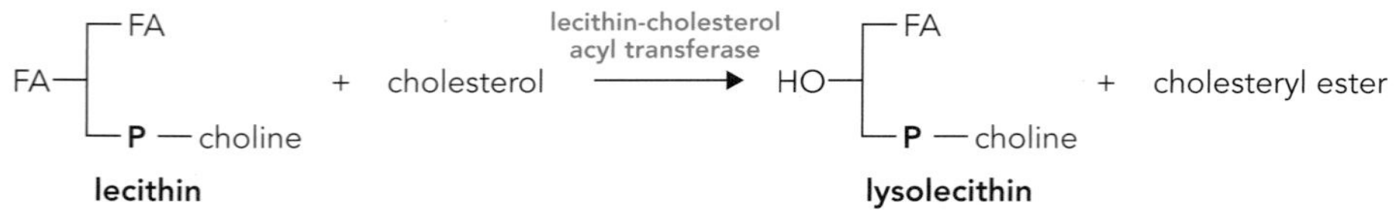

FA(FA)(P — choline) **lecithin** + cholesterol —lecithin-cholesterol acyl transferase→ HO(FA)(P — choline) **lysolecithin** + cholesteryl ester

Figure 18.18 Role of LCAT. The enzyme lecithin-cholesterol acyl transferase (LCAT) is produced in the liver and carried on HDL and LDL. It catalyzes the esterification of free cholesterol with a fatty acid from lecithin, leaving lysolecithin. Cholesteryl ester is more hydrophobic than cholesterol and so moves into the core of the lipoprotein particle, preventing the reverse reaction. Genetic deficiency of LCAT results in cholesterol accumulation in the cornea, kidney, and red blood cells. Although HDL levels are low there is no apparent effect on cardiovascular risk.

apoAI-containing particles bind to specific receptors (ABC transporter 1) on the surface of extrahepatic tissues via apoA1, and free cholesterol in the plasma membrane moves down a concentration gradient into the nascent HDL. At this stage, apoE, phospholipid (phosphatidylcholine; lecithin), and some cholesterol are transferred from IDL. Under the action of a hepatically synthesized enzyme, LCAT, the free cholesterol in the discoidal HDL is converted to cholesteryl ester (**Figure 18.18**) and moves to the hydrophobic interior of the particle. The HDL particle now changes shape, sheds its apoE, and acquires apoAII, becoming HDL_3. Further accumulation of cholesterol from tissues and lipid from VLDL decreases the density of the HDL_3 as it becomes HDL_2. It should be remembered that free cholesterol is transferred from tissue membranes to HDL. This implies that as cholesterol is stored in tissue as cholesteryl ester, hydrolysis of the intracellular ester must occur prior to moving to the plasma membrane and leaving the tissue. Finally, the process of reverse cholesterol transport involves the movement of cholesteryl ester from HDL_2 to the liver. Whilst some may be delivered to the liver directly by HDL_2 via a dedicated receptor, it is probable that much of the HDL cholesteryl ester is exchanged for lipids on other lipoproteins such as LDL and remnant particles of CM and VLDL and delivered to the liver by these particles. Transfer of cholesteryl ester between particles is facilitated by CETP, a plasma protein synthesized in the liver. Interestingly, there is evidence that steroidogenic tissues may receive cholesteryl ester by selective uptake from HDL via a specific receptor (SR-B1) to which HDL binds but which is not internalized by the cell.

The exogenous and endogenous lipid pathways outlined above are normal and essential for energy transport and cell homeostasis. Problems arise when the system is overloaded and pro-atherogenic lipoproteins accumulate in blood, as described below. All of the lipoproteins apart from HDL are, to a greater or lesser extent, pro-atherogenic.

Lipoprotein (a)

Known as lipoprotein "little a," Lp(a) is a variant of LDL first described in the 1960s by Berg as a broad class of lipoproteins with pre-β electrophoretic mobility. It differs from LDL in having an additional apolipoprotein, apo(a), covalently bound via one or two disulfide bonds to apoB100. Apo(a) is a glycoprotein with varying degrees of glycosylation (up to 30% carbohydrate by weight) and exhibiting a high degree of size polymorphism (200–800 kDa), which is genetically determined and due to a variable number (10 to 50 or more) of tandem repeats in the apo(a) gene located on the long arm of chromosome 6. These repeats give rise to an equivalent number of identical repeats of sequences of 114 amino acids similar to those present in plasminogen. The repeated amino acid sequences are held together in triple-loop-like structures resembling the pattern of the Scandinavian pastries called kringles (according to Berg, a Norwegian) and are therefore known as kringle IV repeats. The kringles have high sequence homology with plasminogen. The size of an individual's Lp(a) isoform, which is determined by the number of kringle repeats, appears to be inversely related to its plasma concentration, which can vary from <1 to >200 mg/dL. Unsurprisingly, there are ethnic variations in Lp(a) concentration. The heterogeneity of lipoprotein (a) is due to differences in the protein to lipid ratio, the apoB100 to apo(a) ratio, apo(a) polymorphism, and its degree of glycosylation. There is also heterogeneity in the lipid composition and in particular the cholesteryl ester and triglyceride content of the particle core.

Although an association between Lp(a) and ischemic heart disease has been demonstrated, its value as an independent risk factor is open to debate and it is not routinely measured. Standardization of laboratory methods and appropriate cutoff values are unresolved issues. In addition, although some drugs can alter Lp(a) concentration, it is not affected by lifestyle measures such as diet and exercise and there is no good evidence that lowering Lp(a) is beneficial. The main value of measuring it (once in a lifetime) is in individuals at uncertain levels of risk in whom a high level of Lp(a) would indicate treatment of conventional cardiac risk factors, such as cholesterol and blood pressure.

Lp(a) is synthesized in the liver and although cholesteryl ester-rich and triglyceride-rich species are present in plasma, the physiological role of these particles is unknown at present. However, it is probably not a vital one since healthy subjects can have a very low plasma concentration of Lp(a). From epidemiological studies it is claimed that a concentration of Lp(a) in excess of 500 mg/dL is an independent risk factor for coronary heart disease and Lp(a) has been demonstrated in atherosclerotic plaques. It is possible that the similarity between Lp(a) and plasminogen might enable it to compete with plasminogen for binding to fibrin or the plasminogen receptor and thereby promote thrombogenesis by inhibiting thrombolytic mechanisms.

Lipoprotein X

An occasional cause of elevated plasma cholesterol is lipoprotein X (LpX), a complex composed predominantly of phospholipids and unesterified cholesterol with apoC on its surface. It appears to be a normal constituent of bile. As might be expected, the plasma concentration of LpX rises in liver disease where biliary obstruction is a major feature, for example primary biliary sclerosis or prolonged blockage of the common bile duct by a gallstone. Thus, jaundiced patients may have very elevated plasma cholesterol. It is unclear whether or not LpX is atherogenic in the same way as LDL. It is also not known whether treatment to reduce LpX is beneficial, or if the potential hepatotoxicity of drugs used for treatment outweighs any potential benefit.

18.4 MEASUREMENT OF PLASMA LIPIDS

As may be inferred from the above section on lipoproteins, total plasma cholesterol concentration in the fasting state, where CMs are not present, contains contributions from VLDL, LDL, and HDL and is the sum of all three. Most of the cholesterol in plasma is present in the esterified form and total cholesterol is measured after initial hydrolysis by a cholesteryl esterase. The free cholesterol liberated is oxidized using cholesterol oxidase, the products of which include hydrogen peroxide, which can be measured by further steps resulting in a colored product (**Figure 18.19**).

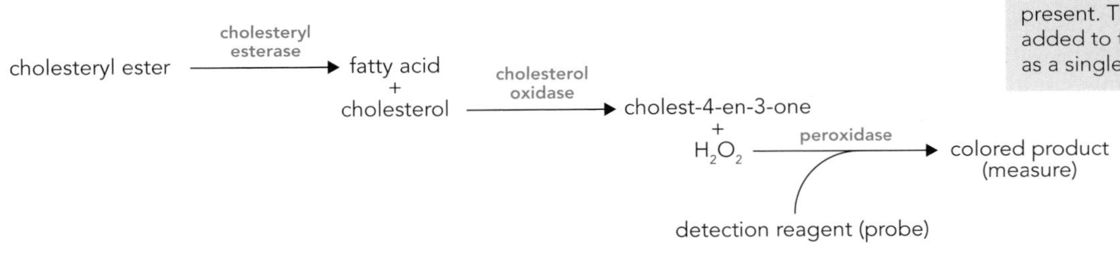

cholesteryl ester —(cholesteryl esterase)→ fatty acid + cholesterol

cholesterol —(cholesterol oxidase)→ cholest-4-en-3-one + H_2O_2

H_2O_2 + detection reagent (probe) —(peroxidase)→ colored product (measure)

Figure 18.19 Assay of cholesterol in plasma.
Total cholesterol is measured by a series of enzymatic steps which results in the production of a colored product, the concentration of which is proportional to the cholesterol present. The enzymes shown are added to the plasma or serum sample as a single reagent.

The contribution of HDL to total plasma lipoproteins is measured indirectly as HDL-cholesterol concentration. As described above, HDL is the only lipoprotein which does not contain apolipoprotein B. Thus, precipitation of apoB-containing lipoproteins with polyanions such as phosphotungstate leaves HDL as the sole cholesterol-containing lipoprotein in the remaining supernatant fraction and allows the measurement of HDL-cholesterol concentration. Such methods are laborious, time consuming, and not suited for the large sample numbers measured routinely in large hospital laboratories. Although in general use until the 1990s, they have been superseded by direct or homogeneous methods, which require no precipitation step and allow high throughput on automated analyzers. Different method principles exist, but one is to modify the enzymes to alter the reaction kinetics and so make them more selective for the cholesterol in HDL compared to LDL and VLDL.

Triglycerides are measured by their glycerol content, using glycerol kinase, after using lipase to hydrolyze the fatty acids (**Figure 18.20**).

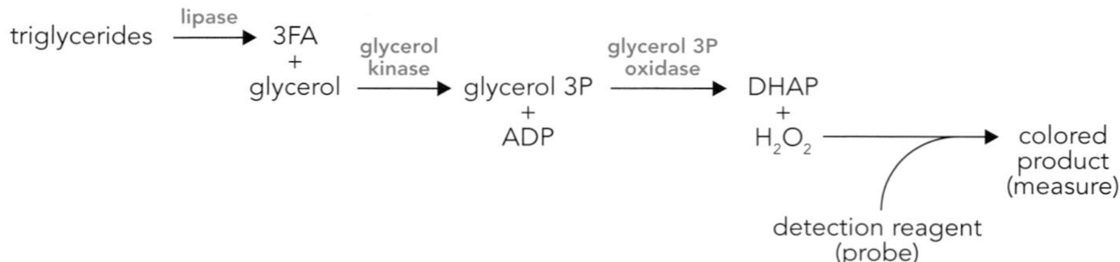

Figure 18.20 Assay of triglycerides in plasma. Triglycerides are measured by a series of enzymatic steps which results in the production of a colored product, the concentration of which is proportional to the glycerol present. Since there is one molecule of glycerol per molecule of triglyceride, this is a measure of the number of triglyceride molecules present. Free glycerol in the sample will give falsely high results, but this is rarely important clinically and can be overcome by running a blank without lipase and subtracting the result. The enzymes shown are added to the plasma or serum sample as a single reagent. FA, fatty acid; glycerol 3P, glycerol 3-phosphate; DHAP, dihydroxyacetone phosphate.

The Friedewald equation, published in 1972, shows the empirical relationship between the three lipoprotein fractions and total cholesterol:

LDL-C = TC − (HDL-C + TG/2.2) (all in mmol/L)

LDL-C = TC − (HDL-C + TG/5) (all in mg/dL)

where LDL-C is low-density lipoprotein cholesterol, TC is total cholesterol, HDL-C is high-density lipoprotein cholesterol, and TG is triglycerides.

The equation holds true for plasma triglyceride concentrations within the normal range and when CMs are absent, hence samples must be taken in the fasting state. Measurement of the total cholesterol and HDL-cholesterol and total triglyceride concentrations allows computation of the LDL cholesterol. This can be measured directly, and there are commercially available methods, but the extra information gained is not usually considered to be worth the cost.

18.5 CLINICAL DISORDERS OF LIPIDS AND LIPOPROTEINS

There has been an explosion in the number of laboratory requests for plasma lipids in the last 20 years. This has been driven by guidelines from expert societies which have translated the outcomes of large trials of lipid-lowering drugs into everyday practice. These guidelines have encompassed ever-larger groups within the population and mandated ever-lower targets for plasma lipids. The aim of treatment is to reduce the number of cardiovascular events, particularly myocardial infarction. The evidence that this is worthwhile has been overwhelmingly provided by a large number of clinical trials carried out since the mid-1990s.

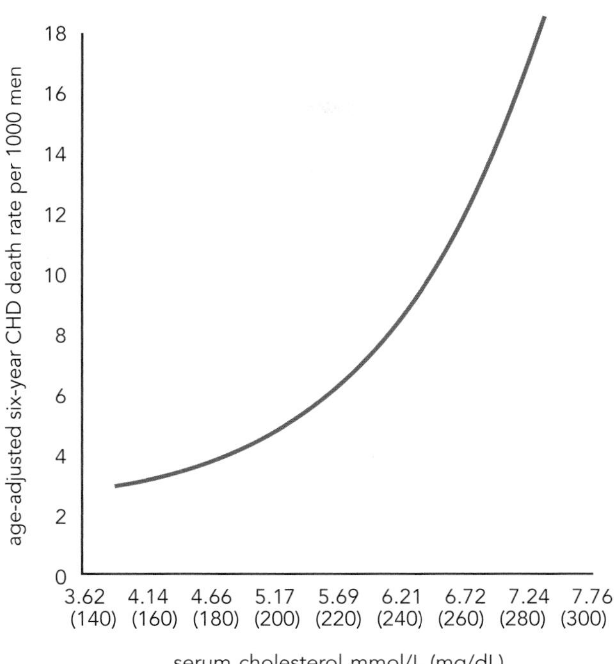

Figure 18.21 Relationship between serum cholesterol and coronary heart disease.
Numerous epidemiological studies have shown a strong correlation between total serum cholesterol and mortality and morbidity from cardiovascular disease, particularly coronary heart disease (CHD). The relationship is nonlinear and elevated cholesterol gives progressively higher risk. Serum cholesterol is predominantly determined by LDL, and risk is better defined by the use of the total cholesterol– HDL cholesterol ratio as this allows for the antiatherogenic effect of high HDL in some individuals. (Data from Martin MJ, Hulley SB, Browner WS et al (1986) *The Lancet*, 328;933–936.)

Plasma lipids and cardiovascular disease

It has been recognized for many decades that the coronary arteries of people dying from myocardial infarction contain fatty material called atheroma (after the Latin for porridge) with a large component of cholesterol. At the individual level, inborn errors of metabolism which cause elevated cholesterol, particularly familial hypercholesterolemia, are associated with ischemic heart disease (IHD) occurring unusually early in life. Studies during the 1970s also showed that populations with higher average plasma cholesterol concentrations had a higher prevalence of IHD than those with lower plasma cholesterol (**Figure 18.21**). Such an association, however, does not prove causation and clinical trials of drugs to lower cholesterol and assess the effect on IHD were required. Initially these were disappointing, due to the lack of truly effective drugs. However, this changed in the 1990s with the advent of inhibitors of the rate-limiting step of cholesterol synthesis, HMG-CoA reductase, and these drugs have become known as the statins. These drugs cause a reduction in plasma LDL-cholesterol of around 30–50%, and their clinical benefits have been demonstrated in long-term clinical trials of up to five years. As statins have become cheaper, they have been widely used to the point where they are the most commonly prescribed drugs in the developed world.

Although the relationship between heart disease and elevated plasma total cholesterol (as a surrogate for LDL particle number) is well established and the benefits of treatment are now undisputed, the situation is less clear cut for triglycerides and HDL-cholesterol. It has been shown by many population studies that high concentrations of HDL are associated with lower rates of IHD; in other words, that HDL is cardioprotective. The exact mechanism by which this effect is mediated is unknown (reverse cholesterol transport and antioxidant effects are both plausible), and trials to increase HDL pharmacologically have given mixed results. The issue of triglycerides as an independent risk factor for IHD is even more controversial. Again, a correlation between raised triglycerides and IHD (this time positive) seen in the population has not translated into reduction in IHD on drug therapy to reduce plasma triglycerides. Very high levels of triglycerides are, however, known to be a cause of acute pancreatitis in a significant minority of patients.

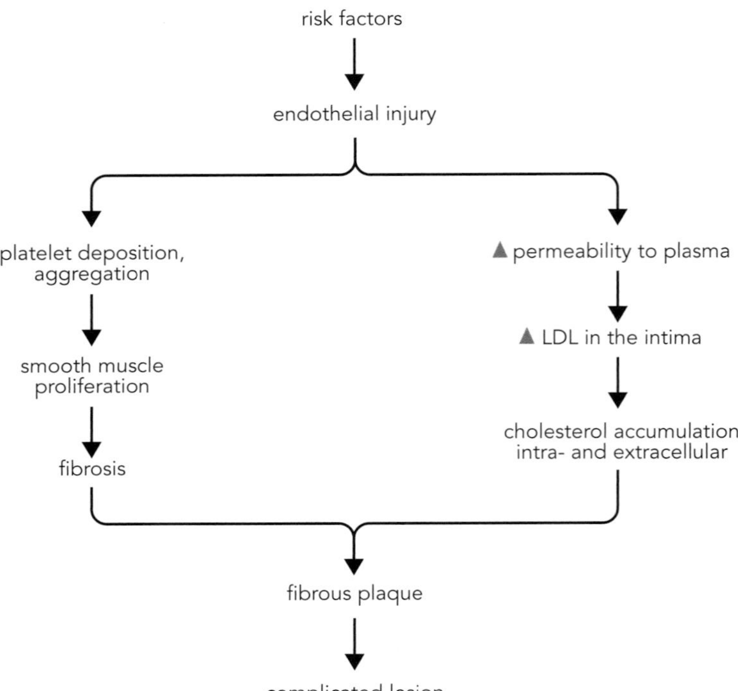

Figure 18.22 Pathogenesis of atherosclerosis.
Atherosclerosis is the process by which atheromatous (porridge-like) plaques form within the walls of medium and large arteries, particularly the coronary and carotid circulations. A complex sequence of events begins with endothelial injury, which is accelerated by the presence of classical risk factors such as smoking, hypertension, and hyperlipidemia. Platelet aggregation releases growth factors which stimulate smooth muscle proliferation and fibrous tissue formation. At the same time, LDL is able to cross the endothelium and enter the intimal layer. The plaques which are eventually formed may be predominantly fibrotic or have a large proportion of lipid material. The latter are particularly likely to rupture, initiating a thrombosis and occlusion of the artery. This will in turn result in ischemia and infarction of the myocardium or brain and cause a heart attack or stroke.

Pathogenesis of atherosclerosis

Atheroma is material which accumulates below the endothelium in medium and large arteries including the coronary, carotid, and lower-limb arteries. An outline of the pathogenesis of atherosclerosis is shown in **Figure 18.22**. When atheroma occludes the blood vessel it precipitates the clinical syndromes of IHD, ischemic stroke, and peripheral vascular disease. A major component of atheroma is cholesterol derived from plasma LDL. These particles can pass through the endothelium where they become oxidized and attracted to macrophages, which endocytose the modified LDL particle, not by the classical LDL/apoB100 receptor, but by another (scavenger) receptor which is not down-regulated by the cholesterol content of the macrophage (**Figure 18.23**). This unregulated accumulation of cholesterol converts the macrophage into a foam cell. Over time the atherosclerotic plaques become bigger and begin to occlude the lumen of the arteries. Ultimately, some plaques may rupture and release their contents into the blood, thereby triggering local platelet aggregation and thrombosis formation. This occludes the lumen entirely causing acute downstream ischemia, which if not reversed within a few hours, results in tissue necrosis. If this happens in the coronary artery it results in a myocardial infarction.

Cholesterol, or more specifically LDL particles, appears to be both necessary and sufficient (if present in high enough concentration) to cause atheroma, but the process is accelerated by the presence of mechanical and chemical damage caused by hypertension, hyperglycemia, and tobacco smoke. In the majority of the population, atheroma is a multifactorial disease and prevention and treatment is directed at all of the risk factors. Cholesterol lowering is particularly effective, especially when mediated by a statin.

Laboratory assessment of plasma lipids

By definition, lipids are insoluble in water but this is easily forgotten when referring to plasma cholesterol and triglycerides. Plasma lipids are rendered miscible with water by being components of lipoproteins and what matters clinically is the relative number and composition of the different lipoprotein particles in

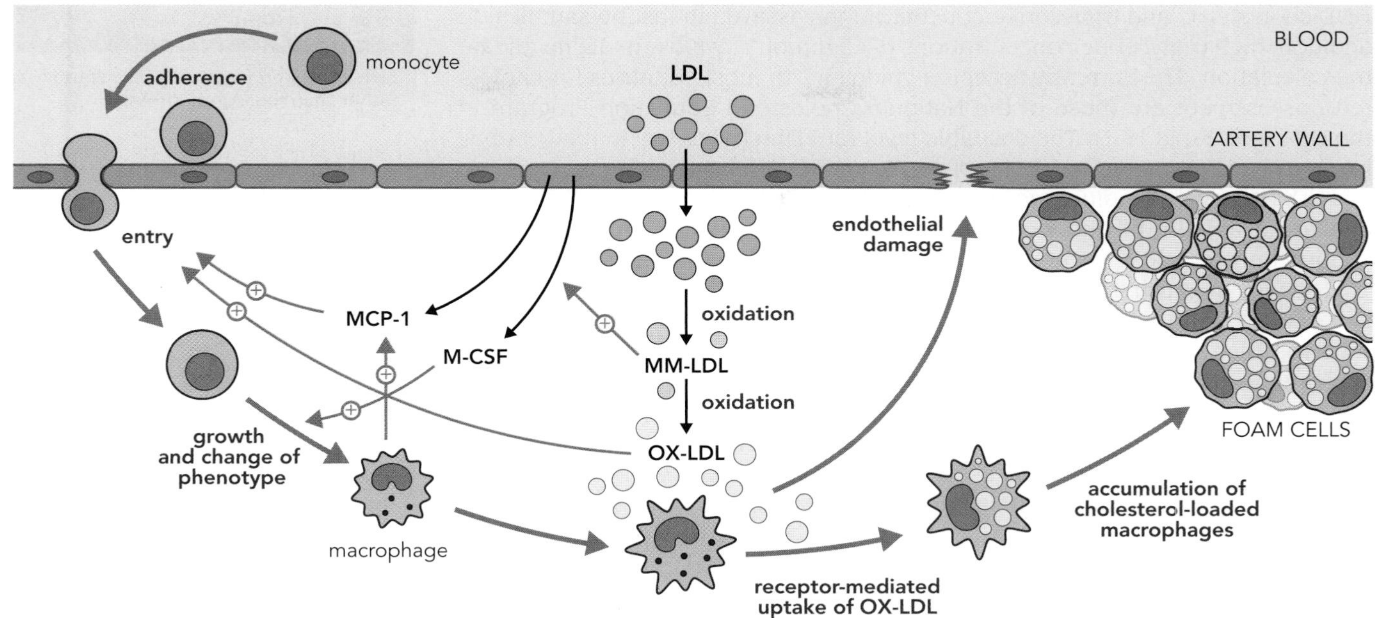

Figure 18.23 A schematic outline of the oxidative modification hypothesis.
The figure indicates the several ways in which oxidized LDL (OX-LDL) may be potentially more atherogenic than native LDL. Monocytes, the major precursor (via tissue macrophages) of foam cells in the fatty streak, are shown adhering to the endothelium and then penetrating to the subendothelial space (top left). Oxidized LDL can stimulate this process directly by virtue of its lysolecithin content, and lightly oxidized LDL (minimally modified LDL; MM-LDL) can stimulate indirectly by increasing the release of monocyte chemotactic factor/protein (MCP-1) from endothelial cells. Oxidized LDL is a ligand for the scavenger receptor that is expressed as the monocyte differentiates to a tissue macrophage and this leads to the accumulation of lipids in the developing foam cells. This monocyte–macrophage differentiation can be facilitated by the release of macrophage colony-stimulating factor (M-CSF) from endothelial cells under the influence of MM-LDL. Finally, OX- LDL can induce endothelial damage and thus facilitate the atherogenic process by allowing both the entry of elements from the blood and the adherence of platelets. Additional properties of OX- LDL, not shown in the figure, which may make it more atherogenic are (i) it is immunogenic and (ii) it interferes with the response of arteries to endothelial-derived relaxation factor.

plasma. However, lipoproteins are rarely measured directly as physical entities and routine laboratory assessment consists of measuring the lipid components as indirect indices. This is due to two factors: history and economics. The concentrations of cholesterol and triglycerides in plasma have been measured for decades and clinical studies have correlated these with cardiovascular events. Thus such measurements have become embedded in the evidence. If research were to start from scratch today, it is likely that direct lipoprotein measurements rather than measurements of individual lipids would be developed. However, such measurements are much more expensive than those of lipids and this precludes their use for the huge number of requests which clinical laboratories have to process daily.

In the laboratory, a standard plasma lipid profile consists of total cholesterol, triglycerides, and HDL-cholesterol. From these data, LDL-cholesterol can be calculated using the Friedewald equation. This calculation assumes that all of the triglyceride-containing particles are VLDL rather than CMs. For this reason, LDL-cholesterol concentration is valid only when calculated using total cholesterol,

Analytical practice point 18.1

Plasma lipid measurements are a surrogate for lipoproteins, which are clinically more important.

HDL-cholesterol, and triglyceride concentrations assayed in fasting samples. In addition, high triglyceride concentrations (>4.5 mmol/L, >400 mg/dL) invalidate the calculation. The currently accepted guidelines in most countries for cholesterol assessment are those of the National Cholesterol Education Program in the USA (see Table 18.1). The desirable level for LDL-cholesterol in the USA has been revised downward to 1.8 mmol/L (70 mg/dL) for those with increased risk of CHD. However, the difference between what is desirable and what is reasonably achievable in the general population as a whole will vary from country to country and depend upon how aggressively lipid-lowering therapy is adopted.

Analytical practice point 18.2

LDL-cholesterol is usually calculated rather than measured directly.

The way in which individual patients with raised total cholesterol are treated will depend upon whether the patient already has IHD or another atherosclerotic vascular condition, such as ischemic stroke or peripheral vascular disease (thus requiring secondary prevention), or is free from disease (primary prevention). Anyone in the secondary prevention category is automatically offered treatment (usually with a statin) and lipid measurements are used to determine response. In recent years, diabetes (particularly type 2) has been considered to be a secondary prevention equivalent, hence treatment is offered to nearly all adult diabetics.

Treatment for the primary prevention of IHD is offered on the basis of 10-year prospective risk and takes into account age, gender, smoking status, lipid levels, and blood pressure as independent risk factors. Additional risk factors are social deprivation, left ventricular hypertrophy, and the presence of inflammatory disease, such as rheumatoid arthritis. The parameters are assigned probability values based on epidemiological data and inserted into one of a number of different calculators, such as Framingham (USA), ASSIGN, and QRISK (both UK) to determine the 10-year risk of CHD. The lipid parameter that is usually used is the total cholesterol–HDL-cholesterol ratio, as this takes into account the protective effect of HDL.

The 10-year risk used to offer treatment has been progressively lowered as results from clinical trials have accumulated and drugs have become cheaper. It currently stands at 7.5–20%, depending on country. By far the most powerful risk factor is age and some authorities argue that the most efficient population strategy would be to treat all people above a certain age, for example 55 years, rather than individually calculating risk.

Some laboratories may measure apolipoproteins A1 and B, but this is usually in a research setting or to investigate unusual lipid profiles. There is increasing interest in using the apoB–LDL cholesterol ratio to detect individuals with small, dense LDL, which is more easily oxidized and so more atherogenic.

Analytical practice point 18.3

Apolipoprotein measurements may become more widely used but at present perform no better than traditional lipid measurements.

Primary hyperlipidemias

The interactions between lipoproteins and their receptors are essential for lipid homeostasis. Genetically determined abnormalities which give rise to overproduction and/or impaired removal of any particular lipoprotein present clinically as primary hyperlipidemias. The clinical consequences of these are determined by the type of lipoproteins present in excess and their concentrations. Since laboratories routinely measure lipids rather than lipoproteins (other than HDL), patients have elevated cholesterol, triglycerides, or a combination of the two. Cholesterol (as a surrogate for LDL) increases cardiovascular risk and moderately raised triglycerides may do so as well. Very high triglyceride levels increase the risk of acute (but not chronic) pancreatitis. Some lipoprotein disorders cause specific clinical signs, particularly in the skin and tendons. For example, in familial hypercholesterolemia the presence of cholesterol accumulation (xanthoma) in the extensor tendons of the hands or the Achilles tendon is diagnostic of the condition. Xanthomas in the palmar creases of the hand may be seen with excess IDL and massive hypertriglyceridemia may cause eruptive xanthomas: crops of itchy raised skin lesions that may occur almost anywhere.

A frequency plot of plasma cholesterol in the population shows a bell-shaped distribution but with a long tail, indicating a small group with very high levels. The majority of people developing IHD have levels of cholesterol which are close to the population average. Because such individuals are common, it is they who are recruited into clinical trials of statins and other lipid-lowering drugs. However, it is important not to overlook the smaller number of individuals who make up the tail of the distribution. They have lipid levels that are greater than the population as a whole, although there is some overlap, due to genetic disorders of lipoprotein metabolism, and individually are at high risk even if other risk factors are absent. Two such disorders that are well characterized are familial hypercholesterolemia and type III (remnant) hyperlipoproteinemia.

Familial hypercholesterolemia

LDL is cleared via LDL (apoB100) receptors, predominantly in the liver. Defects in either apoB100 or the receptor that decrease the interaction of the ligand (apoB100) with the receptor or effectively decrease receptor number will give rise to elevated plasma LDL and hence cholesterol. Familial hypercholesterolemia (FH) is probably the best example of a single-gene lipid disorder. In this condition, the molecular basis of the increased plasma cholesterol concentration is a reduced number (theoretically 50% in heterozygotes) of functional cell surface LDL receptors, due to a dominantly inherited mutation. Since less LDL is taken up into cells there is an accumulation of LDL particles in the circulation, where they may cause atherosclerosis by the mechanisms described above. In FH, the plasma total cholesterol concentration is typically greater than 7.5 mmol/L (290 mg/dL), or LDL-cholesterol >5 mmol/L (190 mg/dL), although it may in some cases be in the range of 10–15 mmol/L (390–580 mg/dL) or LDL 8–12 mmol/L (310–460 mg/dL).

Clinical practice point 18.1

Familial hypercholesterolemia is one of the commonest inborn errors of metabolism and causes premature cardiovascular disease.

The carrier frequency in FH is approximately 1 in 500, making it one of the most common inborn errors of metabolism. Genetic diagnosis is being used increasingly but is complicated by the fact that although hundreds of different mutations have been identified, up to one-third of patients presenting with definite clinical FH have no currently identifiable mutation.

The clinical consequences of FH are those of accelerated atherosclerosis with early onset of IHD in both males and females. By the age of 60 years, 50% of heterozygous males will have developed IHD. However, treatment is usually effective at lowering the LDL levels and, if started early enough, may result in IHD rates at least as low as the general population.

Statistically 1 in 1 million people will carry two defective alleles for the LDL receptor gene and be homozygous (or compound heterozygous) for FH. This results in extremely high plasma cholesterol (for example 20 mmol/L; 780 mg/dL), a tendency to develop IHD in the first two decades of life, and consequent decreased life expectancy.

Type III hyperlipoproteinemia

The term type III refers to the Fredrickson classification of lipid patterns on lipoprotein electrophoresis, an investigation popular in the 1970s but now rarely used. Fredrickson classified lipoprotein patterns from I to V (**Table 18.6**) and these broadly equate to CMs (I), LDL (IIa), IDL (type III), and VLDL (IV and V).

In type III hyperlipoproteinemia there is a defect in the interaction between the LDL receptor and the ligand apoE resulting in increased concentrations of IDL or remnant lipoproteins. There are three forms of apoE termed E2, E3, and E4. Those with type III are homozygous for E2. This genotype makes up 1% of the population, but only around 1% of these develop type III hyperlipoproteinemia, giving a phenotype frequency of 1 in 10,000. It appears that a triggering environmental factor is also required, such as alcohol excess, diabetes, hypothyroidism, or obesity.

TABLE 18.6 World Health Organization/Fredrickson classification of dyslipidemias

Type	Electrophoretic mobility	Lipoproteins raised	Cholesterol	Triglyceride
I	ω	Chylomicrons	–	+++
IIa	β	LDL	++	–
IIb	pre-β and β	VLDL, LDL	++	+
III	broad β	IDL	++	++
IV	pre-β	VLDL	+	++
V	pre-β	VLDL, chylomicrons	+	+++

HDL was not included in this system but has ω mobility. –, Not increased; +, relative degrees of increase.

Severe hypertriglyceridemia

An elevated plasma triglyceride concentration may arise from increased synthesis and secretion of VLDL due to hepatic overproduction of apoB100 or impaired clearance of VLDL and CMs. Very high plasma triglycerides are often a result of both mechanisms occurring simultaneously. Lipoprotein lipase becomes saturated when plasma triglycerides exceed about 5 mmol/L (440 mg/dL), so that further triglyceride loading results in dramatic increases in plasma triglyceride, sometimes as high as 30 mmol/L (2650 mg/dL) and exceptionally up to 100 mmol/L (8850 mg/dL). Severe hypertriglyceridemia is often secondary to an underlying condition, but also appears to require a poorly characterized genetic predisposition. As well as giving plasma a milky appearance (lipemia) due to scattering of light by the large lipoprotein particles, massive hypertriglyceridemia may cause acute pancreatitis. This is probably due to small leakages of pancreatic lipase into the local microcirculation, which reacts with the excess substrate, triggering free radical formation and a chain reaction of lipolysis and acute inflammation.

Clinical practice point 18.2

Severe hypertriglyceridemia may cause acute pancreatitis with a misleadingly normal serum amylase.

One diagnostic pitfall is that plasma amylase, which is usually elevated in acute pancreatitis, may be normal. This appears to be due to an inhibitor of amylase which is only seen in triglyceride-induced pancreatitis. It is therefore important to measure urine amylase or use other tests, such as imaging, before ruling out acute pancreatitis in such patients.

Disorders of HDL metabolism

Since HDL concentration is inversely related to cardiovascular risk, it is an important measurement for interpreting plasma cholesterol. Individuals with low HDL-cholesterol are at higher risk, as reflected in their high total cholesterol–HDL-cholesterol ratios. HDL-cholesterol and triglycerides are also inversely related, and the combination of low HDL-cholesterol and moderately elevated triglycerides most commonly occurs with insulin resistance (metabolic syndrome) and type 2 diabetes mellitus, usually as a consequence of obesity.

High levels of HDL (>2 mmol/L; >80 mg/dL) are seen in some individuals and rarely may be as high as 4 or even 5 mmol/L (160–200 mg/dL). Such high levels may be genetically determined or due to drugs including alcohol. If only the total cholesterol is measured in these subjects, they may be misclassified as being at high cardiovascular risk on the assumption that they have high LDL- rather than HDL-cholesterol. Since individuals with naturally occurring high HDL-cholesterol are at lower cardiovascular risk, it might be logical to assume that drugs which raise HDL-cholesterol will be beneficial. Unfortunately, this has not always been the case. Fibric acid derivatives have been shown to reduce cardiovascular events, whilst torcetrapib—a specific inhibitor of CETP—increased mortality during trials, despite raising HDL-cholesterol. It appears that whereas lowering LDL-cholesterol appears to be beneficial whichever drug is used to

CASE 18.1

A 50-year-old man registered with a new primary care physician. His cholesterol had been measured several times in the previous decade and had always been in the desirable range (<5.0 mmol/L; 190 mg/dL). He had no family history of heart disease and no symptoms himself. His thyroid function tests (TFTs) were normal. After receiving the results below, he was commenced on a statin, but showed little improvement in his cholesterol, despite claiming to take it regularly. On further questioning he reported that he had noticed some swelling of his ankles after prolonged standing and that his urine had been frothy recently. He was referred to a renal physician. He had a number of investigations, including a renal biopsy which showed membranous nephropathy.

	SI units	Reference range	Conventional units	Reference range
Serum				
Total cholesterol	10.1 mmol/L	<5.0 (desirable*)	390 mg/dL	<190 (desirable*)
Triglycerides	1.3 mmol/L	<1.8 (desirable*)	115 mg/dL	<160 (desirable*)
HDL-cholesterol	1.23 mmol/L	>1.00 (desirable*)	48 mg/dL	>40 (desirable*)
LDL-cholesterol	8.2 mmol/L	<3.0 (desirable*)	317 mg/dL	<116 (desirable*)
Albumin	20 g/L	32–52	2.0 g/dL	3.2–5.2
Urine				
Albumin–creatinine ratio	150 mg/mmol	2.5–3.5	1800 μg/mg	30–300

*Reference ranges for serum lipids are not clinically helpful. Desirable ranges are used based on cardiovascular risk and are changed over time according to clinical evidence.

- **What is the significance of high cholesterol with previously normal results?**
- **What does low serum albumin and proteinuria signify?**
- **Why did he not respond to a statin?**

Cholesterol of this level is highly suggestive of familial hypercholesterolemia. However, this causes lifelong raised cholesterol and his increase has been a recent one. Hyperlipidemia of recent onset is likely to be due to an underlying disease or drug; that is, it is secondary rather than primary. His lack of response to a statin is also typical of a secondary hyperlipidemia. Hypothyroidism is excluded by normal TFTs, but his low serum albumin, proteinuria, and edema indicate that he has nephrotic syndrome. Increased protein synthesis by the liver is stimulated by low plasma oncotic pressure, and this includes lipoproteins.

achieve it, the beneficial effects of raising HDL-cholesterol may be drug specific and highly dependent on a particular mechanism.

Secondary hyperlipidemias

A number of medical disorders and drug treatments may disturb lipid metabolism and cause changes in lipoprotein levels. These are termed secondary hyperlipidemias to distinguish them from genetic or primary hyperlipidemias. In most cases when the underlying cause is addressed the lipids return to normal levels.

One way of classifying secondary hyperlipidemias is by the major lipoprotein group affected. Thus, LDL is increased in both hypothyroidism and the nephrotic syndrome. VLDL may be elevated by corticosteroids, β-blockers, and vitamin E analogs used to treat acne. Oral estrogens also may raise VLDL in some individuals, sometimes markedly so, and may even cause acute pancreatitis. This appears to be an effect of first-pass liver metabolism, as it is rarely seen with estrogens when they are delivered as transdermal patches or implants. In some postmenopausal women the main effect of estrogen hormone replacement therapy is to raise the HDL instead. This was previously thought to benefit cardiovascular risk until large studies showed it to be largely ineffective.

Obesity and type 2 diabetes are associated with insulin resistance which tends to moderately raise triglycerides and lower HDL, giving an increased cholesterol–HDL ratio. Alcohol excess, on the other hand, raises both triglycerides and HDL. The latter mechanism may account for lower rates of IHD seen in drinkers compared to abstainers. In type 1 diabetes some individuals tend to have elevated HDL. This appears to be associated with insulin sensitivity and correlates with long-term survival. Some anticonvulsant drugs, for example phenytoin, may elevate HDL. Whether this is cardioprotective is unclear. Causes of secondary hyperlipidemias are given below:

- Hypercholesterolemia
  - Nephrotic syndrome
  - Hypothyroidism
  - Dysgammaglobulinemia
  - Obstructive liver disease
  - Drugs: phenytoin (predominantly HDL)
- Hypertriglyceridemia
  - Diabetes mellitus
  - Uremia
  - Sepsis
  - Obesity
  - Systemic lupus erythematosus
  - Dysgammaglobulinemia
  - Glycogen storage disease type I
  - Lipodystrophy
  - Drugs: estrogens (oral), alcohol, β-blockers

Hypolipidemia

Abetalipoproteinemia

Low plasma total cholesterol (hypocholesterolemia) may result from drug therapy (especially the widespread use of statins), result from malnutrition, or be associated with severe systemic illness such as cancer or following myocardial infarction. A rare cause of a low plasma cholesterol concentration, abetalipoproteinemia, is due to mutations in the gene encoding MTP in both intestine and liver. Although not a component of either CMs or VLDL, its activity is essential in the intracellular assembly of both lipoproteins (see Figure 18.12). Defective MTP activity, arising from a mutant protein, results in no synthesis and hence no export of either lipoprotein into the circulation. Since LDL is a product of VLDL metabolism, this lipoprotein is also absent from plasma and the only cholesterol in the circulation is that associated with HDL. The main clinical consequences are fat malabsorption from the gut resulting in steatorrhea, malnutrition, and failure to thrive. As described earlier, fat-soluble vitamins absorbed from the intestine circulate as components of CMs and are taken up by liver as part of CM remnants. In the absence of CM synthesis, fat-soluble vitamins do not enter the circulation and patients become deficient in vitamins A, D, E, and K. The most important effects on development appear to be due to lack of vitamin E, which results in predominantly neurological symptoms such as ataxia. Vitamin A deficiency causes visual loss, particularly night blindness, and is often the first clinical indication of the condition. The characteristics of abetalipoproteinemia are as follows.

- Rare autosomal disease caused by absence of MTP activity in intestine and liver
- Chronic diarrhea secondary to fat malabsorption
- Fat-filled enterocytes and parenchymal hepatocytes
- Fat-soluble vitamin deficiency
- Acanthocytosis (abnormal erythrocytes)
- Night blindness, loss of color vision
- Ataxia, absence of deep tendon reflex
- Atypical spinal-cerebellar degeneration
- Very low plasma triglyceride due to absence of CMs and VLDL

Hypobetalipoproteinemia

The classical recessive form of abetalipoproteinemia should be distinguished from the co-dominant condition of hypobetalipoproteinemia which presents with similar clinical features. This latter condition arises from mutations in the apoB gene which lead to the production of truncated forms of the apoB protein in intestine and liver. These truncated forms are not of sufficient size to allow the formation of CMs or VLDL. Patients who are homozygous for abetalipoproteinemia or hypobetalipoproteinemia have virtually identical circulating lipid levels. However, heterozygous parents of abetalipoproteinemic patients have lipid levels within the normal range. This contrasts with heterozygous parents of hypobetalipoproteinemic patients, who have an approximately 50% reduction in the plasma concentration of apoB-containing lipoproteins.

Clinical approach to the abnormal lipid profile

Lipids are measured on two groups of patients: those with lipids that are around the average for the population, and a much smaller number of individuals with specific lipid disorders. Both groups may be further divided into primary and secondary prevention on the basis of their clinical history. Lipids are measured to determine risk and to monitor response to and compliance with drug treatment. A classification of primary hyperlipidemias, approximating to the World Health Organization/Fredrickson phenotypes, is given in **Table 18.7**. However, it is rarely used in modern clinical practice. Further investigation and treatment will depend instead on the lipid pattern seen. It is therefore important to assess all the components of the routine profile and to assign individuals to one of the categories described below.

Clinical practice point 18.3

Severe hyperlipidemias may be genetic, in which case family screening may be required, or secondary to another condition.

Clinical practice point 18.4

Plasma lipids, particularly cholesterol, are risk factors for cardiovascular disease, but must be considered with all other risk factors before starting treatment.

Severe hypertriglyceridemia

The major risk in the short term is acute pancreatitis, with the longer-term risk of cardiovascular disease being uncertain. Although such patients invariably have an underlying genetic predisposition to high triglycerides, there is often one or more aggravating factors, such as obesity, alcohol excess, or diabetes mellitus. Addressing these often improves the plasma lipids but rarely normalizes them and drug therapy is usually required. It is important to note that HMG-CoA reductase inhibitors (statins) are rarely effective.

Severe hypercholesterolemia

The first important distinction to make in these patients is between high LDL and high HDL. The former confers a high cardiovascular risk and the latter a low

TABLE 18.7 Genetic/metabolic causes of primary hyperlipidemias

Clinical disorder	Primary disorder	Metabolic disorder	CHD risk	Pancreatitis
Chylomicronemia syndrome (Type I)	Deficiency of lipoprotein lipase or apoCII activity	Impaired clearance of chylomicrons	–	++
Common hypercholesterolemia (Type IIa)	Multiple genetic/environmental factors	LDL overproduction and increased LDL catabolism	+	–
Familial hypercholesterolemia (Type IIa)	Genetic mutation(s) in LDL receptor or apoB	LDL overproduction and impaired LDL catabolism	++++	–
Familial combined hyperlipidemia (Type IIb)	Unknown	Overproduction of VLDL apoB100 and/or LDL apoB100	++	–
Remnant hyperlipidemia (Type III)	Coexistence of nonfunctional apoE isoforms with genetic or acquired disorder of VLDL/LDL metabolism	Impaired conversion of IDL to LDL	+++	?
Familial hypertriglyceridemia, sporadic hypertriglyceridemia, familial combined hyperlipidemia Diabetes (Types IV and V)	Multiple genetic/environmental factors	Increased production of VLDL and/or decreased clearance of VLDL and CM	+	++

The Fredrickson classification is in parentheses.

one. Raised HDL is often found for the first time in the elderly and is a tribute to its protective effect. It does not require drug treatment. It should prompt the consideration of alcohol excess, however, particularly if accompanied by raised triglycerides, gamma-glutamyl transferase (GGT), and mean cell volume (MCV).

Very high LDL is most likely to be due to familial hypercholesterolemia. There are two priorities here: treatment of the index case and identification of affected relatives.

Occasionally, very high LDL may be due to an underlying illness such as hypothyroidism or nephrotic syndrome, and these should be excluded before making the diagnosis of FH.

Raised triglycerides/low HDL

This is the pattern of lipids classically associated with insulin resistance and may be seen in association with obesity and type 2 diabetes. It is increasingly apparent that other disorders and drugs may induce similar changes, for example inflammatory conditions such as rheumatoid arthritis, human immunodeficiency virus (HIV) infection and its treatment, and use of newer (atypical) antipsychotic drugs. Such a pattern of dyslipidemia is considered to present a high risk of cardiovascular disease, even though cholesterol and LDL are not elevated. This is because the LDL that is present is usually small and dense and more prone to oxidation, particularly as the antioxidant effect of HDL is reduced by its low concentration.

CASE 18.2

A 45-year-old man presents with severe abdominal pain worsening over the previous 48 hours and associated with nausea and vomiting. The pain is constant and radiates to his back. He has had several similar but less severe episodes over the previous two years, not requiring hospital admission. On examination he is clinically dehydrated and has a tender, rigid abdomen. The admitting doctor suspects acute pancreatitis. The laboratory reports that the blood sample is lipemic and goes on to measure his lipids. After discussion with the duty biochemist, the doctor requests urine amylase. A computed tomography scan confirms the diagnosis of acute pancreatitis.

	SI units	Reference range	Conventional units	Reference range
Serum				
Total cholesterol	28 mmol/L	<5.0 (desirable*)	1080 mg/dL	<190 (desirable*)
Triglycerides	52 mmol/L	<1.8 (desirable*)	4600 mg/dL	<160 (desirable*)
HDL-cholesterol	1.3 mmol/L	>1.00 (desirable*)	50 mg/dL	>40 (desirable*)
LDL-cholesterol	Not calculated	<3.0 (desirable*)	Not calculated	<116 (desirable*)
Amylase	78 U/L	27–131	78 U/L	27–131
Urine				
Amylase	2000 U/L	No range	2000 U/L	No range

*Reference ranges for serum lipids are not clinically helpful. Desirable ranges are used based on cardiovascular risk and are changed over time according to clinical evidence.

- **Why does the doctor suspect acute pancreatitis?**
- **What is the significance of the serum and urine amylase results?**

The patient has acute abdominal pain with symptoms and signs of acute pancreatitis. One of the less common but well-recognized causes of this is severe hypertriglyceridemia, especially if this is caused by alcohol excess. The first indication that triglycerides are very high is the lipemic appearance of centrifuged serum, due to the presence of large lipoprotein particles (CMs and VLDL) which scatter light. Serum amylase is usually very elevated in acute pancreatitis, but when this is caused by high triglycerides it may be spuriously low, due to interference by an unidentified inhibitor substance. Amylase is a small protein and is cleared by the kidney into the urine where it can be measured. Although there are not well-established reference ranges for urine, levels much higher than serum can be used to indicate pancreatitis as the inhibitor substance appears not to be renally excreted.

DISORDERS OF IMMUNOGLOBULINS AND COMPLEMENT

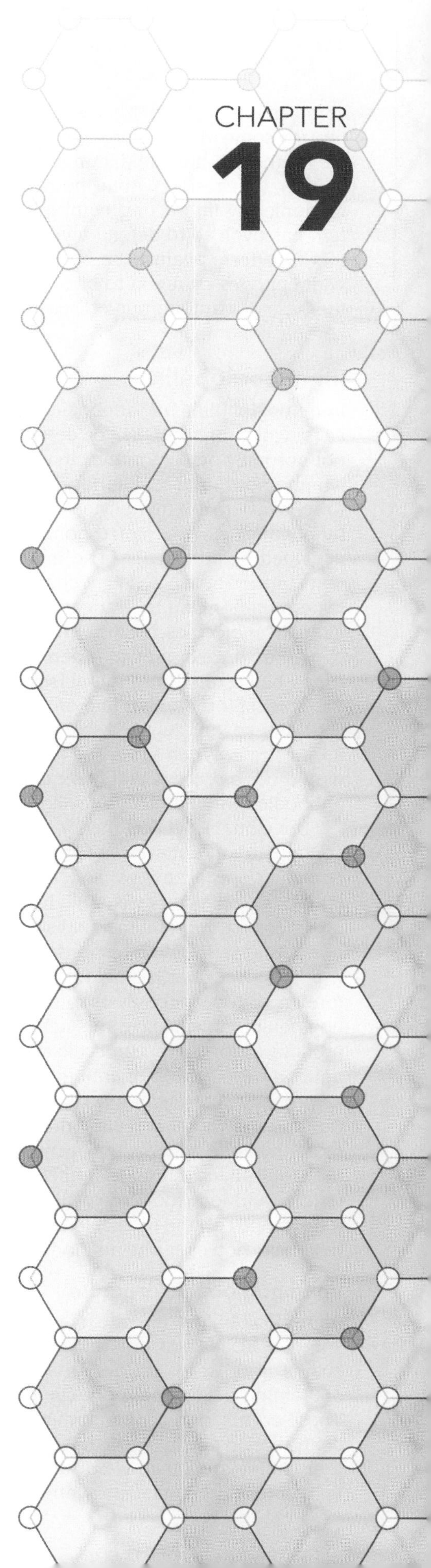

IN THIS CHAPTER

THE IMMUNE SYSTEM

CLINICAL DISORDERS OF IMMUNOGLOBULINS

COMPLEMENT

The human immune system is extremely complex and immunology is a subject in its own right. This chapter does not attempt to be a comprehensive description of the immune system, but concentrates on two particular aspects which are of importance to clinical chemists, namely immunoglobulins and complement. These are proteins that are routinely measured in most clinical chemistry departments to aid in the diagnosis and monitoring of diseases which are usually managed by hematologists and immunologists. In order to understand the role of these measurements in the management of such disorders, some basic understanding of immunology is required, as set out in the first part of the chapter.

19.1 THE IMMUNE SYSTEM

The body has evolved several layers of defense against pathogenic organisms. The first line is the physical barriers of skin and mucous membranes, aided by structures such as cilia in the respiratory tract and acid in the stomach. These form the first part of the innate immune system, components of which are present in all multicellular organisms, which comprises a number of nonspecific responses to infection. In other words, the responses are the same, irrespective of the type of organism. Innate immunity consists of the inflammatory response, the complement cascade, and a number of white blood cell (leukocyte) types that can attack or phagocytose microorganisms. These consist of macrophages, neutrophils, eosinophils, basophils, and natural killer cells. In addition to the innate immune system, vertebrates have evolved a sophisticated system known as adaptive immunity. This has two properties: specificity to antigens found on particular microorganisms; and memory—that is, being able to mount a rapid response to a previously encountered organism. The cells that are responsible for the adaptive immune system are the lymphocytes, which comprise 20–40% of white blood cells. Almost all lymphocytes are within tissues or the lymphatic system, with only around 2% being in the peripheral blood.

The lymphocytes are subdivided into B and T cells. B cells are responsible for humoral immunity, which comprises secretion of antibody molecules (immunoglobulins). These act as markers for phagocytic cells and also activate the complement cascade. This is a series of proteolytic enzymes which are able to destroy some invading organisms, mainly bacteria. T lymphocytes are responsible for the cell-mediated immunity required to deal with organisms, particularly viruses, that reside within the host's cells. T cells also play a role in humoral immunity by interacting with and helping B cells. Although the immune system has evolved to defend against foreign organisms, it may sometime misdirect its effects against the body's own cells, either because of cross-reactivity with epitopes common to both, or because it does not recognize itself. This is the basis for autoimmune disease, which comprises a large number of mainly chronic conditions.

Immunoglobulins

Immunoglobulins are antibodies that are produced in and secreted by plasma cells, which are themselves derived from the B lymphocytes. Plasma cells are not normally present in the circulation, being present in tissues such as spleen, lymph nodes, mucosal surfaces of small intestine, and bone marrow. They are formed when B lymphocytes, synthesized in bone marrow, become activated by contact with the corresponding antigen and factors released by similarly activated helper T cells. The immunoglobulins are proteins made up of subunits termed heavy and light chains. The genes encoding these undergo somatic hypermutation and rearrangement, which results in a vast number of immunoglobulin molecules, each with a slightly different conformation that can recognize a specific antigen. Each B cell produces only one such antibody type. It can be secreted into the plasma or bound to the cell membrane to act as a B-cell receptor. The binding of antibody to antigen is entirely analogous to the lock-and-key model used to explain how the active site of an enzyme binds to its substrate. When the B-cell receptor meets its corresponding antigen, the cell undergoes a process of differentiation that results in a plasma cell that secretes antibodies with identical specificity into the extracellular fluid.

The clonal selection theory, first proposed in the 1950s, states that the body contains a large number of pre-programmed B lymphocytes each able to recognize its specific antigenic determinant (epitope) and, after mitosis and differentiation to plasma cells, able to produce specific (idiotypic) antibody to it. An epitope is a three-dimensional structure on part of a molecule which is capable of eliciting an immune response; any molecule may thus have several epitopes. It is proposed that a single bone marrow-derived B lymphocyte recognizes a foreign epitope through its interaction with a small number of surface immunoglobulin molecules in the cell membrane; this activates the B lymphocyte to produce a clone of plasma cells which secrete large amounts of antigen-specific antibody. Production of antibody to a single epitope by a single plasma cell and its progeny is termed monoclonal. Since foreign bodies such as viruses, bacteria, and individual proteins, nucleic acids, and carbohydrates usually possess more than one epitope, they activate more than one specific B lymphocyte, with the result that a number of different antibodies to the same parent virus, bacterium, and individual molecule are produced. In theory, antibodies to each of the epitopes on the molecule could be produced, yielding a polyclonal antibody response; however, the antibody response to each epitope may differ markedly.

Immunoglobulin structure

Immunoglobulins (Ig) are macromolecular glycoproteins of molecular mass 146–970 kDa. Each immunoglobulin molecule consists of two identical light chains (κ or λ) of 212 amino acids and two identical heavy chains of 450 or 550 amino acids. There are five types of heavy chain termed γ, α, μ, δ, and ε. These correspond to the immunoglobulin classes (in decreasing order of concentration in plasma): IgG, IgA, IgM, IgD, and IgE, respectively (**Table 19.1**). The four chains are linked via disulfide (-S-S-) bridges (**Figure 19.1**). Although the light and heavy chains are synthesized independently, their syntheses in healthy

TABLE 19.1 Heavy-chain types of individual immunoglobulin classes in decreasing order of plasma concentration

Immunoglobulin class	Heavy chain
IgG	γ (gamma)
IgA	α (alpha)
IgM	μ (mu)
IgD	δ (delta)
IgE	ε (epsilon)

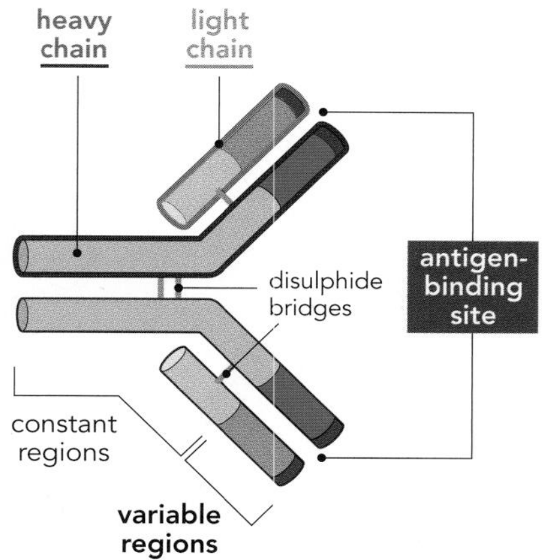

Figure 19.1 Structure of immunoglobulin G. The basic immunoglobulin structure consists of two identical heavy chains (in the case of IgG, these are γ) and two identical light chains (which may be κ or λ), held together by disulfide bridges. A hinge region on the heavy chains allows the molecule to form a Y-shaped configuration. The antigen-binding site is at the N-terminal end and consists of the variable regions of the light (V_L) and heavy chains (V_H), thereby conferring different specificity to different idiotypes. V_L and H_L each contain 110 amino acids. The constant region of the light chain (C_L) also has 110 amino acids, compared to 330 to 340 for the constant region of the heavy chain (C_H). The constant regions of the two heavy chains forming the tail of the Y, and held together by a disulfide bridge, are called F_C. This has an important role in activation of complement. An important function of the constant region is the activation of complement, via the classical pathway, when antigen is bound. There are subclasses of IgG (numbered 1 to 4) which vary in their constant regions.

individuals are tightly coordinated and controlled to prevent overproduction. Each heavy and light chain has a constant and a variable amino acid sequence. The latter gives rise to the antigenic specificity and is at the N-terminal end of the molecule. The differences are in three hypervariable areas of 6–10 amino acids, and in the mature immunoglobulin the three hypervariable areas from each polypeptide (two heavy and two light chains) form the specific antigen-recognition site. The rest of the chains are the constant regions: one in each of the light chains and three in each of the heavy chains. The light chains are identical and may be either κ (κ_2, 60%) or λ (λ_2, 40%) but not κλ. The constant regions of the heavy chains at the C-terminal end are able to bind to receptors on phagocytic cells and to activate complement, thus enhancing the function of the innate immune system. Binding of antigen to antibody causes a change in the three-dimensional shape of the immunoglobulin allowing it to participate in further reactions.

Immunoglobulin classes

The immunoglobulin classes or isotypes have some structural differences, which give rise to specific properties making them better suited to certain functions (**Table 19.2**). They are composed of different numbers of the individual immunoglobulin units. After the B cell is activated, it can switch between different isotypes by altering the constant regions of the heavy chains. Thus the specificity, or idiotype, remains the same but the isotype can change within the daughter cells of the original plasma cell. B cells express IgM and IgD on their surfaces and IgM is the antibody initially secreted in response to antigenic stimulation. Later, IgG, IgA, and IgE can be secreted from the plasma-cell progeny of the original cell.

TABLE 19.2 Some functions of immunoglobulins

Immunoglobulin	Function
IgM and IgG	Prevention and elimination of bacterial infections • IgM: primary response • IgG: secondary response
IgA	Prevention and elimination of bacterial infections at mucosal surfaces, for example gut and lung
IgE	Protection against helminth (worm) infections; mediates immediate (type I) hypersensitivity

IgM

IgM is a large molecule consisting of five basic immunoglobulin units joined together at their constant (C-terminal) ends by a J (joining) chain. Because of its size (800 kDa) it penetrates tissues poorly. This pentameric structure is shown diagrammatically in **Figure 19.2**. Its major role is in intravascular neutralization of organisms, particularly viruses. With five complement binding sites it is particularly suited to complement activation and participates in removal of antibody–antigen complexes, via receptor-mediated endocytosis, via the complement receptor on phagocytic cells.

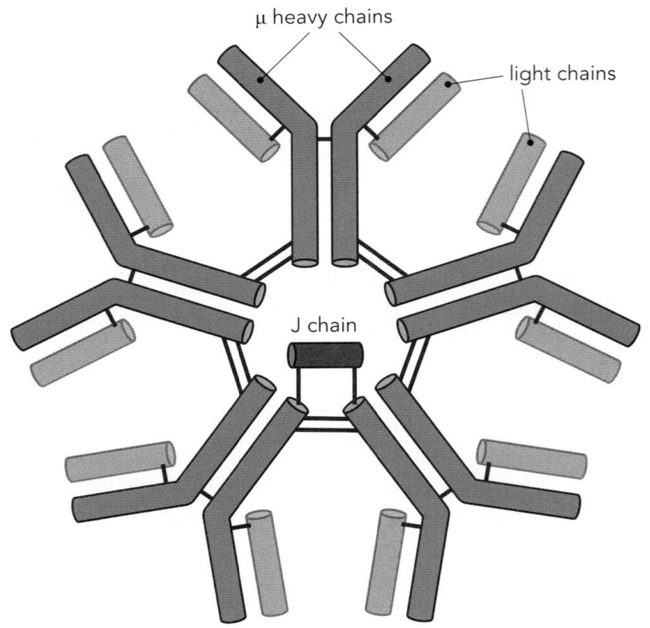

Figure 19.2 Structure of IgM. IgM is usually a pentamer, consisting of five basic immunoglobulin units, but can also exist as a monomer. The individual units differ from IgG in their heavy chains, which are μ rather than γ, although the light chains are the same (κ or λ). In pentameric form, the IgM heavy and light chains are identical to each other and so have the same antigen-binding specificity. The individual units are held together by disulfide bonds (shown in black) and by an additional small protein called the J (joining) chain. Only one is depicted here for clarity. The pentameric structure enables IgM to activate complement more effectively and to agglutinate microorganisms, as several can be attached to the same molecule. A form of monomeric IgM is attached to the surface of B cells where it acts as the antigen receptor. When it meets its complementary antigen, the B cell is stimulated to differentiate into a plasma cell.

IgG

IgG is a monomeric immunoglobulin species (150 kDa; see Figure 19.1). It is the only immunoglobulin which crosses the placenta and can thereby provide the fetus and neonate with immune protection via maternal antibodies. There are four subclasses of IgG (IgG_1–IgG_4): IgG_1 and IgG_3 activate complement very efficiently and are responsible for clearing most protein antigens and microorganisms via receptor-activated endocytosis by phagocytic cells; IgG_2 and IgG_4 interact mainly with carbohydrate antigens.

IgA

IgA is present in gut and respiratory tract secretions as a dimer (385 kDa) in which two basic immunoglobulin molecules are linked by a joining (J) chain plus a second protein (secretory component) which provides protection from proteolysis in intestinal and bronchial secretions (**Figure 19.3**). Maternal IgA may be present in milk. IgA_2 is the predominant subclass in mucosal secretions and it interacts with antigens entering the body via mucous membranes. The concentration of IgA (mainly IgA_1) in serum is very low (**Table 19.3**) and is mainly in the monomeric form. IgA_1 is susceptible to hydrolysis by bacterial proteases.

IgD

IgD is a monomeric immunoglobulin (170 kDa) synthesized by antigen-sensitive B lymphocytes and is present in the cell membrane. Here it acts as an antigen receptor involved in cell activation. Only trace amounts are present in serum.

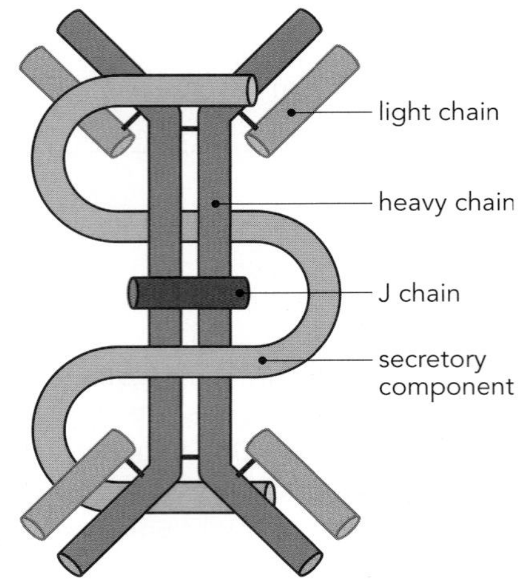

Figure 19.3 Structure of IgA. IgA exists as both a monomer and a dimer; it has unique heavy chains called α and the same κ and λ light chains as the other immunoglobulin classes. The monomeric form is found in the circulation, whereas the dimer is found in body secretions, for example in the gut and the respiratory tract. As with IgM, there is a J (joining) protein that links the monomers together. Secretory IgA also has a second protein attached to it in the epithelial cell called the secretory component. This prevents the immunoglobulin's degradation within secreted fluids such as mucus and saliva.

TABLE 19.3 Immunoglobulin reference ranges in serum

Age	IgG (g/L) (mg/dL)	IgA (g/L) (mg/dL)	IgM (g/L) (mg/dL)	IgD (kU/L)	IgE (kU/L)
Newborn	5.2–18.0 (520–1800)	0–0.02 (0–2)	0.02–0.20 (2–20)		0–10
3 months	2.1–7.7 (210–770)	0.05–0.40 (5–40)	0.15–0.70 (15–70)		0–25
5 years	4.9–16.1 (490–1610)	0.4–2.0 (40–200)	0.5–2.0 (50–200)		0–100
Adult	6.5–16.0 (650–1600)	0.8–4.0 (80–400)	0.5–2.0 (50 – 200)	2–100	0–200

IgE

IgE is also a monomeric immunoglobulin (190 kDa). It is produced by plasma cells and taken up by mast cells and basophils. Only trace amounts are present in serum. However, it has a key role in some allergic reactions as it is able to activate mast cells, which release histamine, in response to specific antigens.

The reference ranges for the immunoglobulins are shown in Table 19.3 and illustrate the development of the immune system from birth to adulthood.

Immunoglobulins as reagents and therapeutic agents

The high specificity of immunoglobulins has been exploited for diagnosis and treatment of disease for several decades. Immunoassays are based on the principle of antibody binding to an epitope on the molecule to be measured, together with a detection system or label. Initially, the label was a radioactive atom (hence the term radioimmunoassay) but in recent years fluorescence and chemiluminescence have largely taken over. Immunoassays are in routine laboratory use for the measurement of many molecules such as hormones, drugs, tumor and cardiac markers, acute-phase proteins, and vitamins. The antibodies are raised in animals against the molecule of interest: these are polyclonal. Techniques to raise monoclonal antibodies were developed in the 1970s using mouse B cells fused with human myeloma cells to give unlimited quantities of highly specific immunoglobulins. Immunoassays may use monoclonal, polyclonal, or a combination of the two antibodies, depending on the required specificity of the assay.

Therapeutic use of monoclonal antibodies has accelerated in recent years. These are directed against key mediators in inflammatory and malignant disease.

19.2 CLINICAL DISORDERS OF IMMUNOGLOBULINS

Both deficiency and excess of immunoglobulin production cause clinical disorders. Immune deficiency may be a primary or congenital disorder, or may arise as a secondary phenomenon due to another disease process which suppresses immunoglobulin production. As would be expected, immune deficiency results in propensity to infections, often by organisms not usually considered to be pathogenic. Such infections are termed opportunistic.

Immunoglobulin excess may be categorized as either polyclonal or monoclonal. In response to an infection or to tissue inflammation, many plasma cells are stimulated to produce more immunoglobulins. This is termed a polyclonal response and is an appropriate part of the body's defenses. The classes of immunoglobulin produced vary according to the stimulus and can be helpful in making a diagnosis of the underlying cause. For example, IgA is predominantly secreted in response to mucosal infections, for example the lungs or the gut. A monoclonal excess of immunoglobulins, however, occurs when a single plasma cell proliferates and produces its own unique antibody idiotype consisting of heavy chains and either κ or λ light chains. This is known as a paraprotein. Commonly, this is due to a malignant disease of plasma cells called multiple myeloma, although other causes of excess monoclonal immunoglobulins also exist.

Laboratory investigation of immunoglobulin disorders

Serum protein electrophoresis

Proteins in serum can be separated by electrophoresis; that is, using an electric current passed through a gel or membrane to move charged molecules. The molecules move at different rates according to their size and charge. After a fixed time interval, the proteins are stained to make them visible. More recently, laboratories have adopted capillary zone electrophoresis (CZE) in response to increasing demand, as it allows a much greater sample throughput. As the name suggests, a narrow capillary replaces the traditional gel plate, and electronic detection and computer software replace the staining step. The pattern obtained by either technique shows a number of bands, usually five or six, depending on the timing used. These are albumin and α_1-, α_2-, β- (or β_1- and β_2-), and γ-globulins (**Table 19.4**).

A typical agarose electrophoresis gel of samples from 30 patients is shown in **Figure 19.4**. **Figure 19.5a–d** shows individual patient samples run on CZE.

TABLE 19.4 Protein components of regions seen on serum electrophoresis

Region	Proteins present
Albumin	Albumin
α_1-Globulins	α_1-Antitrypsin
	α_1-Acid glycoprotein (orosomucoid)
	Alpha-fetoprotein*
α_2-Globulins	α_2-Macroglobulin
	Haptoglobin
	Ceruloplasmin*
β_1-Globulins	Transferrin
β_2-Globulins	Complement C3 and C4
	β_2-Microglobulin*
	C-reactive protein*
γ-Globulins	IgG, IgA, IgM

*Denotes proteins that are not usually visible on electrophoresis due to low concentration.

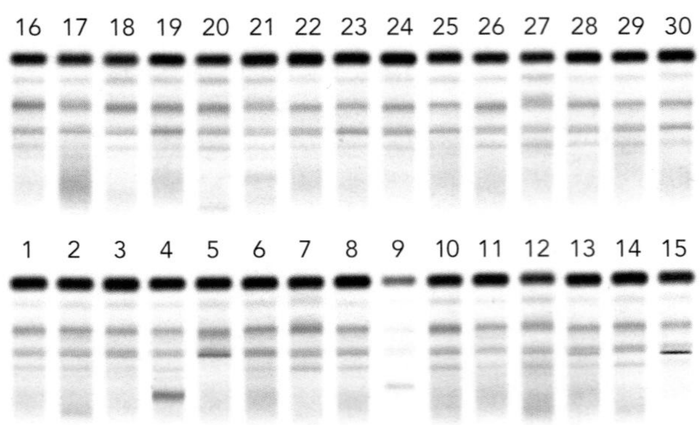

Figure 19.4 An agarose electrophoresis gel showing serum protein electrophoresis for 30 samples. Serum from each patient sample is placed in each lane and an electric current is passed through the gel (anode at the top). After a set time interval the gel is stained to show protein. The band at the top is albumin and the subsequent bands are α_1-, α_2-, β_1-, and β_2-globulin. The diffuse region in the lower part of each lane is the γ region and corresponds to the immunoglobulins. A discrete band here is an abnormal finding and usually represents a paraprotein (see lanes 4, 9, and 20). Lanes 12 and 17 show polyclonal increases in immunoglobulins by darker staining of the γ region. Lane 9 shows generally decreased staining plus a paraprotein. This is caused by the presence of a cryoglobulin, which precipitates at 4°C. The sample had been refrigerated immediately before placing on the gel, hence not all of the proteins were transferred from the sample tube.

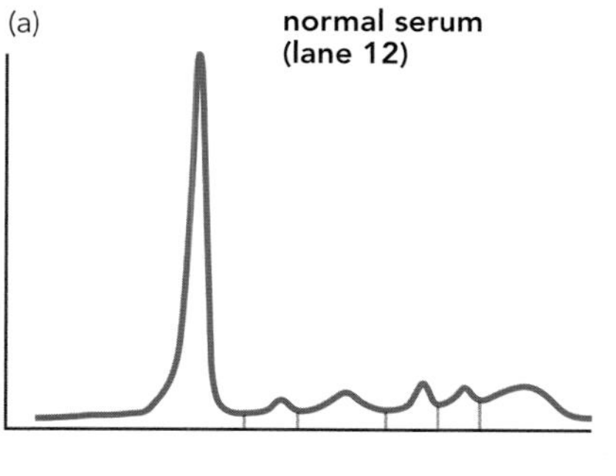

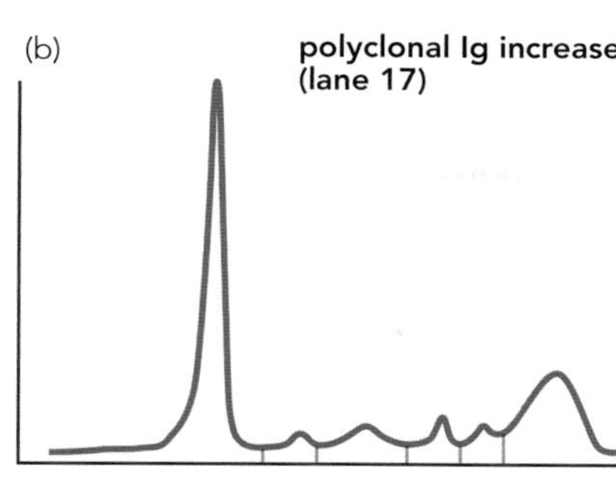

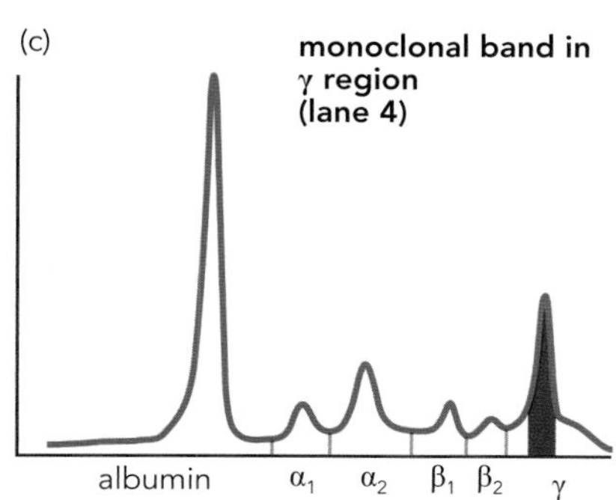

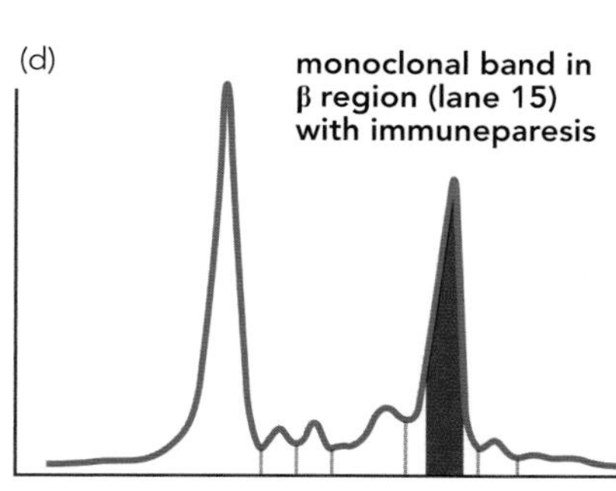

Figure 19.5 Capillary zone electrophoresis of different serum samples.
Serum passes through a thin capillary with an electric potential difference applied along its length (anode on the left). The major peaks from left to right are albumin and α_1-, α_2-, β_1-, β_2-, and γ-globulin, corresponding to the vertical bands of agarose gel in Figure 19.4. The relative concentrations of these are indicated by the areas under each peak. (a) Normal serum. (b) Serum with a polyclonal increase in immunoglobulins (γ region). This is the CZE equivalent of lanes 12 and 17 in Figure 19.4. This appearance would be expected with an immune response to infection or an inflammatory disease. (c) Serum with a monoclonal immunoglobulin (paraprotein) in the γ region. This is the CZE equivalent to lanes 4, 9, and 20 in Figure 19.4. The shaded area represents the concentration of the paraprotein and is calculated as a percentage of the total serum protein or the serum globulins. (d) Serum with a paraprotein in the β region. Paraproteins may sometimes run in the β region. The shaded area represents the concentration of the paraprotein plus the normal β_2 proteins. A small paraprotein may escape detection if it co-migrates with normal β_1 or β_2 proteins.

The largest bands on the gel are albumin and the γ-globulins. The γ region comprises the immunoglobulin molecules IgG, IgA, and IgM and is diffuse compared to the sharp bands of the other regions. The reason for this is that it consists of many thousands of individual immunoglobulin molecules, each with slightly differing charge and size due to their variable regions.

In immune deficiency states, the staining of the γ region is reduced, which is reflected in a decreased amplitude of the region on CZE, as would be expected. In polyclonal immunoglobulin excess, the staining and the amplitude of the γ region are uniformly increased. Monoclonal excess, in contrast, is seen as a single band or spike within the γ region. The other immunoglobulins may be reduced due to crowding out of normal plasma cells by a malignant clone. Sometimes a paraprotein band may be seen in the β region and may co-migrate with normal β-globulins. In other cases, more than one band or spike may be visible, which may be due to polymerization, fragmentation, different heavy-chain production, or biclonal plasma-cell proliferation.

When a potential monoclonal band is identified it is confirmed and characterized by immunofixation. This uses electrophoresis to separate the proteins, as before, but several parallel lanes are used and, before staining, each lane is overlaid with antiserum to each of the heavy- and light-chain molecules. With CZE a similar technique is used to remove the paraprotein before detection. This is called immunosubtraction or immunotyping.

The concentrations of individual immunoglobulin classes (IgG, IgA, and IgM) can be determined separately from electrophoresis, but the assays are not designed to measure monoclonal immunoglobulins and may give falsely high or low results in some cases, particularly with IgA and IgM paraproteins.

Analytical practice point 19.1

The main value of electrophoresis is detection of monoclonal immunoglobulins (paraproteins).

Measurement of light-chain excess

It was recognized in the middle of the nineteenth century that the condition later known as myeloma was associated with abnormalities of the urine. Henry Bence Jones, an English chemical pathologist, first described the protein that now bears his name as a precipitate that appears on heating urine to 40°C, disappears on boiling, and then reappears on cooling. It was not until the 1960s that Bence Jones protein was proved to be identical to the free light chains of immunoglobulins. Unlike intact immunoglobulins, free light chains are sufficiently small to pass through the glomerulus and thus be excreted in the urine.

Excess secretion of monoclonal light chains is a characteristic of multiple myeloma and its finding is an important diagnostic marker. Modern techniques for detecting Bence Jones protein are more sensitive than the original heating and cooling method and use a similar electrophoresis and immunofixation approach as for serum. However, no matter how sensitive the technique used, it has the inherent limitation that light chains are reabsorbed within the renal tubules after filtration and will only be detectable when they exceed a threshold concentration. Serum free light chains are therefore theoretically a better marker, but are technically much harder to measure. However, measurement of serum free light chains has become increasingly popular with the increase in commercially available methods in the last decade. The κ and λ light chains are measured and expressed as a ratio. The normal ratio of κ to λ in serum is between 0.3 and 1.6, reflecting the different synthesis and excretion rates of the two light chains. In polyclonal increases, both κ and λ rise, but the ratio remains normal. In monoclonal disorders, the ratio is abnormal as only one light chain can be secreted by the malignant cell. The corresponding light chain rises and often the other one falls as other plasma cells are crowded out.

Immunoglobulin deficiency disorders

Primary deficiency of immunoglobulins (**Table 19.5**) may occur as part of a recognized syndrome, and several distinct disorders exist with one or more immunoglobulin classes affected.

TABLE 19.5 Conditions causing primary immunodeficiency

Immunoglobulin affected	Clinical condition
IgG	Severe combined immune deficiency (SCID)
	X-linked hypogammaglobulinemia
	Hyper-IgM syndrome
	Common variable immune deficiency
	Transient hypogammaglobulinemia of infancy
IgA	Primary IgA deficiency
	Ataxia telangiectasia
IgM	Selective IgM deficiency
	Wiskott–Aldrich syndrome

In some cases it has been possible to identify specific mutations, which has led to a greater understanding of the pathways involved in B-cell development and immunoglobulin production. However, as the molecular biology becomes better understood, it is increasingly apparent that clinically similar conditions may have different underlying defects. The classification of immune deficiencies is therefore under constant review. The range of clinical presentations is very wide, with life-threatening infections during infancy at one extreme and incidental diagnosis in asymptomatic adults at the other. Interestingly, immune deficiencies are often associated with an increased risk of autoimmune conditions. This may be due to alteration in pathways allowing immune tolerance to occur. This is the term applied to the concept of the immune system ignoring antigens belonging to the host.

Clinical practice point 19.1

Congenital immune deficiency disorders are uncommon. Immune deficiency is more often secondary to another disease.

IgA deficiency

The commonest type of immune deficiency is selective IgA deficiency. It may affect as many as 1 in 700 people of European descent, but its prevalence varies between populations; for example it is much rarer in Japan (1 in 15,000).

Although it is often asymptomatic and detected incidentally, there is a higher rate of mild to moderate infections and autoimmune disease including celiac disease. There is also a risk of blood transfusion reactions due to the presence of anti-IgA antibodies, either IgG or IgE. These can react against IgA in the transfused blood, unless it is specially processed.

Common variable immune deficiency

As its name suggests, this is one of the more common primary immune deficiencies, affecting 1 in 50,000 people, and has a heterogeneous clinical and molecular basis. IgG and IgA are usually both low, but IgM may be either normal or also low. It is often diagnosed after the age of 20 years following a history of recurrent infections, usually respiratory. It is also associated with an increased risk of autoimmune diseases such as rheumatoid arthritis. Immunoglobulin replacement therapy may be given.

Severe combined immune deficiency (SCID)

SCID is a life-threatening condition presenting with severe infection in the first three months of life and is usually fatal by the age of 2 years if untreated. It affects around 1 in 50,000 live births and several molecular defects have been identified. Approximately one-half are X-linked. Definitive treatment requires bone marrow transplant. Although the immunoglobulins are low, IgG may be transiently normal after birth due to transplacental transfer from the mother.

Bruton (X-linked) agammaglobulinemia

This condition affects 1 in 250,000 individuals and results from a defect in a gene involved in maturation of B cells. This leads to a lack of plasma cells and hence immunoglobulins. It usually presents in the first year of life and results in chronic lung infections and inflammatory bowel disease. Treatment usually involves administration of intravenous immunoglobulin.

Wiskott–Aldrich syndrome

Like Bruton agammaglobulinemia, this is an X-linked disorder and has a similar prevalence. It consists of IgM deficiency and low platelets (thrombocytopenia). The latter causes a bleeding tendency, which is usually the presenting condition. Severe infection such as pneumonia and meningitis may also occur. The condition is usually fatal in childhood without bone marrow transplantation.

Hyper-IgM syndrome

This is a very rare condition (1 in 20 million) presenting in childhood. IgM is raised but IgG and IgA are deficient. It can be treated with immunoglobulin replacement.

Secondary immune deficiency

In adult life, immunoglobulin deficiency is more likely to be secondary to another condition (**Table 19.6**). These conditions include malignant disease that invades the bone marrow, increased loss or catabolism of proteins, or drug reactions.

Monoclonal disorders may paradoxically cause decreased immunoglobulins if they predominantly secrete only κ or λ light chains without a corresponding heavy chain. The suppression of IgG, IgA, and IgM production may be accompanied by a very small paraprotein band which may only be detectable by immunofixation, particularly as there is often co-migration with β-band proteins.

TABLE 19.6 Causes of secondary immunodeficiency

Primary cause	Examples
Malignancy	Multiple myeloma
	Lymphoma
Protein loss	Nephrotic syndrome
	Protein-losing enteropathy
	Burns
Drugs	Penicillamine
	Sulfasalazine
	Phenytoin
	Cancer chemotherapy
Marrow impairment	Metastatic infiltration
	Aplastic anemia Severe sepsis

Disorders of immunoglobulin excess

Polyclonal gammopathy

This is a description given to the electrophoretic appearance of increased γ-globulins and implies a generalized stimulation of plasma cells. This would be expected in response to infection, but is also seen in diseases which involve tissue inflammation. Some of these are considered to be autoimmune in nature, but in others the immune system is probably activated as a secondary response. The pattern of immunoglobulins often varies in different diseases and this can be diagnostically useful.

In chronic liver disease, gut bacteria may enter the portal circulation and stimulate the production of IgA. This predominantly runs between the β and γ regions and the appearance on electrophoresis is termed β-γ bridging.

The predominant immunoglobulin classes associated with other diseases are shown in **Table 19.7**.

TABLE 19.7 Predominant polyclonal immunoglobulin increase in clinical conditions

Immunoglobulin(s) increased	Condition
IgG	Connective tissue diseases, for example systemic lupus erythematosis (SLE)
	Chronic active hepatitis
	Sarcoidosis
	Castleman's disease
	Acquired immune deficiency syndrome (AIDS)
	Chronic granulomatous disease
IgA	Respiratory and gastrointestinal infections
	Inflammatory bowel disease
	Cirrhosis
	Rheumatoid arthritis
IgM	Infections: viral, neonatal, tropical, mycoplasma, Q fever, malaria
	Primary biliary cirrhosis
	Sclerosing cholangitis
	Lymphoma
IgG + IgA	Chronic respiratory infection
	Cirrhosis
	Rheumatoid arthritis
	AIDS
IgG + IgM	SLE
	Cirrhosis
	Leprosy
IgG + IgA + IgM	Chronic bacterial infections
	Sarcoidosis
IgD	Hodgkin's disease
	Tuberculosis
	AIDS
	Hyper-IgD syndrome

Monoclonal gammopathy

Monoclonal disorders, or paraproteinemias, arise when a single plasma cell undergoes clonal proliferation. This is often, but by no means always, a malignant condition called multiple myeloma, myelomatosis, or just myeloma. This

Analytical practice point 19.2

Up to 15% of myeloma patients secrete light chains only. Therefore when screening for myeloma, urine Bence Jones protein or serum free light chains should always be determined in addition to undertaking serum electrophoresis.

is a condition characterized by monoclonal immunoglobulin, abnormally large numbers of plasma cells in the bone marrow, and characteristic bone lesions seen on X-ray.

Other causes of monoclonal proteins are:

- Solitary plasmacytoma
- Monoclonal gammopathy of undetermined significance (MGUS)
- Waldenstrom's macroglobulinemia (WM)
- Non-Hodgkin's lymphoma and chronic lymphocytic leukemia
- AL amyloidosis

Multiple myeloma

Myeloma is a common, and currently incurable, malignant disease arising from uncontrolled clonal proliferation of plasma cells. It is usually spread in multiple areas of the skeleton (hence the name) but can occasionally be focal, when it is called a plasmacytoma. Although myeloma is usually associated with excess production of a monoclonal immunoglobulin and/or light chain, as described above, rare cases of nonsecretory myeloma also occur. Infiltration of bone marrow may result in anemia, bleeding due to low platelets, and infection. Bone fractures and hypercalcemia due to local bone resorption are also common. Renal failure is another complication and this may be related, at least in part, to hypercalcemia and the nephrotoxic effects of free light chains. The most common type of paraprotein is IgG (55%) followed by IgA (25%) and light chain only (20%). Overall, around 1% are IgD, IgE, IgM, or nonsecretory. IgM paraproteins are relatively common but usually associated with WM, which is a related but distinct condition arising from clonal proliferation of B lymphocytes. Bone lesions and renal failure are less common in WM than myeloma, but high concentrations of the large IgM molecules can cause hyperviscosity of plasma leading to visual disturbances, bleeding, and seizures due to impaired flow in the microcirculation.

Monoclonal gammopathy of undetermined significance (MGUS)

Monoclonal proteins are increasingly common with increasing age. Whilst they are relatively rare under the age of 40 years, they may be found in 1–2% of people in their sixth decade, 4–5% of those in their eighth decade, and as many as 15% in those over 90. However, only a fraction of these will be due to myeloma. The majority of the remainder is termed MGUS, to denote the uncertainty regarding their prognosis. Around 1% per year will transform into myeloma and so the finding of MGUS mandates regular monitoring. MGUS can be categorized according to the risk of malignant transformation and this can help to determine the frequency of follow up. Criteria for diagnosis of MGUS include the paraprotein concentration and the proportion of plasma cells in the bone marrow. The full criteria are as follows:

- M-protein (paraprotein) in serum <30 g/L
- Bone marrow clonal plasma cells <10% and low-level plasma cell infiltration in a trephine biopsy of the bone marrow
- No myeloma-related organ or tissue impairment
- No evidence of other B-cell lymphoproliferative disorder or light-chain associated amyloidosis, or other light-chain, heavy-chain, or immunoglobulin-associated tissue damage (including bone lesions or symptoms)

Three parameters can be used to separate patients with MGUS into those with low, intermediate, and high risks of developing multiple myeloma (**Table 19.8**). They are the serum M-protein (paraprotein) level, the immunoglobulin isotype, and the free light chain ratio. Furthermore, the intermediate grouping is further subdivided into low-intermediate and high-intermediate.

Clinical practice point 19.2

Monoclonal immunoglobulins are commonly seen in the elderly. It is important to distinguish between clinically benign causes and multiple myeloma.

CASE 19.1

A man, 65 years old, presents with a vertebral fracture. The results of investigations are shown below. Additional investigations showed lytic lesions in several bones on X-ray and 20% plasma cells in the bone marrow on biopsy.

	SI units	Reference range	Conventional units	Reference range
Serum				
Sodium	129 mmol/L	136–142	129 mEq/L	136–142
Potassium	4.2 mmol/L	3.5–5.0	4.2 mEq/L	3.5–5.0
Urea (blood urea nitrogen; BUN)	12.5 mmol/L	2.9–8.2	35 mg/dL	8–23
Creatinine	180 μmol/L	53–106	2.0 mg/dL	0.6–1.2
Calcium (albumin-adjusted)	3.20 mmol/L	2.20–2.60	12.8 mg/dL	8.8–10.4
Total protein	98 g/L	60–80	9.8 g/dL	6.0–8.0
Albumin	28 g/L	32–52	2.8 g/dL	3.2–5.2
Globulin	70	25–40	7.0 g/dL	2.5–4.0
IgG	44.8 g/L	6.5–16.0	4800 mg/dL	650–1600
IgA	0.15 g/L	0.8–4.0	15 mg/dL	80–400
IgM	0.06 g/L	0.5–2.0	6 mg/dL	50–200
Protein electro-phoresis	Abnormal band in γ region	Not present	Abnormal band in γ region	Not present
Immuno-fixation	IgG κ 43 g/L	Not detected	IgG κ 4300 mg/dL	Not detected
Free κ	30.0 mg/L	3.3–19.4	3 mg/dL	0.3–1.9
Free λ	1.00 mg/L	5.7–26.3	0.1 mg/dL	0.6–2.6
κ:λ Ratio	30.0	0.26–1.65	30.0	0.26–1.65

- What is the diagnosis?
- Why is the plasma sodium low?

The patient has multiple myeloma. The three diagnostic criteria are serum paraprotein and/or evidence of monoclonal free light chains in urine or serum, characteristic lytic lesions in the bones, and increased plasma cells in the bone marrow. Two criteria are required for diagnosis; this patient has all three. Note the raised IgG, which corresponds to the measured paraprotein concentration on electrophoresis. This is not always the case, particularly with IgA and IgM paraproteins, and there may be a discrepancy between measured immunoglobulins and the paraprotein estimation. The reason is that monoclonal proteins may not behave as expected in assays designed to measure polyclonal immunoglobulins.

Complications of myeloma include hypercalcemia (due to bone destruction), kidney impairment (due to the nephrotoxic effect of light chains), and immune paresis with reduced levels of the other immunoglobulins. This increases the risk of infection.

The reduced plasma sodium is due to the high protein concentration causing pseudohyponatremia. This is explained in detail in Chapter 5.

TABLE 19.8 Risk groups showing 20-year risk of progression to myeloma (%)

Risk	Paraprotein features
Low (5%)	Serum M-protein (paraprotein) <15 g/L
	IgG isotype
	Appropriate free light chain (FLC) ratio
Low-intermediate (21%)	Presence of an IgA or IgM isotype (note: must be less than 10 g/L) OR Inappropriate FLC ratio
High-intermediate (37%)	Presence of an IgA or IgM isotype (note: must be less than 10 g/L) AND Inappropriate FLC ratio
High (58%)	IgG, M-protein >15 g/L
	IgA or IgM, M-protein >10 g/L
	Inappropriate FLC ratio

CASE 19.2

A woman of age 75 years was investigated for tiredness. The results of investigations are shown below. Further investigations showed no anemia (normal hemoglobin), normal skeletal X-rays, and 8% plasma cells on bone marrow biopsy.

	SI units	Reference range	Conventional units	Reference range
Serum				
Sodium	138 mmol/L	136–142	138 mEq/L	136–142
Potassium	4.2 mmol/L	3.5–5.0	4.2 mEq/L	3.5–5.0
Urea (BUN)	6.2 mmol/L	2.9–8.2	17 mg/dL	8–23
Creatinine	77 μmol/L	53–106	0.9 mg/dL	0.6–1.2
Calcium (albumin-adjusted)	2.40 mmol/L	2.20–2.60	9.6 mg/dL	8.8–10.4
Total protein	79 g/L	60–80	7.9 g/dL	6.0–8.0
Albumin	36 g/L	32–52	3.6 g/dL	3.2–5.2
Globulin	43	25–40	4.3 g/dL	2.5–4.0
IgG	18.0 g/L	6.5–16.0	1800 mg/dL	650–1600
IgA	3.6 g/L	0.8–4.0	360 mg/dL	80–400
IgM	1.2 g/L	0.5–2.0	120 mg/dL	50–200
Protein electrophoresis	Abnormal band in γ region	Not present	Abnormal band in γ region	Not present
Immunofixation	IgG κ 10 g/L	Not detected	IgG κ 1000 mg/dL	Not detected
Free κ	10.15 mg/L	3.3–19.4	1.0 mg/dL	0.3–1.9
Free λ	7.24 mg/L	5.7–26.3	0.72 mg/dL	0.6–2.6
κ:λ Ratio	1.40	0.26–1.65	1.40	0.26–1.65

- **How do these results differ from Case 19.1?**
- **What is the diagnosis?**

This patient has MGUS. Although there is a paraprotein present, it is relatively low concentration and there is no evidence of end-organ damage. There is a risk of malignant transformation with any MGUS and so patients are kept under periodic review to detect increasing paraprotein concentrations and evidence of myeloma.

19.3 COMPLEMENT

As with the immune system as a whole, the complement system is highly complex. The following description is designed as an overview of complement to allow the reader to understand the rationale for measuring complement proteins. It is not intended to be entirely comprehensive; the reader is advised to consult specialist texts for this.

In addition to the synthesis and secretion of antibodies by B cells, the immune response is enhanced by the action of the complement system. This comprises a series of serum proteins, some of which are pro-enzymes, which are sequentially activated, leading ultimately to the formation of a membrane attack complex (MAC) on the surface of an invading organism which initiates lysis of the cell. In addition, various products of the complement pathways described below are involved in a number of immune mechanisms such as opsonization and recruitment of phagocytic cells.

Although complement proteins are numbered C1 to C9, there are over 20 proteins involved in the cascade as a whole. Some of these are regulatory proteins which prevent excessive activation, as this is itself hazardous to the host. Inappropriate activation of the complement system results in the clinical syndrome of anaphylaxis, which may cause fatal edema of the upper airways.

Complement proteins, all of which are acute-phase proteins, are synthesized by hepatocytes and macrophages and constitute a cascade somewhat analogous to the proteins involved in blood clotting, where an activated member of the cascade activates the next member by proteolysis. Indeed, there are such similarities between the two systems of serine proteases that they are believed to have a common ancestral pathway. Recent research has identified interaction between the two cascades that makes them much more of an integrated network than previously thought. The numbering of complement proteins (C1–C9) indicates the order in which they were discovered rather than their order in the cascade. Activated complement proteins have a bar over the number (for example C5, native protein; C$\overline{5}$, activated protein) and proteolytic fragments of activated complement proteins have letters after the number (for example C$\overline{3}$ $\rightarrow$ C3a + C3b). It was proposed originally that activation of the complement cascade required the presence of IgM- or IgG-containing immune complexes (classical pathway), but this has since been shown not to be the case and two other activation pathways, termed alternative and lectin, have been demonstrated. The feature common to all three pathways is the final lytic pathway which results in formation of the MAC on the surface of the invading organism and promotes cell death. The three pathways converge at the proteolytic cleavage of C3 by complement-derived proteases (C3 convertases) to C3a and C3b (**Figure 19.6**). C3b is the key component of further proteolytic enzymes assembled from other complement fragments earlier in the pathways. The C3b-containing complex in turn activates C5 and initiates the final pathway resulting i n MAC formation. The MAC forms a porous channel or pore through the membrane of the invading organism and contains the activated fragment C5b plus C6, C7, C8, and up to 19 molecules of C9. The pore allows water and salts to flow freely between the cell and the extracellular fluid along osmotic and concentration gradients, causing lysis and cell destruction.

Other fragments of complement proteins released during the proteolytic cascade have other important actions in the immune response:

- Vascular dilation and increased permeability
- Cell lysis
- Opsonization (coating of the invading organism with complement proteins to increase phagocytosis by leukocytes)
- Chemotaxis (increased activation and attraction of leukocytes)

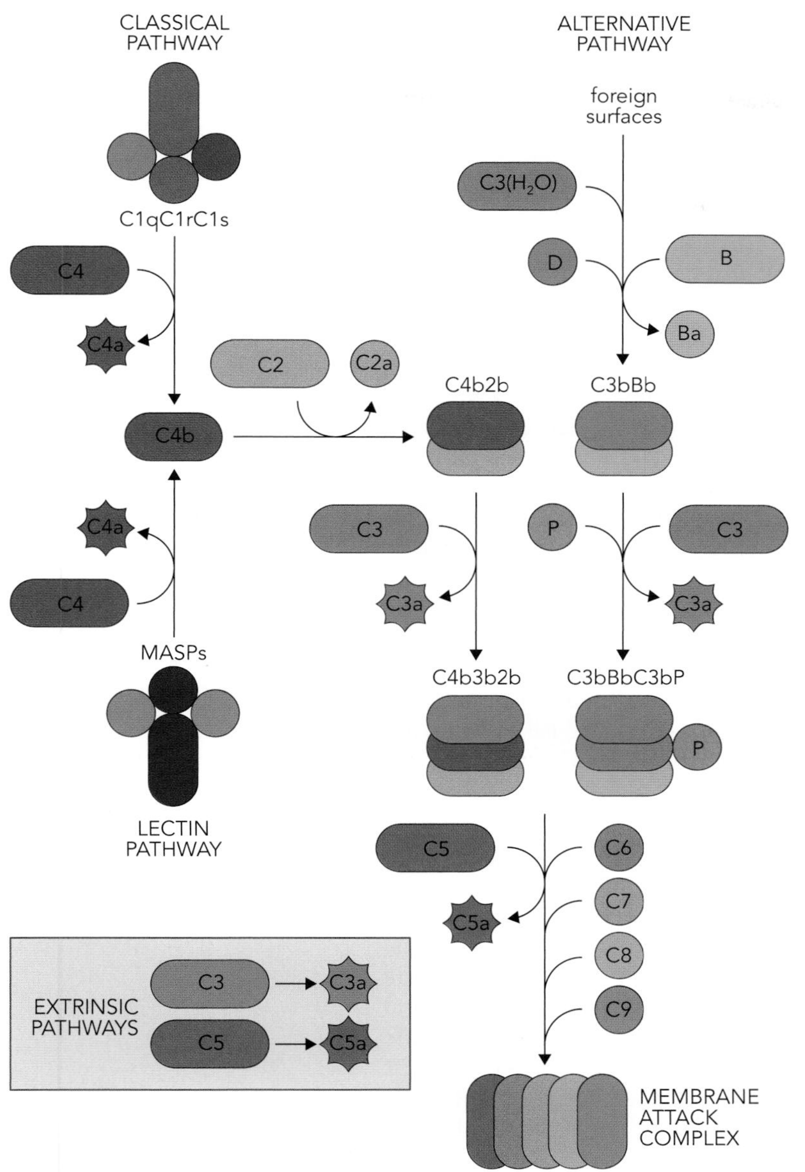

Figure 19.6 The complement cascade.
Complement consists of a number of plasma proteins (numbered C1 to C9 in order of discovery) which are sequentially activated to produce a membrane attack complex that results in lysis of invading microbial cells. The three pathways—classical, alternative, and lectin—are activated by interaction with different molecules, but all result in the activation of C5, the first common step in the final pathway. Activation of a complement component releases fragments such as C4a and C4b. These combine with each other to form a new protease which can activate the next component. The fragments in stars are bound to invading cell membranes and have additional effects such as opsonization. There is overlap between the clotting cascade and complement (inset box) allowing C3 and C5 to be activated. MASP, mannan-binding lectin-associated serine protease; D, factor D; B, factor B; P, factor P (properdin).

The individual components of the three activation pathways are shown in Figure 19.6. In theory, all components of the pathways can be assayed from C1 to C9 to give an indication of the activity of each of the pathways. However, in practice only C3 and C4, and occasionally C1, are measured routinely. These are usually sufficient to give an indication of the pathway most likely to have been activated. The reference ranges for these are shown in **Table 19.9**.

As mentioned earlier, all components of the complement cascade can act as acute-phase proteins and their rate of synthesis is increased in any inflammatory condition such as trauma or infection. For this reason, low levels are usually more relevant clinically as they indicate complement consumption and imply activation of one or more pathways (**Table 19.10**).

TABLE 19.9 Reference ranges for complement proteins

Protein	Reference range
C1	160–330 mg/L (16–33 mg/dL)
C3	880–2520 mg/L (male) (88–252 mg/dL)
	880–2060 mg/L (female) (88–206 mg/dL)
C4	120–720 mg/L (male) (12–72 mg/dL)
	130–750 mg/L (female) (13–75 mg/dL)

Classical pathway

The classical pathway involves activation of C1 by antigen–antibody complexes. There are specific complement-binding domains on the F_C regions of IgM and IgG (see Figure 19.1). Complement protein C1 is multimeric, consisting of C1q,

TABLE 19.10 Causes of changes in serum concentrations of complement proteins C3 and C4

Change	Likely cause
Low C3 and C4	Activation of the classical pathway by antigen–antibody complexes (for example in infection or systemic lupus erythematosis)
Low C3, normal C4	Predominant activation of the alternative pathway (for example by bacterial toxins or in nephritis)
Normal C3, low C4	Lack of C1 esterase inhibitor

a hexameric unit of two trimers, and two copies of two inactive precursor proteases (zymogens), C1r and C1s. The monomeric chains of C1q have been described as resembling a tulip in appearance, with proline-rich collagen-like stalks and globular heads. The two zymogens C1r and C1s bind to the globular domains of the C1q protein. The activation of C1 is initiated by binding to the F_C region of several IgG or IgM molecules bound to the surface of an organism (nonself cell). This binding causes hydrolysis of a peptide bond in the C1r, somewhat akin to the activation of pepsinogen to pepsin by acid in the stomach, and the now-activated C1r hydrolyzes and activates C1s. The hydrolytic cascade process is now set in motion with sequential hydrolysis and activation of C4, C2, C3, and C5, as shown in Figure 19.6. Thus, in this classical pathway, the essential activation of C5 to C5 convertase (C5b) requires participation of surface-bound C3b, C4b, and C2b. A serine protease inhibitor of C1 (C1-INH) blocks the actions of C1q, C1r, and C1s by dissociating the subunits, and is an important regulator of the classical pathway.

Lectin pathway

The lectin pathway is initiated by mannose-binding lectin attaching itself to the surface of a microorganism. The lectin, a serum protein related structurally to C1q, binds with great affinity to mannose-containing carbohydrates (mannans) on the surface of microorganisms and activates the complement cascade via two serine proteases, mannan-binding lectin-associated serine proteases 1 and 2 (MASP-1 and MASP-2), in a manner similar to C1r and C1s as described in the classical pathway. The trimolecular complex so formed cleaves both C4 and C2 to form the C3 convertase, which leads to the lytic pathway. Genetic mutations leading to a deficiency of the lectin are associated with frequent bacterial infections in childhood.

Alternative pathway

As shown in Figure 19.6, activation of the alternative pathway at C3, and subsequently of complement proteins C5 to C9, occurs in the absence of antigen–antibody complex and without involvement of C1, C2, or C4. In evolutionary terms, this pathway probably developed as a nonspecific defense mechanism prior to the classical pathway. The major activators are endotoxins and bacterial cell walls. It is a relatively inefficient pathway requiring high concentrations of the participating components. In contrast to the classical pathway, in this instance, activation of C5 to C5b requires a different C3b complex (C5 convertase). Activation occurs when low levels of C3b, produced intrinsically during normal protein turnover, bind to a surface such as a bacterial cell and, in the presence of magnesium ions, attracts the protein factor B. The C3Bb complex is hydrolyzed by another serine protease, factor D, to produce a short-lived, active C5 convertase (C3bBb). This active protease (C3bBb) in turn hydrolyzes more C3 to produce increased amounts of C3b (positive feedback), and addition of properdin (factor P), a 53 kDa protein, stabilizes the convertase which now consists of two molecules of C3b and one each of Bb and properdin (C3bBbC3bP). The binding of properdin protects the membrane-bound C5 convertase complex (C3bBbC3b) from rapid deactivation. This C5 convertase activates the final lytic cascade.

Clinical use of complement measurement

Most clinical laboratories are able to measure C3 and C4 as indicators of complement consumption. Differential changes in these two tests point toward particular disease processes. The possible patterns are shown in Table 19.10. There may be occasions when C3 or C4 levels are lowered by reduced synthesis rather than increased consumption and these would include genetic and acquired causes, the latter sometimes being seen with severe liver disease. Sometimes complement deficiency may present similarly to immune deficiency with recurrent infections.

Clinical conditions where complement measurements are useful are kidney disease, arthritis, and vasculitis. These are all conditions which have several possible causes and low levels of complement indicate that an immunological cause is likely. This may help to direct other investigations and treatment, particularly immune suppressant drugs.

C1 esterase inhibitor

This is a protease inhibitor which stops activated C1 and prevents uncontrolled complement activation. Deficiency of C1 esterase inhibitor (C1-INH) may occur as a genetic condition in around 1 in 50,000 individuals and is also called Osler's disease. It presents clinically as episodes of edema, which may be externally on the limbs or trunk or within the abdomen or larynx. The latter two may cause abdominal pain or breathing difficulties, which may be fatal. The initial assessment of complement shows low C4, but normal C3, implying activation of the classical pathway. C1-INH itself can be measured to confirm the diagnosis. Rarely, acquired forms of deficiency may also occur, usually in association with a B-cell disorder.

TUMOR MARKERS

CHAPTER
20

IN THIS CHAPTER

USES OF TUMOR MARKERS

TYPES OF TUMOR MARKERS

THE USE OF TUMOR MARKERS IN DETECTION AND MANAGEMENT OF SPECIFIC CANCERS

THE FUTURE

It is perhaps worthwhile to define the terms tumor, neoplasia, cancer, benign, and malignant at the beginning of this chapter to help in understanding the role of tumor markers in clinical chemistry. Tumors, or neoplasia, are abnormal new growths of tissues and may be benign or malignant; cancer indicates a malignant tumor. Benign tumors are formed by accumulation of abnormal cells which remain in the tissue of origin. Malignant tumors (cancers) occur as a result of a combination of genetic, epigenetic, and environmental events that ultimately results in uncontrolled cellular proliferation. This enables cancer cells to invade adjacent tissue and elsewhere in the body, a process known as metastasis. The histological appearance of tissue is suggestive of a malignant state and is often the basis upon which diagnosis, prognosis, and treatment is determined. The histological indications for malignancy include:

- Pleomorphism
- Nuclear aberrations
- Increased mitoses
- Invasion into surrounding cells and tissues

The rationale for searching for a tumor marker specific for a particular tumor is that its detection might lead to early diagnosis of the malignant state when it is the most treatable and before it has had a chance to grow and metastasize. Early treatment hopefully will lead to prevention of spread of the tumor and to its ultimate elimination. In practice, tumor markers are molecules which are produced by cancer cells and sometimes by other cells in the body, including normal cells, in response to a cancerous state. They are usually proteins or cell-specific metabolites which may be present in abnormally high concentration in body fluids, ideally blood or urine, but also in saliva, feces, or in the cancer tissue itself. The release of a specific protein by a malignant tissue is due to a differential pattern of gene expression between the normal and malignant tissue, and this difference has led in some cases to the use of changes in DNA as tumor-specific markers. It is not appropriate in the current text to list the growing number of tumor markers in use throughout the world. Rather, this chapter will describe some general features of tumor markers. Specific examples will be used to illustrate how tumor markers can be used.

Cancers or malignant tumors may be differentiated histologically into three major categories:

1. Carcinomas (endodermal and ectodermal tissues)
2. Sarcomas (mesodermal)
3. Leukemias and lymphomas (white blood cells and the monocyte–macrophage system)

Different subgroups have also been identified based on the tissue type that the tumor originates from. This includes for example adenocarcinoma, which may uniquely express a specific protein. Tumor markers are often associated with a tissue type but may also be identified with a cell type. For example, carcinoembryonic antigen (CEA) is associated with colonic cancer (carcinoma) but may also be expressed in other tumor sites of similar cell origin, for example breast cancer.

20.1 USES OF TUMOR MARKERS

Tumor markers are analytes produced by a tumor, or in response to a tumor, which can help in the diagnosis of disease, can provide prognostic information, and can be used to identify appropriate treatment as well as to monitor treatment. There are five broad uses for tumor markers:

1. Screening
2. Diagnosis
3. Prognosis and appropriate treatment
4. Monitoring therapy
5. Relapse

Screening

Total population screening for detection of a specific cancer is often proposed on the grounds that early diagnosis and treatment will provide a better prognosis. However, this is the exception rather than the rule. Tests need to be diagnostically sensitive and specific. Current screening methods vary widely and include mammography (breast cancer), cytology (cervical cancer), and fecal occult blood test (colon cancer) as well as analysis of body fluids for specific tumor markers. The positive results of screening programs for breast, cervical, and colonic cancers have been widely reported in the literature and are not the subject of this text. However, serum tumor markers are rarely sensitive or specific enough to be used for general population screening. Lack of sensitivity leads to the reporting of a high proportion of false negatives; that is, in patients with an early stage cancer, the tumor marker is negative. Lack of specificity may be as problematic as lack of sensitivity, with false positives leading to overdiagnosis, identifying patients as having cancer when they do not have the disease.

Rather than population screening, an alternative approach is targeted screening in which at-risk individuals may be identified; for example, those with particular genetic abnormalities associated with increased risk of cancer (such as thyroid cancer in patients with multiple endocrine neoplasia). Sometimes the higher risk is simply a matter of age and sex. Screening programs for detection of ovarian, colonic, and prostatic cancers are discussed in more detail later in this chapter.

Clinical practice point 20.1

Tumor markers are never 100% specific and sensitive in the diagnosis of cancer. The higher the tumor marker value, the greater the likelihood of a cancer being present. However, patients with cancer may have normal tumor marker values and, conversely, patients may not have cancer when they have raised tumor marker levels.

Diagnosis

Ideally, for tumor markers to be useful diagnostically, their specificity and sensitivities should approach 100%. Unfortunately, most tumor markers that have been proposed do not match this level of diagnostic performance and are poor diagnostic indicators, even when used in conjunction with other parameters. Commonly, the final diagnosis of the presence of a tumor is made by histology.

Prognosis and treatment

In some cases, the circulating level of a particular tumor marker can give an indication of prognosis. The level reflects the size of the tumor, which in turn may be used to predict survival. For example, human chorionic gonadotropin (hCG) and alpha-fetoprotein (AFP) are prognostic indicators for testicular teratoma (**Figure 20.1**). Similarly, tissue markers such as P53 (the product of expression of a tumor suppressor gene involved in many cancers), E-cadherin (expression of a gene product involved in making cells adherent to one another), and mm23H1 and MMP-2 (products of two genes involved in cell invasion) all aid in the prediction of outcome of node-negative breast cancer.

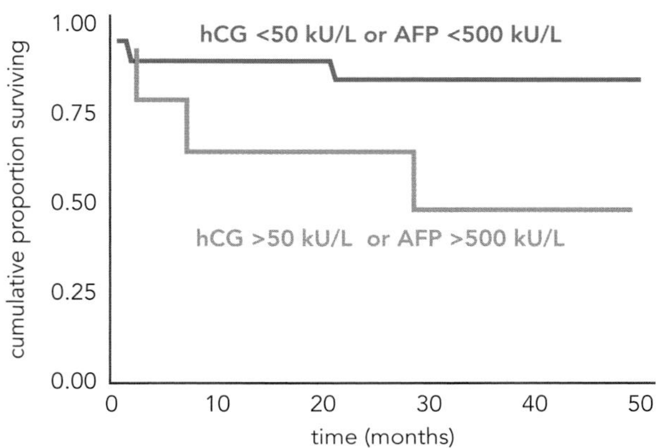

Figure 20.1 Using tumor markers as prognostic indicators. A graph showing survival of patients with testicular teratoma, based on the cumulative number of patients surviving against time. Patients with lower values of the serum tumor markers human chorionic gonadotropin (hCG) and alpha-fetoprotein (AFP) have much better survival rates than those patients with higher values. Reference ranges for hCG and AFP are <2.0 kU/L and <2.5 kU/L, respectively. (Note that kU/L is equivalent to U/mL.)

The presence of tumor tissue markers may also be used to decide on treatment; for example, in breast cancer, the presence of protein receptor HER-2-positive tissue indicates treatment with herceptin (receptor antagonist) and the presence of estrogen receptor-positive tumor tissue indicates treatment with receptor antagonists such as tamoxifen.

Monitoring therapy

Monitoring the effects of treatment in patients with cancer is probably the most useful role for the measurement of tumor markers. This success of monitoring is dependent on there being a direct quantitative relationship between the tumor load and tumor marker concentration. When such a relationship exists it is possible to assess the efficacy of treatment, highlight potential drug resistance, and, in some cases, establish the elimination of the tumor. The criteria used for assessing effectiveness of treatment are:

- No change (tumor marker >50 % of value at time zero, t0)
- Improvement (tumor marker <50% t0)
- Response (tumor marker <10% t0)
- Complete response (tumor marker undetectable or within reference range)

By assessing the rate of decrease of a tumor marker, it is possible to predict the complete response time to when the tumor marker will be undetectable.

While in the ideal situation a complete response may imply total elimination of the tumor, this is not necessarily so. The following scenarios contradict this. First, a tumor may become quiescent or dormant and therefore may not secrete (or even synthesize) the tumor marker. Second, since the success of monitoring treatment will depend on the sensitivity of measurement of the tumor marker, occasionally a tumor may be present but be of inadequate size to produce sufficient marker to allow its detection in plasma (that is, the method employed cannot detect the few molecules of tumor marker per milliliter present in the plasma). In such cases, it may be possible to assess a complete response time (no active tumor cells present) by using the half-life of the tumor marker during treatment and extrapolating down to the presence of zero cells. A theoretical example is given in **Figure 20.2**.

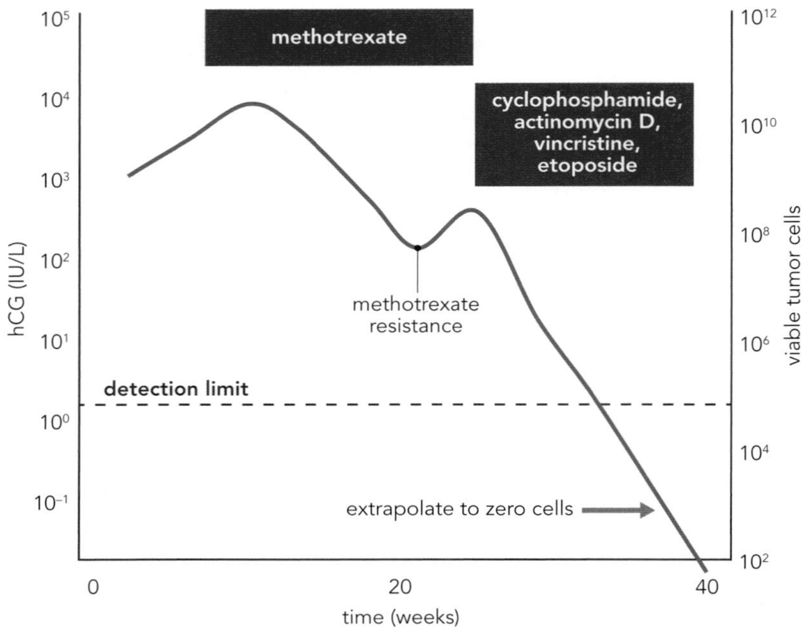

Figure 20.2 Determining when all tumor cells are destroyed. The graph shows the monitoring of a patient with choriocarcinoma using the tumor marker human chorionic gonadotropin (hCG). The patient was initially treated with methotrexate, until they developed methotrexate resistance, and were then treated with the mixture of anti-cancer drugs called CAVE (Cyclophosphamide, Actinomycin D, Vincristine, Etoposide). By relating the serum concentration of hCG to the number of viable tumor cells, and determining the rate of fall of serum hCG, the time needed to continue treatment until all cells are destroyed can be estimated. During this period the concentration of hCG in serum will fall well below the detectable limit of measurement.

Relapse

Another situation in which good analytical sensitivity of tumor marker measurement is critical is monitoring for possible relapse in patients who have been treated successfully for elimination of a tumor. In the ideal scenario, where the tumor marker is undetectable in the treated state, measurement of the marker is extremely valuable in determining re-emergence and/or spread of the tumor. In practice, other factors including frequency of measurement, cost of measurement, and clinical effectiveness have to be considered.

An example of the usefulness of monitoring a marker for assessing re-emergence of an active tumor is in monitoring prostate cancer using prostate specific antigen (PSA). The diagnostic specificity of PSA is limited; values of 4 μg/L (ng/mL) may be found in normal healthy males and even higher values may occur in patients with benign hypertrophy of the prostate gland. However, in patients who have undergone surgery and chemotherapy for treatment of prostatic cancer, plasma PSA is undetectable (<0.01 μg/L). Modern, high-sensitivity PSA assays, with acceptable interbatch coefficients of variation of <10% at the bottom end of the dynamic range, allow early detection of recurrence of the cancer and thus early treatment and increased positive outcomes. The measurement of a tumor marker in this instance obviates the use of alternative, usually more invasive techniques such as biopsy or exposing the patient to radiation (radionuclide or computerized tomography scans). These techniques, which include magnetic resonance imaging, are also considerably more expensive to perform. In addition, where the assay of a tumor marker has a high analytical sensitivity, it can be far more sensitive diagnostically than imaging techniques.

20.2 TYPES OF TUMOR MARKERS

Although most tumor markers are assayed in plasma or serum, they may also be measured on occasion in whole blood, urine and other fluids, feces, sputum, cell scrapes, washings, and tissues. They may be categorized into one of four different types:

1. General nonspecific markers
2. Functional markers
3. Classical tumor markers and hematological malignancies
4. Molecular markers

Nonspecific markers

There are a number of metabolic consequences associated with certain types of malignancies which may lead to abnormalities in routinely measured analytes. This may be illustrated by four examples:

1. Hypercalcemia arising from production of parathyroid hormone related peptide (PTHrP) by squamous cell carcinoma (see Chapter 9)
2. Raised alkaline phosphatase in serum as a consequence of metastasis of primary tumors of breast and prostate to bone
3. Raised alkaline phosphatase and gamma-glutamyl transferase in serum when colonic adenocarcinoma metastasizes to liver
4. Measurement of lactate dehydrogenase in monitoring tumor load in patients with lymphoma

Obviously, calcium, alkaline phosphatase, gamma-glutamyl transferase, and lactate dehydrogenase are not tumor markers in the strict sense but they can indicate a possible malignancy in patients that have no prior history of cancer. In addition, they may provide prognostic information in patients with known malignancies, and may highlight cancer spread and development during monitoring of several types of tumor. While most of these markers are nonspecific and limited in sensitivity, they are used regularly as diagnostic and prognostic indicators in the management of patients. Examples of these nonspecific markers include:

- Calcium
- Erythrocyte sedimentation rate (ESR)
- Sodium
- Lactate dehydrogenase (LDH)
- β_2-Microglobulin
- Alkaline phosphatase
- Tissue polypeptide antigen (TPA)
- Neopterin
- Thymidine kinase
- Tumor-associated trypsin inhibitor (TATI)

Functional tumor markers

Many tumors, often benign, produce excessive amounts of a normal metabolite or hormone in a poorly or sometimes uncontrolled manner. In the case of hormone production, these tumors often arise at the same site from which the hormone is normally synthesized, but sometimes the tumor may arise at a novel site. Examples of such tumors and their markers include:

- Pituitary
  - Prolactin
  - Adrenocorticotropic hormone (ACTH)
  - Growth hormone
  - Thyroid-stimulating hormone (TSH)
- Parathyroid
  - Parathyroid hormone (PTH)
- Adrenal cortex
  - Aldosterone
  - Cortisol
- Adrenal medulla
  - Catecholamines and metabolites
- Ovary
  - Estrogens
  - Androgens
- Gastrointestinal tract
  - Insulin
  - Glucagon
  - Vasoactive intestinal polypeptide (VIP)
  - Gastrin
  - Serotonin and metabolites (5-hydroxyindoleacetic acid; 5-HIAA)

Some of the metabolic and clinical consequences of these tumors have been discussed in detail in other chapters in this book. Measurements of the metabolites listed above are made routinely in clinical practice to detect and monitor many tumors. In some cases, they are very sensitive in the initial detection of a tumor and exquisitely sensitive in monitoring treatment; for example, the measurement of prolactin in the diagnosis and monitoring of prolactinoma (**Figure 20.3**).

Classical tumor markers

Classical tumor markers are proteins which have been used routinely in the diagnosis and monitoring of cancers. Very few of them are specific for a tumor

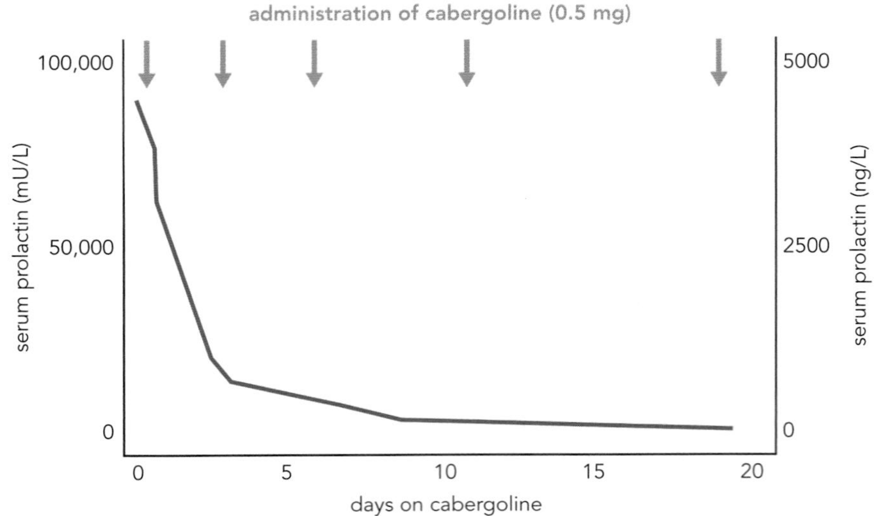

Figure 20.3 Measuring serum prolactin to monitor treatment of prolactinoma.
The graph shows the rapid fall in serum prolactin in a male patient who has been successfully treated with cabergoline. The prolactin falls from very high levels back into the reference range in a matter of days. The reference range for this assay for these prolactin results in men was <400 mU/L, which is equivalent to <20 ng/L or <870 pmol/L.

site and none are 100% specific or sufficiently sensitive to allow diagnosis in the absence of any other clinical information. A list of some widely available markers is given in **Table 20.1**. Whilst each of the tumor markers listed in the table is specific for a cell type, it is easier to discuss them in more detail in relation to the tumor site.

TABLE 20.1 Classical tumor markers

Primary cancer	Tumor marker
Prostate	Prostate specific antigen (PSA)
Ovarian	CA 125
	Inhibin A
Liver	Alpha-fetoprotein (AFP)
Testicular	AFP
	Human chorionic gonadotropin (hCG)
	Placental alkaline phosphatase (PLAP)
Colonic	Carcinoembryonic antigen (CEA)
Pancreatic	CA 19-9
Breast	CA 15-3
Bladder	Nuclear matrix protein 22 (NMP22)
Neuroendocrine	Chromogranin A
	Neuron-specific enolase (NSE)
Thyroid (medullary)	Calcitonin
Thyroid (follicular)	Thyroglobulin
Melanoma	S100

20.3 THE USE OF TUMOR MARKERS IN DETECTION AND MANAGEMENT OF SPECIFIC CANCERS

The incidence of the common cancers in developed countries is shown in **Figure 20.4** and reflects some of the research and development of using tumor markers.

Prostate cancer

Prostatic cancer (**Figure 20.5** and **Figure 20.6**) is the second most common cause of death from cancer in Europe and North America. While in the early stages of the disease patients may be asymptomatic, disease progression is accompanied by classical symptoms of urinary frequency, reduced flow, and blockage,

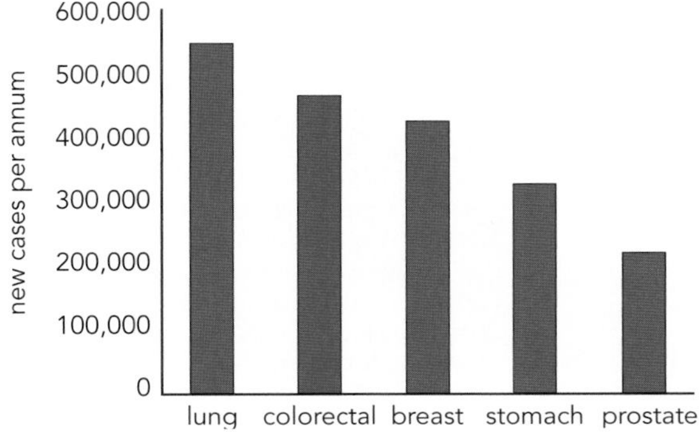

Figure 20.4 Estimates of incidence of the five commonest cancers in developed countries.

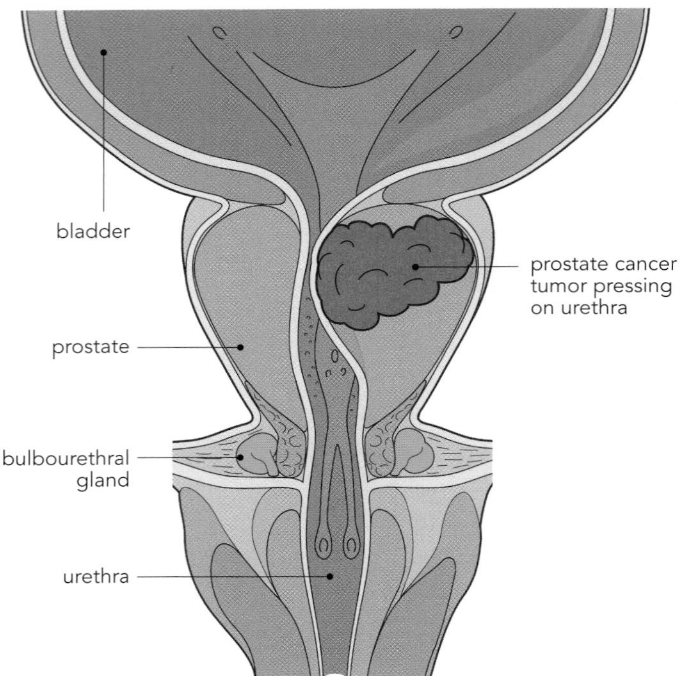

Figure 20.5 Growth of cancer cells in prostate gland.
The figure shows that as the cancer cells grow within the prostate gland they usually start to press on the urethra, causing the classic symptoms of difficulty in urinating.

and later by bone pain as the tumor metastasizes to bone. Measurement of PSA is used extensively in the screening, diagnosis, and monitoring of prostate cancer. The basic features of PSA are:

- Comprised of 240 amino acids
- A 34 kDa glycoprotein (93% peptide and 7% carbohydrate)
- High degree of homology with serine proteases of the kallikrein family
- Circulating PSA is largely complexed with α_1-antichymotrypsin
- Produced by prostate epithelial cells
- Involved in liquefying semen

Being synthesized only by prostate epithelial cells, PSA is usually undetectable in women. The circulating concentration of PSA in healthy males (100% of healthy males <40 years and 97% of healthy males >40 years) is less than 4 μg/L (4 ng/mL). PSA is present both in the free form (minor) and bound form (predominant), where it is complexed with serum proteins, mainly α_1-antichymotrypsin, which inhibits the proteolytic activity of PSA. The ratio of free to bound forms decreases in prostatic disease.

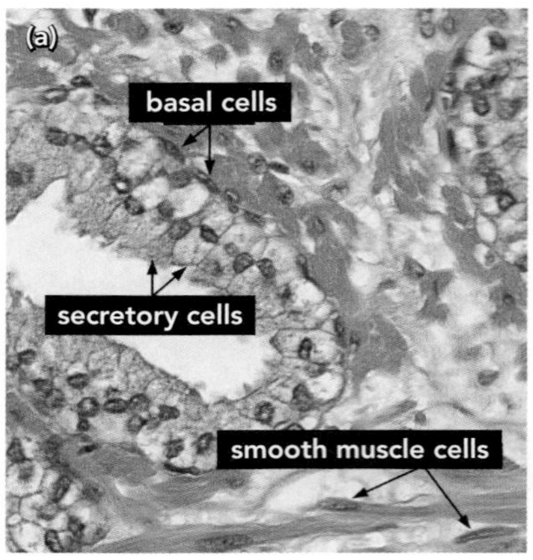

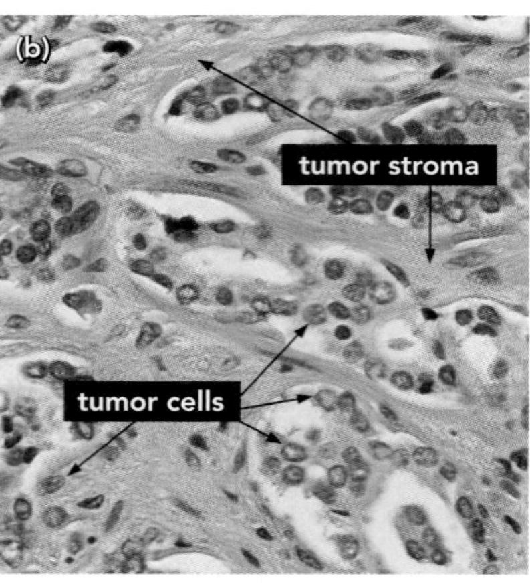

Figure 20.6 Histology of normal prostate and prostate cancer.
(a) The prostate consists of many microscopic glands surrounded by a stroma made up from smooth muscle and fibrous tissue. The glandular cells consist of secretory cells around a central duct (seen as white in the figure) and a layer of basal cells which separate them from the stroma. (b) Prostate cancer (or carcinoma) is graded according to the microscopic pattern of growth, which is correlated with the risk of spread. The malignant cells arise from the glandular component. There is a spectrum of changes ranging from uniform proliferation of well-differentiated glands, through irregular-sized glands, to poorly differentiated glandular cells with invasion into the surrounding stroma. The figure shows the highest grade of cancer. Note that the tumor cells are not differentiated into the columnar epithelium, which makes up the normal secretory cells, and the basal cell layer is missing. Individual tumor cells can also be seen within the stroma, indicating invasion.

The measurement of PSA is used widely in North America to screen for prostate cancer in males over the age of 50 years but this is not currently general practice in the UK. Some screening programs use a single measurement while others measure changes in serum PSA over a period of time before deciding on further investigations. A serum PSA >4 μg/L (or >4 ng/mL) is deemed abnormal (above the reference range) and in patients positive for prostate cancer, the subject might undergo the clinical path shown in **Figure 20.7**.

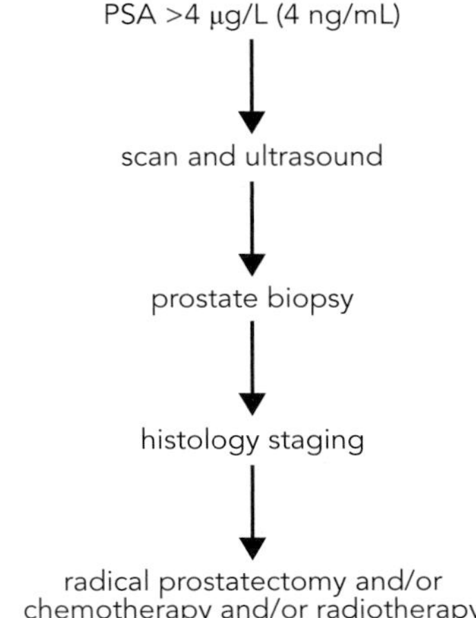

Figure 20.7 Sequence of events in a patient shown to be positive for prostate cancer.
The sequence of events that would happen for a male patient with a serum prostate specific antigen (PSA) level greater than 4.0 μg/L (4 ng/mL), when the histology was positive for prostate cancer. Treatment options, for example surgery, would depend on the histological findings plus other clinical factors.

The final procedure of radical prostatectomy is often combined with chemotherapy and sometimes radiotherapy, the decision for such surgery being dependent on the age and fitness of the patient. Early detection of prostate cancer is usually associated with good prognosis.

Patient preparation is critically important before taking blood for the measurement of PSA as a number of factors may give rise to increased serum PSA, such as:

- Prostatitis, both acute and chronic
- Prostatic infarction
- Acute urinary retention
- Catheterization
- Renal failure
- Riding a bike
- Ejaculation

Digital rectal examination causes only minimal increases in serum PSA in most males, but may occasionally cause a marked increase and yield a false positive result. Any form of prostate massage can cause an appreciable rise in serum PSA in a small number of patients, and since the half-life of PSA in serum is approximately 3 days it may take 2–3 weeks for the serum PSA to return to baseline values after prostatic manipulation or diagnostic intervention in such patients.

PSA as a marker of malignancy

Although PSA is associated in the mind of the public with prostate cancer, in reality it is not 100% specific for the prostate gland, and raised values are certainly not specific for prostate cancer. It is found in small amounts in other tissues (breast, salivary gland, endometrium, and urethral glands) and body fluids (breast milk and amniotic fluid). It may also be present in the serum of patients with cancers of tissues other than prostate, including lung, breast, kidney, and uterus. However, in the vast majority of cases, some abnormality (benign or malignant) or stimulus of the prostate is responsible for the raised plasma PSA values. While not all patients with prostate cancer have an elevated serum PSA, a serum PSA >4 μg/L (>4 ng/mL) requires further investigation.

Benign prostatic hypertrophy

Of major concern in the context of the current chapter, however, is the increase in serum PSA in some patients with benign prostatic hypertrophy (BPH). It is obviously of paramount importance to distinguish between BPH and prostatic cancer to prevent the unnecessary sequence of events shown in Figure 20.7. While serum PSA may be raised in both BPH and prostate cancer, and the higher the serum value the more likely it is that the patient has cancer, there is considerable overlap in the initial serum PSA concentrations in the two conditions, particularly at concentrations <10 μg/L (<10 ng/mL) (**Table 20.2**).

It is only at concentrations above 10 μg/L (10 ng/mL) that there is great divergence between the benign and cancerous conditions. Because of this overlap in serum total PSA concentrations, some centers measure the free and bound forms of PSA to help differentiate between BPH and prostate cancer, since the ratio of free to bound PSA is usually lower in prostate cancer (that is, more PSA is bound to α_1-antichymotrypsin and other plasma proteins). This

TABLE 20.2 Comparison of the percentage of patients with benign prostatic hypertrophy or prostate cancer with increasing serum PSA

Serum PSA	Benign prostatic hypertrophy	Pancreatic cancer
0–4.0 μg/L (ng/mL)	75%	43%
4.1–10 μg/L (ng/mL)	22%	37%
>10 μg/L (ng/mL)	3%	20%

difference in ratio has been suggested to allow differentiation between BPH and prostate cancer at values of serum total PSA <4 μg/L (<4 ng/mL). The ambiguity in interpretation of PSA values illustrates its deficiency as a tumor marker and has prompted the search for other markers of prostate cancer. However, to date, none is being used routinely.

Screening for prostate cancer with PSA in the USA

Measurement of PSA as a screening procedure for prostate cancer has been in place in North America for a number of years, but the expected reduction in deaths due to prostate cancer has not been evident. Some patients with malignant cells in their prostate glands may not develop progressive disease for many

CASE 20.1

A 51-year-old man is concerned about his risk of prostate cancer as his father died of the condition at the age of 65. He has a PSA level measured and digital rectal examination (DRE) performed, which showed some firmness in the gland. He is advised by his urologist to have a biopsy, but is uncertain whether to proceed in view of his PSA.

	SI units	Reference range
Plasma		
PSA (prostate specific antigen)	3.5 μg/L (3.5 ng/mL)	Age dependent: 50–60 years: 0–3 60–70 years: 0–4 >70 years: 0–5

- **Does the PSA level exclude prostate cancer?**
- **What is the difference between a reference range and an action limit?**
- **What conditions can raise PSA other than prostate cancer?**

In common with other tumor markers, PSA is not always raised in cancer. The sensitivity of the test depends on the level set as the action limit and is a trade off against the diagnostic specificity. A low action limit will detect more cancers, but at the expense of more false positives; that is, overdiagnosis. PSA is known to increase with age, due to benign prostatic hypertrophy, and age-specific ranges are used by some authorities to account for this. A man under 60 years of age should have a PSA level below 3 μg/L (3 ng/mL), whilst the limit is below 4 μg/L (4 ng/mL) for men between 60 and 70 years of age. In this case the family history and abnormal DRE are sufficient to advise a prostate biopsy, irrespective of the PSA. Only about one in three men with raised PSA will have prostate cancer, although the likelihood increases with the PSA level. It is important to remember that PSA is prostate specific, not cancer specific, and levels may be raised by urinary infection, prostatic trauma (including catheterization of the bladder, but not routine DRE), and by ejaculation in the previous 48 hours.

years, whilst others may develop metastatic prostate cancer very quickly. Indeed, various studies have shown a high prevalence of cancerous cells in the prostate glands of men over the age of 60 years, where an autopsy had taken place, but the men had no known symptoms of prostate disease. In other words, most men with malignant cells in their prostate gland die from causes other than prostate cancer. The real challenge is to find screening procedures to identify patients who will develop and die from aggressive, malignant prostate cancer, thus allowing early intervention, either therapeutic or surgical.

The use of PSA measurement in staging and monitoring of prostate cancer

The serum concentration of PSA tends to correlate with the clinical stage of prostate cancer, and very high levels are most likely indicative of advanced disease. However, PSA lacks the diagnostic accuracy to distinguish between localized and extra-prostatic (metastatic) disease. It is most useful in the detection of residual tumor and recurrence of cancer following definitive treatment (radical prostatectomy, radiotherapy, and anti-androgen therapy).

Ovarian cancer

Although the incidence of ovarian cancer (approximately 7000 new cases per year in the UK) is much lower than breast cancer, the death rate from ovarian cancer is high in the absence of early detection. The early symptoms of ovarian cancer—persistent pelvic and stomach pain, abdominal bloating, difficulty in eating, and increased frequency of micturition—are somewhat vague, but much has been done to improve early diagnosis and subsequent treatment of this malignancy. Three separate etiologies related to the cell of origin of the tumor have been identified (**Table 20.3**).

TABLE 20.3 Origins of ovarian cancer by cell type

Cell origin	Percentage of ovarian cancer patients
Ovarian epithelium	>85%
Stromal cells	5%
Germ cells	<1%

Ovarian epithelial tumors

Tumor cells originating from the ovarian epithelium constitute the majority of ovarian cancers. They are usually hormone inactive and are often described as serous or mucinous adenocarcinomas. Assay of a high-molecular-mass (200 kDa) glycoprotein in plasma, CA 125, is used to screen and monitor treatment in patients with this type of ovarian cancer. CA 125 is a carbohydrate differentiation antigen secreted by ovarian epithelia and is also found in fallopian epithelia, endometrium, endocervix, mesothelial cells, and peritoneal and pleural cavities. Although dependent on the particular assay used, the upper limit of the normal reference range is between 30 and 35 kU/L (30–35 U/mL); plasma concentrations >35 kU/L (>35 U/mL) are suggestive of a tumor. The plasma concentration of CA 125 increases with the size of the tumor and values in excess of 100 kU/L (100 U/mL) have been measured in 80% of patients in advanced stages of the disease. The plasma concentration of CA 125 is, however, raised in a variety of other conditions (**Table 20.4**) and these must be excluded before a definitive diagnosis is made.

TABLE 20.4 Conditions in which CA 125 may be raised in the absence of ovarian cancer

Condition	Possible serum CA 125 values
Pregnancy	<100 kU/L (U/mL)
Inflammation of the peritoneal and pleural epithelia, pleural effusion, and ascites	>500 kU/L (U/mL)
Peritonitis	>500 kU/L (U/mL)
Some benign ovarian cysts and fibromata	>2000 kU/L (U/mL)
Congestive cardiac failure	>200 kU/L (U/mL)
Pulmonary edema	>100 kU/L (U/mL)

A single measurement of CA 125 is not very useful in determining the cause of an abdominal mass. However, when used in conjunction with abdominal ultrasound, measurement of plasma CA 125 may be used for screening of ovarian cancer, and it is extremely valuable in monitoring response to treatment for the disease.

Other markers that are sometimes used in monitoring ovarian cancer (and also testicular cancers) are inhibins A and B, the structures of which have been described in Chapter 11. The inhibins are dimeric proteins with one common (α) and one variable (β) subunit. Inhibin A is used increasingly to monitor ovarian granulosa cell tumors and inhibin B is used to monitor Sertoli cell testicular cancers. The normal reference ranges for the inhibins are shown in **Table 20.5**.

Clinical practice point 20.2

CA 125 is a mesothelial marker. It is raised by pathology affecting the peritoneum and pleura. Very high levels may occur in ascites and pleural effusions of any cause, even in men.

TABLE 20.5 Normal reference ranges for inhibins in adults (pg/mL)

	Males	Females	
		Pre-menopausal	Postmenopausal
Inhibin A	<2.0	<97.5	<2.1
Inhibin B	<399	<139* <92**	<10

*Follicular phase; **luteal phase.

Stromal cell tumors

Stromal cell ovarian tumors (which are far less common than ovarian epithelial tumors) synthesize and secrete excessive amounts of estrogens and androgens and may be detected by analysis of plasma hormone concentrations. Increased levels of androgens may cause masculinization of the patient. Although not all female patients with slightly raised testosterone levels greater than 1.5 nmol/L but less than 5.0 nmol/L, (>43 ng/dL but <144 ng/dL) will have an androgen-secreting tumor, when the testosterone level is greater than 5 nmol/L (144 ng/dL), the presence of an androgen-secreting tumor of the ovary or adrenal gland is very likely.

Germ-cell tumors

These rare tumors secrete hCG and AFP and changes in the concentration of plasma hCG and AFP are used to monitor treatment.

Breast cancer

Breast cancer is second only to skin cancer in the common causes of cancer in women. Almost 200,000 new cases of invasive breast cancer and approximately 7000 cases of *in situ* breast cancer are found each year in the USA. Thus, detection of breast cancer is a major concern. Most breast tumors are detected via incidental findings on self-examination and mammography screening programs. There are currently no reliable markers for detection of breast cancer through screening. Adenocarcinoma markers and carcinoembryonic antigen in serum have been used in detection of breast tumors, but currently the only serum-based marker specific for tumors of the breast is cancer antigen CA 15-3, the product of the *MUC1* gene. This protein, MUC1, is a large transmembrane glycoprotein with intra- and extracellular domains linked by a membrane-spanning domain. It is synthesized in healthy as well as cancerous breast tissue. It is shed from the cells during turnover and appears in the blood. Its role in the cell is not known with any degree of certainty, but it has been implicated in cell–cell adhesion, immunity, and metastasis. Breast tumor cells contain higher amounts of a less glycosylated MUC1, CA 15-3, and consequently higher levels of CA 15-3 are found in the serum of patients with breast cancer. The serum concentration of CA 15-3 is increased in only approximately 30% of patients with early stage, localized tumors, but this figure rises to 50–90% of patients with metastatic tumors. Patients with a serum concentration >30 U/L have significantly shorter overall survival than those with lower concentrations. The use of CA 15-3 as a prognostic indicator appears to be independent of tumor size, axillary node status, and age of patient. It is widely used in monitoring response to treatment (chemo- and/or radiotherapy) in patients with established breast cancer.

A number of other tests are being used increasingly in some centers in an effort to tailor therapy to the individual patient. These tests are assayed in tumor biopsy tissue (formalin-fixed paraffin-embedded or fresh-frozen specimens) to help establish the nature of the particular cancer cells, and include measuring the expression of genes associated with malignancy; the rationale is that each tumor has a unique genetic makeup which determines the likelihood of it responding to therapy or undergoing metastasis. Some tests assay for the expression of single genes such as estrogen or human epidermal growth factor receptors, but recently, in larger research centers, the number of genes being studied has increased dramatically. Some markers that have been determined by immunohistochemistry in tumor sections are listed here:

- Androgen receptor
- Estrogen/progesterone receptor
- E-cadherin (cell adhesion marker)
- MIB1 (cell proliferation marker)
- Ki67 (cell proliferation marker)
- P53 (tumor suppressor gene)
- KS1 antigen (tumor suppressor gene)
- BRCA1 and BRCA2 (breast cancer susceptibility genes)
- Cathepsin D
- HER-2 protein receptor

In normal tissue, the processes of cell proliferation and apoptosis are tightly coupled so that tissue mass remains constant; that is, the genes governing these two opposing processes are tightly regulated. Unregulated growth such as is seen in cancer results from an uncoupling of proliferation from apoptosis and this can occur through mutations in these genes. Mutations in oncogenes which increase proliferation and in tumor suppressor genes which decrease apoptosis are features of tumors. Identification of such mutations, and mechanisms in which they can be overcome, is a feature of current cancer research. Multiparameter gene expression techniques using reverse transcriptase polymerase chain reaction (RT-PCR), DNA microarray, or RNA expression microarray allow the expression of a battery of genes in a particular tumor to be determined simultaneously, with the results fed into a scoring system which helps in the prediction of prognosis, response to therapy, and likelihood of metastasis. The results from such tests are used to determine the likely response of a particular tumor to therapy and help in the choice of treatment. Multiparameter gene expression assays are restricted to cancers which have not yet spread to neighboring lymph nodes. Once the efficacy of these techniques for the diagnosis and management of individual patients has been established, it is likely that they will become part of the routine armory for breast cancer management in general hospitals. Similar assays may be developed for other cancers but, in terms of numbers, routine demand may not be very high.

Gastrointestinal cancers

Colonic cancer

Cancer of the colon, one of the most common cancers in the developed world, is a good example of where early detection and treatment, in this case surgery, leads to improved outcomes. There are two main staging systems for colonic cancer: Duke's classification and TMN staging. Duke's classification is shown

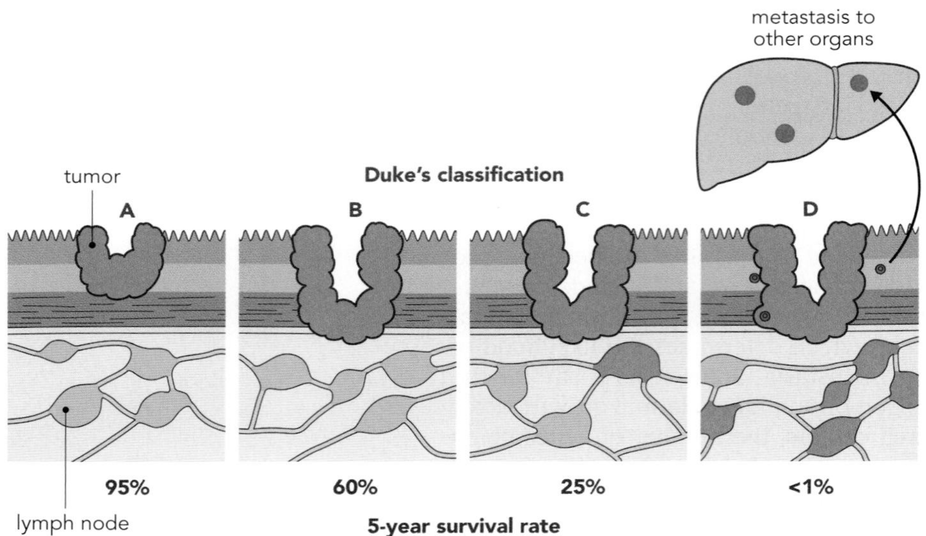

Figure 20.8 Duke's classification of colonic cancer.
In the earliest stage (A) of colonic cancer, the growth is mainly restricted to the mucosa, with only a small degree of penetration into the muscle wall. The next stage (B) is where the cancer has gone through the muscle wall but there is no local or widespread lymph involvement. Stage C is where there is local lymphatic involvement, and finally at stage D there are widespread, distant metastases, often involving the liver and lungs. The earlier the stage at detection, the better the prognosis.

in **Figure 20.8**, where the staging changes from A (mainly *in situ* within the mucosa, with no or very small infiltration into the muscle wall), to B (infiltration through the muscle wall), to C (involvement of at least one local lymph gland), through to D (distant spread, metastases to liver and lung usually). The TMN staging refers to the Tumors, Nodes, and Metastases, and ranges from stage 0 (where the tumor is confined to the mucosa and there is no infiltration of the muscle and no local or distant spread) all the way through to stage IV where there is both local and distant spread.

At the early stage of cancer (Duke's stage A or TMN stage 0), where there is no spread of the tumor, there is a greater than 95% chance of 5 years survival post-surgery, compared to less than 1% once the tumor has metastasized widely. Because of this and the relatively high incidence of the disease, many countries have adopted screening programs targeted primarily at people over the age of 50 years (where the disease is more common) and for high-risk groups.

There is at present no suitable serum-based marker for colon cancer. CEA was proposed but proved to be neither sufficiently sensitive nor specific to screen for colon cancer or other adenocarcinomas. Currently, most screening programs look for increased occult blood in feces, since many colonic cancers cause increased blood loss in the bowel. Fecal occult blood (FOB) is assayed on simple test cards containing the fecal samples and, through home testing, allows for large-scale screening programs. The frequency of testing subjects over the age of 50 years varies from country to country and is normally between 1 and 5 years (annually in the USA; every 3 years in pilot schemes in the UK). A positive FOB result is followed up by colonoscopy and, where necessary, a biopsy for definitive diagnosis by histology. Regular colonoscopy is recommended for higher-risk groups, and some centers use rigid sigmoidoscopy in addition to FOB testing. The advantages of FOB screening are (i) it is relatively inexpensive to screen large numbers of people and (ii) it can identify patients with less advanced (early stage) cancer and provoke early intervention. However, it has relatively low specificity, detecting less than 50% of Duke's stage A cases.

Testing for FOB as a screen for colon cancer is not without its problems, however. Bleeding into the gastrointestinal tract is not unique to colon cancer, and many screen-positive patients will not have cancer. Furthermore, the FOB test has a low sensitivity for early stage cancers (Duke's stage A) since many of these may not have eroded through the gut wall and will not have initiated blood loss. In these cases, the test will be negative despite the presence of a tumor, albeit small.

CASE 20.2

A 67-year-old woman presents with altered bowel habit, weight loss, and anemia. Investigations, including a colonoscopy and biopsy, indicate carcinoma of the colon. She undergoes surgery (day 0) to remove the tumor and CEA (carcinoembryonic antigen) is measured before and at intervals afterward.

	SI units (conventional units)	Reference range
Plasma		
CEA (day 0)	253 μg/L (ng/mL)	0–2.5
CEA (day 60)	55 μg/L (ng/mL)	0–2.5
CEA (day 120)	1.5 μg/L (ng/mL)	0–2.5
CEA (day 365)	10.1 μg/L (ng/mL)	0–2.5

- **What do the changes at the different dates signify?**

Following successful resection of the tumor, CEA levels should fall to normal after 4 to 6 weeks, in line with the half-life of CEA (3 days) and its initial level. The persistently high result at day 60 therefore indicates residual disease. This may be a metastatic deposit in the liver, for example. Following further treatment (such as chemotherapy) the result at day 120 indicates successful eradication of the tumor. However, at day 365 the CEA has risen again, indicating recurrent disease. This may pre-date clinical signs or symptoms by up to 6 months. It is important to note that up to one-third of advanced colorectal cancers do not have raised CEA, which limits its usefulness for monitoring known disease and makes it unsuitable for diagnosis.

A number of alternatives to FOB testing have been suggested, including assays of serum and fecal proteins such as CD44, calprotectin, decay-accelerating factor (DAF), CEA, and cancer antigen 199 (CA 199). Gene expression assays similar to those mentioned for breast cancer have also been proposed using K-ras, P53, adenomatous polyposis coli (APC), and microsatellite instability markers such as Bat-26. None is in routine use and many markers (such as calprotectin) are increased in many nonmalignant states.

Whilst the measurement of the acidic glycoprotein CEA is not suitable as a screening test for colon cancer because of lack of specificity, it is used widely in monitoring response to treatment. The upper end of the normal reference range varies somewhat with the method used, but is usually approximately 2.5 μg/L (2.5 ng/mL); with this value, there is an overall positivity of 67% in the detection of colon cancer. Values >10 μg/L (>10 ng/mL) suggest that malignancy is probable, whilst values >50 μg/L (>50 ng/mL) predict likely metastasis. In addition to colon cancer and other adenocarcinomas, serum CEA is also increased in pregnancy, inflammatory bowel disease (ulcerative colitis, Crohn's disease), pancreatitis, alcoholic cirrhosis, and in smokers.

Other gastrointestinal tumors

Pancreatic cancer

Although several markers of general gastrointestinal (GI) tumors are available, only one—a large 600 kDa sialylated glycoprotein, cancer antigen 19-9 (CA 19-9)—is used routinely in the diagnosis of pancreatic cancer. However, since CA 19-9 is part of the Lewis A blood group, A-negative and AB-negative patients (5–10% of population) who have pancreatic cancer will not express the protein and will be missed on screening. Reference ranges for CA 19-9 vary slightly with the individual assay method but a value <35 kU/L (<35 U/mL) is considered to be within the normal range. Grossly elevated concentrations are found in >80% of patients with advanced pancreatic cancer and, in these cases, measurement of CA 19-9 is of use in the diagnosis and the monitoring of patients undergoing treatment. Serum concentrations of CA 19-9 are usually only slightly increased (<60 kU/L or <60 U/mL) in patients with chronic pancreatitis, a disease which needs to be excluded in making the diagnosis of pancreatic cancer. On the other hand, serum CA 19-9 is increased in other tumors including colon cancer, and very high values (>500 kU/L or >500 U/mL) may be found in obstructive jaundice.

Gut hormone tumors

Islet-cell tumors of the pancreas

Gut hormone tumors synthesize and secrete excessive amounts of different hormones and give rise to characteristic symptoms. For example, insulinomas and glucagonomas—rare causes of hypo- and hyperglycemia (**Table 20.6**)—have been described in Chapter 2.

Other gut hormone tumors include those secreting gastrin, which causes excessive production of acid, and vasoactive intestinal polypeptide, which causes excessive intestinal motility and massive diarrhea. Both are discussed in more detail in Chapter 14.

TABLE 20.6 Gut hormone tumors

Tumor	Hormone produced excessively	Effects
Insulinoma	Insulin	Hypoglycemia
Glucagonoma	Glucagon	Hyperglycemia
Gastrinoma	Gastrin	Excessive acid production
VIPoma	Vasoactive intestinal polypeptide (VIP)	Excessive intestinal motility and massive diarrhea

Carcinoid syndrome

Carcinoid tumors produce excessive amounts of serotonin (5-hydroxytryptamine) which in some cases leads to pronounced flushing and diarrhea in patients (carcinoid syndrome). Measurement of serotonin in plasma, or of its metabolite, 5-HIAA, in urine, may be useful in the diagnosis and monitoring of such tumors, which may be found in the GI tract as well as the lung.

Liver cancer

Primary liver cancers, hepatoma and hepatoblastoma, are associated with elevated plasma concentrations of AFP, a 63 kDa protein with α_1 mobility on protein electrophoresis. The normal reference range for AFP is usually <10 μg/L (<10 ng/mL). As the name implies, AFP is present in fetal liver but it is also synthesized by primary tumors of the liver, being present at high concentrations (>50 μg/L or >50 ng/mL) in >80% of hepatocellular cancers (**Figure 20.9**) and 100% of hepatoblastomas. It is a very useful marker for monitoring primary liver cancers, but its use diagnostically is limited since plasma AFP is also raised in other tumors (for example testicular and ovarian) and in nonmalignant liver disease. Very high plasma AFP levels (>10,000 μg/L or >10,000 ng/mL) may also be seen in patients with tyrosinosis.

Secondary cancers of the liver, often arising from metastatic colon cancer, are generally more common than primary liver tumors. Unfortunately, to date there are no markers specific for secondary liver tumors.

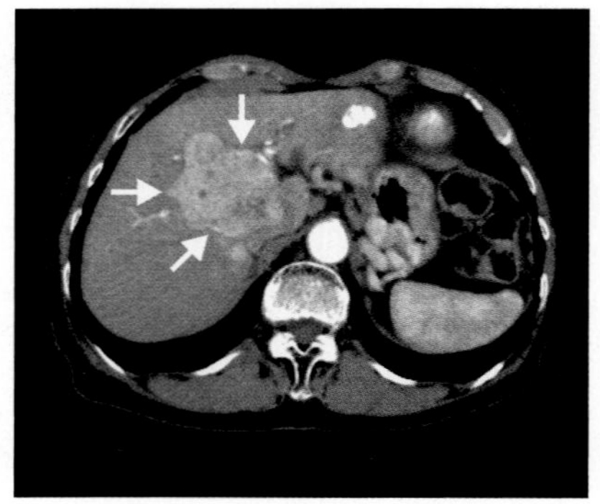

Figure 20.9 Hepatocellular carcinoma.
An enhanced computerized tomography scan showing multiple carcinoma (arrows). (From Kimura T, Umemura T, Ichijo T et al (2010) *IJCRI* 1:16-18. With permission from Edorium.)

Testicular cancer

Testicular cancer is relatively common in males under the age of 40 years. Germ-cell cancers account for approximately 95% of the cases of testicular cancer; roughly half are seminomas and half are nonseminomatous (choriocarcinoma, embryonal carcinoma, and yolk sac tumors). Seminomas are usually treated with orchidectomy and radiotherapy, whilst nonseminomatous tumors are usually treated by resection and chemotherapy. Both types respond well to the different treatments, with 10-year survival rates in excess of 95%.

Over 80% of seminomas, during and after treatment, can be monitored using a combination of measurements of plasma hCG, placental alkaline phosphatase, and lactate dehydrogenase. Nonseminomatous germ-cell tumors are monitored by measuring hCG and AFP. As described in Chapter 12, hCG is a 39.5 kDa glycoprotein with a plasma half-life of 24 hours. The normal reference range for hCG in plasma is <5 IU/L (<5 mIU/mL). The plasma hCG concentration is increased in 75% of teratomatous and combined tumors and in 30% of seminomas and dysgerminomas. It is also increased in hydatidiform moles and gestational choriocarcinomas. A serum–cerebrospinal fluid hCG ratio >60:1 is indicative of brain

CASE 20.3

A 53-year-old woman with severe cardiac failure due to ischemic heart disease is admitted to hospital with worsening shortness of breath. She has noticed weight gain and increasing abdominal girth. On examination she has evidence of ascites and a pleural effusion, as well as pitting edema of her ankles. Amongst the tests requested is a tumor marker panel.

	SI units (conventional units)	Reference range	Conventional units	Reference range
Plasma				
CEA (carcinoembryonic antigen)	1.3 μg/L	0–2.5	1.3 ng/mL	0–2.5
CA 125 (cancer antigen 125)	468 kU/L	0–35	468 U/mL	0–35
CA 19-9 (cancer antigen 19-9)	17 kU/L	0–27	17 U/mL	0–27
AFP (alpha-fetoprotein)	5 μg/L	0–10	5 ng/mL	0–10
hCG (human chorionic gonadotropin)	7 IU/L	0–5	7 mIU/mL	0–5

- **What is the indication for measuring CA 125?**
- **What are the explanations for the elevated CA 125 and hCG?**
- **What are the pitfalls of tumor marker panels?**

CA 125 is used as a marker of ovarian cancer, although its diagnostic sensitivity and specificity are poor. Abdominal bloating in a woman may be caused by ovarian cancer and some guidelines do recommend that it is measured before pelvic ultrasound. CA 125 is a molecule of mesothelial origin and may be greatly raised by conditions that involve the pleural and peritoneal membranes, including effusions and ascites of any cause, even congestive heart failure. In postmenopausal women, hCG levels may be slightly raised due to pituitary secretion. The measurement of tumor markers in patients without known cancers is likely to produce false positive and false negative results and should be avoided.

metastases. Since AFP has a plasma half-life of approximately 5 days, it may be measured after excision to establish residual tumor and/or metastasis. It is essential to measure liver function tests concurrently to eliminate false positives. In current practice, with both types of tumor, it is recommended that markers (AFP, hCG, LDH, and placental alkaline phosphatase) be measured before treatment, 5 days after treatment, and thereafter twice weekly for 8 weeks or until values return to the normal range. After this time point, measurements should be made weekly for six months, monthly for 2 years, and finally quarterly for an additional 2 years. Values within the normal range after this time period indicate elimination of the tumor in the majority of cases.

A pretreatment plasma concentration of hCG is used as an indicator of prognosis, where hCG >100,000 IU/L (>100,000 mIU/mL) predicts a 1-year survival rate of <60% whilst hCG <50,000 IU/L (<50,000 mIU/mL) predicts a 1-year survival rate of >90%. Measurement of hCG has also been used in the diagnosis and monitoring of hydatidiform moles and gestational choriocarcinoma.

Bladder cancer

Traditionally, urinary cystoscopy has been the method of choice in the diagnosis and monitoring of bladder cancer. However, more recently, several candidate urinary markers have been developed, and one—nuclear matrix protein 22 (NMP22)—is now in routine use. It is available as a point-of-care device for the diagnosis and monitoring of bladder cancer, with a better detection rate of recurrence than urinary cystoscopy. Nuclear matrix proteins, derived from cytoskeletal filaments of the nucleus, are released from the cell during apoptosis as part of cell turnover, appearing initially in blood and eventually in urine. NMP22 forms the head domain of the nuclear mitotic protein NuMA. The concentration of the NuMA antigen is tenfold higher in bladder cancer tissue than normal tissue. Complexed and fragmented forms of the protein, including NMP22, are found in urine. Measurement of NMP22 has been invaluable in diagnosing and monitoring bladder cancers. High levels of NMP22 are also present in the urine of patients with transitional cell carcinoma of the urinary tract.

Lung cancer

Although lung cancer is one of the most common causes of death from cancer, tumor markers are not used extensively in its diagnosis or in the monitoring of treatment. Adenocarcinoma and squamous cell carcinomas are usually detected by chest X-ray and treated, where appropriate, with radical surgery to remove the tumor. Measurement of plasma CEA is used only rarely in monitoring outcomes in adenocarcinoma of the lung. Two markers, squamous cell antigen (SCA) and cancer antigen 211, have been used to monitor squamous cell carcinomas of the lung, but are not in widespread routine use.

In lung cancers that respond to chemotherapy rather than surgery (small-cell and oat-cell carcinomas) the neuroendocrine marker neuron-specific enolase (NSE) may be used to monitor treatment. Other neuroendocrine markers, such as chromogranin A—an acidic, monomeric, 52 kDa glycoprotein which is used to monitor pheochromocytoma (see Chapter 8)—can also be used to monitor these lung cancers.

Thyroid cancer

The five general types of thyroid cancer and their relative incidence are shown in **Table 20.7**.

The normal reference range for thyroglobulin, a high-molecular-mass protein (800 kDa) stored in colloid in thyroid tissue (see Chapter 11), in serum is <60 μg/L (<60 ng/mL), but this is exceeded in patients with papillary and follicular thyroid cancer. Patients are treated with surgery and/or radio-iodine to remove the tumor, and their serum thyroglobulin is monitored subsequently.

TABLE 20.7 Relative incidence of the general types of cancer of the thyroid	
Type	**Relative incidence as percentage**
Papillary	60–70%
Follicular	Approximately 15%
Mixed Papillary-Follicular	5–10%
Medullary	Approximately 5%
Anaplastic	Less than 1%

Typically, the serum thyroglobulin concentration post-surgery should be <2 μg/L (<2 ng/mL). A serum thyroglobulin >2 μg/L (>2 ng/mL) suggests a recurrence of the tumor. Analytically, the assay of thyroglobulin is affected by endogenous antibodies to thyroglobulin itself; because of this, these antibodies are also measured routinely in these patients.

Calcitonin (molecular mass 3600 kDa) is routinely measured in the monitoring of medullary carcinoma of the thyroid (MCT). Although it is not recommended for general screening for MCT, its measurement is of great use in screening at-risk groups, such as patients with multiple endocrine neoplasia II.

Plasma cell dyscrasias

Plasma cell dyscrasias, such as multiple myeloma, MGUS (monoclonal gammopathy of undetermined significance), and Waldenstrom's macroglobulinemia, were described in Chapter 19. They are characterized by the appearance of increased amounts of paraproteins in both serum and urine. Paraproteins appear as an abnormal, dense, narrow band on electrophoretograms, and although they usually have γ mobility, they may be present anywhere between the α_2 and γ regions. In multiple myeloma the major immunoglobulins are IgG and IgM with an abundance of free light chains. The appearance of paraproteins is not always indicative of malignancy.

The synthesis of immunoglobulin light and heavy chains by plasma cells is not stoichiometric and an excess of free polyclonal light chains is made. These excess light chains are filtered in the kidney and any that are not metabolized in the renal tubules appear in urine, such that minute, but detectable amounts of free light chain are present even in urine of normal subjects. Patients with renal tubular damage excrete larger amounts of free light chains. Only low-molecular-weight proteins, such as light chains (22 kDa) can be filtered by the kidney; whole IgG (150 kDa) and IgM (800 kDa) molecules are too large to filter and appear in urine only where there is severe damage to the glomerulus. In patients with significant glomerular damage, leakage of all serum proteins, including paraproteins, may occur and may give false positives for myeloma unless the free light chain nature of the M band seen on electrophoresis is confirmed.

Bence Jones proteins are monoclonal κ or λ free light chains found in urine of patients with hematological malignancies. Detection of Bence Jones proteins involves protein electrophoresis of concentrated urine, detection of the M band, and confirmation of monoclonal κ or λ free light chains by immunofixation. Light-chain myeloma is often associated with renal tubular damage due to overload of the capacity of the renal tubules to metabolize and excrete the excessive production of light chains. In this situation, the renal tubules become dilated and accumulate aggregates of protein, which, unless removed by diuresis, leads to atrophy and eventual necrosis of the tubular cells.

Estimates of tumor cell mass in myeloma may be made by measuring the turnover rates of the monoclonal proteins and, in conjunction with measurements of C-reactive protein and β-macroglobulin, may allow prediction of survival before and after therapy.

20.4 THE FUTURE

Although tumor markers are now widely used to help to diagnose and monitor cancer, significant improvements are required in the performance of most of the markers available today. There is a need for candidate markers with better diagnostic sensitivity and specificity, which will detect cancers at an early stage, identify those patients where treatment will be of benefit, be able to monitor the cancer progress readily, but minimize the number of false positives. In some cases it will be necessary to improve the analytical sensitivity and specificity, to detect the presence of just a few molecules of marker per milliliter of blood, and to be sure the assay does not cross-react with other proteins in the blood. It is to be hoped that continued progress in genomics and proteomics may enable some of these candidate markers to be uncovered.

Technical developments in analysis platforms, such as powerful but affordable time-of-flight tandem mass spectrometry, and developments in computer analysis are highlighting candidate markers. These techniques may be used on large blood and tissue banks collected over many years from numerous patients, who have been part of screening programs for both malignant and nonmalignant disease, but who may have developed cancer(s) different to the ones being investigated in the original study. From such longitudinal studies, potential tumor markers may be identified and this may lead to the development of simple assays for routine use.

However, it is essential to be mindful of the heterogeneity of cancer, both in its initial appearance and development, and therefore the holy grail of 100% diagnostic sensitivity and specificity may never be achieved.

THERAPEUTIC DRUG MONITORING

CHAPTER

21

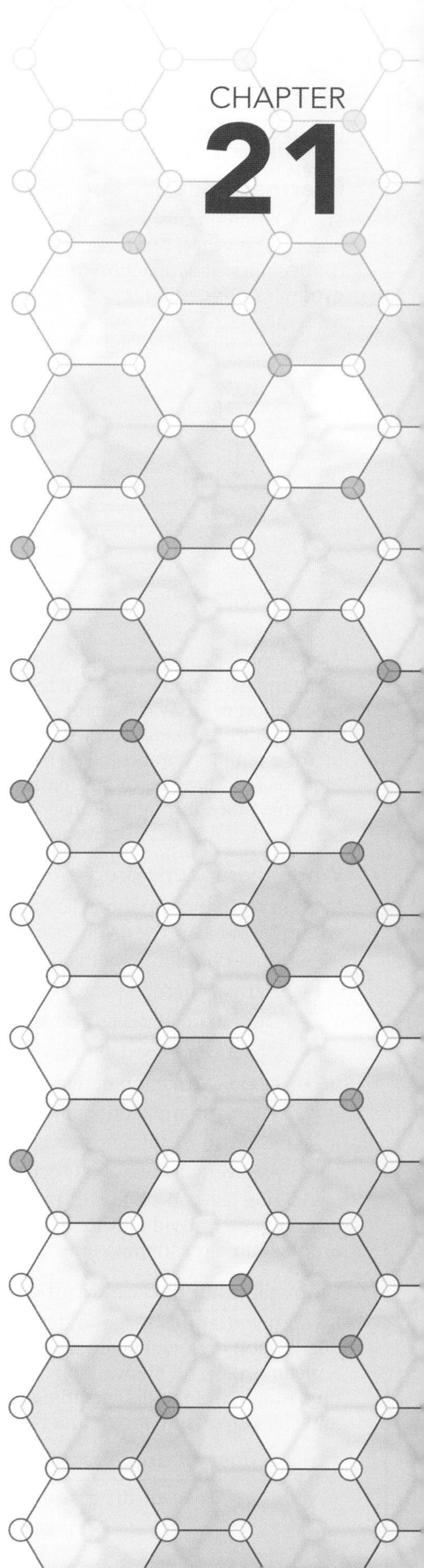

IN THIS CHAPTER

GENERAL ASPECTS OF THERAPEUTIC DRUG MONITORING

DRUG MONITORING IN PRACTICE

Therapeutic drug monitoring (TDM) refers to the assessment of tissue drug concentrations to optimize the dosage for patients receiving drug therapy for treatment (prevention, cure, or alleviation of symptoms) of disease. In practice, the blood concentration is measured as a surrogate for tissue levels. The rationale for its use is that for any particular therapeutic effect there is an optimal range of tissue concentration for the drug; below this range the drug is less effective, while above the range the drug might be toxic. TDM may also be used to monitor patient compliance to a particular therapy. Perhaps surprisingly, however, despite the large number of drugs that are prescribed daily, very few are measured in routine practice as part of a TDM program. There may be a number of reasons why this is so and unless knowledge of a drug concentration fits certain criteria, its routine measurement may offer little help in the management of a patient. Paradoxically, it is of interest to consider when TDM may not be of value (**Table 21.1**) even though an assay of drug concentration may be readily available.

TABLE 21.1 Clinical situations where therapeutic drug monitoring (TDM) would be inappropriate

Clinical situation	Examples
The clinical response is easy to assess	β-Blockers and blood pressure
	Insulin and glucose
	Statins and cholesterol
The drug–dose response is predictable	Most analgesics
The drug has low toxicity at higher concentrations	Gabapentin
Pharmacokinetics are not related to pharmacodynamics (that is, the action of the drug does not correlate well to its measured concentration)	Prednisolone
	Valproate

21.1 GENERAL ASPECTS OF THERAPEUTIC DRUG MONITORING

Pharmacokinetics and pharmacodynamics

From the final point listed in Table 21.1, it may be inferred that TDM can only be of value when there is a relationship between the tissue concentration of a drug and its clinical action (pharmacokinetics must be related to the pharmacodynamics; **Figure 21.1**).

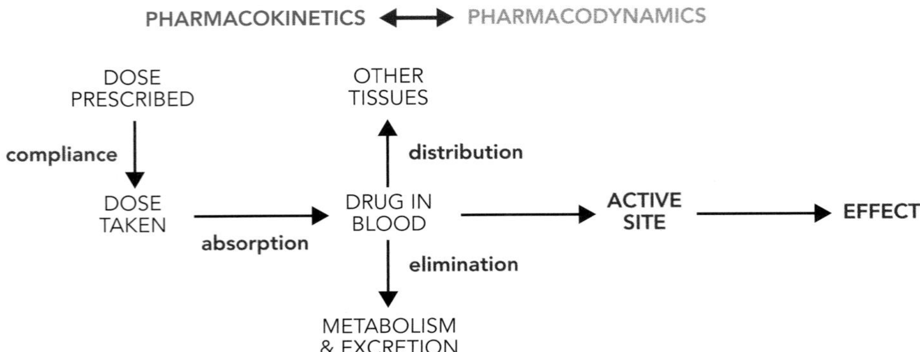

Figure 21.1 For therapeutic drug monitoring to be useful, the pharmacokinetics and pharmacodynamics of a drug need to be dependent.
The figure shows that all of the aspects of the pharmacokinetics of a drug–the absorption, distribution, metabolism, and excretion (or elimination) (ADME)—need to be related to the pharmacodynamics.

For most drugs for which TDM is used, the absorption, distribution, metabolism, and excretion (ADME) of the drug have been well established to enable the calculation of clear ranges of therapeutic effectiveness. Such an understanding of the pharmacokinetics of a drug also allows prediction of the consequences of, for example, compromised liver or renal function on blood levels of the drug. Indeed, TDM may be of particular value in indicating whether a patient is in an unstable state through the use of more frequent assays when the patient presents with renal, hepatic, or gastrointestinal problems.

When therapeutic drug monitoring is useful

In contrast to the situations listed in Table 21.1, there are a number of situations where TDM has proved very useful, including:

- Poor correlation between dose and effect
- Narrow therapeutic window
- Drug toxic at doses above therapeutic window
- Absence of good clinical markers
- Good correlation between plasma level and effect
  - Low pharmacodynamic variability
  - Active metabolites can be measured
  - Reversible action at receptor

To this list can be added the assessment of patient compliance to a given drug regime, providing that the blood concentration of the drug can be measured within a reasonable time of administration of the drug.

Bioavailability and distribution

It is prudent at this stage to define two terms, bioavailability (F) and volume of distribution (V_d), which are used regularly in discussing the pharmacokinetics of a given drug. The bioavailability (F) of a drug administered orally is dependent on a number of factors including formulation, first-pass metabolism in the liver, and salt conversion factor and is defined by the equation

$$\mathbf{F} = \frac{\textbf{dose of drug reaching circulation}}{\textbf{dose of drug administered}}$$

The distribution of an administered drug in the body is also dependent on a number of factors, including:

- Plasma protein binding
- Tissue and receptor uptake
- Localized blood flow
- Body weight and surface area
- Body fat

Many drugs bind to a greater or lesser extent to plasma proteins and it is important to remember that it is the concentration of free drug that is important therapeutically. It is usually the free drug that is involved in tissue and receptor uptake, and this in turn will be affected by the local blood flow to that particular organ or tissue. Although there is an interrelationship between body weight, surface area, and body fat, the variable amount of fat in a patient's body will have a very marked effect upon the distribution of lipophilic drugs.

The volume of distribution is defined by the equation

$$V_d = \frac{\textbf{amount of drug in the body}}{\textbf{plasma concentration of drug}}$$

Drugs that have predominantly intracellular distribution (for example lithium) or are highly lipid soluble (for example amiodarone) have a large V_d and remain within the body for prolonged periods after treatment is stopped.

Metabolism and excretion

The rate of metabolism and excretion of a drug is also dependent on body weight and surface area as well as liver and kidney function. What is critical in terms of TDM is the plasma half-life of a drug ($t_{½}$) which is related to both bioavailability and volume of distribution. Knowledge of the pharmacokinetics of a drug, including $t_{½}$, enables the frequency of dosing and the timing of blood sampling to be determined. Measurements are made at peak and trough levels, the latter being the lowest steady-state value to enable optimum effect of the drug. A sufficiently high peak concentration is required to achieve maximal effect, whilst a low trough concentration minimizes tissue exposure and toxicity. What does this mean in practice? Most, but not all, drug concentrations are measured at troughs; that is, at the lowest steady-state value and just before the next planned dose. However, it is also useful for some drugs with unpredictable pharmacokinetics to measure the highest concentration; that is, the peak level. For optimum treatment, both the peak and trough level should be within the drug's therapeutic window.

For most drugs there is a concentration difference between peak and trough levels in the plasma (**Figure 21.2**). This variation varies between drugs; for example, in some cases, cyclosporine levels may differ fivefold. In such cases, timing of sampling is critical.

In addition, knowledge of $t_{½}$ allows the calculation of when a new steady-state concentration will be reached following a change in dose of the drug. Usually, the longer the $t_{½}$ for a given drug, the longer it takes to reach a new steady state (see Figure 21.2). It is not usually recommended to repeat a drug measurement after a change in dosage until a new steady state has been reached. This requires a length of time equal to five times the $t_{½}$. Knowledge of the pharmacokinetics and pharmacodynamics of any drug enables a therapeutic range to be determined. If the plasma concentration falls below this range there is a reduction in clinical effectiveness of the drug, while above this range there is a danger of the drug becoming toxic with unwanted side effects. Of necessity, such a therapeutic range, which elicits an optimal clinical response in a patient, is only an estimate and may vary from patient to patient. In addition,

Analytical practice point 21.1

Always note the time of the last dose of the drug, and the time the blood sample was taken. For the majority of drugs there is a large difference between trough and peak levels, and failure to accurately time sample collection in relation to dose can lead to misinterpretation and inappropriate clinical management.

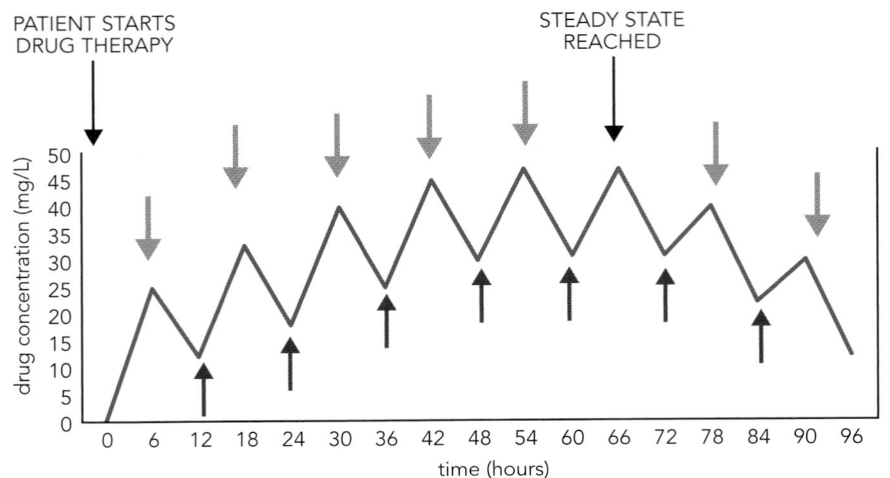

Figure 21.2 Peaks and troughs of drug concentration in plasma. The graph shows the concentration of an orally administered drug measured in plasma over several days. The drug was commenced at 0 hours and given every 12 hours at the same dose. The chart shows the peaks (yellow arrows) and troughs (red arrows) in concentration over this period. The peaks and troughs for the drug concentration rise until a steady state is reached, which is marked by the black arrow. At this time, the dose was decreased, and the peaks and troughs can be seen to fall again.

other factors may affect bioavailability, including interactive effects through co-administration of a second drug. This is particularly important when the co-administered drug competes for binding to plasma proteins, displacing the first drug and leading to increased plasma levels of the free drug with the potential for toxicity, even though the total plasma concentration of the first drug is still within the therapeutic range.

Some drugs for which TDM is appropriate are listed in **Table 21.2**, which is intended not to be exhaustive but to reflect those drugs commonly measured in most large pathology laboratories.

In addition, therapeutic monitoring of methadone is used in assessing addiction patients, and this is discussed in Chapter 22.

TABLE 21.2 Some drugs that are commonly measured in clinical practice

Drug group	Drug
Aminoglycoside and glycopeptide antibiotics	Gentamicin
	Tobramycin
	Amikacin
	Vancomycin
	Teicoplanin
Anticonvulsants	Phenytoin
	Lamotrigine
	Carbamazepine
Immunosuppressants (transplant drugs)	Cyclosporine
	Tacrolimus
	Sirolimus
	Mycophenolic acid
Others	Digoxin
	Lithium
	Theophylline
	Caffeine
	Methotrexate

21.2 DRUG MONITORING IN PRACTICE

Antibiotics

The majority of prescribed antibiotics are administered orally and, in most cases, there is little need, if any, to monitor blood levels. Clinical effectiveness can be determined by alleviation of symptoms and although overdosing may in some cases cause serious side effects (hepatitis with some penicillins, for example), the measurement of drug levels in the blood does not help in predicting these effects. However, there are two small groups of antibiotics, the aminoglycosides (gentamicin, tobramycin, and amikacin) and the glycopeptides (vancomycin and teicoplanin), that are given parenterally and for which the measurement of blood levels is an extremely important aspect of clinical management. Aminoglycoside antimicrobials are derived from moulds of the species *Streptomyces* (streptomycin, kanamycin, tobramycin, and neomycin) and *Micromonospora* (gentamicin and netilmicin). Examples of their structures are shown in **Figure 21.3**.

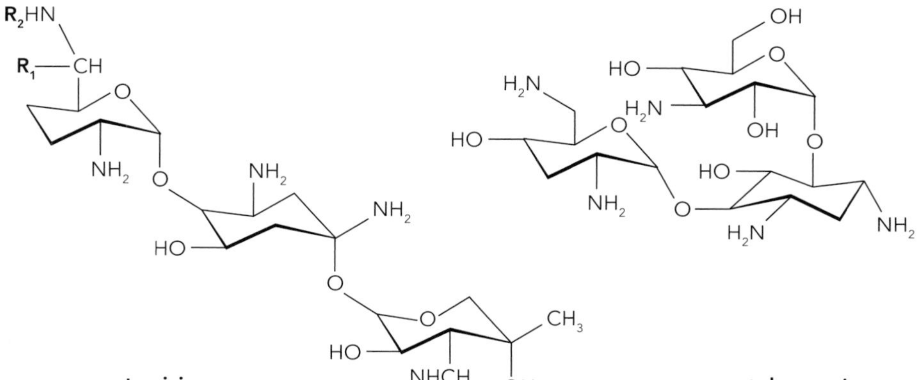

Figure 21.3 Chemical structures of gentamicin and tobramycin. There are three main types of gentamicin, which have different R_1 and R_2 groups:

Gentamicin C_1; R_1 = $-CH_3$ and R_2 = $-CH_3$

Gentamicin C_1a; R_1 = $-CH_3$ and R_2 = $-H$

Gentamicin C_2; R_1 = $-H$ and R_2 = $-H$

These molecules are very effective at killing bacteria but are also very toxic, particularly to the inner ear. Side effects thus include deafness and loss of balance. Neomycin is only used topically and is not routinely measured. Kanamycin can be given orally, intravenously or intramuscularly, whilst netilmicin is only given intravenously. These two aminoglycosides are only usually measured in specialist laboratories. Aminoglycosides are water soluble with a very low bioavailability (<5%) and volume of distribution (0.15–0.4 L/kg). Their short half-life of approximately 2–3 hours is markedly increased in renal failure and may then exceed 100 hours. Particular care is therefore required when administering these drugs to patients with renal impairment. The therapeutic ranges for these drugs are given in **Table 21.3**.

TABLE 21.3 Therapeutic ranges for some aminoglycoside and glycopeptide antibiotics

Antibiotic	Peak level*	Trough level
Gentamicin	5–10 mg/L	<2 mg/L
Tobramycin	5–10 mg/L	<2 mg/L
Amikacin	20–30 mg/L	<10 mg/L
Kanamycin	20–30 mg/L	<10 mg/L
Netilmicin	5–12 mg/L	<4 mg/L
Vancomycin	20–40 mg/L	5–10 mg/L

*The peak level is usually taken 30 minutes after the end of the intravenous infusion of the drug, except for vancomycin, which is usually taken at 60 minutes. All values are given in mg/L which is equivalent to μg/mL; that is, 5 mg/L = 5 μg/mL.

There are a number of reasons why TDM of these drugs is essential, including their variable pharmacokinetics, the adverse effects of prolonged therapy, the impact of renal impairment on $t_{½}$, their toxicity, and the potential catastrophic effects to the patient of not recognizing early therapeutic failure.

Antiepileptic drugs

Assessment of the efficacy of treatment of patients receiving medication for prevention of epileptic fits (seizures) can be particularly challenging. This is because fits may occur intermittently and unpredictably, hence it is difficult to know whether a drug is having an effect. Many antiepileptic drugs have pharmacokinetic properties which are related to their pharmacodynamic properties which is important in therapeutic monitoring. Furthermore these drugs are relatively easy to measure in the laboratory. These two factors mean that these drugs are routinely measured in clinical practice. However, other drugs in common use have variable pharmacokinetic behavior and are toxic at high concentration in blood. In addition, many novel antiepileptic drugs have become available in recent years and the therapeutic ranges for these are not well established. For those with low toxicity, TDM is of lesser importance. From a long list of antiepileptic drugs that are used, only the levels of a few are measured regularly in routine practice (**Table 21.4**). Some specialist centers offer TDM services for some of the less commonly used drugs or drugs with less well characterized pharmacokinetics.

TABLE 21.4 Classes of antiepileptic drugs and where they are measured

Class	Drug
Aldehydes	Paraldehyde[3]
Barbiturates and related compounds	Phenobarbital (phenobarbitone)[1,3]
	Primidone[1]
Benzodiazepines	Clobazam[2]
	Clonazepam[2,3]
	Diazepam[3]
	Midazolam[3]
	Lorazepam[3]
Carboxamides	Carbamazepine[1]
	Oxcarbazepine[2]
Fatty acids and GABA analogs	Valproic acid[1]
	Vigabatrin[2]
	Tiagabine[2]
	Gabapentin[2]
	Pregabalin[2]
Fructose derivatives	Topiramate[2]
Hydantoins	Phenytoin[1,3]
	Fosphenytoin[3] (a pro-drug of phenytoin)
Pyrrolidines	Levetiracetam[2]
Succinimides	Ethosuximide[2]
Sulfonamides	Zonisamide[2]
Triazines	Lamotrigine[2]

[1]Routinely measured in most laboratories; [2]measurement available in specialist laboratories;[3]primarily used in status epilepticus. GABA, gamma-aminobutyric acid.

Therapeutic drug monitoring is usually used in the long-term treatment of epilepsy but it may also be used in the acute treatment of status epilepticus where drugs are given intravenously. However, in the vast majority of these latter, acute interventions, plasma measurement of the administered drugs is not required since the effectiveness is immediately apparent. Occasionally, a drug may be administered as part of both the chronic and acute management of a patient and in such cases TDM is helpful to avoid overdose. For example, it may be useful to measure phenytoin in a patient on long-term treatment where intravenous phenytoin is being considered for status epilepticus. There would be no benefit in giving this if the phenytoin level was already high.

Carbamazepine and phenytoin

As may be inferred from Table 21.4, carbamazepine and phenytoin (**Figure 21.4**) are the two antiepileptic drugs most often measured routinely as part of TDM in the majority of clinical laboratories. They have been in clinical use for many years. These two drugs can be assayed rapidly on most modern analytical platforms and this plays an important role in the monitoring of patients. The properties of the two drugs are listed in **Table 21.5**.

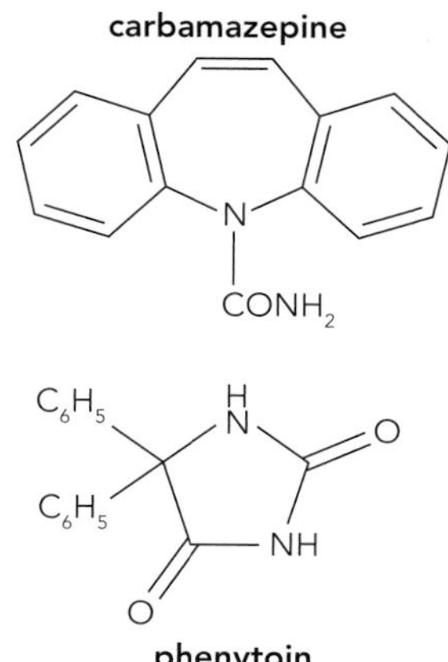

Figure 21.4 Chemical structures of carbamazepine and phenytoin.

TABLE 21.5 Properties of carbamazepine and phenytoin

Property	Carbamazepine	Phenytoin
Bioavailability	75%	90%
Protein binding	75%	93%
Half-life	25–45 hours (chronic 8–24 hours)	9–22 hours* (chronic 20–40 hours)
Target ranges	4–12 mg/L (µg/mL)	10–20 mg/L (µg/mL)

*Can be >100 hours at high doses and <7 hours in some children.

Phenytoin

Phenytoin can be considered as the prototypical drug for TDM because of its unpredictable dose–response characteristics, which are due to its saturable (zero-order) pharmacokinetics.

Although not the drug of first choice in the treatment of epileptic seizures, phenytoin is still commonly used, especially as an anticonvulsant in the elderly. The bioavailability of phenytoin is ~90% and approximately 93% of the drug present in plasma is bound to proteins (see Table 21.5). The drug has a variable $t_{1/2}$ (9–22 hours) and is metabolized mainly in the liver. Hepatic impairment lengthens the $t_{1/2}$. Chronic administration of phenytoin may extend the $t_{1/2}$ to 20–40 hours due to an inability of the hepatic enzymes to cope with a saturating concentration of the drug (that is, phenytoin has dose-dependent pharmacokinetics); extremely high doses may further extend the $t_{1/2}$ to over 100 hours. The saturation of the hepatic enzyme system can lead to a sudden rise in the plasma concentration of the drug and consequent toxicity. Monitoring of the plasma concentration of phenytoin is essential when changing doses, particularly in patients with concomitant conditions which affect hepatic metabolism. The therapeutic plasma concentration range for phenytoin is 10–20 mg/L (10–20 µg/mL), but although trough concentrations are usually measured (as with many drugs), the difference between peak and trough concentrations is not very great once a patient is on a stable dose. It is worth remembering, however, that phenytoin toxicity may occur even within the therapeutic range if a second drug, such as valproic acid, is administered simultaneously and displaces the large protein-bound fraction of phenytoin (inactive) into plasma as free, active drug. Similar signs of toxicity may occur when an increase in dosage of phenytoin saturates the number of plasma protein binding sites and increases the amount of free drug present in plasma.

The plasma concentration of free phenytoin may also be affected by interactions with other co-administered antiepileptic drugs in complex and unpredictable ways. The major problems arise when the co-administered drug alters the rate of hepatic metabolism of phenytoin. In a similar manner, phenytoin itself can have profound effects on the circulating concentrations of other antiepileptic drugs. It is essential therefore to monitor plasma concentrations regularly when moving from monotherapy to multidrug therapy.

Carbamazepine

Carbamazepine is commonly used in the treatment of epilepsy, especially grand mal (tonic-clonic seizures) and focal seizures. It is also frequently used to treat bipolar disease and trigeminal neuralgia. The bioavailability of carbamazepine (75%) is less than that of phenytoin and a smaller fraction (75%) is protein-bound in plasma. Carbamazepine is metabolized slowly in the liver and has a somewhat long $t_{½}$ of 25–45 hours. With chronic use, however, the $t_{½}$ may shorten to 8–24 hours due to induction of the hepatic enzymes. Thus in chronic therapy, the dose of carbamazepine has to be increased to remain effective.

While there is a good relationship between plasma concentration and patient response, trying to predict plasma concentration from a given dose is difficult due to variation between patients. For this reason, TDM is often required in successful management of patients. Measurement of trough concentration (therapeutic range 4–12 mg/L or μg/mL) is used in these cases since peak levels may vary widely after each dose. As is the case with phenytoin, combination therapy of carbamazepine with other drugs may have important effects on the effective plasma concentration of carbamazepine. For example, phenytoin and phenobarbitone can increase carbamazepine concentration, whilst valproic acid and lamotrigine can decrease it. Such effects are variable between patients and are the result of actions of the co-administered drugs on the rate of hepatic metabolism of carbamazepine; this again highlights the necessity of monitoring patients on combination therapy.

The first step in the metabolism of carbamazepine is conversion to a short-lived ($t_{½}$ <5 hours) epoxy derivative which also has considerable pharmacological activity. This product is not measured routinely. Interestingly, carbamazepine also has an antidiuretic effect in many patients and may lead to significant hyponatremia. There is no good clinical evidence to support therapeutic monitoring of the structurally similar drug oxcarbazepine, and assay of this drug is restricted to specialist centers.

Valproic acid (valproate)

Valproic acid is commonly prescribed for the treatment of epilepsy and, to a lesser extent, bipolar disease. Many laboratories offer measurement of the drug but because of a poor relationship between its pharmacokinetic and pharmacodynamic properties, it is not included in routine TDM programs apart from determining patient compliance, therapeutic failure, and possible overdose.

Phenobarbital (barbiturates)

Although phenobarbital was once used frequently in controlling epileptic fits, it is not as commonly used today due to its side effects and the emergence of more effective drugs, and its pro-drug form primidone is rarely if ever used. The drug has a very long half-life of up to 5 days, with a poor correlation of pharmacokinetics and pharmacodynamics. The quoted therapeutic range of 10–40 mg/L (10–40 μg/mL) should be treated with caution, with many patients having good clinical control with concentrations well below 10 mg/L (10 μg/mL). Similarly, some patients exhibit overdose symptoms (for example drowsiness) at concentrations below the therapeutic range.

Since many patients develop a tolerance to phenobarbital, much higher plasma concentrations of the drug may be required (>60 mg/L; 60 μg/mL) to have optimal effect, or produce signs of toxicity. Plasma concentrations are also increased by the co-administration of valproate.

Immunosuppressive drugs

Cyclosporine

Cyclosporine is a cyclic peptide of 11 amino acids (**Figure 21.5**) isolated from the fungus *Tolypocladium inflatum* Gams. It has potent immunosuppressive properties and was one of the first drugs used to prevent rejection of donor organs in a wide range of transplants, including kidney, heart, lung, and liver. It is a calcineurin inhibitor and inhibits activation of T cells, thereby reducing the chances of organ rejection. The drug has a very narrow therapeutic window, high concentrations of the drug being extremely toxic particularly to the kidney. This, coupled with the catastrophic consequences of inadequate treatment (that is, rejection of a donor organ), makes the continued monitoring of blood concentration of the drug vital. In addition to its use in the management of organ transplant patients, cyclosporine is also used to prevent graft rejection in bone marrow transplants. More recently it has been used in the treatment of a wide range of autoimmune disorders including ulcerative colitis, rheumatoid arthritis, and nephrotic syndrome and in some skin complaints including psoriasis.

Cyclosporine may be administered either orally or intravenously. The pharmacokinetic properties of the drug are shown in **Table 21.6**.

Cyclosporine is a highly lipophilic molecule binding avidly to plasma proteins and is taken up rapidly into tissues. The concentration of the drug in erythrocytes is at least three times that in plasma, and whole blood concentrations (using EDTA as a preservative) should be assayed rather than plasma concentrations.

Figure 21.5 Chemical structure of cyclosporine.

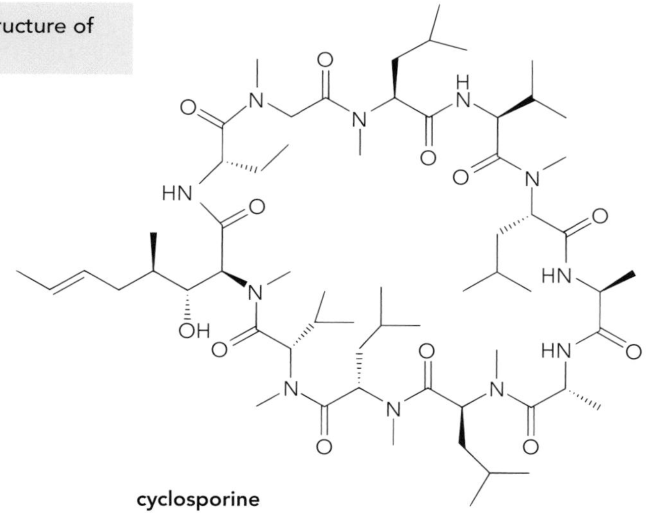

cyclosporine

TABLE 21.6 Properties of cyclosporine

Property	Value
Bioavailability	20–60%
Protein binding	98%
Half-life	5–50 hours
Target ranges	Variable

The pharmacokinetics of cyclosporine are quite complex because of variations in the rate of absorption between patients and the degree to which the drug is metabolized in both the gastrointestinal tract and the liver. Metabolism of cyclosporine in the liver is via the inducible CYP3A system and it is thus affected by co-administered drugs which may modulate this system and also by genetic variation in expression of CYP3A.

When administered orally, less than 50% of the dose appears in the circulation (range 20–60%), with peak concentration achieved within 2 hours in the majority of patients. Because of this wide individual variation in the pharmacokinetics of cyclosporine, the $t_{½}$ of the drug may vary more than tenfold (5–50 hours). For this reason, the effects of the drug correlate far better with the blood concentration than the given dose of the drug. Apart from when the drug is administered intravenously, trough concentrations of whole blood cyclosporine are used to monitor most patients. However, since peak and trough concentrations of cyclosporine may differ by tenfold, and peak concentrations provide a better indication of toxicity, many transplant centers also measure whole blood cyclosporine concentration 2 hours after an oral dose, in order to minimize the toxic effect.

The therapeutic ranges for cyclosporine have been quoted for patients who have undergone transplantation but these may vary somewhat between transplant centers. The ranges may differ with the particular organ being transplanted (heart, liver, and so on) and the time after surgery; for example, typical trough values for early post-renal transplant patients are in the range 200–400 μg/L (ng/mL) and fall to 100–200 μg/L (ng/mL) in patients who have stabilized after surgery. The blood cyclosporine concentration 2 hours after an oral dose should be in the range 1200–1700 μg/L (ng/mL) falling to 1000 μg/L (ng/mL) with time. These are generalized figures and for specific data in any region it is recommended that reference be made to the center providing the analytical service.

It is essential to be aware that many of the metabolites of cyclosporine, while biologically inactive, have the potential to interfere with the immunoassay of the parent drug. The same is true of tacrolimus, described below. It is therefore recommended that both cyclosporine and tacrolimus be assayed by more specific and sensitive technique such as high performance liquid chromatography (HPLC)-tandem mass spectrometry to eliminate such interference.

Tacrolimus

Tacrolimus (FK506) has a similar complex structure to cyclosporine (**Figure 21.6**). It also has a similar mode of action and route of metabolism as cyclosporine but is less toxic. Originally employed as a second-line treatment, it is now used increasingly as a first-line drug in many patients who have undergone transplantation surgery and also in the treatment of autoimmune diseases. While it may be administered intravenously it is usually given orally. A topical skin cream containing tacrolimus is available for treatment of some skin disorders but measurement of blood concentration is not warranted in these cases.

The highly lipophilic nature of tacrolimus ensures that the drug is taken up rapidly by tissues, such that less than 25% of an administered oral dose is present in the circulation. Trough, whole blood tacrolimus concentrations are measured in practice with a target range of 5–10 μg/L (ng/mL) in stable post-transplant patients; a slightly higher range (up to 15 μg/L or ng/mL) may be applied in early post-transplant patients. As with the measurement of cyclosporine, results from measurement of tacrolimus by immunoassay and by tandem mass spectrometry may differ by more than 10 μg/L (ng/mL), especially in patients in early stages of recovery from surgery, in generally unwell patients, in patients with hepatic disorders, and where co-administered drugs have interactive effects on the action and metabolism of tacrolimus. Great care is thus required in interpreting results and, where possible, currently available immunoassays should be avoided.

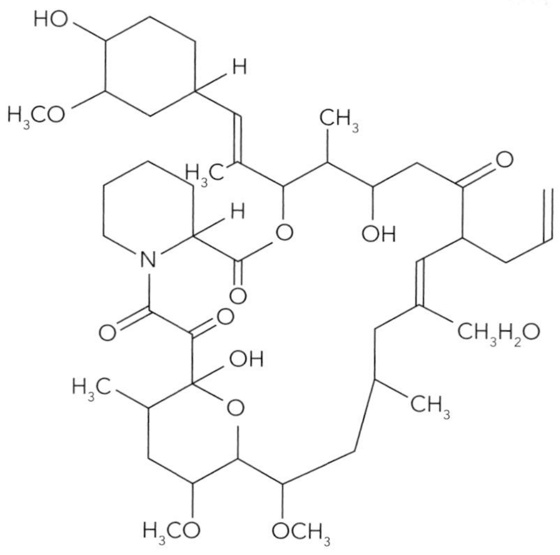

mycophenolic acid

tacrolimus

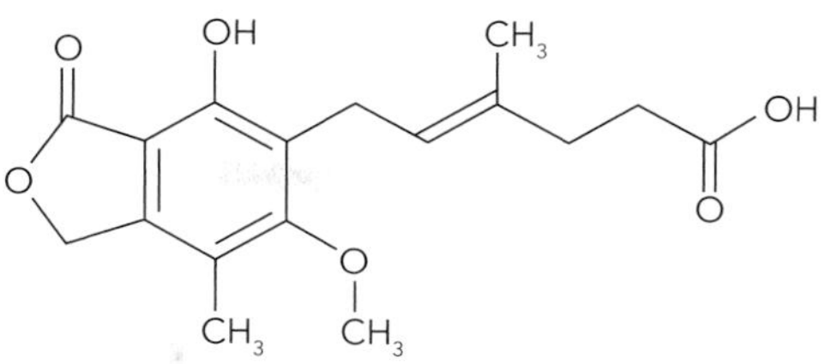

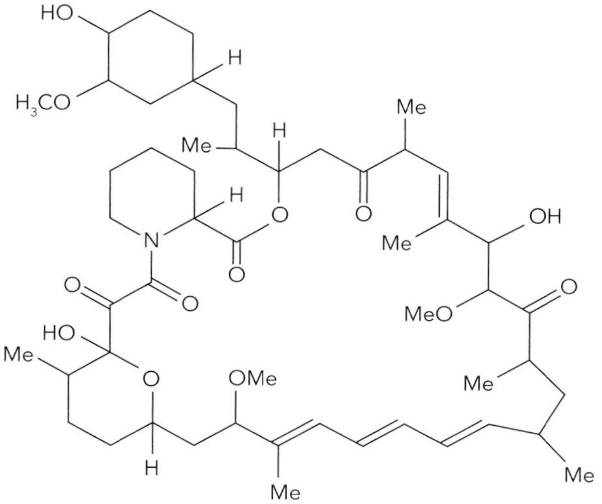

sirolimus

Figure 21.6 Chemical structures of tacrolimus, sirolimus, and mycophenolic acid.

Sirolimus

Sirolimus (see Figure 21.6) is usually only administered orally to relatively low risk transplant patients, mainly after renal transplant. It has a slightly different mode of action from both cyclosporine and tacrolimus and is sometimes given in conjunction with cyclosporine. However, its absorption, distribution, metabolism, and elimination is similar to the two other immunosuppressive drugs described above. Being highly lipophilic, it is rapidly taken up by tissues such that <20% of an oral dose is present in the circulation. Sirolimus is measured ideally by tandem mass spectrometry of whole blood samples, but the quoted therapeutic range of 5–15 µg/L (5–15 ng/mL) is dependent on a variety of clinical factors and may vary from center to center. As with tacrolimus and cyclosporine, local specialist centers should be referred to for this information.

Mycophenolic acid

Mycophenolic acid (see Figure 21.6) is administered as a pro-drug, mycophenolate mofetil, or as its sodium salt and is often used in conjunction with other immunosuppressive drugs. Its structure, mode of action, and pharmacokinetics are quite different from the immunosuppressive drugs discussed so far. It helps to prevent organ rejection in many transplant patients by impeding the proliferation of T and B lymphocytes. Almost the entire orally administered dose of the pro-drug is present in the circulation as the active form, mycophenolic acid, bound to plasma proteins. Its polar nature allows it to be conjugated in the liver and excreted by the kidneys. It has been difficult to illustrate a clear relationship between plasma concentration and therapeutic effect or toxicity. Unlike the situation with cyclosporine, the relationship between dose and toxicity is stronger than that between blood concentration and toxicity. Thus the measurement of a single trough concentration of mycophenolic acid in plasma is of limited value. However, plasma therapeutic ranges have been quoted as 1–3 mg/L.

Other drugs

Lithium

Lithium is commonly used in the treatment of a number of mental disorders, particularly bipolar (manic depressive) illness. It is water soluble with a bioavailability of >98% and does not bind to plasma proteins. It is excreted predominantly by the kidneys and has a plasma $t_{1/2}$ of 10–35 hours, which is increased in patients with renal failure (**Table 21.7**). Therapeutic monitoring is used to confirm compliance, to assess adequate control of dose (which may be difficult in some patients), and to ensure an optimal plasma concentration for individual patients.

TABLE 21.7 Properties of lithium

Property	Value
Bioavailability	>98%
Protein binding	<1%
Half-life	10–35 hours
Target ranges	0.4–0.8 mmol/L (mEq/L) for prophylaxis; up to 1.3 mmol/L for acute mania

Lithium can cause many side effects, including nephrogenic diabetes insipidus, renal impairment, and hypercalcemia; at high plasma concentration, acute toxic effects include circulatory failure, renal failure, and coma. For these reasons, measurement of lithium in suspected overdose cases should be available around the clock.

Lithium has a very narrow therapeutic range with variable individual pharmacokinetics. In long-term therapy, the therapeutic target trough range 12 hours post-dose is 0.4–0.8 mmol/L (0.4–0.8 mEq/L) for prophylaxis, although higher levels of up to 1.0 mmol/L (1.0 mEq/L) may be relevant in younger patients. This target therapeutic range may be even higher (up to 1.3 mmol/L or 1.3 mEq/L) in the treatment of patients with acute mania. Plasma lithium concentrations are sometimes seen above this range and values >2 mmol/L (>2 mEq/L) constitute a medical emergency. Mild toxicity may simply require cessation of treatment and/or administering saline intravenously until lithium plasma concentration falls to within the target range, before placing the patient on a lower-dose regime. Severe toxicity, however, usually requires renal dialysis. Although this can effectively reduce plasma lithium concentration, there is a rebound phase after a dialysis session as lithium leaves the intracellular space where it is mainly distributed. Several sessions of dialysis may therefore be required. It is worthwhile stressing that blood samples from patients on lithium therapy should not be collected into lithium-heparin tubes. Results from blood collected into such tubes often fall within the range of 2–4 mmol/L (2–4 mEq/L), the sort of values seen in patients with lithium overdose!

Analytical practice point 21.2

For obvious reasons, lithium heparin is an unsuitable anticoagulant for plasma lithium measurement. Sometimes cross-contamination can occur if samples are mixed between tubes before reaching the laboratory. The lithium result produced may be high but clinically credible, for example up to twice the therapeutic upper limit.

Digoxin

The use of a crude extract from *Digitalis purpurea* (foxglove) as a treatment for dropsy (cardiac failure) was first described by Sir William Withering in 1785. The active ingredient of this extract, digitalis, and derivatives from it including digoxin (see **Figure 21.7**), have been used since that time in the therapeutic management of patients with cardiac failure.

CASE 21.1

A 35-year-old man with chronic bipolar affective disorder is reviewed by his psychiatrist. He has been on lithium for two years and is well with stable mood. He usually takes his tablets at 11.00 pm. A blood sample is taken at midday. After reviewing the results below he is advised to reduce the dose.

	SI units	Reference range	Conventional units	Reference range
Plasma				
Sodium	142 mmol/L	136–142	142 mEq/L	136–142
Potassium	4.6 mmol/L	3.5–5.0	4.6 mEq/L	3.5–5.0
Urea (blood urea nitrogen, BUN)	6.1 mmol/L	2.9–8.2	17 mg/dL	8–23
Creatinine	84 μmol/L	53–106	1.0 mg/dL	0.6–1.2
Estimated GFR (eGFR)	>90 mL/min/1.73 m^2	>90	>90 mL/min/1.73 m^2	>90
Serum				
Lithium	1.1 mmol/L	0.6–1.2	1.1 mEq/L	0.6–1.2

- **Why is the dose reduced even though the lithium is in the therapeutic range?**
- **Why has his renal and electrolyte profile (urea and electrolytes) also been requested?**
- **What other metabolic complications may be detected by biochemical tests?**

The therapeutic range used for lithium is usually lower in chronic treatment (0.4–0.8 mmol/L) than in acute mania (up to 1.3 mmol/L). Some laboratories do not make this clear on their reports (like the one used in this case), and just put a non-specific range on the report. It is therefore important to know the clinical indication when interpreting the result. Higher concentrations are more likely to cause toxicity. Lithium is excreted by the kidneys, hence it is important to reduce the dose if the estimated GFR is decreased. Acute toxicity may occur with dehydration as well as deliberate overdosing. Nephrogenic diabetes insipidus is a common complication of lithium treatment, even if levels are always therapeutic. This is manifested as polyuria, polydipsia, and hypernatremia with inappropriately dilute urine. Thyroid disease and hypercalcemia are also possible side-effects of lithium therapy and can be detected by thyroid function tests and plasma calcium monitoring, respectively.

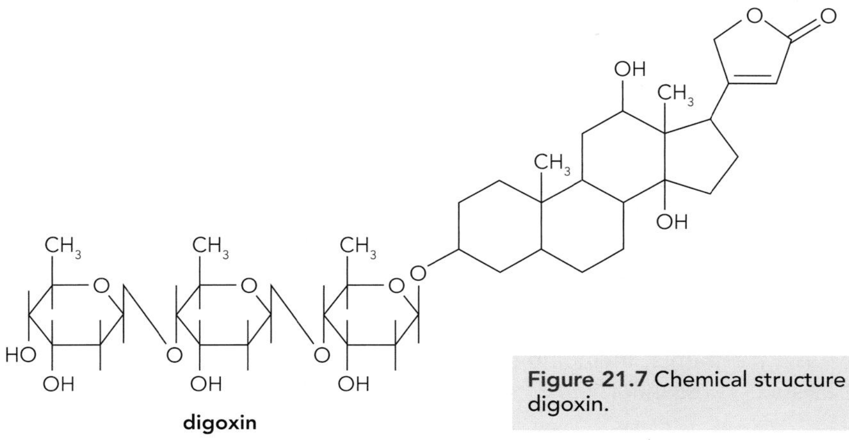

Figure 21.7 Chemical structure of digoxin.

As shown in Figure 21.7, digoxin consists of three sugar residues (glycone) and a noncarbohydrate, aglycone moiety, digoxigenin.

Despite the number of other cardioactive drugs used currently in practice, only digoxin is measured for TDM purposes in most routine clinical laboratories. Digoxin inhibits the Na^+/K^+-ATPase (sodium pump) in cardiac muscle membranes, thereby increasing the force of contraction of the heart muscle and improving heart output. In addition, the action of the drug to improve electrical conduction within the heart explains its usefulness in treating various arrhythmias, especially atrial fibrillation.

Usually administered orally, digoxin is absorbed rapidly from the gut and reaches a peak concentration in blood within an hour. It has a plasma half-life of 24–48 hours, being excreted primarily by the kidneys (**Table 21.8**). Monitoring is usually undertaken on trough plasma concentration or on samples taken at least 6 hours post-dose. The therapeutic range is usually given as 0.5–2.0 ng/mL (0.6–2.6 nmol/L), but this should be used with caution. Recent studies have shown that for most patients the range of 0.5–1.0 ng/mL (0.6–1.3 nmol/L) may be more appropriate; patients with values >1.0 ng/mL (1.3 nmol/L) show poorer outcomes and, of course, have a greater incidence of toxicity.

TABLE 21.8 Properties of digoxin

Property	Value
Bioavailability	>40–100% Marked variation between patients
Protein binding	<20–30%
Half-life	24–48 hours
Target ranges	0.5–2.0 µg/L or ng/mL (0.6–2.6 nmol/L)

Small changes in the dose of digoxin may result in major changes in blood concentration and lack of effect (decreased dose) or toxicity (increased dose). In addition, since the drug is excreted via the kidneys, toxicity can become a problem in patients with decreasing renal function. Co-administration of other drugs has been shown to modify the effectiveness of a given dose of digoxin; for example, decreased absorption (with antacids), increased non-renal elimination (with phenytoin), and increased bioavailability (with erythromycin). For these reasons, care should be taken when changing treatments, when additional drugs are given, and in patients with rapidly changing renal function.

In most patients, symptoms of digoxin toxicity, including nausea, vomiting, bradycardia, and the classically described yellow vision, may present when the blood concentration is >2.0 ng/mL (>2.6 nmol/L). It is essential to measure plasma potassium when administering digoxin, since the threshold at which digoxin becomes toxic is lowered in cases of hypokalemia. In such cases, toxic symptoms may present at digoxin concentrations of less than 2.0 ng/mL (2.6 nmol/L).

In severe cases of digoxin overdose, the antidote Digibind®, which contains digoxin-specific antibody fragments, may be given intravenously to remove digoxin rapidly from the circulation. The downside of this from a clinical chemistry perspective is that the antidote interferes with the commonly used immunoassay for digoxin and it then yields unreliable results. An additional problem with most immunoassays for digoxin is their limited specificity; some assays cross-react with other digoxin-like molecules and fragments, including potentially toxic compounds from other plant species such as oleander.

CASE 21.2

A 72-year-old woman is taking digoxin for atrial fibrillation. She usually takes her tablets at 8.00 am. At her routine medication review she is well and her heart rate is 72 beats per minute. A blood sample is taken at 9.30 am.

	SI units	Reference range	Conventional units	Reference range
Plasma				
Sodium	140 mmol/L	136–142	140 mEq/L	136–142
Potassium	4.1 mmol/L	3.5–5.0	4.1 mEq/L	3.5–5.0
Urea (BUN)	5.2 mmol/L	2.9–8.2	15 mg/dL	8–23
Creatinine	72 μmol/L	53–106	0.8 mg/dL	0.6–1.2
Estimated GFR (eGFR)	>90 mL/min/1.73 m^2	>90	>90 mL/min/1.73 m^2	>90
Digoxin	3.5 nmol/L	0.6–2.6	2.7 ng/mL	0.5–2.0

- **Why is her digoxin level high?**
- **Why has her renal and electrolyte profile (urea and electrolytes) also been requested?**

Digoxin has a large volume of distribution and a long distribution phase. After an oral dose of digoxin, the peak plasma concentration is seen at 1 to 3 hours, but the concentration that correlates with the clinical effects is not seen until the end of the distribution phase after 6 to 8 hours. This is the therapeutic range quoted in the table above. The sample in this case was taken too soon after the dose to be interpretable. It would be important that the physician did not inappropriately reduce the dose on the basis of this result. Measurement of renal function is important as digoxin is excreted by the kidneys and doses should be reduced in kidney disease. Digoxin toxicity is enhanced by hypokalemia—and toxicity may even occur with digoxin within the therapeutic range—hence potassium should also be measured at the same time.

Theophylline and caffeine

Although not a first-line treatment for asthma, theophylline (**Figure 21.8**) is still widely prescribed for many patients where conventional inhaler therapy is not acceptable.

Theophylline is readily absorbed orally (>95% bioavailability), has a relatively short half-life (6–8 hours, but much longer in patients with liver disease), and has a quoted therapeutic range of 10–20 mg/L or μg/mL (trough levels; **Table 21.9**). Theophylline can also be administered intravenously, usually in its ethylene diamine salt form (aminophylline).

Efficacy of treatment and trough drug levels correlate well for theophylline. Side effects of the drug commonly occur at high concentrations, and therapeutic monitoring of theophylline levels is recommended when commencing treatment, when changing dose, or in cases of suspected overdose. Monitoring is also recommended in cases where plasma levels can be affected by such factors as co-administration of drugs which affect liver metabolism, liver disease, smoking, and changes in diet. Because approximately 50–60% of the theophylline present in blood is bound to plasma proteins (mainly albumin), care should be taken when interpreting results from patients with low albumin concentration. Ideally, corrected results should be used.

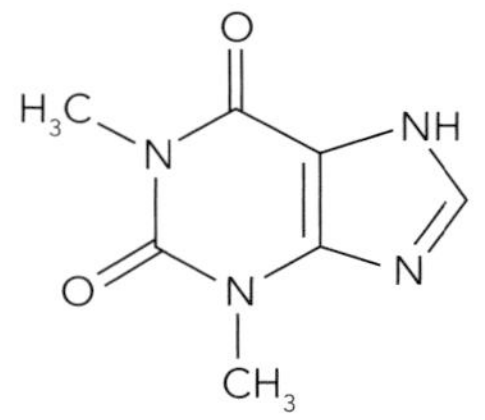

theophylline
(1,3-dimethylxanthine)

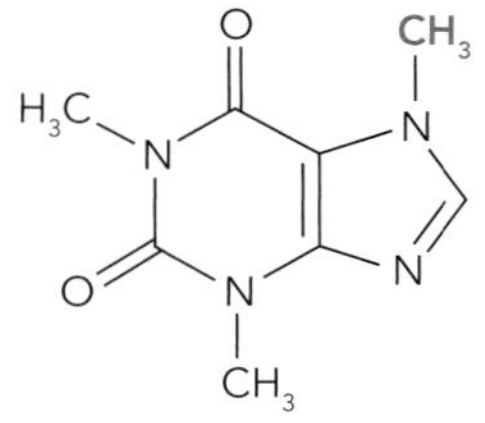

caffeine
(1,3,7-trimethylxanthine)

Figure 21.8 Chemical structures of theophylline and caffeine.
The figure shows the similar structures of the two compounds; an extra methyl group is present on the caffeine molecule.

TABLE 21.9 Properties of theophylline and caffeine

Property	Theophylline	Caffeine
Bioavailability	>95%	>95%
Protein binding	50–60%	<50%
Half-life	6–8 hours (neonates or liver disease >30 hours)	40–230 hours in neonates
Target ranges	10–20 mg/L (μg/mL)	5–20 mg/L (μg/mL)

Caffeine (see Figure 21.8) is predominantly used in premature neonates with apnea. It is not as toxic as theophylline, has a longer half-life (days rather than hours), and is excreted mainly via the urine. It displays a more predictable dose response than theophylline and quoted target ranges are given as 5–20 mg/L or μg/mL (see Table 21.9).

Methotrexate and other drugs

Methotrexate, a derivative of aminopterin, is used at relatively low doses to treat disorders such as rheumatoid arthritis. The major problem from long-term usage in these cases is cirrhosis of the liver. Acute toxicity is seldom a problem and therapeutic monitoring of the drug is not usually carried out. However, therapeutic monitoring of the drug may be employed in cases such as cancer chemotherapy where high doses of methotrexate are used. Here, care needs to be taken to avoid toxicity, which may present when the plasma methotrexate concentration rises above 1000 nmol/L (0.45 mg/L). Rapid intervention in the form of calcium folinate administered intravenously is used to counter the toxic effects of methotrexate.

Many other drugs are measured as part of a therapeutic monitoring service in specialist laboratories but have limited clinical use in routine practice.

POISONS AND DRUGS OF ABUSE

CHAPTER
22

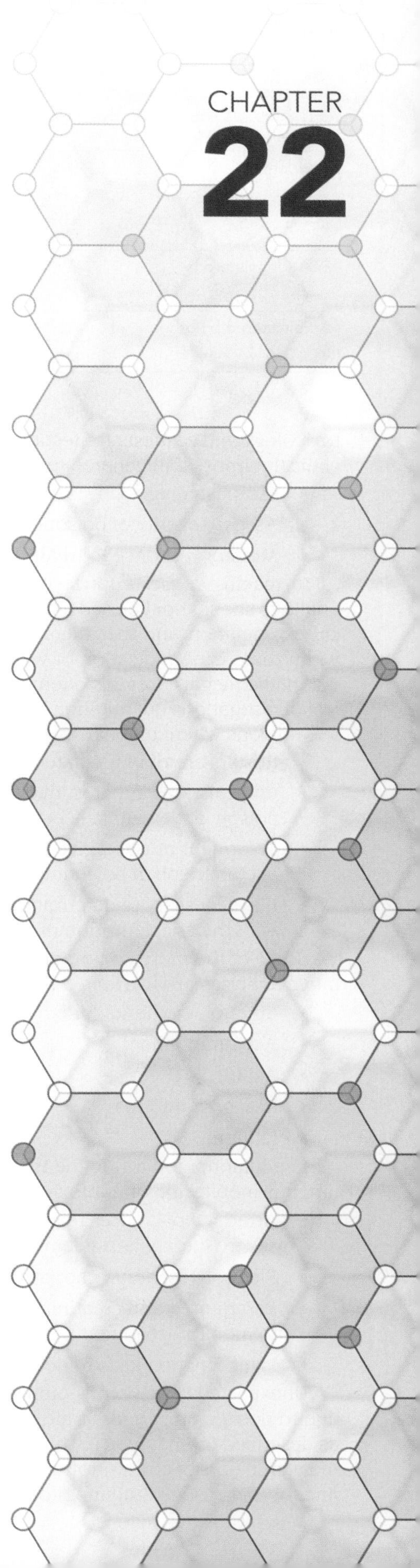

IN THIS CHAPTER

CLINICAL INVESTIGATION INTO POISONING
OVERDOSE OF THERAPEUTIC DRUGS
ESSENTIAL METALS
NONESSENTIAL METALS
ARSENIC
CARBON MONOXIDE POISONING
DRUGS OF ABUSE
PLANT POISONS

A poison is defined in the Oxford English Dictionary as "a substance that, when introduced into or absorbed by a living organism, causes death or injury, especially one that kills by rapid action even in small quantities." In the context of the current chapter, this is broadened to include those substances which are essential in small amounts for life, but which when absorbed/administered in large amounts are toxic; for example, iron and vitamin A. Chemical and biochemical tests are available for a wide range of potential poisons, with the more common being:

- Therapeutic drugs, for example acetaminophen, aspirin, antidepressants, tranquilizers, and iron preparations
- Ethanol, ethylene glycol, and methanol
- Drugs of abuse
- Carbon monoxide
- Metals, especially lead (includes essential and nonessential metals)
- Poisonous plants

Other cases of poisoning occur, for example those involving household and garden products, where laboratory testing is rarely required in the management of the patient.

22.1 CLINICAL INVESTIGATION INTO POISONING

The circumstances in which poisoning occurs may be accidental or, of course, intentional where administration of a poison is deliberately intended to cause harm (**Table 22.1**).

Suspicion of poisoning may develop from circumstantial evidence surrounding the subject and/or the clinical condition of the suspected poisoned individual. Such clinical conditions may include stupor, coma, abnormal behavior, gastrointestinal irritation, and damage to vital organs (liver and kidney). The investigation into poisoning may thus involve forensic rather than clinical

TABLE 22.1 Circumstances in which poisoning occurs

Main areas	Examples
Accidental	Drugs 1. Errors in patient noncompliance, incorrect dosage 2. Effects of disease or treatment
	Exposure to toxic substances at home, work, or elsewhere
Self-administered	With intention to harm or gain attention to problem(s)
	Abuse of drugs or of other substances
Administered by others	Homicide
	Nonaccidental injury

toxicology and requests for testing may be made for medico-legal as well as for clinical purposes. In general, tests are requested to:

- Establish diagnosis
- Assess severity of poisoning and the need for specific treatment
- Monitor response to treatment

In making requests for testing, it is important to remember that chemical analyses should not be used as an alternative to proper history taking and physical examination and that there are practical difficulties in screening for drugs when there is no indication of which drug is suspected as the poisoning agent. For medico-legal cases it is often sufficient to identify the poison, whereas in clinical investigations quantitation of the poison may also be necessary, particularly in situations where urgent treatment is required. These situations include where:

- Results are of immediate use in patient management
- Clinical assessment is inadequate, often due to delayed effects of a poison, for example acetaminophen or iron salts
- Treatment must be given promptly, for example *N*-acetylcysteine (for acetaminophen poisoning) or desferrioxamine (for iron poisoning)
- It is necessary to determine whether the clinical condition is consistent with the concentration of identified poison

Some of the more commonly encountered poisons for which urgent quantitative analyses are warranted include:

- Ethanol, methanol, or ethylene glycol
- Acetaminophen
- Aspirin
- Iron preparations
- Lithium

In addition, monitoring the treatment of poisoning often requires the simultaneous monitoring of nontoxicological parameters, for example:

- Arterial blood gases to monitor patients with respiratory depression or severe metabolic acidemia
- Electrolytes, anion gap, osmolality, and osmolar gap
- Serum urea and creatinine to detect renal damage and serum potassium to monitor treatment by forced alkaline diuresis
- Liver function tests to detect and monitor hepatocellular injury

Unexpected toxicity can complicate treatment with certain therapeutic drugs due to dosage error and/or altered pharmacokinetics and pharmacodynamics, as described in Chapter 21. Thus there may be some degree of overlap between measurements of concentration of drugs in serum for investigation of poisoning and for therapeutic drug monitoring.

Analytical practice point 22.1

Poisons and drugs are usually rapidly cleared from the plasma and samples should be taken as early as possible. Admission samples may only be retained by the laboratory for a few days, due to limited storage space. If poisoning is suspected after admission, the laboratory should be contacted to retrieve the earliest available samples.

Analytical practice point 22.2

Clinical history is vital to direct decisions on which poisons or drugs should be tested for. There is no toxicological screen which can detect all possible chemical entities that are able to cause illness in humans.

22.2 OVERDOSE OF THERAPEUTIC DRUGS

Quantitative measurement of a drug which has been taken to excess may not always be necessary when clinical observation is sufficient to determine a course of action; for example, in overdoses of common opiates such as morphine, the opiate antagonist naloxone is given as a rapidly effective antidote. However, there are common drugs where quantitative measurement is useful, if not essential; these include acetaminophen and aspirin (acetylsalicylic acid).

Acetaminophen

The clinical consequences of an overdose of acetaminophen include:

- Symptoms of lassitude, nausea, and vomiting; these are relatively minor and are due initially to the drug itself and later to liver injury
- Evidence of liver injury
- Jaundice
- Raised serum alanine transaminase, bilirubin, and plasma prothrombin time

Patients who recover have no permanent liver damage.

It is important to remember that in cases of poisoning with co-proxamol tablets (Distalgesic®), which contain d-propoxyphene in addition to acetaminophen, the toxicity of d-propoxyphene (depression of consciousness and respiration) predominates. Furthermore, many patients who deliberately overdose may take a cocktail of drugs and alcohol, and since the antidote to acetaminophen poisoning should ideally be given within 10 hours of ingestion (up to 18 hours after ingestion), measurement of acetaminophen should be considered when there are no presenting features associated with the drug.

The metabolism of acetaminophen and its effects are shown diagrammatically in **Figure 22.1**. The toxicity of acetaminophen depends on the accumulation of

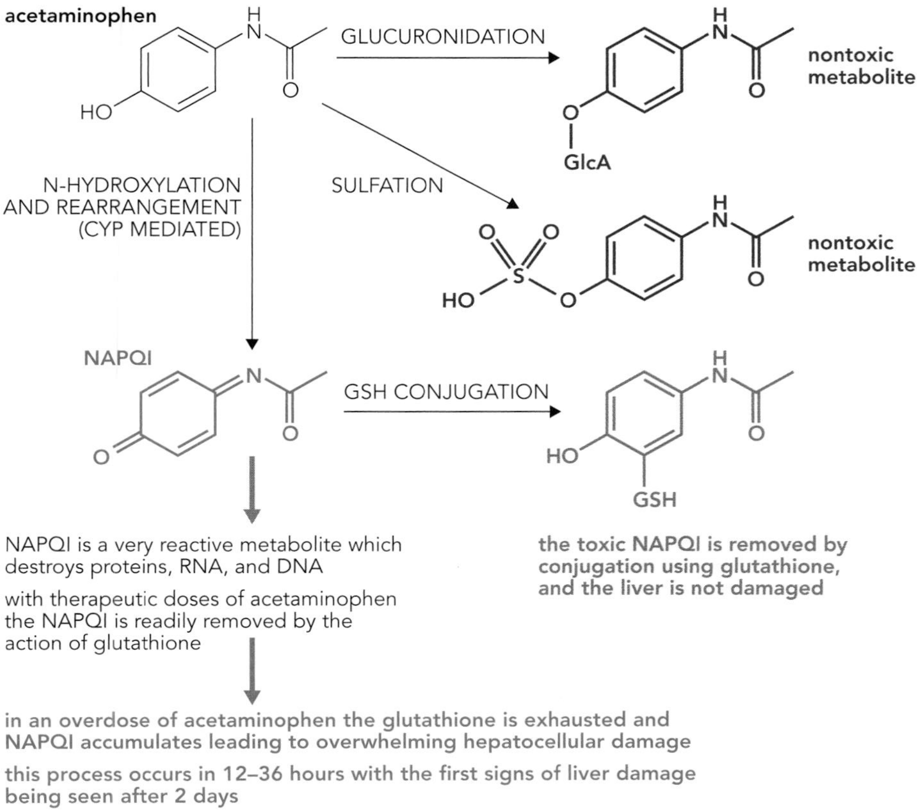

Figure 22.1 Metabolism of acetaminophen by the liver and the sequence of events that lead to liver damage.
NAPQI: *N*-acetyl-*p*-benzoquinone imine.

reactive metabolites secondary to exhaustion of reduced glutathione (GSH), and cell injury occurs only in cases of significant acetaminophen overdose when the supply of reduced GSH has been exhausted. In these cases, severe hepatocellular damage occurs after two days, if untreated. In severe cases of acetaminophen overdose, the rise in alanine transaminase (ALT) can be several hundredfold due to the simultaneous damage of nearly every hepatocyte within the liver. Such huge increases in ALT are very characteristic of untreated acetaminophen overdose, and are associated with acute centrilobular liver necrosis. Treatment for acetaminophen overdose is oral administration within 10 hours of thiol (-SH) donors such as *N*-acetylcysteine (Parvolex®) or methionine, as indicated by a chart based on serum drug concentration. The plasma acetaminophen decay curve and treatment decisions are shown in **Figure 22.2**. Supportive treatment for liver and kidney failure may be required if administration of thiol is delayed beyond 18 hours. In cases of severe, untreated overdoses of acetaminophen, many patients will die from liver failure within a couple of weeks.

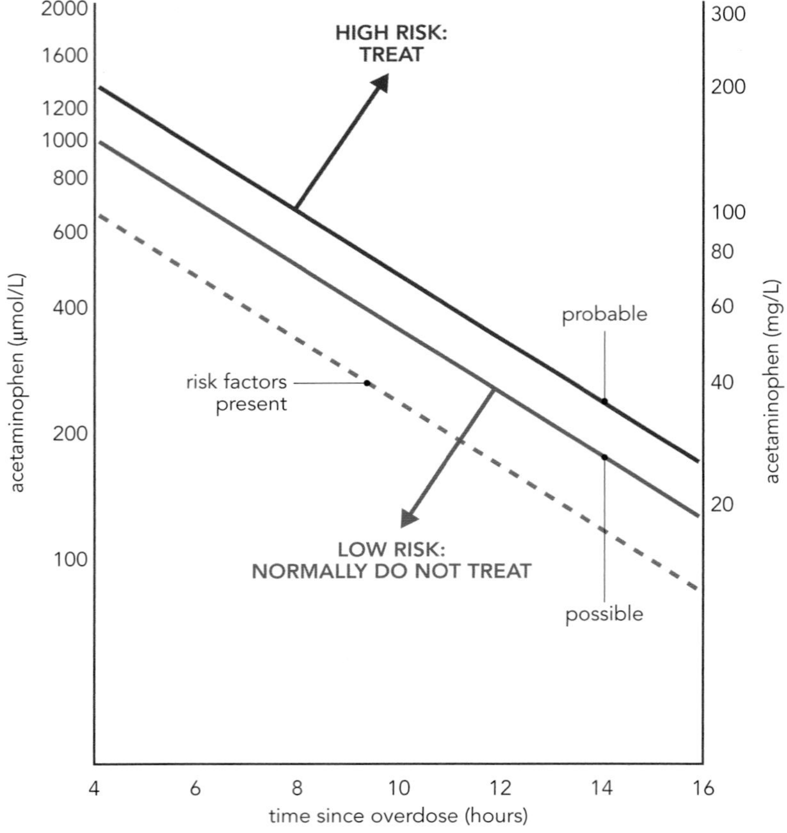

Figure 22.2 Treatment decision graph for acetaminophen overdose. The graph shows the disappearance of plasma acetaminophen after an overdose. The concentrations are given in both SI (μmol/L) and conventional units (mg/L), and are logarithmic scales. If the concentration of acetaminophen is above the red line (probable line) at the correct time after the overdose, treatment should be given. Evidence has shown that if treatment is not given above this line liver damage will occur. If the timings of the suspected overdose and the time the sample was taken are correct, treatment is not given below the blue line. There is a "gray area" between the definite treatment, and no-treatment lines (the probable, red line and the possible, blue line), where there is no good evidence that liver damage will occur if treatment is not given. However, as an extra level of assurance, in practice all acetaminophen-overdose patients are treated above the probable line. If other risk factors are present, for example preexisting liver disease, treatment may well be given at lower levels (dashed line). Since the effects of acetaminophen can be very toxic, if there is doubt about timings treatment may be given even if the result is below the green dashed line.

Salicylate

Salicylate poisoning is most commonly due to aspirin overdose and may be willful or accidental. Aspirin is hydrolyzed to salicylic acid and acetic acid in a first-pass effect in the liver. The clinical features of overdose, which may present within 6 hours of ingestion, occur usually when serum salicylate concentration exceeds 2.2 mmol/L (300 mg/L) and at lower concentrations in children. These features include:

- Nausea, vomiting, and hematemesis
- Tinnitus and deafness
- Sweating

- Acid–base disturbances—initially respiratory alkalemia followed (in severe poisoning in adults and in less severe cases in children) by metabolic acidemia. In both cases there is hyperventilation
- Coma (uncommon in adults unless poisoning is severe; more common in children)
- Hypoglycemia may occur in children

Assessing the severity of poisoning and monitoring response to treatment involve measuring serum salicylate. Arterial blood for acid–base status is only required in very ill patients. Serum potassium is also measured during treatment with forced alkaline diuresis, which is indicated when serum salicylate exceeds ~3 mmol/L (400 mg/L). The rationale for alkaline diuresis is:

- In aqueous solution, salicylic acid is in equilibrium with H^+ and the salicylate anion, the degree of dissociation being proportional to pH
- Salicylate anions are reabsorbed by renal tubules less readily than salicylic acid
- Increasing urinary pH by infusion of a solution containing glucose, sodium chloride, potassium chloride, and sodium bicarbonate increases the dissociation of salicylic acid to salicylate anion, thereby increasing the elimination of the drug into the urine. Potassium chloride is included to prevent potassium depletion secondary to the increased urinary excretion caused by alkalemia.

22.3 ESSENTIAL METALS

Many metals are essential for normal physiological function and they range from the more abundant metals—sodium, potassium, calcium, magnesium, and iron—which have been described in earlier chapters to those whose concentrations in the body are very low and are referred to as trace metals, which include:

- Zinc
- Copper
- Selenium
- Cobalt
- Manganese
- Molybdenum
- Nickel
- Chromium

The role of these trace metals is usually as co-factors for a variety of enzymes.

Abundant metals

The consequences of hyperkalemia, hypernatremia, and hypercalcemia have been described in earlier chapters. Such cases arise from other pathology and are rarely due to overdose, although cases of deliberate poisoning of children by administration of large doses of salt have appeared in the press. Iron toxicity is uncommon but may occur from accidental ingestion of iron tablets by a child mistakenly thinking that the red pills are sweets, for example. It is treated by chelation therapy using desferrioxamine if the plasma iron concentration exceeds 40–50 μmol/L (220–280 μg/dL).

Trace metals

The physiological roles of zinc, copper, and selenium are described in Chapter 24. Toxicity ensues when the plasma concentrations of these metals rise above the homeostatic reference ranges.

Zinc

The serum reference range for zinc is 11–24 μmol/L (72–157 μg/dL) reflecting a diurnal variation with a peak at 10.00 am. It is important to remember when measuring serum zinc concentration that

1. It exhibits this diurnal variation, and
2. Sample contamination may arise from sources such as creams, powder in rubber gloves, and zinc stearate on stoppered blood tubes.

Whereas the signs and symptoms of zinc deficiency have been well described (anemia, growth retardation, hypogonadism, rashes, and failure to heal wounds), zinc poisoning is extremely rare.

Copper

The serum copper concentration range in adults is 11–20 μmol/L (70–127 μg dL) although this increases to 27–40 μmol/L (171–255 μg/dL) in pregnancy. Defects of copper metabolism may arise as a result of uncommon inborn errors of metabolism (for example Wilson's disease or Menkes' kinky hair syndrome). In Wilson's disease, for example, serum copper levels are <15 μmol/L (<95 μg/dL) yet patients develop symptoms of copper toxicity. This is because these patients have low plasma concentrations of the copper-binding protein ceruloplasmin, which results in copper being deposited in various organs including the liver. Diagnosis of Wilson's disease is made by demonstrating a low plasma ceruloplasmin concentration and raised urinary copper output. Measurement of plasma copper is unreliable in the diagnosis. Many patients with Wilson's disease also exhibit raised liver concentrations of copper as demonstrated on a liver biopsy. The acute and chronic toxic effects of copper poisoning are shown in **Table 22.2**. Chronic toxicity may develop from continued use of copper cooking vessels and from chronic consumption of betel nuts.

TABLE 22.2 Acute and chronic effects of copper poisoning

Acute	Chronic
Nausea	Hepatic cirrhosis
Vomiting	Kayser–Fleischer rings in the iris
Intravascular hemolysis	

Selenium

The serum reference range for selenium is 950–1650 nmol/L (74–130 μg/L), much lower than for copper and zinc. Because of its role as a co-factor for both glutathione peroxidase and 5′-iodothyronine deiodinase, selenium is thought to be a powerful antioxidant and, by extension, it has been suggested that an increased risk of cancer and coronary heart disease may be associated with chronic deficiency of selenium. The soil in a number of geographical regions, including parts of China, New Zealand, and Finland, is deficient in selenium, and an endemic cardiomyopathy (Keshan disease) was found in rural areas of Keshan province in China where the selenium content in foods was low.

The low threshold of toxicity for selenium means that symptoms of poisoning (**Table 22.3**) may occur after moderate intake of foodstuffs,

TABLE 22.3 Acute and chronic effects of selenium poisoning

Acute	Chronic
Nausea	Hair and nail loss
Vomiting	Poor dental health
Diarrhea	Liver problems

TABLE 22.4 Members of the *Lecythidaceae* family of trees

Tree	Fruit
Bertholletia excelsa	Brazil nut
Lecythis ollaria	Paradise or sapucaia nut
Couroupita guianensis	Cannonball fruit

particularly nuts, which have a high selenium content. The fruit of trees of the *Lecythidaceae* family (**Table 22.4**) are enriched in selenium, and a major source of the metal in Western diets is the Brazil nut from *Bertholletia excelsa*. Poisoning from consumption of Brazil nuts is uncommon. However, cases of selenium poisoning from consumption of the nuts of the selenium accumulator plant (*Lecythis ollaria*) have been described. These nuts have a very high concentration of selenium (7–12 g per kg dry weight) present as a seleno-cystathionine complex.

Other essential metals

Serum reference ranges for other essential metals are shown in **Table 22.5**. Toxicity results when the serum concentration of any of these metals rises appreciably above the upper limit of the normal range. Most of the causes for toxicity of these metals are due to occupational exposure or to the effects of wear in prosthetic joints, where some of these metals are employed in their construction. These details are summarized in **Table 22.6**.

TABLE 22.5 Plasma reference ranges for some essential metals

Metal	Reference range (all values refer to serum, except where stated)	
	SI units (nmol/L)	Conventional units (ng/mL)
Cobalt	<15	<0.9
Manganese	86–333 (whole blood)	4.7–18.3 (whole blood)
Molybdenum	3–20	0.3–2.0
Nickel	<34	<2.0
Chromium	<5.8	<0.3

TABLE 22.6 Causes of toxicity and pathology for some trace metals

Metal	Cause of toxicity	Pathology
Cobalt	Glass and pigment industries Prosthetic joint wear	Not usually very toxic
		Wide range of findings including pulmonary edema, diarrhea, vomiting, and renal failure
Manganese	Mining industries, inhalation of manganese	Predominantly neurological
		Often irreversible and very disabling with chronic exposure to high doses
Molybdenum	Mainly due to prosthetic joint wear	Toxicity rare
		Effects due to copper deficiency as molybdenum in excess inhibits copper uptake
Nickel	Occupational exposure in glass, ceramics, and battery industries	Industrial-use nickel carbonyl, which is exceedingly toxic, stops oxygen binding to hemoglobin and causes mutagenesis
		Nickel metal is not very toxic
Chromium	Electroplating industry Prosthetic joint wear	Exposure can lead to skin, nasal, and lung irritation
		Chromium exposure can also cause lung cancer

The measurement of these trace elements is usually only carried out by very specialist laboratories. Contamination of samples is a major problem, and great care must be taken with the selection of blood- and urine-collection bottles. The advice of the laboratory should always be sought in the first instance.

22.4 NONESSENTIAL METALS

Other metals are not required for human physiological processes and are toxic when ingested (**Table 22.7**). Of these, the more common poisonous metals are aluminum, arsenic, cadmium, lead, and mercury, but even so, cases of severe toxicity are rare.

TABLE 22.7 Toxic nonessential metals

Aluminum	Cadmium	Silver
Antimony	Germanium	Strontium
Arsenic	Gold	Tellurium
Barium	Lead	Thallium
Beryllium	Lithium	Tin
Bismuth	Mercury	Titanium
Boron	Platinum	Vanadium

Aluminum

Aluminum was, for many years, considered to be an inert metal but, more recently, particularly after a disastrous event in Camelford in the UK where aluminum waste was emptied into the water supply, it has been shown to cause major problems after ingestion by humans. Laboratory indices of aluminum toxicity are shown in **Table 22.8**.

TABLE 22.8 Laboratory indices of aluminum toxicity

Serum aluminum concentration		Relevance and action
SI units (μmol/L)	**Conventional units (μg/L)**	
>2.2	>60	Increased body burden
>3.7	>100	Health surveillance required
>7.4	>200	Requires urgent medical attention

Aluminum toxicity was also a problem in patients undergoing renal dialysis, as the aluminum in the dialysis water passed into the blood during the dialysis process, leading to the buildup of toxic levels. However, since the introduction of reverse osmosis to remove aluminum ions from the dialysis water, aluminum toxicity in renal dialysis is currently a rare finding. The pathological consequences of aluminum intoxication are encephalopathy encompassing speech disorder, dementia, convulsions, and even death, and osteodystrophy characterized as an osteomalacia intractable to vitamin D. The speech disorder was described as dialysis dementia when it was found in chronic renal failure patients undergoing hemodialysis. Furthermore, occupational exposure to aluminum dust causes pulmonary fibrosis.

Cadmium

Normally, whole blood cadmium concentrations are less than 27 nmol/L (<3 ng/mL), but smokers can have values of up to 54 nmol/L (6 ng/mL). Acute toxicity usually only occurs when cadmium concentrations are greater than 450 nmol/L (50 ng/mL). Cadmium poisoning is rare but may occur as a result of industrial exposure to cadmium fumes or ingestion of cadmium salts. Cadmium is used extensively in industry (several thousand tons per year in the USA and Europe) as a component in alloys in the manufacture of nickel-cadmium batteries, as cadmium pigments in ceramics, glass, and paints, as cadmium stabilizers of plastics such as polyvinylchloride, as cadmium coatings of metal surfaces, and in cadmium electronic compounds such as cadmium telluride.

Inhalation of cadmium fumes leads acutely to pulmonary edema and chronically to emphysema, while ingestion of cadmium salts causes nausea, vomiting, and diarrhea. Chronic low exposure to cadmium is associated with symptoms of an acquired Fanconi syndrome (see Chapter 4) with renal dysfunction and osteomalacia. The condition is extremely painful and has been called Itai-Itai or "ouch-ouch" disease following the first reports of the cadmium poisoning described in 1955 in the Toyama Prefecture in Japan. Since one of the toxic effects of cadmium is proximal tubular nephropathy, urinary β_2-microglobulin or retinol-binding protein can be measured. These two low-molecular-weight proteins are normally reabsorbed in the proximal tubule. In cadmium toxicity, however, this reabsorption is inhibited and an early indicator of cadmium toxicity is the measurement of raised levels of β_2-microglobulin and retinol-binding protein in urine.

Lead

Lead has no known function in humans but exposure to the metal causes both acute and chronic problems because it is absorbed through both the gut and lungs. It is absorbed via the gut more rapidly in children than in adults. Once absorbed, lead is sequestered in bone by exchange with calcium in hydroxyapatite. The exchange is slow and easily overloaded. When intake of lead is high, it accumulates in soft tissues, especially liver, kidneys, and brain. Toxicity of lead is due to inactivation of thiol groups in enzymes.

Inorganic lead

As listed in **Table 22.9**, situations associated with exposure to excess inorganic lead (lead salts) may be industrial or domestic.

Various guidelines are available to restrict exposure of industrial workers to inorganic lead. In the UK, for example, there are the Health and Safety Executive guidelines (last updated in 2012; http://www.hse.gov.uk/pubns/indg305.pdf) which prescribe surveillance of all lead workers. This involves:

- A medical examination before employment and subsequently at least once a year

TABLE 22.9 Exposure to excess inorganic lead

Industrial exposure	Domestic exposure
Rarely in smelting or manufacturing in regulated premises	Chiefly in children living in old houses where there is lead paint
Usually found in unsupervised workers, for example battery salvage or demolition	Ceramics for food and drink glazed with lead-containing pigments; water pipes
Jewelry making	Rarely from lead shot or bullets
Ceramics	Asian eye cosmetics including surma
Building	Exhaust fumes from vehicles using leaded petrol (much less common than formerly)

TABLE 22.10 Action and suspension limits for industrial lead workers

Category	Blood lead level (μg/dL)	
	Action level	Suspension level
General employees	50	60
Women of childbearing age	25	30
Young people under 18 years	40	50

Data from UK HSE guidelines 2012 (http://www.hse.gov.uk/pubns/indg305.pdf).

- Blood lead monitoring, the frequency of which depends on results
- Suspension from working with lead if the blood lead level rises above set action limits (**Table 22.10**)
- Suspension from work if a female worker is pregnant

It is essential to identify lead poisoning early, particularly in children, because of the devastating pathology arising from a lack of detection. The hematological changes in inorganic lead poisoning, occurring at all ages, are due to inhibition of heme synthesis and red cell maturation. Inhibition of 5-aminolevulinic acid (ALA) dehydratase and ferrochelatase results in accumulation of heme precursors, decreased heme synthesis (see Chapter 16), and consequent microcytic, hypochromic anemia, while inhibition of pyrimidine 5′-nucleotidase causes retention of RNA in maturing red cells, which is manifest as basophilic stippling. Indices used in the diagnosis of inorganic lead poisoning are shown in **Table 22.11**.

TABLE 22.11 Diagnosis of lead poisoning

Area of damage	Laboratory finding
Indices of increased body lead content	Increased blood lead (95% in red blood cells)
	Increased urinary lead excretion after EDTA infusion
Indices of biological effects	Hypochromic, microcytic anemia with basophilic stippling of red blood cells
	Decreased red blood cell 5-aminolevulinic acid (ALA) dehydratase activity
	Increased urinary excretion of ALA and coproporphyrin III
	Increased red blood cell zinc-protoporphyrin

EDTA, ethylenediaminetetraacetic acid.

In addition to the anemia associated with lead poisoning, neurological changes and renal tubular damage also occur. In children, early signs may be headaches and hyperactivity, but more serious cases can lead to blindness, convulsions, and coma. Increased levels of protein can be found in the cerebrospinal fluid. The renal tubular damage produces generalized aminoaciduria, glycosuria (without a raised plasma glucose), and hypophosphaturia. In adults, patients often present with abdominal pain and constipation. The renal nephropathy can lead to chronic renal failure. Finally, hyperuricemia may also occur which can lead to the clinical features found in gout (for example tophi).

Treatment of inorganic lead poisoning obviously includes removal of the patient from exposure to lead and modification of the working environment if that is the source of lead. In addition, administration of one of a number of drugs which increase urinary excretion of lead serves to decrease the total body load of the metal.

Organic lead

Lead alkyls were once used routinely as antiknock agents in leaded fuel but with the advent of lead-free petroleum, poisoning by organic lead compounds is exceptionally rare nowadays. Such lead alkyls were lipid soluble and accumulated in the brain, causing toxic psychosis (anxiety and insomnia) and encephalopathy (convulsions and coma).

Mercury

Mercury has been commonly used for several centuries and is still present in many thermometers, sphygmomanometers, and dental amalgams. It is toxic at blood levels >20 nmol/L (>4 μg/L). Humans may be exposed to organic mercury such as methylmercury chloride, inorganic mercury salts, and metallic mercury and the symptoms and treatment for each form of mercury varies. Organic mercury is extremely toxic, reacting spontaneously with thiol groups on proteins or glutathione, for instance, to form organo-mercury adducts (**Figure 22.3**).

Organo-mercurials are neurotoxic and teratogenic. Cationic mercury, such as is present in dental amalgams, does not appear to be toxic and there is no evidence that leaching of mercury salts from fillings causes any major problem apart from discomfort. Interestingly, the same appears to be true for ingestion of metallic mercury per se, but inhalation of metallic mercury vapor is extremely toxic, prolonged exposure giving rise to respiratory failure and renal disease.

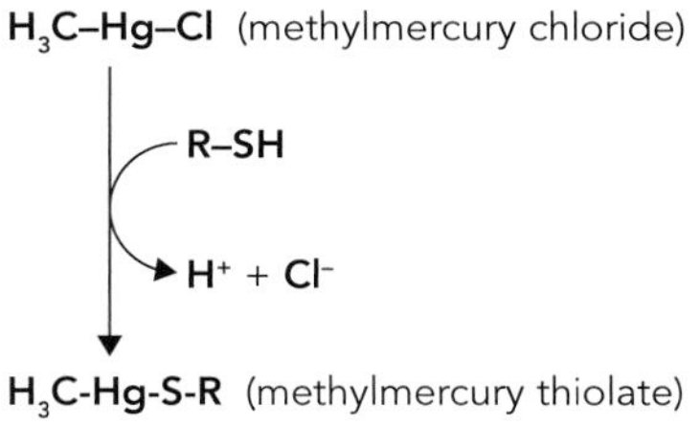

$H_3C{-}Hg{-}Cl$ (methylmercury chloride)

R–SH

$H^+ + Cl^-$

$H_3C{-}Hg{-}S{-}R$ (methylmercury thiolate)

Figure 22.3 Reaction of an organo-mercurial with a thiol residue to form an organo-mercury adduct.

22.5 ARSENIC

Arsenic is a ubiquitous element and is extremely toxic. Serum concentrations are normally very low, less than 130 nmol/L (10 ng/mL). Humans may be exposed to arsenic through eating contaminated seafood (fish and shellfish) harvested from rivers with runoffs from mineral workings, working in the smelting industry, and exposure to insecticides, herbicides, and wood preservatives. In addition, some traditional Asian remedies contain arsenic. Acute exposure may result in abdominal pain, muscle weakness, trembling, and diarrhea and vomiting while chronic exposure leads to debilitating disease. Treatment of arsenic poisoning involves administration of compounds with reactive thiol groups (for example dimercaprol, dimercaptosuccinic acid) which complex with the arsenic and promote its elimination from the body.

22.6 CARBON MONOXIDE POISONING

The toxicity of carbon monoxide (CO) is due to its ability to bind to heme residues of both red blood cell hemoglobin and mitochondrial cytochromes. CO has 200 times the affinity to bind to heme groups compared to molecular oxygen. Carbon monoxide thus competes for oxygen binding to hemoglobin and decreases the delivery of oxygen to tissues. It also has an allosteric effect on oxygen binding to hemoglobin, shifting the oxygen dissociation curve to the left. In mitochondria, carbon monoxide competes for oxygen binding to the terminal cytochrome, cytochrome oxidase, of the electron-transport chain on the inner membrane, and impairs oxidative metabolism in the cell such that severe carbon monoxide poisoning causes lactic acidemia.

The binding of CO is reversible, with an elimination half-life of approximately 5 hours when patients breathe pure air and approximately 1 hour when breathing pure oxygen.

The major sources of carbon monoxide are mainly from the incomplete combustion of organic matter and include:

- Tobacco smoke
- Inefficient gas appliances
- Motor vehicle exhaust
- Smoke (including hydrogen cyanide and irritants)

Exposure to dichloromethane is another source of CO and very small amounts of carbon monoxide are produced in humans *in vivo* during the catabolism of heme to bilirubin (see Chapter 3).

Clinical aspects of carbon monoxide poisoning

The effects of chronic, low-grade exposure to carbon monoxide poisoning, such as may occur in properties with inefficient gas heating, are headache, dyspnea on exertion, and insomnia. Erythrocytosis may also be apparent due to increased secretion of erythropoietin. In relation to carbon monoxide poisoning, it is important to bear in mind that:

- There is a poor correlation between the clinical condition and the carboxyhemoglobin (HbCO) concentration
- Toxicity is potentiated by anemia, vascular disease, drugs, other poisons, and old age
- Headache is the earliest complaint, experienced when the HbCO level is ~ 20%
- Patients are usually pale due to shock but tissues appear cherry red at autopsy
- Acute complications include cerebral edema, hypothermia, rhabdomyolysis, and renal failure
- Chronic mental disturbance is common after severe poisoning

While measurement of the concentration of carbon monoxide in blood may be pertinent for medico-legal purposes, it is not required or essential for establishing or confirming the diagnosis for clinical purposes. Quantitative measurements of HbCO concentration are of value only if the blood sample is taken immediately at the scene of poisoning and carbon monoxide exposure may be determined by the proportion of HbCO in whole blood. Historically, this was difficult to measure, but modern co-oximeters are readily available as point-of-care instruments and HbCO levels can be measured as easily as blood gases. Whilst the finding of an elevated concentration of HbCO in blood may be useful to confirm poisoning, it is not always high if some time has elapsed following exposure or if oxygen has been given. It should also be borne in mind that smokers may have HbCO levels of up to around 10%—twice as high as nonsmokers.

Treatment of carbon monoxide poisoning requires immediate removal of the patient from exposure and, in cases of acute poisoning, administration of oxygen. This may involve the use of hyperbaric facilities, if available. Since CO poisoning causes tissue ischemia, complications include lactic acidosis, myocardial infarction, and rhabdomyolysis. The latter may cause acute kidney injury.

22.7 DRUGS OF ABUSE

Overview

A wide range of drugs may cause harm to the individual and unsupervised use of these is considered a criminal offence in many countries. Some are prescription drugs while others have no established therapeutic role and are used solely for recreational purposes. Of the latter, many derived from plants or fungi have been used for centuries. More recently, there has been an increase in the use of

TABLE 22.12 Examples of drugs of abuse	
Drug class	**Examples (not an exhaustive list)**
Alcohols	Ethanol, ethylene glycol, methanol
Cannabinoids	Marijuana (tetrahydrocannabinol) and metabolites
Opiates	Heroin, morphine, methadone, buprenorphine, naloxone, naltrexone
Tranquilizers	Barbiturates: phenobarbitone
	Benzodiazepines: benzodiazepoxide (Librium®), temazepam, diazepam (Valium®), lorazepam (Ativan®), clonazepam
Hallucinogens	Indoleamines: lysergic acid diethylamide (LSD), dimethyltryptamine, psilocybin (magic mushrooms)
	Phenylethylamines: 3,4-methylenedioxyamphetamine (MDA), 3,4-methylenedioxymethamphetamine (MDMA), mescaline
	Related drugs: phencyclidine (PCP, angel dust), nitrous oxide, amyl and butyl nitrites, and plant compounds (nutmeg, seeds of morning glory, catnip)

designer drugs, specifically synthesized for illegal recreational purposes. All of these different drugs are classed as drugs of abuse and many laboratories screen for them in various settings such as for:

- Accidental poisoning
- Workplace screening
- Drugs administered illegally (especially to children)
- Unconscious and/or severely ill patients where there is suspected toxicity from a drug of abuse
- Drug abuse by athletes
- Monitoring drug abuse by addicts

Drug abuse is an important health concern in both developed and developing countries and, besides the health consequences for the individual consumer, the financial consequences of casual drug use as well as drug dependency are major drains on the health budgets of these countries. The principal classes of drugs that fit the category of drugs of abuse are listed, with examples of each, in **Table 22.12**. Alcohol (ethanol) is by far the most widely used drug worldwide in a growing list of drugs of abuse for which screening now exists.

Detection and measurement of drugs of abuse

The detection of drug abuse in athletes is a very specialized area and beyond the scope of most routine laboratories. However, many laboratories are involved in the routine testing of drug addicts to assess compliance with their medication and detection of illegal substances.

The measurement of drugs of abuse is often a two-step process; the first step is a simple screening test, in which an individual's urine (or sometimes saliva) is assessed against a panel of specific and generic immunoassays for qualitative analysis of the more common drugs of abuse. The second step would be a more specific, often quantitative analysis of body fluids including blood.

Thus, simple immunoassays are used to detect the different drugs, either on automated platforms or using point-of-care testing systems. Some of these immunoassays are quite specific for the drug in question, for example for cannabis and methadone metabolites, whilst others have a wide range of cross-reactivity with a family of drugs, for example opiates and sympathomimetics. In such a case, for example, a patient taking codeine would produce a positive opiate

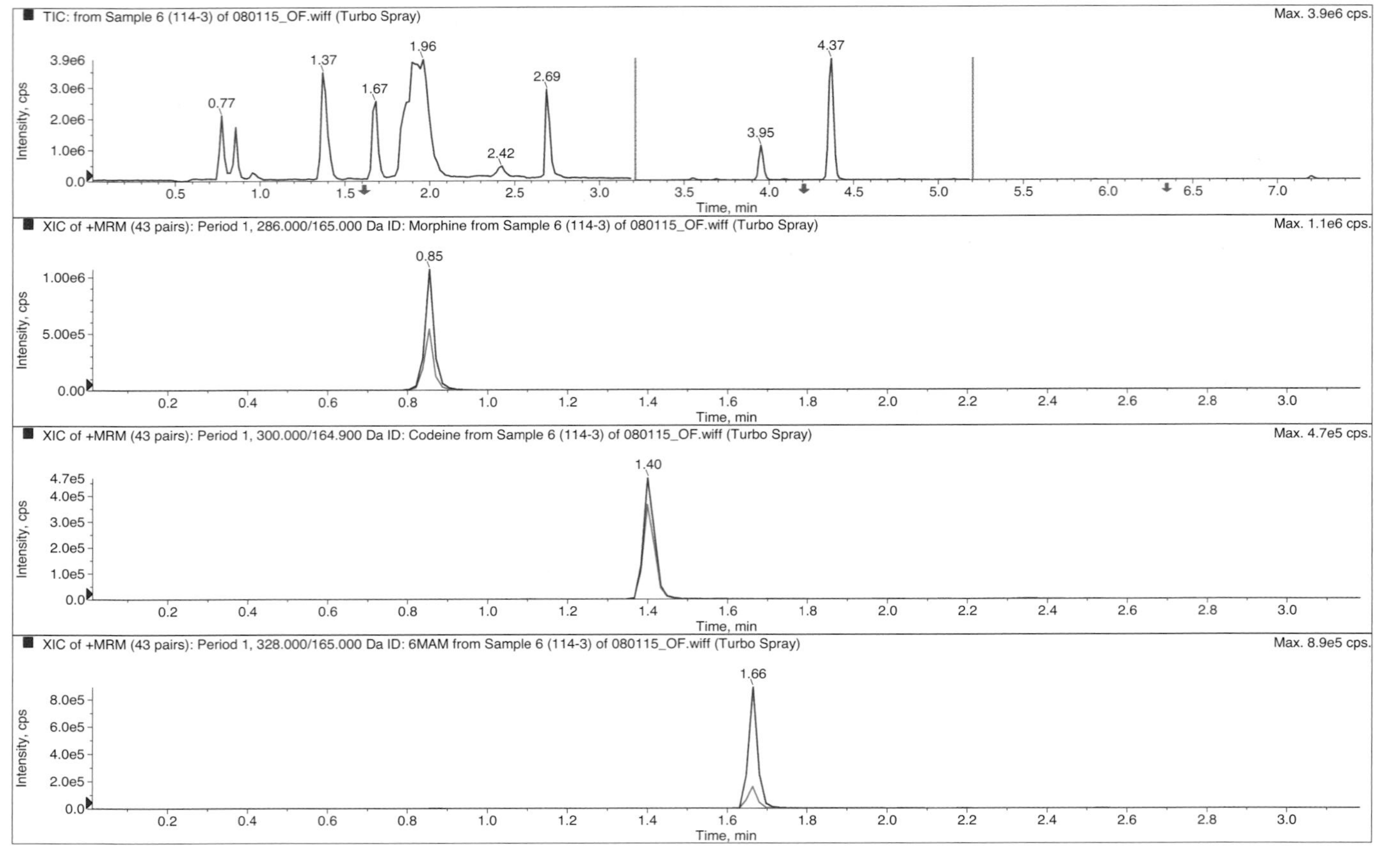

Figure 22.4 Analysis for drugs of abuse on oral fluid using liquid-liquid extraction followed by liquid chromatography tandem mass spectrometry (LC-MS/MS).
The top panel shows the total ion chromatogram (TIC) with the extracted ion chromatograms (XIC) shown below for morphine, codeine, and 6-monoacetylmorphine (6MAM). Heroin (diacetylmorphine) has a very short half-life and is metabolized sequentially to 6MAM and then to morphine. The detection of 6MAM is conclusive evidence of heroin use. Street heroin often also contains acetylcodeine which is metabolized to codeine. Therefore the chromatogram shows a characteristic profile for heroin abuse including 6MAM, morphine, and codeine. It must be noted that 6MAM has a very short detection window (approximately 6 hours in oral fluid, 24 hours in urine) and the presence of morphine and codeine without 6MAM cannot be used to identify heroin use. (Courtesy of Alex Lawson, Heartlands Hospital, UK.)

screen and further testing, usually by gas chromatography–mass spectrometry (GC-MS) or tandem MS, may be required to determine the precise drug of abuse (**Figure 22.4**). Clearly, this is vital in workplace testing, where the employment of an individual may be dependent upon the result. Many of the newer designer drugs show up as negative in current screening tests, and full chromatographic analysis is always required for their detection.

Most drugs of abuse and their metabolites are excreted via the kidneys into urine, and urine is usually the source material for analysis. In most cases urine testing should be performed within 3 days of ingestion of the drug. The times of appearance in urine of ingested drugs and their major metabolites are shown in **Table 22.13**.

TABLE 22.13 Time after ingestion during which abused drugs and/or metabolites may be tested for in urine

Drug	Compound detected in urine	Time compound may be detected in urine after ingestion
Heroin	Morphine 6-Monoacetylmorphine	1–3 days
Codeine	Codeine Morphine	1–3 days
Methadone	Methadone	2–4 days
Cocaine	Benzoylecgonine	1–3 days
Amphetamine	Amphetamine	2–4 days
Methamphetamine	Methamphetamine Amphetamine	2–4 days
Marijuana	Tetrahydrocannabinol	1–3 days (casual use); up to 30 days (chronic use)
Phencyclidine	Methamphetamine	2–7 days (casual use); up to 30 days (chronic use)
Benzodiazepines	Oxazepam, diazepam, other benzodiazepines	Up to 30 days
Barbiturates	amobarbital, secobarbital, other barbiturates	2–4 days (short acting); 30 days (long acting)

CASE 22.1

A 28-year-old woman with a history of opiate addiction is enrolled on a methadone program. She is required to take methadone under supervision, to undertake not to use illicit drugs, and to submit to random drug tests. After one month, a random urine sample is tested.

Urine	Result
Amphetamine	Negative
Barbiturates	Negative
Benzodiazepines	Negative
Cocaine metabolites	Positive
Methadone metabolite	Positive
Opiate screen	Negative
Cannabinoids	Positive
Appearance	Normal
Creatinine	7.9 mmol/L (89 mg/dL)
pH	7

- **What is the value of urine drug screening in methadone programs?**
- **Why are creatinine and pH measured?**

Methadone is used as an orally administered opiate to control withdrawal symptoms and avoid the risks of intravenous injections. Methadone metabolite in the urine is an indicator that methadone is actually being taken, rather than some being added to the urine sample and some being sold to other drug abusers. The presence of cocaine metabolites and cannabinoids indicates the use of other illicit drugs and may be grounds to terminate her participation in the program. To avoid detection of other drug use, some drug users manipulate their urine samples, for example by dilution with water or addition of other chemicals. Integrity checks by the laboratory are usually carried out before analysis. These include visual inspection and pH measurement. Creatinine is a marker of urinary concentration: low levels (for example <1 mmol/L) may indicate dilution of a sample with tap water.

CASE 22.2

A 17-year-old adolescent boy is admitted to hospital having taken some recreational drugs. He is initially confused but becomes comatose with muscle tremors and sweating. On examination he has increased temperature, heart rate, and blood pressure. He is admitted to the intensive therapy unit with a clinical diagnosis of serotonin syndrome due to overdose of unknown drugs. A urine sample is sent to the laboratory and the full report is available three days later. With supportive therapy the patient recovers and is sent home after one week in hospital.

Urine	Result
Amphetamine	Negative
Barbiturates	Negative
Benzodiazepines	Negative
Cocaine metabolites	Negative
Methadone metabolite	Negative
Opiate screen	Negative
Appearance	Normal
Creatinine	3.4 mmol/L (38 mg/dL)
pH	7
Toxicology by LC-MS QTOF*	α-Methyltryptamine and mephedrone detected

* Liquid chromatography–mass spectrometry quadrupole time of flight

- **What is the value of urine drug analysis in this patient?**

The serotonin syndrome is caused by drugs which increase the neurotransmitter serotonin in the central and peripheral nervous system. It presents as a triad of cognitive, autonomic, and muscular effects and may be fatal. Drugs used in high doses or in combination that may cause serotonin syndrome may be prescribed, over-the-counter, or illicit substances. In this case two illicit drugs were found: α-methyltryptamine and mephedrone. Neither was detected in the routine screen for drugs of abuse done initially. The LC-MS QTOF analysis done at the referral laboratory was able to detect the drugs by comparison of the sample with a library of known molecules. Unless there is a specific antidote (which is rare), the value of drug screening is to confirm the clinical diagnosis and it may be useful if the same patient presents again or other patients are admitted with similar symptoms, due to the local availability of the same illicit drug supply. It is worthwhile noting that although a simple urine drug screen is usually available 24 hours a day, seven days a week in most large hospitals, and can be performed within an hour, complex drug analyses are usually only performed in specialist toxicology centers and, by their very nature, take a long time to perform.

Ethanol

Moderate consumption of ethanol is said to have beneficial effects on health through increased plasma high-density lipoprotein (HDL) and increased rates of fibrinolysis. However, the wide availability of cheap alcoholic drinks has contributed to the situation where ethanol (alcohol) has become probably the most widely used drug of abuse in Western societies, and its chronic excessive intake is the most common cause of liver disease. The progression of alcoholic liver disease follows a familiar pattern:

1. Excessive intake
2. Alcoholic hepatitis

3. Steatosis (fat deposition in the liver)
4. Liver fibrosis leading to cirrhosis
5. Liver failure

Excessive consumption of alcohol is also a major cause of morbidity and mortality in the West, being the third most preventable cause of death after smoking and obesity. This small, water-soluble molecule is absorbed predominantly in the small intestine. The rate of absorption is affected by simultaneous ingestion of carbohydrate and fizzy (carbonated) drinks, the food content of the stomach, and the rate of ingestion of alcohol. The physical nature of ethanol ensures it has a wide volume of distribution and once in the blood it equilibrates rapidly across the blood–brain barrier. Ethanol is a sedative-hypnotic and acute ingestion has immediate effects on the brain, where it binds to a number of receptors [including gamma-aminobutyric acid (GABA), *N*-methyl D-aspartate, glycinergic, and serotoninergic] and produces a euphoric state. More chronic excessive intake is associated with damage to other tissues in addition to the liver, as mentioned above, and also with an increased risk for a number of cancers. Tissues susceptible to damage by alcohol include the:

- Gastrointestinal tract (esophagus and stomach)
- Immune system (bone marrow, immune cell function)
- Endocrine system (pancreas, gonads)
- Muscle (cardiac and skeletal)
- Nervous system (brain, nerves)
- Liver

In addition, excessive alcohol intake is associated with nutritional and other disorders including:

- Obesity
- Wasting
- Protein malnutrition
- Vitamin deficiency
- Psychological disorders
- Infections

Metabolism of ethanol

The stepwise metabolism of ethanol to acetic acid (ethanoic acid) via acetaldehyde (ethanal) is shown in **Figure 22.5**. Alcohol dehydrogenase (ADH) has a K_m for ethanol of 1.74 mM and at low doses approximately 90% of the ingested ethanol is metabolized by this ADH/aldehyde dehydrogenase (ALDH) system. However, chronic high doses of ethanol induce a hepatic microsomal ethanol oxidizing system (MEOS) with a much higher K_m for ethanol; CYP2E1, a cytochrome P450-linked enzyme, has a K_m for ethanol of 8.7 mM, and this system uses $NADP^+$ as an electron acceptor. Less than 10% of the ingested dose is excreted unchanged in the urine via the kidneys. In adults, ethanol is cleared from the blood at a rate of approximately 100 mg/kg/h such that blood ethanol concentration falls by approximately 150 mg/L/h. In addition to ethanol, the MEOS is also induced by various other drugs. The effects of ethanol on the toxicity and clearance of other drugs is shown in **Table 22.14**.

1. ethanol is oxidized to acetaldehyde by alcohol dehydrogenase (ADH)

$$CH_3CH_2OH \xrightarrow[NAD^+ \quad NADH]{ADH} CH_3CHO$$

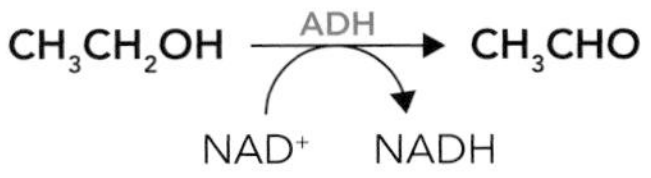

2. acetaldehyde is oxidized to acetic acid by aldehyde dehydrogenase (ALDH)

$$CH_3CHO \xrightarrow[NAD^+ \quad NADH]{ALDH} CH_3COOH$$

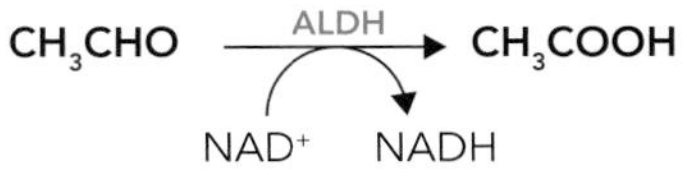

Figure 22.5 Metabolism of ethanol to acetic acid.

TABLE 22.14 Examples of the effects of ethanol on toxicity and clearance of drugs

Increases toxicity	Increases clearance
Acetaminophen	Phenytoin
Isoniazid	Phenobarbitone
Halothane	Rifampicin

The overall effects of metabolism of chronic doses of ethanol in the liver are to increase the intracellular concentration of acetaldehyde and to change the redox state of the cell by increasing the pyridine nucleotide ratios ($NADH:NAD^+$ and $NADPH:NADP^+$) in favor of the reduced state. In metabolic terms, this means that acetaldehyde will form Schiff bases with free amino groups of lysine residues on proteins (compare with glycated proteins) and may thereby change the physical nature and function of the proteins and contribute to tissue injury. These changes may manifest as:

- Decreased formation of microtubules and protein export
- Decreased enzyme activity
- Changes in cell surface and mitochondrial membranes
- Decreased DNA repair processes
- Increased formation of free radicals
- Increased formation of collagen

The change in redox state will affect the activities of pathways requiring NAD^+ (glycolysis, tricarboxylic acid cycle, fatty acid oxidation, and gluconeogenesis). Under these circumstances, the liver attempts to regenerate NAD^+ by converting pyruvate to lactate (**Figure 22.6**). This may produce a lactic acidosis and inhibit gluconeogenesis (NAD^+ is required by malate dehydrogenase and glycerol 3-phosphate dehydrogenase; see Chapter 2) in poorly nourished alcoholics. A raised intra-mitochondrial $NADH:NAD^+$ ratio inhibits β-oxidation of fatty acids and reduces the ability of mitochondria to use fatty acids as an energy source. Furthermore, induction of the MEOS yields increased amounts of NADPH which can be used to promote fatty acid synthesis and the formation of triglyceride. Initially this leads to hyperlipidemia as the liver exports the excess triglyceride as a component of very-low-density lipoprotein (VLDL). However, eventually, protein synthesis becomes compromised due to raised levels of acetaldehyde and VLDL production falls, producing a fatty liver. Some of the metabolic sequelae of the excessive consumption of ethanol are thus:

Figure 22.6 Reduction of pyruvate to lactate and regeneration of NAD^+.

- Hypoglycemia
- Hypertriglyceridemia
- Hyperlactatemia
- Ketosis
- Hyperuricemia
- Hypomagnesemia
- Hypophosphatemia
- Precipitation of acute episodes in acute intermittent porphyria
- Male hypogonadism
- Pseudo-Cushing's syndrome
- Osteoporosis

The possible link between chronic ethanol ingestion and tumor formation may lie in the fact that the action of CYP2E1 gives rise to radical oxygen species such as superoxide and perhydroxyl, which are postulated to be teratogenic agents.

Ethanol-induced hypoglycemia

Ethanol-induced hypoglycemia arises due to inhibition of gluconeogenesis. The change in redox state in the liver yields a deficiency of NAD^+ that inhibits the reactions mediated by three key dehydrogenases of the gluconeogenesis pathway: lactate dehydrogenase, glycerol 3-phosphate dehydrogenase, and malate dehydrogenase). Hypoglycemia occurs during fasting when liver glycogen has been exhausted. The condition is seen in adults who drink heavily and in children who drink surreptitiously. It is an important clinical condition, often causing a coma which does not respond to glucagon.

Ethanol-induced hypoglycemia may also arise as a result of potentiation by ethanol of glucose-induced secretion of insulin. This is sometimes referred to as the "gin and tonic lunch syndrome" and the resulting hypoglycemia is rarely severe.

Measurement of ethanol

In most routine laboratories, ethanol level is determined by enzymatic methods using alcohol dehydrogenase, but when it is measured for legal reasons (for example to assess drink-driving blood levels) it is usually measured by head-space gas–liquid chromatography. The presence of alcohols in the blood may be inferred when the patient has a high osmolar gap. Breath tests using dichromate, infrared absorbance, or fuel cells are used in the detection of drink-driving. The detection of exposure to ethanol may be seen from measurements of ethanol in blood, urine, and breath (**Table 22.15**).

TABLE 22.15 Relative ratios of ethanol concentrations in blood, urine, and breath

Comparison	Ratio
Blood:urine	1:1.3 to 1:20
Blood:breath	2000:1

Currently, in the UK, it is an offence to be in charge of a motor vehicle when whole blood ethanol concentration is >80 mg/dL (17 mmol/L) and/or breath ethanol concentration is >35 μg/dL (7.6 mmol/L) although this is likely to be revised downward. The tolerance limits vary throughout the world, with Scandinavian countries, for example, exhibiting a zero tolerance policy.

Markers of chronic exposure to ethanol are raised serum concentrations of gamma-glutamyl transferase, carbohydrate-deficient transferrin, and folate-dependent macrocytosis.

Ethylene glycol (ethanediol)

Ethylene glycol poisoning arises mainly from accidental or deliberate ingestion of the antifreeze used in car radiators. It is detected in blood samples by gas–liquid chromatographic analysis. Ingestion of ethylene glycol causes immediate, serious problems due to its rapid metabolism to toxic metabolites, as outlined in **Figure 22.7**. Most if not all of these reactive products are toxic. Oxidation of ethylene glycol, in addition to changing the redox state of hepatocytes with

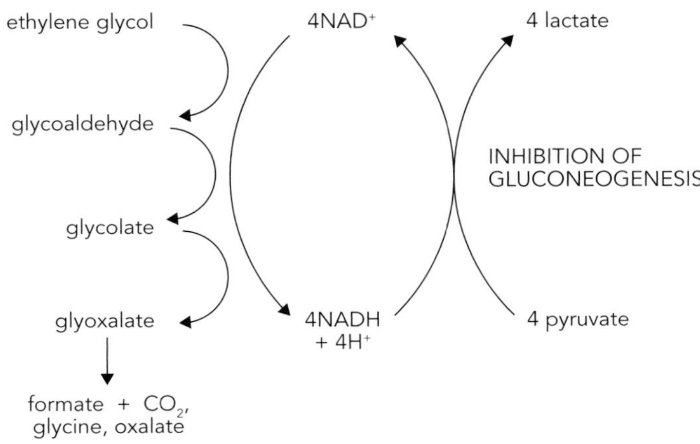

Figure 22.7 Metabolism of ethylene glycol to form toxic products. Sequential conversion of ethylene glycol to glyoxalate, via glycoaldehyde and glycolate, involves the reduction of four molecules of NAD^+. The glyoxalate will then break down to glycine, oxalate, formate, and CO_2. Glyoxalate and formate are both very toxic to cells and can cause considerable tissue damage, especially in the kidneys. The generation of NADH leads to pyruvate being converted to lactate, which inhibits gluconeogenesis.

the consequent metabolic changes described for ethanol, produces toxic products including oxalic acid, a powerful chelator of calcium and glycolic acid. As such, calcium oxalate crystals in the urine are a common finding in ethylene glycol poisoning. Treatment needs to be immediate and involves, paradoxically, administration of ethanol in an attempt to swamp the alcohol and aldehyde dehydrogenases to promote urinary excretion of non-metabolized ethylene glycol. Treatment may also include administration of 4-methyl pyrazole and dialysis to remove the glycol and its metabolites. Toxic effects of ethylene glycol poisoning include metabolic acidemia, renal failure, and hypocalcemia.

Other alcohols including methanol, isopropanol, and butanol may also be ingested, either deliberately (drinking methylated spirits) or as products from distillation in homemade wooden stills (wood spirit). These alcohols are metabolized by the same enzyme system as described for ethanol. Thus methanol is metabolized to methanal (formaldehyde) and methanoic (formic) acid producing a metabolic acidemia due to methanoic and lactic acids. Toxic effects of methanol include visual disturbance and pancreatitis. As for ethylene glycol poisoning, treatment involves ethanol infusion or administering a specific antidote (fomepizole), as well as correction of the acidemia. In severe cases hemodialysis is required.

22.8 PLANT POISONS

Fortunately, deaths from ingestion of poisonous plants are not common, and the clinical laboratory is rarely asked to perform analyses to identify the possible toxins in live patients. Poisoning is often accidental, arising from children eating berries or seeds, adults mistakenly identifying poisonous plants (especially fungi) as edible, or occasionally by deliberate administration of poisonous plants. The cause of the poisoning is usually ascertained from a combination of symptoms and collected evidence (seeds and flowers for example) brought in by or with the patient's caregivers. The analyses are performed post mortem to identify potential causes of deaths and are usually undertaken in specialist toxicology laboratories.

The cardiac glycosides digoxin and digitoxin, found in numerous foxglove (*Digitalis*) species, have been known as plant poisons for centuries and their actions as Na^+/K^+-ATPase channel antagonists are well described in the literature. Cardiac glycosides are also present in a number of other plant species including *Helleborus* and *Adenium*. Oleander (*Nerium oleander*) contains the cardiac glycoside oleandrin, one of the most common causes of plant poisoning in the USA. Various attempts have been made to measure digoxin-like substances in cases of plant poisoning, by making use of the cross-reactivity with immunoassays for digoxin. Patients can be actively treated using the digoxin antidote, Digibind®.

Poisoning from consumption of flowers or berries of *Laburnum anagyroides* has also been well described. The toxin in this case, cytisine, mimics the action of nicotine and the toxic effects are mediated by agonistic stimulation of some of the acetylcholine receptors. Many other plants contain similar nicotine-type toxins, including hemlock (*Conium maculatum*) which has been attributed to be the poison that killed Socrates in 399 BC. Angel's trumpets (*Brugmansia* and *Datura* species) and deadly nightshade (*Atropa belladonna*) are amongst several plants which contain high concentrations of atropine, which antagonizes muscarinic-type acetylcholine receptors.

Several plants, in particular members of the buttercup family (*Ranunculaceae*), contain poisons which act as sodium channel activators and cause inadequate repolarization of nerves and cardiac tissue, leading to heart and respiratory failure. For example, monkshood (*Aconitum napellus*; also known as friar's cap or garden wolfsbane) is believed to be one of the most poisonous plants known; it contains aconitine, an extremely toxic compound which can cause death within a couple of hours of ingestion. Various *Rhododendron* species (including the common wild species *Rhododendron ponticum*) also contain a sodium channel activator, acetylandromedol (also known as grayanotoxin I, a diterpene compound), which has been shown to pass from the nectar into bees' honey, and to be the cause of mad honey disease.

The yew tree (*Taxus baccata*) contains an arsenal of toxins, amongst which is the potentially fatal compound taxane, which acts as both a calcium and sodium channel blocker. Taxane is found throughout the plant except for the flesh of the bright red berries. The berries are toxic only if the seeds, which have a high concentration of the toxin, are also eaten. Recently, taxane has been used as a cytotoxic agent in the treatment of cancer.

Several other plants are particularly toxic, including water henbane (*Cicuta virosa*) which contains the poison cicutoxin. A single bite of the plant's root can be fatal within an hour. Arguably the most toxic compound derived from plants is ricin extracted from the castor oil plant (*Ricinus communis*). Injection of as little as 1 mg can be lethal.

Finally, one particular form of poisoning—Jamaican vomiting sickness—mimics the features of an inborn error of metabolism, glutaric acidemia type II. This poisoning is caused by eating unripe ackee fruit (*Blighia sapida*) which contains high concentrations of hypoglycin A, an irreversible inhibitor of acyl-CoA dehydrogenase.

INBORN ERRORS OF METABOLISM

CHAPTER 23

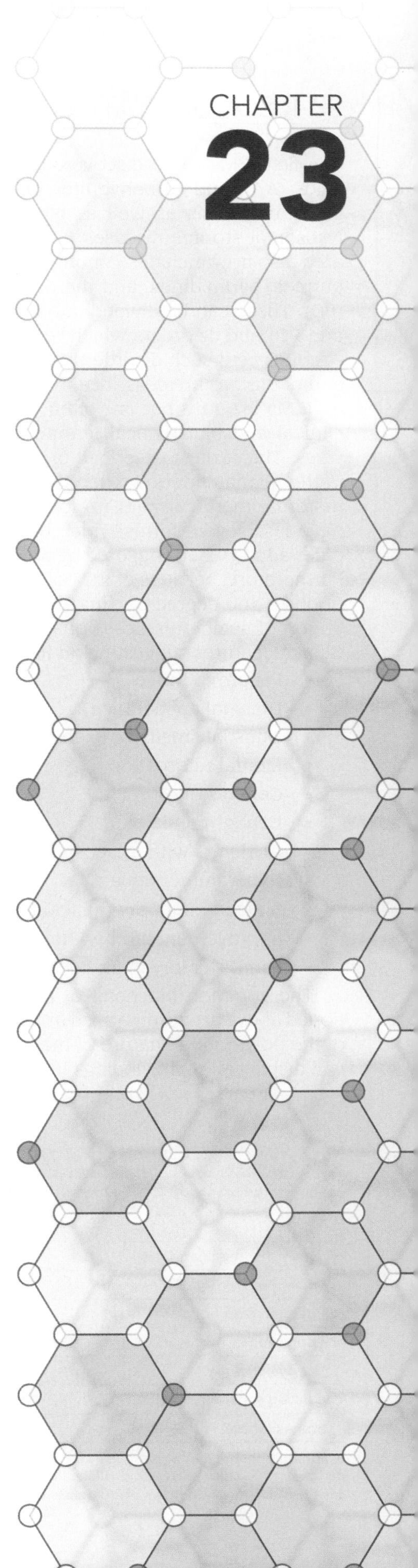

IN THIS CHAPTER

The term "inborn error of metabolism" (IEM) was introduced by Sir Archibald Garrod following his pioneering work at the end of the nineteenth and in the early part of the twentieth century on the inheritance of alkaptonuria, albinism, cystinuria, and pentosuria in affected families. Garrod postulated correctly that such disorders arose from the absence of a particular enzyme in a metabolic pathway and opened the way for a field of study leading to the development of molecular biology.

A single base change or deletion in the coding sequence of a gene, and the subsequent translation of a protein with a single amino acid difference in its primary structure from the wild-type protein, will often not cause any problem for the cell if the function of the variant protein is the same or similar to the wild-type protein. Examples of such polymorphisms are seen for lactate dehydrogenase, hemoglobins, and collagens. However, when a base change gives rise to a translated protein whose structure and function are grossly different from the wild-type protein an IEM may be the result. IEMs comprise a large number of individually rare diseases characterized in most cases by a defective or absent enzyme in a metabolic pathway. The underlying cause in each case is a single gene defect and these monogenetic disorders, while somewhat rare, have often given valuable information allowing the elucidation of biochemical pathways. Such monogenetic disorders may be (i) autosomal dominant, in which a subject heterozygous for the mutation (single defective allele) exhibits the clinical phenotype of the disorder; (ii) autosomal recessive, in which the subject is homozygous for the mutation (two defective alleles); or (iii) X-linked, where the mutation may be dominant or recessive. Mutations in a given gene may give rise to a gene product (enzyme/protein) that may have increased or decreased function compared to the wild-type product. The definition of an IEM has widened since the first disorders were recognized and now many authorities would consider genetic defects of other functional proteins, such as transmembrane ion transporters, to be no different in terms of the underlying causation of the disorder. For this reason, conditions not traditionally considered to be IEMs have been included, for example cystic fibrosis.

23.1 CLASSIFICATION OF INBORN ERRORS

The first IEMs to be discovered were disorders of organic acids (OA), amino acids (AA), and carbohydrates (CD; these include disorders of glycolysis and gluconeogenesis as well as glycogen storage diseases), and porphyrias and lysosomal storage diseases (LSD). As new defects have been discovered, the area also now includes, amongst others, disorders of fatty acid oxidation (FAO), purines, pyrimidines, and the organelles mitochondria (MD) and peroxisomes (PD). These can be divided into disorders of energy metabolism (FAO, CD, MD, and PD) and disorders which result in the accumulation of toxic products (OA, AA, LSD, and such disorders as galactose intolerance). Disorders of porphyrins and of purines are described in Chapters 16 and 17, respectively.

IEMs are rare but are often disabling or lethal conditions with a range of clinical and biochemical features, usually presenting in a limited number of ways. The earliest described or classical presentations of IEMs generally occur in the neonatal period when the baby becomes metabolically independent and maternal metabolism is no longer able to compensate. However, many IEMs have less severe forms which become apparent later in childhood or even in older adults when triggered by metabolic stress or other environmental factors. Furthermore, because of the polymorphic nature of individual disorders, the age of onset and degree of clinical presentation may vary between cases, with initiation of symptoms occasionally precipitated by dietary factors and infection. Clinical features suggesting an IEM in childhood include:

- Failure to thrive
- Unusual appearance
- Unusual smell, for example of maple syrup or cat urine
- Mental retardation
- Convulsions
- Hypoglycemia
- Acidemia with increased anion gap
- Hyperammonemia
- Sudden infant death including unexplained deaths in infancy
- Improvement after withdrawal of protein or a specific sugar
- Family history of an IEM

The common biochemical presentations of the major inborn errors mentioned above are summarized in **Table 23.1**. It is important to note that, because of the polymorphic nature of many IEMs, not all cases of an individual disorder will display all of the findings listed.

Clinical practice point 23.1

IEMs are usually thought of as being exclusively disorders of infants, but can present at any age. The neonatal forms of IEMs are usually the most severe and are often rapidly fatal.

TABLE 23.1 Laboratory findings in inborn errors of metabolism

Laboratory finding	Amino acid disorders	Organic acid disorders	Urea cycle disorders	Carbo-hydrate disorders	Glycogen storage disorders	Fatty acid disorders	Lysosomal storage disorders	Peroxisome disorders	Mitochondrial disorders
Primary metabolic acidosis	1	3	2	2	1	2	0	0	1
Primary respiratory alkalosis	0	0	2	0	0	0	0	0	0
Hyperammonemia	1	2	3	1	0	2	0	0	1
Hypoglycemia	1	1	0	2	1	2	0	0	1
Liver dysfunction	1	1	1	2	1	2	1	1	1
Reducing substances (urine)	1	0	0	2	0	0	0	0	0
Ketones (urine)	A	H	A	A	L/A	L	A	A	H/A

0, absent; 1, sometimes present; 2, usually present; 3, always present; H, inappropriately high; A, appropriate; L, inappropriately low. Not all diseases within a particular group may present with all the laboratory findings for that group. In disorders of an episodic nature the findings may present only during an episodic crisis, while in those disorders of a progressive nature the findings may not present in early stages of the disorder.

Individual IEMs are relatively rare conditions presenting usually at a frequency of between 1 in 1500 and 1 in 5000 live births. Overall, however, the more than 1000 IEMs so far documented constitute a group of disorders with a cumulative incidence of approximately 1 in 800 live births and contribute significantly to worldwide morbidity and mortality.

Treatment depends on the particular enzyme defect. Occasionally, this will consist simply of the administration of the missing end product of the affected pathway; for example, of cortisol in 21-hydroxylase deficiency (see Chapter 11). Sometimes therapeutic replacement enzymes may be available, such as glucocerebrosidase for the storage disorder Gaucher's disease. The missing enzyme may also be replaced in some disorders by tissue transplantation; for example, kidney in cystinosis, or bone marrow in some other lysosomal storage disorders. For many years there has been hope that gene replacement therapy will be the ultimate cure for these disorders, but progress has been frustratingly slow. Nevertheless, work in this therapeutic area continues and has been encouraged by some recent successes.

The study of IEMs is a large and expanding area with new disorders being recognized all the time. The current chapter gives an overview of IEMs using some examples of the more commonly recognized disorders, but does not attempt in any way to be a comprehensive text. The reader is referred to the excellent facility provided by the National Library of Medicine of the US National Institutes of Health (http://ghr.nlm.nih.gov) for extensive coverage of the subject. The clinical laboratory plays a key role in the diagnosis and management of IEMs. Traditionally, the major role has been played by clinical chemistry, but molecular biology has become increasingly important in recent years.

23.2 GENETICS

IEMs may arise from gene duplication, gene deletion, or commonly from a single point mutation in the genes encoding proteins which are essential for metabolic homeostasis, including enzymes of metabolic pathways, proteins of transport processes, and structural proteins. They are usually inherited as autosomal recessive diseases but a gene dosage effect may be present in rare instances. In addition, occasionally autosomal dominant disorders with variable expression occur, for example with porphyrias (except congenital erythropoietic porphyria) and hyperinsulinism with hyperammonemia. There are, however, also a few X-linked conditions including glucose 6-phosphate dehydrogenase (G6PD) variants and Lesch–Nyhan syndrome, both of which are X-linked recessive, and deficiency of the urea cycle enzyme ornithine transcarbamylase, hypophosphatemic rickets, and Hunter syndrome, which are X-linked dominant disorders. In addition, expression of a genetic defect may be variable due to genetic and allelic heterogeneity, an effect of diet or of metabolic stress, and genetic background.

23.3 PATHOPHYSIOLOGY

The effect of a mutation may manifest itself in different ways in terms of the protein expressed from the mutant gene. Only single examples from the many known are given for each of the situations listed below.

Causes

Absence of enzyme/protein

In the most extreme situation, the mutation is such that the protein is not synthesized at all or the protein which is synthesized is recognized as faulty and destroyed by the quality control mechanisms in the cell. For example, mutations in the gene encoding apolipoprotein B100 lead to the absence of a functioning apolipoprotein B100 in abetalipoproteinemia (see Chapter 18).

Synthesis of an abnormal enzyme

An active enzyme is synthesized but the enzyme has properties which differ from the wild-type enzyme; for example, altered affinity for its substrate. Two examples are (1) argininosuccinic acid (ASA) synthase mutations leading to decreased affinity (increased K_m) for its substrate, and (2) 5-phosphoribosylpyrophosphate synthase mutations leading to an increased affinity (decreased K_m) for substrate.

Synthesis of a less stable enzyme

An active enzyme is synthesized but is recognized by the quality control system of the cell as not normal and is turned over more rapidly than the wild-type enzyme/protein. For example, some variants of G6PD (see Chapter 2) have a shorter half-life than the wild-type enzyme.

Synthesis of a protein with decreased affinity for co-factor (vitamin responsive)

The three-dimensional structure of the variant enzyme is such that it does not bind its co-factor with the same affinity as the wild-type enzyme, with a consequent decreased enzyme activity. Examples include propionyl-CoA carboxylase (biotin co-factor), cystathionine synthase (pyridoxyl phosphate), and methylmalonyl mutase (adenosylcobalamin).

Compromised enzyme activity due to deficiency of proteins involved in co-factor biosynthesis

In this case the mutation is not in the gene encoding the enzyme itself but in that for an enzyme involved in the synthesis of a co-factor essential for the enzyme activity. This mutation may result in the absence of this protein or synthesis of a less functional enzyme which produces insufficient co-factor. For example, there are variants of phenylketonuria which are due to a deficiency of biopterin reductase (dihydropteridine reductase) rather than of phenylalanine hydroxylase activity (**Figure 23.1**).

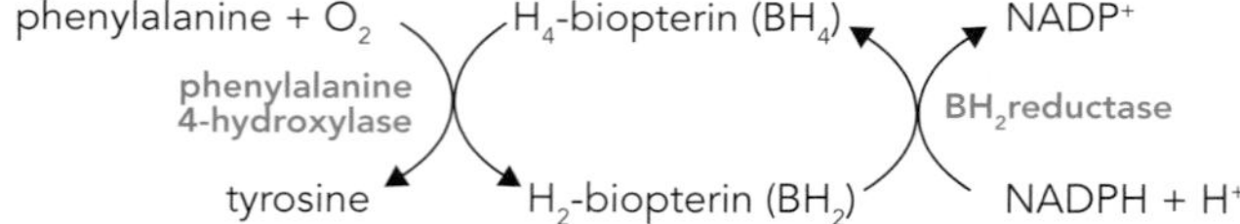

Figure 23.1 Hydroxylation of phenylalanine to tyrosine by phenylalanine hydroxylase. The hydroxylation of phenylalanine to tyrosine (4-hydroxyphenylalanine) by phenylalanine hydroxylase requires molecular oxygen and reducing equivalents from the reduced co-factor tetrahydrobiopterin (H_4-biopterin), which is oxidized to dihydropterin (H_2-biopterin) in the reaction. Tetrahydrobiopterin is regenerated through the action of dihydropteridine reductase (BH_2 reductase), which uses NADPH as a co-factor.

Metabolic consequences of abnormal enzyme activity

A hypothetical metabolic pathway is shown in **Figure 23.2** where substrate X is converted, via intermediate Y, to Z and beyond by a sequence of enzymes (E). X may also be converted to by-products (p) when metabolism to Y is compromised. Possible metabolic consequences of the abnormal activity of enzyme E1 are:

- Accumulation of substrate (precursor) X or intermediate Y by the absence of E1 or E2 activity
- Overproduction of side product(s) p1, p2, and/or p3 through absence of E1
- Deficiency of end-product Z
- Accumulation of intermediate or end-product Z
- Increased turnover of precursor, intermediate, and/or end product

As stated earlier, the overall consequences of such defects may occur due to the accumulation of a toxic product or due to a decrease in energy production in cells in which the defect is present.

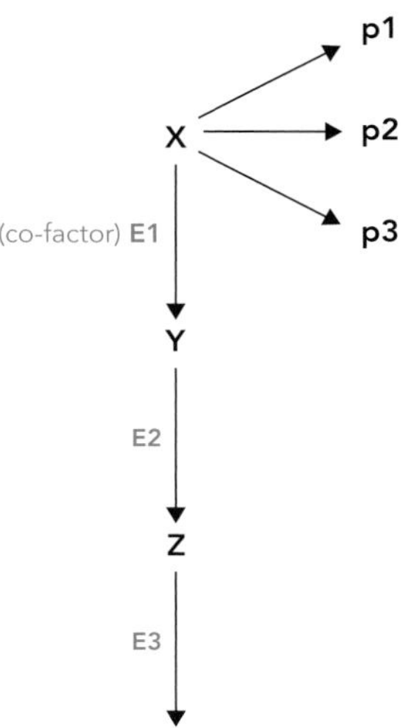

Figure 23.2 A hypothetical metabolic pathway.
Sequential enzymes are shown as E1, E2, and E3; possible by-products as a result of compromised activity of E1 are shown by p1, p2, and p3; and the product of E1 activity is indicated as Y and the product of E2 activity as Z.

Accumulation of precursor/intermediate

Example: alkaptonuria

The catabolic pathway of tyrosine to CO_2 and water involves opening of the aromatic ring in a step, catalyzed by homogentisic acid oxidase, in which homogentisic acid is converted to maleylacetoacetate (**Figure 23.3**). The normal plasma concentration of phenylalanine or tyrosine is very low and is testament to the rapid and efficient catabolism of these aromatic amino acids. Furthermore, homogentisic acid is not normally found in urine because of the action of homogentisic acid oxidase, particularly in liver, kidney, and prostate. Alkaptonuria is a rare autosomal recessive disorder (affecting <1 per 250,000 live births) in which the activity of homogentisic acid oxidase is deficient. In this disorder, homogentisic acid is produced in large amounts, with plasma concentrations rising to 175–200 μmol/L and between 4 and 8 g/day of the acid being excreted in urine. Interestingly, the urine of patients with this disorder turns black on standing as homogentisate oxidizes spontaneously to a dark pigment under the alkaline conditions of urine. With time, however, homogentisic acid is deposited in cartilage and connective tissue where it undergoes oxidation, catalyzed by a homogentisate polyphenyloxidase, to an achromatic pigment which causes joint stiffness, leading to degenerative arthritis particularly of the spine and larger peripheral joints.

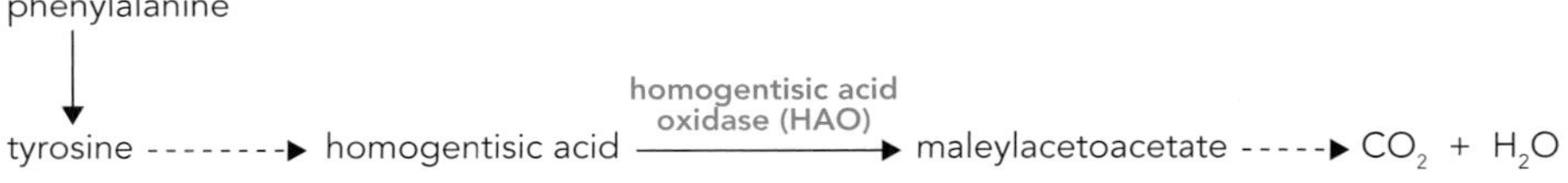

Figure 23.3 The role of homogentisic acid oxidase (HAO) in the catabolism of phenylalanine and tyrosine.
This key step in tyrosine metabolism involves the opening of the aromatic ring of the original amino acid. In the absence of HAO (alkaptonuria), homogentisic acid accumulates and appears in the urine. Longer term, homogentisic acid accumulates in the joints. Broken arrows indicate several enzymatic steps.

Overproduction of side products

Example: disorders of galactose metabolism

Absence of the activity of any of the three enzymes of galactose metabolism leads to accumulation of the precursor galactose and presents clinically as galactosemia (**Figure 23.4**). Galactitol accumulates as galactose is metabolized through an alternate pathway by aldose reductase.

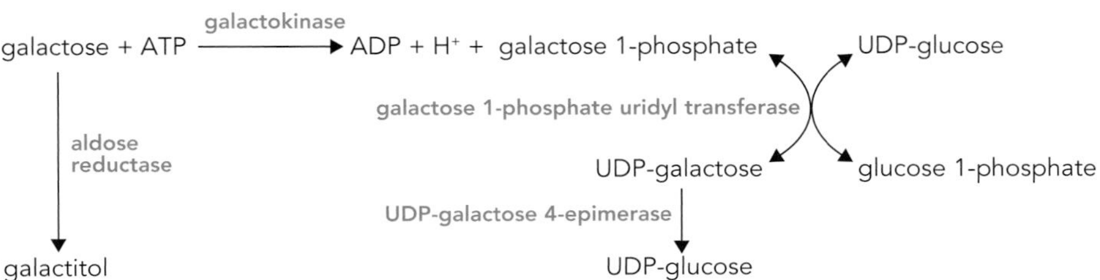

Figure 23.4 Metabolism of galactose.
Under normal conditions, galactose is metabolized to UDP-glucose through the sequential actions of galactokinase, galactose 1-phosphate uridyl transferase, and UDP-galactose 4-epimerase. In the second reaction, galactose 1-phosphate is exchanged with the glucose 1-phosphate of UDP-glucose to yield UDP-galactose. UDP-galactose undergoes epimerization at the 4-hydroxyl group of the galactose moiety to form UDP-glucose, which can donate the galactose-derived glucose to glycogen. If the metabolic route from galactose to UDP-glucose is compromised, galactose accumulates and is converted to the metabolic dead-end molecule galactitol by aldose reductase.

CASE 23.1

A week-old newborn male has poor feeding, vomiting, and is jaundiced. A urine sample is positive for reducing substances on a qualitative test, but is negative for glucose. Further tests were performed and whilst waiting for the results the newborn was switched from breast milk to soya milk.

	SI units	Reference range	Conventional units	Reference range
Plasma				
Glucose	1.9 mmol/L	3.9–6.1	34 mg/dL	70–110
Bilirubin	220 μmol/L	<21	13 mg/dL	<1.3
Alanine transaminase (ALT)	360 U/L	5–40	360 U/L	5–40
Red blood cell				
Galactose 1-phosphate uridyl transferase (GALPUT)	Absent activity			

- What do urine reducing substances indicate?
- Why is the change in milk made?
- What might cause a falsely normal enzyme result?

"Urine reducing substances" is a term used to describe a long-established but nonspecific colorimetric test for sugars, based on the reduction of copper ions. Glucose, galactose, and fructose give positive results and glucose can be specifically measured by an enzymatic method. The clinical picture is of galactosemia due to inability to metabolize galactose from the lactose in milk: a disaccharide of glucose and galactose. The commonest enzyme defect is of galactose 1-phosphate uridyl transferase (GALPUT), with deficiencies of galactose kinase and galactose 6-phosphate epimerase being less common. Demonstration of reduced GALPUT enzyme activity in red cells is used to diagnose classical galactosemia, but this enzyme activity will be falsely normal in patients who have received a recent blood transfusion. Accumulation of galactose within the cell results in depletion of phosphate, which reduces formation of ATP and causes multiorgan dysfunction including liver disease, poor growth, cataracts, seizures, and muscle hypotonia. Restriction of dietary lactose is essential to prevent complications and is usually continued for life. In some countries, including the USA, detection of galactosemia is included in the neonatal screening program.

Deficiency of end product

Example: congenital adrenal hyperplasia

This disorder has been described in Chapter 11. A lack of 21-hydroxylase is responsible for the failure to synthesize cortisol.

Accumulation of intermediate product

The process of turnover of molecules in cells requires their continual synthesis and degradation. Thus if the catabolic enzymes required for degradation are absent, the macromolecule accumulates if synthesis is unimpaired.

Example: lysosomal storage diseases

These disorders occur when lysosomal hydrolases are absent, such as in Gaucher's disease (deficiency of glucocerebrosidase) and Tay–Sachs disease (deficiency of hexosaminidase A). Both of these are described in more detail later in this chapter.

Increased turnover of precursor, intermediate(s), and end product

Example: hyperuricemia

This disorder is due to an overactivity of 5-phosphoribosylpyrophosphate (PRPP) synthase as described in Chapter 17.

23.4 EXAMPLES OF INBORN ERRORS OF METABOLISM

IEMs are associated with each organelle of the cell including the plasma membrane (transport functions), cytosol, mitochondria, lysosomes, endoplasmic reticulum, and nucleus. It is not possible within the remit of the current text to provide a comprehensive review of them all. The examples listed are the more common ones but most are still extremely rare.

Deficiencies of lysosomal enzymes: lysosomal storage diseases

Lysosomes are cellular organelles containing a battery of acid hydrolases which are involved in the catabolism of a number of the structural molecules of the cell. In homeostasis, the rate of breakdown of these molecules is balanced by their rate of synthesis so as to maintain a steady-state level of the molecule. In the absence of one of the lysosomal hydrolases, the substrate of the hydrolase accumulates within the lysosome. In most cases the defect is in a single specific enzyme and the common outcome is that eventually the lysosome becomes damaged and releases its other acid hydrolases into the cell with pathological consequences (**Figure 23.5**). Increased activity of non-prostatic acid phosphatase (ACP) in serum is a marker for lysosomal damage.

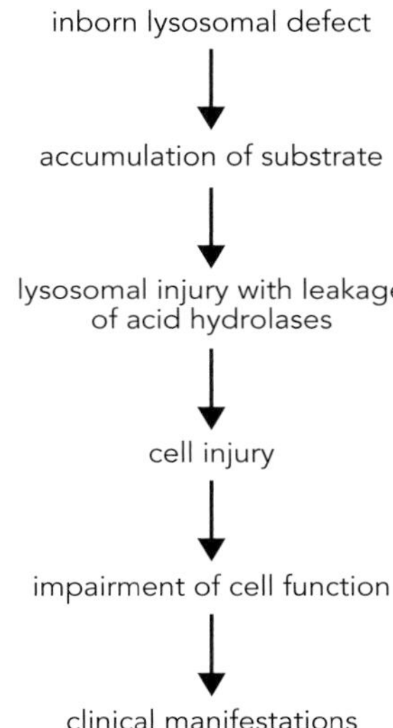

Figure 23.5 Pathogenesis of lysosomal storage diseases. The gradual accumulation in lysosomes of macromolecules from the extracellular matrix and connective tissue (substrates of lysosomal enzymes) eventually causes rupture of the organelles and leakage of acid hydrolases into the cell cytosol and eventual cell death.

The lysosomal storage diseases are classified by the nature of the particular molecule which accumulates, for example:

- Sphingolipidoses
- Mucopolysaccharidoses
- Mucolipidoses
- Glycogenosis type II
- Cystinosis
- Cholesterol ester storage disease

The inheritance of these diseases is usually autosomal recessive and they occur rarely apart from in close-bred ethnic groups. For example, Tay–Sachs disease was relatively common in Ashkenazi Jewish communities, such as in Baltimore in the USA, prior to the introduction of genetic counseling programs. In some conditions, expression of the genetic defect varies markedly between individuals with severe, early-onset and less severe, late-onset variants. Lysosomal storage diseases are rare due to the autosomal recessive inheritance and because the severity of illness usually precludes reproduction. The most common lysosomal storage diseases in the UK are Gaucher's disease (sphingolipidosis; 1 in 50,000 live births) and cystinosis (1 in 200,000 live births).

Eight of the lysosomal acid hydrolases are glycoproteins which are translated in the rough endoplasmic reticulum and undergo post-translational modification as they traverse through the smooth endoplasmic reticulum and Golgi apparatus, before being transferred to the lysosome. A key post-translational step which ensures that the hydrolases are targeted to the lysosome is the formation, in the *cis*-Golgi, of a covalently bound mannose 6-phosphate residue on the enzyme. This requires the participation of *N*-acetylglucosamine 1-phosphotransferase (**Figure 23.6**) which catalyzes the formation of a phosphodiester linkage between a mannose residue on the core oligosaccharide of the

mannose — protein backbone — *N*-acetylglucosamine 1-phosphotransferase (UDP-GlcNAc → UMP) → intermediate — phosphodiesterase (H_2O → GlcNAc) → mannose 6-phosphate

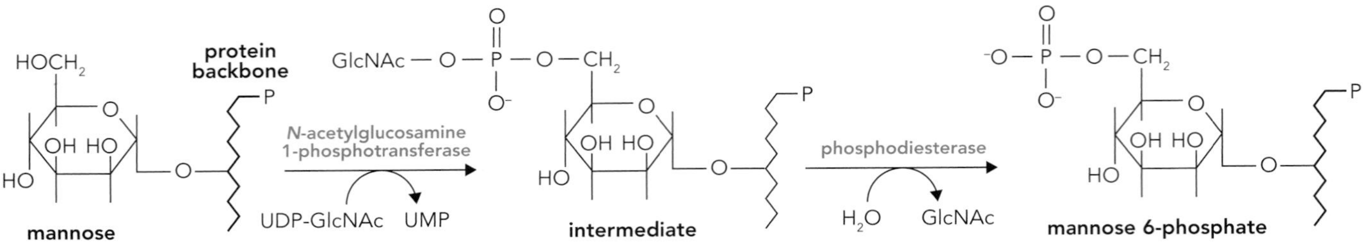

Figure 23.6 The action of *N*-acetylglucosamine 1-phosphotransferase in the formation of the mannose 6-phosphate label for targeting lysosomal enzymes to the lysosome.
N-acetylglucosamine 1-phosphotransferase, an enzyme of the *cis*-Golgi, catalyzes the addition of *N*-acetylglucosamine (GlcNAc) to a mannose residue on the newly synthesized lysosomal enzyme via a phosphodiester bond at C6 of the mannose. Hydrolysis of the *N*-acetylglucosamine residue by a phosphodiesterase leaves a phosphate at C6 of the mannose of the lysosomal enzyme. This mannose 6-phosphate is the label that ensures transfer of the enzyme into the lysosome.

hydrolase enzyme and acetylglucosamine from the activated donor, UDP-*N*-acetylglucosamine, with the release of UMP. A phosphodiesterase hydrolyzes this bond to leave a mannose 6-phosphate residue on the hydrolase enzyme.

Sphingolipidoses

This group of diseases presents with the lysosomal accumulation of sphingolipids and, apart from the common form of Gaucher's disease, affects neural tissue. Sphingolipids contain the lipophilic molecule sphingosine (an 18-carbon molecule) as a core structure; a ceramide has a fatty acid in amide linkage with this sphingosine base (**Figure 23.7**).

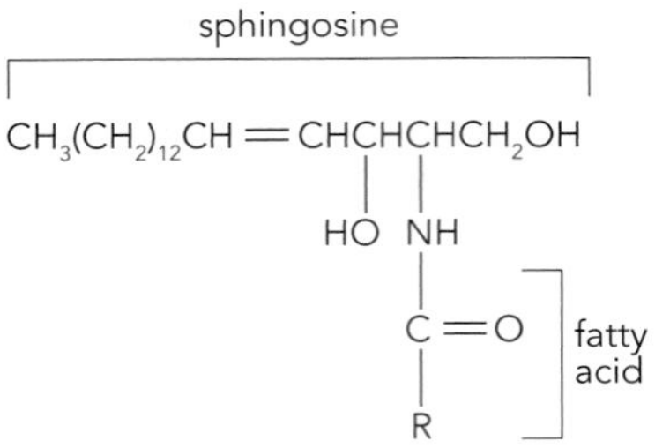

sphingosine: $CH_3(CH_2)_{12}CH{=}CHCHCHCH_2OH$ (HO, NH; fatty acid: $C{=}O$, R)

cerebrosides: CER-sugar(s)
gangliosides: CER-sialic acid-sugar
sphingomyelin: CER-phosphorylcholine

Figure 23.7 The chemical structure of a ceramide.
Sphingosine, synthesized from palmitoyl-CoA and serine, is the backbone of sphingolipids. In all sphingolipids the amino group of sphingosine is acylated with a long-chain fatty acyl CoA to form ceramide (CER). Thus ceramides are *N*-acylsphingosines and form part of sphingomyelin (CER-phosphorylcholine), cerebrosides (CER-sugar), and gangliosides (CER-sialic acid-sugar).

Gaucher's disease

Glucocerebrosides are formed during the breakdown of erythrocytes, white blood cells, and platelets and are normally degraded by macrophages of the reticuloendothelial system. The initial step in their catabolism is hydrolysis of the bond between ceramide and glucose by a glucocerebrosidase (**Figure 23.8**).

In the absence of glucocerebrosidase (Gaucher's disease), glucocerebroside accumulates in macrophages so that the cells become enlarged, and patients present with splenomegaly, hypersplenism, and lesions in bone, lung, liver, and occasionally brain. In the usual form of Gaucher's disease (type I), an autosomal recessive disorder, patients are homozygous for the mutation N370S; rarer neuropathic forms arise from the homozygous mutations L444P (type II) and 84insG (type III) (**Table 23.2**). Over 80 mutations have been reported. Gaucher's disease is detected in an affected fetus and heterozygous carrier by analysis of glucocerebrosidase activity in white blood cells and more recently by molecular biological techniques. The incidence in developed countries is approximately 1 in 50,000 live births (carriers, about 1 in 100) but it is much higher in the Ashkenazi Jewish population, where almost 10% of individuals are carriers and the birth incidence of the disease is 1 in 450.

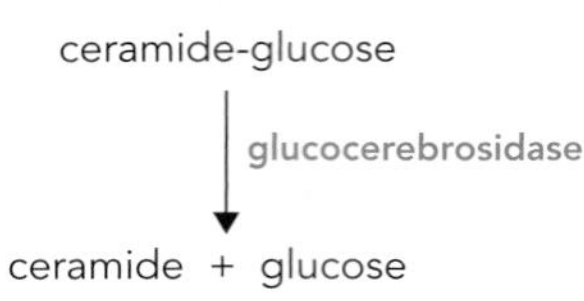

Figure 23.8 The action of glucocerebrosidase.
β-Glucocerebrosidase (also called acid β-glucosidase, D-glucosyl-*N*-acylsphingosine glucohydrolase, or GCase) is a lysosomal enzyme that hydrolyzes the β-glycosidic linkage of glucocerebroside. Mutations in the glucocerebrosidase gene cause Gaucher's disease, a lysosomal storage disease characterized by an accumulation of glucocerebrosides.

TABLE 23.2 Clinical presentation of the three types of Gaucher's disease

Type	Clinical features
I	Most common form, occurring at any age; liver and spleen enlargement, bone pain and fractures, and sometimes lung and kidney problems. No brain involvement
II	Severe brain damage, appears in infants; death by age 2 years
III	Liver and spleen enlargement; gradual appearance of signs of brain involvement

Laboratory findings in Gaucher's disease

The following pathological findings are present in the general presentation of Gaucher's disease:

- Elevated serum angiotensin-converting enzyme (ACE), non-prostatic acid phosphatase (tartrate-resistant acid phosphatase, TRAP), and chitotriosidase activity
- The presence of Gaucher cells in marrow
- A lack of glucocerebrosidase activity in white blood cells

Treatment of Gaucher's disease

While there is no cure for the type II disorder, treatment for Gaucher's disease types I and III can be very effective and may include:

- Enzyme replacement therapy
- Splenectomy
- Bone marrow transplantation

Gangliosidoses

Gangliosidoses are diseases in which there is an accumulation of gangliosides due to a lack of the glycosidases β-galactosidase and hexosaminidase, which are required to initiate their degradation.

GM1 gangliosidosis

In this disorder, which affects 1 in 100,000–200,000 live births, β-galactosidase is deficient and the nondegraded ganglioside accumulates to toxic levels in lysosomes leading to the progressive destruction of tissues, particularly neurons in the brain and spinal cord. Although infants are normal at birth, their development slows and muscles weaken. Continued accumulation of ganglioside causes hepatosplenomegaly, skeletal deformation, seizures, and low intellectual ability. GM1 gangliosidosis is an autosomal recessive disorder. The mutations responsible (>80 have been reported) are present in the *galactosidase beta 1* (*GLB1*) gene on the short arm of chromosome 3. Variations in the activity of the expressed β-galactosidase give rise to GM1 type 2 (late infantile) and type 3 (juvenile), in which the severity of the disease is somewhat less than the early infantile GM1 type 1 condition. Different mutations in *GLB1* are responsible for mucopolysaccharidosis IV, where GM1 ganglioside does not accumulate; in this case, keratin sulfate accumulates.

GM2 gangliosidosis

Hexosaminidase is a lysosomal enzyme which hydrolyzes the terminal *N*-acetyl-D-hexosamine residues in complex lipids (**Figure 23.9**). It is a dimeric enzyme (αβ) coded by two separate genes, *HEXA* and *HEXB*, for the α and β chains, respectively. GM2 ganglioside, a key component of the central nervous system, has a terminal *N*-acetylgalactosamine and is a substrate for the enzyme. Mutations in *HEXA* lead to Tay–Sachs disease, while mutations in *HEXB* are found in Sandhoff disease.

ceramide-Glu-Gal(NANA)-GalNAc

↓ hexosaminidase

ceramide-Glu-Gal + GalNAc

Figure 23.9 Hexosaminidase is a lysosomal enzyme that hydrolyzes the terminal *N*-acetyl-D-hexosamine residues in complex lipids. NANA, *N*-acetylneuraminic acid; GalNAc, *N*-acetylgalactosamine.

Tay–Sachs disease

In this condition the absence of hexosaminidase A activity results in the accumulation of ganglioside GM2, mainly in neural tissues, which is seen as a cherry-red spot in the macula. Although rare in Caucasian groups, it occurs in the Ashkenazi Jewish population with a gene frequency of 1 in 30. Typically the disease is manifest at about age 6 months with impairment of mental, motor, sight, and hearing functions and death occurs by about age 3 years.

Sandhoff disease

This is a variant form of gangliosidosis which presents with visceral as well as neural involvement and appears to be due to a deficiency of hexosaminidase B.

Gangliosidoses are detected by measurement of hexosaminidase or β-galactosidase activity in serum and white blood cells. Subjects who are heterogeneous for the mutation have decreased levels of activity. Genetic screening techniques are also available.

Mucopolysaccharidoses

Proteoglycans, integral components of connective tissue and extracellular matrix, are huge, complex, polyanionic molecules consisting of a core protein (5% of structure) to which multiple polysaccharide chains (glycosaminoglycans; GAGs, 95%) are covalently linked. The GAGs consist of disaccharide repeating units in which one of the sugars of the repeat is an amino sugar (glucosamine or galactosamine) and at least one of the sugars is negatively charged with carboxylate (glucuronic acid or its isomer iduronic acid) or sulfate groups (at the 4- and/or 6-hydroxyls on the amino sugar). The major glycosaminoglycans, the repeating disaccharide structures of which are shown in **Figure 23.10**, are

- Chondroitin sulfate (glucuronic acid:*N*-acetylgalactosamine-sulfate)
- Dermatan sulfate (iduronic acid:*N*-acetylgalactosamine-sulfate)
- Keratan sulfate (galactose:*N*-acetylglucosamine-sulfate)
- Heparin (iduronic acid sulfate:*N*-acetylglucosamine sulfate)
- Hyaluronic acid (glucuronic acid:*N*-acetylglucosamine)

Heparan sulfate is similar in structure to heparin but has a lower degree of sulfation. The sulfates of chondroitin, heparan, and dermatan are widely distributed in connective tissue whilst dermatan sulfate is synthesized only in cartilage. Healthy connective tissue requires continual turnover of these macromolecules.

The catabolism of proteoglycans requires a battery of lysosomal enzymes, the initial step being the hydrolysis of the GAGs from the core protein by lysosomal proteases. This is followed by the sequential action of a number of acid exoglycosidases and sulfatases to release individual monosaccharides from the GAG. If any one of these hydrolytic enzymes is missing, the whole degradation process is impaired and nondegraded GAG accumulates in the lysosome. Eventually, the lysosome becomes leaky and its contents, including acid hydrolases, appear in the cytosol causing cellular damage and compromising cellular activity. Glycosaminoglycans are also released into blood and excreted in urine and more than a dozen mucopolysaccharidoses have been identified by urinalysis of GAGs.

The lysosomal enzymes required for degradation of the major GAGs include:

- *N*-acetylglucosamine (GlcNAc) sulfatase
- Iduronate sulfatase
- Galactose (Gal) sulfatase
- *N*-sulfatase
- Iduronidase
- Hexosaminidase

CHONDROITIN SULFATE

D-glucuronic acid (GlcA) **N-acetyl-D-galactosamine** (GalNac)

DERMATAN SULFATE

L-iduronic acid (IdoA) **N-acetyl-D-galactosamine** (GalNac)

HEPARAN SULFATE

D-glucuronic acid (GlcA) **D-glucosamine** ($GlcNH_2$)

HEPARIN

L-iduronic acid (IdoA) **D-glucosamine** ($GlcNH_2$)

HYALURONIC ACID

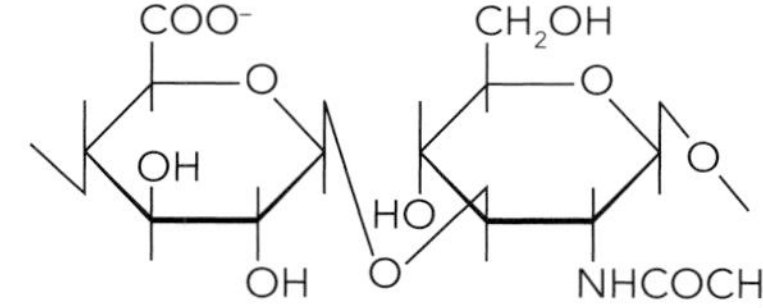

D-glucuronic acid **N-acetyl-D-glucosamine**

KERATAN SULFATE

D-galactose **N-acetyl-D-glucosamine-6-sulfate**

Figure 23.10 Chemical structures of disaccharide repeats of glycosaminoglycans. All monosaccharides in these structures are D-isomers. Chondroitin sulfate A and dermatan sulfate (chondroitin B) are copolymers with linked repeats of a β-(1,3)-uronic acid and β-(1,4)-*N*-acetylgalactosamine 4-sulfate; it is β-glucuronic acid in chondroitin sulfate A and iduronic acid in dermatan sulfate. Heparan and heparin are repeats of a β-(1,4)-uronic acid (glucuronic acid or iduronic acid) and either β-(1,4)-glucosamine or β-(1,4)-*N*-acetylglucosamine. Sulfation (zero to three residues per monosaccharide) may be present at oxygen or nitrogen groups. Heparin is more sulfated than heparan. Hyaluronic acid consists of β-(1,3)-glucuronic acid and β-(1,4)-*N*-acetylglucosamine repeats and is not sulfated. Keratan sulfate is a copolymer of β-(1,3)-galactose and β-(1,4)-*N*-acetylglucosamine 6-sulfate. R, sulfate.

- Galactosidase
- Heparin sulfamidase
- Glucosaminidase

The mucopolysaccharidoses represent a group of disorders characterized by the accumulation of GAGs (also known as mucopolysaccharides) in the lysosomes of tissues of mesenchymal origin, endothelium, and neurons. Some of the named mucopolysaccharidoses and the associated genetic defects are listed in **Table 23.3**. They are somewhat rare conditions, occurring at a frequency of 1 in 100,000 live births and all are inherited as autosomal recessive disorders apart from Hunter syndrome, which is X-linked. The lysosomal accumulation of polysaccharide is due to a deficiency of one of the enzymes required for the sequential degradation of GAGs, and the polysaccharide which accumulates is the substrate for the particular missing enzyme. The age of presentation of the different mucopolysaccharidoses varies from very early in development in MPS IH to adulthood in MPS IV, but symptoms invariably worsen with age of the patient. In most cases patients present with coarse facial features, cardiac valve disease, hepatosplenomegaly, bone dysostosis, corneal opacities, and intellectual disabilities.

If a mucopolysaccharidosis is suspected, diagnosis may be made initially by identification of urinary GAGs and confirmed by assay for the specific hydrolase in serum, white blood cells, or fibroblasts. Prenatal screening is often carried out in at-risk pregnancies.

TABLE 23.3 Mucopolysaccharidoses

Mucopolysaccharide type	Trivial name	Enzyme defect	Tissues affected	
			Connective tissue	CNS
MPS IH MPS IS	Hurler syndrome Scheie syndrome	α-L-iduronidase α-L-iduronidase	Yes Yes	Yes No
MPS II	Hunter syndrome	Iduronate sulfatase	Yes	Yes
MPS IIIA MPS IIIB MPS IIIC MPS IIID	Sanfilippo A Sanfilippo B Sanfilippo C Sanfilippo D	Heparan sulfate sulfatase *N*-acetylglucosaminidase Acetyl-CoA:α-glucosamide acetyltransferase *N*-acetylglucosamine 6-sulfatase	Mild Mild Mild Mild	Yes Yes Yes Yes
MPS IVA MPS IVB	Morquio A Morquio B	Galactosamine 6-sulfate sulfatase β-Galactosidase	Yes Yes	No No
MPS VI	Maroteaux–Lamy syndrome	*N*-acetylgalactosamine 4-sulfatase	Yes	No
MPS VII	Sly syndrome	β-Glucuronidase	Variable	Variable
MPS IX	–	Hyaluronidase	Yes	Yes

All are autosomal recessive disorders, apart from Hunter syndrome (MPS II) which is X-linked, and all are due to the deficiency of a single lysosomal enzyme of the ten required for complete degradation of glycosaminoglycans. Pathology is apparent early in development and worsens with age but there is a wide spectrum of clinical severity. Although death by the time of adolescence may occur in MPS IH and MPS II patients, those with other types of mucopolysaccharidoses may survive into adulthood. Intelligence appears to be unaffected in MPS IS, MPS IV, and MPS VI. CNS, central nervous system.

Mucolipidoses

Mucolipidosis I (sialidosis I)

Mucolipidosis I (sialidosis I) is due to the absence or decreased activity of lysosomal neuraminidase 1 due to mutations in the *NEU1* gene. This enzyme hydrolyzes *N*-acetylneuraminic acid (sialic acid) residues from complex sugar molecules and in its absence sialic acid-containing molecules accumulate in lysosomes.

Mucolipidoses II and III

In these disorders, substrates of lysosomal enzymes accumulate in lysosomes and as such they are considered to be lysosomal storage diseases. However, unlike the other lysosomal storage diseases, the gene for the hydrolytic enzyme required for breakdown of the mucolipids is expressed but the enzyme is secreted from the cell rather than incorporated into the lysosome. As described below, the enzyme deficiency here is of a glycosyltransferase activity in the Golgi, responsible for targeting the lysosomal enzyme to its destination in the lysosome.

Glycogenosis type II (Pompe's disease)

Pompe's disease is characterized by an accumulation of glycogen in skeletal and cardiac muscle due to a lack of lysosomal α-1,4-glucosidase activity. The major site of glycogen accumulation depends on the age at which the disease presents (**Table 23.4**). Death occurs from heart failure in infancy and from respiratory

TABLE 23.4 Major sites of accumulation of glycogen with age in patients with Pompe's disease

Age at presentation	Major tissue involved
Infancy	Heart
Childhood	Skeletal muscle
Adulthood	Skeletal muscle

failure in the childhood type. In contrast with type I glycogenosis, there are no abnormalities of plasma glucose, lactate, urate, or triglycerides since the hepatic enzyme activity is normal. It is diagnosed primarily by enzyme testing in blood where there is a marked increase in serum creatine phosphokinase. Enzyme replacement therapy is now available for this disorder and its use has changed dramatically the prognosis for the disease.

Cystinosis

Cystinosis is a lysosomal storage disease which is not due to a deficiency of a lysosomal enzyme but to a defect in the transporter protein mediating export of the amino acid cysteine out of the lysosome. Cysteine thus accumulates in the organelle usually in its oxidized form, cystine. Like many lysosomal storage diseases, cystinosis shows a spectrum of severity. In the mildest cases the accumulation of cystine is asymptomatic, while in severe cases with the accumulation of large amounts of cystine, death from renal failure often occurs in childhood. It is the commonest lysosomal storage disease in the UK and has a frequency of 1 in 200,000 births in developed countries. The pathophysiological progress of cystinosis is shown in **Figure 23.11**. Cystine accumulates in the lysosomes of most tissues, but especially in kidneys, thyroid, bone marrow, and cornea. However, in the clinical setting, the most important site of accumulation is the kidneys, where it results in symptoms of Fanconi syndrome where the tubules are affected and progressive renal failure where the glomerulus is affected. The manifestations of Fanconi syndrome, often seen in cystinosis, include

- Renal glycosuria
- Hypophosphatemia; hypouricemia
- Generalized leakage aminoaciduria
- Renal tubular acidosis
- Failure to concentrate urine (that is, high volume and dilute)
- Polyuria and increased thirst due mainly to impairment of water reabsorption

In addition, hypothyroidism is relatively common in patients with advanced disease. Crystals of cystine (the oxidized form of cysteine) may be observed in the cornea by slit lamp examination and in bone marrow by microscopy of marrow aspirate. Excessive accumulation of ^{35}S-Cys into cells cultured from amniotic fluid provides a means of prenatal diagnosis in conjunction with molecular analysis whenever available.

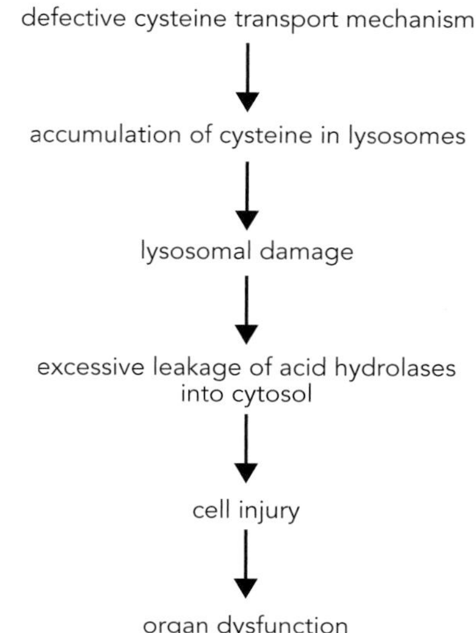

Figure 23.11 Pathophysiology of cystinosis.
Cystinosis is a disorder in which cysteine derived from lysosomal hydrolysis of proteins is unable to leave the lysosome due to a defect in the transporter system required for exit of the amino acid from the organelle. Cysteine thus accumulates, eventually damaging the lysosome and causing leakage of hydrolytic enzymes into the cell cytosol. Cell physiology is then compromised and tissue dysfunction follows.

Deficiency of an enzyme of the endoplasmic reticulum/Golgi apparatus

Phenylketonuria

Phenylalanine is an essential amino acid which is incorporated into proteins during *de novo* synthesis as part of general protein turnover. As described earlier in this chapter, excess dietary phenylalanine is converted to tyrosine, and in healthy individuals the hydroxylation of phenylalanine to tyrosine is a very efficient process such that little free phenylalanine is present in plasma (<125 μmol/L). This hydroxylation of phenylalanine to tyrosine is catalyzed by phenylalanine 4-hydroxylase (PAH), a mixed function oxidase which requires stoichiometric amounts of tetrahydrobiopterin (BH_4) as a co-factor. In the hydroxylation reaction, BH_4 (tetrahydrobipterin) is oxidised to BH_2 (dihydrobiopterin) in the PAH reaction and is regenerated by the action of BH_2 reductase to allow it to participate in the hydroxylation reaction once more. BH_2 can be synthesized *de novo* from H_2-neopterin by the BH_2 synthase complex and reduced to BH_4 for participation in the PAH reaction (**Figure 23.12**). The biochemical defect in phenylketonuria (PKU) is the inability to hydroxylate phenylalanine to tyrosine, with the result that phenylalanine accumulates in all body

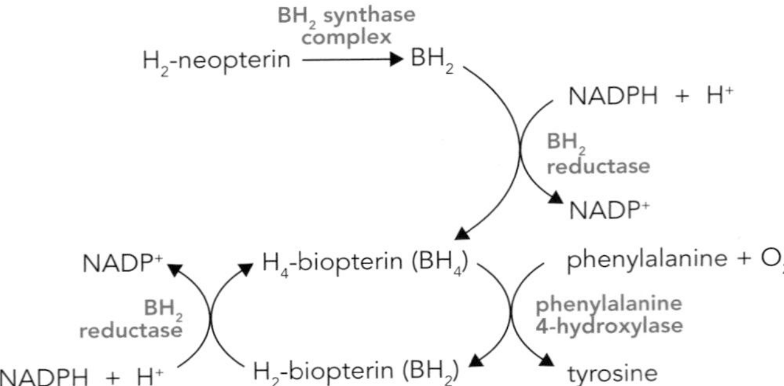

Figure 23.12 *De novo* **synthesis of dihydrobiopterin (BH_2) and the hydroxylation of phenylalanine to tyrosine.**
BH_2 is synthesized *de novo* from H_2-neopterin by the BH_2 synthase complex and reduced by BH_2 reductase to BH_4 for participation in the hydroxylation of phenylalanine to tyrosine by phenylalanine 4-hydroxylase. BH_4 is also regenerated by the action of BH_2 reductase on BH_2 formed in the phenylalanine hydroxylase reaction.

fluids; there it is metabolized by alternative pathways leading to production of acidic metabolites which are normally only minor metabolites of phenylalanine (**Figure 23.13**).

In the classical form of PKU (98% of cases) there is a decrease in PAH activity (0–25% of normal activity) and more than 400 mutations in the PAH gene have been identified. Approximately 1.5–2% of Western populations are heterozygous for a mutation and while such individuals appear normal, they have lowered blood cell phenyalanine hydroxylase activity and raised plasma phenylalanine levels. The defect is inherited as an autosomal recessive disorder with a prevalence in the UK population of approximately 1 in 10,000 and in the USA of 1 in 30,000. In variant forms of PKU (2% of cases; ~50 mutations) there is a reduced availability of BH_4 due to defects in the activity of dihydropteridine reductase or BH_2 synthase complex. In such cases there is also a concomitant decrease in the syntheses of both dihydroxyphenylalanine (DOPA) and 5-hydroxytryptamine.

Detection and diagnosis of PKU

Phenylketonuria is usually detected by routine neonatal blood spot screening (Guthrie test) for raised phenylalanine on day six after birth. Babies born to families with a positive history of PKU are tested on day 3 in addition to day 6. Should the baby have not been feeding for 3 days prior to blood collection, a further sample is taken as soon as feeding has been reestablished.

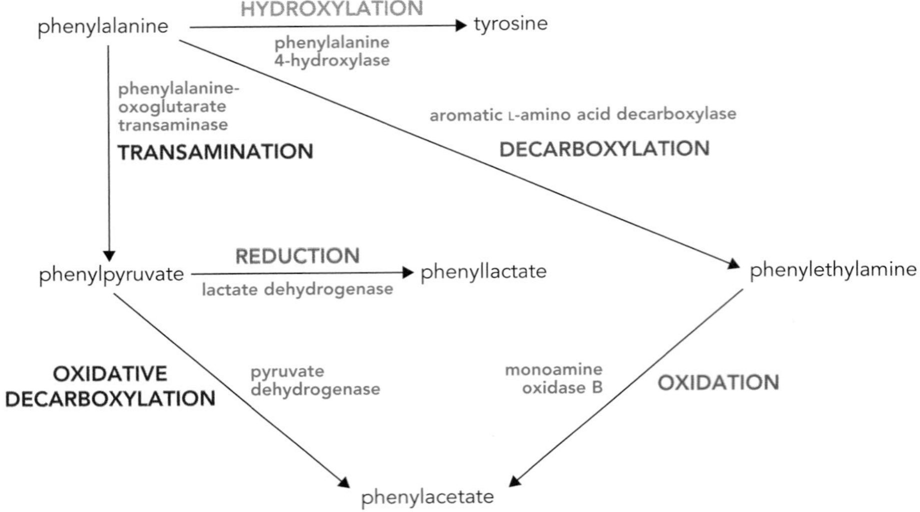

Figure 23.13 Products of phenylalanine metabolism in the absence of phenylalanine 4-hydroxylase activity.
The fate of phenylpyruvate is similar to that of pyruvate at the last stages of glycolysis and involves the same enzymes. Only trace amounts of phenylethylamine are detectable due to its efficient conversion to phenylacetate by hepatic monoamine oxidase B.

Interpretation of results

In many classical cases of PKU, blood phenylalanine concentration may be in excess of 1000 μmol/L (16.5 mg/dL) at day 6 after birth. It is important to repeat the measurement if it is in the equivocal range of 200–400 μmol/L (3.3–6.6 mg/dL). It is also important to differentiate PKU from other causes of raised blood phenylalanine, such as transient hyperphenylalaninemia and where increased phenylalanine is a component of generalized raised plasma amino acid concentration due to liver disease, for example.

Investigations for variant PKU

Investigations for variant PKU should be performed in all PKU patients. These tests should include:

- Measurement of BH_2 synthase and BH_2 reductase in red blood cells
- Measurement of the plasma neopterin–biopterin ratio, which is increased in BH_2 synthase complex deficiency

Clinical consequences of PKU

Untreated or inadequately treated PKU results in progressive mental retardation of affected individuals and also of non-PKU fetuses in affected mothers with poor metabolic control during pregnancy. Mental retardation in the untreated newborn is progressive and severe at 1 year; life expectancy is concomitantly reduced, with 50% mortality by age 20. Mental retardation can be prevented in affected individuals by early and sustained treatment with low-phenylalanine diets; that is, diets with a high tyrosine content and just sufficient phenylalanine to allow growth. Fetal development is extremely sensitive to the toxic effects of raised phenylalanine and its metabolites, and injury *in utero* to the brain in non-PKU offspring is much reduced when PKU mothers eat a diet low in phenylalanine throughout pregnancy to maintain blood phenylalanine at <200 μmol/L (<3.3 mg/dL). It is important to screen for PKU in mothers of retarded non-PKU children. The clinical manifestations of PKU are particularly severe in patients with variant PKU due to tetrahydrobiopterin (BH_4) deficiency as these patients respond little, if at all, to diets low in phenyalanine. Since the BH_4 system is also involved in reactions catalyzed by tyrosine hydroxylase and tryptophan hydroxylase, mutations in BH_2 synthase complex or dihydropteridine reductase will also affect tyrosine catabolism and tryptamine synthesis. Patients have a high incidence of epilepsy, light skin pigmentation (due to decreased melanin synthesis), and abnormal posture.

While PKU was formerly a significant clinical problem, early diagnosis and implementation of a strict dietary regime in which phenylalanine is limited has virtually overcome this condition.

Clinical practice point 23.2

It is important for females with PKU to follow a low-phenylalanine diet before conception and during pregnancy. They may have discontinued the diet after the critical stage of their own brain development and not be aware of the risk to their own offspring.

I-cell disease (mucolipidosis II, ML II) and pseudo-Hurler polydystrophy (mucolipidosis III, ML III)

In these diseases the hydrolytic enzymes (acid hydrolases) required for the degradation of mucolipids and glycosaminoglycans are expressed but they are not delivered to the lysosomes; rather, they are secreted from the cell. This has two major consequences:

1. The substrates of these enzymes—mucolipids and glycosaminoglycans—accumulate in the lysosome and eventually compromise cellular activity. These lysosomal inclusions give rise to the trivial name of I-cell disease for ML II
2. Inappropriate hydrolytic activity present in plasma causes hydrolysis of extracellular matrix and connective tissue

The translation of enzymes destined for the lysosomes, like all proteins, occurs on the rough endoplasmic reticulum with post-translational modifications taking place during movement of the protein through the smooth endoplasmic reticulum and Golgi apparatus. A key step in the synthesis of a lysosomal enzyme is the post-translational addition of a mannose 6-phosphate tag, the signal which ensures its entry into the lysosome. The enzyme that donates the phosphate group to a mannose residue on the newly synthesized lysosomal protein is an *N*-acetylglucosamine 1-phosphotransferase present in the smooth *cis*-Golgi. In this reaction, *N*-acetylglucosamine becomes attached to the mannose via a phosphodiester bond; release of the *N*-acetylglucosamine residue by a phosphodiesterase leaves a mannose 6-phosphate on the core oligosaccharide part of the protein (see Figure 23.6). The *N*-acetylglucosamine 1-phosphotransferase appears to be specific for lysosomal transferases, recognizing their particular three-dimensional structure, and does not act on other mannose residues present in the cell. The enzyme is a hexameric ($\alpha_2\beta_2\gamma_2$ subunits) protein and is the product of two genes—*GNPTG* on chromosome 16 (γ subunits) and *GNPTAB* on chromosome 12 (α and β subunits). In the absence of the mannose 6-phosphate tag, nascent lysosomal enzymes are no longer targeted to the lysosome but follow the default pathway of exocytosis and secretion from the cell into extracellular fluid and eventually into plasma. It is the enzyme required for tagging the eight nascent lysosomal enzymes which is deficient in ML II and ML III, rather than the lysosomal enzyme activity itself. As a consequence, macromolecules which enter the lysosomes for degradation accumulate and give rise to the inclusion vesicles characteristic of I-cell disease (**Figure 23.14**).

The presentation of ML II is similar to but more severe than MPS IH, with coarse facial features, skeletal and connective tissue abnormalities, and mental retardation. Death usually occurs by early adolescence. Patients with ML III are of short stature and suffer joint stiffness characteristic of connective tissue damage. They may survive into adulthood.

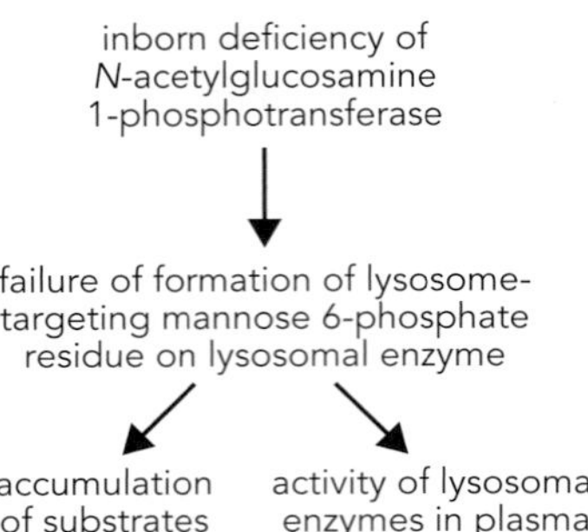

Figure 23.14 Pathogenesis of mucolipidoses. Targeting of lysosomal enzymes (glycoproteins) to their site of action in the cell (that is, the lysosome) requires attachment of a signal mannose 6-phosphate residue to an *N*-acetylglucosamine residue on the enzyme by *N*-acetylglucosamine 1-phosphotransferase during the lysosomal enzyme's passage through the endoplasmic reticulum. In the absence of this signal the hydrolases follow the default secretory route of exocytosis and the enzymes are found outside of the cell. The consequent lack of specific hydrolytic activity in the lysosome results in accumulation of the enzyme substrate in the lysosome, which becomes enlarged.

Deficiency of a cytosolic enzyme

Glucose 6-phosphate dehydrogenase in the red blood cell

G6PD is an enzyme of the pentose phosphate pathway, described in Chapter 2. A key role of this reaction is the generation of NADPH, essential for the reduction of oxidized glutathione. Since mature red blood cells lack mitochondria, the pentose phosphate pathway is the only source of NADPH, and a deficiency of NADPH reduces the ability of red cells to accommodate oxidative stress induced by the use of antimalarial drugs such as pamaquine. This is discussed more fully in the section on red cell enzymopathies.

Disorders of specific transport mechanisms in the plasma membrane

As mentioned earlier in this chapter, mutations in genes encoding specific transport systems may also lead to disease states. These include disorders of specific transport mechanisms that affect absorption from the intestine such as:

- Deficiency of disaccharidases
- Glucose–galactose malabsorption
- Cystinuria, neutral aminoaciduria (Hartnup disease)
- Metal malabsorption: copper (Wilson's disease), zinc, magnesium
- Folate- or cobalamin-deficiency anemias
- Triglyceride malabsorption; abetalipoproteinemia
- Familial chloride diarrhea

and disorders of specific transport mechanisms affecting renal tubules such as:

- Renal glycosuria
- Renal tubular acidosis

- Hypophosphatemic rickets
- Cystinuria, neutral aminoaciduria (Hartnup disease)
- Nephrogenic diabetes insipidus

The following examples of plasma membrane transporter defects have been chosen to illustrate the problems arising from a lack of transport activity. The list is in no way exhaustive and other examples in the kidney and gut are described in Chapters 4 and 14.

Cystinuria

Cystinuria is an autosomal recessive disorder in which the transporter protein of cysteine and other dibasic amino acids (arginine, ornithine, and lysine) in the plasma membrane of the kidney cells is defective, thereby impairing their reabsorption by renal tubules. Three mutations giving rise to the same clinical phenotype have been reported (cystinuria types I, II, and IIII). In this disease only the renal tubular defect is important clinically, although transport into intestinal mucosal cells is also defective (**Table 23.5**). Decreased tubular reabsorption of dibasic amino acids leads to increased urinary excretion of cystine (the oxidized form of cysteine), ornithine, arginine, and lysine. Urinary excretion of cystine in adults with cystinuria is 300–1000 mg/day compared with 30–100 mg/day in a healthy adult. Since cystine is relatively insoluble in acidic urine, patients are at risk of forming stones in the urinary tract. Cystine stones are often large and staghorn in shape, moderately radio-opaque, and may also contain calcium and magnesium phosphate. Such cystine stones may cause renal colic and/or obstruction and be associated with urinary tract infection.

TABLE 23.5 Classification of cystinuria

Clinical finding		Type I	Type II	Type III
Relative increase in urinary excretion of cystine, ornithine, arginine, and lysine	Homozygote	+++	+++	+++
	Heterozygote		++	++
Intestinal absorption of cysteine		Decreased	Decreased	Variable

Diagnosis of cystinuria

The following tests should be carried out to confirm the diagnosis of cystinuria:

- Analysis of stone if available
- Cyanide-nitroprusside screening test for excess urinary cystine
- If screening test is positive, quantitative thin-layer chromatography (TLC) of urinary amino acids showing increased cystine, ornithine, arginine, and lysine
- Radiological or ultrasound examination for stones
- Cyanide-nitroprusside screening test on urine of relatives and follow up if necessary

Treatment of cystinuria

Treatment of cystinuria includes high fluid intake, urine alkalinization, and administration of compounds which will help reduce the urinary concentration of cystine to below its solubility product. These compounds include D-penicillamine and more recently α-mercaptopropionyl glycine, which form cysteine adducts.

Hypophosphatemic rickets

Hypophosphatemic rickets is an X-linked disorder in which renal tubular reabsorption of phosphate is defective. The resulting decrease in intracellular phosphate in the distal tubule reduces the activity of the 1α-hydroxylase required to form the active metabolite of vitamin D, 1,25-dihydroxy vitamin D. Patients with this condition present with bowed legs and growth retardation and radiological changes associated with rickets. Serum analysis shows a normal calcium, decreased phosphate, and increased alkaline phosphatase. Urinary phosphate excretion is increased and urinary amino acids are normal. Treatment includes dietary phosphate supplementation and calcitriol.

Cystic fibrosis

Cystic fibrosis (CF) is the most common fatal inherited disease in Caucasian populations with a carrier frequency of between 1 in 20 and 1 in 30 and a disease rate of 1 per 10,000 live births. It is somewhat less common in other ethnic populations. It is not classically an IEM as there is no defective enzyme in a metabolic pathway. However, the underlying cause is similar: an abnormal gene product (protein) with reduced function leading to clinical disease. The condition is inherited in an autosomal recessive manner and is due to a point mutation in the gene encoding the cystic fibrosis transmembrane (conductance) regulator (CFTR), a transmembrane protein in the apical membrane of epithelial cells. The CFTR protein contains a chloride channel and regulates chloride movement across the membrane. The most common mutation (ΔF508), accounting for >70% of cases, involves deletion of a 3-base-pair codon for phenylalanine at amino acid 508 in the native protein. This deletion has profound effects on the three-dimensional structure of the mutant protein such that it fails to open its chloride channel in response to a stimulus (cyclic adenosine monophosphate; cAMP) in the epithelial cell. The physiological effect of this is to decrease chloride excretion into the epithelial lumen and increase reabsorption of sodium into epithelial cells, with a consequent failure to hydrate mucus at epithelial secreting surfaces. Thus the viscosity and tenacity of secretions are increased. The major tissues affected are the lungs, pancreas, and genitourinary system. In addition, compromised sweat glands secrete excessive amounts (three times normal) of sodium and chloride.

In the lungs, the more viscous nature of secretions makes it difficult for cilia to move them upward toward the pharynx, leading to congestion and increased susceptibility to infection. Progressive respiratory complications (for example bronchiectasis) and eventually respiratory failure are features of the disease. Greater than 90% of CF patients experience pancreatic failure, the majority in the perinatal period, due to obstruction of the pancreatic ducts causing damage to acinar cells and impaired production of pancreatic enzymes. Deficiency of pancreatic enzyme activity results in a failure to hydrolyze completely dietary components, and the malabsorption of sugars, amino acids, fats, and fat-soluble vitamins leads to stunting of growth.

Treatment of cystic fibrosis

Treatment of CF patients includes intense daily physiotherapy to move mucus in the airways and provision of high-calorie, high-fat diets containing enteric-coated pancreatic enzyme to hydrolyze dietary protein and lipid. Heart–lung transplantation has proved successful in a number of cases.

Deficiency of a mitochondrial enzyme activity

Long-chain hydroxyacyl dehydrogenase deficiency

Fatty acids, stored as triglycerides, represent the major energy source in the body. Energy in the form of ATP is produced from β-oxidation of fatty acids and oxidative phosphorylation in mitochondria as described in Chapter 18. The trifunctional protein (TFP) in β-oxidation, as its name implies, has three enzyme activities: hydroxyacyl-CoA dehydrogenase, 3-ketothiolase, and enoyl-CoA hydratase. Mutations in the gene encoding the α-subunit of the TFP, the hydroxyacyl dehydrogenase activity (the *HADHA* gene on the short arm of chromosome 2), prevent β-oxidation such that long-chain fatty acids cannot be processed and, as a consequence, they accumulate in cells causing damage to tissues, particularly the liver, muscles, and retina. Deprived of a major energy source, tissues have to rely on glucose for energy production and patients become hypoglycemic very quickly, especially in periods of fasting or illness such as infection. The signs and symptoms—difficulty in feeding, lethargy, weak muscle tone, and liver problems—appear in infancy and continue into early childhood.

CASE 23.2

A girl aged 2 years is brought to the emergency department having had a first seizure. She has been mildly unwell with symptoms of a viral illness for 2 days, but had no other medical history. A point-of-care blood glucose measurement showed hypoglycemia. Blood and urine samples sent to the laboratory showed the following.

	SI units	Reference range	Conventional units	Reference range
Plasma				
Ammonia	70 μmol/L	11–32	98 μg/dL	15–45
Glucose	1.2 mmol/L	3.9–6.1	21 mg/dL	70–110
Free fatty acids (qualitative)	Increased			
Ketones measured as 3-hydroxybutyrate (qualitative)	Decreased			
Acylcarnitines (qualitative)	Increased C6, C8, and C10			
Urine				
Organic acids (qualitative)	Increased dicarboxylic acids C6, C8, and C10			
Acylglycines (qualitative)	Increased hexanoylglycine, suberylglycine, and phenylpropionylglycine			

- **How does the plasma glucose relate to the presentation?**
- **What is the significance of the raised free fatty acids and low ketones?**
- **Why has the patient presented now?**

Hypoglycemia causes acute neurological symptoms ranging from lethargy to seizures and coma, reflecting a reduced energy supply to the brain. The alternative energy source—fatty acids from adipose tissue—requires β-oxidation in the mitochondria. The excess acetyl-CoA produced is metabolized to the ketones acetoacetate and 3-hydroxybutyrate. Raised free fatty acids in the plasma is therefore appropriate in the face of hypoglycemia, but hypoketonemia is not. This finding points to a defect of β-oxidation, confirmed by the finding of raised acylcarnitines and increased excretion of dicarboxylic acids and acylglycines. The carbon chain length corresponds to a defect in the mitochondrial acyl-CoA dehydrogenase responsible for the metabolism of medium-chain fatty acids (C6 to C10). The enzyme is referred to as MCAD: medium-chain acyl-CoA dehydrogenase. The affected individual may appear well until they have an extended fast or increased metabolic stress, which depletes glucose supply. Normally, this would be supplemented as an energy source by fatty acid oxidation, but in MCAD deficiency this cannot happen and glucose is rapidly exhausted, causing hypoglycemia which may be fatal. MCAD deficiency is the commonest fatty acid oxidation defect, but disorders preventing the oxidation of long-chain and very-long-chain fatty acids (LCAD and VLCAD deficiencies) also cause hypoketotic hypoglycemia as well as myopathy.

23.5 RED CELL ENZYMOPATHIES

Metabolism in the red blood cell is severely restricted because of the absence of a nucleus and other organelles including an endoplasmic reticulum and mitochondria. Glucose derived from plasma is the red cell's sole energy source. In health, approximately 90% of the glucose in the red cell is metabolized anaerobically to lactate with the generation of two molecules of ATP per molecule of glucose metabolized. Red cells require ATP for operation of ion pumps that regulate their intracellular Na^+ and K^+ concentrations to help maintain cell volume and elasticity. A decrease in the rate of ATP production in the red cell increases its rate of destruction (hemolysis). Approximately 10% of glucose in the red cell is oxidized via the hexose monophosphate shunt (pentose phosphate pathway, see Chapter 2), which is essential for the production of NADPH by reduction of $NADP^+$. The continued production of NADPH is required to reduce glutathione, which is being continually oxidized during the maintenance of free –SH groups on cell proteins, both in the membranes and cytosol and including hemoglobin. The oxidative stress on the red cell in the circulation drives oxidation of cysteine residues to cystine in polypeptide chains, and reduced glutathione (GSH) reduces these residues back to cysteine while itself being oxidized to GSSG (**Figure 23.15**). A deficiency in the generation of NADPH impairs the ability of red cells to resist this oxidative stress, with the result that cell proteins (particularly hemoglobin) become denatured and accumulate in Heinz bodies with subsequent damage to the cells. Such damaged cells are destroyed more rapidly than normal cells by the fixed macrophages of the reticuloendothelial system.

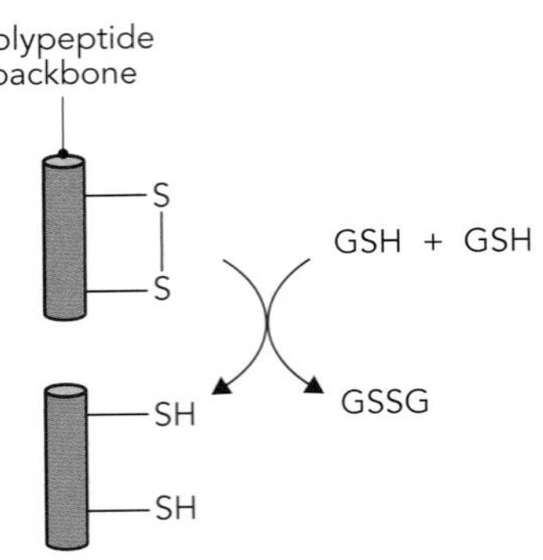

Figure 23.15 Oxidation of reduced glutathione to maintain cysteine –SH groups in proteins. Oxidative stress in the red cell drives the oxidation of cysteine residues in cellular proteins to cystine which, if unchecked, might lead to altered three-dimensional structure and loss of function. A redox reaction involving reduced cytoplasmic glutathione (GSH) allows the reduction of cystine residues in proteins to cysteines and continued functionality of the protein. Oxidized glutathione (GSSG) can be reduced by NADPH generated in the pentose phosphate pathway.

Inherited enzyme defects causing hemolysis

Some inherited disorders which give rise to hemolysis are listed in **Table 23.6**.

TABLE 23.6 Inherited enzyme defects causing hemolysis

Defect	Incidence	Hemolysis
Glucose 6-phosphate dehydrogenase (G6PD)	Common in certain ethnic groups	Episodic with G6PD variants
Pyruvate kinase (PK)	Rare	Persistent with PK variants
Others	Exceptionally rare	Common finding

Glucose 6-phosphate dehydrogenase variants

More than 200 genetic mutations (variants) of G6PD have been reported in the literature, and while some have no effect on enzyme activity or half-life, others cause severe disease states. The most common variants with normal activity are the A+ and B variants which have differential ethnic distributions (**Table 23.7**).

Mutation in the G6PD gene is the most common enzymatic cause of hemolytic anemia. Both of the common abnormal variants—$G6PD^{A-}$ found in the ethnic Black population in the USA and Africa, and $G6PD^{Med}$ found in indigenous peoples of the Mediterranean region—differ from the normal variants by only a single amino acid but this is sufficient to change significantly the kinetic properties of the enzyme or its half-life. For example, the A– variant has normal activity but a half-life of only 12–15 days compared with 60 days for the B and A+ variants. The red cell cannot compensate for this shortened half-life by increasing production of new enzyme because, lacking a nucleus, it is unable to carry out protein synthesis. New enzyme can only appear in new red cells, and reticulocytosis is a feature of acute hemolytic anemia by any cause.

TABLE 23.7 Distribution of G6PD variants in Caucasian and Black ethnic groups		
G6PD variant	**Caucasians**	**US Blacks and Black Africans**
Most common normal variants		
A+	<1%	20%
B	>99%	70%
Most common abnormal variant*		
A–	<1%	10%

*Other common variants are the Med variant associated with people of the Mediterranean region, including Arabs, Greeks, Italians, and also Indians, Southeast Asians, and Sephardic Jews, and the Canton variant associated with ethnic Chinese populations.

The G6PD gene is located on the long arm of the X chromosome and thus G6PD deficiency is an X-linked recessive trait. Males (XY) will be either normal or abnormal, inheriting the disease in an X-linked dominant fashion, while females may be normal, heterozygous, or homozygous for the deficiency. Female carriers (heterozygotes) have a mixed population of red blood cells, some with normal and others with abnormal enzyme.

The consequence of a reduced capacity for production of NADPH, particularly in response to oxidative stress, is oxidative damage to the red cell (membrane lipid peroxidation and protein denaturation) and hemolysis, with the release of hemoglobin into the circulation. Such hemolytic episodes occur mainly in affected males (and homozygous females). Patients are often asymptomatic until exposed to oxidative stress precipitated by:

- Intercurrent illness or severe infection
- Drugs including antimalarials, sulfonamides, and dapsone
- Fava beans

The hemoglobin released from the breakdown of red cells is metabolized in the liver to bilirubin as described in Chapter 3. In severe cases of hemolytic anemia the excessive production of bilirubin may cause jaundice as bilirubin accumulates in tissues. Laboratory findings for blood, serum, and urine during acute hemolytic episodes are listed in **Table 23.8**.

TABLE 23.8 Laboratory findings during acute hemolytic episodes in glucose 6-phosphate dehydrogenase deficiency	
Tissue	**Laboratory findings**
Blood	Decreased hemoglobin
	Distorted red blood cells due to presence of Heinz bodies
	Reticulocytosis
Serum	Increased unconjugated bilirubin
	Decreased haptoglobins
	Hemoglobinemia if hemolysis is very severe
Urine	Increased urobilinogen
	Bilirubin negative
	Hemoglobinuria if hemolysis is very severe

Pyruvate kinase variants causing hemolysis

Pyruvate kinase variants are relatively uncommon autosomal recessive disorders (affecting 1 per 10,000 of the population). Red cells containing the defective

TABLE 23.9 Laboratory findings in chronic hemolysis in individuals with pyruvate kinase variants

Tissue	Laboratory finding
Blood	Red blood cells appear normal
	No Heinz bodies
	Reticulocytosis
Serum	Increased unconjugated bilirubin
	Decreased haptoglobins
Urine	Increased urobilinogen
	Bilirubin negative

enzyme are unable to synthesize sufficient ATP for the maintenance of ion gradients across the red cell membrane and thus cell shape. Pyruvate kinase deficiency is associated with a chronic hemolytic state, often first manifested by prolonged neonatal jaundice. Symptoms vary between individuals and include anemia, jaundice, and splenomegaly; hemolysis may be increased by intercurrent illness. The laboratory findings with this disorder are shown in **Table 23.9**. Interestingly, in the absence of pyruvate kinase activity, the concentration of 2,3-diphosphoglycerate is increased in the red blood cell, decreasing the affinity of oxygen for hemoglobin and promoting oxygen delivery to muscle during exercise and to the fetus during pregnancy.

23.6 SCREENING AND DIAGNOSIS OF INBORN ERRORS OF METABOLISM

Neonatal screening for metabolic disorders

Because of the devastating metabolic effects of some genetic disorders, many countries have initiated screening programs to detect such disorders in newborn infants. This has resulted in one of the most successful public health initiatives of the last century. The rationale for screening is that it allows the early identification of affected babies and introduction of appropriate therapy and specialist treatment to improve their quality of life. The traditional sample collection method, introduced in the 1960s in the USA, UK, and Ireland, has been to obtain blood by heel prick and to collect the blood sample on to a Guthrie card (fine-grade chromatography paper) for analysis. The introduction of the dried blood spot provided a robust method of sample collection with ease of processing and storage of samples. Initially, the screen was for phenylketonuria by bacterial inhibition assay, but screening has now been expanded to include a number of other disorders. With the advent of the more sophisticated techniques such as tandem mass spectrometry (TMS), immunoassay, enzyme/microchemical assay, electrophoresis, high performance liquid chromatography (HPLC), and molecular/DNA analysis, screening programs have been broadened to include many less common disorders.

The criteria for screening for diseases, such as breast cancer, in the general population need to be modified somewhat for screening of the newborn; for example, the neonate is unable to give consent. However, general criteria for the introduction of an analyte into a newborn screening program can be defined, for example:

- The disorder is a rare but significant public health problem
- The metabolic defect and development of the disorder are known

- There is a reliable test for detection of the disorder, with high specificity and sensitivity
- There is a benefit to the newborn from early diagnosis
- Benefit to the patient is balanced against financial and other costs; for example the impact on the family
- Satisfactory support systems are in place for
  (i) Clinical management of the newborn
  (ii) Counseling of family and, later on, the patient
  (iii) Patient follow-up and therapy throughout their life

The nature of the screening program, even within an individual country, may vary in areas where there is a specific ethnic population; for example Afro-Caribbean and Asian populations in urban areas of the UK. The incidence of genetic disorders in these ethnic groups will differ from a Caucasian group and the screening program may be tailored to reflect this. In addition, screening for hemoglobinopathies and thalassemias occurs in some countries but, apart from sickle cell anemia, these disorders are not discussed here. The following tables contain data from the screening programs in the UK (UK Newborn Screening Programme Centre) and the Republic of Ireland (National Newborn Bloodspot Screening Programme in the Republic of Ireland). The data have been chosen to illustrate the differences in (i) the screening in specific countries as a result of the incidence of the more common IEMs in those countries; and (ii) the approach to testing—selective in Europe and much broader in the USA. The incidence of metabolic disorders that are screened for in the UK, the Republic of Ireland, and worldwide is shown in **Table 23.10**.

TABLE 23.10 Incidence of screened inherited metabolic disorders in the UK and Irish Republic compared with the worldwide incidence of the disorders

Metabolic disorder	UK	Irish Republic	Worldwide
Sickle cell anemia	1 in 2000	Rare	*
Cystic fibrosis	1 in 2500	1 in 1350	1 in 3000
Phenylketonuria	1 in 10,000	1 in 4600	1 in 12,000
Congenital hypothyroidism	1 in 4000	1 in 3500	1 in 4000
MCAD deficiency	1 in 10,000		1 in 14,600
Galactosemia type 1		1 in 19,500	1 in 45,000
Maple syrup urine disease		1 in 125,000	1 in 225,000
Homocystinuria		1 in 65,000	1 in 200,000

Five disorders are currently screened in the UK while six form the Irish screening program. The differences in the disorders screened for reflect the different ethnic backgrounds of the two populations. Incidence is quoted per number of live births and worldwide incidence is shown for comparison. * The incidence of sickle cell anemia varies widely with ethnic background, being high in Afro-Americans (1 in 500) and Hispanic-Americans (1 in 1000–1400), and is of similar incidence in areas of the world where malaria is or was common. MCAD, medium-chain acyl-CoA dehydrogenase.

Sickle cell anemia, for instance, is screened for in the UK, where it affects approximately 350 of 700,000 newborn babies per year, but is not screened for in the Irish Republic, where there has been no appreciable Afro-Caribbean immigration. On the other hand, the Republic of Ireland has the highest incidence of cystic fibrosis in the countries of the European Union, which probably reflects the genetic makeup of the indigenous population. These screening programs are being revisited, with pilot studies expanding the number of serious conditions tested for to include those with an incidence of 1 in 100,000 live births (approximately seven babies per year in the UK). These conditions include:

- Maple syrup urine disease
- Homocystinuria
- Glutaric acidemia type 1
- Isovaleric acidemia
- Fatty acidemias: deficiency of very-long-chain (VLCAD), long-chain (LCAD), and medium-chain (MCAD) acyl-CoA dehydrogenase

In addition to the likely incidence of a particular disorder in a population, the cost of screening, particularly with techniques using expensive equipment, may be a major deterrent to the introduction of new tests. This may be exacerbated by the cost of treatment for an affected child, particularly with some enzyme replacement therapies which are vastly more expensive than conventional supportive treatment. The expanded newborn screening program proposed for the Republic of Ireland (**Table 23.11**) takes advantage of the increased throughput and analytical capacity of modern techniques such as TMS to cover a wide range of disorders of amino acids, fatty acids, and organic acids. The availability of modern analytical systems, particularly TMS, and funding for screening has driven the widespread neonatal screening program in the USA, as may be seen on the website for the US National Newborn Screening and Global Resource Center (NNSGRC; http://genes-r-us.uthscsa.edu/). Despite the different ethnic mixes in states such as California (>50% Hispanic), Mississippi (46% African American), and Vermont (>95% White Caucasian), for example, screening for 29 IEMs, including the core conditions listed in Table 23.11, is required by law in all fifty US states; testing for further (>50) rarer disorders, seen as secondary targets, is offered to select populations. It is not appropriate within the scope of this text

TABLE 23.11 Expanded newborn screening program in the Republic of Ireland

Generic class of disorders	Specific disorder screened
Amino acid disorders	Phenylketonuria
	Maple syrup urine disease
	Homocystinuria
	Tyrosinemia
Urea cycle defects	Argininosuccinic acidemia
	Citrullinemia
Fatty acid defects	Long-chain hydroxyacid dehydrogenase deficiency
	Medium-chain acid dehydrogenase deficiency
	Very-long-chain acid dehydrogenase deficiency
	Trifunctional protein defects
Carnitine uptake defects	Defect in plasma membrane carnitine transporter
Organic acid disorders	3-Methylcrotonyl-CoA carboxylase deficiency
	3-Hydroxy 3-methylglutaric acidemia
	β-Ketothiolase deficiency
	Isovaleric acidemia
	Glutaric acidemia type 1
	Methylmalonic acidemia
	Propionic acidemia
	Multiple carboxylase deficiency

to list these here. A cost–benefit analysis precludes such a blanket screening program in most other countries and, to date, there is no consensus on what should be offered routinely in European countries.

Diagnosis

Despite neonatal screening, IEMs may present clinically in the ways described earlier in the chapter. This is because some conditions present earlier than the age at which screening is performed, the condition is not covered by the local screening program, or the child may have missed screening altogether, especially if they were born before it was introduced. Classical presentations occur within the first few days of life with metabolic decompensation and may be rapidly fatal. The most common reasons for considering an IEM in the newborn are lethargy, feeding difficulties, vomiting, and failure to thrive. Additional pointers would be a family history (usually a sibling) or known consanguineous parents.

Initial investigations

Initial biochemical tests include measurement of serum urea and electrolytes and liver function tests, with blood gases, bicarbonate, and anion gap to establish acid–base status. Plasma glucose (low), lactate (raised), ammonia (raised), and ketones (low or raised) may be affected in a number of disorders of intermediary metabolism. Urine and plasma are required for second-line testing at a specialist pediatric laboratory. In particular, urinary and plasma amino acids may be quantitated and patterns of urinary organic acid excretion determined. Plasma acylcarnitines are measured in suspected fatty acid oxidation disorders. Urinary sugars and glycosaminoglycans can also be detected. Biochemical investigations are usually sufficient to indicate the presence of an IEM and to make a provisional diagnosis. Additional confirmation may be required, but is not always necessary.

Urinalysis

Initial urinary tests will include measurement of urinary pH and ketones and also analysis of reducing substances in the urine to eliminate obvious causes. Reducing substances react with hot alkaline copper solutions (Benedict's solution) or in a Clinitest® tablet test, but glucose is the only reducing substance which reacts with GOD-POD (glucose oxidase–peroxidase) strip tests.

Glucose and other reducing substances appear in urine due to renal tubular injury, in galactosemia, and in hereditary fructose intolerance. Renal glycosuria occurs with normal plasma glucose concentration. In Fanconi syndrome there is generalized leakage aminoaciduria, phosphaturia, and hypophosphatemia and hypouricemia due to a urine concentration defect (see above).

Galactose is excreted in urine after ingestion of lactose (milk sugar) in two IEMs: galactokinase deficiency and galactose 1-phosphate uridyl transferase deficiency. Simple screening tests (ultraviolet fluorescence on filter paper; dye decoloration tube test) are available to detect the deficiencies, as well as a quantitative assay for a definitive diagnosis based on assessment of the activity of galactose 1-phosphate uridyl transferase in red blood cells.

Fructose is excreted in urine after ingestion of sucrose in the inborn deficiencies of both fructose 1-phosphate aldolase in hereditary fructose intolerance and fructokinase in essential fructosuria. Diagnosis of hereditary fructose intolerance was made originally by measurement of the activity of fructose 1-phosphate aldolase in liver tissue obtained by biopsy but this has been superseded by molecular analysis of the gene encoding fructose 1-phosphate aldolase.

Homogentisate (found in urine of patients with alkaptonuria, due to deficiency of homogentisic acid oxidase) is identified by TLC by reduction of cold ammoniacal silver nitrate solution (silver mirror test) and atypical reaction with alkaline copper solutions.

CASE 23.3

A 5-day-old newborn male has poor feeding and weight gain with vomiting. On examination he has increased muscle tone and is irritable, then becomes floppy and difficult to rouse. His urine is noted to smell of maple syrup. Blood gases show a metabolic acidosis. Further tests show the following.

	SI units	Reference range	Conventional units	Reference range
Plasma				
Ammonia	102 μmol/L	11–32	142 μg/dL	15–45
Glucose	1.6 mmol/L	3.9–6.1	28 mg/dL	70–110
Ketones measured as 3-hydroxybutyrate (qualitative)	Increased			
Amino acids (qualitative)	Increased valine, leucine, and isoleucine			

- **What do the elevated amino acids have in common?**
- **What gives the urine its characteristic smell?**

The patient has maple syrup urine disease (MSUD) in which there is an enzymatic block in the pathway responsible for the metabolism of the branched-chain amino acids leucine, isoleucine, and valine. The enzyme affected is branched-chain α-ketoacid dehydrogenase, which consists of four subunits, each coded by a separate gene. There is accumulation of the three amino acids and their corresponding ketoacids, which are toxic to the central nervous system. There are several phenotypes (which correspond poorly with the genotype) with varying ages of presentation, but the classical form affects neonates in the first week of life. Biochemical features include ketoacidosis, hyperammonemia, and hypoglycemia. Early diagnosis and restriction of the branched-chain amino acids prevents neurological damage and detection of MSUD is incorporated into neonatal screening programs. The characteristic smell is due to the presence of sotolone, a small lactone molecule, which is produced spontaneously in MSUD and is also found in a number of aromatic foodstuffs.

Confirmatory investigations

In many cases further investigations are needed to make a firm diagnosis. These may include measurement of enzyme activity in cells, most often white blood cells, fibroblasts, or liver cells. In some conditions the molecular biology is sufficiently well characterized that genetic testing can be a useful investigation. This is likely to be the case if:

- Few specific mutations have been shown to cause the disease;
- A family mutation has previously been characterized; or
- Several deleterious mutations have already been described.

Genetic testing is likely to be less informative if most reported mutations are private mutations and if the presenting infant is the index case. There are also examples of IEMs where the genetic variant correlates poorly with the clinical severity and is unhelpful in predicting outcome. If a genetic diagnosis is made, this can be utilized for prenatal or even preimplantation diagnosis in subsequent pregnancies.

Principles of treatment

If the IEM is known, specific treatment is often available for the patient. However, if an IEM is suspected but not yet identified, one or more of the following treatments may be appropriate in the short term:

- Elimination of nutrients other than glucose
- Administration of large doses of water-soluble vitamins
- Management of intercurrent illness, especially infections

CASE 23.4

A boy aged 12 years is brought to the emergency department with confusion and vomiting, which had begun 12 hours previously. On examination he was disoriented and breathing rapidly. He had increased muscle tone and a tremor. He was noted to be small for his age. Urgent biochemical investigations showed very high plasma ammonia. Treatment was initiated and samples were referred to a specialist pediatric laboratory.

	SI units	Reference range	Conventional units	Reference range
Plasma				
Ammonia	400 μmol/L	11–32	560 μg/dL	15–45
Glucose	4.2 mmol/L	3.9–6.1	76 mg/dL	70–110
Urea	0.5 mmol/L	2.9–8.2	1.4 mg/dL	8–23
Bilirubin	10 μmol/L	<21	0.6 mg/dL	<1.3
Alanine transaminase (ALT)	30 U/L	5–40	30 U/L	5–40
Amino acids (qualitative)	Increased ornithine, glutamine, and alanine Low citrulline			
Urine				
Organic acids (qualitative)	Increased orotic acid			

- **How does the raised ammonia relate to his symptoms?**
- **What type of defect is likely to cause severe hyperammonemia?**
- **What is required to make a definitive diagnosis?**

Ammonia is toxic only to the central nervous system and his neurological symptoms are the manifestation of this. Urgent treatment to lower the ammonia is the priority. Hypoglycemia may cause confusion, but his plasma glucose was normal. Severe hyperammonemia may be caused by a defect of the urea cycle, and low plasma urea is supportive evidence of this. Lesser degrees of hyperammonemia may be seen in infants, particularly if they are premature or sick, and in liver failure (hepatic encephalopathy). In this case, the defect is ornithine transcarbamylase (OTC) deficiency. The enzyme is responsible for condensation of carbamoyl phosphate and ornithine to form citrulline (see Chapter 3), hence the increase in amino acids proximal to the block and decreased citrulline. Excess carbamoyl phosphate is converted to orotic acid, which is an important diagnostic finding. Definitive diagnosis requires enzyme analysis on liver tissue or DNA analysis to show the mutation. The OTC gene is carried on the X chromosome, but females may sometimes be affected. The age at presentation is extremely variable, from neonatal to adulthood, indicating some cases have residual enzyme function.

More specific management may be possible where the IEM has been identified and this includes:

- Elimination or control of dietary intake of the substrate of a defective enzyme
    - For example, phenylalanine in phenylketonuria; galactose in classic galactosemia; fructose in hereditary fructose intolerance
- Administration of large doses of water-soluble vitamins
    - For example, biotin in biotin-responsive propionic acidemia; pyridoxine in pyridoxine-responsive homocystinuria; cobalamin in cobalamin-responsive methylmalonic acidemia
- Inhibition of overproduction of toxic side products
    - For example, cortisol administration in 21-hydroxylase deficiency; allopurinol administration in urate overproduction
- Replacement of the defective enzyme, which is produced by recombinant technology
    - For example, α-galactosidase for Fabry disease; glucocerebrosidase for Gaucher's disease
- Replacement of the defective gene in the target tissue (gene therapy)
    - For example, *ABCD1* for the peroxisomal disorder adrenoleukodystrophy; lipoprotein lipase for the lipid disorder lipoprotein lipase deficiency
- Replacement of the defective tissue by organ transplantation
    - For example, liver transplantation for tyrosinemia and urea cycle defects

VITAMINS, ESSENTIAL FATTY ACIDS, AND TRACE ELEMENTS

IN THIS CHAPTER

FAT-SOLUBLE VITAMINS

WATER-SOLUBLE VITAMINS

POLYUNSATURATED FATTY ACIDS (PUFAS)

TRACE ELEMENTS

Vitamins and trace elements are micronutrients which are essential dietary components required for the maintenance of health. The dietary requirements for individual micronutrients vary and are subject to a number of different factors. A window of the amount to be ingested daily exists for each, and clinical disorders may arise from deficiency or excess. The most commonly used guideline values quoted for this are the recommended daily amounts (RDA) published regularly by the United States National Academy of Sciences Food and Nutrition Board (**Table 24.1**), the recommended nutritional intake (RNI) published by the Committee on Medical Aspects of Food and Nutrition Policy (COMA) in the UK, and the recommended daily intake (RDI) published by the World Health Organization. These values define amounts which are adequate to meet known nutrient needs of practically all healthy persons. It should be stressed, however, that this is adequate and may not be optimal in individual cases.

24.1 FAT-SOLUBLE VITAMINS

Absorption and transport

Fat-soluble vitamins in the diet are incorporated into the mixed micelles formed during digestion of dietary fats in the jejunum by pancreatic lipases (see Chapter 14). After absorption from the micelles into enterocytes, they are incorporated into chylomicrons and transported via the lymph into the peripheral circulation, eventually reaching the liver as a component of chylomicron remnants. As described below, some fat-soluble vitamins are stored in the liver itself, while others are stored in adipose tissue. A failure to synthesize chylomicrons in the enterocytes leads to fat-soluble vitamin deficiency (see Chapter 18).

TABLE 24.1 Recommended daily amount (RDA) and references ranges for micronutrients

	RDA	Reference range (SI units)	Reference range (conventional units)
Fat-soluble vitamins			
Vitamin A	900 μg (M) 700 μg (F)	1.1–2.8 μmol/L	30–80 μg/dL
Vitamin D	5 μg (19–50 years) 10 μg (51–70 years) 15 μg (>70 years)	See Chapter 9 for vitamin D and its metabolites	See Chapter 9 for vitamin D and its metabolites
Vitamin E	15 mg	12–42 μmol/L	5.0–18.0 μg/mL
Vitamin K	120 μg (M) 90 μg (F)	0.22–4.88 nmol/L	0.10–2.20 ng/mL
Water-soluble vitamins			
Vitamin B_1	1.2 mg (M) 1.1 mg (F)	50–220 nmol/L (measured in whole blood)	2–6 μg/dL (measured in whole blood)
Vitamin B_2	1.3 mg (M) 1.1 mg (F)	210–410 nmol/L (measured in whole blood)	8–15 μg/dL (measured in whole blood)
Vitamin B_3	16 mg (M) 14 mg (F)	0.9–8.2 μmol/L (measured in whole blood)	0.2–1.8 μg/mL (measured in whole blood)
Vitamin B_6*	1.3 mg (19–50 years) 1.7 mg (M >50 years) 1.5 mg (F >50 years)	20–121 nmol/L	5–30 ng/mL
Vitamin B_{12}	2.4 μg	146–616 pmol/L	200–835 pg/mL
Folate	400 μg	7–36 nmol/L 317–1422 nmol/L (RBC)#	3–16 ng/mL 140–628 ng/mL (RBC)#
Vitamin C**	90 mg (M) 75 mg (F)	28–85 μmol/L	0.5–1.5 mg/dL
Biotin	30 μg	0.82–2.05 nmol/L	200–500 pg/mL
Pantothenic acid	5 mg	168–671 nmol/L	37–147 μg/L

All assays are performed on serum samples apart from * (plasma/EDTA) and ** (plasma/oxalate, heparin, or EDTA). #These values are for packed erythrocytes (red blood cells; RBC) isolated in EDTA-containing medium. M, male; F, female; EDTA, ethylenediaminetetraacetic acid.

Vitamin A

Vitamin A is the collective term used to include retinol and retinyl esters, sometimes referred to as preformed vitamin A. Many population groups obtain vitamin A through metabolism of the plant-derived precursor provitamin β-carotene. Primary sources of β-carotene are oily fish, eggs, milk products such as cheese and yoghurt, and fortified margarine. It is stored in the liver of animals and this tissue is a rich source of the vitamin, particularly when served as pâté. However, because of the high levels of vitamin A in liver, consumption of liver-containing foodstuffs during pregnancy should be avoided.

Transport and storage

Dietary-derived vitamin A and β-carotene, brought to the liver as components of chylomicron remnants, are stored as retinyl esters in the stellate cells (fat cells) in liver. These retinyl esters are transported to extrahepatic tissues bound to two proteins synthesized by the liver, a specific retinol-binding protein and transthyretin.

Chemistry

The structures of retinol and β-carotene are shown in **Figure 24.1**. Retinol is an all-*trans* molecule consisting of a β-ionone ring with a conjugated isoprenoid side chain and a polar terminal group.

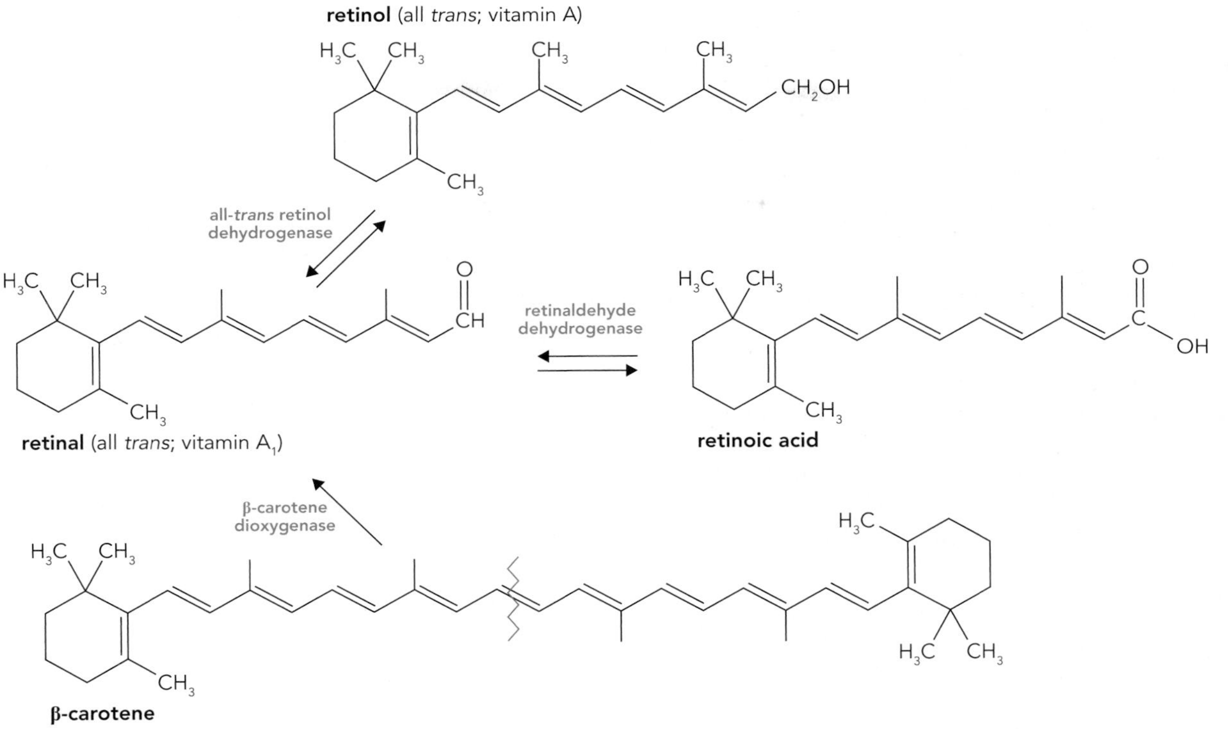

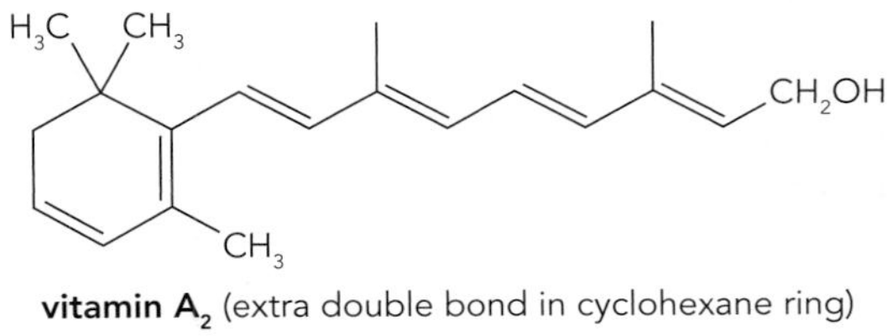

retinol (all *trans*; vitamin A)

all-*trans* retinol dehydrogenase

retinal (all *trans*; vitamin A_1)

retinaldehyde dehydrogenase

retinoic acid

β-carotene dioxygenase

β-carotene

vitamin A_2 (extra double bond in cyclohexane ring)

Figure 24.1 The structures of vitamin A (all-*trans* retinol) and related molecules.
The figure shows the enzymes involved in the metabolic pathway from β-carotene, the precursor molecule present in the diet, which is split in the first reaction. Vitamin A_2 is shown for comparison. It differs from vitamin A_1 by an additional double bond in the cyclohexane ring.

Functions

The major physiological roles of vitamin A are:

- Vision in dim light; it is a component of rhodopsins and iodopsins in rod and cone cells in the retina
- Maintenance of cell-mediated immunity
- Induction and maintenance of differentiation in some tissues, for example skin and mucous linings
- Signal for morphogenesis in developing embryos

Besides being an essential component of the visual process, it also acts at the genetic level to control gene expression of differential growth through its binding to the retinoid nuclear receptors RAR and RXR.

Deficiency

The vital importance of vitamin A to vision is illustrated by the problems caused by its deficiency. Early signs of vitamin A deficiency are night blindness and hyperkeratosis followed by conjunctival keratosis (dry eye) and degeneration of the cornea (keratomalacia). This may eventually lead to blindness due to destruction of the cornea and retinal dysfunction if not treated with vitamin A supplements. Prolonged deficiency also results in dedifferentiation of rapidly growing tissues such as gastric epithelium.

Excess

Symptoms of vitamin A excess are rare but may occur in cases of chronic overdose. Such cases may exhibit skin thickening, cracked lips, conjunctivitis, erythematous eruptions, chronic headache, painful joints, and damage to the liver and eyes. Damage to the liver, eyes, and bone (decreased bone mineral density) may be permanent. Single high doses of vitamin A (100,000 IU in infants; 10,000 IU in adults) give rise to abdominal pain, anorexia, irritability, vomiting, blurred vision, and headaches (1 IU is equivalent to 0.3 μg of vitamin A [retinol]). High intake of vitamin A during pregnancy, through consumption of liver and liver-derived products such as pâté, is discouraged due to an association with birth defects.

Clinical chemistry

Vitamin A can be measured in serum as retinol. Since levels of retinol are under homeostatic control, they do not fall until late in vitamin A deficiency. Levels are also affected by infection, liver disease, and other nutrients. Measurement of carotene may be useful in suspected carotenemia where excessive ingestion of foods such as carrots leads to orange coloration of the skin due to deposition of β-carotene. This is otherwise a harmless condition.

Vitamin D

The structures, sources, and clinical chemistry of vitamin D and its derivatives are described in detail in Chapter 9.

Clinical chemistry

Vitamin D status is assessed by measurement of total 25-hydroxy vitamin D (25-OHD; D_2 plus D_3). Laboratories measuring vitamin D have experienced huge annual increases in workload in recent years due to the recognition that deficiency and insufficiency are very common, especially at northern latitudes where sunlight is limited. Even in countries where sunlight is abundant, deficiency may still be seen due to use of high-factor sun block and campaigns to avoid sunburn. Vitamin D is found in few foods, except oily fish or where fortification is carried out. Many chronic diseases have been associated with low levels of vitamin D and these are unrelated to its effect on calcium and bone metabolism. However, the precise relationships remain controversial.

Vitamin E

Vitamin E is the collective name given to a group of at least eight naturally occurring plant tocopherols and tocotrienols which may be metabolized in the body to the biologically active form of the vitamin, α-tocopherol. Derivatives of both tocopherols and tocotrienols have been given Greek prefixes: α, β, δ, and γ. The most biologically active is d-α-tocopherol (1 IU is equal to 67 mg) and the activities of other derivatives are related to this. Primary sources of the vitamin are leaves and green parts of higher plants (that is, plants other than algae) while plant oils are the richest source (soybean oil, 560–1600 mg/kg; corn oil, 530–1620 mg/kg; olive oil 50–150 mg/kg). Generally, foodstuffs rich in polyunsaturated fatty acids (PUFAs) contain large amounts of vitamin E. Much smaller amounts are found in the fat of animal tissues and the vitamin E content may be reduced considerably during processing and packaging of food.

Transport and storage

After absorption from the gut, vitamin E is incorporated into chylomicrons, enters the circulation, and is transported to liver as part of the chylomicron remnant. Unlike vitamins A and D, there is no specific transport protein for the vitamin and it is distributed to peripheral tissues by hepatically synthesized lipoproteins. There appears to be a long-term, stable storage pool of vitamin E in adipose tissue, in contrast to the more labile, rapidly depleted pools in plasma and liver. Most vitamin E (up to 70%) is excreted as a glucuronide into feces via

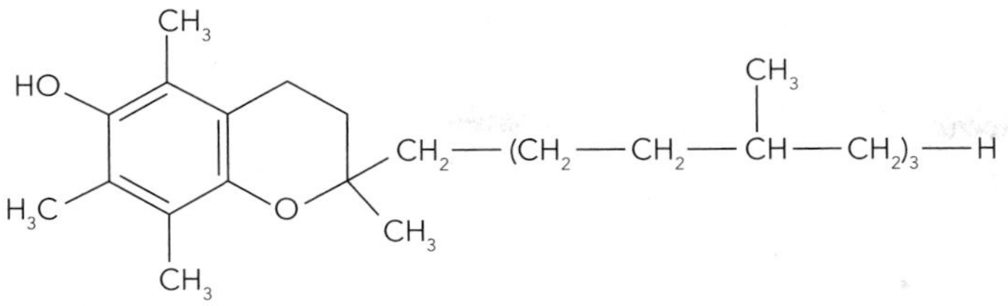

α-tocopherol

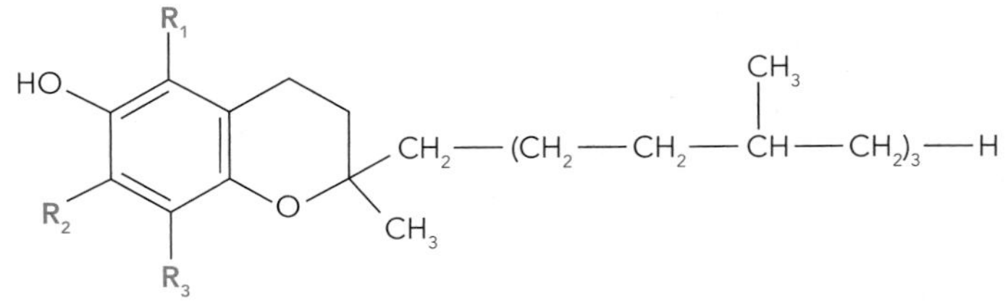

tocopherol	R_1	R_2	R_3
β	CH_3	H	CH_3
γ	H	CH_3	CH_3
δ	H	H	CH_3

Figure 24.2 Structures of the E vitamins (tocopherols).
The side chain usually consists of three saturated, isoprene-derived, 5-carbon units as shown, but may occasionally be more than this. Methyl substituents on the aromatic ring may vary in position yielding other members of the vitamin E family: β-, γ-, and δ-tocopherols.

the bile. Very little (<1%) is excreted in urine while some vitamin E may be eliminated via sloughed-off skin.

Chemistry

Structures of the active form of the vitamin, α-tocopherol, and tocotrienols are shown in **Figure 24.2**. Vitamins in both classes are designated by the methyl groups on the aromatic ring.

Function

The major, and maybe only, role of vitamin E in adults is as an antioxidant and free-radical scavenger in a lipid environment, where it helps to maintain the integrity of all cell membranes. It may also have an anti-atherosclerotic action by protecting against oxidation of low-density lipoproteins.

Deficiency

Deficiency of vitamin E results in oxidative destruction of unsaturated bonds in fatty acids of membrane phospholipids with consequent deleterious effects on cell function. Clinically this presents as red blood cell fragility, sometimes leading to hemolytic anemia, and peripheral neuropathies due to neuronal degeneration. Further changes may present in the vascular and reproductive systems. Genetic abnormalities associated with vitamin E metabolism have been described; for example, the Friedreich type of spinocerebellar ataxia is associated with defects in hepatic α-tocopherol transferase. The major clinical manifestations of abetalipoproteinemia (see Chapter 18) are due to inadequate vitamin E delivery to the tissues.

Excess

Vitamin E has low toxicity. Humans and animals appear to be able to tolerate levels of the vitamin two orders of magnitude above nutritional requirements (for example 1000–2000 IU/kg diet) without untoward effects (1 IU of alpha-tocopherol is equivalent to 0.67 mg of the natural form of the vitamin or 0.45 mg of the synthetic form). At very high doses, however, vitamin E can produce signs indicative of antagonism to the function of the other fat-soluble vitamins (vitamins A, D, and K). Isolated reports of sporadic adverse effects in humans consuming up to 1000 IU (670 mg) of vitamin E per day have included the following symptoms: headache, fatigue, nausea, double vision, muscle weakness, mild creatinuria, and gastrointestinal distress. Vitamin E also has an antiplatelet and anticoagulant effect.

Clinical chemistry

Measurement of serum or plasma concentration of α-tocopherol provides the simplest and most direct evidence of vitamin E status, and reference ranges for adults and children are established. Values of 5–18 μg/mL for adults and children of twelve years or older, and values of 3–15 μg/mL for children under twelve years, indicate acceptable levels of intake. Since vitamin E is transported by lipoproteins, many laboratories express results as lipid-adjusted concentrations, taking into account high and low serum lipids. Previously, a widely used indicator of vitamin E status was the extent of hemolysis of red blood cells in the presence of hydrogen peroxide. A high degree of hemolysis accompanies a vitamin E deficiency of greater than 20%, but hemolysis is not specific to vitamin E deficiency and this method is no longer used extensively.

Vitamin K

Vitamin K is the generic term for members of a family of naphthoquinones which have similar biological activity: namely, the ability to promote the post-translational γ-carboxylation of proteins, thereby increasing the capacity of the protein to bind calcium. Such proteins include procoagulant proteins of the clotting cascade (see Chapter 3), including prothrombin and Factors VII, IX, and X, anticoagulant proteins S and C, bone matrix proteins including osteocalcin (see Chapter 9), and proteins of the renal epithelium. The primary sources of vitamin K_1 are green leafy vegetables (for example broccoli and spinach), vegetable oils, and cereals. Small amounts are present in meat, fish, and dairy products such as cheese. Synthesis of vitamin K_2 by Gram-positive bacteria in the jejunum and ileum also makes a significant contribution to the daily requirement. Excess vitamin K is stored in the liver.

Chemistry

Members of the vitamin K family are derived from 2-methyl 1,4-naphthoquinone. The major dietary vitamin K is phylloquinone, or vitamin K_1 (2-methyl 3-phytyl 1,4-naphthoquinone; **Figure 24.3**), which is found predominantly in plants. Vitamin K_2, or menaquinone, is synthesized by Gram-positive bacteria,

(a)

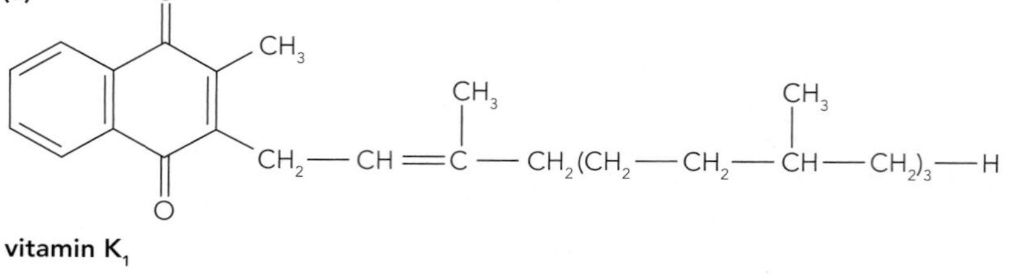

CH_3 ; $CH_2—CH{=}C(CH_3)—CH_2(CH_2—CH_2—CH(CH_3)—CH_2)_3—H$

vitamin K_1

(b)

CH_3 ; $(CH_2—CH{=}C(CH_3)—CH_2)_n—H$

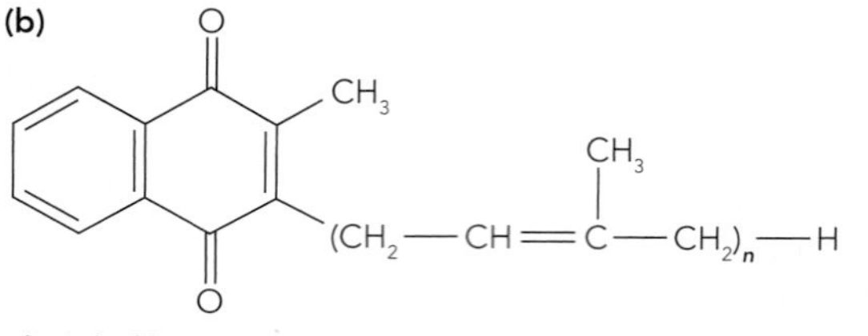

vitamin K_2

Figure 24.3 Structures of the K vitamins.
(a) Vitamin K_1 (phylloquinone). (b) Vitamin K_2 (menaquinones; menaquinones vary in the number of isoprene units (n = 6, 7, or 9) in the polyisoprenoid side chain.

including those present in the human gut. Synthetic vitamin K is also available commercially as menadione (K_3) and menaquinol (K_4).

Function

Vitamin K is involved in the γ-carboxylation of proteins, as described in Chapter 3 as one of the functions of the liver.

Deficiency

Since gut bacteria synthesize vitamin K, and the daily requirement for vitamin K is low anyway, deficiency of the vitamin is uncommon. It occurs occasionally in breast-fed infants and in adults secondary to malabsorption, such as in abetalipoproteinemia, for example, or impaired synthesis in the gut. However, newborn infants are susceptible to vitamin K deficiency due to the relative lack of the vitamin in breast milk and the time taken for bacterial flora to become

CASE 24.1

A 46-year-old homeless man is admitted to hospital with confusion. He is unable to give any history, but later confirms extreme self-neglect, heavy alcohol use, and poor diet. Two years previously he had a hip fracture following a minor fall. On examination he is extremely thin with low blood pressure and low body temperature. He has widespread abnormalities of his skin and both eyes show corneal ulceration. His immediate treatment includes intravenous water-soluble vitamins (B_1, B_2, B_3, B_6, and C). Blood is taken to assess his fat-soluble vitamin status.

	SI units	Reference range	Conventional units	Reference range
Serum				
Vitamin A	<0.1 μmol/L	1.1–2.8	<5 μg/dL	30–80
Vitamin D	2.7 nmol/L	37–104	6.9 ng/mL	15–42
Vitamin E	6.1 μmol/L	12–42	2 μg/mL	5–18
Cholesterol	1.3 mmol/L	3.5–6.5	50 mg/dL	135–250
Triglycerides	0.52 mmol/L	<1.8	46 mg/dL	<160

- **Which vitamin deficiencies are likely to be responsible for his clinical features?**
- **Why are the water-soluble vitamins not measured?**
- **Why are the plasma lipids low and how might this affect the other results?**

In this unusually severe case of malnutrition there is deficiency of all of the fat-soluble vitamins. His skin and eye conditions are reflections of vitamin A deficiency, which worldwide is a major cause of blindness, although rarely in industrialized countries. In the skin there is loss of goblet cells and excessive keratin formation. Low-trauma hip fractures are unusual in young men and may be caused by vitamin D deficiency due to a lack of sunlight exposure and inadequate intake. Vitamin E deficiency may cause neurological abnormalities, but these can be difficult to distinguish from other causes.

It was not necessary to measure the water-soluble vitamins as it was assumed that they would be low and were treated anyway. There would be no value in measuring them after treatment.

Acquired hypolipidemia may be seen with severe restriction of dietary fat. Vitamin E is transported by lipoproteins and low levels can be difficult to interpret in the presence of very low plasma lipids. Some laboratories express vitamin E as a molar ratio to plasma lipids to account for this.

established. Deficiency of vitamin K in this instance results in hemorrhagic disease of the newborn. Hence an injection of vitamin K is routinely given at birth.

Clinical chemistry

Unlike other fat-soluble vitamins, vitamin K is not routinely measured in serum. Vitamin K functional status is instead assessed indirectly by its effect on blood coagulation. Vitamin K deficiency causes a prolongation of the prothrombin time, which is routinely measured in hematology laboratories. Anticoagulant drugs (warfarin and other coumarins) interfere with the function of vitamin K and thereby increase the prothrombin time as a therapeutic effect. The prothrombin time and the index derived from it (international normalized ratio or INR) are used to adjust dosage of these drugs.

24.2 WATER-SOLUBLE VITAMINS

Unlike the fat-soluble vitamins, and apart from vitamin B_{12}, there is no body store of water-soluble vitamins and thus these have to be supplied regularly via the diet.

Clinical practice point 24.1

It is usually easier and cheaper to replace the water-soluble vitamins B_1, B_2, B_3, B_6, and C than to measure them in suspected deficiency states.

Vitamin B_1 (thiamine)

Thiamine is present in unrefined grain products such as cereals and unpolished rice; other major dietary sources of thiamine are vegetables, meat and dairy products, fruit, and eggs. The vitamin is lost during the refining of flour and polishing of rice and it is mandatory in the UK to fortify white and brown flour with thiamine to a minimum concentration of 0.24 mg per 100 g of flour. At concentrations normally present in the diet, thiamine is absorbed rapidly via a saturable specific transporter system in the intestine. A slower, passive absorption system also exists and comes into play at high dietary doses of the vitamin. Thiamine has a half-life of 10–20 days in the body and deficiency may develop rapidly. Severe deficiency of thiamine presents as beriberi.

Chemistry

The structure of thiamine is shown in **Figure 24.4**. It consists of substituted pyrimidine and thiazole rings linked via a methylene bridge. A hydroxyethyl side chain is attached to the thiazole ring.

Function

Thiamine is present in tissues and blood in free and phosphorylated forms, predominantly as thiamine pyrophosphate (TPP) intracellularly, with the highest tissue concentration in the liver. Erythrocytes transport free and phosphorylated thiamine, although only thiamine monophosphate (TMP) and free thiamine are found in plasma and cerebrospinal fluid. Thiamine undergoes activation to its pyrophosphate (TPP) derivative before participating as a co-factor of a

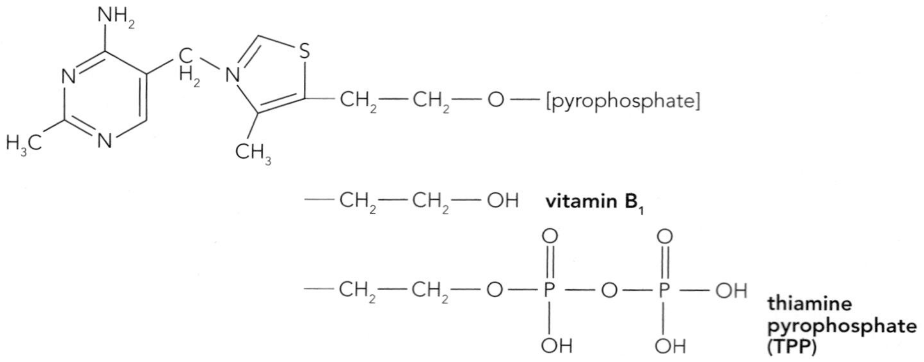

NH2 — N — C H2 — N — S — H3C — N — CH3 — CH2—CH2—O—[pyrophosphate]

—CH2—CH2—OH **vitamin B1**

—CH2—CH2—O—P(=O)(OH)—O—P(=O)(OH)—OH **thiamine pyrophosphate (TPP)**

Figure 24.4 Structures of vitamin B_1 (thiamine) and thiamine pyrophosphate (TPP). Thiamine pyrophosphate is the active form of the vitamin which participates in metabolic reactions such as the oxidative decarboxylation of α-ketoglutarate and pyruvate.

number of enzymes (decarboxylases and transketolases). This activation process requires the donation of phosphate from adenosine triphosphate (ATP). Thiamine pyrophosphate is an essential co-factor of the decarboxylation activity of α-ketoacid dehydrogenase complexes including, for example, pyruvate dehydrogenase in the glycolytic pathway and α-ketoglutarate dehydrogenase in the tricarboxylic acid (TCA) cycle (see Chapter 2). It is also a co-factor of transketolases of the pentose phosphate pathway. There is evidence for a role in neuronal cells and excitable tissues including skeletal muscle. Excess thiamine and its metabolites are excreted in urine.

Deficiency

Deficiency of thiamine results in an inability to carry out the biochemical reactions outlined above; thus there is a decreased ability to synthesize ATP from glucose and to metabolize glucose via the pentose phosphate pathway. Clinically, a mild deficiency leads to irritability, fatigue, headaches, appetite loss, constipation, nausea, and unsteadiness. Prolonged deficiency of thiamine (a daily intake of less than 0.2 mg per 1000 kilocalories) presents as beriberi, a condition characterized by peripheral neuropathy with cardiovascular and cerebral dysfunction. In some cases, an accompanying fluid retention (edema) leads to congestive heart failure and low peripheral resistance (wet beriberi). Chronic deficiency

CASE 24.2

A 52-year-old man is admitted to hospital with acute confusion. He is known to have a long history of excessive alcohol intake, but has not drunk that day. On examination he has abnormal eye movements and is unsteady on his feet with ataxia. He is given an injection of thiamine, and immediately before this a blood sample is taken. The results return a few days later.

	SI units	Reference range	Conventional units	Reference range
Whole blood				
Thiamine	3 nmol/L	50–220	1 μg/dL	2–6
Transketolase activation test	130% of baseline	100–115 of baseline	130% of baseline	100–115 of baseline

- **What is the diagnosis?**
- **Why is transketolase measured?**
- **What other biochemical abnormalities might occur in this patient?**

The patient has Wernicke's encephalopathy, which is a manifestation of thiamine deficiency. It is usually, but not exclusively, seen in individuals who abuse alcohol and have a deficient diet. It causes confusion, eye movement abnormalities (ophthalmoplegia), and signs of cerebellar damage. Diagnosis is made clinically as treatment must be given immediately. However, confirmation by demonstrating low blood thiamine can be helpful. An older, and generally superseded test, is the measurement of red-cell transketolase before and after addition of thiamine pyrophosphate (TPP) *in vitro*. The enzyme is a key part of the pentose phosphate pathway and is thiamine dependent. In normal individuals there is no increase in transketolase activity after TPP, but in deficiency the enzyme activity increases by more than 20%.

Alcoholics often have magnesium and phosphate deficiencies, which may require intravenous replacement.

If the patient's lifestyle is unchanged there is a high risk of developing Korsakoff syndrome consisting of memory loss, confabulation, and chronic confusion.

may also present with central nervous system problems such as those seen in Wernicke's encephalopathy, which is characterized by confusion, ataxia, nystagmus, ophthalmoplegia, and coma. Occasionally, Wernicke's encephalopathy is accompanied by psychosis (Korsakoff psychosis), and both conditions are found typically in alcoholics whose dietary vitamin intake is severely compromised (Wernicke–Korsakoff syndrome). A further complication is that alcohol itself can impair the uptake and metabolism of thiamine and chronic alcoholism is the major cause of thiamine deficiency in developed countries. Consumption of high quantities of polished rice is a major cause of deficiency in developing countries.

Clinical chemistry

Traditional assessment of thiamine status was by measurement of the activity of red cell transketolase before and after the addition of thiamine pyrophosphate. However, this has been replaced in many laboratories by direct measurement of red cell thiamine, which is 80% thiamine pyrophosphate, or whole blood thiamine. Measurements are usually requested for patients with clinical features of Wernicke–Korsakoff syndrome, although treatment with thiamine should be given if this is suspected and not be withheld until blood results are available.

Vitamin B_2 (riboflavin)

Riboflavin is a relatively unstable water-soluble vitamin which decomposes in both acidic and alkaline solution. All living cells have an absolute requirement for riboflavin for ongoing metabolic processes and small amounts are present in most foodstuffs. The major dietary sources of riboflavin are eggs, milk, cereals, grain, green vegetables, liver, and some lean meats. Riboflavin is light sensitive and degrades in sunlight and ultraviolet (UV) light. About 20% is lost during milk pasteurization, and exposure to alkali during baking also destroys the vitamin. It is a permitted coloring agent in food preparations and may also be added to salad dressings, some cakes, and powdered drinks. Riboflavin is absorbed from the intestine via a specific transport system which involves its phosphorylation to flavin mononucleotide (FMN). Approximately 10–20% of the normal daily intake is excreted unchanged in urine.

Structure

Dietary riboflavin, 7,8-dimethyl 10-ribitylisoalloxazine (**Figure 24.5**), is metabolized in the body to the flavin nucleotides FMN and flavin adenine dinucleotide (FAD). The isoalloxazine ring exists in oxidized and reduced forms and allows the two nucleotides to participate in redox reactions.

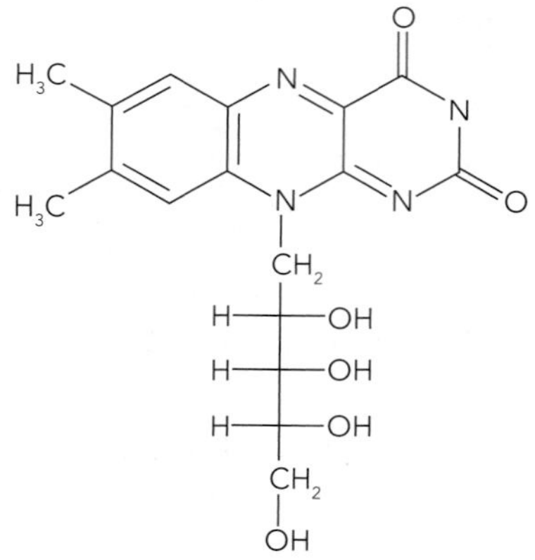

Figure 24.5 Structure of riboflavin (vitamin B_2).
Riboflavin is a constituent of the flavin nucleotide FMN (flavin mononucleotide), where OH is replaced by phosphate, and FAD (flavin adenine dinucleotide), where FMN is in a phosphodiester linkage with adenosine monophosphate. FAD is a co-factor of dehydrogenases such as succinate dehydrogenase, working through reduction and oxidation of the double bond in the isoalloxazine ring.

Function

FMN and FAD are coenzymes in a number of redox reactions, both anabolic and catabolic. This includes reactions of fatty acid oxidation and the TCA cycle (catabolic), and steroid, fatty acid, and epinephrine synthesis (anabolic). Generally, riboflavin supports normal growth in children and erythropoiesis. Riboflavin is an effective lipophilic antioxidant, preventing oxidation of unsaturated bonds in membrane lipids, and this property helps to maintain the integrity of mucous membranes, skin, eyes, and the nervous system.

Deficiency

Groups at risk of riboflavin deficiency are poorly nourished children in developing countries, undernourished elderly individuals, and individuals with problems of malabsorption. Since thyroid hormones stimulate the formation of flavin nucleotides from riboflavin, while major tranquilizers such as phenothiazines and possibly also tricyclic antidepressants inhibit the synthesis of FAD, deficiency may also occur in hypothyroid patients and patients on prescribed tranquilizers. Riboflavin deficiency impairs the intestinal absorption of iron, zinc, and calcium, possibly due to the effects of loss of membrane integrity on transporter systems.

Although photophobia, angular stomatitis, glossitis, scaly dermatitis, and anemia are associated with riboflavin deficiency, isolated cases are rare and usually occur in conjunction with a generalized deficiency of all B vitamins. The anemia in this case is normochromic and normocytic. It may present as photophobia in newborn babies undergoing phototherapy for neonatal jaundice where UV light photolyzes the vitamin.

Clinical chemistry

Riboflavin measurements are rarely required for clinical purposes, it usually being safer and more cost effective to give replacement vitamin doses and assess for improvement. This is true for several of the water-soluble vitamins since unnecessary over-replacement is unlikely to do harm. Measurements that have been used are urinary, blood, or red cell riboflavin and glutathione reductase activity.

Vitamin B3 (niacin)

Niacin is a collective term for nicotinic acid and its amide (nicotinamide). Only small amounts of these molecules are present in the free form in cells and the principal niacin-containing molecules in animal cells are the pyridine nucleotides: nicotinamide adenine dinucleotide (NAD^+) and its phosphorylated derivative nicotinamide adenine dinucleotide phosphate ($NADP^+$). Like vitamin D, niacin is not strictly a vitamin since it can be synthesized in humans from tryptophan and thus is not essential given an adequate daily intake of this amino acid. However, the synthesis of 1 mg of niacin requires 60 mg of dietary tryptophan and the participation of B vitamins riboflavin, thiamine, and pyridoxine. Niacin, present as the pyridine nucleotides, is found in all cells, but good dietary sources are meat (beef and pork), wheat and cornflour, eggs, and milk. Interestingly, human milk is richer in niacin than cow's milk. Preparation of food may lead to enzymatic hydrolysis of the pyridine nucleotides and, in the UK, this has prompted the mandatory fortification with nicotinic acid of all flour apart from wholemeal and other specified types.

Structure

The structures of nicotinic acid, nicotinamide, and the oxidized form of the cofactor NAD^+ are shown in **Figure 24.6**.

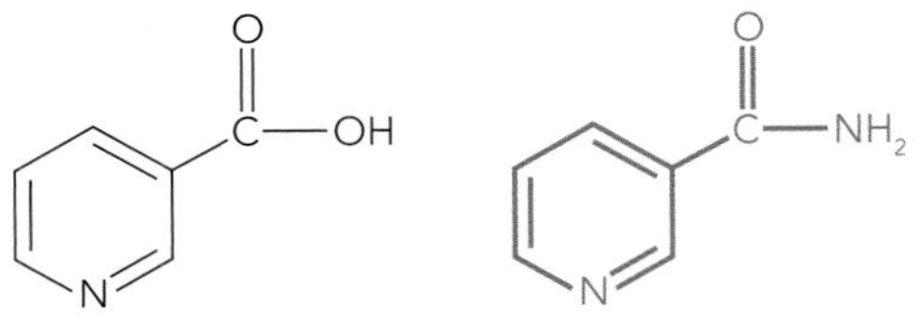

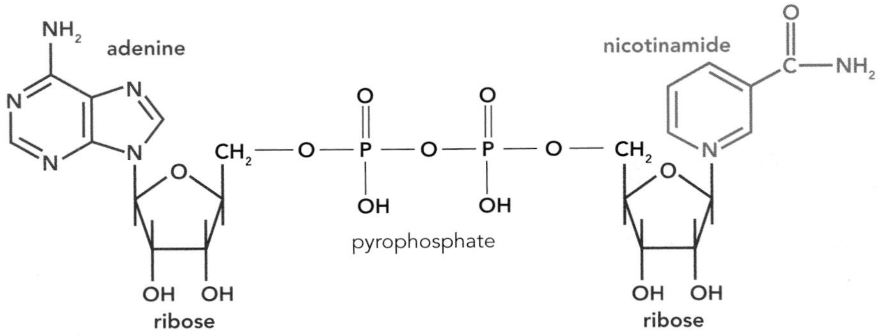

Figure 24.6 Structures of the B3 vitamins.
(a) Nicotinic acid and nicotinamide and (b) nicotinamide adenine dinucleotide, NAD^+. The active form of the vitamin, NAD^+, is a co-factor for many catabolic dehydrogenases such as lactate dehydrogenase and the dehydrogenases in the TCA cycle (see Chapter 2). $NADP^+$ has an additional phosphate substituent on C2 of the ribose moiety of the adenosine component.

Function

As mentioned above, niacin is a component of the enzyme co-factors NAD^+ and $NADP^+$ which participate in over 200 redox reactions in the cell. In general, NAD^+-linked reactions are catabolic, for example in fatty acid oxidation, glycolysis, and the TCA cycle, while $NADP^+$-linked enzymes are involved in anabolic reactions of reductive biosynthesis, for example fatty acid and cholesterol synthesis. $NADP^+$ is also an important co-factor of the pentose phosphate pathway.

Deficiency

The classical presentation of niacin deficiency is pellagra, which is often seen in populations where corn is the major dietary energy source. This condition presents with the following:

- Changes in skin ("pellagra" means raw skin); pigmented dermatitis is a feature of areas of skin exposed to sunlight or pressure
- Inflammation of the epithelial surfaces of the mouth and gastrointestinal (GI) tract
- Diarrhea is a common consequence of inflammation of the GI tract
- Neurological disturbances may eventually lead to dementia

If untreated, prolonged deficiency of niacin may be fatal.

Clinical chemistry

The plasma half-life of nicotinic acid is about 1 hour and the vitamin may be stored in limited amounts in liver, adipose tissue, and kidneys. The major metabolites, *N*-methylnicotinamide, *N*-methyl 2-pyridone 5-carboxamide and *N*-methyl 4-pyridone 5-carboxamide, are excreted via the kidneys along with small amounts of unchanged nicotinic acid, nicotinamide-*N*-oxide, and 6-hydroxynicotinamide. These can be measured and expressed as ratios which vary with niacin status. However, as with riboflavin, laboratory assessment of niacin is rarely required for clinical purposes (as opposed to nutritional research). Methods for measurement include fluorimetric assays and high performance liquid chromatography (HPLC), which measure *N*-methyl-nicotinamide concentration in urine.

Vitamin B_6 (pyridoxine)

The three members of the B_6 vitamin family, which are all present in major foodstuffs, are water-soluble derivatives of pyridine: pyridoxine itself, pyridoxal, and pyridoxamine. They are sensitive to light and long-term storage may lead to breakdown of the vitamin. Boiling of food also results in the loss of the vitamin from the food preparation into water. B_6 vitamins are found in all cells but are enriched in the liver and kidneys of animals and fish. High concentrations are present in wheatgerm and brewer's yeast while other good, non-meat sources are brown rice, oats, wholewheat grains, soybeans, peanuts, and walnuts. Smaller concentrations are present in milk and eggs.

Structures of vitamin B_6

The three forms of the vitamin, which are interconvertible, are shown in **Figure 24.7**.

Function

Members of the B_6 family undergo activation to pyridoxal 5-phosphate in the liver, a reaction which requires riboflavin and zinc. Pyridoxal 5-phosphate is a co-factor for a wide range of transaminases essential in amino acid metabolism and metabolic homeostasis. Some of the metabolic processes which require pyridoxal 5-phosphate include the following:

- Gluconeogenesis
- Glycogenolysis

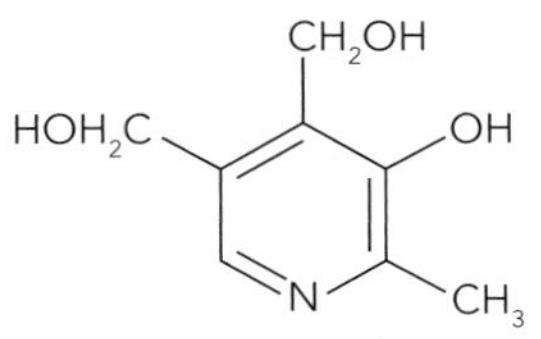

pyridoxine

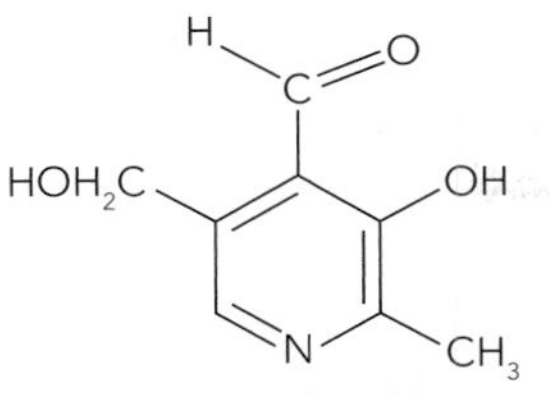

pyridoxal

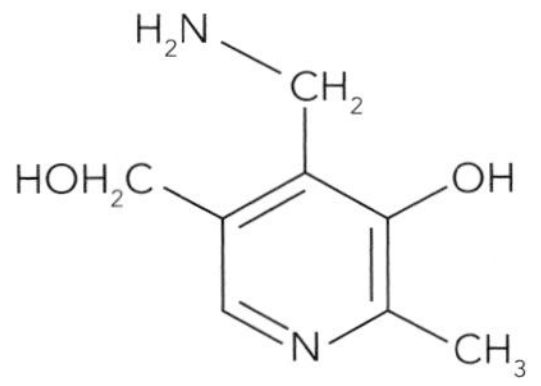

pyridoxamine

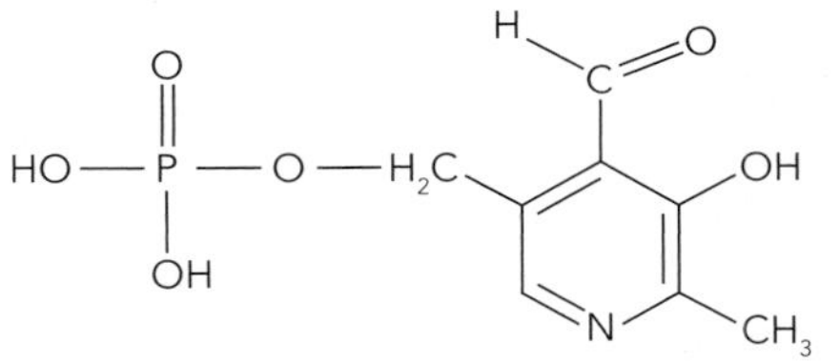

pyridoxal 5-phosphate

Figure 24.7 Structures of the B_6 vitamins.
Pyridoxine, pyridoxamine, and pyridoxal are interconvertible. Pyridoxine and pyridoxamine must be converted to pyridoxal before hepatic activation to pyridoxal 5-phosphate, the form in which the vitamin participates in metabolic reactions.

- Sphingolipid synthesis
- Synthesis of neurotransmitters [serotonin, gamma-aminobutyric acid (GABA), dopamine, and norepinephrine]
- Heme synthesis
- Niacin synthesis from tryptophan
- Absorption of vitamin B_{12}
- Production of gastric acid
- Prostaglandin synthesis
- Erythropoiesis

Deficiency

Deficiency of pyridoxine is rare and occurs usually in association with other water-soluble vitamin deficiencies. However, a number of inborn errors of metabolism involving vitamin B_6 function (the pyridoxine-responsive disorders) have been described. Mild deficiency leads to irritability, nervousness, and depression. More prolonged deficiency may result in peripheral neuropathy, convulsions, and coma as well as skin changes and anemia.

Clinical chemistry

Both urine and blood assays have been developed for laboratory assessment of vitamin B_6 levels, but measurement of the vitamin is rarely required for clinical purposes. Indirect measures include blood transaminases and urinary tryptophan metabolites before and after an oral load of tryptophan.

Vitamin B_{12} (cobalamin)

Like the other water-soluble vitamins, vitamin B_{12} constitutes a family of structurally related molecules, in this case the corrinoids. The key feature of these molecules is the corrin nucleus consisting of a tetrapyrrole ring structure to which an atom of cobalt is chelated. This is the only known role for cobalt in the body. Since vitamin B_{12} is not synthesized by plants, the absolute requirement of all animal cells for the vitamin is satisfied by that synthesized by bacteria, fungi, and algae. Meat and fish (particularly salmon and tuna) are the common dietary sources of cobalamins. Liver is the tissue with the highest content. Commercial preparations used in food fortification and as dietary supplements contain cyanocobalamin and/or hydroxycobalamin. Absorption of vitamin B_{12} requires the presence of intrinsic factor (IF), which is secreted by the stomach. This is a

50 kDa glycoprotein which binds to vitamin B_{12} and is bound by epithelial receptors when it reaches the distal part of the ileum. The complex is internalized and vitamin B_{12} is released to be transported by the portal circulation to the liver.

Structure

The structure of vitamin B_{12} is based on a planar tetrapyrrole ring not unlike that seen in heme (see Chapter 16), except that in this case the metal ion at the center of the molecule is cobalt rather than iron (**Figure 24.8**). The tetrapyrrole ring system is also more saturated (rings A and D) than in heme. There are six coordination sites for the cobalt ion; four are filled by nitrogens of the four pyrrole rings and one by a nitrogen of dimethylbenzimidazole. This dimethylbenzimidazole is part of a bridging system consisting of ribose 5-phosphate and 2-amino 1-methylethan-1-ol (aminoisopropanol) which is linked via an amide bond to a propionyl side chain of ring A in the corrin structure. Occupation of the sixth coordination site by a hydroxyl ion yields an inactive form of the vitamin where cobalt is present as Co^{3+}. In the active form of the vitamin, where the cobalt ion has been reduced by flavoprotein reductases to Co^{+}, this site is occupied by deoxyadenosylcobalamin.

The other active form is methylcobalamin where the sixth coordination site is occupied by a methyl residue. The two active forms of vitamin B_{12} are the only known examples of a carbon–metal bond in natural biological systems.

Function

The functions of cobalamin should be considered in conjunction with those of folate. Deoxyadenosylcobalamin is a co-factor for methylmalonyl-CoA mutase, an essential enzyme in the oxidation of odd-numbered fatty acids (see Chapter 18). It is involved in the conversion of propionyl-CoA to succinyl-CoA, thereby providing succinyl-CoA for carbohydrate and lipid metabolism.

Methylcobalamin is a co-factor for homocysteine methyl transferase (methionine synthase), a key enzyme of one-carbon metabolism and important

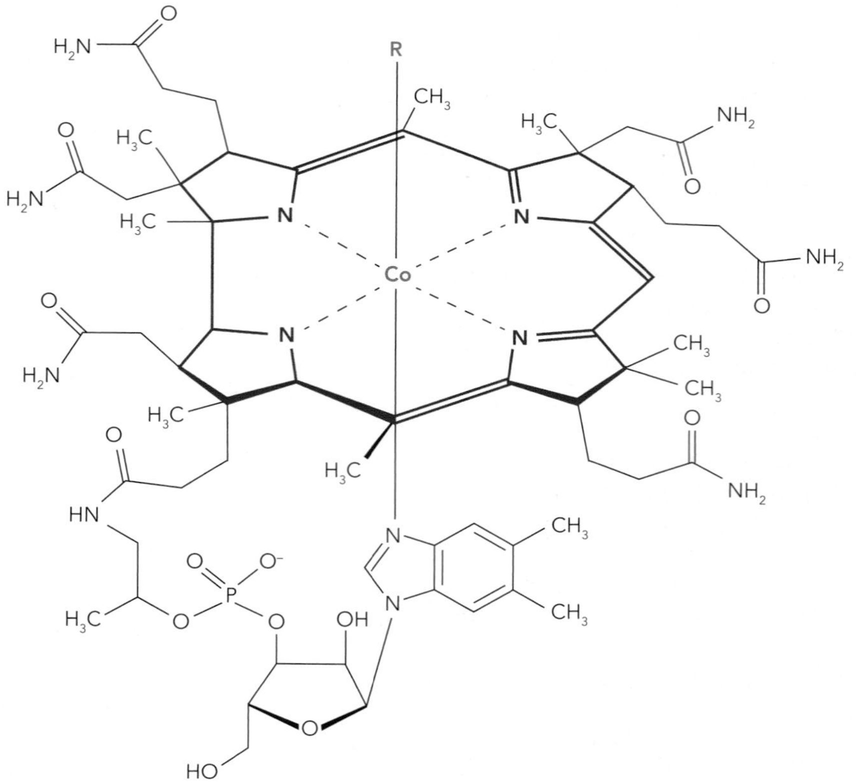

Figure 24.8 Structure of vitamin B_{12} (5-deoxyadenosylcyanocobalamin). The complex structure of vitamin B_{12} is based on a tetrapyrrole (corrinoid) structure shown in outline in blue. The central cobalt ion is coordinated with a nitrogen from each of the pyrroles and a nitrogen of 5,6-dimethylbenzimidazole. The sixth coordination site R may vary; in the parent vitamin B_{12}, R = 5-deoxyadenosine, while R = methyl (CH_3) in methylcobalamin and R = cyanide (CN–) in cyanocobalamin. A detailed description of the structure is beyond the scope of this text.

in the synthesis of the universal methyl donor *S*-adenosylmethionine, which is required for nucleotide synthesis, for example, and cellular transport and recycling of folate.

Deficiency

Dietary deficiency of cobalamin is rare but may occur in subjects on a strict vegan diet and occasionally in the elderly in institutions. Clinical conditions which may present with deficiency are due to a lack of intestinal absorption and include:

- Pernicious anemia
- Pancreatic insufficiency
- Atrophic gastritis
- Small bowel bacterial overgrowth
- Ileal disease

Pernicious anemia is the name given to the autoimmune condition resulting in the destruction of the gastric cells that produce IF. Without intrinsic factor, vitamin B_{12} in normal dietary doses cannot be effectively absorbed, although very high oral doses can be absorbed through non-IF routes.

Deficiency of vitamin B_{12} affects both erythropoiesis and the function of the nervous system (megaloblastic anemia, and demyelination of peripheral nerves, posterior and lateral columns of the spinal cord, and nerves in the brain). The clinical effects on the spinal cord are termed subacute combined degeneration of the cord.

Clinical chemistry

Serum vitamin B_{12} is routinely measured by automated immunoassay, usually at the same time as folic acid and ferritin as part of a hematinics panel for the investigation of anemia. However, a significant proportion of vitamin B_{12}-deficient patients do not have anemia or a raised mean cell volume, and B_{12} should be measured in patients with neuropathy and other neurological symptoms. When B_{12} is found to be low, further investigation may include the measurement of antibodies to gastric intrinsic factor. These are present in pernicious anemia and indicate the need for regular B_{12} injections or high-dose oral B_{12} supplementation. The Schilling test, which measures the absorption of radiolabeled B_{12} with and without co-administration of IF, is rarely used today since it is easier and cheaper to give B_{12} treatment.

Folic acid (folate)

Folate is the generic name for derivatives of folic acid (pteroylglutamic acid), the most stable of the vitamins comprising the folate group. By convention, folic acid refers to the parent molecule, pteroylglutamic acid, while folate is used for individual or mixtures of other members of the group. Folates consist of three components—a substituted pteridine ring, *p*-aminobenzoate, and glutamic acid (**Figure 24.9**). Humans and mammals in general are unable

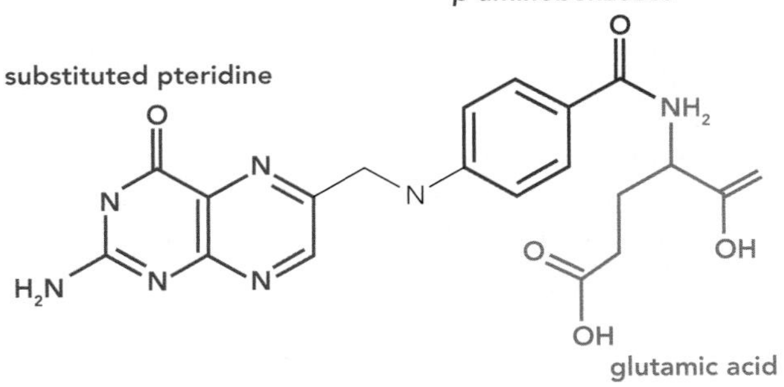

Figure 24.9 Structure of folic acid. Folic acid consists of three regions: a substituted pteridine ring (red), *p*-aminobenzoate (blue), and glutamic acid (green).

to synthesize the pteridine ring and thus this has to come from the diet or from intestinal flora. Little of the parent, fully oxidized folic acid is present in food but more than 35 derivatives are found naturally, mainly as the reduced folates 5,6,7,8-tetrahydrofolate and 7,8-dihydrofolate. Major sources of folates include yeast extract, leafy vegetables such as broccoli and Brussels sprouts, peas, asparagus, chickpeas, and brown rice. Some fruits, particularly oranges and bananas, also contain significant amounts. The parent molecule, pteroyl-glutamic acid, is used pharmaceutically and for fortification of some breads and breakfast cereals.

Structure

As shown in Figure 24.9, the structure of folates contains a pteridine ring linked to the amino group of *p*-aminobenzoate, which is in turn linked via an amide bond to glutamic acid. Nitrogen atoms N5 and N10 provide the active part of the molecule, being involved in one-carbon metabolism.

Function

All the roles of members of the folate family relate to their ability to transfer one-carbon groups (methyl, methenyl, formyl, and formamino, but not CO_2) in biochemical reactions. Thus, they are essential for the synthesis of nucleotides and the amino acid methionine and for replenishing the methyl donor *S*-adenosylmethionine. The essential role of folate in nucleic acid synthesis has been the rationale for the development of drugs which inhibit the metabolism of folate in bacteria (sulfonamides) and cancer cells (methotrexate and 5-fluorouracil).

Deficiency

Deficiency of folate may lead to dramatic consequences, as illustrated by three examples:

- Women of childbearing age are likely to be folate deficient, and deficiency during pregnancy has been associated with increased incidence of neural tube defects (spina bifida) in the fetus
- Defective DNA synthesis in erythropoiesis as a result of folate deficiency manifests as megaloblastic anemia, where erythrocyte precursor cells become enlarged and have underdeveloped nuclei
- Other rapidly proliferating epithelial tissue, such as buccal mucosa and cells of the GI tract, may also present with megaloblastic changes of glossitis and diarrhea, respectively.

Absorption of folate is inhibited by sulfathiazines and diphenylphenytoin and use of these compounds may lead to deficiency of the vitamin.

Clinical chemistry

Folate status may be assessed by its measurement in serum and red blood cells. Both can be performed by high-volume immunoassay analyzers in modern clinical laboratories, although measurement of serum folate is technically easier. Red-cell folate is considered to be a better reflection of tissue stores and is less susceptible to short-term rises following recent folic acid intake.

Folate function is dependent upon adequate levels of vitamin B_{12} being available because B_{12} is required for the demethylation of 5-methyltetrahydrofolate to the active tetrahydrofolate. In cases of vitamin B_{12} deficiency, high doses of folate can bypass this block (the folate trap) and reverse the megaloblastic anemia. It will not, however, improve the deleterious neurological effects of vitamin B_{12} deficiency and indeed may accelerate them. Thus folate administration alone in this situation can precipitate the dreaded complication of subacute combined degeneration of the spinal cord.

Vitamin C (ascorbic acid)

Vitamin C is the trivial name for L-ascorbic acid, a strong reducing agent, and its oxidized derivative dehydroascorbic acid. Most animals are able to synthesize ascorbic acid but, unfortunately, this does not include primates and guinea pigs. Thus an adequate daily supply of vitamin C must be present in the human (and guinea pig) diet. Green leafy vegetables and citrus and soft fruits are good sources of vitamin C. These include Brussels sprouts, broccoli, peppers, sweet potatoes, oranges, lemons, limes, and kiwi fruit. Unfortunately, much vitamin C is destroyed by oxidation during cooking.

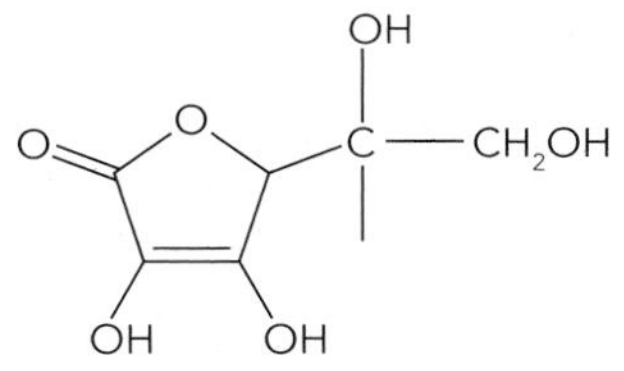

Structure

Ascorbic acid is a six-carbon molecule not dissimilar in structure to glucose (**Figure 24.10**). It undergoes oxidation to dehydroascorbic acid by loss of hydrogen atoms across carbons 2 and 3.

ascorbic acid

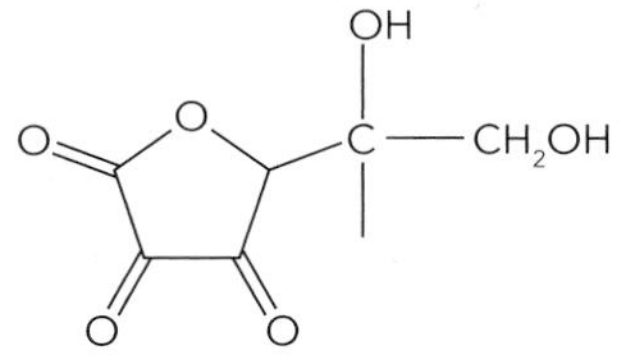

dehydroascorbic acid
(oxidized)

Figure 24.10 Structures of the oxidized and reduced forms of ascorbic acid (vitamin C).

Function

Ascorbic acid is a powerful, water-soluble antioxidant active in cell water and plasma. It is required for the biosynthesis of a number of molecules including collagen, bile acids, carnitine, and norepinephrine. It is also essential for the function of the microsomal P450-linked mixed function oxidase system. Dietary vitamin C increases the rate of absorption of dietary non-heme iron from the gut by reducing it from Fe^{3+} to Fe^{2+}.

Deficiency

Vitamin C deficiency is uncommon in developed countries. Perhaps the best-known presentation of deficiency is scurvy, which is well-described in accounts of long-distance voyages in the eighteenth century. Early symptoms of deficiency include depression, fatigue, and aches and weakness in muscles and joints. More prolonged deficiency, associated particularly with defective synthesis of collagen, presents with petechiae, inflamed gingivae, perifollicular hemorrhage, impaired wound healing, coiled hairs, hyperkeratosis, and defects in ossification and bone growth. Hemorrhage into body cavities may be present in infants.

Ascorbic acid is absorbed actively from the small intestine via a saturable transporter system and is widely distributed in the body, particularly in the adrenal glands, pituitary, and retina. Lesser amounts are present in kidney and muscle. It is metabolized via dehydroascorbic acid and ketogulonic acid to oxalic and threonic acids; unchanged ascorbic acid and its metabolites are excreted in urine.

Clinical chemistry

Measurement of vitamin C may be performed on plasma or white blood cells, the latter being more indicative of tissue levels. Vitamin C may also be measured in urine before and after an oral load. However, measurements are not performed in most routine laboratories and plasma samples must be stabilized rapidly using strong acids, which present a health and safety risk. For these reasons, it is often more straightforward to give vitamin C supplements on the basis of a clinical diagnosis, especially in obvious cases of malnutrition. Vitamin C can be measured using spectrophotometric, colorimetric, enzymatic, capillary electrophoresis, and microdialysis methods. Challenges in measurement involve interference, sample instability, and sensitivity issues. HPLC assay is the mainstay of measurement of vitamin C.

It has been suggested that very high doses of vitamin C are effective for preventing and treating the common cold. Although this is disputed, some people may be convinced to take megadoses of over-the-counter vitamin C. The risks of this are small but do include a greater likelihood of renal stones and excessive iron absorption. These may be significant in individuals with genetic susceptibilities such as hyperoxaluria and hemochromatosis.

Biotin

Biotin (vitamin B_7), sometimes known as vitamin H, is a water-soluble vitamin synthesized by bacteria, including those in the intestine, and primitive eukaryotes such as yeasts, algae, and molds. Biotin synthesized by intestinal bacteria may constitute a high proportion of the daily intake. Although widely distributed in food, biotin is present at a much lower concentration compared to other water-soluble vitamins. However, yeast extracts, red meats including muscle, liver, and kidney, and egg yolks are useful sources. The levels of biotin in vegetables and fruits are much lower than in liver, by as much as a factor of two. Biotin in foodstuffs is usually linked via an amide bond to a lysine residue in dietary proteins and must be hydrolyzed to free biotin in the intestine prior to absorption. This reaction is catalyzed by an intestinal biotinidase.

Structure

Biotin is a bicyclic molecule consisting of an ureido ring fused to a heterocyclic substituted tetrahydrothiophene ring with a valeric acid side chain (**Figure 24.11**).

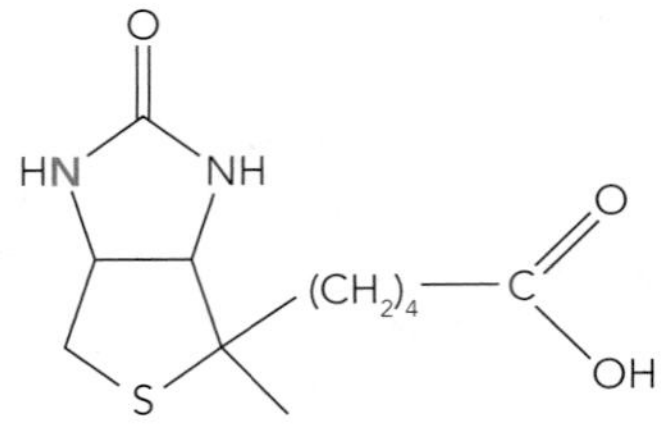

HN NH O $(CH_2)_4$ C O OH S

vitamin B_7
(biotin)

Figure 24.11 Structure of biotin (vitamin B_7).
Biotin is found in nature in an amide linkage to ε-amino groups of lysine residues in proteins. The vitamin participates in carboxylation reactions such as those catalyzed by acetyl-CoA carboxylase and pyruvate carboxylase. The nitrogen, shown in red, in the ureido ring binds a molecule of CO_2, forming a carboxybiotin intermediate, during the carboxylation.

Function

Biotin is a co-factor in a number of carboxylation reactions, being linked covalently via an amide bond to a specific lysine residue on the carboxylase enzyme. A nitrogen atom in the ureido ring is involved in the transfer of CO_2 in the substrate, as shown in Figure 24.11. Thus, it is a co-factor for four carboxylases: (i) acetyl-CoA carboxylase in fatty acid synthesis; (ii) propionyl-CoA carboxylase, a late step in the catabolism of odd-numbered fatty acids; (iii) methylcrotonyl-CoA carboxylase in branched-chain amino acid metabolism; and (iv) pyruvate carboxylase, a key enzyme of gluconeogenesis. It has also been suggested that biotin may exert some action on gene expression via interaction with nuclear histone proteins.

Deficiency

In the absence of a general dietary deficiency of water-soluble vitamins, an isolated deficiency of biotin is rare. However, biotin deficiency has been observed in patients maintained on total parenteral nutrition (TPN) and also in individuals consuming large quantities of raw egg white. Egg white contains a high level of the protein avidin which binds biotin with extremely high affinity rendering it unavailable for biological use. Biotin deficiency has also been reported in patients on hemodialysis and in some subjects undergoing prolonged anticonvulsant therapy.

Apparent biotin deficiency arises from inborn errors of metabolism associated with biotinidase, holocarboxylase synthetase, and the specific carboxylases mentioned earlier and presents clinically in the same manner as true vitamin deficiency.

The clinical symptoms of biotin deficiency are a fine, scaly, desquamating dermatitis; a characteristic rash around the eyes, nose, and mouth; alopecia; changes in mental status; and anorexia.

Clinical chemistry

Assays for biotin have been developed but are not routinely used since isolated biotin deficiency is vanishingly rare. However, an assay for biotinidase is offered by pediatric laboratories to diagnose the inborn errors of metabolism listed above.

Pantothenic acid

Pantothenic acid (vitamin B_5) is a water-soluble molecule present in practically all living cells—animal, plant, and microbial—where it is present as part of a co-factor, coenzyme A. It is found in virtually all foodstuffs and good sources

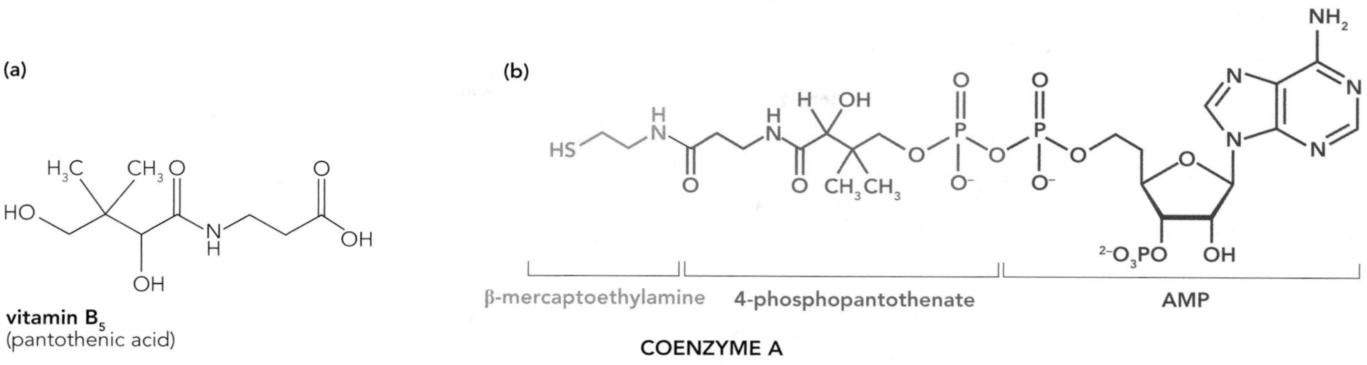

(a)

vitamin B_5
(pantothenic acid)

(b)

β-mercaptoethylamine 4-phosphopantothenate AMP

COENZYME A

Figure 24.12 Structure of pantothenic acid (vitamin B_5). (a) Pantothenic acid is a component of (b) coenzyme A (CoA). In the formation of CoA, pantothenic acid is phosphorylated by ATP to 4-phosphopantothenate, which reacts with cysteine to form 4-phosphopantothenoyl cysteine. Decarboxylation of this molecule yields 4-phosphopantotheine to which AMP is transferred from ATP to form coenzyme A, which has essential roles in both fatty acid synthesis and oxidation.

include beef, chicken, kidney, eggs, broccoli, tomatoes, whole grains, and brown rice. Royal bee jelly and ovaries of tuna and cod are particularly rich in the vitamin. Up to 50% of the vitamin may be destroyed during cooking or processing.

Structure

Pantothenic acid (**Figure 24.12a**) consists of a molecule of pantoic acid linked covalently via an amide bond to β-alanine (3-aminopropionate).

Function

Pantothenic acid is a component of coenzyme A (**Figure 24.12b**), which consists of adenosine monophosphate (AMP) linked via a phosphodiester bond to 4-phosphopantothenate. Coenzyme A is an essential co-factor in the β-oxidation of fatty acids and the syntheses of cholesterol, steroid hormones, and other isoprenoid molecules and some amino acids. It is also involved in the synthesis of δ-aminolevulinic acid, a precursor of the porphyrin rings of heme and cytochromes, and in the acetylation and fatty acylation of proteins.

Deficiency

Although it has been possible to induce pantothenic acid deficiency experimentally by dietary vitamin exclusion, natural deficiency of pantothenic acid has not been described clinically. For this reason there is no clinical necessity for laboratory measurements.

24.3 POLYUNSATURATED FATTY ACIDS (PUFAS)

The structure of fatty acids is described in Chapter 18. Mention was made there of the inability of humans to desaturate a fatty acid chain beyond carbon 9. However, there is a need for PUFAs in order to maintain whole-body homeostasis and thus such fatty acids, known as essential fatty acids, or their precursors have to be provided by the diet. Important dietary sources of PUFAs are vegetable and fish oils. PUFAs are present in mammalian cells as components of membrane phospholipids and, on release, serve as precursors for families of key signaling molecules known as the eicosanoids: prostaglandins, prostacyclins, thromboxanes, and leukotrienes. Arachidonic acid, a fatty acid of 20 carbons and 4 double bonds (see Chapter 18), is the common precursor for these molecules. In the absence of dietary α-linolenic acid, arachidonic acid, like vitamins, is an essential dietary component.

Metabolism of arachidonic acid to eicosanoids

Arachidonic acid is a minor component of membrane phospholipids, being esterified to carbon 2 of the glycerol backbone. The initial stage of its metabolism is its release from the membrane (usually endoplasmic reticulum) by the action of phospholipase A_2. There are a number of tissue phospholipase A_2 enzymes, but the type IV cytosolic phospholipase A_2 ($cPLA_2$) has a high specificity for arachidonic acid. The enzyme is activated by a number of hormones which increase the concentration of intracellular calcium, causing calcium-dependent movement of the enzyme to membranes of the nucleus and endoplasmic reticulum and the subsequent release of arachidonic acid from membrane phospholipids. The $cPLA_2$ is inhibited by a cytosolic calcium- and phospholipid-binding protein, lipocortin, which is activated by steroids, including synthetic steroids (for example betamethasone), and this provides a rationale for the use of steroids as anti-inflammatory agents.

Once released by the action of $cPLA_2$, arachidonic acid becomes the precursor for a number of 20-carbon molecules known collectively as eicosanoids (from the Greek word for "twenty"). Which precise product is formed in any given tissue is dependent on the expression of particular pathways in that tissue (**Figure 24.13**); thus leukotrienes are formed from the 5-lipoxygenase pathway, prostaglandins from the cyclooxygenase pathway, and epoxyeicosatrienoic acids via a cytochrome P450-linked system.

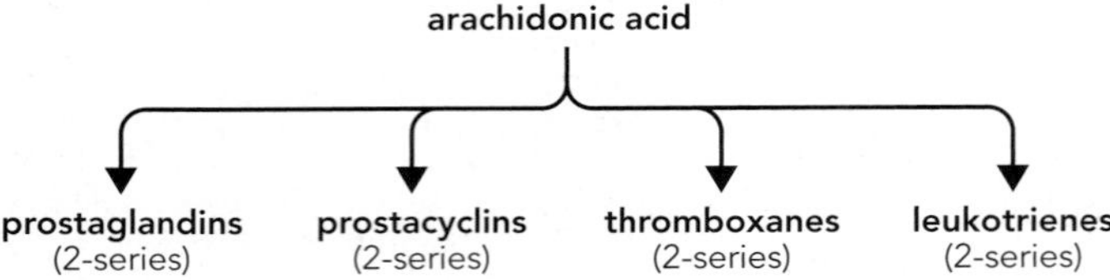

Figure 24.13 Overview of the metabolism of arachidonic acid. Arachidonic acid, also known as eicosatetraenoic acid (C20:$\Delta^{5,8,11,14}$), is a precursor of eicosanoids (20-carbon derivatives).The formation and proposed role of other metabolic products from arachidonic acid, including the hydroxyeicosatetraenoic acids (HETEs), are beyond the scope of the current text. The "2" denotes the number of double bonds in the product.

Cyclooxygenase pathway

This metabolic pathway is responsible for the production of precursors of classical prostaglandins, prostacyclins, and thromboxanes. The products of this pathway were named from the saturated 20-carbon acid prostanoic acid (**Figure 24.14**) which has an additional bond between carbons 8 and 12 to form a five-membered (cyclopentane) ring in the molecule. The prostaglandins and prostacyclins derived from arachidonic acid retain 20 carbons and the five-membered ring.

Cyclooxygenase or prostaglandin G/H synthase

Cyclooxygenases (COXs) are dimeric proteins that sit in the membrane of the endoplasmic reticulum. They have both cyclooxygenase and hydroperoxidase activities that form, sequentially, two short-lived intermediates: first, PGG_2 and then PGH_2 (**Figure 24.15**). The letter after PG (prostaglandin) defines the functional groups on the cyclopentane ring, while the subscript number denotes the number of double bonds in the molecule. Two of the four double bonds of arachidonic acid are used in production of PGH_2. The cyclooxygenase activity of the enzyme is involved in the addition of two molecules of oxygen to the arachidonic acid, leading to the formation of the cyclopentane ring between carbons 8 and 12, and peroxidation at carbon 15. The PGG_2 thus produced has two double bonds less than the substrate. Hydroperoxidase activity of the COX enzyme then converts the hydroperoxy residue at carbon 15 to a hydroxyl group

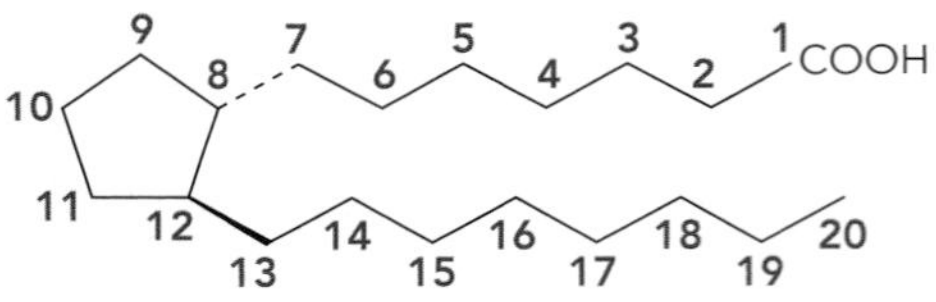

Figure 24.14 The structure of prostanoic acid. The prostaglandins are named from the trivial name for this molecule. Arachidonic acid, the precursor of the prostanoids, is shown in Figure 24.15 and has been drawn to illustrate its structural similarity with prostanoic acid.

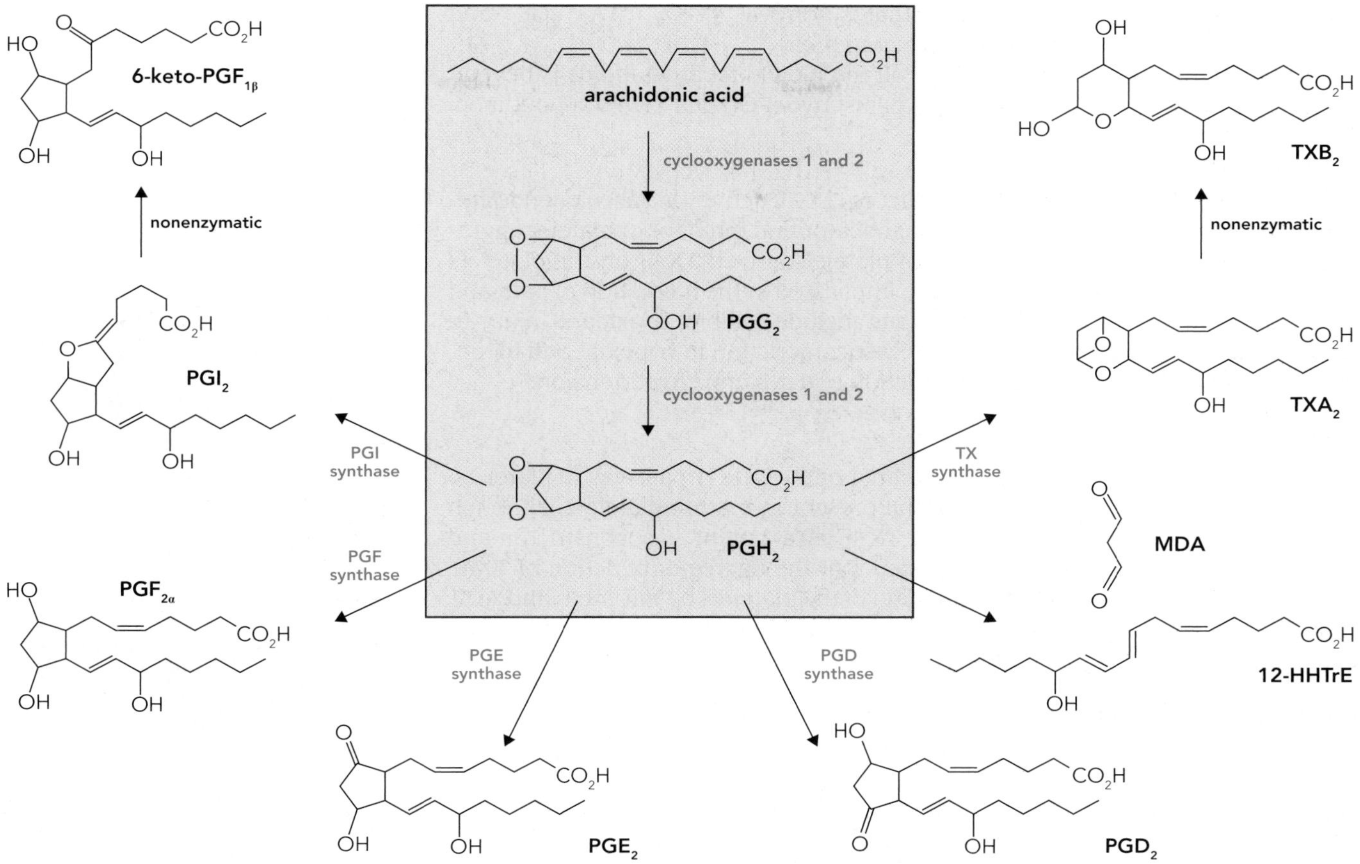

Figure 24.15 Detailed metabolism of arachidonic acid showing the formation and structures of derived prostanoids.
TX, thromboxane; PG, prostaglandin; PGI, prostacyclin; 12-HHTrE, 12-hydroxyheptadecatrienoic acid; MDA, malonyldialdehyde.

to form PGH_2. If the substrate were eicosapentaenoic acid (five double bonds), the products would have been PGG_3 and PGH_3, respectively. Thus the product of COX will have two double bonds less than the substrate fatty acid.

Two COX genes, *COX-1* and *COX-2*, have been described and they code for enzymes which have similar molecular structures. Both have a hydrophobic tunnel into which the substrate polyunsaturated fatty acid can enter and which provides access to the catalytic site. The tunnel is slightly larger in COX-2 and this has implications in the design of drugs which might selectively inhibit the enzymes. The intracellular distribution of the enzymes is also similar (endoplasmic reticulum) but, as mentioned earlier, they are linked to different downstream enzymes, dependent on the tissue-specific expression of these enzymes. For example, thromboxane A_2 is the major product from COX-1 activity in the platelet.

A major difference between COX-1 and COX-2 is their expression; COX-1 is constitutively expressed in most cells while COX-2 is induced by such factors as shear stress, cytokines, and tumor promoters. This suggests that COX-1 is involved in routine, day-to-day maintenance of cellular homeostasis, hemostasis, and cytoprotection (particularly of gastric epithelia), with COX-2 being the major source of prostanoids in inflammation and cancer. However, the finding of both enzymes in synovial tissue of patients with rheumatoid arthritis and also in atherosclerotic plaques suggests that both enzymes may be involved in the inflammatory response.

The products of the cyclooxygenase pathways—prostaglandins, prostacyclins, and thromboxanes—have very short half-lives, on the order of seconds to minutes, implying that they act on cells close to their site of synthesis or even on the cells in which they are synthesized (autocoids). Specific receptors for each class of products have been identified (that is, for PGI_2, $PGF_{2\alpha}$, TXA_2, PGE_2, and PGD_2) and have similar heptahelical structures.

Discussion of the precise actions of all the individual prostanoids is beyond the remit of this chapter, and the effects of just two (PGI_2 and TXA_2) are outlined.

Prostacyclin I_2 (PGI_2)

PGI_2 (see Figure 24.15) is the major product of COX-2 activity in vascular endothelium in healthy subjects. It is a potent vasodilator and inhibitor of platelet aggregation, and thereby limits the effects of platelet-derived TXA_2 on the platelets themselves and the vessel wall. It is also implicated in the mediation of pain and the inflammatory response. Other actions include inhibiting oxidant injury to cardiac myocytes, opposing pulmonary vasoconstriction in hypoxia, and blocking angiotensin II-induced renal constriction and systemic hypertension.

Thromboxane A_2 (TXA_2)

TXA_2 (see Figure 24.15) is the major product of the COX-1 pathway in platelets. COX-1 is the only isoform of the enzyme present in mature platelets, although COX-2 is present in megakaryocytes. TXA_2 is a potent vasoconstrictor and platelet aggregator. It has been suggested that the aggregatory action of TXA_2 occurs subsequent to the initial stimulation of the platelets by thrombin and ADP. COX-2 activity in macrophages also produces mainly TXA_2 and it is thought that this might contribute to atherogenesis.

Inhibition of eicosanoid synthesis

Eicosanoid synthesis can be decreased by inhibiting the $cPLA_2$ which liberates arachidonic acid and also by inhibiting the cyclooxygenases. As mentioned above, lipocortin is an endogenous inhibitor of $cPLA_2$ and glucocorticoids and other synthetic steroids activate lipocortin either directly or by increasing its rate of transcription. This provides a rationale for the use of steroids as anti-inflammatory agents.

Aspirin

Aspirin (acetylsalicylic acid) causes irreversible inactivation of cyclooxygenases by acetylation of a specific serine residue (residue 529) near the active site in the hydrophobic tunnel of the enzyme (**Figure 24.16**). Acetylation of this serine prevents access of the arachidonic acid to the active site. This irreversible process results in termination of COX activity in the platelet since the platelet, having no protein synthetic machinery, cannot replace the inhibited enzyme. Platelet thromboxane synthetic activity can only come from new platelets.

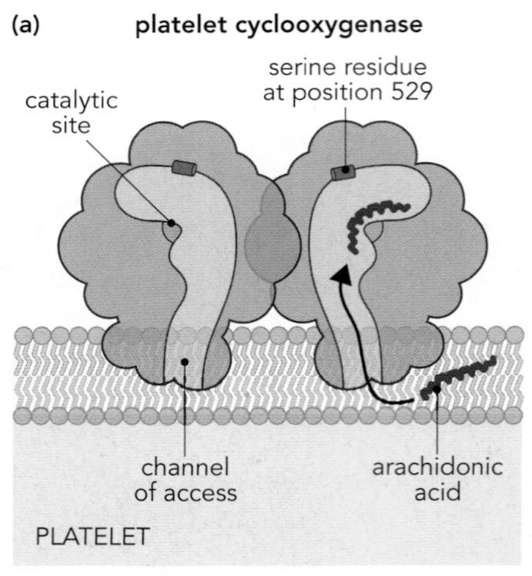

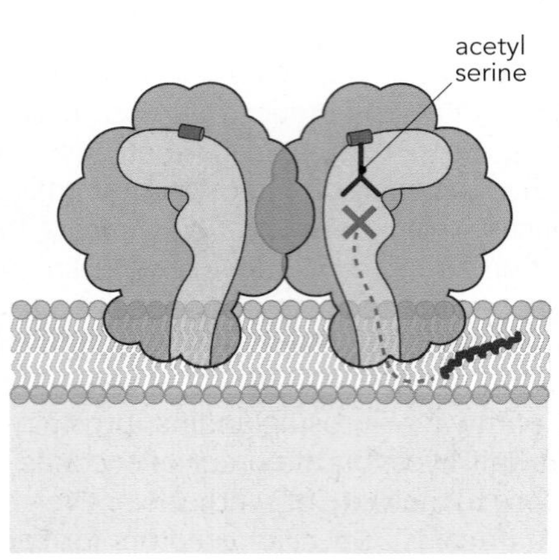

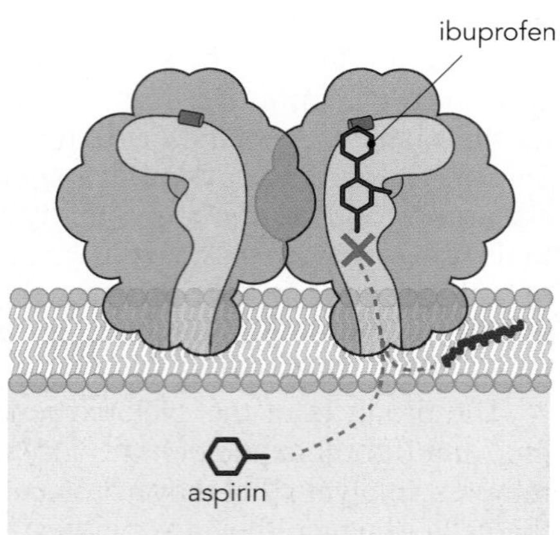

Figure 24.16 The action of platelet cyclooxygenase (prostaglandin G/H 1 synthase) and its inhibition by aspirin and nonsteroidal anti-inflammatory drugs (NSAIDs). In these diagrams the enzyme is shown as a dimer in the platelet membrane. (a) The formation of prostaglandin G_2 requires access of arachidonic acid, released from membrane phospholipids by the action of a phospholipase A_2, to the active site of the enzyme via a hydrophobic channel in the interior of the enzyme. (b) Aspirin (acetylsalicylic acid) enters the hydrophobic core of COX and transacetylates irreversibly a serine residue at position 529 in the polypeptide chain. While this serine residue is not in the active site of the COX, its acetylation prevents access of the arachidonic acid substrate to the active site. Because the inhibition is irreversible, the enzyme cannot be reactivated in the cell. (c) NSAIDs such as ibuprofen enter the hydrophobic channel and bind reversibly at the active site of the enzyme thereby preventing access for the arachidonic acid substrate. The enzyme becomes active once more when drug administration is ceased. Prior binding of ibuprofen to the active site also prevents access of aspirin to its target serine residue and the irreversible inactivation of the enzyme.

Although similar inactivation of COX can occur in the vascular endothelial cells, being nucleate, such cells can replace the inhibited enzyme through protein synthesis. This is the principle of aspirin use in the treatment of unstable angina and stroke. Aspirin is nonselective and inhibits both COX-1 and COX-2. Salicylic acid is only a very mild reversible inhibitor, illustrating the key effect of acetylation.

Nonsteroidal anti-inflammatory drugs (NSAIDs)

The various members of the NSAID family are shown in **Table 24.2**. These hydrophobic molecules are nonselective and sit in the hydrophobic tunnel in both COX-1 and COX-2 (see Figure 24.16). They bind through hydrophobic interactions at the active site. They do not chemically modify the enzyme and their action is reversible, such that when their concentration in the circulation falls, their action decreases. By binding to the enzyme at the active site, NSAIDs prevent access of the substrate polyunsaturated fatty acid to the site and thereby block its metabolism. NSAIDs have anti-inflammatory and analgesic effects probably due to suppression of PGE_2 and PGI_2 synthesis. However, these two prostaglandins also confer cytoprotection to gastric epithelia and inhibition of COX-1 in the GI tract may lead to gastric bleeding. It has also been reported recently that continued high doses of NSAIDs may contribute to cardiovascular problems.

TABLE 24.2 Nonsteroidal anti-inflammatory drugs (NSAIDs): inhibitors of arachidonic acid metabolism

NSAID class	Examples
Arylpropionic acids	Ibuprofen, naproxen
Indole acetic acids	Indomethacin
Heteroaryl acetic acids	Diclofenac
Enolic acids	Piroxicam
	Phenylbutazone
Alkanones	Nabumetone

COXIBs

The COXIBs are selective inhibitors of COX-2, the inducible COX. They were designed based on knowledge of the slight differences in molecular structure between COX-1 and COX-2, particularly in the size of the hydrophobic tunnel. COX-2 is a major source of prostaglandins involved in the mediation of pain and inflammation. COXIBs are effective anti-inflammatory and analgesic agents, similar to NSAIDs, but, on a dose-for-dose basis, they cause less gastric ulceration. However, some have been withdrawn due to adverse cardiovascular effects associated with their use.

Leukotrienes (LTs) and the 5-lipoxygenase pathway

The synthesis of leukotrienes represents an alternative fate of arachidonic acid. Again, production of LTs depends on the particular complement of downstream enzymes expressed in an individual tissue. LTs are produced mainly by inflammatory cells such as mast cells, neutrophils, and macrophages. The synthesis of LTA_4 requires the release of arachidonic acid from phospholipids in the nuclear membrane by $cPLA_2$ and the actions of a 5-lipoxygenase and a 5-lipoxygenase-activating protein (FLAP). LTB_4 is produced from LTA_4 by the action of a specific hydrolase.

Alveolar macrophages, eosinophils, and mast cells express enzymes for the synthesis of the cysteine-linked leukotrienes: LTC_4, LTD_4, and LTE_4. This requires conjugation of LTA_4 with glutathione and further metabolism.

Leukotrienes are important mediators of the asthmatic response and the use of inhaled steroids in treating asthma (betamethasone, for example) causes direct inhibition of $cPLA_2$ activity in the inflammatory cells responsible for the condition, thereby preventing supply of arachidonic acid to the 5-lipoxygenase pathway.

24.4 TRACE ELEMENTS

A small number of elements are present in the body in trace amounts yet play an essential role in homeostasis. These elements, which include metals such as chromium, copper, molybdenum, manganese, selenium, and zinc and halogens such as iodine and fluorine, are found in soil and rocks. They become available in the diet when accumulated into plants and also occur in meat products derived from animals feeding on plants. In the case of trace metals, the elements may be present in their various cationic forms as well as in anions such as chromate, molybdate, manganate, and selenate. Selenium is also found in amino acid complexes, particularly selenomethionine and selenocysteine. In addition, iodine as iodide is present in seawater and may enter the food chain via plants grown near the sea, while fluorine, as fluoride, is added to drinking water (and toothpaste) in many countries. The roles and recommended allowances for the trace elements are shown in **Table 24.3**.

TABLE 24.3 Recommended daily amounts and functions of some trace elements

Element	RDA	Function
Chromium	35 μg (M) 25 μg (F)	Potentiates insulin action on carbohydrate, fat, and amino acid metabolism
Manganese	2.3 mg (M) 1.8 mg (F)	Co-factor in metalloenzymes, for example superoxide dismutase in mitochondria, glycosyl transferases
Molybdenum	45 μg	Co-factor in metalloenzymes, for example oxidoreductases (xanthine oxidase, sulfite oxidase); DNA synthesis and repair
Selenium	55 μg	Immune system; tetra-iodothyronine diodinase ($T_4 \rightarrow T_3$); antioxidant (superoxide dismutase) with vitamin E prevents oxidative and free radical damage; thioredoxin reductase
Iodine	150 μg	Component of thyroid hormones
Fluorine (children)	150 μg	Protects against dental caries

M, male; F, female; T_4, tetra-iodothyronine (thyroxine); T_3, tri-iodothyronine.

An adequate daily supply of trace elements is found in a normal balanced diet of cereals, fruit, vegetables, eggs, fish, and meat. Apart from copper, zinc, and iodine, in most cases symptoms of deficiency have not been described and reference ranges have not been established; laboratory measurements are thus not really meaningful clinically.

The role of iodine in the biosynthesis of thyroid hormones has been described in Chapter 11 and the clinical chemistry of only copper and zinc are described here.

Trace elements can be measured in biological fluids to diagnose deficiency or excess. High levels may come from iatrogenic or industrial sources. Not all trace elements are routinely measured in the clinical laboratory, however, since the concentrations of some elements are too low to be assayed reliably. These also tend to be the elements where deficiency almost never occurs. The major issue with trace element analysis is environmental sample contamination which may come from the skin, venipuncture needles, or the sample tubes.

Copper

Approximately 70 mg (1 mmol) of copper is found in adult humans and is stored mainly in liver, heart, kidneys, and brain. Copper is excreted from the body via the bile into the gut. Its main role is as a co-factor for a number of enzymes including:

- Ferrochelatase
- Monoamine oxidase
- Lysyl oxidase
- Cytochrome *c* oxidase
- Dopamine β-hydroxylase
- Superoxide dismutase

Approximately 80% of the copper in blood is found in erythrocytes. Of the copper present in serum (10–25 μmol/L), 95% is bound to ceruloplasmin, a copper transporter protein synthesized in the liver (see Chapter 3), while the remainder is ultrafilterable and bound mostly to amino acids.

Copper deficiency

Copper deficiency may be acquired or be due to a severe absorptive defect such as in Menkes' disease. Acquired copper deficiency occurs mainly in patients with intestinal disease and occasionally in patients on TPN. It is manifest in these cases by iron-unresponsive hypochromic, microcytic anemia. Menkes' disease is an X-linked recessive disorder presenting as postnatal copper deficiency secondary to a severe absorptive defect. Again it is manifest by iron-unresponsive hypochromic, microcytic anemia, and kinky hair, mental retardation, and early death, especially from infection. Very low concentrations of copper and ceruloplasmin in serum and high urinary copper excretion provide a laboratory diagnosis.

Wilson's disease is an autosomal recessive disorder characterized by chronic copper toxicity, with accumulation of the metal arising secondarily to impaired secretion. Deposition in tissues leads to a variety of manifestations of the disease as detailed in **Table 24.4**. In childhood, the condition presents with liver disease. Neurological manifestations do not occur prior to adolescence. Typical laboratory diagnosis of Wilson's disease is shown in **Table 24.5**. In addition, there would be evidence of Fanconi syndrome, including:

- Generalized aminoaciduria
- Renal glycosuria
- Hypophosphatemia
- Hypouricemia

Wilson's disease is usually treated with oral penicillamine and complications include rash, nephritic syndrome, and aplastic anemia.

TABLE 24.4 Clinical sequelae of copper deposition in different tissues in Wilson's disease

Tissue	Manifestation
Brain (basal ganglion lesions)	Tremor, rigidity, ataxia
Liver	Hepatitis, cirrhosis
Kidneys	Fanconi syndrome
Blood	Hemolytic crises

TABLE 24.5 Typical laboratory diagnosis of Wilson's disease

Analyte	Finding			
	SI units	Reference range	Conventional units	Reference range
Serum ceruloplasmin	<100 mg/L	200–400 mg/L	<10 mg/dL	20–40 mg/dL
Serum copper	Variable	11–22 μmol/L	Variable	70–140 μg/dL
Urine copper	39 μmol/L	0.2–0.9 μmol/L	250 μg/L	15–60 μg/L
Liver copper	>15 μmol/g dry weight	<0.5 μmol/g dry weight	>1000 μg/g dry weight	<35 μg/g dry weight

Zinc

The adult human body contains approximately 2 g (30 mmol) of zinc stored mainly in prostate, skin, liver, and pancreas. Its major role is as a co-factor for enzymes such as alkaline phosphatase, several dehydrogenases, and RNA and DNA polymerases. Approximately 80% of the zinc in blood is found in erythrocytes, while 90% of that in serum (10–18 μmol/L) is protein bound, mainly to albumin and to a lesser degree to α_2-macroglobulin.

Zinc deficiency

The effects of severe deficiency of zinc include:

- Diarrhea
- Rash
- Irritability and depression
- Loss of sensation of taste
- Impaired wound healing
- Male hypogonadism; decreased serum testosterone, increased serum follicle-stimulating hormone, luteinizing hormone, and prolactin

Chronic deficiency may also be associated with other causes of nutritional failure, such as dwarfism, and anemia. A diagnosis of zinc deficiency is based on finding a low serum zinc concentration, allowing for serum albumin concentration.

The major causes of zinc deficiency are:

- Inadequate intake: anorexia, deficient diet, TPN
- Impaired absorption: intestinal disease, binding to dietary phytate, acrodermatitis enteropathica, eating clay (particularly in younger children)
- Excessive loss: gut disease, TPN where zinc is lost into urine bound to amino acids

Analytical practice point 24.1

Zinc contamination is often a problem with standard blood tubes. Falsely raised results can be obvious, but there is a risk of missing true deficiency unless tubes certified for trace metal analysis are used.

CLINICAL CHEMISTRY IN THE ACUTELY ILL PATIENT

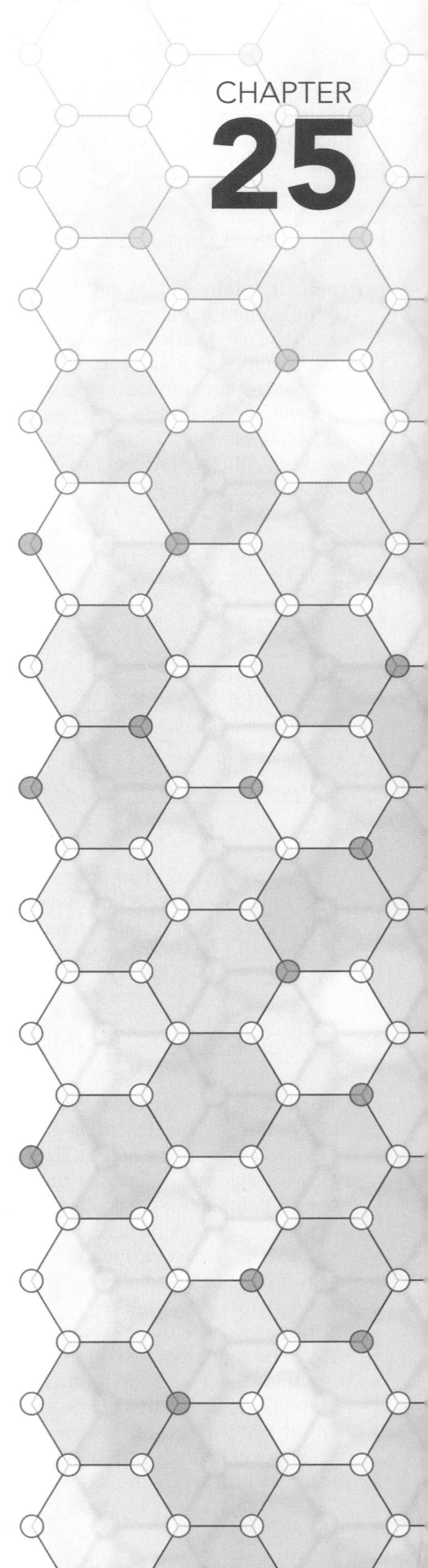

IN THIS CHAPTER

When patients present to the emergency department with acute illness, clinical chemistry investigations play an important role in their assessment and management. Presenting conditions may be primarily metabolic with disturbed biochemistry as the predominant feature or, more commonly, may be disorders where biochemical changes are secondary to another pathology. Examples of the first category would include hypoglycemia and electrolyte disorders, whilst examples of the second category would include myocardial infarction and acute pancreatitis. Common disorders which present as acute severe illness and where clinical chemistry plays an important role are listed below.

- Acute abdominal pain
- Overdoses and poisoning
- Acute chest pain
- Altered consciousness
- Shortness of breath (dyspnea)
- Diabetic emergencies
- Acute kidney injury
- Sepsis

However, in all cases, it must be emphasized that the most important part of the assessment starts with a good clinical history and thorough examination of the patient. The choice of investigations, which include blood tests and imaging, is directed from this initial assessment.

Clinical practice point 25.1

Clinical chemistry tests may be used:

To diagnose predominantly metabolic disorders, for example diabetic ketoacidosis

To diagnose non-metabolic conditions with specific biomarkers, for example myocardial infarction

To monitor secondary metabolic disturbances, for example renal failure

25.1 ACUTE ABDOMINAL PAIN

There are many causes of the acute abdomen, most of which are considered to be the dominion of the surgeon, including appendicitis, cholecystitis, and peptic ulcer disease. In such cases, the role of clinical chemistry tests lies in helping to assess fluid losses and renal function by measurement of urea, creatinine, and electrolytes. Liver function tests may point to cholestasis. In other cases of acute abdominal pain (**Table 25.1**) more specific tests may be appropriate. For example, acute pancreatitis is diagnosed by elevated levels of amylase or lipase and biochemical tests are used to decide on severity and prognosis using validated scoring systems.

TABLE 25.1 Biochemical tests in the diagnosis of acute abdominal pain

Cause	Biochemical test
Acute pancreatitis	Diagnosis: amylase and/or lipase Severity: PO_2, glucose, urea, albumin, calcium Cause: calcium, triglycerides
Acute porphyria	Urine porphobilinogen and porphyrins
Diabetic ketoacidosis	Blood glucose, blood/urine ketones
Hereditary angioedema	C1 esterase inhibitor
Lead toxicity	Blood lead

Less commonly, abdominal pain may be due to metabolic diseases and these must be identified to avoid the patient undergoing unnecessary surgery. Such causes include diabetic ketoacidosis, acute porphyria, lead poisoning, and C1 esterase inhibitor deficiency. Each of these is described in detail in earlier chapters.

25.2 OVERDOSES AND POISONING

One of the most common presentations to the emergency department is deliberate overdose of medicines as either an attempt at suicide or, more commonly, as a so-called "cry for help" following failure to cope with emotional stress. The type of drugs taken depends on their availability and most often are over-the-counter painkillers, particularly acetaminophen. Less often, patients may take overdoses of drugs prescribed to them or members of their household. Most of these are not toxic. In other cases, patients may take non-drug chemicals, such as antifreeze or weedkiller, or poisonous plants. These cases may include accidental ingestion by children.

In many cases there is no role for the clinical laboratory since the ingested substance is nontoxic. In others, the laboratory may provide basic investigations to monitor liver and renal function and blood glucose where the substance is potentially toxic but there is no specific antidote. In a smaller, well-defined group of poisoned patients the laboratory may measure levels of the toxic substance in order to decide on the use of antidotes, monitor progress, and predict prognosis. Examples of some of the latter group are described in greater detail below.

Examples of drugs and other substances where biochemical measurement is of value in the treatment of overdose include:

- Acetaminophen
- Salicylates
- Cardiac glycosides
- Iron
- Ethylene glycol (antifreeze)
- Lithium
- Paraquat
- Theophylline

Acetaminophen

Overdose of acetaminophen is an extremely important and common cause of poisoning. Although only a small proportion of patients have serious toxicity, it can cause acute liver failure, which may be fatal or require an immediate organ transplant. Liver toxicity does not usually develop for 2–3 days and patients may not, therefore, realize how severely ill they may become. Acetaminophen levels measured 4 hours or more after a single overdose can predict the severity of the poisoning and are used to determine the need for the antidotes *N*-acetylcysteine or methionine. A nomogram (see Figure 22.2) is used with treatment lines for normal and high-risk patients. Risk factors include chronic alcoholism and the use of anticonvulsant drugs and rifampicin. All of these induce liver enzymes and can accelerate the formation of the toxic metabolite of acetaminophen. It is standard practice to measure acetaminophen levels on all poisoned patients, irrespective of their history, since it is potentially fatal, can be readily treated, and overdose patients often take multiple substances together.

Salicylates

Poisoning with acetylsalicylic acid (aspirin) has declined in recent years but remains relatively common. Other sources of salicylates are teething gels and ointments. Overdose causes a metabolic acidosis with a raised anion gap as well as a respiratory alkalosis, due to stimulation of the respiratory center. Other biochemical effects include hypoglycemia and hypo- or hyperkalemia. There is no direct correlation between the plasma salicylate concentration and clinical effects, but levels are used to guide treatment with bicarbonate to alkalinize the urine and keep the acid dissociated. Very high levels are an indication for hemodialysis.

Cardiac glycosides

The most commonly prescribed cardiac glycoside is digoxin, which is used to treat cardiac arrhythmias and heart failure. It is relatively toxic in overdose, with a high mortality rate. The drugs inhibit the sodium–potassium pump (Na^+/K^+-ATPase), causing hyperkalemia and a metabolic acidosis. As well as use of supportive treatment and cardiac drugs that can counter the effects of digoxin, there is a specific antidote available in the form of antibody fragments (Fab) directed against digoxin. Serum digoxin levels are used to confirm the diagnosis and also to calculate the required dose of the antidote. As the Fab interferes in the immunoassay for digoxin, further levels after its administration cannot be reliably measured.

Analytical practice point 25.1

Plasma digoxin levels cannot be reliably measured after the administration of antidote Fab (antibody fractions).

Iron

Poisoning with iron salts is potentially serious as iron is very toxic at high concentrations. It has a direct irritant effect on the upper gastrointestinal tract and is also systemically toxic to the liver, heart, and vascular system. Serum iron levels are used to determine the need for treatment with the antidote desferrioxamine, which chelates iron.

Ethylene glycol

Poisoning with ethylene glycol (the major component in antifreeze) is discussed in detail in Chapter 7. Measurement of ethylene glycol levels is required to decide on appropriate treatment with ethanol, fomepizole, or hemodialysis.

Lithium

Lithium is a commonly prescribed mood-stabilizing drug used in psychiatry to treat depression and bipolar disorder. It has a narrow therapeutic window and is potentially toxic in overdose. More commonly, poisoning occurs due to gradual overdosage rather than a single overdose. Treatment may require hemodialysis to remove lithium ions, but the toxicity is related to intracellular concentrations and several dialysis sessions may be needed. Between these sessions, lithium may rise as it leaves the intracellular space and enters the extracellular fluid.

Paraquat

Paraquat is one of the most widely used herbicides in the world, although it was banned by the European Union in 2007 and its use is restricted to licensed individuals in the USA. It is highly toxic and even as little as 10 mL may be fatal, primarily through lung injury and fibrosis, although death may take several days to occur. Once absorbed, there is no antidote. There are qualitative urine tests and quantitative plasma assays available, but their use is only to confirm poisoning and prognosis, not to direct treatment.

Theophylline

Theophylline is a phosphodiesterase inhibitor used as a long-acting bronchodilator in the treatment of asthma. Toxicity may occur with either unintended chronic overdosage or deliberate massive overdose. In high concentration it stimulates the sodium–potassium pump (Na^+/K^+-ATPase) and the release of insulin and catecholamines, with the effect of causing acute hypokalemia. The major clinical effects are cardiac arrhythmias, hypotension, seizures, and sometimes rhabdomyolysis. Measurement of plasma theophylline concentration can be helpful in confirming the diagnosis, but treatment is directed toward correcting potassium, arrhythmias, and seizures, rather than a specific antidote. However, it is important to measure plasma levels in patients known to be on theophylline before giving intravenous doses in acute asthma, to avoid iatrogenic overdose.

25.3 ACUTE CHEST PAIN

The onset of acute chest pain is always taken seriously as a potential heart attack, and increased awareness in recent years amongst the general population has resulted in earlier presentation. Whilst this allows early intervention with thrombolytic drugs and angioplasty in those with coronary artery occlusion, it also results in many patients being kept in hospital for serial electrocardiograms (ECGs) and blood tests. The early diagnosis and the treatment of heart attack have evolved together and clinical chemistry plays an important role in the rule-in and rule-out strategies used in emergency care. Terminology has also evolved to encompass the concept of myocardial injury, which would not have been detectable using insensitive tests in the past.

Patients presenting with chest pain are assessed by their history, ECG, and biochemical markers, the most important in terms of cardiac specificity being troponin. Both troponin T and I appear to be equally useful, the choice being determined by the manufacturer of the analytical platform used. Other biochemical markers which may be used are the MB isoenzyme of creatine kinase (CKMB) and myoglobin. These are released from injured myocardium more quickly than troponins and may be used as an early rule-out test, allowing patients with a normal ECG and an unconvincing history to be rapidly discharged from hospital. Elevated troponin indicates myocardial damage, but does not indicate the cause. Many illnesses are associated with acute and chronic release of troponin, not just ischemia from occlusion of the coronary arteries, and serial measurement of troponin over several hours is used to help differentiate the possible causes.

Non-cardiac causes of chest pain include pulmonary embolism, pneumonia, esophageal spasm, and musculoskeletal disease.

25.4 ALTERED CONSCIOUSNESS

A reduced conscious level, including acute confusion, has many potential causes, with the commonest metabolic cause being hypoglycemia. This most often occurs in previously known diabetics who may deliberately or inadvertently take too much insulin or sulfonylurea drug. Hypoglycemia occurring in a non-diabetic individual raises the possibility of either misuse of insulin (by the patient or their carer) or an insulin-secreting tumor of the pancreas (insulinoma). Samples taken at the time of hypoglycemia for measurement of insulin and C-peptide are invaluable to distinguish between endogenous and exogenous sources.

Acute hyper- or hyponatremia may cause changes in brain cell volume leading to symptoms of confusion and altered mental state. In severe acute hyponatremia, coma may occur. This is rarely seen unless a large amount of water has been consumed in a short time or, in the more common scenario, an excess of hypotonic fluid has been given intravenously.

25.5 SHORTNESS OF BREATH

Patients presenting with difficulty breathing may require assessment with arterial blood gas analysis to determine their oxygenation and acid–base status. The condition may vary from a respiratory alkalosis due to anxiety-induced hyperventilation to metabolic acidosis due to lactic acid formation in hypoxic tissues. Blood gases play an important role in assessing the severity of respiratory emergencies. For example, a rapid rise in PCO_2 in acute asthma is an indication of inadequate gas exchange and the need for mechanical ventilation.

25.6 DIABETIC EMERGENCIES

There are three major diabetic metabolic emergencies: hypoglycemia, diabetic ketoacidosis (DKA), and hyperglycemic hyperosmolar state (HHS), previously called hyperosmolar non-ketotic state (HONK). Hypoglycemia has already been discussed above. In hyperglycemic emergencies there is significant fluid loss due to osmotic diuresis, and hypernatremia is common in HHS. Clinical chemistry plays a central role in the diagnosis of these states and in the monitoring of treatment. Full descriptions of these conditions are given in Chapter 2 and Chapter 5.

25.7 ACUTE KIDNEY INJURY

In recent years, hospital mortality surveys have shown the prognostic importance of even small rises in plasma creatinine during admission. This has led to the concept of acute kidney injury being detected by serial measurement of

TABLE 25.2 Acute kidney injury (AKI) scoring

AKI score	Defining features
Stage 1: Risk	GFR decrease >25%, serum creatinine increased 1.5 times OR urine production of <0.5 mL/kg/hr for 6 hours
Stage 2: Injury	GFR decrease >50%, doubling of creatinine OR urine production <0.5 mL/kg/hr for 12 hours
Stage 3: Failure	GFR decrease >75%, tripling of creatinine, or creatinine >354 μmol/L (>4 mg/dL) [with a rise of >44 μmol/L (>0.5 mg/dL)] OR urine output below 0.3 mL/kg/hr for 24 hours

plasma creatinine and scored as acute kidney injury (AKI) 1 to 3, with 3 being the most severe (**Table 25.2**). AKI has multiple causes, including reduced renal blood flow in dehydration, infection, and drug toxicity. The clinical laboratory plays a key role in the identification of AKI and alerting of the medical team when it is present. This is because previous creatinine results may already be in the laboratory database and can be used as a baseline. What may appear, at the clinician's first glance, to be an unremarkable creatinine result takes on new significance when it is realized to be a threefold increase from a week previously, as it would indicate AKI stage 3.

25.8 SEPSIS

The modern concept of sepsis is that of an extreme inflammatory response to systemic bacterial infection. The host response is termed systemic inflammatory response syndrome (SIRS). This may also occur as a response to major non-infectious illnesses such as trauma, poisoning, acute pancreatitis, and burns. Sepsis is graded as severe if organ dysfunction or systemic hypoperfusion is present. The term septic shock is used when tissue perfusion remains low despite many liters of intravenous fluid therapy. The scoring systems for SIRS and sepsis use several laboratory parameters, including blood glucose, PCO_2, white blood cell count, bilirubin, alanine transaminase (ALT), lactate, and creatinine.

The definitions used in cases of infection are as follows.

- SIRS: two or more of:
  - Temperature >38°C or <36°C
  - Heart rate >90 beats/min
  - Respiratory rate >20 breaths/min or PCO_2 <32 mmHg
  - White blood cell count >12,000/mm^3, <4000/mm^3, or >10% immature (band) forms
- Sepsis: SIRS plus evidence of infection
- Severe sepsis: sepsis plus evidence of organ hypoperfusion:
  - General: lactic acidosis
  - Kidney: oliguria or rise in serum creatinine
  - Brain: mental changes
  - Liver: raised bilirubin or ALT
  - Heart: raised troponin
- Septic shock: severe sepsis despite adequate fluid resuscitation

Lactate is used clinically as a marker of poor tissue perfusion, since reduced delivery of oxygenated blood results in anaerobic metabolism and an increase in plasma lactate. Lactate may be measured in the laboratory or at the bedside using a specific electrode in a blood gas analyzer.

25.9 POINT-OF-CARE TESTING

In the case of the acutely ill patient, the more quickly that test results are available, the more rapidly the clinician can make important decisions about diagnosis, treatment, and discharge. In extremely busy emergency departments, the ability to discharge a patient quickly can be as important to the efficient running of the hospital as it is to the care of the patient. In order to reduce the turnaround time of laboratory tests—which includes transport to a central laboratory, centrifugation, analysis, and reporting—it may be appropriate to do these tests at the bedside using point-of-care or near-patient devices. A number of such tests have been made commercially available in the last decade and are capable of performing as well as their laboratory counterparts in terms of accuracy and precision. For many years, emergency departments have had meters to measure blood glucose and blood gas analyzers. The latter may have additional electrodes to measure sodium, potassium, chloride, ionized calcium, and lactate. More recently, tests for cardiac markers, particularly troponin, and D-dimers have been used to rapidly rule out myocardial infarction and pulmonary embolism. Other tests are finding their way from the laboratory to the bedside.

The major downside of point-of-care testing (POCT) is that of ensuring all of the potential operators—nurses, doctors, and support staff—are properly trained and that the equipment is adequately maintained, calibrated, and subjected to quality control procedures. Unless there is strict supervision, generally by the local clinical laboratory, POCT equipment can become unreliable or be used inappropriately in patients with clinical contraindications. For example, capillary blood glucose is unreliable in patients with poor peripheral circulation, such as in shock. It is, therefore, very important before introducing POCT that the reduced turnaround time is balanced against the clinical risk of tests being done by non-laboratory personnel. The additional cost per test also needs to be taken into account. Finally, there must be a full audit trail which allows any POCT result to be traced back to the operator, the analyzer, and the batch number of the reagent strip used to generate it. There have been several instances in recent years of manufacturers or regulators withdrawing batches of reagent strips, POCT analyzers, or entire tests due to unacceptable performance. These issues usually relate to manufacturing defects. It is vitally important to be able to identify individual patients who might have been affected by such recalls, hence the need for a complete audit trail for every POCT analysis.

Analytical practice point 25.2

Point-of-care testing offers the advantage of more rapidly available results, but must be properly managed, ideally by the clinical laboratory.

ABBREVIATIONS

Abbreviation	Definition
(25-OH)D 1α-hydroxylase	25-hydroxy vitamin D_3 1α-hydroxylase
α-MSC	α-melanocyte-stimulating hormone
$1,25(OH)_2D$	1,25-dihydroxy vitamin D
16-OH DHEAS	16-hydroxy-dehydroepiandrosterone sulfate
17-OHP	17-hydroxyprogesterone
2,8-DHA	2,8-dihydroxyadenine
25-OHD	25-hydroxy vitamin D
5-HIAA	5-hydroxyindole acetic acid
AA	amino acids
AAT	α_1-antitrypsin
ACAT	acylcoenzyme A:cholesterol acyl transferase
ACE	angiotensin-converting enzyme
acetyl-CoA	acetyl-coenzyme A
ACh	acetylcholine
ACP	acyl carrier protein
ACP	acid phosphatase
ACR	albumin–creatinine ratio
ACS	acute coronary syndrome
ACTH	adrenocorticotropic hormone
AD	alcohol dehydrogenase
ADA	adenosine deaminase
ADH	antidiuretic hormone
ADHR	autosomal dominant hypophosphatemic rickets
ADME	absorption, distribution, metabolism, excretion
ADP	adenosine diphosphate
ADP	aminolevulinic acid dehydratase deficiency porphyria
AFP	alpha-fetoprotein
AGE	advanced glycation end product
AIDS	acquired immune deficiency syndrome
AIP	acute intermittent porphyria
AKI	acute kidney injury
ALA	5-aminolevulinic acid
ALAD	aminolevulinic acid dehydratase
ALDH	aldehyde dehydrogenase
ALP	alkaline phosphatase

ALT	alanine transaminase
AMP	adenosine monophosphate
ANP	atrial natriuretic peptide
APC	adenomatous polyposis coli
apoB	apolipoprotein B
APRT	adenine phosphoribosyl transferase
APRTD	adenine phosphoribosyl transferase deficiency
ARB	angiotensin receptor blocker
ARG	arginase
ARHR	autosomal recessive hypophosphatemic rickets
ASL	argininosuccinate lyase
ASS	argininosuccinate synthetase
AST	aspartate transaminase
ATP	adenosine triphosphate
ATPase	adenosine triphosphatase
AVP	arginine vasopressin
BH_4	tetrahydrobiopterin
BNP	brain natriuretic peptide
bone GLA protein	bone gamma-carboxyglutamic acid protein
BP	blood pressure
BPG	bisphosphoglycerate (aka DPG)
BPH	benign prostatic hypertrophy
C1-INH	C1 esterase inhibitor
CAH	congenital adrenal hyperplasia
cAMP	cyclic adenosine monophosphate
CASR	*calcium sensing receptor* (gene)
CAT	computerized axial tomographic
CBFA1	core-binding factor, Runt domain, alpha subunit 1 (gene)
CCK	cholecystokinin
CD	carbohydrates
CEA	carcinoembryonic antigen
CEH	cholesteryl ester hydrolase
CEP	congenital erythropoietic porphyria
CER	ceramide
CETP	cholesteryl ester transfer protein
CF	cystic fibrosis
CFTR	cystic fibrosis transmembrane (conductance) regulator
CHD	coronary heart disease
CK	creatine kinase
CKD	chronic kidney disease
CLIA	Clinical Laboratory Improvement Amendments
CM	chylomicron
CMOAT	Canalicular Multispecific Organic Anion Transporter
CMR	chylomicron remnant
CO	carbon monoxide

CO	cardiac output
CO_2	carbon dioxide
CoA	coenzyme A
COMA	Committee on Medical Aspects of Food and Nutrition Policy
COMT	catechol-*O*-methyl transferase
COX	cyclooxygenase
CPA	Clinical Pathology Accreditation
CPK	creatine phosphokinase
$cPLA_2$	cytosolic phospholipase A_2
CPM	central pontine myelinolysis
CPO	coproporphyrinogen oxidase
CPSI	carbamoyl phosphate synthase I
CPT	carnitine palmitoyl transferase
CRH	corticotropin-releasing hormone
CRP	C-reactive protein
CSF	cerebrospinal fluid
cTnI	cardiac troponin I
cTnT	cardiac troponin T
Cv	coefficient of variation
CVS	chorion villus sampling
CYP	cytochromes P450
CZE	capillary zone electrophoresis
DβH	dopamine-β-hydroxylase
DAF	decay-accelerating factor
dATP	deoxyadenosine triphosphate
DBP	diastolic blood pressure
DCCT	Diabetes Control and Complications Trial
DCT	distal convoluted tubule
Dcyt*b*	duodenal cytochrome *c*
DHA	docosahexaenoic acid
DHEA	dehydroepiandrosterone
DHEA-S	dehydroepiandrosterone sulfate
DI	diabetes insipidus
DIT	di-iodotyrosine
DKA	diabetic ketoacidosis
DMD	dystrophin (gene)
DMP1	dentin matrix acidic phosphoprotein 1 (gene)
DMT1	divalent metal (ion) transporter 1
DNA	deoxyribonucleic acid
DOPA	dihydroxyphenylalanine
DPG	diphosphoglycerate(aka BPG)
eCCr	estimated creatinine clearance
ECF	extracellular fluid
ECG	electrocardiogram
EDTA	ethylenediaminetetraacetic acid

eGFR	estimated glomerular filtration rate
EMA	endomysial antibody
EPA	eicosapentaenoic acid
EPO	erythropoietin
EPP	erythropoietic protoporphyria
EQA	external quality assurance
EQC	external quality control
ESR	erythrocyte sedimentation rate
F	bioavailability (of a drug)
FA	fatty acid
Fab	antibody fragments
FAD	flavin adenine dinucleotide
$FADH_2$	flavin adenine dinucleotide (reduced)
FAO	fatty acid oxidation
FASCoA	fatty acyl coenzyme A
FBC	full blood count
FDA	Food and Drug Administration
FGF	fibroblast growth factor
FH	familial hypercholesterolemia
FHH	familial hypocalciuric hypercalcemia
FLAP	5-lipoxygenase-activating protein
FMN	flavin mononucleotide
FOB	fecal occult blood
FSH	follicle-stimulating hormone
fT_3	free T_3
fT_4	free T_4
G6PD	glucose 6-phosphate dehydrogenase
GABA	gamma-aminobutyric acid
GAG	glycosaminoglycan
GC	gas chromatography
GC-MS	gas chromatography–mass spectrometry
GERD	gastro-esophageal reflux disease
GFR	glomerular filtration rate
GGT	gamma-glutamyl transferase
GH	growth hormone
GHRH	growth hormone-releasing hormone
GI	gastrointestinal
GIP	gastric inhibitory peptide
GlcNAc	*N*-acetylglucosamine
GLP-1 and GLP-2	glucagon-like peptides
GLUT	glucose transporter
GMP	guanosine monophosphate
GnRH	gonadotropin-releasing hormone
GOD-POD	glucose oxidase–peroxidase
GRPP	glicentin-related peptide

GSH	glutathione (reduced)
GSSG	glutathione (oxidized)
GTP	guanosine triphosphate
GTPase	guanosine triphosphate hydrolytic enzyme
Hb	hemoglobin
HbA	adult hemoglobin
HbA_{1c}	glycated hemoglobin
HbCO	carboxyhemoglobin
HbF	fetal hemoglobin
hCG	human chorionic gonadotropin
HCl	hydrochloric acid
HCM	human chorionic somatomammotropin
HCP	hereditary coproporphyria
HDL	high-density lipoprotein
HELLP	hemolysis, elevated liver enzymes, and low platelets
HEP	hepatoerythropoietic porphyria
HETEs	hydroxyeicosatetraenoic acids
HGPRT	hypoxanthine-guanine phosphoribosyl transferase
HH	hereditary hemochromatosis
HHS	hyperglycemic hyperosmolar state
HIV	human immunodeficiency virus
HMB	hydroxymethylbilane
HMBS	hydroxymethylbilane synthase (aka PBG deaminase)
HMG-CoA	3-hydroxy 3-methylglutaryl-CoA
HONK	hyperosmolar non-ketotic state
HPG axis	hypothalamus–pituitary–gonad axis
HPL	human placental lactogen
HPLC	high performance liquid chromatography
HPT axis	hypothalamus–pituitary–thyroid axis
HTGL	hepatic triglyceride lipase
ICF	intracellular fluid
IDL	intermediate-density lipoproteins
IEM	inborn error of metabolism
IF	intrinsic factor
Ig	immunoglobulin
IgA	immunoglobulin A
IGF	insulin-like growth factor
IGF1	insulin-like growth factor 1
IGF2	insulin-like growth factor 2
IGFBP4	insulin-like growth factor binding protein 4
IgG	immunoglobulin G
IHD	ischemic heart disease
IMP	inosine monophosphate
INR	international normalized ratio
IP	intervening peptide

IP3	inositol trisphosphate
IQC	internal quality control
IRIDA	iron-refractory iron-deficiency anemia
ISE	ion-selective electrodes
ISO	International Organization for Standardization
JH	juvenile hemochromatosis
KDIGO	Kidney Disease: Improving Global Outcomes
LCAD	long-chain acyl-CoA dehydrogenase
LCAT	lecithin-cholesterol acyl transferase
LDH	lactate dehydrogenase
LDL	low-density lipoprotein
LDLR	LDL receptor
LES	lower esophageal sphincter
LFT	liver function test
LH	luteinizing hormone
LMP	last menstrual period
LNS	Lesch–Nyhan syndrome
Lp(a)	lipoprotein (a)
LPL	lipoprotein lipase
LpX	lipoprotein X
LRP	LDL receptor-related protein
LSD	lysosomal storage disease
LT	leukotriene
MABP	mean arterial blood pressure
MAC	membrane attack complex
MALT	mucosa-associated lymphoid tissue
MAO	monoamine oxidase
MASP	mannan-binding lectin-associated serine protease
MBP	major basic protein
MCAD	medium-chain acyl-CoA dehydrogenase
MCH	mean cell hemoglobin
MCHC	mean cell hemoglobin concentration
MCP-1	monocyte chemotactic factor/protein
M-CSF	macrophage colony-stimulating factor
MCT	medullary carcinoma of the thyroid
MCV	mean cell volume
MD	mitochondria
MDRD	Modified Diet in Renal Disease
MEN	multiple endocrine neoplasia
MEOS	microsomal ethanol oxidizing system
MGUS	monoclonal gammopathy of undetermined significance
MI	myocardial infarction
MIT	mono-iodotyrosine
MM-LDL	minimally modified LDL
MoM	multiples of the median

MRI	magnetic resonance imaging
MS	mass spectrometry
MS	multiple sclerosis
MSUD	maple syrup urine disease
MTP	microsomal triglyceride transport protein
NAD	nicotinamide adenine dinucleotide
NAD^+	nicotinamide adenine dinucleotide (oxidized)
NADH	nicotinamide adenine dinucleotide (reduced)
NADP	nicotinamide adenine dinucleotide phosphate
NADPH	nicotinamide adenine dinucleotide phosphate (reduced)
NAFLD	non-alcoholic fatty liver disease
NAGS	*N*-acetyl glutamate synthetase
NAPQI	*N*-acetyl-*p*-benzoquinone imine
NASH	non-alcoholic steatohepatitis
NCEP	National Cholesterol Education Program
NET	neuroendocrine tumors
NMP	nucleoside monophosphate
NMP22	nuclear matrix protein 22
NNSGRC	National Newborn Screening and Global Resource Center
NSAIDs	nonsteroidal anti-inflammatory drugs
NSE	neuron-specific enolase
OA	organic acids
OAT	organic anion transporters
OGTT	oral glucose tolerance test
OTC	ornithine transcarbamylase
OX-LDL	oxidized LDL
P	properdin
PABA	*p*-aminobenzoic acid (or 4-aminobenzoic acid)
PAH	phenylalanine 4-hydroxylase
PAH	para-aminohippuric acid
PAPPA	pregnancy-associated plasma protein A
PBC	primary biliary cirrhosis
PBG	porphobilinogen
P_{BS}	Bowman's space hydrostatic pressure
P_{cap}	glomerular capillary hydrostatic pressure
PCOS	polycystic ovary syndrome
PCR	polymerase chain reaction
PCR	protein–creatinine ratio
PCT	porphyria cutanea tarda
PCT	proximal convoluted tubule
PD	peroxisomes
PDH	pyruvate dehydrogenase
PepT1	peptide transporter 1
PFK	phosphofructokinase
PG	prostaglandin

PHEX	phosphate-regulating endopeptidase (gene)
PI	Protease Inhibitor (gene)
PKU	phenylketonuria
PLAP	placental alkaline phosphatase
PlGF	placental growth factor
PNETs	pancreatic NETs
PNMT	phenylethanolamine-*N*-methyl transferase
PNP	purine nucleoside phosphorylase
PO_2	partial pressure of oxygen
POCT	point-of-care testing
POF	premature ovarian failure
POMC	pro-opiomelanocortin
PON	paraoxonase
PPI	proton pump inhibitor
PPO	protoporphyrinogen oxidase
PRL	prolactin
proMBP	precursor of major basic protein
PRPP	phosphoribosylpyrophosphate
PSA	prostate specific antigen
PTH	parathyroid hormone
PTHR1	parathyroid hormone receptor 1 (gene)
PTHrP	parathyroid hormone related peptide
PUFA	polyunsaturated fatty acid
qBH4	quinonoid isomer of tetrahydrobiopterin
QTL	quantitative trait locus
R	resistance to flow
RAADP	routine antenatal anti-D prophylaxis
RAAS	renin–angiotensin–aldosterone system
RBC	red blood cells
RDA	recommended daily amount
RDI	recommended daily intake
RhD–ve	Rhesus D negative
RhD+ve	Rhesus D positive
RNA	ribonucleic acid
RNI	recommended nutritional intake
RPF	renal plasma flow
rT_3	reverse T_3
RTA	renal tubular acidosis
RT-PCR	reverse transcriptase polymerase chain reaction
S	solubility constant
SAH	subarachnoid hemorrhage
SBP	systolic blood pressure
SCA	squamous cell antigen
SCID	severe combined immune deficiency
SD	standard deviation

SERPINA1	Serpin Peptidase Inhibitor A1 (also known as PI)
SGLT	sodium–glucose linked transporter
SHBG	sex-hormone-binding globulin
SHH	syndrome of hyporeninemic hypoaldosteronism
SI	Standard International
SIADH	syndrome of inappropriate ADH
SIRS	systemic inflammatory response syndrome
SLE	systemic lupus erythematosus
SOB	shortness of breath
SSRI	selective serotonin reuptake inhibitor
SST	short Synacthen test
STR	soluble transferrin receptor
STR	short tandem repeat
sVEGFR-1	soluble vascular endothelial growth factor receptor 1 (also known as sFlt-1)
$t_{1/2}$	plasma half-life (of a drug)
T_3	tri-iodothyronine
T_4	tetra-iodothyronione (thyroxine)
TATI	tumor-associated trypsin inhibitor
TBG	thyroid hormone-binding globulin
TBPA	thyroid hormone-binding prealbumin
TCA	tricarboxylic acid
TDM	therapeutic drug monitoring
TFP	trifunctional protein
TFT	thyroid function test
TG	thyroglobulin
TGF-β	transforming growth factor-β
Th	helper T (cells)
TIBC	total iron-binding capacity
TLC	thin-layer chromatography
T_m	transport maximum
TMP	thiamine monophosphate
T_mP	renal threshold for phosphate
TMS	tandem mass spectrometry
TPA	tissue polypeptide antigen
TPN	total parenteral nutrition
TPO	thyroid peroxidase
TPP	thiamine pyrophosphate
TPR	total peripheral resistance
TRAP	tartrate-resistant acid phosphatase
TRH	thyrotropin-releasing hormone
TSAT	transferrin saturation
TSH	thyroid-stimulating hormone
TTG	tissue transglutaminase
UCD	urea cycle disorder

UDP	uridine diphosphate
UDP-glucose	uridine diphosphate glucose
UES	upper esophageal sphincter
UGT1A1	UDP-glucuronyl transferase gene
UKPDS	United Kingdom Prospective Diabetes Study
UMP	uridine monophosphate
UP	ultrafiltration pressure
UROD	uroporphyrinogen decarboxylase
UROS	uroporphyrinogen synthase
UV	ultraviolet
V_d	volume of distribution (of a drug)
VDDR1	vitamin D-dependent rickets type 1
VIP	vasoactive intestinal polypeptide
VLCAD	very-long-chain acyl-CoA dehydrogenase
VLDL	very-low-density lipoproteins
VLDLR	very-low-density lipoprotein remnant
VMA	vanillylmandelic acid
VP	variegate porphyria
WM	Waldenstrom's macroglobulinemia
π_{BS}	Bowman's space colloid osmotic pressure
π_{cap}	blood colloid osmotic pressure

INDEX

Readers are reminded that page references may indicate only a relevant Table, Figure caption, Case description or practice point, not necessarily coverage in the main text on that page.